ASSESSMENT METHODS FOR SOIL CARBON

Advances in Soil Science

Series Editor: *B. A. Stewart*

Published Titles

ASSESSMENT METHODS FOR SOIL CARBON

Edited by

R. Lal
J.M. Kimble
R.F. Follett
B.A. Stewart

CRC Press
Taylor & Francis Group
Boca Raton London New York

CRC Press is an imprint of the
Taylor & Francis Group, an **informa** business

CRC Press
Taylor & Francis Group
6000 Broken Sound Parkway NW, Suite 300
Boca Raton, FL 33487-2742

First issued in paperback 2019

© 2001 by Taylor & Francis Group, LLC
CRC Press is an imprint of Taylor & Francis Group, an Informa business

No claim to original U.S. Government works

ISBN-13: 978-1-56670-461-8 (hbk)
ISBN-13: 978-0-367-39768-5 (pbk)
Library of Congress Card Number 00-050666

Library of Congress Cataloging-in-Publication Data

Assessment methods for soil carbon / [edited by] R. Lal ... [et al.].
     p. cm.
Includes bibliographical references and index.
ISBN 1-56670-461-8 (alk. paper)
    1. Soils—Carbon content—Analysis—Congresses. I. Lal, R.
S592.6.C35 A88 2000
631.4′17—dc21

                                          00-050666

Visit the Taylor & Francis Web site at
http://www.taylorandfrancis.com

and the CRC Press Web site at
http://www.crcpress.com

Preface

The international workshop on "Assessment Methods for Soil C Pools" was held at The Ohio State University, Columbus, Ohio, in November 1998. Approximately 50 participants from 10 countries presented research information on recent advances in methods for assessment of different carbon (C) pools in soil. The participants were drawn from Australia, Brazil, Canada, Ecuador, France, Germany, Kenya, Russia, Sweden, and the United States. This workshop was the fifth in a series of workshops (the arid regions: Tunis, Tunisia, October 1997; tropical ecosystems: Belem, Brazil, December 1997; cold regions: Columbus, Ohio, March 1998; rangeland: Las Cruces, New Mexico, September 1998) that addressed specific topics related to carbon sequestration in these ecoregions. Organization of these workshops emerged as a recommendation of the second international symposium on "Carbon Sequestration in Soils" held at The Ohio State University in July 1996.

The workshop deliberations included presentation of approximately 50 papers under seven general themes: (i) assessment of C pool (13 papers); (ii) soil sampling and sample preparation (8 papers); (iii) analytical techniques (8 papers); (iv) soil erosion and sedimentation (4 papers); (v) remote sensing, GIS and modeling (4 papers); (vi) scaling procedures (10 papers); and (vii) economic analysis (3 papers). All papers published in this volume were reviewed by the editorial committee and revised by the authors.

The 41 contributor chapters of this volume are divided thematically into seven sections, with introductory and concluding chapters (Section I, Introduction, and Section VIII, Synthesis) written by the editors. Section II, Soil Sampling and Sample Preparation, contains 7 chapters. Section III, Assessment of Carbon Pools, comprises 4 chapters. Section IV, Assessment and Analytical Techniques, consists of 15 chapters. Section V, Soil Erosion and Sedimentation, consists of 4 chapters. Section VI, Modeling and Scaling Procedures, contains 8 chapters. Section VII, Economics and Policy Issues, comprises 3 chapters.

The organization of the symposium and publication of this volume were made possible by the cooperation and funding of the USDA Natural Resources Conservation Service, the Agricultural Research Service, and The Ohio State University. The editors thank the authors for their outstanding efforts to document and present their information on the current understanding of soil processes and the carbon cycle in a timely fashion. Their efforts have contributed to enhancing the overall understanding of pedospheric processes, and how to better use soils as a sink for carbon while also managing soils to minimize pedosphere contributions of carbon dioxide and other greenhouse gases to the atmosphere. These efforts have advanced the frontiers of soil science and improved the understanding of the pedosphere into the broader scientific arena of linking soils to the global carbon cycle, soil productivity, and environment quality.

Thanks are also due to the staff of Lewis Publishers/CRC Press for their timely efforts in publishing this information to make it available to the scientific community. In addition, valuable contributions were made by numerous colleagues, graduate students, and OSU staff. We especially thank Ms. Lynn Everett for her efforts in organizing the conference and for handling the flow of papers to and from the authors throughout the review process. We also offer special thanks to Ms. Brenda Swank for her help in preparing this material and for her assistance in all aspects of the symposium. The efforts of many others were also very important in publishing this relevant and important scientific information in a timely manner.

The Editors

About the Editors

Dr. R. Lal is a professor of soil science in the School of Natural Resources at The Ohio State University. Prior to joining Ohio State in 1987, he served as a soil scientist for 18 years at the International Institute of Tropical Agriculture, Ibadan, Nigeria. Professor Lal is a fellow of the Soil Science Society of America, American Society of Agronomy, Third World Academy of Sciences, American Association for the Advancement of Sciences, Soil and Water Conservation Society, and the Indian Academy of Agricultural Sciences. He is a recipient of the International Soil Science Award, the Soil Science Applied Research Award of the Soil Science Society of America, the International Agronomy Award of the American Society of America, the International Agronomy Award of the American Society of Agronomy, and the Hugh Hammond Bennett Award of the Association of the Soil and Water Conservation Society. He is past president of the World Association of the Soil and Water Conservation and the International Soil Tillage Research Organization.

Dr. John M. Kimble is a Research Soil Scientist at the USDA Natural Resources Conservation Service National Soil Survey Laboratory in Lincoln, Nebraska. Dr. Kimble manages the Global Change project of the Natural Resources Conservation Service, and has worked more than 15 years with the US Agency for International Development projects dealing with soils-related problems in more than 40 developing countries. He is a member of the American Society of Agronomy, the Soil Science Society of America, the International Soil Science Society, and the International Humic Substances Society.

Dr. R.F. Follett is Supervisory Soil Scientist, USDA-ARS, Soil Plant Nutrient Research Unit, Fort Collins, CO. He previously served 10 years as a National Program Leader with ARS headquarters in Beltsville, MD. Dr. Follett is a Fellow of the Soil Science Society of America, American Society of Agronomy, and the Soil and Water Conservation Society. He was twice awarded the USDA Distinguished Service Award (USDA's highest award). Dr. Follett organized and wrote the ARS Strategic Plans for both "Ground-Water Quality Protection-Biogeochemical Dynamics." Dr. Follett has been a lead editor for several books and a guest editor for the Journal of Contaminant Hydrology. His scientific publications include topics about nutrient management for forage production, soil-N and -C cycling, ground-water quality protection, global climate change, agroecosystems, soil and crop management systems, soil erosion and crop productivity, plant mineral nutrition, animal nutrition, irrigation, and drainage.

Dr. B.A. Stewart is Distinguished Professor of Agriculture, and Director of the Dryland Agriculture Institute at West Texas A&M University. Prior to joining West Texas A&M University in 1993, he was Director of the USDA Conservation and Production Research Laboratory, Bushland, Texas. Dr. Stewart is a past president of the Soil Science Society of America, and was a member of the 1990-93 Committee on Long Range Soil and Water Policy, National Research Council, National Academy of Sciences. He is a Fellow of the Soil Science Society of America, American Society of Agronomy, Soil and Water Conservation Society, a recipient of the USDA Superior Service Award, and a recipient of the Hugh Hammond Bennett Award of the Soil and Water Conservation Society.

Contributors

W. Amelung, Institute of Soil Science and Soil Geography, University of Bayreuth, 95440 Bayreuth, Germany.

O. Andrén, Department of Soil Science, P.O. Box 7014, SLU, S-750 07, Uppsala, Sweden.

M.J. Apps, Natural Resources Canada, Canadian Forest Service, Northern Forestry Center, 5320 122nd Street, Edmonton, Alberta T6H 3S5, Canada.

J. Balesdent, IRA-CNRS/CEA, Laboratoire d'Ecologie Michobienne de la Rhizosphere DEVM, CEA, Cadarache, France.

J. Bauhus, Department of Forestry, ANU, Canberra A.C.T. 0200, Australia.

M. Beare, University of Georgia, Athens, Georgia 30602, U.S.A.

P. Becker-Heidmann, Institute of Soil Science, University of Hamburg, Allendeplatz 2, D-20146, Hamburg, Germany.

J.S. Bhatti, Canadian Forest Service, Northern Forestry Center, 5320, 122nd Street, Edmonton, Alberta T6H 3S5, Canada.

G. Blair, Division of Agronomy and Soil Science, University of New England, Armidale NSW 2351, Australia.

N. Blair, Division of Agronomy and Soil Science, University of New England, Armidale NSW 2351, Australia.

N. Bliss, Raytheon Stx Corporation, USGS Eros Data Center, Sioux Falls, SD 57198, U.S.A.

S. Böhm, Department of Crop and Soil Sciences, Michigan State University, 286 Plant and Soil Science, East Lansing, MI 48824-1325, U.S.A.

P.V. Bolstad, University of Minnesota, Department of Forest Resources, 115 Green Hall, 1530 N. Cleveland Avenue, St. Paul, MN 55108, U.S.A.

C.A. Cambardella, National Soil Tilth Laboratory, USDA-ARS, 2150 Pammel Drive, Ames, IA 50011-0001, U.S.A.

R.J. Candler, University of Alaska-Fairbanks, Palmer Research Station, 533 E. Fireweed Avenue, Palmer, AK 99645, U.S.A.

C. Cerri, Centro de Energia Nuclear na Agricultura, Universidade de São Paulo, Campus "Luiz De Queiroz," Avenida Centenario 303, Caixa Postal 96, Cep 13400-970, Piracicaba SP, Brazil.

H.H. Cheng, Department of Soil Science, Water and Climate, University of Minnesota, 1991 Upper Buford Circle, St. Paul, MN 55108-6028, U.S.A.

A. Conteh, Division of Agronomy and Soil Science, University of New England, Armidale NSW 2351, Australia.

X.Y. Dai, University of Alaska-Fairbanks, Palmer Research Station, 533 E. Fireweed Avenue, Palmer, AK 99645, U.S.A.

J.W. Doran, USDA-ARS, Soil and Water Conservation Research Unit, 116 Keim Hall, East Campus University of Nebraska, Lincoln, NE 68583-0915, U.S.A.

L.R. Drees, Texas A&M University, Soil and Crop Sciences, College of Agriculture and Life Sciences, College Station, TX 77843-2474, U.S.A.

B.H. Ellert, P.O. Box 3000, Lethbridge, Alberta T1J 4B1, Canada.

E.T. Elliott, Natural Resource Ecology Laboratory, Colorado State University, Fort Collins, CO 80523, U.S.A.

M. Eve, USDA-ARS, Soil Plant Nutrient Research Unit, P.O. Box E, Fort Collins, CO 80522, U.S.A.

C. Feller, CENA, Caixa Postal 96, CEP 13400-970, Piracicaba, São Paulo, Brazil, or IRD (EX-ORSTOM) (L.C.S.C.), 34032 Montpellier Cedex, France.

R.F. Follett, Soil Plant Nutrient Research, Room 407, USDA-ARS, P.O. Box E, 301 S. Howes, Fort Collins, CO 80522-0470, U.S.A.

C.D. Franks, USDA-NRCS, Federal Building, Room 152, 100 Centennial Mall North, Lincoln, NE 68508-3866, U.S.A.

A.M. Gajda, Institute of Soil Science and Plant Cultivation, Department of Agricultural Microbiology, 24-100 Pulawy, Poland.

J.A. Galantini, Departmento de Agronomia, Universidad Nacional Del Sur, 8000 Bahia Blanca, Argentina.

J.C. Gasparoni, Departmento de Agronomia, Universidad Nacional Del Sur, 8000 Bahia Blanca, Argentina.

R.J. Gilkes, Soil Science and Plant Nutrition, Faculty of Agriculture, The University of Western Australia, Nedlands, Western Australia 6907, Australia.

R.F. Grant, Department of Renewable Resources, University of Alberta, 4-42 Earth Sciences Building, Edmonton, Alberta T6G 2E3, Canada.

R.B. Grossman, National Soil Survey Center, USDA-NRCS, Federal Building Room 152, 100 Centennial Mall North, Lincoln, NE 68508-3866, U.S.A.

Y. Hao, The Ohio State University, School of Natural Resources, 210 Kottman Hall, 2021 Coffey Road, Columbus, OH 43210, U.S.A.

M.E. Harmon, Oregon State University, Department of Forest Science, 020 Forestry Sciences Laboratory, Corvallis, OR 97331-501, U.S.A.

D.S. Harms, USDA NRCS, National Soil Survey Center, Federal Building, Room 152, 100 Centennial Mall North, Lincoln, NE 68508-3866, U.S.A.

R.J. Harper, Department of Conservation and Land Management, Locked Mail Bag 104, Bentley Delivery Center, Western Australia 6983, Australia.

D. Harris, Stable Isotope Facility, University of California, Davis, CA 95616, U.S.A.

U. Hartwig, Institute of Plant Science, Swiss Federal Institute of Technology, 8-92 Zurich, Switzerland.

K.H. Haugen-Kozyra, Conservation and Development Branch, Alberta Agriculture Food and Rural Development, Edmonton, Alberta, Canada.

J.C. Hiley, Agriculture and Agri-Food Canada, Edmonton, Alberta, Canada.

J. Hopkins, Department of Agriculture, Environment, and Development Economics, The Ohio State University, 2120 Fyffe Road, Columbus, OH 43210, U.S.A.

W.R. Horwath, Department of Land, Air, and Water Resources, One Shields Avenue, University of California at Davis, Davis, CA 95616-8627, U.S.A.

R.C. Izaurralde, Integrated Earth Studies, Battelle, Pacific Northwest Laboratory, 901 D Street, S.W. Suite 900, Washington, D.C. 20024-2115, U.S.A.

P. Jacinthe, The Ohio State University, School of Natural Resources, 210 Kottman Hall, 2021 Coffey Road, Columbus, OH 43210, U.S.A.

D.C. Jans, Agrium Fertilizers, Saskatoon, SASK, Canada.

H.H. Janzen, P.O. Box 3000, Lethbridge, Alberta T1J 4B1, Canada.

A.B. Jenkins, USDA NRCS, Soil Survey Information Center, P.O. Box 1,449 Water Street, Room 209, Summerville, WV 26651, U.S.A.

H. Jiang, Department of Renewable Resources, University of Alberta, Edmonton, Alberta T6G 2H1, Canada.

N.K. Karanja, Department of Soil Science, University of Nairobi, P.O. Box 29053, Nairobi, Kenya.

T. Kätterer, Department of Soil Science, P.O. Box 7014, SLU, S-75007, Uppsala, Sweden.

T.A. Kettler, USDA-ARS, Soil and Water Conservation Research Unit, 116 Keim Hall, East Campus, University of Nebraska, Lincoln, NE 68583-0915, U.S.A.

P.K. Khanna, Division of Forest Resources, CSIRO, P.O. Box E4008, Canberra, A.C.T. 2604, Australia.

J.M. Kimble, National Soil Survey Center, USDA-NRCS, Federal Building, Room 152, 100 Centennial Mall North, Lincoln, NE 68508-3866, U.S.A.

D.F. Kingsbury, USDA NRCS, MLRA 13, 75 High Street, Room 301, Morgantown, WV 26505, U.S.A.

O.N. Krankina, Oregon State University, Department of Forest Science, 020 Forestry Sciences Laboratory, Corvallis, OR 97331-501, U.S.A.

R. Lacelle, ECORC, Research Branch, Agriculture Canada CEF, Room 1135 K.W. Neatby Building, Ottawa, Ontario K1A OC6, Canada.

R. Lal, The Ohio State University, School of Natural Resources, 210 Kottman Hall, 2021 Coffey Road, Columbus, OH 43210, U.S.A.

R. Lefroy, IBSRAM, P.O. Box 9-109, Phaholyotin Road, Jatujak, Bangkok 10900, Thailand.

B. Ludwig, Institute of Soil Scinece and Forest Nutrition, University of Göttingen, Büsgenweg 2, 37077 Göttingen, Germany.

A. Manale, USEPA, Office of Planning and Evaluation, 401 M St. S.W., Washington, D.C. 20460, U.S.A.

E. Matthews, Oregon State University, Department of Forest Science, 020 Forestry Sciences Laboratory, Corvallis, OR 97331-501, U.S.A.

G.W. McCarty, USDA ARS Environmental Chemistry Laboratory, BARC-West Building 007, Room 201, Beltsville, MD 20705, U.S.A.

B. McConkey, P.O. Box 1030, Swift Current, Saskatchewan S9H 3X2, Canada.

W.B. McGill, Department of Renewable Resources, University of Alberta, 4-42 Earth Sciences Building, Edmonton, Alberta T6G 2E3, Canada.

G.J. Michaelson, Agriculture Forestry Experiment Station, University of Alaska - Fairbanks, 533 East Fireweed, Palmer, AK 99645-6629, U.S.A.

S.J. Morris, Department of Crop and Soil Sciences, Michigan State University, 286 Plant and Soil Science, East Lansing, MI 48824-1325, U.S.A.

E.W. Murage, Kenya Agricultural Research Institute, P.O. Box 57811, Nairobi, Kenya.

B. Nicolardot, INRA, Unité d'Agronomie de Châlons-Reims, Reims, France.

L.C. Nordt, Texas A&M University, Soil and Crop Sciences, College of Agriculture and Life Sciences, College Station, TX 77843-2474, U.S.A.

C. O'Hara, Department of Forestry, Australian National University, Canberra A.C.T. 0200, Australia.

F. Orozco-Chavez, ECORC, Research Branch, Agriculture Canada, C.E.F., Neatby Building, Ottawa K1A 0C6, Canada.

L. Owens, North Appalachian Experimental Watershed, P.O. Box 478, SR 621, Coshocton, OH 43812-0478, U.S.A.

E.A. Paul, Department of Crop and Soil Sciences, Michigan State University, 286 Plant and Soil Science, East Lansing, MI 48824-1325, U.S.A.

K. Paustian, Natural Resources Ecology Laboratory, Colorado State University, Fort Collins, CO 80523-1499, U.S.A.

E.M. Pfeiffer, Institute of Soil Science, University of Hamburg, Allendeplatz 2, D-20146 Hamburg, Germany.

J.L. Pikul, Jr., USDA-ARS, Northern Grain Insect Research Laboratory, Brookings, SD, U.S.A.

C.L. Ping, University of Alaska-Fairbanks, Palmer Research Station, 533 E. Fireweed Avenue, Palmer, AK 99645, U.S.A.

E.G. Pruessner, Soil Plant Nutrient Research, Room 407, USDA-ARS, PO Box E, 301 S. Howes, Fort Collins, CO 80522-0470, U.S.A.

G. Rapalee, University of California, Irvine and NASA/Goddard Space Flight Center, Code 923, Greenbelt, MD 20771, U.S.A.

J.B. Reeves III, USDA ARS Environmental Chemistry Laboratory, BARC-West Building 007, Room 201, Beltsville, MD 20705, U.S.A.

J.C. Ritchie, USDA ARS Hydrology Laboratory, BARC-West Building 007, Beltsville, MD 20705, U.S.A.

R.A. Rosell, Departmento de Agronomia, Universidad Nacional Del Sur, 8000 Bahia Blanca, Argentina.

S.E. Samson-Liebig, USDA-ARS, Northern Great Plains Research Laboratory, P.O. Box 459, Mandan, ND 58554-0459, U.S.A.

H.W. Scharpenseel, Institute of Soil Science, University of Hamburg, Allendeplatz 2, D-20146 Hamburg, Germany.

R.K. Shaw, USDA NRCS, North Jersey Field Support Office, Box 3, Suite 21, 1322 Route 31, Annandale, NJ 08801, U.S.A.

G.R. Smith, Environmental Resources Trust, 209 NW 58th Street, Seattle, WA 98107-2030, U.S.A.

T.M. Sobecki, USDA Natural Resources Conservation Service, Box 2890, Washington, D.C. 20013, U.S.A.

B. Sohngen, Department of Agriculture, Environment and Development Economics, The Ohio State University, 2120 Fyffe Road, Columbus, OH 43210, U.S.A.

G.C. Starr, USDA ARS, Walnut Gulch Experimental Watershed, P.O. Box 213, Tombstone, AZ 85638-0213, U.S.A.

D. Sunding, Department of Agricultural and Resource Economics, 207 Giannini Hall, University of California - Berkeley, Berkeley, CA 94720-3310, U.S.A.

L. Tweeten, Department of Agriculture, Environment and Development Economics, The Ohio State University, 2120 Fyffe Road, Columbus, OH 43210, U.S.A.

S. Waltman, USDA Natural Resources Conservation Service, Box 2890, Washington, D.C. 20013, U.S.A.

C. van Kessel, Department of Agronomy and Range Science, University of California at Davis, Davis, CA 95616, U.S.A.

J.M. Vose, USDA Forest Service, Coweeta Hydrologic Laboratory, 3160 Coweeta Laboratory Road, Otto, NC 28763, U.S.A.

A. Whitbread, Division of Agronomy and Soil Science, University of New England, Armidale NSW 2351, Australia.

B.J. Wienhold, USDA-ARS, Soil and Water Conservation Research Unit, 116 Keim Hall, East Campus, University of Nebraska, Lincoln, NE 68583-0915, U.S.A.

L.P. Wilding, Department of Soil and Crop Science, Texas A&M University, College Station, TX 77843-2474, U.S.A.

P.L. Woomer, Tropical Soil Biology and Fertility Programme, P.O. Box 30592, Nairobi, Kenya.

M. Yatskov, Oregon State University, Department of Forest Science, 020 Forestry Sciences Laboratory, Corvallis, OR 97331, U.S.A.

D. Zilberman, University of California - Berkeley, 207 Giannini Hall, Berkeley, CA 94720-3310, U.S.A.

Contents

Section I

Introduction

Section 1

Introduction

CHAPTER 1

Methods for Assessing Soil C Pools

J.M. Kimble, R. Lal and R.F. Follett

I. Introduction

Because of the 1997 Kyoto Protocol, carbon (C) sequestration in soils and in aboveground biomass and its effect on C fluxes is receiving much more attention. Compared to the base established in 1990, this protocol requires a major reduction in emissions by the next decade. Figure 1 shows strategies that can help meet the reduction in greenhouse gas fluxes. The Committee of the Parties (COP) for the United Nations Framework Convention on Climate Change (FCC), which is working to implement the Kyoto Protocol and to make needed changes, asked the IPCC (Intergovernmental Panel on Climate Change) for a special report addressing the protocol's major points related to C. Article 3.4 deals with methods to establish baselines of C stocks and modalities, rules and guidelines, and how and which additional human-induced activities related to changes in greenhouse gas emissions by sources and removals by sinks in the agriculture soils and land use and forestry categories shall be added to, and subtracted from, the assigned amounts of C sequestration. This is on top of forests and land use change (Article 3.3) that may be accepted by the COP and is of special interest in relation to soil C as this is the only place other practices may be added.

Past meetings and their proceedings have shown the need for standard procedures to measure the different C pools and flux rates, and the need to scale point data to farm, regional, national, and even international levels. There is an additional need for procedures to measure the C pool changes caused by anthropogenic impacts (e.g., standard tillage vs. no till, and other management methods). These past meetings were as follows: International Symposium on Soils and Greenhouse Gasses, The Ohio State University, July, 1993; International Symposium on Soils and C Sequestration, The Ohio State University, July, 1996; Global Climate Change and Pedogenic Carbonates, October, 1997, Tunis, Tunisia; C Pools and Dynamics in Tropical Ecosystems, December, 1997, Belem, Brazil; International Workshop on C Pools and Cold Ecosystems, April, 1998, Columbus, Ohio; and C Sequestration Meeting on U.S. Grazing Lands, September, 1998, New Mexico; the proceedings of these meetings (1994, 1995a, 1995bc, 1997, 1998a, 1998b, and 2000) are available.

Several years ago, Dr. Pedro Sanchez (Director General of ICRAF) asked, "What do we really know about C dynamics and its sequestration in soils?" (Kimble et al., 1994). No clear answer could be given then but many advances have been made and we understand much more about Soil Organic Carbon (SOC) and its dynamics. Standard methodologies are in place and more are being developed. In many cases, information has been put into a format that will help policy makers develop programs for sequestering C. Estimates as to the value of SOC are developed and were addressed by authors who attended. Numbers quoted ranged from very small to quite large and in time the C market would fall into place. Some scientists consider the benefits of C only to the fertility status of the soil, while

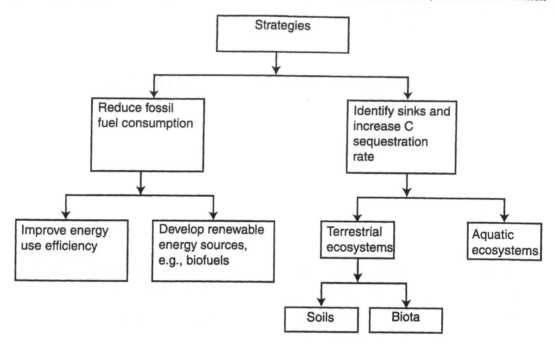

Figure 1. Strategies for reductions in greenhouse gases to help reduce the risks of global warming.

others try to include societal benefits (reduced erosion by both water and wind, better water quality, better wildlife habitat, improved yields, etc.). These societal benefits are important and must be considered in any policies developed.

We know that SOC and soil inorganic carbon (SIC) have value. For the farmer and land manager, SOC and SIC contribute to the soil's capacity to supply nutrients, buffer the movement of nutrients, pesticides, and herbicides, improve the soil's water holding capacity, and increase soil aggregation. For society in general, SOC and SIC improve water quality, reduce soil erosion, reduce sedimentation, and improve wild life habitat. However, we need standardized methods for measuring C pool dynamics so that we can determine these values accurately.

Anthropogenic activities have caused large losses of C from the soil and from aboveground biomass. In many cases, these activities – especially those involving poor planning, development of marginal lands, and use of tillage methods that sacrifice SOC and SIC – have caused land degradation and the failure of the overall system (Figure 2). If we continue with business as usual, soil C will continue to be lost both as SOC and SIC.

Agriculture and forestry have been part of the problem. Until about the 1950s, in fact, agricultural and forestry activities were the major contributors to the increase of atmospheric CO_2 (Lal et al., 1998). We now recognize this but also know that agricultural and forestry practices instead are becoming part of the solution. Agricultural practices cannot by themselves solve the greenhouse gas flux problem, but adoption of Best Management Practices (BMPs) and other agricultural strategies can decrease the overall net flux and can stash a significant part of the currently emitted C into the soil and into the above- and belowground biomass.

Why, then, has the rate of adoption of BMPs slowed and in some areas even reversed? We as scientists know that reduced tillage and no-till reduce erosion; we have conducted many studies showing this, but the old tillage practices are coming back. Did we not answer all of the land managers' questions? If not, what is missing? If the land use managers and farmers perceive prob-

Figure 2. The net effects of poor land management on systems' sustainability.

lems with procedures and practices we developed, why are we not addressing their concerns? We tend to move on to the next new exciting problem or question and forget the mundane work needed to see that our good scientific ideas are adopted. This must change if we are to succeed in helping reduce the rate of increase in greenhouse gases and in making agriculture a sustainable system that is part of the solution.

II. About the Workshop

With these ideas in mind, the Workshop on Methods for Assessing Soil C Pools was organized to begin to:
1. Provide some standardized methods to measure different C pools
2. Address problems with C and related data
3. Provide means to scale point data for use at many different levels (farm, regional, national, and international)
4. Provide methods for sampling to measure the different pools and flux rates
5. Develop schemes for C trading
6. Provide methods for verifying changes in C stocks and rates of sequestration

Our initial goal was to make advances toward the following:
1. Providing standard methods for measuring C pools and C dynamics
2. Developing increased awareness of the problems and creating a better dialogue among scientists in many different specialties
3. Creating a better understanding of policy options by increasing the dialogue between scientists and policy makers (another meeting in July 1999 also addressed this topic)

The workshop thus was organized into eight broad areas, which the chapters in this volume cover in detail:
1. Overviews, to set the stage and provide background information on what we know and do not know
2. C pools, what they are and which are important
3. Soil sampling and sample preparation
4. Assessment and analytical technique
5. Soil erosion and sedimentation
6. Scaling procedures
7. Scaling and C estimates
8. Policy and economic issues

The workshop's objectives were to
1. Consider standardized procedures for:
   a. Sampling depth in relation to land use and soil management
   b. Sample preparation and handling for different C pools
   c. Measurement of soil bulk density
   d. Analytical techniques for different C pools
2. Describe procedures to evaluate total system C
3. Describe and standardize scaling procedures
4. Identify procedures for developing C credits
5. Assess monetary/societal/economic value of soil C

III. The Need for Standardized Methods

Many publications describe methods for measuring SOC, SIC, C flux rates, and fractionation of C pools (Black, 1965), and this in itself is one of the problems. Everyone has a method and feels it is the best. Many methods were developed for limited areas (e.g., for one soil type, which has a specific set of chemical, mineralogical, and physical properties), and extensions to other areas may not apply. The solution is to test methods on a broad basis, to define the parameters under which they work, and to make sure they are described in enough detail for others to use them and get the same results.

Specific chapters in this book address these related problem areas: (1) analytical methodology for SOC; (2) general data problems; (3) microbial biomass; (4) bulk density; (5) scaling the data; and (6) sampling.

A. Analytical Methodology for SOC

1. Problems with the Walkley-Black Procedure

For many years, the standard method for measuring SOM was the Walkley-Black procedure (WB) (Walkley, 1935). Most scientists recognized limitations to this procedure and used a standard correction factor (of 0.7) because of the incomplete measurement of all the C in the sample. This correction factor was known to vary with depth and soil type, yet only the one factor was used. In addition, in soils with high levels of Mn, the Mn might act as an electron donor and bias the results (Cheng and Kimble, 2000). Nor did this procedure address the measurement of charcoal in many soils. However, the procedure was usable in most laboratories, although many modified it in ways that may have affected the results and, unfortunately, no one was very concerned.

Today, we seldom use the WB procedure, not because of scientific concerns but because of environmental ones. A serious question here is why we disregarded our own scientific concerns and simply reverted to an easy and long-used method.

2. Problems with Measuring CO_2 Output

We now measure SOC by heating the sample in a furnace and measuring the CO_2 given off. The results are very reproducible on a given sample, but even here, the temperatures used may vary. Some laboratories measure SOC based on the weight change of a sample and not on CO_2 emitted. However, hydroscopic water in the different clays can affect weight changes, as can high amounts of amorphous material in water. We must address problems such as these if we are to compare samples and methodology over large areas.

3. Problems with Measuring Charcoal and Carbonates

Another problem with the evolution and measurement of CO_2 is the presence of charcoal and carbonates. Some laboratories leach the samples with HCl; others treat with H_2PO_4 to remove the carbonates; and others measure the carbonates and make a correction for the C in them based on the weight of calcium carbonate. All work to an extent and all have problems. The carbonates we measure may not all be $CaCO_3$, so the weight used for correction may be wrong. Acid treatments do not answer the question of charcoal, either, when carbonates are removed or when adjustments are made for the total carbonates present. In many areas, charcoal may not be a problem but, in areas of frequent fires (tropics, many grazing lands in all ecoregions, boreal forests), large buildups of charcoal may occur.

When acids are used to leach or remove carbonates, do we also remove soluble C products from the soil, particularly from the active pool? We talk about C leaching down into the profile, and this implies that some of the C is soluble. We thus should expect some of the active or mobile pool to be removed by acid treatment, but we do not report it in our results. This may seem minor but, in a C-trading scheme, it will affect verification and measurement. This soluble C may be in the active pool and small when compared to the large soil release pools, but it may be of interest when we are concerned with short-term changes (five to ten years, for which C credits may be paid). A land manager will not be happy to know that part of this sequestered C is leached away and no credit given for it. Methods to account for this C must be developed, tested, and standardized.

All analytical work must use procedures that adjust for carbonates and for charcoal and that address the soluble SOC pool. For data published to date, it is often hard to tell if and how such adjustments were made or even considered. Thus, databases are compiled from many different sources, but no one knows if the data were run with the same procedures or need adjustments; and modelers at regional and global scales take data from many different sources and assume the data are the same, but in many cases, they are not. We must remedy this problem if we are to standardize measurement and verification for C trading and if we are to determine if soils are a sink or source of greenhouse gases or, if managed properly, soils can help alleviate the greenhouse effect.

B. General Data Problems

The database problem just mentioned highlights our general data problem. How can we create regional, national, and international databases when we are not sure of the methodologies used? We need reference samples, agreed-upon methodologies, standardization of terms, etc. We also must highlight problems in our databases and make policy makers aware of potential problems and pitfalls. How can we correctly interpret data developed with different procedures when we are not even sure what methods generated it?

Some data uncertainties we need to be aware of are:
1. Problems with methods used for C analysis
2. Lack of standardized procedures for measurement
3. Problems with estimates of total system C versus SOC and SIC
4. A general lack of data on SIC
5. No assessment of SOC and SIC to a uniform depth, at least 2 meters
6. Difficulties in scaling the data (i.e., point data to large areas)
7. Difficulties in assessing or estimating historical C levels and changes in them, including:
   a. Lack of data to at least a depth of 2 meters
   b. Lack of reference or benchmark sites
   c. Missing data on soil bulk density
   d. Problems of establishing the "cause-and-effect" relationship between land use and C pools.

To answer some of these questions and concerns, we must (1) set priorities in data collection and research; (2) give estimates of error; (3) provide complete and accurate documentation; and (4) rigorously test the methods and the model outputs from the input data. Samples have been collected at many points in the past, yet the land use history, the actual sample site, or even the date when sampled are not known. We cannot develop a 1990 base map with all these known limitations. Do we really need a baseline map when most likely it can be developed only by the use of models? Alternatively, are we interested only in the rate of change or actual change for a specific time? If the latter is the case and it seems to be the most practical way to go, then all we need is to measure soil C, especially SOC, at time (x) and at x+1, where 1 is the time interval of interest.

C. Microbial Biomass

Microbial biomass is used as an indicator of the active C pool. The procedure commonly used (Jenkins, 1988) has been modified several times over the years. The original procedure used fumigation and incubation. The incubation step is sometimes not used and an extraction phase follows the fumigation.

Scientists disagree about running a control that is subtracted from the incubated results, specifically, about the correction for the control. Some have used dried samples, re-wet them, and then let them stabilize, but how long they should sit before the fumigation and incubation steps is not clear.

The temperature at which the incubation is run also varies. Is the use of a 25 or 30°C incubation relevant in a soil in a frigid or cryic temperature regime where the soil temperature never rises above 5 or 10°C? Does the higher temperature stimulate extra microbial growth that would not be found in natural soils? What steps are needed in sample preparation and storage? Do the samples need to be kept cold as soon as they are sampled?

Several chapters in this volume address the microbial biomass. Some present new methods that may be simpler to run and give equivalent results, e.g., the $KMnO_4$ extraction, hot water extraction, fumigation by microwaving, formaldehyde fumigation, etc. Until standardized procedures are accepted, we will have problems saying what the active pool in the soil really is.

D. Bulk Density

In any C trading/sequestration system, we will need to know the bulk density of the soil. The proposed payments for C sequestration are based on how many grams, kilograms, or tons of C are stored. These are all volume measurements. Methods used to measure C give percentages, and changing percentages to volumes requires knowing the bulk density of the soil.

Bulk density changes with time and land use. Immediately after tillage, the surface may be very loose and fluffy, i.e., a lower bulk density would be expected then than at the end of a growing season. Should the time when we measure bulk density vary with land use and management?

There are also several methods for measuring bulk density, e.g., with undisturbed soil clods, or standard cores, or, in unconsolidated soils, with compliant cavity procedures. All these methods have their proponents and all have their strengths and weaknesses, so it is hard to settle on a single procedure, but we must have some kind of standardization. Do we report the results using a standard moisture content (33 kPa), or on an oven dry basis? A further difficulty is that the same method seldom is used both to develop the baseline and to determine the change.

E. Scaling the Data

Holmgem (1988) discussed the concept of "the Point Pedon." He pointed out that we sample points and that some types of analysis and property measurements can be applied to a larger area than can others. This concept needs to be thought out completely for any C sequestration scheme for, once we have data for C, how should we scale the data about that point in time and space to larger areas? In essence, we want to take a single point in time and space and scale it to larger and larger land areas (point < field < regional < national < global). Every move to a larger area introduces more uncertainty and variability.

Modelers use many procedures to develop information for large and larger areas, but we must remember that a model is no better than the data from which we develop and verify it. Modeling to scale data is very useful and will produce pretty maps and figures, but they must have a measure of error, and this measurement is hard to determine. In addition, the error varies with the scale of the model and the property modeled.

We can look at any scale system as a spider web (Figure 3). If we measure something at one point within the web or modify another point, how does this relate to some other point on the web? Is the web (field) uniform? Did we make the measurement at the correct or representative spot? We never really know, yet we base what we do on these single point measurements and then scale them up and up.

Figure 3. An example of the interconnections of the many different parts of systems is illustrated by this spider web; changes in any one part will cause changes in all the other parts.

F. Sampling

One of the greatest challenges in soil science is sampling. We can have all kinds of methods to "slice and dice" the sample to produce varieties of information about the different C pools. However, if the sample collection is not done in a way that can be repeated, that will give a representative sample, the other tests are in vain. If sampling is done incorrectly, nothing else really matters.

In sampling soils, we need to determine what we want to sample. Soil including or excluding roots? If roots are included, what size? Living or dead? How do we deal with coarse fragments? Do we sample by depth increments or by genetic soil horizon?

For a C-trading scheme, do we also need to standardize the time of the year we sample. Do we sample at the start of a growing season, the end, or some other time? How is this taken care of in areas where crops grow year-round? How do we randomize the sampling? How deep do we sample? Some crops have very shallow roots; others may have roots to several meters; what is the total C change with depth? All these questions need answers. The answers may vary for the type of crop and the climatic region but, once you start with a sampling system in a given area, you must stick with it.

IV. Conclusion

The diversity of topics covered in this meeting and presented in this volume shows that scientists indeed want to solve the many problems associated with assessing soil C pools and ensure that sound management practices are adopted and rewarded as universally as possible. We are now meeting with the modelers who use the data and the policy makers who need information on the C pools to develop national and international policies.

Part of the uncertainty in methodology is our desire to see what is on the other side of the hill. We like to modify procedures and develop new and better ones, not just settle into routines but push forward and learn. Nevertheless, policy makers must develop such programs as C trading schemes to allow payments for positive impacts on SOC and penalties for negative impacts. We must help by developing methods that can be applied broadly, to many different areas and conditions; and, although the data we use must come from many laboratories, they must be based on the same standard methods.

We also must answer policy makers' questions with the knowledge we have now. Later we may have a better answer, but at any given moment we must be able to say, "This is the best procedure now; this level of error is associated with it; this is how we can use it now." And we must make sure that policy makers, land managers, and people who will pay for the sequestration of C can also understand and use the information we provide to measure and verify SOC and SIC changes.

Policies and programs developed to enhance C sequestration must have positive environmental and societal impacts that are understandable and measurable and have appeal to the farmer/land manager and to the public. We thus must ensure not only that policy makers know and understand and can use the results that we develop, but also that the public in general sees the value of sequestering C and is willing to help pay for it.

References

Black, C.A. 1965. Methods of Soil Analysis. Part 2. Chemical and Microbiological Properties Monog. 9. V. 2.

Cheng, H.H. and J.M. Kimble. 2000. Methods of analysis for soil carbon: an overview. p. 333-339. In: R. Lal, J.M. Kimble and B.A. Stewart. *Global Climate Change and Tropical Ecosystems*. Lewis Publishers, Boca Raton, FL.

Holmgren, G.G.S. 1988. The point representation of soil. *Soil Sci. Soc. Am. J.* 52:712-716.

Jenkinson, D.S. 1988. Determination of microbial biomass C and nitrogen in soils. p. 368-386. In: J.R. Wilson (ed.), *Advances in Nitrogen Cycling in Agriculture Ecosystems*. C.A.B. International, Wallingford, U.K.

Kimble, J.M. and E.R. Levine. 1994. The Nairobi Conference: Topics, Results, and Research Needs. p. 151-162. The International Society of Soil Science and The Mexican Society of Soil Science, Volume 9.

Lal, R., J. M. Kimble, H. Eswaran and B.A. Stewart. 2000. *Global Climate Change and Pedogenic Carbonates*. Lewis Publishers, Boca Raton, FL.

Lal, R., J. Kimble and E. Levine. (eds.) 1994. *Soil Processes and Greenhouse Gas Emissions*. USDA-SCS-NSSC, Lincoln, NE.

Lal, R., J. Kimble and R. Follett (eds.). 1997. *Soil Properties and Their Management for C Sequestration*. USDA-NRCS-NSSC, Lincoln, NE.

Lal, R., J. Kimble, E. Levine and B.A. Stewart (eds.). 1995a. *Soil Management and Greenhouse Effect*. Advances in Soil Science, Lewis Publishers, Boca Raton, FL.

Lal, R., J. Kimble, E. Levine and B.A. Stewart (eds.). 1995b. *Soil and Global Change*. Advances in Soil Science, Lewis Publishers, Boca Raton, FL.

Lal, R., J. Kimble, R. Follett and B.A. Stewart (eds.). 1998a. *Soil Processes and the C Cycle*. Lewis Publishers, Boca Raton, FL.

Lal, R., J.M. Kimble, R.F. Follett and C.V. Cole. 1998. *The Potential of U.S. Cropland to Sequester C and Mitigate the Greenhouse Effect*. Ann Arbor Press, Chelsea, MI.

Lal, R., J. Kimble, R. Follett and B.A. Stewart (eds.). 1998b. *Management of C Sequestration*. Lewis Publsihers, Boca Raton, FL.

Lal, R., J.M. Kimble and B.A. Stewart (eds.). 2000. *Global Climate Change and Tropical Ecosystems*. CRC Press, Boca Raton, FL.

Walkley, A. 1935. An examination of methods for the determining organic C and nitrogen in soils. *J. Agr. Sci.* 25:598-609.

Section II

Soil Sampling and Sample Preparation

CHAPTER 2

Methodology for Sampling and Preparation for Soil Carbon Determinations

J.M. Kimble, R.B. Grossman and S.E. Samson-Liebig

I. Introduction

Errors made in sampling may result in unusable data. Proper field sampling and then handling of samples from the field and their preparation are the first steps for ensuring data of the highest quality. This needs to be considered regardless of the planned measurements to be made. Well-executed sample plans and proper sampling techniques can still result in data of little value if the samples are not properly prepared, stored, and/or recorded. This chapter covers methodology for the sampling of soils for carbon determinations. The procedures discussed may be applied to other soil analytical determinations.

The use of standardized protocols in all aspects of sampling, preparation, and analyses is very important. No simple set of rules can be given to cover every scenario, but general ones are needed and will be reviewed. If care is taken to document what protocols were followed, others will be able to understand and duplicate the procedures used. Documentation of procedures cannot be overemphasized. When developing standard protocols for sampling, the following need to be considered: (1) Type of sampling. Issues may be the sampling design and if it is to be statically sampled. (2) Site selection. This is an important aspect for all types of sampling. Good site selection is important to ensure the sample is representative of the area. In some cases, a representative site is sampled completely for characterization with satellite samples for the surface horizon. Another option would be to take a composite sample of the surface horizons from a number of points for carbon determinations. (3) Size of the sample. New analytical equipment may use very small sample weights; however, large samples are required to allow for proper representation of the material sampled.

The Intergovernmental Panel on Climate Change (IPCC) is currently completing a report requested by the UN Framework Convention on Climate Change (UF-FCCC) to respond to questions raised by the Kyoto Protocol. Article 3.3 of the protocol deals with forestry as a sink for greenhouse gases. Article 3.4 leaves the door open to methods to establish baselines of carbon stocks and modalities, and rules and guidelines as to how and which additional human-induced activities might be included to reduce greenhouse gas emissions. One avenue to reduce atmospheric CO_2 is through the process of photosynthesis. CO_2 transformed into plant carbon is sequestered either in above- or belowground biomass and/or as soil carbon. The carbon may be in the form of either soil organic carbon (SOC) or soil inorganic carbon (SIC). Questions remain as to the feasibility of the ability to measure changes in SOC and SIC and to what level of precision over defined times.

The concept of buying and selling of carbon is becoming a reality and is referred to as trading "carbon credits." In general, a contract is made to pay for sequestering carbon within or above the soil. A degree of certainty is needed that the amount of carbon contracted can be measured and verified. If X_o is the baseline carbon and the level at the beginning of the accounting period is T_o, then the change is from T_o to T_{o+x}, where x is any selected length of time. The carbon level (X_{o+x}) measured at T_{o+x} is $(X_{o+x}) - (X_o)$ and is the amount that represents the change in soil carbon over a set period of time T_o to T_{o+x} which is usually not less than 5 years.

Year-to-year variability in climatic conditions can have a major influence on carbon sequestration. Carbon sequestration needs to be measured for a time period in which the values can be normalized, and dry and/or wet years averaged out. The yearly change is not the main concern. The change over the selected time is the question. As far as carbon credits, it is hoped that there will be a gain in sequestered carbon.

The purpose of this chapter is to discuss concerns in sampling in the context of measurements of carbon to establish changes in carbon stocks.

II. Sampling Concerns

Concerns that will be addressed with respect to sampling for SOC are: (1) errors in field sampling that can result in unreliable or useless data; (2) the type of sampling; (3) the size of sample needed; (4) what should be sampled; and (5) what does the sample represent. Figure 1 represents the idea of sampling a complex system, with soils being like this in many ways. How you sample the dishes on the table will allow you to determine what is in all of them. Soils vary in many ways both with depth and spatially within the landscape. What is sampled at a given point may only represent that one point. A soil property (organic carbon, nitrogen for example) may vary within meters or less distance, depending on the soil property, and other properties may vary across larger areas. The concept of a point pedon is discussed by Holmgren (1988). It defines different sizes of area dependent on the soil properties. Soil carbon in a native site may be uniform within a given soil. However, it may vary greatly within a complex landscape that is made up of several soils occurring on different parts of the landscape. SOC gains or losses may also be influenced by factors such as soil management (tillage systems, fertilization, and types of crops grown, etc.) and wind and water erosion. These factors affecting soil carbon in cropland are discussed by Lal et al. (1998) and for grazing lands by Follett et al. (2000).

The ultimate use of the sample must be considered when sampling. Measurement of the change in the amount of carbon at point "x" over time "y," is different from sampling to determine the carbon in a specific soil type. The reason for this is that, in many cases, one may want to extrapolate the data from a point to a larger area.

Most soil taxonomic systems require sampling complete pedons for detailed characterization analyses. Many properties are analyzed when doing a complete characterization; the cost can be quite substantial. Therefore, it is important to be sure that what is sampled is representative of the field or area of interest. Before sampling a soils map of the area will need to be made. The sites to be sampled must be representative of the soils in the area. Sites should be located in areas that represent mappable and extensive soils. Special care should be taken to ensure the site is away from roads, fencerows, old living areas, and any other feature that may cause aberrant properties. The site should be located so that a pit at least 1 to 2 m across and 1 m wide can be dug. The reliability and usefulness of the data gathered could be enhanced through selection of paired sites (pedons), transect sites, and satellite sites. In addition, when sampling for carbon changes, the site must be located where it can resampled. Most of the considerations for taxonomic purposes apply to point sampling for carbon.

Figure 1. An illustration of a complex system with several properties that can be sampled in various ways to account for variability.

Soil variability has been studied and discussed for years. Cline (1944) discussed soil sampling and the points he made are still true. The concept of soil variability is discussed by Boone et al. (1999) and will not be covered in detail here. The main point to remember regarding soils is that their properties vary in both time and space. Space can be defined as both depth and location within a landscape. Soils vary with depth as a result of differences in horizon formations. Chemical and physical weathering of the soil help create the different horizons found in the soil profile. Some horizons are zones of accumulation and others of loss. Carbon in soil varies with depth; usually the concentration decreases with depth because of fewer roots. The distribution is irregular in many alluvial soils (Fluvents) because layers are composed of materials with different particle sizes. In addition, the root distribution, amount of litter returned, the microbial and macro soil fauna population have a major influence on the distribution of carbon within the soil profile.

Anthropogenic impacts may influence the carbon distribution and can change the soil variability significantly. When a soil is cultivated (disked, plowed, etc.), the surface layer is mixed and in areas of intense cultivation, is greatly homogenized. This mixing is reduced under minimum tillage or no-till. In these circumstances, the changes in carbon concentration tend to be more pronounced in the top few centimeters, at least for the short term. Different sampling protocols may be needed for no-till versus tilled fields. With different tillage practices, there are also changes in the density of the soil over time. Because of this, sampling equal amounts of soil is critical. For example, if a layer is more compact in a given soil and one samples an equal depth layer in another soil that is less compact, the percent carbon obtained may differ. However, the difference may be an artifact of the sampling because the sample for the denser soil is effectively from a greater depth, as discussed later in the section on sampling strategies.

A. Sampling Protocol

Are standard sampling protocols needed? The answer is yes. The soil variability and the agronomic practices used need to be considered. Sampling strategies will be discussed later in this chapter; however, a few points that need to go into a standard sampling protocol will be discussed here. The same sampling protocol may not necessarily be used in all places. However, the same protocol must be used to determine the baseline (T_o) as is used for $T_{(o+x)}$. If this is not done, there will be no way to compare the data obtained.

Issues that need to be addressed when sampling for soil carbon are: (1) should samples be taken with cores, by profile sampling, or taken from several points in a field and composited; (2) the time of year the samples are collected; (3) the depth to sample; (4) the number of samples to take within a map unit (this is what is included within a delineation made in soil mapping which may be a single soil or a group of similar or even dissimilar soils) or field; and (5) how will the data be used for scaling to larger areas. This list could contain many more items, but these are the immediate questions that need to be considered in any sampling protocol. These, and other aspects, are discussed by Boone et al. (1999) and will not be considered here in detail.

A question often asked is which sampling technique to use: is it a core sample, sampling a complete soil profile, or compositing a number of samples from the surface layer, as is done for fertility measurements? The approach should be based on the objectives, the type of soil to be sampled, and the initial soil variability expected. To make the determination, a good understanding of the landscape and soil properties to be sampled are necessary. Figure 2 shows a soil with horizons mostly parallel to the soil surface. In this type of soil, core sampling would present few problems. Figure 3 represents a highly turbated (extensively mixed) soil, which cannot be sampled using the core method (Kimble et al., 1992).

A major factor is the time of sampling. For example, it is difficult to determine the change if the baseline sample (X_o at T_o) is collected at a different time of year than the sample collected at the end of the period (X_{o+1} for T_{o+1}). If the first sampling was performed immediately after harvest, the residue in the soil from the previous year will be at the maximum state of breakdown. If the next sampling was conducted at the beginning of a growing season, the residues will be at their maximum input level. Therefore, a standard time of sampling needs to be established. It need not be the same for all areas, but at a given point, it must be the same.

So, what protocol is needed when considering the depth to sample? If observing short-term changes in agricultural lands (5 to 10 years), the greatest change in most crops will be in the top 20 to 30 cm. In studies on CRP sites (Follett and Kimble, unpublished data), the greatest change observed was in the near surface. They found that samples obtained from 0 to 5, 5 to 10, and 10 to 25 cm depths showed the largest change. This can differ by the tillage system used; in a 0 to 30 cm plow layer, unless completely mixed, the changes will be more evident in the top 0 to 5 cm. Changes may be detected much deeper in the profile in many cases. This may be controlled by the species being grown the soil type, etc. Under trees, the sampling depth may need to be much deeper. As a rule, it is suggested that samples be taken 0 to 5, 5 to 10, 10 to 25 cm and then by genetic soil horizon with no sample layer > 25 cm thick to a depth of 2 m. For most carbon studies, determining changes to 50 cm should be sufficient. Finally, the sampling depth needs to be determined when the first samples are taken and continued at this depth in subsequent sampling.

Figure 4 represents two different range conditions within the same map unit and both having the same soils. However, because of diverse grazing management, which results in different conditions, the carbon inputs would be expected to be quite different. The sampling protocol may need to be somewhat different for each of these conditions, because of vastly differing amounts of woody biomass and other organic inputs.

Figure 2. A soil with horizon boundaries parallel to the soil surface where core sampling would pose no problem.

Figure 3. A highly turbated soil on the North Slope of Alaska.

Figure 4. Range managed differently, which may confound sampling.

Many variables need to be considered when sampling. Some general points that must be considered are: (1) always sample a fresh exposure, i.e., not one that has been open for a long time; (2) auger or probe to assess the local variability in the area to be sampled to ensure a representative sample; (3) make sure the sample represents a uniform layer; and (4) thick horizons need to be split.

III. Pedon Sampling

The pit should be 1 by 2 m across and at least 2 m in depth, or to bedrock if the soil is less than 2 m. Supplemental auguring or excavations may be needed to assess and describe pedons larger than the pit. In laterally uniform pedons, sampling is done in a horizontal pattern 30 to 50 cm wide. Each sample should represent the entire cross section of each soil feature or soil horizon in the pedon. A general rule is that the lateral area of exposure should be 100 to 200 times the size of the feature of interest. This cannot always be met but should be strived for whenever possible.

When sampling a soil pedon, equal volumes of soil should be taken from the top to bottom of the layer being sampled, taking care not to sample too deep into the wall of the soil pit. Sample laterally to increase the volume of material rather than removing a greater depth. What is deeper in the soil pit face or wall is not described and therefore less well known. A sample area of 20 cm high by 10 cm wide and 20 cm deep has a volume of 4000 cm^3, which is commonly sufficient. However, a sample 20 cm high, 40 cm wide, and only 5 cm deep will provide a more representative sample and still provides the same volume of material.

Horizons thicker than 25 cm should be split for sampling. They may look alike in the field, but once the material is mixed, you will never be able to separate the results. As a general rule of thumb, if there is a question about the thickness, split into two sections for sampling. The more data you collect, the more you will know about the site.

Figure 5. A soil high in coarse rock fragments that must be measured in sampling so adjustments can be made to the <2 mm material.

The amount of sample will vary depending on what measurements need to be performed. However, it should not be smaller than 3 to 5 kg. If rock fragments are present, usually the 20 to 75 mm fraction is weighed; this requires perhaps as much as 50 to 75 kg. Usually volume estimates are made of the 75 to 250 mm material. Return the <20 mm to the laboratory and obtain the weight percent 2 to 20 mm. Combine the estimated volume >75 mm from records for the map unit component, the field weights 20 to 75 mm and the 2 to 20 mm from the laboratory. The volume of the total coarse fragments is used to make adjustments to a volume basis. It is best to sample a large area, place the material on a clean canvas tarp, mix it well, and then subsample this material. An improper sampling technique is to take soil material from only one place on the face of the pit.

Figure 5 shows a profile that is high in coarse fragments. The volume of coarse fragments must be determined so that the analytical results can be adjusted to reflect the non-soil material. When this is done, a general rule of thumb is to take a sample at least 100 times as large as the sample to be collected. Figure 6 shows the volume of rocks that need to be weighted to obtain a 2-kg soil sample.

The sample should contain a sampling of all features of the horizon in question in proportion to their areal percentage. If there are special features of interest, a subsample should be taken. For example, a horizon that is 50% red and 50% yellow would be sampled by taking a mixture of both colors and then two subsamples, one of each color. Or, if the horizons of the pedon are discontinuous

Figure 6. Coarse fragments are removed and measured to adjust for the <2-mm material.

or vary greatly in thickness or degree of expression, samples that represent the dominate characteristic are taken along with subsamples to characterize major differences.

Sampling generally starts at the bottom of the soil pit. However, each person has a different preference in sampling. Some people start from the top and work down while others work from the bottom up. It is usually easier to sample from the bottom up since you would not need to excavate as you sample nor would you contaminate the lower samples with materials from above. If the surface is very soft and loose, the top may need to be sampled first.

IV. Sample Handling

The handling of the samples once they are collected is very important. Each sample needs to be placed in a separate bag that is clearly labeled. Plastic bags are preferable because samples are less easily contaminated and field-state moisture is maintained. Cloth bags allow the loss of moisture and the movement of dust between the samples resulting in contamination. Each bag should be labeled with the depth sampled, horizon number, identification and date sampled. Tags should be attached to the outside of the bag unless they are made of materials that will not break down when moist. A consecutive number on each bag makes sorting the samples at the laboratory a much easier task. Samples that are collected for edaphic properties (for example, organic carbon) may not represent all horizons and/or depths represented in taxonomic sampling. Performing the analyses on the edaphic samples that would be made on the taxonomic samples increases the opportunity to place the edaphic samples in a taxonomic framework.

V. Type of Sampling for Carbon

The type of soil sampling needs to be considered before starting. One sample is not enough per field. Different soils within the field where the experiments are located need to be sampled individually. The number of samples will vary depending on the size of the field (several hectares versus one or two hectares will need more samples). If the same type of soil with different surface textures occurs in the same field, each texture should be sampled as if it were a different soil. A maximum area would be in the order of 10 ha.

When taking cores for a composite sample, one can sample 0 to 25 cm and 25 to 50 cm, resulting in two samples for each field with each being made up of 10 to 20 cores as discussed earlier. These depths may vary. When taking cores, it is important to understand the past soil management practices. Samples should not be taken in areas where more fertilizer was used because greater plant growth and more biomass may have been returned giving possible erroneous results.

The question arises frequently when sampling for edaphic properties regarding the depth samples should be taken. As stated previously, the general rule is to sample to 50 cm. However, deep sampling is also valuable as many crops root deeper than 50 cm. The greater the variability, the more sampling is needed when the samples are composited and an average sample used for analysis.

The depth of sampling should be dependent on the plants, the expected rooting depth, and depth of available water during the growing season. There is little need to sample below the expected wetting front in a dry area since roots will not generally extend below the wetting front. It is difficult to apply the same rule to all crops and all fields. Each farmer and/or researcher will need to take into consideration all variables and make a decision.

VI. Carbon Analysis

For a discussion on the different types of soil carbon analysis and the potential problems, see Cheng and Kimble (2000), Cheng and Kimble (Chapter 9, this volume), and Robertson et al. (1999), which describes standard methods for LTER Sites. Soil carbon sequestration measurements are similar to the measurements needed for the LTER sites.

VII. Site and Pedon Descriptions

A pedon/site description should be completed. Figure 7 shows the recording of site and profile data during sampling. Descriptions should include: what was sampled, where the sample was taken, and how the sample was taken. Some simple rules as to what is needed are: (1) enough detail to create a mental picture of what was sampled; (2) enough information to classify the soil sampled (so that the data can be scaled to larger areas using soil maps); and (3) enough information to build a database for future use. This is all part of monitoring and verification that will be needed for carbon credits. Locations can be determined by the use of a GPS so that one can return later within a matter of a few meters of the selected site. If a GPS unit is unavailable, a metal object can be buried and the exact site found with a magnetometer. Keep in mind that you would not want to sample the exact same point, as it will be changed by the prior sampling. A method to follow may be to sample at a spacing of 1 m from a fixed point sampling each time at a different compass bearing.

Figure 7. Site information is needed and must be collected when samples are collected.

VIII. How Many Samples to Collect

The numbers of samples required depends on what is being sampled. For field sampling, composites by the core method is attractive. This is the procedure used for fertility sampling. If using control points in conjunction with a model and/or experimental studies for model development and validation, then samples need to be taken at specific points on the landscape. When sampling a large area, which in carbon trading will be the case, a determination needs to be made of how to sample a soil map unit component. The complexity of the map unit will need to be considered. For example, should all inclusions (soils found within the map unit but not named) or just the major soils from the map unit name be sampled? This will vary from field to field, but needs to be part of the sampling protocol.

Figure 8 is a typical landscape located in the U.S. grain belt. Figure 9 shows the different parts of such a landscape. It is to be expected that each part of the landscape will have different carbon levels and each will need to be sampled separately if we are going to determine the total carbon in the area. If only the rate of change within a given area is of interest then not all parts of the map unit will need to be sampled but only selected ones to determine the rate of change. The sampling can vary if only specific fields are included for carbon measurements, but in many cases, many different areas will be combined to reach a level of carbon for emission trading.

IX. Sampling Strategies

The following section covers in more detail a technique used in carbon sampling, called Time Separated Pairs. In this sampling, one returns to the same site after a set period. To accomplish this

Figure 8. Typical Midwest landscape showing different soil patterns.

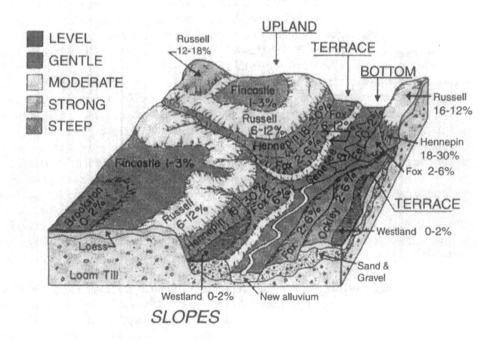

Figure 9. Schematic of a soil landscape in which each part would be expected to have different carbon levels.

you must know the location of the site and the date it was originally sampled. Locations can be obtained using a GPS or buried metal object such as a cow magnet to find your control point. Specific considerations (Figure 1) in sampling are:

Interfield:
1. Must know the map unit component to be sampled;
2. Use generalization, i.e., years of no-till.

Intrafield
1. Treatment covers to a field (i.e., fertilization, tillage, etc.);
2. Was application to the whole field to be characterized or were treatments only to parts of a field.

You must combine "Interfield" and "Intrafield" concerns so that their interactions or lack of interactions can be used in sample site selection as well as the actual sampling.

The following is a detailed sampling procedure for consideration for sampling for carbon determinations. The purpose is to present ideas for sampling for the assessment of carbon sequestration to relatively shallow depths. First, two general remarks:

1. There seem to be three possible bases for payments for carbon sequestration:
   a. Comparison of measured values for a length of time (5 years?);
   b. Comparison of measured values to average expected values to establish carbon credits. Credit rate would increase relative to an expected or normal level. This would permit credit for prior good management. Without some such provision, good managers with relatively high initial carbon levels would be penalized. Note that here the credit is less for future sequestration, and more for maintenance of existing levels.
   Even in well-managed fields, there is an expectation that more carbon can be added. It is estimated that up to 50% of the original carbon (Lal et al., 1998) has been lost from many of our cultivated soils. In addition, with a few years of good management, it is not expected that these soils will not have reached their new equilibrium. Moreover, with better nutrient management and higher biomass returns, they may actually exceed the native levels. The main idea is to reward good managers for their past practices and also for future gains;
   c. Award for an assumed increase in soil organic carbon based on the management practice and measured initial soil carbon levels. It is assumed that the effect of a practice is related to the initial soil carbon. This could be based on rates of increase using a model such as Century (Paustian et al., 1997). The amount of biomass returned could be based on a grain to biomass ratio. With the advent of yield monitors, the use of this type of modeling should increase.

2. Many standard procedures for organic carbon report to 0.1 percent absolute. If a surface layer is assumed to be 15 cm thick with a bulk density of 1.30 g cc^{-1}, then for 1 m^2, the weight is about 200,000 g. Therefore, 0.1% absolute of this 200,000 g is 200 g. A sequestration rate of 20 g m^2 yr^{-1} is reasonable (Lal et al., 1998). Over 5 years, 100 g would accumulate. Strictly speaking, 100 g is not detectable if measurements are reported to 0.1% absolute. However, new analytical equipment measures soil carbon to 0.01% absolute, which would be 20 g, and that may well be an easily obtainable increase.

A map unit-use (MUU) concept is evaluated. Example: Miami silt loam, 5 to 9% slopes, eroded, no-till > 5 years. Additionally intrafield (IFD) differences in use may be recognized. An example would be a difference in fertilizer application over several years previous that resulted in more grass growth in some areas (Step 1 in Figure 10). Any difference that may be considered relative to carbon sequestration that is not part of the use concept assigned to a map unit may be recognized (Step 2). Both MUU and IFD separations may extend across geographic separated fields or management areas. MUU and IFD would be combined (MUU/IFD) and be the identifier for an area of land (Step 3). Up to 15% of the area may be excluded. This would remove small MUU/IFD areas.

Transects are established in the direction that divides the MUU/IFD into approximately two equal parts (Step 4). Points for organic carbon sampling are placed along the transects. One point at least is established for every 5 to 10 ha (the density of the points can vary depending on the uniformity/complexity of the soils to be sampled) in the MUU/IFD area. Note that this is total area, which may be separate parcels of land. At least one point is established in each MUU/IFD:MUU/IFD that in aggregate are $\leq$ 15% of the area may be excluded. This point should be at least 15 m from field boundaries. Figure 10 shows these relationships and their interactions. The use of soil survey information to determine areal estimates of organic carbon is described by Grossman et al. (1992).

Location of the points can be established with a GPS to within a circle of error of about 10 m. Cow magnets could be buried to get exact locations of prior sampling. The initial sample would be taken to the base of the plow layer, but not less that 15 cm or 20 cm if no Ap horizon is present. The subsequent sample would be taken to the depth so that weight per unit area is the same as for the first (this will be discussed in detail later).

Three samples a meter apart (sub-point) would be obtained. The samples would be taken at points where the bulk density with depth is very similar. Commonly, only a single sample would be taken for each of the three sub-points. Samples would be collected in such a manner that the proportion of the whole sample contributed by each layer equals the proportion that the layer is of the total thickness. For one of the three samples, the weight (one would composite and compute the weight per unit area) per unit area (g cm^{-2}) is determined.

The sample depths for the initial and subsequent sampling should be such that the weight of soil removed per unit area is very similar. Thickness of the zone to be sampled is inversely related to the depth-weighted bulk density. The sub-point samples could be composited or analyzed separately. Bulk density measurements are unnecessary for the initial sampling if the weight of soil removed from various zones is directly proportional to its thickness. This could be achieved most easily by removal of a core of soil over the full depth sampled. The weight per unit area is calculated for the initial sample. A second core is taken after the period of carbon accumulation to the depth that the weight per unit area is equal to that for the initial core.

For the subsequent sampling, one approach requires measurement or estimation of the bulk density of the constituent layers and then calculation of the adjusted length of the bottom layer (Lx) to make the weight per unit area (M/A) the same as the initial sampling. Commonly M/A would be in g cm^{-2}.

This is the calculation: $M/A = L_1\rho_1 + L_2\rho_2 + L_3\rho_3 + L_x\rho_{x...}$

Where M/A is the total weight per unit area for the initial sample L_1 to L_x are the thickness of the contributing sublayers for the subsequent sampling, and ρ_1 to ρ_x are the respective bulk densities. The subscript x indicates the bottom zone.

The equation can be rearranged to: $L_x = \dfrac{M/A - (L_1\rho_1 + L_2\rho_2 + L_3\rho_3)...}{\rho_x}$

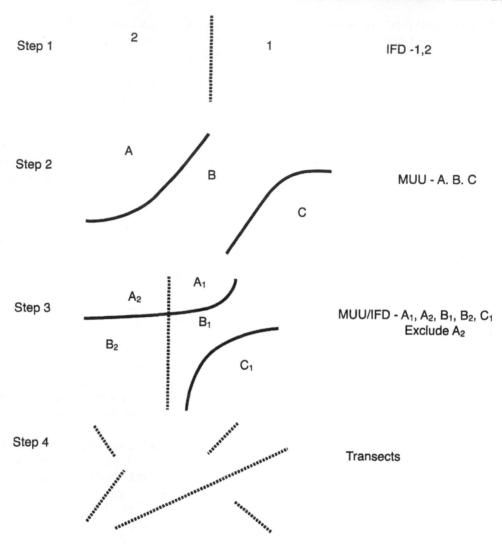

Step 1   2   1   IFD -1,2

Step 2   A   MUU - A. B. C
                            B
                                      C

Step 3   A₂   A₁   MUU/IFD - A₁, A₂, B₁, B₂, C₁
                            B₁   Exclude A₂
        B₂
                    C₁

Step 4   Transects

Figure 10. The concepts that would be sampled for agricultural fields for sequential organic carbon evaluation; MUU is Map Unit-Use and IFD is Intrafield Difference.

Note that the adjustment may be positive or negative.

We have had experience with several bulk density procedures. The clod method, 4A1 in Soil Survey Staff (1996), is very robust but it is not suitable for fragile layers or layers high in larger rock fragments. We also use a partially filled core method where the volume not filled is subtracted from the total as well as several methods dependent on the change in volume on excavation. Details are found in Grossman et al. (Chapter 7, this volume).

Constant mass sampling may be achieved by determination of the bulk density of the sublayers through a depth that would yield the same M/A as the initial sample, and therefrom calculate the depth increment L_x to sample. Alternatively, one could during the second sampling remove segments, measure the M/A separately, and combine to give an M/A for the original sampling. Heinonea (1960) has designed a slotted core sampler that may have promise for constant M/A sampling. Another alternative is to sample by core to more than one depth and select the core that gives an M/A closest to the initial.

Some general comments to consider in this type of sampling just described are as follows.

There may be merit in wetting the soil as follows. A ring 30 cm in diameter is mounted on the soil surface to a 5-cm depth and sufficient water is placed in it to bring the soil to field capacity to 20 to 30 cm. The amount of water to add is about 3 to 12 cm. Note that a 1-gal jug contains enough water to wet 5 cm of soil when the 30-cm ring is used. The next day, if the near surface is loose from cultivation, standard impact energy is applied to increase the density appreciably in order to make it easier to measure bulk density. The wetting usually would facilitate all of the measurements mentioned.

A question that remains is if it is reasonable to detect the small differences expected over a 5-year period. Izaurralde et al. (1998) found that 108 samples would need to be collected twice for a statistically significant increase at the 0.1% absolute level. This raises the question of how many samples are needed if the level measured is reduced to 0.01%. Further, would there be the same variability in pairs of samples taken within 1 m of each other, as described compared to random sampling.

Strategies may be considered to make the sampling and analysis effort more concentrated. One option is to rent the use of parts of fields that are representative of large areas. The initial soil OC levels would be specified as within the expected range. The land manager would be required to keep a certain use over 5 years and permit sampling, at the beginning and at the end of the sequestration period. The sampling intensity would be raised at least an order of magnitude. The MUU/IFD would have to be applicable for a considerable area.

X. Bulk Density

Bulk density measurements are necessary when collecting data for carbon sequestration if the data are to be expressed on a volume basis. For discussions of bulk density, see Lal and Kimble and Grossman et al. (Chapters 3 and 7, respectively, this volume). Weight percent of carbon must be converted to a volume basis using the calculated bulk density. This is required for the trading of "Carbon Credits" and for the overall determination of the change in the soil carbon stock. Many concerns discussed for soil carbon sampling also apply to soil bulk density. In addition, rapid changes can be expected in bulk density; this is known as a temporal property of the soil and does change over the seasons, thus emphasizing the importance of sampling both the baseline and the end of sequestration at the same time of the year. The bulk density can be affected by traffic, i.e., the creation of compacted layers. This traffic can be caused by tillage and harvesting or by grazing of livestock. The management practices over a given time period need to be recorded. This is even more important when the results are modeled. Note bulk density is not needed for different layers if constant mass sampling is used.

XI. Conclusion

A soil is a three-dimensional body that varies in time and space. This is important with regard to sampling methodology. The laboratory data generated is only as accurate as the sample collected. A sample not representative of the area it was collected from will have laboratory data of little value regardless of precision. Documentation as to how and what was sampled is important in order to return in 5 or 10 years and sample. Finally, in addition, all steps used must be able to be duplicated by others, which requires that a defined protocol that should be followed.

References

Boone, R.D., D.F. Grigal, P. Sollins, R.J. Ahrens and D.E. Armstrong. 1999. Soil Sampling, Preparation, Archiving, and Quality Control. p. 3-28. In: G.R. Robertson, D.C. Colman, C.S. Bledsoe and P. Sollins (eds.), *Standard Soil Methods for Long-Term Ecological Research.* Oxford University Press, New York.

Cheng, H. H. and J. M. Kimble. 2000a. Methods of Analysis for Soil Carbon: An Overview. In: R. Lal, J.M. Kimble and B.A. Stewart (eds.), *Global Climate Change and Tropical Ecosystems.* Lewis Publishers, Boca Raton, FL.

Cheng, H.H. and J.M. Kimble. 2000b. Characterization of Soil Organic Carbon Pools. In: R. Lal, J.M. Kimble, R. Follett and B.A. Stewart (eds.), *Assessment Methods for Soil Carbon.* Lewis Publishers, Boca Raton, FL.

Cline, M.G. 1944. Principles of soil sampling. *Soil Sci.* 58:275-288.

Follett, R.F., J.M. Kimble and R. Lal. 2000. The potential of U.S. grazing lands to sequester soil carbon. p. 401-403. In: R.F. Follett, J.M. Kimble and R. Lal (eds.), *The Potential of U.S. Grazing Lands to Sequester Carbon and Mitigate the Greenhouse Effect.* Lewis Publishers, Boca Raton, FL. 442 pp.

Grossman, R.B., E.C. Benham, J.R. Fortner, S.W. Waltman, J.M. Kimble and C.E. Branham. 1992. A demonstration of the use of soil survey information to obtain aerial estimates of organic carbon. ASPRS/ACSM/RX92. Tech. Papers, Vol. 4. p. 457-465. In: Am. Soc. Photogrammetry and Remote Sensing and Am. Cong. Surveying and Mapping, Bethesda, MD.

Heinonen, R. 1960. A soil core sampler with provisions for cutting successive layers. *Maatalou-stieteellinen Aikakauskirja* 32:176-178.

Holmgren, G.G.S. 1988. The Point Representation of Soil. *Soil Sci. Soc. Am. J.* 52:712-716.

Kimble, J.M., C. Tarnocai, C.L. Ping, R. Ahrens, C.A.S. Smith, J. Moore and W. Lynn. 1993. Determination of the amount of carbon in highly cryoturbated soils. Russian Academy of Sciences. Joint Russian-American Seminar on Cryopedology and Global Change November 15-16, 1992, Pushchino. Post-Seminar Proceedings.

Paustian, K., E.T. Elliott, and K. Killian. 1997. Modeling Soil Carbon in Relation to Management and Climate Change in Some Agroecosystems in Central North America. In: R. Lal, J.M. Kimble, R. Follett and B.A. Stewart (eds.), *Soil Process and the Carbon Cycle.* Lewis Publishers, Boca Raton, FL.

Soil Survey Staff. 1996. Soil Survey Laboratory Methods Manual. SSIR No. 42. Version 3.0. USDA-NRCS-NSSC, Lincoln, NE. 693 pp.

CHAPTER 3

Importance of Soil Bulk Density and Methods of Its Importance

R. Lal and J.M. Kimble

I. Introduction

Soil bulk density (ρ_b), the weight per unit volume, is an important soil physical property that influences biomass productivity and environment quality. The biomass productivity effects of soil bulk density influence root growth and proliferation, which influence uptake of plant nutrients and water. The environmental effects of soil bulk density are due to its effects on aeration, soil-water regime, runoff and erosion. Anaerobiosis can lead to emissions of CH_4 and N_2O, whereas accelerated runoff and erosion can cause pollution of natural waters. Runoff and erosion are the principal causes of non-point source pollution. Soil bulk density is influenced by natural and anthropogenic factors (Figure 1). Important among natural factors are soil properties, particle size distribution, degree of aggregation, soil organic carbon content, cation exchange capacity (CEC), exchangeable cations, flora and fauna including soil biotic activity, and climatic factors. The clay content and nature of clay strongly interact with soil organic carbon (SOC) and other cations to form aggregates. Degree of aggregation and stability of aggregates strongly influence soil bulk density. High activity clays (HACs, 2:1 minerals) tend to have higher bulk densities than soils low in clay and SOC contents. Low activity clays (LACs, 1:1 minerals) tend to have lower bulk densities and behave like low clay soils. Soils dominated by amorphous materials (e.g., Andisols) have very low bulk densities, as do organic soils (e.g., Histosols, Mollisols).

Soil bulk density is a dynamic property with wide temporal variations that are more pronounced for the surface than sub-surface horizons. The magnitude of temporal variations in ρ_b of the surface layer depends on the changes in soil moisture and temperature regimes, and is governed by the amount and nature of clay minerals, and the SOC content. For soils containing large contents of HACs, ρ_b of the surface layer can change with moisture and temperature regimes. Repeated cycles of freezing and thawing, and wetting and drying can also alter ρ_b, soil aggregation, porosity and pore size distribution.

There are at least five anthropogenic factors that affect both magnitude and temporal variations in ρ_b (Figure 1). Land use is an important factor because it determines the ground cover and the degree and frequency of soil disturbance. In general, ρ_b is in the order arable > pastoral > silvicultural > natural land use. In many cultivated fields, ρ_b varies with season. The ρ_b is usually low immediately after tillage. In some cases, the ρ_b of the plowed layer soon after tillage may be lower than that of the natural/undisturbed soil. Tillage can loosen the compacted soil and make it extremely friable and loose. Yet, the bulk density increases at the harvest time due to compaction caused by vehicular traffic and climate. If the stocking rate is high, ρ_b in pastoral system may exceed that in the

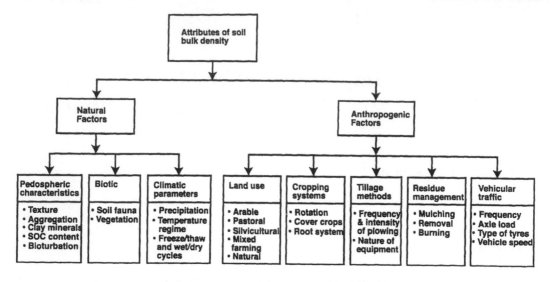

Figure 1. Natural and anthropogenic factors affecting soil bulk density.

arable land use. Tillage methods, residue management, and cropping systems affect ρ_b within an arable land use through alterations of ground cover, incorporation of biomass in soil, and nature of the root system. Vehicular traffic is another important factor that influences ρ_b because of some compaction. In some cases, ρ_b under natural ecosystems is high because of soil properties and ecoregional factors (Barber et al., 1989).

The objective of this chapter is to highlight the importance of ρ_b, indicate the importance of natural and anthropogenic factors, and highlight the relative merits and limitations of most commonly used methods. For details of different techniques, readers are referred to other comprehensive reviews (Culley, 1993; Blake and Hartge, 1986; Erbach, 1987).

II. Edaphologic and Ecologic Impacts of Soil Bulk Density

Both magnitude and the degree of temporal variations in ρ_b play an important role in plant growth and other edaphological effects, and in moderating the soil environment (Figure 2). Edaphological effects of ρ_b are well documented and related to root growth and proliferation in the sub-soil. There is a critical range of ρ_b above which root growth is drastically curtailed. This critical range varies among soils (texture) and crops (tap root vs. fibrous root system). Both water and nutrient use efficiencies are influenced by root growth and proliferation. Soil densification can influence economic yield by 10 to 20% (Soane and Van Ouerkerk, 1994).

Environmental effects of ρ_b are manifested in at least three ways (Figure 2): (i) alterations in water balance influencing infiltration, runoff and soil erosion; (ii) changes in transport (and diffusion) of plant nutrients and chemicals into groundwater; and (iii) influence on soil aeration including gaseous exchange leading to exchange of gases with the atmosphere, anaerobiosis with effects on mineralization of SOC content leading to emissions of CO_2, methanogenesis leading to emission of CH_4, and denitrification leading to emission of N_2O and NO_x.

Both edaphological and environmental effects of ρ_b are the basis of choosing appropriate strategies of soil surface management for optimizing crop growth and minimizing adverse effects on the envir-

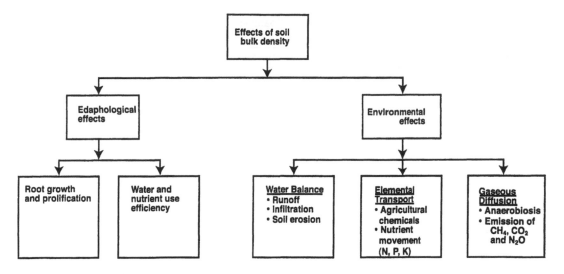

Figure 2. Edaphological and environmental effects of soil bulk density.

onments. Techniques of ρ_b management include tillage methods, crop residue management, and crops with appropriate root system (e.g., tap root) for loosening the sub-soil (Hulugalle and Lal, 1986). Soil fauna also plays an important role in influencing ρ_b. Bioturbation through activity of earthworms, termites and other soil fauna improves soil structure and lowers ρ_b (Lavelle, 1983; Lavelle et al., 1992; Lal, 1988; 1991).

III. Pedospheric Characteristics and Soil Bulk Density

Soil forming factors have a strong influence on ρ_b. Under natural conditions ρ_b may range from 0.1 to 0.5 Mg m^{-3} for organic soils, to 0.4 to 0.9 Mg/m^3 for volcanic soils and 0.8 to 1.6 Mg m^{-3} for non-volcanic mineral soils. The data in Table 1 for a Histosol show a low ρ_b of 0.13 Mg m^{-3} to 0.36 Mg m^{-3} for the organic horizon with SOC content ranging from 16% to 40%. The ρ_b of the mineral horizon, below the 60-cm depth, increases abruptly from 0.36 Mg m^{-3} to 1.61 Mg m^{-3}. However, with a high SOC content of > 50%, it is often difficult to estimate ρ_b of the surface horizon (Table 2). Data in Tables 3 and 4 show low ρ_b of 0.4 Mg m^{-3} to 1.0 Mg m^{-3} for Andisols from Japan (Table 3) and Hawaii (Table 4).

The ρ_b profile of a permafrost (Cryosol) soil from Alaska is shown in Table 5. Because of high SOC content, ρ_b is generally low and often < 0.5 Mg m^{-3}. Soil horizons being frozen, however, it is difficult to obtain reliable measurements of ρ_b below the permafrost layer. Hence, variability in the measurement of ρ_b is often high.

Spatial variation in ρ_b with depth of several mineral soils from diverse ecoregions are shown in Figures 3 to 6. The data in Figure 3 show a range of ρ_b from 0.96 Mg m^{-3} to 1.15 Mg m^{-3} for a Mollisol from Wyoming. The surface horizons (0 to 8 cm and 8 to 18 cm depths) have a higher SOC contents and lower ρ_b than the deeper depths. However, not all Mollisols have a low ρ_b. The data in Figure 4 for a Mollisol from Oregon show that even surface horizons (0 to 15 cm and 15 to 23 cm depths), with higher concentration of SOC can have a high ρ_b of 1.46 Mg m^{-3} to 1.49 Mg m^{-3}. Soils

Table 1. Soil bulk density profile of a Histosol (loamy-mixed, eutic, mesic, terric sulfihemist) from Dorchester County, Maryland

Depth (cm)	Bulk density (Mg m^{-3})	SOC content (%)
0–18	0.19	28.4
18–36	0.13	35.4
36–53	0.24	39.9
53–61	0.36	15.7
61–81	1.61	1.4
81–122	1.62	0.5
122–152	1.64	0.2
152-165	1.65	0.2

Table 2. Soil bulk density profile of a Histosol (fine, mixed, acid, isomesic, histic, placaquept) from Hawaii County, Hawaii

Depth (cm)	Bulk density (Mg m^{-3})	SOC content (%)
0–7	Not determined	54.6
7–15	Not determined	55.0
15–29	0.31	57.0
29–36	0.25	52.6
36–54	0.35	28.8
54–63	Not determined	24.4

Table 3. Soil bulk density profile of an Andosol (mixed, mesic type dystrandep) from Memuro town, Japan

Depth (cm)	Bulk density (Mg m^{-3})
0–23	0.53
23–35	0.42
35–49	0.75

of arid regions, Aridisols, are often characterized by high ρ_b and low SOC content. The data in Figure 5 show a ρ_b range of 1.47 Mg m^{-3} to 1.71 Mg m^{-3} for SOC content range of 0.35% to 0.14%. The data in Figure 6 for an Aridisol from Nevada show high ρ_b of 1.60 to 1.70 Mg m^{-3} for SOC content of < 0.1%.

In general, mineral surface horizons in humid and sub-humid regions have low ρ_b under the native vegetation cover (Moormann et al., 1975). However, deforestation and conversion of these soils to an arable land use leads to densification with a corresponding increase in ρ_b (Lal and Cummings, 1979; Hulugalle et al., 1984).

Table 4. Soil bulk density profile of a Typic Eutrandopt from Hawaii

Depth (cm)	Bulk density (Mg m^{-3})
0–10	0.97
10–25	0.81
25–41	0.84
41–63	1.09

Table 5. Bulk density profile of an organic soil from Alaska

Layer	Bulk density (Mg m^{-3})	SOC content (%)
1	0.21	56.8
2	1.08	31.5
3	0.48	41.6
4	0.31	31.1
5	1.44	46.3

IV. Importance of Soil Bulk Density

Knowledge of ρ_b is relevant to understanding several pedospheric processes of edaphologic and ecologic importance. Assessment of soil's water and nutrient holding capacity, and changes in water and elemental pools in natural and managed ecosystems require the knowledge of ρ_b and its spatial and temporal variations. The importance of ρ_b in estimating water and SOC pool of the 0 to 30 cm layer of two soils is explained by the data in Table 6. On % basis, soil A contains more SOC and has a higher AWC than soil B. However, considering the mass of soil within a 30-cm depth, which is 50% more in soil B than soil A, soil B contains 20% more SOC and available water. The data in Table 7 show that even a slight change in ρ_b, due to natural and anthropogenic factors, can have a strong impact on pools of water and SOC. Therefore, pools of water and elements and their fluxes cannot be accurately determined without the knowledge of ρ_b. In addition to understanding its impact on plant root growth and nutrient and water uptake, knowledge of ρ_b is also needed to study pools and fluxes of water, C, and plant nutrients. Studies of biogeochemical cycles, in both natural and managed ecosystems and their changes due to anthropogenic activities, cannot be done without reliable information on ρ_b and its temporal and spatial variations.

V. Measurement of Soil Bulk Density

Measurement of ρ_b involves assessment of two parameters: mass of soil and total volume of soil (Equation 1).

$$\rho_b = M_s/V_t \qquad\qquad\qquad\qquad\qquad\qquad\qquad \text{(Eq. 1)}$$

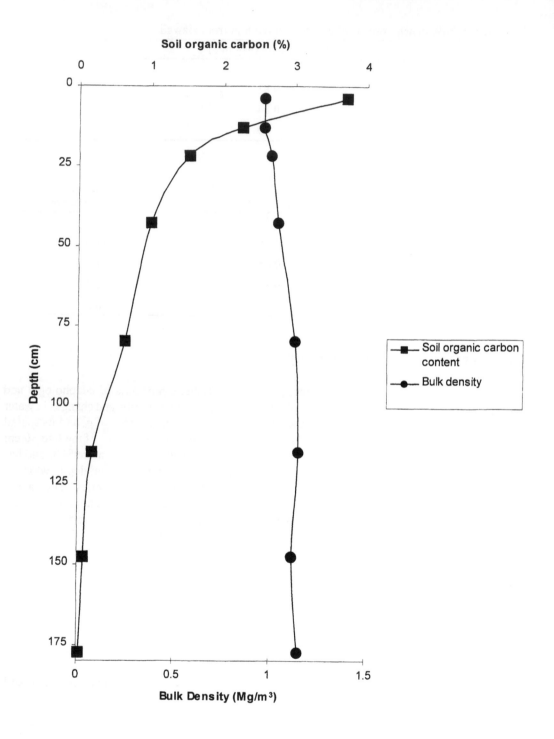

Figure 3. Soil bulk density of a Mollisol in Laramie County, Wyoming.

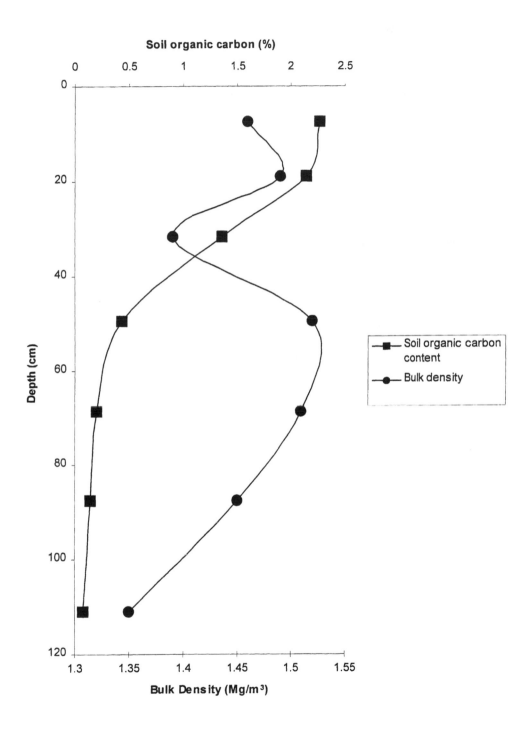

Figure 4. Soil bulk density of a Mollisol from Benton County, Oregon.

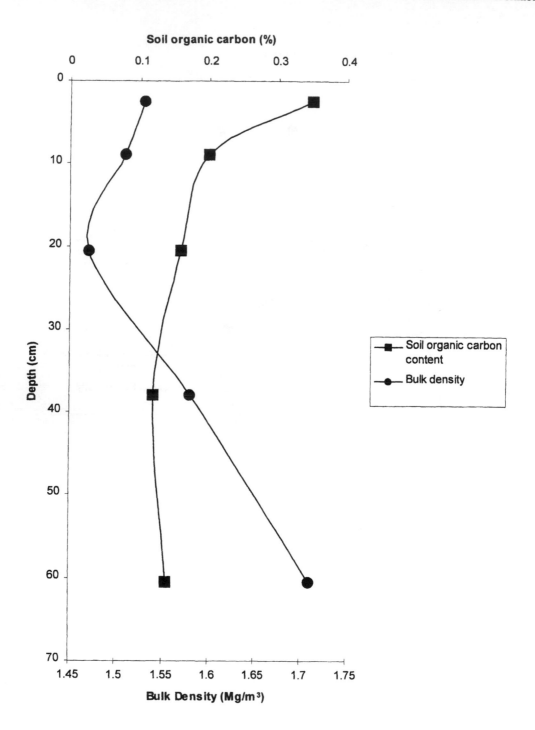

Figure 5. Soil bulk density of an Aridisol from Nye County, Nevada.

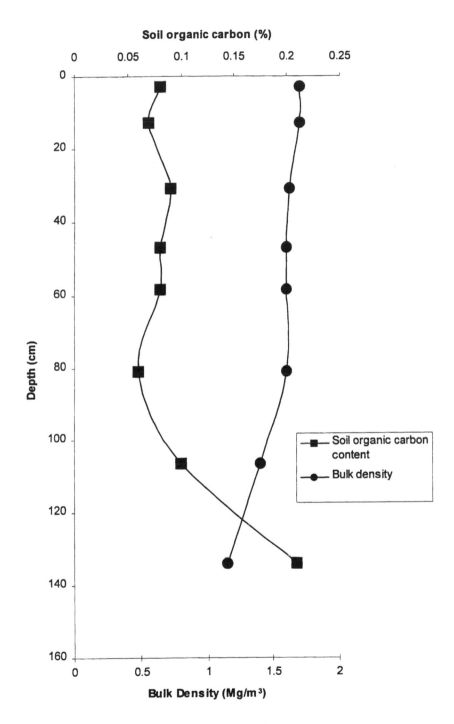

Figure 6. Soil bulk density of an Aridisol from Lyon County, Nevada.

Table 6. Impact of soil bulk density on SOC pool and water holding capacity of 0 to 30 cm depth for two hypothetical soils

Parameter	Soil A	Soil B
Bulk density (Mg m^{-3})	1.0	1.5
SOC content (%)	2.0	1.6
SOC pool (Mg ha^{-1})	60	72
Available water capacity (%)	10	8
Available water capacity (cm/30 cm)	3.0	3.6

Table 7. Effects of slight change in ρ_b on H$_2$O and SOC pool in a hectare-meter of a hypothetical soil containing 20% moisture content by weight and 2% SOC content

ρ_b (Mg m^{-3})	Water pool (kg m^{-3})	SOC pool (Mg ha^{-1})
1.100	220	220
1.105	221	221
1.110	222	222
1.115	223	223
1.120	224	224

Where ρ_b is in Mg/m^3, M$_s$ is mass of oven dry soil (Mg) and V$_t$ is total volume of soil in (m^3). Dry ρ_b can also be estimated from wet soil bulk density (ρ'_b) by making corrections for the gravimetric water fraction (w) as shown in Equation 2.

$$\rho_b = \rho'_b/(1+w)$$ (Eq. 2)

In theory, therefore, measurement of ρ_b is apparently simple and involves assessment of weight and volume of soil expressed on oven dry basis. While gravimetric measurement of soil weight is a routine procedure, accurate determination of soil volume remains a major challenge. Direct measurement of ρ_b involves a range of techniques to measure the weight of a known volume of soil. These direct methods include core method, clod method, rubber balloon method, sand replacement method, and excavation method. The most commonly used among these are the core and clod methods. The core method involves measuring the weight of a known volume of soil sampled by the cylindrical device of 75 to 100 mm diameter (Brasher et al., 1966). In contrast, the clod method involves measuring the volume and weight of a natural clod. Assessing the volume of a clod involves coating with an impermeable material such as saran. Limitations, precautions, and proposed modifications of the core and clod methods are described in Table 8. Pros and cons of direct methods are described in detail by Campbell and Henshall (1990) and Culley (1993). Sources of variability in ρ_b due to the core method are described by Raper and Erbach (1987).

There are also numerous indirect methods of measuring ρ_b. Important among these are those involving soil's effects on radiation. These techniques based on soil's interaction with γ-radiation are described in detail by Campbell and Henshall (1990), and are based on two different interactions:
  (a) Attenuation or transmission of γ-radiation through the soil, and
  (b) Scattering of γ-radiation by the soil.

Table 8. Limitations and proposed modifications of the core and clod methods of measuring ρ_b

Method	Limitations		Modifications
1. Core	(i) Compaction during sampling	(i)	Thin wall cylinders
		(ii)	Lubrication
		(iii)	Core diameter ≥ 75 mm
		(iv)	Cylinder height $\leq$ diameter
		(v)	Two nested/concentric cylinders
	(ii) Profile samples for different depths	(i)	Hydraulically driven probes
		(ii)	Hand-driven samplers
2. Clod	(i) Non-representative samples	(i)	Sample naturally occurring clods
		(ii)	Clods on or near the surface are unnatural
	(ii) Penetration of plastic lacquer into the voids	(i)	Spraying water on the clod and immediately dipping in the plastic lacquer
		(ii)	Use more viscous plastic lacquers
		(iii)	If bubble appears, the seal is poor; use another sealant (saran)
	(iii) High variability	(i)	Use larger clods
		(ii)	Weigh up to 1 mg precision
		(iii)	Use a large number of sub-samples
	(iv) High experimental error	(i)	Weigh separately wire and the mesh
		(ii)	Make correction for the weight and volume of plastic lacquer
		(iii)	Make corrections for rock fragments

Design of scattering and transmission gages are different. In back-scatter gages, γ-ray source and detector are fixed relative to each other and shielded from each other in an assembly designed to prevent measurement of directly transmitted photons. The assembly either rests on the soil surface or is lowered into an access hole. The relation between count rate (ratio) and ρ_b is complex and soil-specific calibration is essential (Lal, 1974).

In transmission gages, the sample to be evaluated for ρ_b is located between the source and the detector. The detected photons observe Beer's Law (Equation 3).

$$I = I_o e^{\mu \rho'_b x} \tag{Eq. 3}$$

Where I is transmitted radiation, I_o is the incidence radiation, μ is the mass absorption coefficient, ρ'_b is wet soil bulk density, and x is soil thickness. Some merits and limitations of radiation gages are outlined in Table 9. A principal merit of these techniques is non-destructive sampling, ability to make repeated measurements overtime for the same site, and large soil volume in use for making the measurements. These devices are, however, expensive and also pose health hazards (Table 9).

Table 9. Merits and constraints of radiation techniques for measuring bulk density

Merits	Limitations
Non-destructive or *in situ* measurement	Soil chemical composition affects μ
Repeated measurements over time	Gravel concentration may affect μ
Saving in labor and time	Requires independent measurement of soil wetness
	Soil or site specific calibration is needed
	The technique measures ρ_b of bulk soil comprising large volume
	The device is expensive, both initial investment and maintenance
	There is some degree of health hazard

VI. Problem Soils for Measuring Soil Bulk Density

While ρ_b is an important parameter, reliable measurement remains a challenge for several soils including the following:

Gravelly and stoney soils:
Two densities are commonly required for understanding the physical behavior of soils with a high proportion of coarse fragments: total bulk density (ρ_{bt}) and bulk density of the fine earth (ρ_{bfe}). The core/clod sampling and radiation techniques are difficult to use for soils with high gravel and stone contents. The excavation method may be more appropriate for skeletal soils (Lal, 1979). The ρ_{bt} must be corrected for gravel and stone fraction to obtain ρ_{bfe}, and relationship established between the skeletal fraction and ρ_{bt} and ρ_{bfe} (ρ_{bfe} = (mass of fine earth < 2mm)/(total volume – volume of gravels)). Poesen and Bunte (1996) developed relationship between ρ_{bt} and ρ_{fe} and the gravel fraction (G_f) for a soil in the Mediterranean region. The ρ_{bt} increases with increase in G_f to a maximum at about 40% beyond which it decreases. In comparison, the ρ_{bfe} decreases monotonically with increase in G_f:

$$\rho_{bfe} = a - b(e^{cG_f})$$

where a, b, and c are empirical constants. Poesen and Bunte attribute the decrease in ρ_{bfe} with increase in G_f to three factors: (i) insufficient fine earth to fill the voids, (ii) open rather than close packing arrangements, and (iii) change in the nature of the fine earth itself due to increase in G_f.
In addition to ρ_b, the soil organic carbon (SOC) content should also be corrected for the G_f, because SOC is mixed only in the fine earth fraction.

$$SOC_e = \frac{SOC}{1-G_f}$$

where SOC_e is the effective proportion of SOC measured in the sieved sample, and G_f is the gravel fraction (Kirkby et al., 1996).

In addition to accuracy and reliability, the question remains whether inter-gravel bulk density is relevant to evaluating edaphologic and ecologic processes.

Cracking soils:
Soils containing predominantly HACs have large swell-shrink capacity, and develop wide and deep cracks on drying. Whereas bulk density of such soils is computed at field moisture capacity, *in situ* measurements at the time when it has numerous cracks poses a major challenge. If clods are taken, the bulk density of the inter-crack clods can be determined.

Organic soils:
Measuring bulk density of organic soils can be a challenge. In some cases, large blocks can be cut to measure weight and volume. Muck and peat that are partly decomposed can be sampled by excavation and also by the clod method. The excavation method is useful for the partially decomposed biomass that does not have a cohesive mass to form clods. Despite the difficulties involved, measuring ρ_b of organic soils is extremely important for assessing the SOC pool, CEC, fertility status, and the available water capacity.

Wet soils:
Similar to organic soils, there is a serious problem of obtaining an accurate sample of wet soils. The relative proportion of the solid fraction is rather small. Therefore, at what moisture content should the ρ_b be expressed for wetlands or even rice paddies?

Frozen soils:
The conventional techniques of clod and core methods do not apply to measuring the ρ_b of frozen soils. The procedure involves thawing the sample, which, depending on moisture content, may drastically alter the results and their interpretations. Many frozen soils have layers of ice that may make up much of the matrix. This needs to be determined and then the volume adjusted in a similar manner as used for rock fragments. Frozen soils can also be sampled with a coring drill, and some can also be sampled as clods. Determining soil-water content and making appropriate corrections are extremely important.

Single-grained soils:
It is difficult to obtain reliable estimates of the volume of a sample comprising cohesionless material of coarse texture by the conventional clod or core methods. Both sand cone and air balloon methods, and compliant cavity measurements can provide the data on volume of the cavity.

Moisture content:
A general problem is using field moisture, air-dried or a pre-defined moisture content to measure soil bulk density that can vary widely. There can be 10 to 20% change in ρ_b with change in moisture potential from 0.03 MPa to 1.5 MPa. When ρ_b is measured at field moisture content, it can give erroneous information. Similarly, ρ_b measured at air dry moisture content can give an overestimation. Expressing ρ_b to a standardized moisture content (0.03 MPa) is an important procedural consideration.

VII. Conclusions

Reliable data of ρ_b is needed to assess pool size and fluxes of H_2O, C and other elements. Although it is a simple property, its precise measurement using simple and routine methods remains a challenge. There are several problem soils that pose a special challenge including sandy soils, gravelly and stony soils, cracking soils, organic soils, frozen soils, and wet or hydromorphic soils. There is a need to develop and standardize methods of measuring ρ_b of these soils. Simple methods of measuring V_t *in situ* need to be developed.

References

Barber, R.G., C. Herrera and O. Diaz. 1989. Compaction status and compaction susceptibility of alluvial soils in Santa Cruz, Bolivia. *Soil & Tillage Res.* 15:153-167.

Blake, G.R. and K.H. Hartge. 1986. Bulk density. p. 363-375. In: A. Klute (ed.), *Methods of Soil Analyses. Part 1.* American Society of Agronomy, Monograph 9, ASA, Madison, WI.

Brasher, B.R., D.P. Franzmeier, V. Valassis and S.E. Davidson. 1966. Use of saran resin to coat natural soil clods for bulk density and water retention measurements. *Soil Sci.* 101:108-110.

Campbell, D.J. and J.K. Henshall. 1990. Bulk density. p. 329-366. In: K.A. Smith and C.E. Mullins (eds.), *Soil Analysis: Physical Methods.* Marcel Dekker, New York.

Culley, J.L.B. 1993. Density and compressibility. p. 529-539. In: M.R. Carter (ed.), *Soil Sampling and Methods of Analysis.* Lewis Publishers, Boca Raton, FL.

Erbach, D.C. 1987. Measurement of soil bulk density and moisture. *Trans. ASAE* 30:922-931.

Hulugalle, N.R., R. Lal and C.H.H. ter Kuile. 1984. Soil physical changes and crop root growth following different methods of land clearing in western Nigeria. *Soil Sci.* 138:172-179.

Hulugalle, N.R. and R. Lal. 1986. Root growth of maize in a compacted gravelly tropical Alfisol as affected by rotation with a woody perennial. *Field Crops Res.* 13:33-44.

Kirkby, M.J., A.J. Baird, S.M. Diamond, J.G. Lockwood, M.L. McMahon, P.L. Mitchell, J. Shao, J.E. Sheehy, J.B. Thornes and F.I. Woodward. 1996. The MEDALUS slope catena model: A physically-based process model for hydrology, ecology and land degradation interactions. p. 303-354. In: C.J. Brandt and J.B. Thornes (eds.), *Mediterranean Desertification and Land Use.* John Wiley & Sons, Chichester, U.K.

Lal, R. 1974. Effect of soil texture and density on the neutron and density probe calibration for some tropical soils. *Soil Sci.* 117:183-190.

Lal, R. 1979. Physical characteristics of soils of the tropics: determination and management. p. 7-44. In: R. Lal and D.J. Greenland (eds.), *Soil Physical Properties and Crop Production in the Tropics* John Wiley & Sons, Chichester, U.K.

Lal, R. 1988. Effects of macrofauna on soil properties in tropical ecosystems. *Agric. Ecosystems and Env.* 24:101-116.

Lal, R. 1991. Soil conservation and biodiversity. p. 89-103. In: D.L. Hawksworth (ed.), *The Biodiversity of Microorganisms and Invertebrates: Its Role in Sustainable Agriculture.* CAB International, Wallingford, U.K.

Lal, R. and D.J. Cummings. 1979. Changes in soil and microclimate after clearing a tropical forest. *Field Crops Res.* 2:91-107.

Lavelle, P. 1983. The soil fauna of tropical savannas. II. The earthworms. p. 465-504. In: F. Bourliere (ed.), *Tropical Savannas.* Elsevier Scientific Publishing Co., Amsterdam.

Lavelle, P., A.V. Spain, E. Blanchart, A. Martin and S. Martin. 1992. Impacts of soil fauna on the properties of soils in the humid tropics. p. 157-185. In: R. Lal and P.A. Sanchez (eds.), *Myths and Science of Soils of the Tropics*, SSSA Special Publication N. 29, Madison, WI.

Moormann, F.R., R. Lal and A.S.R. Juo. 1975. Soils of IITA. Tech. Bull. 3, IITA, Ibadan, Nigeria.

Poesen, J. and K. Bunte. 1996. The effects of rock fragments on desertification processes in Mediterranean environments. p. 247-269. In: C.J. Brandt and J.B. Thornes (eds.), *Mediterranean Desertification and Land Use.* John Wiley & Sons, Chichester, U.K.

Raper, R.L. and D.C. Erbach. 1985. Accurate bulk density measurements using a core sampler. Paper 85-1542, American Society Agricultural Engineers, St. Joseph, MI.

Soane, B.D. and C. Van Ouverkerk (eds.). 1994. *Soil Compaction in Crop Production.* Elsevier Science Publishers, Amsterdam, Holland. 662 pp.

CHAPTER 4

The Effects of Terrain Position and Elevation on Soil C in the Southern Appalachians

P.V. Bolstad and J.M. Vose

I. Introduction

Soil carbon (C) content affects biogeochemical cycles, and varies with climate, vegetation, soil moisture, and other factors. These factors vary in predictable patterns in many regions, including the southern Appalachians (Lindsay and Sawyer, 1970; Whittaker, 1956). For example, temperature decreases and precipitation increases from low to high elevations in the southern Appalachians (Meiners et al., 1984; Bolstad et al., 1998a), and soil moisture decreases from cove to ridge and increases from low to high elevations (Helvey et al., 1972; Yeakley et al., 1998).

Soil carbon changes predictably across an undisturbed landscape, generally increasing from warmer to cooler and from drier to wetter locations. Disturbance generally changes the amounts and spatial patterns of soil and/or vegetation C, particularly when forested lands are converted to agricultural uses. Warmer soil temperatures, lower organic matter inputs, and thorough mixing generally result in decreasing soil C, and the reduction in soil C is often proportional to the duration of disturbance (Schlesinger, 1986). Forest harvests or fire also reduces soil C, both by direct loss and by longer-term reduction in inputs, at least until root and aboveground biomass recovers and litterfall nears pre-disturbance conditions.

While the general relationships between environmental variables and soil C are known, there is uncertainty regarding the specific, quantitative relationships between soil C and environmental variables. There are many instances where landscape estimates of soil C are required. For example, C pools are important determinants of total soil C flux, and so are inputs in comprehensive carbon and biogeochemical cylcling models (Burke, et al., 1995). Spatially explicit, process-driven models are the focus of much current research; however, data are often of unknown accuracy, casting doubt on results and limiting our ability to base policy decisions on model predictions.

Currently, soil maps are most commonly used when estimates of soil properties are required over large areas (Aber et al., 1995; Lathrop et al., 1995), but many regions of the country have old or non-existent county soil surveys, and so must rely on coarser data. Work over the past decade has produced a national digitial soils data layer organized by state boundaries, identified as STATSGO data (SCS, 1991). While quite general, STATSCO data have been investigated and used in many modeling and analyses initiatives (Davidson, 1995), because they are at times the best data available, and because they are often considered to be at an appropriate sampling grain for regional or continental analyses.

An alternative to soil maps involves using soils, geologic, and/or environmental relationships to predict soil properties on a landscape basis. Soils properties are determined by climate, parent material, vegetation, disturbance history, and time. This approach identifies the relationships between all or some of these soil-forming factors and specific soil properties, and develops specific predictive relationships. Developed equations may be inaccurate and hence quite useless for point-specific predictions, given the complex set of factors controlling soil properties. Factors that can be easily mapped or determined with current data or field measurements (e.g., precipitation, temperature, vegetation, elevation, aspect) may control soil C less than factors that are more difficult to determine for each site (e.g., parent material, disturbance history), and thus use in a predictive model. This chapter describes observed relationships between soil C, elevation, and terrain position in the southern Appalachian Mountains.

II. Sample Collection and Analysis

Soil samples were collected between November 1995 and April 1998 at 72 sites in the southern Appalachian mountains (Figure 1). Samples were collected across a range of elevations, latitudes, vegetation types, and terrain positions (Table 1). Soil samples were collected with a 3.5-cm diameter probe to a depth of 30 cm. The probe could not penetrate roots larger than approximately 2.5 cm, the effective upper limit on root samples. Litter was removed, and the probe inserted perpendicular to the soil surface and a randomly chosen spot. At least 15 samples were collected per site. Large stones and high coarse fraction at 6 of the 72 sites required frequent re-selection of sampling locations within the site. In these cases, every attempt was made to sample to the full 30-cm depth, resulting in from 8 to 14 samples at these sites. Samples were placed in a plastic bucket, mixed for at least 5 min, and a 2-kg sub-sample was placed in a plastic bag for transport and storage until laboratory analyses. Site location and elevation were determined from differentially corrected GPS measurements, typically accurate to within 5 m. Overstory species composition was noted for all trees greater than 30 cm within 15 m of plot center, and basal area determined using a 10 factor prism. Intact soil cores were removed at a subset of sites (22) to measure bulk density, dry weight basis. Cores were taken with an 8-cm sampling tube. Soils were stored and dried at 70°C for 48 h and weight determined. Soil coarse fraction was determined by passing soils through a 2-mm sieve, and weight and volume of the coarse fraction determined. Sieved soil was dried at 70°C for 48 h, and soil C and N determined with the CHN analyzer. Fine and coarse fraction mass per mass sampled were determined, and combined with bulk density and carbon content data to estimate carbon per unit volume.

III. Spatial and Statistical Analyses

Plot locations were entered into a spatial data layer in a GIS. Digital elevation data (DEMs) were assembled from 1:24,000 scale USGS data (USGS, 1990). Quadrangles were joined, and terrain shape data were derived from this mosaic DEM, using a 9-cell kernal to calculate a modified version of McNab's (1989) Terrain Shape Index. Elevation and terrain shape were extracted for each field sampling location through a cartographic overlay of GPS-determined plot centers. Plots were categorized into terrain position (cove, sideslopes, and ridges) and elevation classes (high, mid, and low). Linear regression models were fit between soil C, elevation and terrain shape, and common regression tests and diagnostics calculated. Regressions of soil C were fit for the total data set, and then grouped by parent material, (a) meta-sedimentary, or (b) igneous origin (Hatcher, 1988). All statistical analyses were performed using SAS™. Regression diagnostics included scatter diagrams to test the assumption of uniform variance, regression significance via F-tests on Type IV sums of squares, R^2, Cook's-D, and K-S tests for normality.

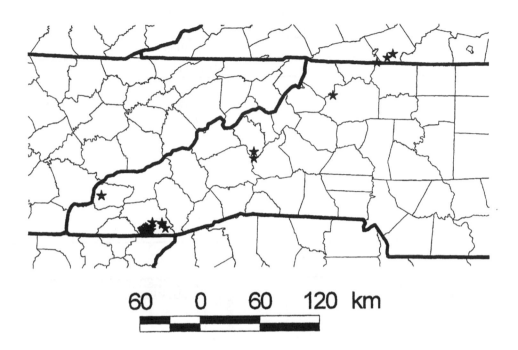

Figure 1. Sample locations in the southern Appalachian Mountains.

Table 1. Characteristics of sampling sites

	Elevation (m)	Terrain shape (deg.)	Basal area (m² ha⁻¹)	Root biomass (kg ha⁻¹)
Mean	978.84	1.71	22.2	4160.01
Maximum	2005.69	32.90	41.1	6539.49
Minimum	632.70	−17.53	14.7	974.17
n	84	84	72	18

IV. Soil C, Elevation and Terrain

Soil C was related to both elevation and terrain shape (Table 2). Carbon content was lowest at low elevation ridges, and highest at high elevation cove sites. Differences among mean values were significant ($p < 0.05$, t-test) for high vs. low elevation sites and cove vs. ridge sites when pooled. High elevations (>1150 m) had significantly higher soil C than low elevations (<900 m) at sideslope and ridge, but not at cove terrain positions ($p < 0.05$). Cove sites were significantly different than ridge sites at low and intermediate elevations, but not at high elevations.

Table 2. Average soil C % (and standard error) by elevation and terrain position, mature forested sites (> 70 years old); means are based on from 5 to 9 samples

Terrain position	Elevation		
	<900 m	901 to 1150 m	> 1150 m
Cove/flat			
TS< -5, or	3.94	4.58	6.54
-2< TS <2	(0.89)	(0.57)	(1.72)
Sideslope			
-5≤ TS ≤ -2, or	1.49	2.90	7.60
2≤ TS ≤ 5	(0.21)	(0.18)	(1.20)
Ridge	1.64	1.61	5.00
TS> 5	(0.31)	(0.93)	(1.68)

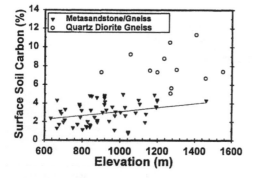 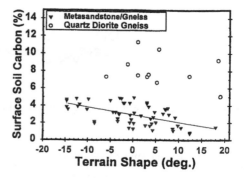

Figure 2. Relationships between soil carbon, elevation, and terrain shape. Soil carbon increased with elevation and decreased at more convex (ridge) terrain shapes. There were also strong differences associated with soil parent material, with lower average soil C in mixed metasandstone/gneiss derived soils. Lines are significant regression ($p < 0.05$, linear, first-order) models, fit for each parent material type. Only metasandstone/gneiss subset significant, $R^2 = 0.19$ for elevation, $R^2 = 0.39$ for terrain shape.

Soil C increased with elevation (Figure 2, left), and linear regression between soil C and elevation showed an almost four-fold increase in soil C, from 2.1 to 8.0% (mass-based) between 600 to 1600m. However, bedrock geology had a significant impact on soil C, which was somewhat confounded with elevation. Mixed metasandstone/gneiss parent materials are more common at low elevations in the study region, while gneiss is more typical at upper elevations. The former also had significantly lower soil C content, once adjusted for elevation. Thus, separate regression lines of soil C vs. elevation were warranted for the lower elevation mixed metasandstone and higher-elevation gneiss-derived soils. There were significant relationships between soil C and elevation for the metasandstone-origin sample sites, but not for the gneiss-derived samples, even though the gneiss subset showed an upward trend (Figure 2, left). Average soil C approximately doubled, from 2.1 to 3.9% over the sampled 900m elevation range for metasandstone-origin soils. Lack of significance for the gneiss samples may be

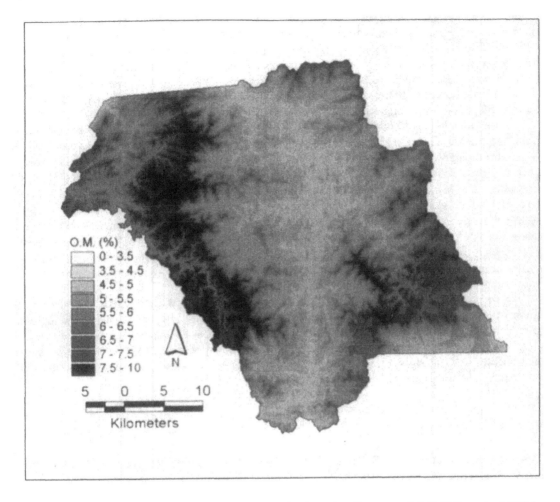

Figure 3. Estimated soil C using elevation and terrain shape in a linear model. Soil C content (%) in the top 30 cm.

due in part to a smaller sample size, and in part to apparently higher variability in soil C at sites with gneiss parent materials.

Soil C decreased significantly and linearly as terrain shape increased (Figure 2, right), at least on metasandstone-derived soils. Soil C averaged 4% in deep coves (terrain shape −15°) and fell to 1.7% on well-defined ridges (terrain shape +15°). There were no apparent or significant terrain-caused differences in soil C in gneiss-derived soils.

Higher Ca, Mg, and other base concentration may be the cause of higher soil C. Gneiss-derived soils are more base-rich than metasandstone, which is likely to aid higher gross primary production and may lead to greater accumulation of organic matter over long time frames. Some of the increased organic matter may be due to segregation of vegetation and higher litter quality in the northern hardwood at higher elevations. However, evidence against this hypothesis include a high proportion of species that produce lower quality litter at high elevations, including *Q. rubra, Q. alba,* and *Q. prinus* (Bolstad et al., 1998b), and because northern hardwood types occur only in the upper half of the richer gneiss-derived soils.

Increases in soil C with increasing elevation are probably related to decreases in decomposition relative to production, and hence higher long-term accumulation of C in the forest floor and C. Leaf area and biomass, and hence annual leaf litter production, is similar at high and low elevations, although wood production is lower (Vose and Bolstad, 1998). Soil temperatures at a 5-cm depth average about 4.5°C lower at 1600 than 600 m (Bolstad et al., 1998a), and because decomposition is an exponential function of temperature, reduced temperature may retard soil decomposition relative to production, leading to long-term C accumulation.

Landscape estimates of soil C were derived, based on the linear regression models based on terrain shape and elevation. Our field sites were skewed slightly towards mid- to higher elevations, perhaps oversampling high-elevation sites, so a simple sample mean is not appropriate for a landscape-wide mean C estimate. A landscape map was produced based on a linear terrain/elevation/soil C model (Figure 3). The linear model was then used with 30-m grain digital elevation and terrain shape data to predict soil C, and summed by cell for the study area. Total soil C in the top 30 cm for the upper Little Tennessee River watershed was estimated at 18.1 T/ha based on the linear model. While these values initially appear quite low when compared to other published data (up to 72 T/ha C), here soil C is reported, not organic matter, found in the top 30 cm. When adjusting appropriately for depth and to organic matter, the current observations are comparable to those previously published for the region (Monk and Day, 1988).

We conclude that there are large differences in soil C across the landscape, and that these differences are most strongly related to parent material, temperature, terrain position, in the southern Appalachian Mountains. Soil C varied from less than 1 to almost 12% of the top 30 cm at forested sites, and significantly increased with elevation. Soil C decreased from cove to ridge sites. These relationships were strongest on soils derived from a mixture of metasandstone and gneiss, and were not observed at higher elevation gneiss-derived soils.

Acknowledgments

This research was support by NSF Grants BSR9011661 and DEB-9596191. We wish to thank Cread Brown, E.H. Pape and Charles Marshall for their efforts in collecting the field data.

References

Aber, J.D., S.V. Ollinger, C.A. Federer, P.B. Reich, M.L. Goulden, D.W. Kicklighter, J.M. Melillo and R.G. Lathrop. 1995. *Clim. Res.* 5:207-222.

Bolstad, P.V., L. Swift, F. Collins and J. Regniere. 1998a. Measured and predicted air temperatures at basin to regional scales in the Southern Appalachian Mountains. *Ag. For. Met.* 91:167-176.

Bolstad, P.V., W.T. Swank and J. Vose. 1998b. Predicting Southern Appalachian overstory vegetation with digital terrain data. *Land. Ecol.* 13:271-283.

Burke, I.C., E.T. Elliott and C.V. Cole. 1995. Influence of macroclimate, landscape position, and management on soil organic matter in agroecosystems. *Ecol. App.* 5:124-131.

Davidson, E.A. 1995. Spatial covariation of soil organic carbon, clay content, and drainage class at a regional scale. *Land. Ecol.* 10:349-362.

Hatcher, R.D., Jr. 1988. Bedrock geology and regional geologic setting of Coweeta Hydrologic Laboratory in the Eastern Blue Ridge. In: W.T. Swank and D.A. Crossley, Jr. (eds.), Forest Hydrology and Ecology at Coweeta. Springer-Verlag, Berlin. 469 pp.

Helvey, J.D., J.D. Hewlett and J.E. Douglass. 1972. Predicting soil moisture in the southern Appalachians. *Soil Sci. Soc. of Amer. Proc.* 36:954-959.

Hunt, E.R., S.C. Piper, R. Nemani, C.D. Keeling, R.D. Otto and S.W. Running. 1996. Global net carbon exchange and intra-annual atmospheric CO_2 concentrations predicted by an ecosystem process model and three-dimensional atmospheric transport model. *Glob. Biogeochem. Cycles* 10:431-456.

Lathrop, R.G., J.D. Aber and J.A. Bognar. 1995. Spatial variability of digital soil maps and its impact on regional ecosystem modeling. *Ecol. Model.* 82:1-10.

Lindsey, A.A. and J.O. Sawyer. 1970. Vegetation-climate relationships in the eastern United States. *Proc. Indiana Acad. Sci.*, 80:210-214.

McNab, W.H. 1989. Terrain shape index: quantifying effect of minor landforms on tree height. *For. Sci.* 35:91-104.

Meiners, T.M., D.W. Smith, T.E. Sharik and D.E. Beck. 1984. Soil and plant water stress in an Appalachian oak forest in relation to topography and stand age. *Plant Soil* 80:171-179.

Monk, C.D. and F.P. Day, Jr. 1988. Biomass, primary production, and selected nutrient budgets for and undisturbed watershed. In: W.T. Swank and D.A. Crossley, Jr. (eds.), *Forest Hydrology and Ecology at Coweeta*. Springer-Verlag, Berlin. 469 pp.

Schlesinger, W.H. 1986. Changes in soil carbon storage and associated properties with disturbance and recovery. p.194-220. In: J.R. Trabalka and D.E. Reichle (eds.), *The Changing Carbon Cycle: A Global Analysis*. Springer-Verlag, New York.

Soil Conservation Service. 1991. *State Soil Geographic Database (STATSCO): Data User's Guide*. USDA. SCS Miscellaneous Publication No. 1492. U.S. Government Printing Office, Washington, D.C.

USGS, 1990. Digital Elevation Models Data User's Guide 5., Reston, VA. 51 pp.

Vose, J.M. and P.V. Bolstad. 1998. Biotic and abiotic factors regulating forest floor CO_2 evolution across a range of forest age classes in the southern Appalachians. *Can. J. of For. Res.* (submitted).

Whittaker, R.H. 1956. Vegetation of the Great Smoky Mountains. *Ecol. Mon.* 26:1-80.

Yeakley, J.A., W.T. Swank, L. Swift, G.M. Hornberger and H.H. Shugart. 1998. Soil moisture gradients and controls on a southern Appalachian hillslope from drought through recharge. *Hyd. and Earth Sys. Sci.* 2:31-39.

CHAPTER 5

Approaching "Functional" Soil Organic Matter Pools through Particle-Size Fractionation: Examples for Tropical Soils

C. Feller, J. Balesdent, B. Nicolardot and C. Cerri

I. Introduction

The notion of soil organic matter (SOM) or soil organic carbon (SOC) functional pool is often quoted in literature (Tiessen et al., 1984; Duxburry et al., 1989; Theng et al., 1989; TSBF, 1989; Bonde et al., 1992; Christensen, 1992; Cambardella and Elliott, 1993, 1994; Feller, 1993; Woomer, 1993; Herrick and Wander, 1997; Monreal et al., 1997; Sternberg, 1998), but this notion is generally poorly defined, never quantified and is more generally restricted to the dynamics of SOM than applied to the different and numerous functions that SOM plays in the soils or in the soil-plant-atmosphere system. In fact, total SOM (expressed as total carbon Ct) exerts essential and different *functions* in soil:

- biological functions, such as easily mineralizable carbon or nitrogen (Cm, Nm), microbial nitrogen immobilization, enzymatic activities,
- exchange and sorption functions such as cation exchange capacity or sorption of pesticides,
- function of aggregation,
- functions of medium- to long-term storage ("sequestration") of elements and/or nutrients for plants and soil organisms, such as total organic carbon, nitrogen, phosphorus, sulfur, non-exchangeable bases associated to SOM.

It is important to identify, for a given function, which part(s) of the total SOM represent(s) the majority of the considered function. This is the notion we shall refer to when functionality of SOM pools is discussed throughout this chapter. These functions follow from *properties* of organic entities or molecules, and these properties are quantified owing to static or dynamic *descriptors* such as mineralizable carbon Cm or nitrogen Nm (expressed e.g., in g kg^{-1} C), organic CEC (in cmole(+) kg^{-1} C), sorption coefficient Koc of organic molecules M (in mg Mg^{-1} C), mean residence time (MRT) or half-life of SOC.

Total SOM is characterized by average values of such descriptors, but there exists a wide distribution of properties within different SOM pools. This distribution can be approached and simplified into sub-classes by SOM fractionation of the whole soil, without affecting the properties of the separates. Identification of functional pools through fractionation procedures would also be valuable for the modelling of soil organic matter properties and behavior (Balesdent, 1996).

The objectives of this chapter are (1) to propose a general approach for the quantification of the functionality of different soil organic fractions; (2) to propose a definition of functional SOM pools;

and (3) to illustrate the conceptual approach in the case of particle-size fractionation of SOM in relation to three different functions: short-term mineralization of carbon *(Cm)*, short-term mineralization of nitrogen *(Nm)*, medium- to long-term SOC sequestration.

II. Materials and Methods

A. Sites, Soils and Land Use

The results reported here were obtained from the pedological situations summarized in Table 1. With the exception of the Entisol Ps1 the clay fraction of which was rich in smectite, all the soils were low activity clay (LAC) soils with a mineralogy of the clay fraction dominated by kaolinite or halloysite associated with iron and/or aluminium oxyhydroxides. The soils belonged to the following orders of the U.S. Soil Taxonomy: Oxisol (Fr4, Fr7), Ultisol (Fr2, Fr3), Inceptisol (Fi6), Entisol (Ft1). The selected LAC soils covered a wide range of texture, from sandy (Ft1) to clayey (Fr4, Fi6, Fr7), with SOC contents ranging from 1.9 (Ft1) to 41.2 (Fr4) $gC.kg^{-1}$ soil.

For some of these situations, comparisons were made between plots under continuous annual cropping (situations A) and plots corresponding to alternatives for soil carbon sequestration (situations B):

- situations A: groundnut (*Arachis hypogaea*)-millet (*Pennisetum typhoïdes*) rotation, food crops, corn *(Zea mays)*, market gardening and sugarcane (*Saccharum* spp.);
- situations B: grass or bush fallows and artificial meadow.

Results for the sole 0 to 10 cm layers will be presented in this chapter.

Some analytical data are reported in Table 1, others are detailed in Feller (1995). All the sites chosen were apparently not eroded.

B. Soil Sampling

Each soil sample (0 to 10 cm) was constituted from 6 to 12 replicates. Each replicate was measured for total SOC content. The coefficients of variation ranged from 3 to 18% with a mean value of 11% (Feller, 1995).

C. Particle-Size Fractionation Method

The particle-size fractionation method used in this study was described in Feller et al. (1991, "method R/US"). Briefly, it consists of shaking for 2 to 16 h (duration depending upon the soil texture) the 0 to 2 mm soil sample (40 g) in water (300 ml) in presence of a cationic resin (R) saturated with Na^+ to improve the soil dispersion. This was followed by wet sieving at 200 and 50 μm to separate the coarse (200 to 2000 μm) and fine (50 to 200 μm) sand fractions. An ultrasonic treatment (US) of the 0 to 50 μm suspension ($100J\ ml^{-1}$) improved the clay dispersion. The coarse silt fraction (20 to 50 μm) was obtained by sieving. The fine silt (2 to 20 μm) was separated from clay (0 to 2 μm) by repeated centrifugation.

C and N analysis were performed by dry combustion with a CHN Analyser (Carlo Erba, Mod. 1106). SOM solubilized during the fractionation procedure (less than 4% of total SOM) was not considered in this study.

The above method provided a high dispersion of the soil constituents even with no application of ultrasonic treatment to the whole 0 to 2 mm soil sample. Balesdent et al. (1991) showed that an ultra-

Table 1. Some soil, climatic, and land use characteristics of the studied sites

Location[a]	Site[a]	Climate P (mm)	Climate T°C	Soil order	Samples[a]	Vegetation of crops	Horizon 0–10 cm Clay	Carbon	C/N
							— (g kg^{-1} soil) —		
Sites studied for C and N mineralization									
Senegal	Ft1	700	29	Entisol	Mi6	A - millet	51	1.9	10.6
Martinique	Fi6	1820	26	Inceptisol	Ca50	A - sugarcane	493	21.8	12.0
Sites studied for C sequestration									
Senegal	Ps1	700	29	Entisol	Am6	A - groundnut-millet	87	5.4	10.0
					Ja21	B - grass fallow	84	7.3	12.7
Ivory Coast	Fr2	1360	26	Ultisol	Rv10	A - rice-corn-manioc	207	9.7	13.9
					Ja12	B - bush fallow	186	15.9	17.8
Togo	Fr3	1040	27	Ultisol	Ms14	A - corn	89	5.3	13.3
					Ja6	B - bush fallow	64	12.4	12.4
Guadaloupe	Fr4	3000	25	Oxisol	Rm10	A - corn-market gardening	670	19.1	9.8
					Pr10	B - artificial meadow[b]	639	41.2	13.3
St. Lucia	Fr7	2700	25	Oxisol	Rv10	A - corn-yam-market gardening	522	18.6	12.9
					Jh10	B - grass fallow	539	29.6	14.3

[a]The symbols refer to nomenclature used by Feller (1995). For the sample symbol the number refers to the last duration (years) of the agricultural system.
[b]Planted with *Digitaria decumbens*.
(Adapted from Feller, 1995; Feller et al., 1996.)

sonic treatment of the whole soil may lead to an artificial transfer (about 50%) of OM associated with sands (plant debris) into the fine fractions (< 50 μm).

By simplification, and according to previous studies on SOM (Feller, 1995; Feller et al., 1996; Feller and Beare, 1997) and particle-size fractions (morphology, C/N, xylose/mannose ratios and dynamics), we shall only consider here the three following fractions:

- fraction 20 to 2000 μm (f20 to 2000), the "plant debris fraction": predominance of plant debris at different stages of decomposition, with carbon to nitrogen (C/N) ratios ranging between 12 to 33 (mean value 20.4);
- fraction 2 to 20 μm (f2 to 20), the "organo-silt complex": consisting of very humified plant and fungi debris associated with stable organomineral microaggregates which have not been destroyed during the fractionation. C/N ratios vary from 11 to 17 (mean value 14.5);
- fraction < 2 μm (f0 to 2), the "organo-clay fraction": with predominance of amorphous OM acting as a cement for the clay matrix. Sometimes, under forest or savanna, presence of plant cell walls occur in the coarse clay fraction but usually not in the fine clay. Very often, bacterial cells or colonies at different stages of decomposition can be observed in both fractions. C/N ratios vary from 8 to 12 (mean value 9.8).

D. Soil Carbon and Nitrogen Mineralization (Whole Soil and Particle-Size Fractions)

For the whole soil (0 to 2 mm), 25 g was moistened at 80% of its field capacity (pF 2.5) and incubated in 125-ml flasks for 28 days at 28° C. Mineral N (sum of $N-NH_4^+$, $N-NO_3^-$ and $N-NO_2^-$) was extracted at 0 and 28 days with $1 M$ KCl and was determined according to Nicolardot (1988). Net mineralization in 28 days was defined as Nm. Evolved CO_2 was measured at 0, 2, 7, 14 and 28 days (Nicolardot, 1988) and cumulated CO_2 evolved after 28 days was defined as Cm. For the size fractions, the fractions larger than 20 μm were incubated alone, but each of the 2 to 20 and 0 to 2 μm fractions were mixed (1/1, w/w) with coarse commercial sand. The incubation conditions were similar to those applied to the whole soil. All determinations were conducted in triplicate.

III. Results and Discussion

A. Theoretical Approach and Definitions

For a given property "x" and a given fraction "i" or the total soil "t," we define:

- a "descriptor value" $DV-x_i$ or $DV-x_t$:
  $DV-x_i$ and $DV-x_t$ are expressed on the basis on the C (or N) concentration of the fraction or of the soil ($DV-x_i$ in g (or other units) $\cdot$ g^{-1} C fraction, $DV-x_t$ in g (or other units) $\cdot$ g^{-1} C soil),
- a "functionality index" $FI-x_i$ given by the formula:
$$FI-x_i = 100 \cdot (DV-x_i / DV-x_t) \cdot (C_i/C_t) \qquad (1)$$
with C_i and C_t the respective carbon amounts of the fraction and soil expressed in g C kg^{-1} soil.

$FI-x_i$ represents the participation (in % of the fraction) to the total property expressed by the whole soil.

- a "functionality index variation" ΔFI_{A-B}. The objective is to quantify the participation of each fraction to the total variation observed for a given property when there is a change in the soil use or land management from a previous situation A to a new situation B.

ΔFI_{A-B} is given by the formula:

$$\Delta FI_{A-B} = 100 \cdot (C_{iA} \cdot DV_{iA} - C_{iB} \cdot DV_{iB}) / (C_{tA} \cdot DV_{tA} - C_{tB} \cdot DV_{tB}) \qquad (2)$$

where :
* DV_{iA} and DV_{iB} are the descriptor values for the fractions i of the respective situations A and B,
* DV_{tA} and DV_{tB} are the descriptor values for the total soil of the respective situations A and B,
* C_{iA} and C_{iB} are the carbon amounts (g C kg^{-1} soil) for the fractions i of the respective situations A and B,
* C_{tA} and C_{tB} are the total soil carbon content (g C kg^{-1} soil) for the respective situations A and B,

ΔFI_{A-B} represents the participation in % of the fraction "i" to the total variation Δ of the property "x" expressed by the whole soil, when there is a change in soil use (or management) between a situation A and B.

Both FI-x_i and ΔFI_{A-B} are expressions of the functionality of a fraction i. For a number of fraction n equal or higher than 2, we define the following "functionality scale":

- low functionality, when FI_{ix} or $\Delta FI_{A-B} = 75/n$ (3)
- medium functionality, when $75/n < FI_{ix}$ or $\Delta FI_{A-B} = 125/n$ (4)
- high functionality, when FI_{ix} or $\Delta FI_{A-B} > 125/n$ (5)

Although the index ΔFI_{A-B} will not be used in this chapter, this definition remains important for other potential studies.

Definition of a "functional SOM pool":
A SOM pool P_i will be considered as a "functional" pool for a given function F_x, or a given variation of that function with changes in soil use if its functionality index FI-x_i, or its functionality index variation ΔFI_{A-B}, is "high" according to the functionality scale.

A virtual example of the "descriptor values" DV-x_i, and the "functionality index" FI – x_i is presented in Figure 1 for a fractionation procedure involving three (n = 3) fractions f1, f2, f3. In this example, the fraction f1 can be considered as a functional pool for sample A but not for sample B. For this one, the fraction f3 is the functional pool.

B. Application to the Short-Term Mineralization of Soil C and N

The study concerns the 0 to 10 cm layer of two low activity clay (LAC) soils: the sandy soil Ft1 cultivated with millet and the clayey soil Fi6 cultivated with sugarcane.

The concentrations and the distributions in the particle size fractions are given in Table 2 for total carbon *(C)*, total nitrogen *(N)*, mineralized carbon *(Cm)* and mineralized nitrogen *(Nm)*.

1. Mass, Carbon and Nitrogen Balances of the Fractionation Procedure

The mass balance of the fractionation procedure (Sum of fractions / non-fractionated soil) was 99.9 and 100.4% for samples Ft1 and Fi6, respectively. The corresponding balances for the total carbon (C) were 88.2 and 108.9% and those for mineralized carbon (Cm) were 78.3 and 78.4%. The N and Nm balances for sample Fi6 were 105.5 and 84.2%, respectively.

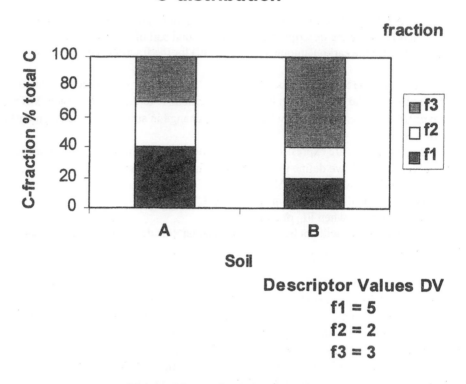

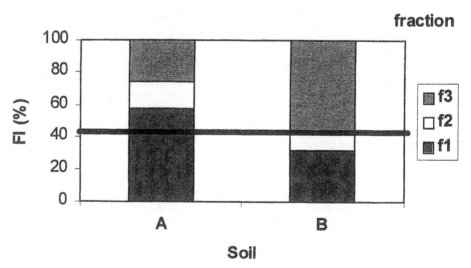

Figure 1. A virtual example of the calculation and expression of the functionality index (FI) from C-distribution and descriptor values data.

Table 2. Characteristics of the total (C,N) and mineralized (Cm, Nm) carbon and nitrogen if the particle-size fractions and their descriptor value (DV) and functionality index (FI) NF soil = not fractionated soil

Site	Characteristics	Fraction (µm)				Sum	NF soil
		20–2000	2–20	0–2	H_2O^a		
Ft1	mass (g 100g^{-1} soil)	954.9	11.1	27.5	5.0	998.5	1000.0
	C (g kg^{-1} fraction)	0.9	27.2	28.5	nd		
	C (g kg^{-1} soil)	0.9	0.3	0.8	nd	1.9	2.2
	Cm (mg kg^{-1} fraction)	85.8	281.8	392.4	nd		
	Cm (mg kg^{-1} soil)	81.9	3.1	10.8	nd	95.8	122.5
	DV-Cm	95.3	10.4	13.8	nd	49.3	55.7
	FI-Cm (%)	85.5	3.3	11.3		100.0	
	C/N	12.3	12.1	8.9	nd		12.7
Fi6	mass (g 100g^{-1} soil)	200.3	164.5	601.9	37.0	1003.7	1000.0
	C (g kg^{-1} fraction)	21.4	24.0	21.1	nd		
	C (g kg^{-1} soil)	4.3	3.9	12.7	nd	20.9	19.2
	Cm (mg kg^{-1} fraction)	540.0	202.0	330.2	nd		
	Cm (mg kg^{-1} soil)	108.2	33.2	198.7	nd	340.1	434.0
	DV-Cm	25.2	8.4	15.6	nd	16.2	22.6
	FI-Cm (%)	31.8	9.8	58.4	nd	100.0	
	N (g kg^{-1} fraction)	0.9	1.5	2.2	nd		
	N (g kg^{-1} soil)	0.2	0.2	1.3	nd	1.7	1.6
	Nm (mg kg^{-1} fraction)	9.4	42.1	86.8	nd		
	Nm (mg kg^{-1} soil)	1.9	6.9	52.3	nd	61.1	72.5
	DV-Nm	11.0	28.3	39.8	nd	35.3	44.2
	FI-Nm (%)	3.1	11.3	85.6		100.0	
	C/N	25.2	16.1	9.7			11.7

^{a}Water content (105°C) of the air-dried sample.

The Cm and Nm balances were systematically lower than C and N, with values varying from 78 to 84%. This can be due to the loss of water soluble C and N (but these were not taken into account in this work), and to environmental differences in the incubation of bulk versus fractionated samples, including interaction between fractions within the bulk sample. Therefore, we shall now only consider the three particle-size fractions and their sums in the study on functionality.

2. Functional Pools for Short-Term Carbon Mineralization

The descriptor value (DV) symbolized by DV–Cm (DV–Cm = Cm_i / C_i) was expressed in g Cm kg^{-1} C fraction. The DV–Cm value of the whole soil was higher for the sandy sample Ft1 than for the clayey one, Fi6, but both samples varied in the order : f20 to 2000 > f0 to 2 > f2 to 20. The two soils differed more in the value of the f20 to 2000 fraction (higher for Ft1) than in that of the f0 to 2 and f2 to 20 fractions. This trend toward higher DV-Cm values (or an equivalent index) for the "sand-

C. Feller, J. Balesdent, B. Nicolardot and C. Cerri

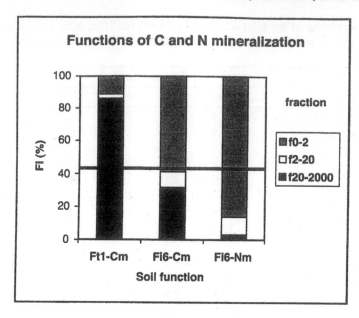

Figure 2. Functionality Index (FI) for the C and N mineralization function.

size fractions" with regard to silt- and clay-size fraction was also observed by Christensen (1987), Gregorich et al. (1989), Hassink (1995) but not by Bernhardt-Reversat (1987, 1988) in the case of tropical sandy soils under savannah or tree plantation.

The functionality index (FI–Cm, Equation (1)) of the particle size fractions of the two soils are presented in Figure 2. The plant debris fraction (f20 to 2000) appeared to be the functional pool for the sandy soil Ft1 (FI% = 85.5) but not for the clayey soil Fi6 (FI% = 31.8). For this soil the functional pool was the organo-clay fraction f0 to 2 (FI% = 58.4). The difference in the functionality of the same fraction according to the type of soil might be attributed to the distribution of the total C within the fractions (textural effect) rather than to the differences of the descriptor values DV–Cm. For example, even if we applied the corresponding DV–Cm values of sample Fi6 to the Ft1 fractions, the plant debris fraction (f20 to 2000) would have remained the functional pool for soil Ft1. Hassink (1995) also observed that DV–Cm did not differ significantly among size fractions in soils of different texture for temperate grasslands.

This emphasizes the fact that the only descriptor value seems insufficient to evaluate the role played by a given organic compartment for a given soil property; distribution of the property within the soils fractions must also be considered.

3. Comparisons between Nitrogen and Carbon Mineralization Functions

We only considered sample Fi6 for net N mineralization. The descriptor value DV–Nm varied in the order: f0 to 2 > f2 to 20 > f20 to 2000. This order differed from that of DV–Cm. The DV–Nm value increased as the C/N ratio of the fraction decreased (Table 2). A significant correlation between the N mineralization coefficient (equivalent to DV–Nm) and the C:N ratios of different particle size fractions, was observed for tropical sandy soils by Bernhardt-Reversat (1981), the low values of the sand-size fractions being attributed to nitrogen immobilization by presence of plant debris slightly decomposed. Similar observations were done by Sollins et al. (1984) and Barrios et al. (1996) with the existence of a negative relationship between "light fractions" (density <1.6 or 1.7) and equivalents of DV–Nm.

For sample Fi6, the comparison of the functionality index for Cm (FI–Cm) and Nm (FI–Nm) showed (Figure 2) that the FI% of the organo-clay fraction was largely more important for the nitrogen mineralization function than for the carbon mineralization one.

Finally, these two examples for LAC tropical soils showed that:

- the same fraction did not exhibit the same level of functionality according to the soil characteristics,
- the same fraction, for a given soil, did not exhibit the same level of functionality according to the type of the studied function.

C. Application to the Function of SOC Sequestration

In relation to the global greenhouse effect, more and more studies are concerned with the SOM potential for soil organic carbon sequestration (Cseq). The main purpose is to identify, for different bio-physical and socio-economical environments, the alternatives in soil use or land management that allow an increased storage of OC in soil. Another purpose is to identify which SOC pool is involved in the process of carbon sequestration. Those pools could be considered as functional pools when one discusses the function of carbon sequestration.

To illustrate this function, we studied by means of a synchronic approach in sub-Saharan Africa and in the Lesser Antilles, low activity clay (LAC) soils formerly cultivated under continuous annual crops (situation A) and later cultivated with a potential SOC sequestering system (situation B) during variable durations. For each site we considered a pair of plots.

The potential SOC sequestering systems studied (Table 1) were the following:
- spontaneous herbaceous or shrub fallows during 6 (site Fr3), 10 (site Fr7), 12 (site Fr2) and 21 (site Ps1) years after cultivation during 10 years or more with cereal, root crops and/or market gardening,
- artificial meadow during 10 years after 10 years of market gardening (site Fr4).

The relative increase in total SOC content in the 0 to 10 cm layer following these systems varied from 34 to 134% of the initial value under cultivation.

In order to study the SOM forms involved in the SOC variations observed, we conducted a particle-size fractionation of the different samples. The detailed results are presented in Table 3.

For the function of C-sequestration (Cseq) we defined the following descriptor values DV–Cseq:
- for each fraction i , DV–Cseq$_i$ = $\Delta C_{i\,A\text{-}B}$ / C_{iA}

with $\Delta C_{iA\text{-}B}$ representing the difference in the amount of carbon (g C kg^{-1} soil) in the fraction i between the situation A and B,
- for the sum of the fractions, DV–Cseq$_t$ = $\Delta C_{tA\text{-}B}$ / C_{tA}

with $\Delta C_{tA\text{-}B}$ representing the difference in the carbon amount (g C kg^{-1} soil) of the sum of the fractions between the situation A and B.

Therefore, the functionality index, FI–Cseq calculated with equation (1), is given for each fraction by the formula:

FI–Cseq$_i$ = 100 · $\Delta C_{iA\text{-}B}$ / $\Delta C_{tA\text{-}B}$

Table 3. Characteristics (C,N) of the particle-size fractions and their descriptor values (DV) and functionality index (FI) for the function of C sequestration.

Fraction (µm)	Characteristics	Fr3 A	Fr3 B	PS1 A	PS1 B	Fr2 A	Fr2 B	Fr7 A	Fr7 B	Fr4 A	Fr4 B
	Clay content (g 100 g⁻¹ soil) (mean value of A and B)	7.6		8.5		19.7		53.1		65.5	
20–2000	mass (g 100 g⁻¹ soil)	85.0	86.9	86.0	86.4	69.4	71.1	23.7	27.2	10.7	13.5
	C (g kg⁻¹ fraction)	13.0	62.5	12.9	33.0	29.7	55.8	145.6	291.9	143.9	565.7
	C (g lg⁻¹ soil)	1.1	5.4	1.1	2.8	2.1	4.0	3.5	7.9	1.5	7.6
	C/N	19.7	14.5	12.0	13.1	19.5	20.1	24.6	33.1	19.3	31.8
	DV−CseqC (g ΔC kg⁻¹ CsoilA)		3.9		1.6		0.9		1.3		3.9
	FI−Cseq (%)		72.9		59.1		29.4		45.1		31.6
2–20	mass (g 100 g⁻¹ soil)	4.4	4.6	3.0	3.6	7.7	8.8	17.2	13.9	13.5	18.4
	C (g kg⁻¹ fraction)	41.1	64.6	22.3	33.3	38.4	61.1	13.9	22.8	47.5	41.6
	C (g kg⁻¹ soil)	1.8	3.0	0.7	1.2	3.0	5.4	2.4	3.2	6.4	7.7
	C/N	15.2	10.9	10.6	10.6	18.1	19.6	13.5	16.5	13.5	16.6
	DV−Cseq		0.7		0.8		0.8		0.3		0.2
	FI−Cseq (%)		20.1		17.6		37.2		7.9		6.4
0–2	mass (g 100 g⁻¹ soil)	8.9	6.4	8.7	8.4	20.7	18.6	52.2	53.9	67.0	63.9
	C (g kg⁻¹ fraction)	23.1	38.6	23.1	32.0	20.6	34.5	23.9	31.8	15.7	35.1
	C (g kg⁻¹ soil)	2.0	2.5	2.0	2.7	4.3	6.4	12.5	17.1	10.5	22.4
	C/N	9.2	8.4	9.9	9.0	10.7	11.8	10.7	10.6	8.5	10.7
	DV−Cseq		0.2		0.3		0.5		0.4		1.1
	FI−Cseq (%)		7.0		23.3		33.4		46.9		61.9
H₂O	mass (g 100 g⁻¹ soil)	0.5	0.7	0.8	0.8	1.2	1.2	7.5	5.8	11.3	5.6
Sum	mass (g 100 g⁻¹ soil)	102.5	98.0	98.5	99.2	99.1	99.7	100.6	100.8	102.5	101.4
	C (g kg⁻¹ soil)	5.0	10.9	3.8	6.7	9.3	15.8	18.3	28.2	18.5	37.7
	C/N	14.0	12.7	10.6	10.7	13.9	15.5	12.3	13.8	10.3	13.4
	DV−Cseq		1.2		0.8		0.7		0.5		1.0
	FI−Cseq (%)		100.0		100.0		100.0		100.0		100.0

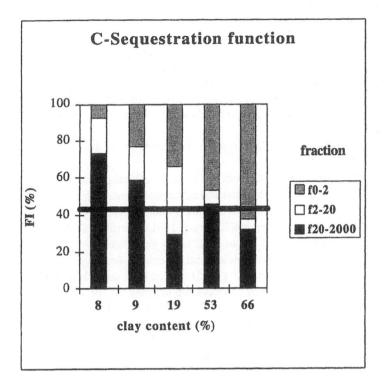

Figure 3. Functionality Index (FI) for the C sequestration function.

As the soil carbon content of the surface horizons of tropical LAC clay soil is generally strongly correlated with the clay content (Feller and Beare, 1997), we shall present the results in relation to soil texture.

1. Mass and Carbon Balances of the Fractionation Procedure

The mass balance (Sum × 100 / NF Soil) of the fractionation procedure varied between 98.0 and 102.5%. With the exception of sample PS1-Am6 with a low carbon balance of 70.2%, the other samples exhibit an acceptable carbon balance in the range of 84.4 to 103.9%.

2. Functional Pools for Carbon Sequestration

Results are presented in Table 3 and summarized for FI in Figure 3. The descriptor value DV−Cseq for the sum of the fractions varied from 0.5 (grass fallow 10 years) to 1.2 (bush fallow 6 years). The value for the 10-year artificial meadow (site Fr4) was relatively high (1.0). For each sample, the DV−Cseq for the fractions generally varied in the order : f20 to 2000 > f2 to 20 = f0 to 2, except for the artificial meadow of site Fr4, which displayed a relatively high value for f0 to 2. All situations gathered, the DV–Cseq of the 20 to 2000 μm fraction was 1.8 (site Fr2) to 5.9 (site Fr3) times higher than that of the 0 to 2 μm fraction.

However, if we consider the functionality index FI, it appeared (Figure 3) that the type of functional pools (FI > 50%) depended also on the texture.

- the plant debris compartment (20 to 2000 μm fraction) for sandy to sandy-clay soils (sites Fr3 and PS1) with clay content less than 20% ,
- the organo-clay (0 to 2 μm fraction) for clayey soils with a clay content higher than 50%.

But none of the soil fractions with an intermediate clay content of ca. 20% (site Fr2) exhibited an FI% higher than 50.

Finally, whatever the samples, the organo-silt complex never appeared as a functional pool for C-sequestration.

Different results were published about the benefits of tree plantation, agroforestry or planted pastures on C-sequestration for the tropical and subtropical zones, which seem to confirm this trend. In the Congo the positive effect of a eucalyptus plantation on SOC content of a sandy soil was attributed to >50 μm fractions (Bernhard-Reversat, 1991). Harmand (1998) and Harmand and Nitji (1998) showed in North Cameroon for an LAC sandy soil (% clay = 5) that the positive effect of different legume or non-legume tree species, and especially *Acacia polyacantha*, was mainly due (64%) to > 50 μm fractions. Lehman et al. (1998) studied for a sandy loam plinthic acrisol with 12% clay and with a litter bag experiment of 120 days, the increase in C content for different agroforestry systems including tree legume leaves inputs). For *Calliandra calothyrsus*, the increase was mainly due (near 100%) to the 20 to 2000 μm fraction. However, with *Senna siamea*, this increase only represented 21 % for the 20 to 2000 μm against 49 and 28% for the 2 to 20 and 0 to 2 μm fraction, respectively. Quiroga et al. (1996) showed that for different sandy to loamy soils (Argentina, 0 to 20 cm horizon, % clay from 5 to 25%) the positive effect of a crop-pasture (4 years crops with conventional tillage and 4 years pasture) rotation on C-sequestration as compared to continuous cropping cultivation was mainly due (ca. 66%) to fraction 50 to 2000 μm. Guggenberger et al. (1995) observed that pastures with *Digitaria decumbens* following native savannah in Columbia allowed an increase of C content in the A horizon of an oxisol with 40% clay. The increase due to 20 to 2000 μm fraction was only 21% versus 52 and 27% for the 2 to 20 and 0 to 2 μm fraction, respectively.

IV. Conclusion

The approach of SOC functional pools and quantification of the functionality was only applied here to three types of function: short-term mineralization of carbon, short-term mineralization of nitrogen and pluriannual term of C-sequestration. These examples already showed that:

- for a given soil, the functional pool and the functionality intensity will depend on the function studied,
- for a given function, a particle-size fraction can be considered as a functional pool for a specific soil but not for another, and the importance of soil texture was emphasized,
- the plant debris fraction (20 to 2000 μm) seems to play an important role in the functioning of coarse textured soils, compared to the organo-clay fraction in fine textured soils. For the functions studied, the organo-silt complex (2 to 20 μm fraction) does not appear as an important functional pool.

Therefore, the particle-size fractionation seems to be an interesting approach to identify functional SOC pools, as long as the texture is taken into consideration for the quantification of the functionality. This approach has to be extended to other SOM functions (cationic and anionic exchange, organic molecules sorption, aggregation, etc.). And different land uses, potentially interesting for C-sequestration, must also be studied in different tropical environments.

Acknowledgments

We thank the Organizing Committee of the International Workshop "Methods of Assessment of Soil Carbon Pool" (November 2–4, 1998) for financial support in the preparation of this chapter.

References

Balesdent, J. 1996. The significance of organic separates to carbon dynamics and its modelling in some cultivated soils. *European J. Soil Sci.* 47:485-493.

Balesdent, J., J.P. Petraud and C. Feller. 1991. Effets des ultrasons sur la distribution granulométrique des matières organiques des sols. *Science du Sol* 29:95-106.

Barrios, E., R.J. Buresh and J.I. Sprent. 1996. Organic matter in soil particle size and density fractions from maize and legume cropping systems. *Soil Biol. Biochem.* 28:185-193.

Bernhard-Reversat, F. 1981. Participation of light and organomineral fractions of soil organic matter in nitrogen mineralization in sahelian savanna soil. *Zbl. Bakt. II Abt.* 136:281-290.

Bernhard-Reversat, F. 1987. Litter incorporation to soil organic matter in natural and planted tree stands in Senegal. *Pedobiologia* 30: 401-417.

Bernhard-Reversat, F. 1988. Soil nitrogen mineralization under a Eucalyptus plantation and a natural Acacia forest in Senegal. *Forest Ecology and Management* 23: 233-244.

Bernhard-Reversat, F. 1991. Evolution of the soil litter interface under Eucalyptus plantation on sandy soil in Congo. *Acta Oecologica* 12: 825-828.

Bonde, T.A., B.T. Christensen and C.C. Cerri. 1992. Dynamics of soil organic matter as reflected by natural ^{13}C abundance in particle size fractions of forested and cultivated oxisols. *Soil Biol. Biochem.* 24:275-277.

Cambardella, C.A. and E.T. Elliott. 1993. Methods for physical separation and characterization of soil organic matter fractions. *Geoderma* 56: 449-457.

Cambardella, C.A. and E.T. Elliott. 1994. Carbon and nitrogen dynamics of soil organic matter fractions from cultivated grassland soils. *Soil Sci. Soc. Am. J.* 58:123-130.

Christensen, B.T. 1987. Decomposability of organic matter in particle-size fractions from field soils with straw incorporation. *Soil Biol. Biochem.* 19:429-435.

Christensen, B.T. 1992. Physical fractionation of soil and organic matter in primary particle size and density separates. p. 1-90. In: *Advances in Soil Science.* Vol. 20. Springer-Verlag, New York.

Duxbury, J.M., M.S. Smith and J.W. Doran. 1989. Soil organic matter as a source and sink of plant nutrients. p. 33-67. In: D.C. Coleman, J.M. Oades and G. Uehara (eds.), Dynamics of soil organic matter in tropical ecosystems. NifTAL Project, University of Hawaii.

Feller, C. 1993. Organic inputs, soil organic matter and functional soil organic compartments in low activity clay soils in tropical zones. p. 77-88. In: K. Mulongoy and R. Merckx (eds.), *Soil Organic Matter Dynamics and Substainability of Tropical Agriculture.* John Wiley-Sayce, Chichester.

Feller, C. 1995. La matière organique dans les sols tropicaux à argile 1:1. Recherche de compartiments organiques fonctionnels. Une approche granulométrique. ORSTOM, Coll. TDM N° 144, Paris, 393 p. + Annex.

Feller, C., G. Burtin, B. Gérard and J. Balesdent. 1991. Utilisation des résines sodiques et des ultrasons dans le fractionnement granulométrique de la matière organique des sols. Intérêt et limites. *Science du Sol* 29:77-94.

Feller, C., A. Albrecht and D. Tessier. 1996. Aggregation and organic matter storage in kaolinitic and smectitic tropical soils. p. 309-359. In: M.R. Carter and B.A. Stewart (eds.), *Structure and Organic Matter Storage in Agricultural Soils*. Advances in Soil Science,. CRC Press, Boca Raton, FL.

Feller, C. and M.H. Beare. 1997. Physical control of soil organic matter dynamics in the tropics. *Geoderma* 79: 69-116.

Gregorich, E.G., R.G. Kachanoski and R.P. Voroney. 1989. Ultrasonic dispersion of organic matter in size fractions. *Can. J. Soil Sci.* 68:395-403.

Guggenberger, G.W. and J. Thomas. 1995. Lignin and carbohydrate alteration in particle-size separates of an oxisol under tropical pastures following native savannah. *Soil Biol. Biochem.* 27: 1629-1638.

Harmand, J.M. 1998. Rôle des espèces ligneuses à croissance rapide dans le fonctionnement biog ochimique de la jachère. Effet sur la restauration de la fertilité des sols ferrugineux tropicaux. (Bassin de la Bénoué au Nord-Cameroun). Edition de la thèse de doctorat soutenue en 1997 à l'Université de Paris VI. IRAD (Cameroun) - CIRAD (France), CIRAD-Forêt, Vol. 1, Montpellier, 213 pp.

Harmand, J.M. and C.F. Njiti.1998. Effets des jachères agroforestières sur les propriétés d'un sol ferrugineux et sur la production céréalière. *Agriculture et Développement* 18:21-29.

Hassink, J. 1995. Decomposition rate constants of size and density fractions of soil organic matter. *Soil Sci. Soc. Am. J.* 59:1631-1635.

Herrick, J.E. and M.M. Wander. 1997. Relationships between soil organic carbon and soil quality in cropped and rangeland soils: the importance of distribution, composition, and soil biological activity. p.405-425. In: R. Lal, J.M. Kimble, R.F. Follett and B.A. Stewart (eds.), *Soil Processes and the Carbon Cycle*. Advances in Soil Science, Lewis Publishers, Boca Raton, FL.

Lehmann, J., N. Poidy, G. Schroth and W. Zech. 1998. Short-term effects of soil amendment with tree legume biomass on carbon and nitrogen in particle size separates in Central Togo. *Soil Biol. Biochem.* 30: 1545-1552.

Monreal, C.M., H. Dinel, M. Schnitzer and D.S. Gamble. 1997. Impact of carbon sequestration on functional indicators of soil quality as influenced by management in sustainable agriculture. p. 435-457. In: R. Lal, J.M. Kimble, R.F. Follett and B.A. Stewart (eds.), *Soil Processes and the Carbon Cycle*. Advances in Soil Science, Lewis Publishers, Boca Raton, FL.

Nicolardot, B. 1988. Evolution du niveau de biomasse microbienne du sol au cours d'une incubation de longue durée: relations avec la minéralisation du carbone et de l'azote organique. *Rev. Ecol. Biol. Sol* 25:287-304.

Quiroga, A.R., D.E. Buschiazzo and N. Peinemann. 1996. Soil organic matter particle size fractions of the semiarid Argentinian pampas. *Soil Science* 161: 104-108.

Sollins, P., G. Spycher and C.A. Glassmann. 1984. Net nitrogen mineralization from light and heavy-fraction forest soil organic matter. *Soil Biol. Biochem.* 16: 31-37.

Sternberg, B. 1998. Soil attributes as predictors of crop production under standardized conditions. *Biol. Fertil. Soils* 27:104-112.

Theng, K.G., K.R. Tate and P. Sollins. 1989. Constituents of organic matter in temperate and tropical soils. p. 5-32. In: D.C. Coleman, J.M. Oades and G. Uehara (eds.), Dynamics of soil organic matter in tropical ecosystems. NifTAL Project, University of Hawaii.

Tiessen, H., J.W.B. Stewart and H.W. Hunt. 1984. Concepts of soil organic matter transformations in relation to organo-mineral particle size fractions. *Plant Soil* 76:287-295.

Tiessen, H., J.W.B. Stewart and H.W. Hunt. 1984. Concepts of soil organic matter transformations in relation to organo-mineral particle size fractions. *Plant Soil* 76:287-295.

T.S.B.F. 1989. *Tropical Soil Biology and Fertility: A Handbook of Methods*. J.M. Anderson and J.S.I. Ingram (eds.). CAB International, Oxon, U.K., 171 pp.

Woomer, P.L. 1993. Modelling soil organic matter dynamics in tropical ecosystems: model adoption, uses and limitations. p. 279-294. In: K. Mulongoy and R. Merckx (eds.), *Soil Organic Matter Dynamics and Sustainability of Tropical Agriculture*. John Wiley & Sons, Chichester.

CHAPTER 6

Spatial Variability: Enhancing the Mean Estimate of Organic and Inorganic Carbon in a Sampling Unit

L.P. Wilding, L.R. Drees and L.C. Nordt

I. Introduction

The release (natural and anthropogenic) and sequestration of carbon has taken on increased significance in recent years due to its potential impact on global climate. Soil plays a key role in the global carbon balance because it supports all terrestrial ecosystems that cycle much of the atmospheric and terrestrial carbon (Rosenberg et al., 1999). Soil (pedosphere) also provides the biogeochemical linkage between the other major carbon reservoirs: the biosphere, atmosphere and hydrosphere (rivers, lakes and oceans). Most carbon cycles between these large dynamic reservoirs with soil as a mediator. In the broad issues of management of CO_2 and other greenhouse gases, the role that soil serves in the global carbon cycle and as a long-term reservoir of carbon is not well understood or appreciated. However, that role is changing. The pedosphere is considered to be an active and significant component in global carbon emission and sequestration potential (CAST, 1992; Lal et al., 1998c, 1999; Rosenberg et al., 1999). In fact, soil carbon sequestration is considered to be a bridge to the future in controlling increased levels of atmospheric CO_2 until other direct or indirect technologies for its control are developed (Edmonds et al., 1996 a, b; Cole et al., 1996; DOE, 1999). Because soil is an important source and sink for carbon, several workshops and conferences have addressed the issue of soil and its role in global climate change and carbon sequestration (Lal et al., 1995a, b, 1998a, b, c, 1999).

   In the past, most pedological studies on spatial variability of organic (SOC) and inorganic (SIC) forms of carbon have focused on pedon, soil series, or mapping unit variability (Ball and Williams, 1968; Bascomb and Jarvis, 1976; Beckett and Webster, 1971; Wilding and Drees, 1983; Mahinak-barzadeh et al., 1991). During this time information on spatial variability was germane to soil survey activities (Arnold and Wilding, 1991), while little attention was given to the spatial variability of these carbon reserves in estimating available carbon pools. Lal et al. (1995c) point out that "An accurate assessment of global carbon balance is faced with many uncertainties. These uncertainties may lead to erroneous data and misinterpretations"; and "There is a great deal of uncertainty in the data on global soil carbon pool and fluxes. Information generated can only be as good as the quality of input data." The quality of input data was exemplified by the comparison of the SOC content of about 1100 soil profiles representing five soil orders in temperate and tropical regions (Kimble et al., 1990). Soil organic carbon content (expressed as kg m^{-2}/50 cm) had a coefficient of variation ranging from 42 to 70%. This large variation creates problems in extrapolating data to similar soils and integrating SOC data into models of carbon flux. Thus, our intent is (1) to caption the nature of soil carbon spatial vari-

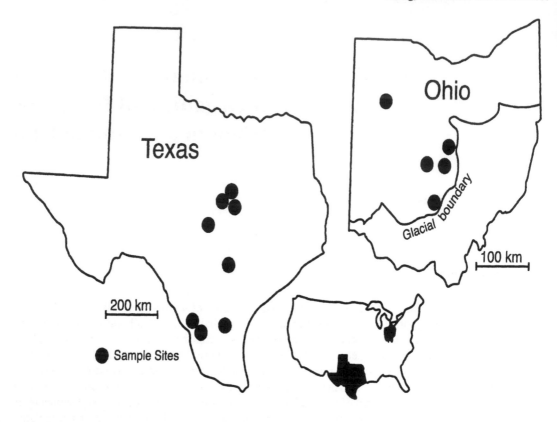

Figure 1. Location of soil sampling pedons in Ohio and Texas.

ability at the pedon level and (2) to indicate the importance of spatial variability on properties of soil carbon stocks, carbon fluxes, and the potential of soil to sequester carbon. It is necessary to understand the spatial variability of soil carbon pools in order to make accurate mean estimates of available carbon and integrate variability into predictive models of soil carbon reserves and sequestration potential.

II. Methods

A. Field

Eight pedons from Texas and five from Ohio were used in this study (Figure 1, Tables 1 and 2). These pedons have a broad range of parent materials, texture, drainage, landform position, SIC and SOC status and hence serve as a generalized data set. Additional information for these pedons can be found in Drees et al. (2000) and Nordt et al. (1999). Within the confines of a pedon (about 1 m²), four to six sampling profiles were selected using a systematic multiple sub-sampling scheme (Figure 2) (Drees and Wilding, 1973; Wilding and Drees, 1983). Within each sub-profile, four to six identical horizons were selected for analysis. Thus, close interval variability within an area of approximately 1 m² could be determined.

Table 1. Soil series and classification for profiles in study

Pedon number	Series	Classification
		Ohio profiles
PT-28	Nappanee	Fine, illitic, mesic Aeric Epiaqualfs
FR-57	Celina	Fine, mixed, active, mesic Aquic Hapludalfs
LC-5	Alford	Fine-silty mixed, superactive, mesic Utic Hapludalfs
LC-12	Ockley	Fine-loamy, mixed, active, mesic Typic Hapludalfs
PY-20	Ockley	Fine-loamy, mixed, active, mesic Typic Hapludalfs
		Texas profiles
S80TX-479-1	Brundage	Fine-loamy, mixed hyperthermic Ustollic Natrargids
S82TX-479-2	Catarina	Fine, smectitic, hyperthermic Typic Torrerts
S84TX-299-1	Voca	Fine, mixed, thermic Typic Paleustalfs
S86TX-249-1	Papalote	Fine, mixed, hyperthermic Aquic Paleustalfs
S86TX-187-1	Branyon	Fine, smectitic, thermic Udic Haplusterts
S90TX-099-1	Lewisville	Fine, smectitic, thermic Udic Calclusterts
S92TX-099-1	Bastsil	Fine-loamy, siliceous thermic Udic Paleustalfs
S97TX-099-1	Lewisville	Fine, smectitic, thermic Udic Calclusterts

Table 2. Properties of soil profiles

Pedon number	Solum thickness (cm)	Organic carbon A horizon (%)	$CaCO_3$ equivalent C horizon
	Ohio profiles		
PT-28	97	2.0	20.1
FR-57	135	1.7	19.7
LC-5	107	1.3	NC
LC-12	112	1.5	25.2
PY-20	152	1.2	41.5
	Texas profiles		
S80TX-479-1	183	0.8	17.2
S82TX-479-2	185	0.9	14.5
S84TX-299-1	154	0.9	NC
S86TX-249-1	300	0.6	25.5
S86TX-187-1	250	2.6	73.9
S90TX-099-1	270	2.3	49.8
S92TX-099-1	263	0.3	NC
S97TX-099-1	210	2.2	42.1

NC = non-calcareous

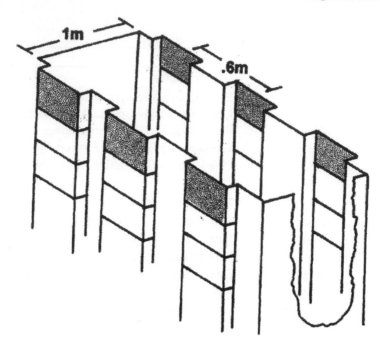

Figure 2. Generalized pedon showing location of six sampling profiles around pedon. Shaded areas represent multiple subsamples of a given horizon within a pedon.

B. Laboratory

Soil organic carbon and SIC were determined for each horizon selected. The SIC was determined using the gasometric procedure of Dreimamis (1962). This procedure determines total carbonate, which is then converted to total inorganic carbon (SIC). Organic carbon was determined by the difference between total carbon and SIC by the dry combustion method (Nelson and Sommers, 1982). Bulk density was determined by the Saran clod method of Brasher et al. (1966).

Quantitative reconstruction (mass balance analysis) techniques (Smeck and Wilding, 1980; Chadwick et al., 1990; Nordt et al., 1999) were used to access the net gain/loss of SIC and SOC for profiles where all horizons and parent material were sampled (Profiles PT-28 and LC-12). Soil profile reconstruction is used to calculate the net gain or loss (flux) on a unit volume or horizon basis of a soil constituent since the initiation of pedogenesis. Additional details on profile reconstruction can be found in Drees et al. (2000), while Nordt et al. (1999) give a sample equation for quantitative reconstruction similar to those used in this study.

The mean, standard deviation (SD) and coefficient of variation (CV) were determined for SOC and SIC for each horizon around the pedon sampling unit. For the profiles with complete profile data, the reconstructed gain of SOC or loss of SIC was also statistically analyzed. Reconstruction calculations can be made over the complete soil profile, or to an arbitrary depth. We have selected a depth of 1 meter as the depth for calculating reconstructed values because this is below the major depth of SOC sequestration in these soils and the flux of SIC during pedogenesis.

III. Significance of Spatial Variability

Spatial variability is the change in a soil property as a function of time and space. Arnold and Wilding (1991) state that "Never before in the history of soil science has the knowledge of soil spatial variability been so germane. It reaches to the heart of the pedology profession and is critical to the success of agronomic practice, agriculture development, land management, and earth science on a global scale." During the last third of the 20th century, pedologists have taken increased interest in the origin, magnitude and implications of spatial variability (Beckett, 1967; Ball and Williams, 1968, 1971; Beckett and Webster, 1971; Wilding and Drees, 1978, 1983; Upchurch et al., 1988; Mausbach and Wilding, 1991; Arnold and Wilding, 1991). The implications of spatial variability affect concepts of soil management, soil survey, soil interpretations and formulating predictive models (Ruark and Zarnoch, 1992). It is particularly germane to predict estimates of SOC and SIC sequestration and flux as these parameters are incorporated into predictive models.

Spatial variability expresses itself in various forms; depending on the parameters measured, variability may be temporal or permanent, but it is a real landscape attribute. Soil variability is anisotropic and unique. Soils are a continuum across the landscape, many properties are not single values, many properties are temporal, many properties result from interactive processes, and properties are systematically and spatially dependent. This leads to both vertical and lateral anisotropic conditions. Thus, spatial diversity within a pedon sampling unit directly affects the accuracy of mean estimates of SOC and SIC sequestration or release potential.

IV. Origin and Magnitude of Spatial Variability

From a true pedological perspective, soil is a natural three-dimensional body systematically related to the landscape. Landscapes are not static, but dynamic, changing in response to internal and external forces. Thus, soils superposed on landscapes will also change in response to local, seasonal and regional perturbations. These perturbations produce systematic and random error sources which influence pedon variability. Systematic variation is governed by processes of soil formation that are in turn conditioned by lithology, climate, biology and relief active through geologic time. Random variation is local and includes site-specific differences in infiltration, erosion, accretion, biological activity, turbation, differential lithology, differential weathering intensities, temporal affects of soil moisture or soil management, and sampling/analytical errors. Soil variability depends on the level of resolution (microscopic to megascopic), property being measured, and size of sampling unit (Figure 3). For the purposes of this discussion the focus will be on horizon and pedon sampling units. For most properties as the size of the sampling unit increases, so does the magnitude of soil variability. As the variability of a soil property increases, the precision of statements that can be made about that soil property decreases.

Within a pedon, soil properties may be divided into dynamic and stable properties. Dynamic properties are ones that are variable over space and time. These properties respond more quickly to changing environmental conditions and may vary over time periods of hours to seasons to years. Examples include pH, iron mobility, organic carbon levels and water content. Properties termed stable, such as soil texture, mineralogy and soil depth, exhibit slower response and greater permanence over time and space (Figure 4). Organic and inorganic forms of carbon may be considered metastable to dynamic as changes in climate, water balance, or land management may alter the horizon and depth distribution of these constituents (Wilding et al., 1994). Pedon variability is also controlled, in part, by parent material with variability in the order of loess < till < fluvial deposits < pyroclastics and tectonic rocks < drastically disturbed materials (Wilding and Drees, 1983). It is no wonder that soils exhibit diversity, considering the dynamic nature of complex interactive systems superposed on processes that are active over short and long time scales.

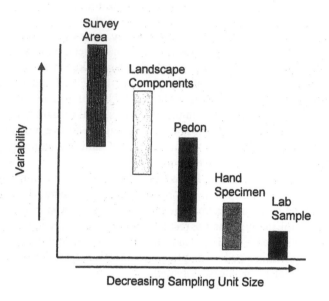

Figure 3. Relative variability of soil properties as a function of size of sampling unit. (Modified from Wilding et al., 1994.)

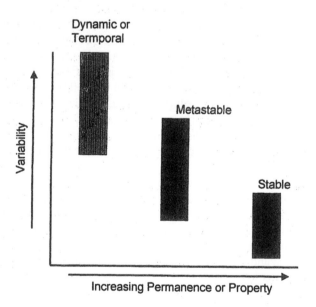

Figure 4. Relative variability of soil properties as a function of permanence. (Modified from Wilding et al., 1994.)

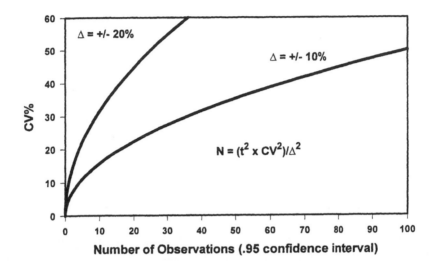

Figure 5. Coefficient of variation (CV) versus number of observations necessary to estimate population mean within limits of $\pm 10\%$ and $\pm 20\%$.

V. Statistical Considerations

It is often convenient to compare variation among parameters, independent of the property mean. Coefficient of Variation (CV = Standard Deviation/Mean x 100) is a useful index to express the magnitude of sample diversity (Trangmar et al., 1985; Wilding and Drees, 1983). Coefficient of variation measures the spread in sample values as a function of the sample mean (Upchurch et al., 1988), is commonly independent of the property mean, and allows comparison of variation among different soil properties. Caution must be exercised in using CV when the variation is covariant with the sample mean or when the sample mean is near zero. Covariations are evident when comparing properties that may vary systematically with depth, such as SOC. For example, horizons with low SOC may exhibit high CV's while samples near the surface with higher amounts of SOC have low CV's. With increasing CV, the number of samples necessary to estimate the mean within given confidence limits increases exponentially (Figure 5). For example, the confidence interval is proportional to the square root of the number of samples collected within the sampling unit. Figure 5 illustrates the exponential relationship of CV with number of observations necessary to make mean estimates at a given confidence interval. It should be noted that to achieve an accuracy of $\pm 10\%$ of the population mean for any given CV requires about four times as many observations as for $\pm 20\%$.

In most investigations, it is desirable to make probability statements about the expected or measured variance of sample mean within a sampling unit (pedon) (Young et al., 1991). Limit of accuracy (LA) curves can be constructed based on the CV, number of samples and the desired confidence level. Limit of accuracy is defined as: $LA = SE \times t$

where $SE = SD/\sqrt{n}$

Table 3. Expected percentage variation about the mean (Limit of Accuracy) for different CV values at 95% and 80% confidence levels for different number of samples

Number of samples	CV = 2%		CV = 5%		CV = 10%		CV = 25%		CV = 50%	
	95%	80%	95%	80%	95%	80%	95%	80%	95%	80%
1	5	3	11	7	23	14	57	35	113	69
2	3	2	8	5	16	10	40	25	80	49
4	2	1	6	4	11	7	28	17	57	35
6	2	1	5	3	9	6	23	14	46	28
8	2	1	4	2	8	5	20	12	40	25
10	1	1	4	2	7	4	18	11	36	22

SE = standard error of the mean
SD = standard deviation about the mean
n = number of samples
t = t distribution for selected confidence interval (t = 2.26 for n = 10 at 95% level).

The sample mean ± the LA provides a value range for a given confidence limit. In these calculations CV is substituted for SD to normalize the variation. Frequently, the 95% confidence limit is selected, but a lower confidence limit may be more appropriate in some situations where the number of samples necessary to achieve a given LA is unrealistic. Accepting a lower confidence limit of 80% will decrease the number of samples required by about half. Such curves permit one to estimate the number of subsamples within a pedon necessary to estimate the mean within certain limits with a given degree of confidence. Accurate mean estimates are essential for determining the quantity of SOC or SIC sequestered during pedogenesis. Table 3 represents limit of accuracy, or expected variation about the mean for different CV values at 95% and 80% confidence levels for different numbers of samples used to estimate the mean. For moderate CV's (15 to 35%) to high CV's (>35%) (Wilding and Drees, 1983) going from a single sample to four subsamples will reduce the variance by half.

VI. Results

A. Variability

Laboratory analysis of SOC and SIC contents were conducted within the confines of a pedon. Figures 6 and 7 illustrate SOC and SIC, respectively, for the pedon sampling scheme. Results for SOC are for Ap and BA horizons of the LC-12 pedon (Figure 6) and for SIC in a Bk and C horizon of the PT-28 pedon (Figure 7). It may be noted that for both SOC and SIC, the CV increases as the mean decreases. The question is thus raised whether SOC and SIC CV's are covariable with magnitude of respective means. The mean, SD and CV for all pedons and all horizons are summarized for SIC and SOC (Tables 4, 5). In these tables the mean and CV's were partitioned into class groups of approximately equal variability, rather than partitioned by horizon, because the population of values differ considerably in SOC or SIC content among horizons. It should be noted that for these parameters the variability does covary with class mean which influences the number of samples required to make mean estimates with a given degree of confidence. Such information is important in planning sampling strategies because there is no given number of observations appropriate to use which is applicable over

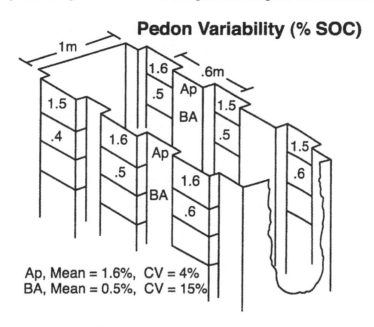

Figure 6. Pedon variability in % SOC for Ap and BA horizons of LC-12 pedon.

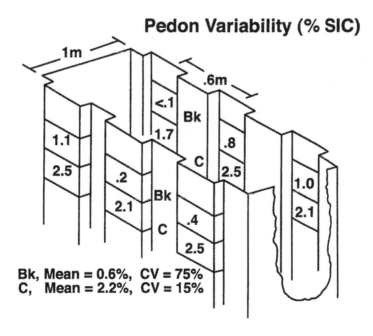

Figure 7. Pedon variability in % SIC Bk for C horizons of PT-28 pedon. Percent SIC originally measured as $CaCO_3$ and converted to SIC equivalent.

Table 4. Horizon variability in SOC (%) for 47 horizons partitioned by SOC content

	Range in SOC (%)			
	<0.2%	0.2 to 0.5%	0.5 to 1.0%[a]	>1.0%[a]
Mean[b]	0.1	0.3	0.6	1.5
CV(%)	44.7	18.1	13.8	11.6

[a]Values >0.5% usually represent A and upper B horizons; [b]4 to 6 subsamples per horizon.

Table 5. Horizon variability in SIC (%) for 28 horizons partitioned by SIC content

	Range in SOC (%)			
	<1%	1 to 2.5%	2.5 to 5%[a]	>5%[a]
Mean[b]	0.6	1.8	3.1	6.0
CV(%)	49.7	17.5	10.2	2.3

[a]Values >2.5% commonly represent Bk horizons; [b]4 to 6 subsamples per horizon.

the whole range of SOC or SIC magnitudes for a predicted probable error. Higher CV's for lower class means commonly reflect more dynamic soil regimes. For example, Figure 7 illustrates the difference in SIC content between a Bk and C horizon. The Bk horizon is more variable because of differential leaching and precipitation within the pedon while the C horizon represents less altered materials reflecting primarily static conditions of parent material composition. In the case of SOC (Table 4), lower organic values are from deeper horizons where differential root proliferation and translocation of organic constituents are largely controlled by differential water movement and plant-available water.

Based on the variability of all the horizons considered, LA curves were constructed for SOC <0.2% (Figure 8), SOC >0.2% (Figure 9), SIC <1% (Figure 10) and SIC >5% (Figure 11). The Limit of Accuracy is expressed as a percentage of the mean for two levels of confidence (80% and 95%). Because the variation has been shown to be covariant with the sample mean for SOC and SIC, many more samples are required for samples of low mean values to make accurate mean estimates. For example, for samples with <1% SIC (Figure 10), if only one sample had been collected and analyzed, one could not estimate the mean closer than about 70% of the mean value at the 80% confidence level. At the 95% confidence level, mean estimates could be expected to vary by 100% of the actual mean based on a single sample analyzed. The analysis of eight samples (either as individual samples or a composite sample) would permit accuracy of the mean estimate to about 25% at the 80% confidence level or 40% at the 95% confidence level. However, with SIC values >5% (Figure 11) analysis of a single sample would permit mean estimates with probable error of 6%, 95% confidence level or 3%, 80% confidence level. Multiple sampling would improve the accuracy, but would not be practical considering the errors of laboratory determination. While the 95% confidence limit is commonly selected, from a practical standpoint a lower confidence may be more appropriate in highly variable populations. For example, in horizons with <1% SIC (Figure 10) or SOC <0.2% (Figure 8), accepting a lower confidence limit of 80% will decrease the number of samples required by about half.

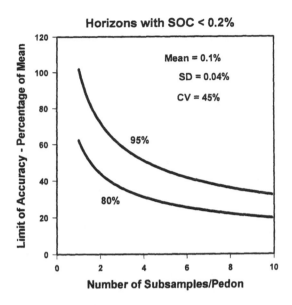

Figure 8. Number of observations needed to estimate population mean for SOC . <0.2% at 80% and 95% confidence level.

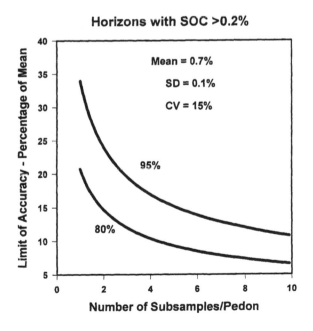

Figure 9. Number of observations needed to estimate population mean for SOC >0.2% at 80% and 95% confidence level.

L.P. Wilding, L.R. Drees and L.C. Nordt

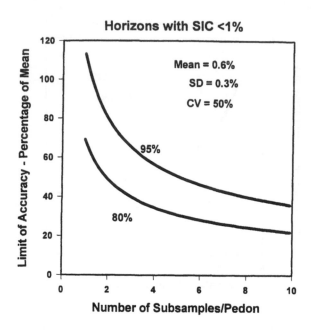

Figure 10. Number of observations needed to estimate population mean for SIC <1% at 80% and 95% confidence level.

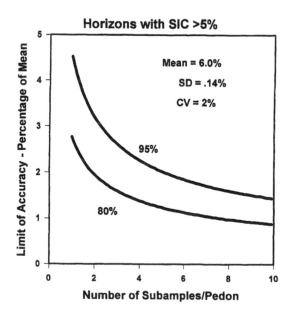

Figure 11. Number of observations needed to estimate population mean for SIC >5% at 80% and 95% confidence level.

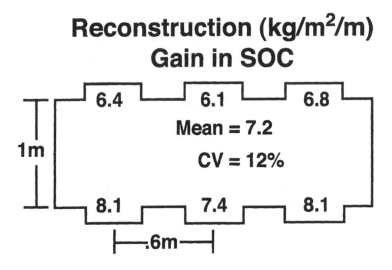

Figure 12. Variability in gain in SOC (kg m^{-2} m^{-1}) for six profiles within PT-28 pedon.

B. Reconstruction

Variability in gravimetric SOC or SIC for an individual horizon for a given parent material is only the initial step in understanding the probable errors in estimating SOC and SIC sample means. Such values are a percentage of the given horizon, but do not indicate quantity per unit volume for a soil profile. For accurate mean estimates of carbon sequestration or loss it is necessary to calculate changes in SOC and SIC per unit volume of soil using a stable reference base (Smeck and Wilding, 1980; Chadwick et al., 1990). However, reconstruction is not a panacea. It cannot indicate the rate of accumulation or loss integrated under variable periods of pedogenesis, only the net gain/loss (flux) over the entire period of soil formation.

The net gain in SOC (Figure 12) and loss of SIC (Figure 13) in kg m^{-2} m^{-1} is illustrated for pedon PT-28 subsampled at 6 profile locations within the pedon. These values represent the net sequestration of carbon within a soil pedon. In this example the values for SIC can vary from 37.8 to 43.8 kg m^{-2} m^{-1} (Figure 13) within a square meter, a difference of 6 kg m^{-2} m^{-1} or about 15% and CV = 5%. For SOC, comparable values range from 6.1 to 8.1 kg m^{-2} m^{-1} (Figure 12) with a CV of 11.5% for the sampled pedon. Again in a pedon where most of the SOC is in the cultivated Ap horizon and morphology indicates pedon uniformity, without multiple sampling schemes mean estimates can be quite variable and misleading. Such examples, especially for SIC, illustrate the variability observed in apparently uniform soils for dynamic properties influenced by pedogenic leaching processes. This also points out the difficulty of making accurate predictions of net sequestration rates in soil systems based on single observations.

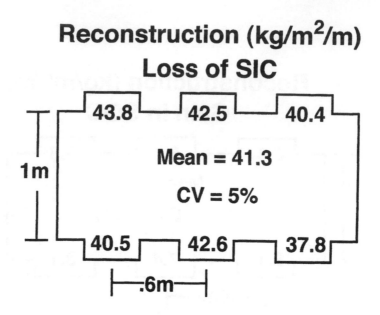

Figure 13. Variability in loss in SIC (kg m^{-2} m^{-1}) for six profiles within PT-28 pedon.

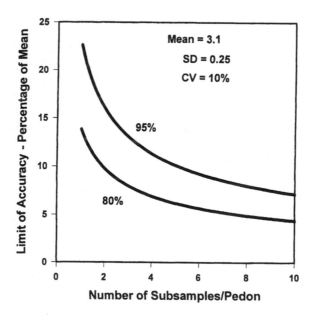

Figure 14. Number of observations needed to estimate volumetric (kg m^{-2}) gain of SOC in Ap horizons of PT-28 pedon at 80% and 95% confidence level.

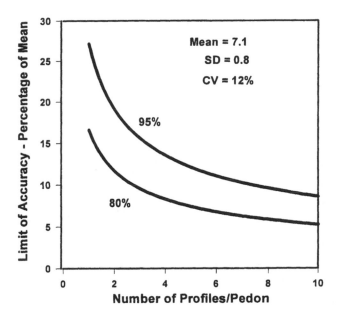

Figure 15. Number of observations needed to estimate volumetric (kg m^{-2} m^{-1}) gain of SOC for complete profile of PT-28 pedon at 80% and 95% confidence level.

Limit of accuracy curves are just as applicable in profile reconstruction estimates as they are for individual horizons. Again, profile PT-28 was used as an example with LA curves for SOC in the Ap horizon (Figure 14) and complete profile (Figure 15). Limit of accuracy curves are also given for SIC for the complete profile (Figure 16). Mean estimates of SOC within a pedon using only one profile/pedon cannot be estimated closer than about ±1.9 kg m^{-2} m^{-1} (LA = 27% times mean of 7.1% = 2.0 kg m^{-2} m^{-1}) using a 95% confidence interval. This is under the most ideal soil conditions (CV's = 10 to 15%). Similar mean estimates for SIC total fluxes are about ±6 kg m^{-2} m^{-1} (LA = 11% times mean of 41 kg m^{-2} m^{-1} = 4.5 kg m^{-2} m^{-1}) under the least variable soil conditions (CV's = 5%) and at the 95% confidence level (Figure 16). However, if four subsamples are taken within a sampling unit and analyzed individually or as a composite sample, the probable error can be halved. Likewise, accepting a lower confidence interval will also reduce the probable error.

Estimates of accuracy of mean values or the number of samples required to achieve a given level of mean confidence can only be obtained once the sample variation is known or estimated based on prior or ancillary information. This requires multiple sample analysis, which can be laborious, time consuming and costly. However, once the variation is known for a given population or can be estimated from other work, composite sampling of *n* number of samples should achieve the same accuracy as statistically analyzing *n* individual samples (Ruark and Zarnoch, 1992). In both cases the mean should be identical but in composite sampling one does not gain the variability knowledge. Composite sampling may take a little longer, but compositing six to ten subsamples around a pedon will have the effect of substantially increasing the accuracy of mean estimates.

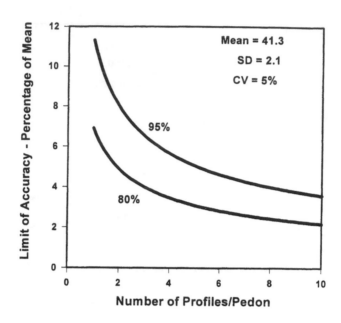

Figure 16. Number of observations needed to estimate volumetric (kg m^{-2} m^{-1}) loss of SIC for complete profile of PT-28 pedon at 80% and 95% confidence limit.

VII. Conclusions

The pedosphere contains major carbon reserves and provides the biogeochemical linkage between other major carbon reserves. Uncertainties in determining the carbon balance in soil may lead to erroneous data and misinterpretations of extrapolated data. Hence, variability information is directly germane to the matter of verification of carbon stocks in soils. Although uncertainties cannot be eliminated, understanding the source and magnitude of variability in SOC and SIC on a horizon or pedon level can help provide mean estimates of these carbon reserves with a given degree of accuracy and confidence. Accurate estimates are vital in any quantitative modeling of carbon release or sequestration. The accuracy of mean estimates is proportional to the quantity of SOC or SIC in a horizon or profile because these two parameters have CV's covariable with mean size (Tables 4 and 5). In some situations, such as low SOC or SIC contents, a lower confidence level of mean estimates may be more appropriate rather than obtaining additional samples to reduce sampling errors (Table 3).

The calculations and figures given herein are for selected soils and should not be interpreted as baseline values for estimating SOC or SIC in all soil resource conditions. Rather, these data are intended to emphasize the need to consider spatial variability in any estimate of soil carbon balance. Composite sampling can significantly enhance accuracy of estimating means of SOC and/or SIC in a pedon (Table 3). Hence, it is strongly recommended that compositing four to six subsamples be conducted as the norm for SOC and SIC measurements, once the magnitude of soil variability of these parameters are determined with a pilot project. This is the most accurate, efficient and cost-effective sampling scheme to verify carbon stocks or carbon fluxes in soils.

References

Arnold, R.W. and L.P. Wilding. 1991. The need to quantify spatial variability. p. 1-8. In: M.J. Mausbach and L.P. Wilding (eds.), *Spatial Variabilities of Soils and Landforms*. Soil Sci. Soc. Am. Spec. Publ. 28. Soil Science Society of America, Madison, WI.

Ball, D.F. and W.M. Williams. 1968. Variability of soil chemical properties in two uncultivated Brown earths. *J. Soil Sci.* 19:379-391.

Ball, D.F. and W.M. Williams. 1971. Further studies on variability of soil chemical properties: efficiency of sampling programmes on an uncultivated Brown earth. *J. Soil Sci.* 22:60-68.

Bascomb, C.L. and M.G. Jarvis. 1976. Variability in three areas of the Denchworth soil map unit. I. Purity of the map unit and property variability within it. *J. Soil Sci.* 27:420-437.

Beckett, P.H.T. 1967. Lateral changes in soil variability. *J. Aust. Inst. Agric. Sci.* 33:172-179.

Beckett, P.H.T. and R. Webster. 1971. Soil variability: a review. *Soils Fert.* 34:1-15.

Brasher, B., D.P. Franzmeier, V.T. Vallassis and S.E. Davidson. 1966. Use of Saran resin to coat natural soil clods for bulk density and water retention measurements. *Soil Sci.* 101:108.

CAST. 1992. Preparing U.S. agriculture for global climate change. Council for Agricultural Science and Technology Task Force Report No. 119.

Chadwick, O.A., G.H. Brimhall and D.M. Hendricks. 1990. From a black to a gray box — a mass balance interpretation of pedogenesis. *Geomorphology* 3:369-390.

Cole, V., C. Cerri, K. Minami, A. Mosier, N. Rosenberg and D. Sauerbeck. 1996. Agricultural options for mitigation of greenhouse gas emissions. p. 745-771. In: R.T. Watson, M.C. Zinowera and R.H. Moss (eds.), Climate Change 1995. *Impacts, Adaptations and Mitigation of Climate Change, Report of IPCC Working Group II*. Cambridge University Press, New York.

DOE. 1999. More tools may be at hand to combat global warming. News release from Pacific Northwest National Laboratory, Jan. 20, 1999.

Drees, L.R. and L.P. Wilding. 1973. Elemental variability within a sampling unit. *Soil Sci. Soc. Am. Proc.* 37:82-87.

Drees, L.R., L.P. Wilding and L.C. Nordt. Reconstruction of inorganic and organic carbon sequestration across broad geoclimatic regions. In: R. Lal and K. McSweeney (eds.), *Soil Management for Carbon Sequestration*. SSSA Special Publication. Soil Science Society of America, Madison, WI (in press).

Dreimanis, A. 1962. Quantitative gasometric determination of calcite and dolomite by using the Chittic apparatus. *J. Sediment. Petrol.* 32:520-529.

Edmonds, J., M. Wise, H. Pitcher, R. Richels, T. Wigley and C. MacCracken. 1996a. An integrated assessment of climate change and the accelerated introduction of advanced energy technologies: An application of MiniCAM 1.0. *Mitigation Strategies for Climate Change* 1(4):311-339.

Edmonds, J., M. Wise, R. Sands, R. Brown and H. Kheshgi. 1996b. Agricultural land-use, and commercial biomass energy: A preliminary integrated analysis of the potential of biomass energy for reducing future greenhouse related emissions. PNNL-11155. Pacific Northwest National Laboratories, Richland, WA.

Kimble, J.M., H. Eswaran and T. Cook. 1990. Organic carbon on a volume basis in tropical and termerate soils. p. 248-253. *Trans. 14th Int. Cong. Soil Sci.*, Volume 5. Kyoto, Japan.

Lal, R., J. Kimble, H. Eswaran and B.A. Stewart. 1999. *Global Climate Change and Pedogenic Carbonates*. Lewis Publishers, Boca Raton, FL.

Lal, R., J.M. Kimble and R.F. Follett. 1998a. *Soil Processes and the Carbon Cycle*. Lewis Publishers, Boca Raton, FL.

Lal, R., J.M. Kimble and R.F. Follett. 1998b. *Management of Carbon Sequestration in Soils*. Lewis Publishers, Boca Raton, FL.

Lal, R., J.M. Kimble, R.F. Follett and C.V. Cole. 1998c. *The Potential of U.S. Cropland to Sequester Carbon and Mitigate the Greenhouse Effect*. Sleeping Bear Press, Chelsea, MI.

Lal, R., J. Kimble, E. Levine and B.A. Stewart. 1995a. *Soils and Global Change*. Lewis Publishers, Boca Raton, FL.

Lal, R., J. Kimble, E. Levine and B.A. Stewart. 1995b. *Soil Management and Greenhouse Effect*. Lewis Publishers, Boca Raton, FL.

Lal, R., J. Kimble, E. Levine and C. Whitman. 1995c. World soils and greenhouse effect: an overview. p. 1-7. In: R. Lal., J. Kimble, E. Levine and B.A. Stewart (eds.), *Soils and Global Change*. Lewis Publishers, Boca Raton, FL.

Mahinakbarzadeh, M., S. Simkins and P.L.M. Veneman. 1991. Spatial variability of organic matter content in selected Massachusetts map units. p. 231-242. In: M.J. Mausbach and L.P. Wilding (eds.), *Spatial Variabilities of Soils and Landforms*. Soil Sci. Soc. Am. Spec. Publ. 28. Soil Science Society of America, Madison, WI.

Mausbach, M.J. and L.P. Wilding (eds.). 1991. *Spatial Variabilities of Soils and Landforms*. Soil Sci. Soc. Am. Spec. Publ. 28. Soil Science Society of America, Madison, WI.

Nelson, D.W. and L.E. Sommers. 1982. Total carbon, organic carbon and organic matter. p. 539-580. In: A.L. Page (ed.), *Methods of Soil Analysis, Part 2*. 2nd ed. Agronomy 9, ASA, Madison, WI.

Nordt, L.C., L.P. Wilding and L.R. Drees. 1999. Pedogenic carbonate transformations in leaching soil systems: implications for the global C cycle. p. 43-64. In: R. Lal, J. Kimble, H. Eswaran and B.A. Stewart (eds.), *Global Climate Change and Pedogenic Carbonates*. Lewis Publishers, Boca Raton, FL.

Rosenberg, N.J., R.C. Izaurralde and E.L. Malone (eds.). 1999. *Carbon Sequestration in Soils: Science, Monitoring and Beyond*. Battelle Press, Columbus, OH.

Ruark, G.A. and S.J. Zarnoch. 1992 Soil carbon, nitrogen and fine root biomass sampling. *Soil Sci. Soc. Am. J.* 56:1945-1950.

Smeck, N.E. and L.P. Wilding. 1980. Quantitative evaluation of pedon formation in calcareous glacial deposits in Ohio. *Geoderma* 24:1-16.

Trangmar, B.B., R.S. Yost and G. Uehara. 1985. Application of geostatistics to spatial studies of soil properties. *Adv. Agron.* 38:45-94.

Upchurch, D.R., L.P. Wilding and J.L. Hatfield. 1988. Methods to evaluate spatial variability. p. 201-229. In: L.R. Hossner (ed.), *Reclamation of Surface-Mined Lands*. CRC Press, Boca Raton, FL.

Wilding, L.P., J. Bouma, and D.W. Goss. 1994. Impact of spatial variability on interpretive modeling. p. 61-75. In: R.B. Bryant and R.W. Arnold (eds.), *Quantitative Modeling of Soil Forming Processes*. Soil Sci. Soc. Am. Spec. Publ. 39. Soil Science Society of America, Madison, WI.

Wilding, L.P. and L.R. Drees. 1978. Spatial variability: a pedologist's viewpoint. p. 1-12. In: *Diversity of Soils in the Tropics*. Soil Sci. Soc. Am. Spec. Publ. 34. Soil Science Society of America, Madison, WI.

Wilding, L.P. and L.R. Drees. 1983. Spatial variability and pedology. p. 83-116. In: L.P. Wilding, N.E. Smeck and G.F. Hall (eds.), *Pedogenesis and Soil Taxonomy I. Concepts and Interactions*. Elsevier, New York.

Young, F.J., J.M. Maatta and R.D. Hammer. 1991. Confidence intervals for soil properties within many units. p. 213-229. In: M.J. Mausbach and L.P. Wilding (eds.), *Spatial Variabilities of Soils and Landforms*. Soil Sci. Soc. Am. Spec. Publ. 28. Soil Science Society of America, Madison, WI.

CHAPTER 7

Assessment of Soil Organic Carbon Using the U.S. Soil Survey

R.B. Grossman, D.S. Harms, D.F. Kingsbury, R.K. Shaw and A.B. Jenkins

I. Introduction

Assessment of the terrestrial soil organic carbon (SOC) pool for the U.S. requires an appreciation of the information available from soil survey. The reason is that soil survey deals with the areal quantification of land at a very pertinent scale. Relatedly, the amount of information in the soil survey is very large; some 70 to 80% of the U.S. has first generation soil surveys with a higher proportion for private than for public land. Soil survey systematics is discussed briefly. Both the estimates of SOC for areas of land (map unit components) and the laboratory data base of point samples of SOC are explored. Bulk density and rock fragment volume needed to calculate SOC on a volume basis are considered. Examples of areal expression of SOC are included. Standing biomass is briefly considered. Methods for successive estimates of organic carbon are discussed.

II. Soil Survey Systematics and Practice

The soil series is the lowest and most detailed category in Soil Taxonomy (Soil Survey Staff, 1999). Soil series share a unique set of chemical and physical properties including temperature and precipitation. Series are defined by ranges of soil properties. The defining range cannot be so wide that large interpretive (use and management) differences occur within a series. Nor can the range be so narrow that the series is difficult or impractical to map.

Soil mapping involves delineating individual soil polygons on landscapes. Landforms, slope inflections, and other landscape relationships mark the soil boundaries. The soil delineations that have common properties form a map unit — the concept that is mapped. The map unit is usually named by both soil series and by other features that pertain to use and management that are not part of the series; slope and erosion are examples. A map unit consists of soil components. There are both named and unnamed components. The components that are contrasting in terms of interpretations are named and discussed.

There are about 20,000 soil series and probably over 50,000 map unit components. The estimated properties stipulated for the map unit components are used to make interpretations. Soil properties for map units of soil surveys are stored in the National Soil Information System (NASIS) (USDA-NRCS, 1998) at the level of the map unit component (http://nasis.nrcs.usda.gov., consult the home page). Structured queries of database relationships may be made about any portion of the database. For example, the average organic matter values for 0 to 10 cm, 10 to 20 cm, and 20 to 30 cm depths can be obtained for a selected component in a map unit, for a selected map unit, or for all map units within a given soil survey area.

NASIS can store point data and permits queries of profile descriptions, laboratory data, field measurements, and transect data. Point data can be queried at various levels to help determine properties of components within a map unit. Laboratory data has not yet been entered. The latter information may, however, be obtained from the National Soil Characterization Database homepage (http://www.statlab.iastate.edu/soils/ssi/natch_data.html).

III. Organic Carbon Database

A. Interpretive Database

For each map unit component estimates of the weight percent organic matter (organic carbon x 1.7) are given by soil layer for the <2mm fraction. Conversion to an areal basis is discussed in Section IV. The estimates of organic matter do not necessarily reflect recent change to no-till nor into or out of the Conservation Reserve Program (CRP). Trees can be harvested by clear cut, replanted, and grown for 50 years without change in the interpretive organic matter database. Section VII A gives an effect of use. Both a range and a so-called Representative Value (RV) are provided. The RV is the most commonly applicable or expected value. The RV commonly is the mean of the range, but need not be so.

B. Point Database

Standardized soil sampling in the soil survey has been carried out for about 50 years. In the federal program, about 11,000 point sites have been analyzed and classified. Agricultural Experiment Stations have also sampled a large number of point sites. Site descriptions include soil use and vegetation. It is difficult, however, to relate soil use and organic carbon (SOC) values because use and vegetation information has not been generalized into classes. Further, sampling has been largely to classify the soil; soil use commonly affects the near surface, which may not be relevant to the classification. Additionally, samples collected a long time ago may not reflect recent changes in agricultural practices. Finally, the federal data collection has been largely request driven and not closely planned nationally. Consequently, data may be lacking for important soils.

The vast majority of organic carbon determinations in the federal soil survey database are by the Walkley-Black method (Walkley and Black, 1934; method 6A1 in Soil Survey Laboratory Staff, 1996). The federal Soil Survey Laboratory has assumed a recovery factor for the Walkley-Black method of 77%. For mineral soils, there is a tendency for the recovery factor to be higher for surface layers than for lower. Reinsch et al. (1987) for samples with less than 3% SOC determined a mean recovery factor of 77% for A horizons and 72% for B horizons leading to an overestimation. For most soils, much of the organic carbon occurs in shallow layers (A horizons); therefore, the lower recovery for subsoils has reduced importance. Walkley-Black gives low recovery for charcoal and very low recovery for partially decomposed plant residue and fresh organic matter.

Recently, dry combustion methods have been used (Method 6A2e in Soil Survey Staff, 1996). In dry combustion methods, carbonate carbon is either removed with acid before combustion or measured separately and subtracted from the total carbon. The Soil Survey Laboratory measures the carbonate manometrically and subtracts it from the total carbon (Method 6E1h). Sherrod (2000) presents a volumetric method for carbonate that has a detection limit of 0.25% $CaCO_3$ (0.03% inorganic carbon).

IV. Areal Calculation

A. General Equation

Summation of SOC by layers on an areal basis, here kg m^{-2}, employs the equation:

$$SOC = \frac{L_1 \, x SOCP_1 x \rho_1 \left(1 - \frac{V_{>2_1}}{100}\right) + L_2 \, x SOCP_2 x \rho_2 x \left(1 - \frac{V_{>2_2}}{100}\right) + ...}{10} \tag{1}$$

Where SOC = Soil organic carbon in kg m^{-2}
SOCP = Soil organic carbon percent
L = Thickness of the layer in cm
ρ = Moist bulk density of the <2 mm fabric; usually at 33 kPa suction
V>2 = Volume percent >2 mm

The 10 in the denominator is the quotient of 10,000 cm^2 m^{-2} and the product of 1000 g kg^{-1} and the 100 necessary to convert from percent organic carbon to a weight fraction.

Both moist bulk density (ρ) of the <2mm and the volume percent rock fragments (V>2) are necessary. For all bulk density methods, both the weight and the calculated volume of the rock fragments are subtracted from the weight and in-place volume, respectively, of the field sample to obtain the bulk density of the <2mm fabric.

B. Bulk Density

In agriculture, bulk density is obtained by dividing the ovendry weight of the <2mm fabric by the volume of <2mm fabric at or near field capacity. If the soil is drier than field capacity, both bulk density and to a lesser extent the field volume estimates of rock fragments may be higher. Bulk density is subject to the most uncertainty at shallow depths where organic carbon is usually the highest. This review does not consider nuclear radiation methods except in passing. Campbell (1994) and Culley (1993) review radiation methods. Campbell (1994) is a particularly fine review of bulk density generally.

If the sample depth is a few decimeters, it is feasible to wet a sample point to near field capacity and subsequently determine the bulk density. A ring approximately 30 cm in diameter is inserted to the shallowest depth that the water can be retained. The objective is to remove the ring without disturbing the soil. The water is allowed to infiltrate overnight, and bulk density is determined by layer.

1. Organic Soils

There are two principal classical approaches (test D4531-86; American Society of Testing Materials, 1999; Parent and Caron, 1993). One is to coat pieces of the fabric with wax and obtain the volume by immersion in water. The other is to remove a core of known volume. Much effort has been expended on development of coring devices. Sheppard et al. (1993) present a thorough review.

2. Clod Method

In the clod method (4A1 in Soil Survey Laboratory Staff, 1996; Blake and Hartge, 1986) pieces of soil 60 to 70 mm on edge are immersed briefly in the field in liquid Saran. On evaporation of the solvent a flexible coating is formed that is impervious to liquid water but is pervious to water vapor. Volume is determined by weight loss on immersion in water. The pieces of soil are equilibrated at 33 kPa (10kPa if coarse) and the volume is determined. The soil is ovendried and the volume again determined. Bulk density of the <2 mm is the ratio of the ovendry weight divided by the volume at 33 (or 10) kPa, both corrected for rock fragments.

The method has the important advantage that samples can be obtained while the soil is below field capacity. Dry samples are equilibrated at a water content near field capacity and bulk density obtained at what is considered a "standard water state" in soil survey.

If the soil is at or near field capacity, it is unnecessary to equilibrate at a given suction and the equipment and facilities become simple. Further, wax may be substituted for Saran because flexibility of the coating to allow for expansion of the soil is not a consideration. Wax can be used in a closed room whereas Saran, because of the solvent, cannot.

The clod method is considered generally to overestimate the bulk density from 0.07 to 0.09 g cm^{-3} (Tisdall, 1951; Van Remortel and Shields, 1993; Blake and Hartge, 1986) because of a tendency to exclude voids between structural units. The clod method has been a source of overestimation of the bulk density of tillage zones. Commonly, a single layer is recognized that contains the recently tilled and loosened upper part of the tillage zone together with the subjacent compacted part (compacted in the operation that loosens and bulks the upper part). Clods commonly can only be taken from the compacted part. Inclusion of the two parts together has been justified on the assumption that composition is the same due to frequent mixing. The differences between the measured and actual bulk densities can be large. The loosened part may have a bulk density of 0.90 to 1.20g cm^{-3} and the compacted part 1.40 to 1.60 g cm^{-3}. The measured bulk density (and hence the interpretive value) would be 1.50 g cm^{-3}. Assuming equal thickness of the two zones and the midpoint in the bulk density ranges, the applicable value for computation of SOC on an areal basis should be 1.28 g cm^{-3} instead of 1.50 g cm^{-3}.

It is feasible to use the clod method to measure the bulk density at or near complete dryness (ρd) and at an intermediate water content (ρi) and then to calculate the bulk density at field capacity (ρf). This makes it possible to obtain the needed bulk density (ρf) without equilibration at 33 (or 10) kPa and hence avoids the need of a sophisticated laboratory. The sample removed must be free of desiccation cracks both at the intermediate water content and completely dry. The bulk density (ρi) at intermediate water content is calculated as follows (Grossman et al., 1992):

$$\rho i = \rho f + (\rho d - \rho f) W^1 \tag{2}$$

Where

$$W^1 = \frac{Wf - Wi}{Wf - (W_{15} x F)} \tag{3}$$

Wf   =   Gravimetric water content at field capacity for <2mm
Wi   =   Gravimetric intermediate water content for <2mm
W_{15} =   Gravimetric water content at 1500 kPa for <2mm
F   =   Factor from 0.4 to 0.6, which decreases as extensibility increases

Water quantities can be obtained as follows:

Wf — This may be obtained from a similar layer that has been characterized.
Wi — Measured
W_{15} — 0.4 times the clay percent plus twice the SOC

Equation (2) is rearranged to obtain the estimate of the bulk density at field capacity (pf):

$$pf = \frac{pi - (pd \times W^1)^1}{1 - W^1} \tag{4}$$

To recapitulate, equations 2 to 4 may be used to calculate the bulk density at field capacity from the bulk density exclusive of dessication cracks at intermediate water content. A major uncertainty is the estimation of the water content at field capacity. In most situations, it is more direct to wet the site (section B).

3. Standard Core Method

In the standard procedure (Blake and Hartge, 1986), cores 8 to 10 cm diameter and about the same height are inserted to full height into the soil, broken out, and the bottom cut flush with the base of the cores. Alternatively, the cores may be inserted 2 or 3 cm, the empty space determined, and subtracted from the core volume to obtain the volume of the soil. This is referred to as the variable height core method. To execute, a piece of slotted U-shaped rod (shelf standard) is placed across the core. The vertical distance to the soil is measured at several points with a piece of retractable ruler dropped through the slots. The average of these depths is subtracted from the distance to the base while the core is empty. The difference is multiplied by the area of the core to obtain the volume of the specimen. Culley (1993) describes the use of glass beads to determine the empty volume. Incomplete filling of the core has the advantage that less compaction occurs than if the core is inserted its complete height. The core methods described cannot be used if the soil is dry and hard, very fragile, or rock fragments precluded insertion.

For the variable height core method and two excavation methods to follow, the same alignment tool is used. It may be useful to describe the tool. A CPVC compression coupler 75 mm long is used. BIBB neoprene washers 19 mm diameter are placed in each end. A length of tape 30 cm long is taken from a retractable ruler. Sectors are cut from each washer, to provide just enough room for the free downward movement of the tape. The tape is dropped through the alignment tool to the surface from which the height is determined. The concave side of the tape faces outward. The flat end of the compression coupler provides a reference plane.

4. Excavation Procedures

The methods discussed here are attractive if the soil layer is fragile or high in larger rock fragments prohibiting insertion of cores. An advantage of the excavation method is that depth may be controlled by soil morphology instead of an arbitrary standard depth.

Blake and Hartge (1986) and Test D 1556-90, American Society of Testing Materials, 1999, describe the sand-funnel method. One problem is that it may be difficult to keep the sand dry and to avoid variable compaction. A second problem is the need to level the surface which changes the specimen. Flint and Childs (1984) describe an excavation procedure that is similar to the sand cone method but use plastic spheres. The procedure is designed for sites that are steep, remote and high in rock fragments. The method makes it unnecessary to flatten the soil on steep slopes. The rubber-balloon method is an alternative (Blake and Hartge, 1986 and Test D 2167-94, American Society of Testing Materials, 1999). For weak soil fabrics such as the mechanically bulked part of the tillage zone, the pressure exerted in the filling of the balloon increases the volume of the hole. The method was designed for evaluation of compaction of fill which usually is much stronger than most near surface layers. Blake and Hartge (1986) describe a method which is similar to the frame method to be described. Shipp and Matelski (1965) describe a procedure in which holes are sprayed with a substance that is impermeable to water and the volume determined by pouring water in the hole up to a guide. Muller and Hamilton (1992) give a method that uses expanding polyurethane foam to form a cast of the hole which is used to obtain the volume. Howard and Singer (1981) present a method that is similar to the compliant cavity method that follows. A difference is that the ground surface must be planned and hence the very near surface disturbed.

To follow are three bulk density methods developed by the first and second writers. Additionally, a method is described for area SOC that does not involve bulk density.

a. Compliant Cavity

An annulus of soft plastic foam, 13 to 15 cm inside diameter, is placed on the soil surface (Bradford and Grossman, 1983). It is covered by a rigid annulus with the same inside diameter. The assembly is mounted on the soil surface with three threaded rods that pass through holes in the rigid annulus. A cavity is formed that because of the foam annulus is compliant with irregularities of the soil surface. The cavity is lined with thin plastic. A bar is placed across the cavity to which is attached a hook gauge. Water is added to the cavity from a graduated cylinder up to the tip of the hook gauge. (A liquid that does not freeze may be used.) The soil is then exposed and excavated 2 to 5 cm, and the volume of water determined again. The difference between the two volumes of water is the volume of the specimen. Successive depths may be removed. The diameter selected was determined by the scale of tillage-determined features, e.g., shoulder versus interrow axis.

b. Non-Removable Ring

A ring 20 cm diameter and 20 cm high is driven into the soil to below the depth to be sampled. A piece of shelf standard is placed across the top of the ring and the distance to the soil surface is measured with the alignment tool described previously at about 16 points in two transects at right angles. Soil is removed to the edge of the ring and the distance to the soil surface measured again. The mean difference between the two distances multiplied by the inside area of the ring gives the volume of soil removed. Successive depths of soil may be removed.

c. Frame

A square frame 0.1 m² (31.6 cm on edge) is mounted on the soil surface with threaded rods. A plastic plate is mounted on the frame. The plate has five parallel equally spaced slots that extend close to the edge. The mean height to the ground surface is obtained from measurements with the alignment tool at 30 to 40 points. The plate is then removed and soil excavated flush to the inside of the frame. The plate is returned and the mean distance again measured. From the change in mean distance to the soil and the area within the apparatus, the volume of the soil is calculated. The area of the sample is probably sufficient to measure up to 20-mm rock fragments.

d. Cord Method

An area square or rectangular commonly about 1m^2 is established with a length of cord. In our experience, 1 m^2 is suitable for Oi horizons, $\frac{1}{2}$ m^2 for Oe, and $\frac{1}{4}$ m^2 for Oa or A horizons. All soil material within a vertical plane marked by the string is removed to the base of the layer of interest. Rock fragments up to 20 mm are included in the field sample. The ovendry weight of the <2 mm fraction is obtained and the organic carbon measured. The product of the percentage of organic carbon and the weight of <2 mm gives the weight of organic carbon for the area sampled. A correction is made for the total volume of rock fragments sampled. The equation for calculation of SOC in kg m^{-2}:

$$SOC = \frac{W_{<2} \, xSOCPx \left(1 - \frac{V_{>2}}{100}\right)}{SA}$$

(5)

Where

$W_{<2}$	—	Ovendry weight of <2mm in kilograms
SOCP	—	Soil organic carbon percent
$V_{>2}$	—	The volume percent of >2mm
SA	—	Surface area from which the sample is taken in square meters

5. Estimation

Adams (1973) gives an equation applicable to both organic and mineral soil materials:

$$\rho a = \frac{100}{\dfrac{OMP}{0.22} + \dfrac{100 - OMP}{\rho m}}$$

(6)

Where

ρa	—	The bulk density adjusted for organic matter
OMP	—	The organic matter percent
ρm	—	The bulk density with zero organic matter

For 20% organic matter and a mineral bulk density of 1.60 g cm^{-3}, the calculated bulk density by the Adams equation is 0.71 g cm^{-3}. For the same inputs, Curtis and Post (1964), and Huntington et al. (1989) both yield values about 0.50 g cm^{-3}.

a. Organic Layers

Organic soil layers as defined by soil taxonomy contain at least 12% organic carbon with no clay which rises to 18% organic carbon at 50% clay (Soil Survey Division Staff, 1998).

Table 1. Estimates of bulk densities for O horizons

Horizon	Jenkins (1998)	Huntington et al. (1989)	Lynn et al. (1974)	Boelter (1969)
	(g cm⁻³)	(g cm⁻³)	(g cm⁻³)	(g cm⁻³)
Oi	0.07	0.15	0.10	<0.075
Oe	0.11	0.15	0.10 - 0.20	0.075 - 0.20
Oa	0.26	0.20	0.10 - 0.30	≥0.20

Histosols are usually wet soils and all have an organic layer ≥40 cm thick. The only exception to the wetness stipulation is soils (Folists) with the organic layer directly on rock or soil very high in rock fragments.

Organic soil layers are commonly described on state of decomposition, which is based on the volume percent of fibers that persist after a prescribed rubbing (Soil Survey Laboratory Staff, 1996: Method 8G1). The volume of fibers would be used to assign bulk density for low mineral material in place of equation 6. Three degrees of decomposition are recognized: fibric (Oi), hemic (Oe), and sapric (Oa). The fiber volume in one study ranged from 80 to 100% for fibric, 20 to 75% for hemic, and 0 to 17% for sapric (Lynn et al., 1974). Boelter (1969) uses unrubbed fiber >0.1mm obtained by a wet sieving procedure after dispersion instead of rubbed fiber. Fibric was defined as having >67% unrubbed fiber, sapric ≤ percent, and hemic intermediate. Boelter's work suggests that unrubbed fiber measurements may be more predictive of bulk density than rubbed fiber.

Decomposition based on fiber volume is predictive of bulk density if the mineral content is low. Table 1 shows bulk density estimations for soil materials relatively low in mineral material by Jenkins (1998) and by Huntington et al. (1989) for soils other than Histosols, and by Lynn et al. (1974) and Boelter (1969) for Histosols. An alternative approach to fiber content is the von Post humification scale measurement. In this test, the specimen is pressed in the palm. Characteristics and properties of expressed material are the basis for placement (Parent and Caron, 1993). Silc and Stanek (1977) obtained a linear relationship between humification on the von Post scale and bulk density with $R^2 = 0.88$. Loss-on-ignition at 450°C to predict bulk density of organic layers has been studied by Griegal et al. (1989) for northern Minnesota. They report a relationship with an $R^2 = 0.89$ for a range in bulk density from 0.02 to 0.19 g cm⁻³. Lynn et al. (1974) also report that mineral content (reciprocal of loss-on-ignition) and bulk density are closely related.

In summary, it would seem practicable for organic soil materials to combine either fiber percentage or the humification index with loss-on-ignition to predict bulk density. The fiber percentage alone and hence the taxonomic horizon designations suffice if mineral material is low. For such estimates, Equation 6 would be used substituting the estimate of the bulk density at 100% OMP for 0.23.

b. Mineral Layers

A procedure has been developed that employs particle size, moist rupture resistance, and clay mineralogy to calculate the mineral bulk density devoid of organic matter (ρm of equation 6'). Table 2 contains estimates of ρm up to 35% clay for combinations of moist rupture resistance and family particle size, classes (Soil Survey Staff, 1995), and grading. Table 3 gives the estimates of ρm for above ≥35% clay based on moist rupture resistance, 2 to 0.1 mm to 2 to 0.002 mm ratio, and the clay mineralogy. The values of ρm assume a particle density of 2.65 g cm⁻³. ρm may be adjusted by multiplying by the ratio of the computed particle density over 2.65. The particle density may be computed based on extractable iron and glass of rhyolitic composition (Soil Survey Division Staff,

Table 2. Bulk density estimates for combinations of moist rupture resistance, family texture classes excluding the >2 mm for < 35% clay, and grading class

Moist rupture resistance	Bulk density[a]							
	Sandy		Coarse-loamy		Coarse-silty		Coarse-silty	Fine-silty
	Well graded	Other	Well graded	Other	Well graded	Other		
	(g cm^{-3})							
Very firm and stronger	2.00	1.90	1.85	1.75	1.75	1.65	1.60	1.55
Firm	1.85	1.75	1.75	1.70	1.70	1.60	1.50	1.45
Friable, very friable	1.65	1.65	1.55	1.55	1.50	1.50	1.30	1.20
Loose	1.60	1.50	1.50	1.40	1.35	1.30	1.10	1.00

[a]Family textural classes applied to the <2 mm only; >2 mm is excluded. Well graded material has 1/4 to 3/4 of the 2 to 0.002 m in the 2 to 0.25 mm range and of this 2 to 0.25 mm less than 2/3 is either 2 to 1 or 0.5 to 0.25 mm. Noncarbonate clay on a carbonate-containing basis Employ.

1993). ρm after adjustment for particle density is entered in Equation 6 with the organic matter percentage and the bulk density adjusted for organic matter (ρa) is calculated. Bulk density of certain horizons defined taxonomically should be considered separately. For example, spodic horizons have lower bulk densities and surface horizons of certain Xeralfs have higher bulk densities. Statistical comparisons of measured and calculated bulk densities (Soil Survey Staff, 1975) show an approximately linear 1:1 relationship with an r^2 of about 0.60.

C. Rock Fragment Volume

The weight of the sample should be 50 to 100 times the weight of the rock fragment size that encompasses about 90% of the volume of the rock fragments. The instructions for engineers (Test D2488-93; American Society for Testing and Materials, 1999) are supportive.

The equations to convert weight to volume are as follows:

$$V_2 = W_2 \times \rho t / \rho_2 \tag{7}$$

$$\rho t = \frac{100}{\dfrac{W_2}{\rho_2} + \dfrac{\left(1 - \dfrac{W_2}{100}\right)}{\rho}} \tag{8}$$

Table 3. Bulk density estimates for soil materials with ≥35% clay[a]

Clay[a]	2-0.1 mm 2-0.002 mm	Clay mineralogy[b]	Bulk density			
			Rupture resistance			
			Very firm or stronger	Firm	Friable or very friable	Loose
(%)	(%)		(g cm⁻³)			
35-45	<20	Swelling	1.40	1.35	1.30	1.25
		Not swelling	1.45	1.45	1.40	1.35
	20-40	Swelling	1.45	1.45	1.40	1.35
		Not swelling	1.55	1.55	1.50	1.45
	40-70	Swelling	1.55	1.55	1.50	1.50
		Not swelling	1.60	1.60	1.55	1.50
	≥70	Swelling	1.65	1.60	1.55	1.50
		Not swelling	1.65	1.65	1.60	1.55
45-55	<30	Swelling	1.40	1.25	1.20	1.10
		Not swelling	1.40	1.30	1.30	1.20
	30-70	Swelling	1.45	1.40	1.35	1.30
		Not swelling	1.50	1.45	1.45	1.40
	≥70	Swelling	1.55	1.50	1.50	1.45
		Not swelling	1.60	1.55	1.55	1.45
55-65	<50	Swelling	1.25	1.20	1.15	1.10
		Not swelling	1.35	1.30	1.25	1.20
	≥50	Swelling	1.40	1.35	1.30	1.25
		Not swelling	1.45	1.40	1.35	1.30
65-75	<50	Swelling	1.20	1.15	1.10	1.05
		Not swelling	1.30	1.25	1.20	1.15
	≥50	Swelling	1.35	1.35	1.30	1.25
		Not swelling	1.40	1.40	1.35	1.25
≥75	NA	Swelling	1.15	1.10	1.05	1.00
		Not swelling	1.25	1.20	1.15	1.10

[a]If 15 bar/clay ≥0.6, use 2.5 times the 15 bar retention, subtract the quantity of organic carbon times 1.7 from the 15 bar retention value.
[b]Swelling clay is assumed if CEC by NH_4OAc minus 2.6 times the organic carbon, the difference divided by clay percentage is ≥ 0.60.

Where

V_2 = Volume of >2 mm
W_2 = Weight of >2 mm
ρt = Bulk density inclusive of >2 mm
ρ_2 = Particle density of >2 mm
ρ = Bulk density <2 mm

ρt is substituted into equation 7 to give equation 9:

$$V_2 = \frac{W_2}{\dfrac{W_2}{100} + \left[\left(\dfrac{100 - W_2}{100}\right) \times \dfrac{\rho_2}{\rho}\right]} \tag{9}$$

Equation 9 contains the particle density of the rock fragments (more nearly a bulk density). Commonly, rock fragments have a particle density of 2.50 to 2.70 g cm^{-3}. The range is from unity or below for volcanic ejecta to exceeding 4.0 g cm^{-3} for ironstone. The volume of rock fragments for a given weight percent (Equation 9) decreases as the particle density of the rock fragments increases and increases as the bulk density of the <2 mm increases.

If the interpretive records only contain weight percentages, equation 10 is used to calculate the total rock fragment weight percentage. The Pass 10 in equation 10 is the <2 mm as a percent of the <75 mm. The weight percentage obtained from equation 10 is introduced into Equation 9.

$$W_2 = W_{>250} + W_{75-250} + \left(100 - Pass\ 10\right)\left(1 - \frac{W_{250} + W_{75-250}}{100}\right) \tag{10}$$

Commonly, in the field the weight 2 to 20 mm or 2 to 75 mm is obtained gravimetrically. Larger rock fragments are estimated visually as the volume percent. The weight and volume estimates need to be combined to obtain an overall volume percentages. Equations 7 to 9 are used to calculate the 2 to 20 or 2 to 75 mm as a percent by volume for the <20 (V") or <75 mm (V'). We then combine, using Equations 11 and 12, the volumes for 2 to 20 or 2 to 75 mm as a percent of the <20 or <75 mm with the volume estimates of 20 to 250 or 75 to 250 mm for the <250mm. Note that to this point the volume of >250 mm is not considered.

$$V_{2-250} = V_{>250} + V''_{2-20}\left(1 - \frac{V_{20-250}}{100}\right) \tag{11}$$

$$V_{2-250} = V_{75-250} + V'_{2-75}\left(1 - \frac{V_{75-250}}{100}\right) \tag{12}$$

Commonly, the volume >250 mm cannot be estimated from the usual point exposure and would be estimated from the description of the map unit component and/or point or line intercept transects. To obtain the total volume of rock fragments (V_2):

$$V_2 = V_{>250} + V_{2-250}\left(1 - \frac{V_{>250}}{100}\right)$$ (13)

For layers so high in rock fragments that they touch, the interstices between the touching rock fragments commonly are only partially filled with <2 mm. The effective bulk density to use in equation 9 is calculated by multiplying the bulk density of the <2 mm fabric by the fraction of the interstices filled with <2 mm fabric. The volume of rock fragments cannot reach 100%. The value where the rock fragments touch is 100 minus the volume percent of interstices between the rock fragments. A value of 10 to 40% for the interstial space may be assumed depending on the packing arrangement of the rock fragments. Hence, the maximum volume of rock fragments is 60 to 90%.

V. Biomass Sampling

Methods for collecting standing biomass and woody debris data in wooded areas of New Jersey have been adapted from methods developed by the USDA Forest Service. Table 4 gives the categories of vegetation reported for an example. Forbs were measured but not reported and woody debris was not measured. Plot size ranges from about 0.1 ha for trees to 1 m^2 for forbs. Measurements involve the number and diameter at breast height (Dbh) for larger-sized categories. Biomass C is calculated in kg m^{-2} using regression equations from the literature. Regressions used in Maine, New York, and West Virginia are similar. Those from Maine were largely used. Site Index (Bell, 1984) was obtained to gain knowledge about using it to approximate values obtained with the biomass methods here described.

SOC for the example is 9.3 kg m^{-2} which is similar to that of the standing biomass. Relatedly, in a study for the northern lake states, SOC and biomass carbon were fairly similar (Rollinger et al., 1998).

VI. Successive Sampling

A. General Considerations

The small absolute percentage change in SOC due to management may pose a problem. Suppose it was desired to measure an increase of 10 g m^{-2} yr^{-1} in organic carbon over a 5-year period or 50 g. Assume that the change occurs in a zone 15 cm thick with a bulk density of 1.30 g cm^{-3}; the weight is 195 kg. The 50g increase in organic carbon would be 0.026% of the hypothetical soil mass. The increase in organic carbon may be measured analytically but spatial variability may be too high to make detection practicable.

Izaurralde et al. (1998) assumed an increase in organic carbon of 30 g m^{-2} y^{-1} or 150 g m^{-2} in 5 years. The study concluded that one sample would be needed per hectare. The cost per sample is estimated at $10 to $30. The writers question if because of cost it would be practicable to sample at this intensity. For lower rates of increase, such as the 10 g m^{-2} yr^{-1} discussed earlier, the necessary number of samples would be higher.

Table 4. Standing biomass for a site of Lakehurst soil

Category	Biomass carbon (kg m^{-2})[a]
Trees ≥13 cm dameter Dbh[b]	7.4
Trees, saplings 3-13 cm Dbh	0.50
Seedlings, sampling 0.3-3 cm Dbh	0.04
Seedling 3 to 10 cm high	0.003
Total	7.9

[a]An organic carbon weight percentage of 50 for dry wood was assumed (Sampson and Winnett, 1972).
[b]Dbh is diameter breast height for trees.

B. Constant Mass Sampling

For comparison of the same cropland site over time the mass per unit area of <2 mm should be kept constant. Thickness alone is not satisfactory because, as a result of tillage, the bulk density may change. Units of g cm^{-2} are convenient.

The mass per unit area (M/A) for the initial sampling is as follows:

$$M/A\left(gcm^{-2}\right) = \left(L_1 x\rho_1\right) + \left(L_2 x\rho_2\right) + \left(L_3 x\rho_3\right) + ...\left(Lx\rho x\right) \tag{14}$$

For the subsequent sampling, equation 12 is rearranged, solving for Lx, which is the deepest layer of the subsequent sampling:

$$Lx = \frac{M/A - \left(L_1\rho_1 + L_2\rho_2 + L_3\rho_3\right)}{\rho_x} \tag{15}$$

Suppose the M/A for the original sampling was 50 g cm^{-2}, the depths for the subsequent sampling were 10, 15 and 5 cm, plus Lx beneath and the bulk density for the subsequent sampling is 1.20, 1.30, 1.35 and 1.40 g cm^{-3}. Lx would be 8.4 cm. There would be four layers with the deepest 8.4 cm thick. The total depth would be 38.4 cm.

C. Field Techniques

Suggestions to follow for cropland are based on limited experience. A possible approach is to sample nearby pairs of points with a time interval between the two samplings. The points would be 1 to 2 m apart and in a known orientation to each other. Statistical procedures to analyze such a sampling plan are available, referred to as matched pair or repeated measures. A cow magnet (placed in stomachs of bovines to sequester metal) would be buried at the initial sampling. At the time of the second measurement geopositioning would be used to establish the general area and a magnetometer to locate the initial point exactly. Constant mass sampling would be maintained between the two sampling dates.

Some experience has been obtained using a constant sampling procedure to follow. Bulk and mix while moist the upper few centimeters, preferably of an interrow. Tamp the ground surface uniformly. At 20- to 30-cm intervals, insert a cylinder with a drive head either hydraulically or by impact energy. Attach spinet levels to the cylinder to help judge vertical orientation. At 120° spacing mount longitudinally three sections of retractable ruler with 1-mm graduation. Measure and record the depth of insertion to the nearest millimeter (not used directly but provides uniformity of procedure). Empty the cylinder quantitatively and obtain the oven dry weight of <2 mm plus the rock fragment volume. Determine the minimum area of the drive head and divide into the weight of <2 mm to obtain the mass per unit area. Reduce for the >2 mm volume.

For the second sampling, the cores may be taken with a cylinder that permits segmentation in the lower part. These cores may extend from the ground surface or below a shallow depth sampled by excavation as described previously. Slots may be cut in the cylinders (Heinonen, 1900). Another approach is to use a driver tube within which segmented liner tubes are placed (commercially available). Segments are combined to give a field sample with near the same g cm^{-2} as the initial sample. An alternative may be to sample to, say, three depths across the depth that would be expected to give the same g cm^{-2} as the first. The core would be selected that gives the closest g cm^{-2} as the initial sample. Finally, the bulk density of the zones within the second sample may be determined and the required depth of the second core calculated.

Gamma radiation attenuation should be explored for successive sampling. The single-probe design in which the source and detector are above a plate that rests on the soil would seem very applicable to measurements at shallow depth in cultivated fields. A question is how constant the depth of measurement remains over changes in near-surface bulk density and whether the precision suffices.

VII. Areal Representation of Laboratory Data

A. New Jersey Matched Pair Study

A study was started in 1991 in New Jersey to document the effects of cultivation practices on soil physical and chemical properties. Matched pairs of soils from wooded and cultivated sites of important agricultural soils have been sampled and analyzed. To date, 30 pairs have been sampled. A variety of parent material types and particle size classes, as well as several types of crops and cultivation practices, are represented. The project has involved the measurement for the woodland of the carbon in the leaf litter, and use of excavation bulk density procedures (see section IVB) for the sequence of 0 horizons and surficial mineral horizons that are highly fragile. Results through 1994 are in Table 5. They show a 49% average soil carbon decrease upon cultivation. The study demonstrates the need for a use-dependent approach to document SOC.

B. Assignment to Map Unit Components

The laboratory database point samples can be assigned to map unit components and thereby provide information on measured soil carbon content on an areal bases. This was done for a county in southwest Iowa (Grossman et al., 1992). Relevant laboratory data were obtained from the federal laboratory files. If no data existed for a soil series, similar series were employed. Laboratory data were assigned to all 125 map unit components. Assignments were made by the soil scientist who had been responsible for the survey. The 150 records assigned were drawn from five states; they were not restricted to the county to which the exercise applied. The same kind of assignment was done for Major Land Resource Area 106 (Soil Conservation Service, 1981), a 2.8 million-ha area in southeast Nebraska and northeast Kansas (Grossman et al., 1998). The legend was arranged by soil series not

Table 5. Areal organic carbon for sites of the same soil series in woodland and cultivated fields in New Jersey

| Series | Classification[a] | Amount soil carbon | |
		Wooded	Cultivated
		——— kg m^{-2} ———	
Aura	fine-loamy, Typic Fragiudults	12.7	8.1
Bath	coarse-loamy, Typic Fragiudepts	24.5	9.1
Chillum	fine-silty, Typic Hapludults	14	7.9
Collington	fine-loamy, Humic Hapludults	11.1	4.4
Evesboro	Typic Quartzipsamments	8.8	3.1
Freehold	fine-loamy, Typic Hapludults	17.3	11.9
Galway	coarse-loamy, Typic Eutrudepts	16.7	5.6
Gladstone	fine-loamy, Typic Hapludults	12.3	6.3
Hazen	coarse-loamy over sandy, Mollic Hapludalfs	9.5	4.2
Holmdel	fine-loamy, Aquic Hapludults	12.8	7.1
Keyport	fine, Aquic Hapludults	20.9	10.5
Klej	Aquic Quartzipsamments	12.7	7.3
Lakehurst	Spodic Quartzipsamments	11.9	11.2
Lakewood	Spodic Quartzipsamments	9.3	8.5
Lordstown	coarse-loamy, Typic Dystrudepts	18.8	9.7
Penn	fine-loamy, Ultic Hapludalfs	11.5	7.6
Quakertown	fine-loamy, Typic Hapludults	16.1	7.0
Ryder	fine-loamy, Ultic Hapludalfs	7.5	6.6
Sassafras	fine-loamy, Typic Hapludults	12.6	4.4
Washington 1991	fine-loamy, Ultic Hapludalfs	15.0	6.9
Washington 1994	fine-loamy, Ultic Hapludalfs	15.4	5.5
Westphalia	coarse-loamy, Ochreptic Hapludults	12.2	3.0
Mean		13.8	7.0

[a]Only particle size of the family criteria is given.

by map unit component. One hundred sixty laboratory records were assigned to 77 series. If more than one record were applicable, average values were employed. Finally, for a 75,000-ha area in south central New Mexico (Grossman et al., 1995) 93 laboratory records were assigned to 509 map unit components. The large number of map unit components is because of the uncommon situation that for each map unit the components were listed down to components of 5% areal percentage.

VII. Overview

This chapter reviews application of the U.S. soil survey to the areal estimation of SOC. The interpretive and laboratory databases are reviewed and limitations considered. Bulk density and volume of rock fragments are needed for SOC areal estimates. Clod bulk density is widely used for mineral soil layers. Core methods are considered. Excavation procedures are presented as an alternative for layers that are fragile or high in larger rock fragment. Alternatively, soil material beneath a specified area can be removed and the SOC reported areally. For organic soil layers, fiber percentage and loss-on-ignition are considered for bulk density prediction. Estimation of bulk density from texture, moist rupture resistance, and SOC is presented. Volume of rock fragments commonly requires the combination of measured weight percentages for the smaller sizes and visual volume estimates for the larger sizes.

Successive sampling over time of pairs of point sites in cropland is explored with emphasis on maintenance of constant mass per unit area between the point samples. Suggestions on the mechanics of the sampling are given. Biomass sampling of wooded areas is given. A study in New Jersey is described that compares SOC for sites in woodland and cropland and examples of assignments of laboratory data to mapping concepts to express SOC areally are provided.

References

Adams, W.A., 1973. The effect of organic matter on the bulk and true densities of some uncultivated podzolic soils. *J. Soil Sci.* 24:10-17.

American Society for Testing and Materials. 1999a. Standard test method for density and unit weight of soil in place by the sand-cone method. D 1556-90. 1999. Annual Book of ASTM Standards (I), Vol. 04.08:121-126. American Society for Testing and Materials, Philadelphia, PA.

American Society for Testing and Materials. 1999b. Standard test method for density and unit weight of soil in place by the rubber balloon method. D 2167-99. 1999. Annual Book of ASTM Standards (I), Vol. 04.08:178-184. American Society for Testing and Materials, Philadelphia, PA.

American Society for Testing and Materials. 1999c. Standard practice for description and identification of soils (visual-manual procedure). D 2488-93. Annual book of ASTM Standards. Vol. 04.08:231-241. American Society for Testing and Materials, Philadelphia, PA.

American Society for Testing and Materials. 1999d. Standard test methods for bulk density of peat and peat products. D 4531-86. Annual Book of ASTM Standards, Vol. 04.08:653-654. American Society for Testing and Materials, Philadelphia, PA.

Bell, J.F. 1984. Timber measurements. In: K.F. Wengen (ed.), *Forestry Handbook*, 2nd ed. John Wiley & Sons, New York.

Blake, G.R. and K.H. Hartge. 1986. Bulk density. In: A. Klute (ed.), *Methods of Soil Analysis. Agronomy*, Series No. 9, American Society of Agronomy, Madison, WI.

Boelter, D.H. 1969. Physical properties of peats as related to degree of decomposition. *Soil Sci. Soc. Am. Proc.* 33:606-609.

Bradford, J.M. and R.B. Grossman. 1982. In-site measurement of near-surface soil strength by the Fall Cone device. *Soil Sci. Am. J.* 46:685-688.

Campbell, D.J. 1994. Determination and use of soil bulk density in relation to soil compaction. In: B.D. Soane and C. Van Ovwerkerk (eds.), *Soil Compaction and Crop Production*. Elsevier Sciences, B.V., Amsterdam.

Culley, J.L.B. 1993. Density and compressibility. In: M.L. Carter (ed.), *Soil Sampling and Methods of Analyses*. Lewis Publishers, Boca Raton, FL.

Curtis, R.O. and B.W. Post. 1964. Estimating bulk density from organic-matter content in some Vermont forest soils. *Soi Sci. Soc. Am. Proc.* 28:285-286.

Flint, A.L. and S. Childs (1984). Development and calibration of an irregular hole bulk density sampler. *Soil Sci. Soc. Am. J.* 48:374-378.

Grigal, D.F., S.L. Brovold, W.S. Nord and L.F. Ohmann. 1989. Bulk density of surface soils and peat in the North Central United States. *Can. J. Soil Sci.* 69:895-900.

Grossman, R.B., E.C. Benham, J.R. Fortner, S.W. Waltman, J.M. Kimble and C.E. Branham. 1992. A demonstration of the use of soil survey information to obtain areal estimates of organic carbon. ASPRS/ACSM/RX 92. Tech. Papers Vol. 4. p. 457-465. In: *American Society Photogrammetry and Remote Sensing and American Congress Surveying and Mapping*, Bethesda, MD.

Grossman, R.B., R.J. Ahrens, L.A. Gile, C.E. Montoya and O. Chadwick. 1995. Areal evaluation of organic and carbonate carbon in a desert area of southern New Mexico. In: R. Lal, J. Kimble, E. Levine and B.A. Stewart (eds.), *Soils and Global Change*. Lewis Publishers, Boca Raton, FL.

Grossman, R.B., D.H. Harms, M.S. Kuzila, S.A. Glaum, S.L. Hartung and J.R. Fortner. 1998. Organic carbon in deep alluvium in southeast Nebraska and northeast Kansas. In: R. Lal, J.M. Kimble, R.F. Follett and B.A. Stewart (eds.), *Soil Processes and the Carbon Cycle*. CRC Press, Boca Raton, FL.

Heinonen, R. 1960. A soil core sample with provision for cutting successive layers. *J. Sci. Agric. Soc., Finland* 32:176-178.

Howard, R.F. and M.J. Singer. 1981. Measuring forest soil bulk density using irregular hole, paraffin clod, and air permeability. *Forest Sci.* 27:316-322.

Huntington, T.G., C.E. Johnson, A.H. Johnson,T.G. Siccama and D.F. Ryan. 1989. Carbon, organic matter, and bulk density relationships in a forested Spodosol. *Soil Sci.* 148:380-386.

Izaurralde, R.C., W.M. McGill, A. Bryden, S. Graham, M. Ward and P. Dickey. 1998. Scientific challenges in developing a plan to predict and verify carbon storage in Canadian prairie soils. In: R. Lal, J.M. Kimble, R.F. Follett and B.A. Stewart (eds.). *Management of Carbon Sequestration in Soil*. CRC Press, Boca Raton, FL.

Jenkins, A. 1998. Personal communication.

Lynn, W.C., W.E. McKenzie and R.B. Grossman. 1974. Field laboratory tests for characterization of Histosols. In: A.R. Aandahl, S.W. Buol, D.E. Hill and H.H. Bailey (eds.), *Histosols: Their Characteristics, Classification, and Use*. SSSA Special Publ. 6, Soil Science Society of America, Madison, WI.

Muller, R.N. and M.E. Hamilton. 1992. A simple, effective method for determining the bulk density of stony soils. *Common. Soil Sci. Plant Anal.* 23:313-319

Parent, L.E. and J. Caron. 1993. Physical properties of organic soils. In: M.L. Carter (ed.), *Soil Sampling and Methods of Soil Analyses*. Lewis Publishers, Boca Raton, FL.

Reinsch, T.G., M.D. Mays and J.M. Kimble. 1987. A comparison of total carbon by dry combustion to organic carbon using the modified Walkley-Black Method. 1987. Agron. Abs., American Society of Agronomy, Madison, WI.

Rollinger, J.L., T.F. Strong and D.F. Grigal. 1998. Forested soil carbon storage in landscapes of the Northern Great Lakes Region. In: R. Lal, J.M. Kimble, R. F. Follett and B.A. Stewart (eds.), *Management of Carbon Sequestration in Soil*. CRC Press, Boca Raton, FL.

Sampson, R.N. and W.M. Winnett. 1992. Trees, forests, and carbon. In: R.N. Sampson and D. Hair (eds.), *Forest and Global Change, Vol. 1. Opportunities for Increasing Forest Cover*. American Forests, Washington, D.C.

Sheppard, M.I., C. Tarnocai and D. H. Thibault. 1993. Sampling organic soils. In: M.L. Carter (ed.), *Soil Sampling and Methods of Analyses*. Lewis Publishers, Boca Raton, FL.

Sherrod, L.A. 2000. Personal communication. Total inorganic carbon analysis by a modified pressure calcimeter method.

Shipp, R.F. and R.P. Matelski. 1965. Bulk-density and coarse-fragment determinations on some Pennsylvania soils. *Soil Sci.* 99:392-397.

Silc, T. and W. Stanek. 1977. Bulk density estimation of several plots in northern Ontario using the Von Post humification scale. *Can. J. Soil Sci.* 57:75.

Soil Conservation Service. 1981. Land resource regions and major land resource areas of the United States. U.S. Dept. Agric. Handbook 296. Washington, D.C.

Soil Survey Division Staff. 1975. *Soil Taxonomy*. USDA Handbook 436. Washington, D.C.

Soil Survey Division Staff. 1993. *Soil Survey Manual*. USDA Agriculture Handbook 18. Washington, D.C.

Soil Survey Division Staff. (1999). *Soil Taxonomy*. Agriculture Handbook 436. Soil Conservation Service, USDA. Washington, D.C.

Soil Survey Laboratory Staff. 1996. In: R. Burt (ed.), *Soil Survey Laboratory Methods Manual*, Soil Survey Invest. Report 42, Version 3.0.

Tisdall, A.L. 1951. Comparison of methods of determining apparent density of soils. *Aust. J. Agric. Res.* 2:349-354.

USDA, NRCS, Soil Survey Division and USDA, Information Technology Center. 1998. National Soil Information System. Version 4.0. National Soil Survey Center, Lincoln, NE.

Van Remortel, R.D. and B.A. Shields. 1993. Comparison of clod and core methods for determination of soil bulk density. *Comm. Soil Sci. and Plant Anal.* 24:2517-2528.

Walkley, A. and I.A. Black. 1934. An examination of the Degtjareff Method for determining soil organic matter and a proposed modification of the chromic acid titration method. *Soil Sci.* 37:29-38.

CHAPTER 8

Organic Carbon Methods, Microbial Biomass, Root Biomass, and Sampling Design under Development by NRCS

C.D. Franks, J.M. Kimble, S.E. Samson-Liebig and T.M. Sobecki

I. Introduction

The Natural Resources Conservation Service (NRCS) National Soil Survey Center (NSSC) has initiated a Soil Biology and Soil Carbon research area within the NRCS Global Climate Change Initiative. The purpose is to examine the role microorganisms play in nutrient and energy cycling and carbon pools in soil systems. Microbes drive nutrient cycling. Thus, it is important for soil scientists to measure microbial activity, and pools of carbon (C), phosphorus (P), sulfur (S), nitrogen (N), and trace elements.

The research effort to date has been directed toward five thrusts: to establish a framework for site selection; to develop sampling protocols; to select, adapt and test existing analytical procedures; to put appropriate methods into a soil biology characterization mode; and to explore content and structure of a database.

Areas of Emphasis
Under the auspices of the Soil Biology and Soil Carbon research area, 10 emphasis areas were selected to help us focus our work. The following areas were selected because they have the potential to meet or enhance an application within an NRCS administrative programmatic area. Areas of emphasis for Soil Biology and Soil Carbon include: Soil and Ecosystem Function, Global Climate Change, Soil Interpretations and Management Impacts, Database Development, Field Sampling and Observations, Soil Quality, Soil Characterization, Resource Inventory and Assessment, Wetland Identification and Function, and Bioremediation.

Applications
Each project within the research emphasis areas has a potential use in one or more of the following applications. The eight NRCS research projects under way, for the Soil Biology and Carbon Program, will generate carbon data that can be used in Marco-nutrient Cycling, Soil Quality Determinations, Resource Assessments, Global Climate Change Predictions, Long-term Soil Fertility Assessments, Impact Analysis for Erosion Effects, Impact Analysis for Conservation Management Practices and Carbon Sequestration.

II. Projects

The projects lead in a natural progression from field sampling to laboratory characterization to database development followed by data analysis and ultimately interpretation and application, with NRCS programmatic needs taken into consideration. Each project was developed in such a way that it supports at least one of the other projects. Our projects are in four major groups in preparation for the primary research: research framework, sampling protocols, analyses and analytical approaches and database development. Current projects include Benchmark Sites – A Research Framework, Field Sampling Protocols, Soil Food Web Database, Expanded Soil Organic Matter Characterization, Microbial Biomass and Activity Analysis, Wetland Ecosystem Project, Macro-Nutrient Cycling, and FAME – Microbial Community Fingerprinting Project. A brief overview of some of these major projects follows.

Research Framework
1. Benchmark Sites – A Research Framework
This project is an outgrowth of previous hydrologic work by the NRCS National Range Study Team (NRST) and the Interagency Rangeland Water Erosion Team (Franks et al., 1998). NRST used statistically valid directed site selection and field sampling protocols. Benchmark Sites are statistically valid because the criteria for selection are determined prior to field selection. We applied these protocols and developed a database of hydrologic measurements from different plant communities on a variety of rangeland soils using paired sites with different vegetative states and known land use history. This database is as comprehensive as possible because we used Benchmark Sites.

Benchmark sites are a major improvement over benchmark soils because they use an interdisciplinary approach that integrates plant communities and land use history, not just soils. We selected soils that meet the criteria for benchmark soils and added benchmark plant communities on a limited number of sites representing more extensive areas within the western U.S. With this approach we were able to project results over wider areas and use the database in the Water Erosion Prediction Project (WEPP). WEPP is the acronym for the project as well as the name of the model developed in the project. WEPP is the new parameterized hydrologic model developed by ARS with BLM, NRCS and FS, which will eventually replace the Universal Soil Loss Equation (USLE) for estimating erosion nationwide.

Using the same statistically valid approach described above, we hope to develop a similar small, but representative database for use in global climate change modeling and carbon sequestration. The primary objectives of this project follow, as does a brief outline of the site selection procedure. This is the template and experimental design we use for selecting soils in the other projects.

Objective 1:
To select and characterize benchmark sites which consist of representative, extensive soil series phases and their associated benchmark plant communities (crop, forest, range, etc.) in combination with known management history.

Objective 2:
To develop a pilot database to be used in modeling efforts. This database will contain micro-spatial and temporal near-surface soil properties, plant community characterization, known management history and complete pedon characterization to a depth of 1.5 m.

Benchmark Site Selection Procedure (Franks et al., 1998):
1. Define a 'benchmark' for the study.
2. Review benchmark soils criteria (USDA NRCS, 1996) for applicability. Select those that apply.
3. Develop new criteria essential for the study.
4. Determine which criteria occur in the National Soils Information System (NASIS).
5. Develop new potential benchmark soils lists.
6. Select soils from these new lists which have sites where:
   — Land use history is known and available.
   — Contrasting vegetation states within an ecological site or potential plant community or cropping systems.
   — Easy access.
7. Identify other limiting factors that affect the site selection process.

Sampling Protocols
2. Field Sampling Protocols
This project was developed to implement standardized sampling methods for the new microbiological and organic carbon analyses we are developing. With this approach, written instructions can be provided to a variety of soil scientists, from field soil survey parties to NSSC staff, who can consistently sample, store and ship samples from many locations.

Analyses And Analytical Approaches
3. Expanded Soil Organic Matter Characterization
This project is designed to lead the National Soil Survey Laboratory, in a focused manner, into a broader range of soil organic matter testing rather than just total organic matter or total carbon as is currently performed. Within this project, organic matter and carbon analyses are tested and adapted in a research setting. The goal is to make as many of these analyses as possible available on a routine basis through the National Soil Survey Laboratory (NSSL) and ultimately build a database of these measurements to support modeling.

The proposed dataset (Gregorich et al., 1994) includes organic C, Total N, light fraction (we selected POM), microbial activity and biomass, potential mineralizable N, as well as carbohydrates and enzymes. For our first cut through this minimum dataset we selected organic C, Total N, Particulate Organic Matter (POM) as our light fraction and Chloroform Fumigation Incubation (CFI) for microbial biomass and activity. Carbohydrates and enzymes will be pursued at a later date.

Objective 1:
Test the minimum dataset for organic carbon, proposed by Gregorich et al. (1994), for applicability to NRCS programs and projects.

Objective 2:
Build a characterization database based on benchmark sites that includes data elements needed in carbon and nutrient modeling.

Objective 3:
Use the pilot database in modeling efforts.

4. Microbial Biomass and Activity Analysis

This project is the first one developed under the Expanded Organic Matter Characterization Project. We selected this method because it can be applied as a standard for comparing different soils (within the same moisture and temperature regimes). It also serves as a measure for soil microbial activity and biomass, which represent a small but significant carbon pool, which, along with climate, largely controls the turn-over rate of the labile carbon pool as well as nutrient cycling within the soil.

Objective 1:

Adapt, develop, and implement National Soil Survey Laboratory (NSSL) capability to analyze fresh soil samples for Microbial Biomass determinations from CO_2 evolved by microbial respiration using Chloroform Fumigation Incubation (CFI) combined with sodium hydroxide titration or gas chromatography.

Objective 2:

Test soil samples for the Conservation Reserve Program (CRP) and International AID projects, in cooperation with the Agricultural Research Service (ARS).

Objective 3:

Test soil samples from selected benchmark sites throughout the U.S., in order to develop a database.

Objective 4:

Use the biomass database to develop new interpretations and assess the impacts of management practices on the soil system.

Our Microbial Biomass and Activity analysis was developed in this project. We use Chloroform Fumigation Incubation followed by sodium hydroxide (NaOH) titration or gas chromatography (GC) to determine CO_2 evolved.

5. Macronutrient Cycling

Soil organic carbon levels and climate are not the only factors that affect carbon cycling. Microbial activity and thus the turn over rate of organic matter is influenced by the available nitrogen, phosphorous and sulfur levels. Because of the nature of microbial metabolism, nitrogen levels are particularly important and microbes operate most efficiently when C to N ratios range from 4 or 5 to 1 (bacteria) but not exceeding 10 to 1 (fungi). Following the 'Law of the Minimum' other macro-nutrients, in particular phosphorus and sulfur, then can become quite important in their influence on both microbes and plants.

In order to address carbon cycling, under this project we have included measurements of the various carbon pools associated with soils. As an essential part of the project we are attempting to quantify the physical fractions of organic matter as well as the labile carbon pool.

Objective 1:

Develop laboratory analyses for evaluating carbon, nitrogen, and phosphorus and sulfur cycling in soils.

Objective 2:

Develop a suite of carbon analyses in order to capture the physical size fractions of organic matter associated with the soil. Include other soil organic carbon measures that separate the labile fraction as well. This will give us a total picture of organic matter pools in the soil.

Objective 3:
Evaluate nutrient cycling on benchmark sites. Evaluate selected management practices for their impact on nutrient cycling.

Analyses developed under this project include Root Biomass, Microbial Biomass and Activity, Particulate Organic Matter (POM), O Horizons and Residue, and Potential Mineralizable N; and NSSL has developed Bray P, Oxalate P, New Zealand P, Olsen P and both sulfate and sulfur analyses (USDA - NRCS, 1996).

6. Wetland Ecosystem Project
All of the microbial processes as well as carbon and macronutrient cycling occur in wetland soil systems as well. Many of the systems are quite productive and store a great deal of carbon. However, microbial processes differ under anaerobic conditions. Gases other than CO_2 are evolved following respiration. Examples of these gases would be methane (CH_4), nitrous oxide (N_2O) and hydrogen sulfide (H_2S). Different analytical methods are needed to measure these products of respiration. To this end we will adapt and develop appropriate methods.

Objective 1:
Develop a set of anaerobic laboratory analyses to measure microbially released nitrogen and carbon.

Objective 2:
Test soil samples from a variety of wetland soil catenas. Concomitantly, test the same samples under aerobic conditions. Analyze the paired data.

Objective 3:
Develop interpretations applicable to wetland sites.

The following analytical methods are under development for use in this project:
Water-logged Microbial Biomass and Activity
Anaerobic Potential Mineralizable Nitrogen

7. FAME – Microbial Community Fingerprinting Project
(in cooperation with Rhae Drijber, University of Nebraska, Lincoln.)
This project is designed to characterize soil microbial communities on a variety of soils. We hope to identify many of the common microbial functions being performed in a variety of soils from around the country using this technique. Phospholipid biomass is paired with this analysis rather than microbial biomass because the soils under study have been dried at 45°C. Phospholipids including fatty acids are less heat labile and thus less likely to be denatured than other compounds found in the soil that could be used to represent microbial biomass.

Objective 1:
Test Fatty Acid Methyl Ester (FAME) microbial analysis for applicability on archive soil samples dried at 45°C. (Archive samples were selected from rangeland benchmark sites.)

Objective 2:
Re-sample benchmark sites and perform the analyses on fresh samples.

Objective 3:
Test the suitability of this method to characterize soil microbial communities, through the FAME signature, on a national basis.

Fatty Acid Methyl Ester (FAME) Analysis and Phospholipid Biomass methods are under development by Dr. Rhae Drijber, University of Nebraska, Lincoln in cooperation with the NRCS National Soil Survey Center.

Database development
8. Soil Food Web Database Pilot
Undertaken as a cooperative project between NRCS and Oregon State University, this project was designed as a pilot to test ways that microbial data from laboratories outside NRCS could be made compatible with the NRCS National Soil Information System (NASIS). In addition, this project tested an external linkage to the NASIS database.

Objective 1:
Using georeferencing, link data such as the microbial Food Web Database developed by Oregon State University's Soil Microbial Biomass Service to the NRCS mapping unit database in the National Soil Information System (NASIS).

Objective 2:
Develop the Logical Data Model for enhancing the Food Web Database, in order to make the linkage.

Objective 3:
Use this data model as a pilot to develop the database structure for new point/site attributes (such as microbial data) for pedons and their associated sites in NASIS. The alpha test of the database structure is complete.

III. Analyses

The following section includes a brief synopsis of each of the microbiological, carbon or carbon-related analyses selected by NRCS under the Global Climate Change Initiative as part of the Soil Biology and Soil Carbon research area.

A. Aerobic Microbial Biomass and Activity

The intent is to measure carbon dioxide evolved through microbial respiration. A soil sample is incubated under standard conditions. Carbon dioxide respired by microorganisms in the soil sample is measured periodically. We selected the Chloroform Fumigation Incubation (CFI) method (Jenkinson and Powlson, 1976). Fresh moist soil samples are fumigated with chloroform. Carbon dioxide levels are determined for the fumigated samples and unfumigated control samples.

Two alternative methods are used to determine the carbon dioxide respired: sodium hydroxide titration and gas chromatography. Each method has advantages. The titration method (Horwath and Clark, 1994: Follett and Schimel, 1989: and Jenkinson and Powlson, 1976) applies to a wider range of soils than the gas chromatography method, although the determination is less precise. The Gas

Chromatography (GC) method (Zibilske, 1994 as modified by Doran and Kettler, 1995) works best when applied to acidic soils and is more precise when no carbonates are present (Horwath and Clark, 1994).

B. Potential Mineralizable Nitrogen

The intent is to measure exchangeable ammonium ions, a major component of the nitrogen pool readily available to plants. Ammonium in soil samples incubated for aerobic Microbial Biomass is replaced by potassium chloride (KCl), and ammonium determined on a Lachet Flow Injection Analyzer (Hart et al., 1994, and Follett and Pruessner, 1997).

C. Labile Carbon

The intent is to measure the labile carbon pool, a component of the soil organic carbon pool that is readily available. We are currently pursuing two potential methods for determining labile carbon contents of soils. One is a modified Loginow Procedure (Blair, et al., 1996) and employs potassium permanganate (KMnO4) extraction of the labile carbon fraction. The second is a hot water extraction, the literature search for which is under way.

D. POM – Particulate Organic Matter (modified Campardella)

The intent is to measure the coarser physical fraction of soil organic matter, which is a seasonally more stable carbon pool than microbial biomass, yet reflects changes in the organic matter pool occurring in one to a few years. This analysis separates the soil organic matter into two fractions. Particulate Organic Matter is found in the particle-size fraction between 53μ and 2 mm in diameter (Campardella and Elliott, 1992 and Follett and Pruessner, 1997). The fraction less than 53μ is typically dominated by soil minerals. Total Carbon and Total nitrogen content are determined for both fractions.

E. Total Carbon

The intent is to measure the total carbon content of the soil (USDA NRCS, 1996 and Nelson and Sommers, 1982). A Leco (combustion) C Analyzer is used for this determination and it is reported as percent. This value contains both organic and inorganic carbon contained in the soil. NSSL includes this method as a standard soil characterization analysis.

F. Total Nitrogen

The intent is to measure the total nitrogen content of the soil (USDA NRCS, 1996 and Bremmer and Mulvaney, 1992). A Leco (combustion) N Analyzer is used for this determination which is reported in percent. NSSL includes this method as a standard soil characterization analysis.

G. Walkley-Black Organic Carbon

Until last year, the NRCS National Soil Survey Laboratory included Walkley-Black organic carbon (Walkley and Black, 1934) as a standard soil characterization analysis. This method employs rapid dichromate oxidation of organic matter. NSSL has chosen to discontinue this analysis because of environmental concerns. In an effort to replace this important analysis with another organic carbon method NSSL is switching to Organic Carbon by Subtraction, which is described in the next section.

H. Organic Carbon by Subtraction

The intent is to measure the total soil organic carbon content. In this case total carbon is determined (described above), inorganic carbon in carbonates is determined manometrically (Nelson, 1982) and total organic C is calculated from the difference (Nelson and Sommers, 1982). This carbonate procedure lacks the precision needed for some carbon pool assessments.

We continue to seek a more accurate method for determining total organic carbon for use in our Global Climate Change Initiative and the Soil Biology and Soil Carbon research area. We are currently testing a gas chromatography method for determining inorganic carbon in carbonates. This value is used to determine organic carbon by subtraction as described above.

In the gas chromatography method, ferrous chloride ($FeCl_2$) is used as an antioxidant, and acidification using chilled hydrochloric acid is employed to release CO_2 from carbonates found in the soil. The carbonate CO_2 is then subtracted from the total carbon determined by combustion thus yielding the total organic carbon value (Smith, 1983 and Doran and Kettler, 1992).

Alternatively, carbonates could be acidified through pretreatment with hydrochloric acid. The treated sample is used for total carbon analysis by combustion (described above), thus yielding a direct measure of the organic carbon. A small amount of organic carbon may be lost through the pretreatment step.

I. Root Biomass

The intent is to measure the biomass of roots in soil samples, and thus capture this larger size fraction of organic matter found in the whole soil. Soil samples are gently rotated through a water bath and sprayed with water. Roots and nonliving organic matter are retained on #30 mesh with 0.084-mm openings (after Brown and Thilenius, 1976). Retained sand is removed and the samples are dried and weighed.

J. Organic Matter Residues (O Horizons and Litter)

The intent is to measure the organic matter residues that occur on the mineral soil surface, in order to capture this important carbon pool. The 'O' horizon (whether continuous or discontinuous in the case of litter) is sampled separately in the field and Total C and Total N are determined for each sample.

IV. Conclusions

The NRCS Soil Biology and Soil Carbon Program is well under way. Several new laboratory analyses are under development, which could potentially join our suite of standard characterization analyses. We have focused in particular on organic matter-related analyses, such as microbial biomass, particulate organic matter, root biomass and labile carbon. The new data generated will be used for Global Climate Change as well Nutrient Cycling modeling efforts. A carbon database is a key objective for this program. We will also develop new interpretations to help predict soil behavior, assess management practice impacts, and measure carbon sequestration.

References

Bremmer, J.M. and C.S. Mulvaney. 1982. Nitrogen – Total. p. 595-624. In: A.L. Page, R.H. Miller, and D.R. Keeney (eds.), *Methods of Soil Analysis. Part 2. Chemical and Microbiological Properties.* Second Edition. Agronomy 9, American Society of Agronomy, Madison, WI.

Brown, G.R. and J.F. Thilenius. 1976. A low-cost machine for separation of roots from soil material. *J. Range Management* 29:506-508.

Campardella, C.A. and E.T. Elliott. 1992. Particulate soil organic matter changes across a grassland cultivation sequence. *Soil Sci. Soc. Amer. J.* 56:777-783.

Doran, J. and T. Kettler. 1995. GC method for determining CO_2 evolution. Agricultural Research Service, University of Nebraska, Lincoln (unpublished).

Doran, J. and T. Kettler. 1992. Determination of soil carbonate concentration by acid decomposition and GC CO_2 analysis. Agricultural Research Service, University of Nebraska, Lincoln (unpublished).

Follett, R.F. and E. Pruessner. 1997a. POM procedure. Agricultural Research Service, Fort Collins, CO (unpublished).

Follett, R.F. and E. Pruessner. 1997b. Mineralizable nitrogen. Agricultural Research Service, Fort Collins, CO (unpublished).

Follett, R.F. and D.S. Schimel. 1989. Effect of tillage practices on microbial biomass dynamics. *Soil Sci. Soc. Amer. J.* 53:1091-1096.

Franks, C.D., F.B. Pierson, A.G. Mendenhall, K.E. Spaeth and M.A. Weltz. 1998 Interagency Rangeland Water Erosion Project Report and State Data Summaries. USDA Interagency Water Erosion Team (ARS and NRCS) and National Range Study Team (NRCS), NRCS National Soil Survey Center, Lincoln, NE.

Hart, S.C., J.M. Stark, E.A. Davidson and M.K. Firestone. 1982. Nitrogen mineralization, immobilization, and nitrification. In: *Methods of Soil Analysis, Part 2 — Microbiological and Biochemical Properties.* Soil Science Society of America, Madison, WI.

Horwath, W.R. and E.A. Paul, 1994. Microbial biomass. p. 753-773. In: *Method of Soil Analysis, Part 2 – Microbiological and Biochemical Properties*, Soil Science Society of America, Madison, WI.

Jenkinson, D.S. and D.S. Powlson. 1976. The Effects of Biocidal Treatments on Metabolism in Soil – V. A Method for Measuring Soil Biomass. *Soil Biology and Biochemistry* 8:209-213.

Smith, K.A. 1983. Gas chromatographic analysis of the soil atmosphere. p. 407-454. In: K.A. Smith (ed.), *Soil Analysis: Instrumental Techniques and Related Procedures.* Marcel Dekker, New York.

USDA Natural Resources Conservation Service. 1996. *National Soil Survey Handbook.* USDA NRCS, National Soil Survey Center, Lincoln, NE, p. 603-1.

USDA Natural Resources Conservation Service. 1996. *Soil Survey Laboratory Methods Manual*, Version 3.0. Soil Survey Investigations Report No. 42, USDA-NRCS, National Soil Survey Center, Lincoln, NE.

Voroney, R.P. and E.A. Paul. 1984. Determination of k_c and k_n in situ for calibration of chloroform fumigation incubation method. *Soil Biology and Biochemistry* 16:9-14.

Walkley, A. and I.A. Black. 1934. An examination of the Degtajareff Method for determining soil organic matter and a proposed modification of the chromic acid titration method. *Soil Sci.* 37:29-38.

Zibilske, L.M. 1994. Carbon mineralization. p. 835-863. In: *Methods of Soil Analysis, Part 2 – Microbiological and Biochemical Properties*, Soil Science Society of America, Madison, WI.

Section III

Assessment of Carbon Pools

Characterization of Soil Organic Carbon Pools

H.H. Cheng and J.M. Kimble

I. Introduction

The term "soil organic carbon (SOC) pools" is commonly used to characterize various organic C components in the soil. For different researchers, the term may have very different meanings. To some, it may mean a physical fraction of plant residues at different stages of decomposition, or SOC associated with different soil particle size fractions; to others, it may be a chemical fraction containing specific chemical structures or functional groups. In recent years, the term has become popular in the literature on modeling of the dynamics of C transformations in the environment, as SOC can be compartmentalized into pools of different reactivities. The model-defined pools may or may not have any resemblance to the physically or chemically defined pools.

Interest of scientists and policy makers in SOC pools and in the overall global carbon budget increased dramatically following the proposed establishment of the Kyoto Protocol in December 1997. The goal of the Kyoto Protocol is to reduce the amount of greenhouse gases going into or already in the atmosphere. Reduction of the net emissions would be compared with the measured emission level using 1990 as the baseline. If the emission level could not be reduced sufficiently, the increased emission could be offset by increasing C sequestration by various means. The potential of providing "C credit" for C sequestration as a means to trade off for increased C emission has stimulated much interest in C sequestration research.

Even though ocean and forests are most often mentioned as having the greatest potential for C sequestration, the role of soil as a net sink for atmospheric carbon dioxide (CO_2) has not yet been explored fully (Lal et al., 1995). Under Article 3.4 of the Protocol, the idea of sequestering C in soils as SOC is implied as a possible means to reduce atmospheric CO_2. This approach is being considered by countries which are signatories of the Protocol.

However, before soil can be credited as a sink for atmospheric CO_2, there is need to understand the C sequestration processes in the soil and develop appropriate methodology for determining how much and for how long C is sequestered in the soil. This means the methods must be able to measure the specific portion of SOC (or the specific SOC pools) involved in C sequestration before they are useful for global C budgeting purposes.

The aim of this chapter is to provide first a brief background on the nature of SOC and its role in the C cycling processes and factors affecting these processes in the soil. This background will provide the conceptual framework for defining the appropriate SOC pools for C sequestration characterization and for C credit assessment. Once the SOC pools are defined conceptually, technical aspects of methodology development can then be examined.

II. Soil Organic Carbon and Carbon Cycling

Carbon (C) is a chemical element existing in nature in many different forms, ranging from the elemental forms (e.g., diamond, graphite, and charcoal) to the combined inorganic (carbonate and bicarbonate ions, salts, and minerals) and innumerable organic forms. It exists under ambient conditions in gas (e.g., carbon dioxide, carbon monoxide, methane), liquid (numerous organic solvents), and solid state in mineral form (carbonates) and as the backbone of biological materials. Other than C stored in the ocean, the soil is the largest container of C in the natural environment. Lal et al. (1998) estimated that the global storage of SOC ranged from 1500 to 2000 Pg and soil inorganic C ranged from 700 to 1000 Pg.

All the different forms of C in the natural environment are linked by the process of C cycling. The atmospheric carbon dioxide (CO_2) is transformed by the photosynthesis process into plant and microbial tissues. The plants, microbes, and animals that have consumed the plants as food in turn convert the biological forms of C into CO_2 by the metabolic processes involved in respiration, completing the direct cycling. However, most of the dead animal and plant materials would enter the soil and undergo more complex pathways before these biological C forms are decomposed to CO_2. The organic C in soil is reactive both as an energy source driving biochemical reactions and as reactants and products of various transformation processes. It exists in soil simultaneously as the relatively undecomposed plant and animal residues, microbial biomass and dead tissues, decomposed and decomposable C-containing chemicals, naturally humified or recalcitrant carbonaceous materials, as well as manufacturing wastes and sludges (biosolids). Plant and animal residues are mostly natural sources of C, whereas manufacturing wastes and sludges are anthropogenic sources. Once these materials are in the soil, they are subject to transformations. Soil microorganisms are the predominant agents in driving the transformation processes, with the readily decomposable organic C fractions, including the dead microbial tissues, as the energy source. The decomposable substances will eventually degrade to CO_2. Other fractions of C materials may become stabilized or resynthesized into more recalcitrant or humified fractions (Figure 1). Although soil C has long been known to be the backbone of soil organic matter and to provide the driving force in biochemical transformations in the soil environment, there has been little direct interest in studying the nature of soil C until relatively recently. In comparison, far more extensive studies have been devoted to understanding the nature of soil nitrogen and the nitrogen transformation and cycling processes in the environment.

Yet to truly understand the soil nitrogen transformation processes, one must first understand the energy forces (or C) that drive the transformation processes. Most of the time, researchers were content with a simple determination of the total amount of C present in a soil as an indicator of the biochemical reactivity of the soil, even though it was recognized that only a small fraction of the total SOC serves as the driving force for biochemical transformations in the soil environment.

However, in recent years, interest in the nature of SOC and C cycling has increased greatly because of the concern about increase of greenhouse gases in the atmosphere and its potential for causing global climate change or global warming. Soil is a major source or sink for atmospheric CO_2. Soil processes affect the amount of CO_2 in the atmosphere. Understanding of the C cycling processes in the environment can help assess the impact of increasing emission of CO_2 into the atmosphere and the mitigating effect of soil on the C status in various environmental compartments.

In most soils, C is organic in nature and constitutes approximately 50% of soil organic matter. It serves a structural and a functional role in soil and as the driving force for transformation processes. Whereas biochemical oxidation of SOC can lead to emission of CO_2 into the atmosphere, making soil a source of atmospheric C, the humification processes lead to sequestration of organic C. Whether soil is a source or sink of terrestrial C depends on the balance between the oxidation and humification processes.

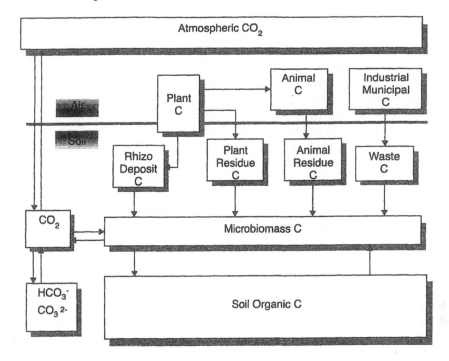

Figure 1. Simplified terrestrial carbon cycle.

III. Fractionation, Analysis, and Modeling of Soil Organic Carbon

The difficulty in understanding the nature of soil organic matter is not only in the complexity and diversity of its physical structure and chemical composition, but also in the dynamics of its varied transformations. To assess the specific contributions of the various components of SOC to the transformation processes, these components may either be isolated from other components by separation methods or be identified selectively by using functional group analysis methods.

A. Fractionation Approaches

A number of fractionation approaches have been developed to separate and isolate various SOC fractions. These approaches can be broadly categorized as physical, chemical, and biochemical separation or fractionation methods. The total C in the fractions obtained is normally used to represent the size of the components.

The physical methods are generally attempting to separate soil organic components by their particle size or by their density (Christensen, 1992; Cambardella and Elliott, 1993; Swift, 1996). Depending on the severity of treatment, size separation can be achieved at aggregate or particle levels. Similarly in density separation, by passing a soil sample through a column of different density-gradient fluids, fractions of organic matter of similar densities can be separated in different layers of density fluids, which can then be extracted and isolated. In general, organic separates are lighter in density than organo-mineral complexes.

Soil organic components can also be separated physically by mechanical means, such by shaking or by using sonic or ultrasonic vibration. The mechanical energy may force the disruption of weaker bonds tying soil C materials to other soil components.

Chemical separation methods are mostly based on the solubility and affinity of certain SOC components in different solvents or extracting solutions (Swift, 1996). The solutions range from water and polar and non-polar solvents (such as alcohols, acetonitrile, acetone, and hexane) to inorganic salt solutions (such as KCl and K_2SO_4), acids and bases of varying strengths, and chelating agents (Cheng, 1990). By far the most effective extracting solution in terms of the amount of SOC extracted is 0.5 M NaOH. There exists a voluminous literature on fractionation of humic substances into fulvic acid, humic acid, and humin components (Stevenson, 1965). Other chemical separation methods may not extract as much C from the soil as NaOH, but provide more definitive fractions related to the strength of their binding to soil particles. A number of chelating agents and resins as well as neutral reagents, such as sodium pyrophosphate, have been used as extractants for metal-organic complexes in the soil (Bremner and Lees, 1949). Once the carbonaceous chemicals are solubilized into solution, further separation by selective precipitation, solvent affinity, chromatographic, electrophoretic, and size exclusion means can be achieved.

Biological or biochemical components of SOC can be separated by physical, chemical, and biochemical means. For instance, fractions separated by density gradient may be related to the stages of decomposition of the organic materials. The light fractions (LF), or particulate organic matter (POM), isolated by density gradient separation are likely partially decomposed plant residues, with the readily decomposable components such as the carbohydrates decomposed and the more resistant lignin complexes remaining (Cambardella and Elliott, 1992; Bremer et al., 1995). Many components of complex biochemical macromolecules or structures are held together by hydrolytic bonds which can be separated by acid, base, or enzymatic hydrolysis (Cheng and Molina, 1995; Leavitt et al., 1997; Xu et al., 1997).

Soil microbial biomass is a component of soil organic matter that has attracted a great deal of attention for the past 25 years (Jenkinson and Ladd, 1981; Parkinson and Paul, 1982; Sparling and Ross, 1993; Beck et al., 1997; Wardle, 1998). This component has most commonly been estimated by destruction of the living matter in the soil sample by fumigation followed by mineralization of the dead tissues to CO_2 by incubation or by solvent extraction of the dead tissues. Other means of measuring soil microbial biomass includes measuring the respiration rates of the biomass treated with an excess of readily available substrate, or measuring such representative microbial constituents as specific enzymes, or ATP, or lipids present.

These approaches were designed for obtaining appropriate fractions to answer specific research questions. Not all of them are appropriate for C transformation dynamics or C sequestration characterization. For instance, although the fulvic acid, humic acid, and humin fractions in the soil organic matter have been associated with different stability in the soil environment, these fractions do not reflect the transformation dynamics of their C contents.

B. Functional Group Analysis Approaches

Instead of measuring the total C in a soil fraction after separation, specific structural components and functional groups of the organic C in a fraction may be detected and measured analytically with specialized techniques (Swift, 1996). Some of these techniques can also be applied directly to the total soil sample without first separating the fractions from the soil.

Although infrared and ultraviolet spectroscopic techniques have long been used for characterizing organic functional groups, the technique that has been used most frequently not only in characterizing the structural components of SOC, but also in monitoring the transformation dynamics of these structural components is the nuclear magnetic resonance (NMR) technique. However, these techniques are more useful in assessing which of the structure components are changed over time rather than measuring how much and how fast these components are changed.

A unique characteristics of certain plants to preferentially using one isotope of C rather than another has provided a useful tool to assess the residence time of C from various plant residues in the soil (Balesdent and Mariotti, 1987). It is now established that C3 plants have different $^{12}C/^{13}C$ isotope ratios than C4 plants. The organic C of the soil under C3 plants also have a different isotopic C ratio from that under C4 plants. When the vegetation changes from a community of predominantly C3 plants to that of C4 plants, or vice versa, the rate of change of the isotopic C ratio can be a measure of the transformation rate of various fractions of SOC in the soil. A combination of fractionation and analytical approaches can further identify these fractions and estimate the age of these fractions in soil.

C. Modeling Approaches

Monitoring the rates of change from one form of C to another and from one environmental compartment to another has provided a basis to model the dynamics of SOC transformation and the C cycling processes. As analytical techniques became more sophisticated, models also became more refined. Jansson (1958), using ^{15}N isotope tracing technique to monitor soil nitrogen transformation, was among the first to propose that soil organic matter can be divided into active and passive phases (or pools). Jenkinson and Rayner (1977) applied the same concept to SOC transformations. It is assumed that C materials in a pool have similar reactivities. Under a set of environmental parameters, the kinetics of transformation of C in each pool can be characterized by a rate constant, which can be determined by a standardized incubation experiment or by a specific set of conditions characteristic of a specific location.

Such pools have been useful for model characterization of the dynamics of soil C transformation. With the advent of techniques for estimating soil microbial biomass, most models have now included at least three pools of different reactivities (Paul et al., 1995; 1997). These are the active or labile pool, with a very fast turnover time (< 10 years), the slow release or intermediate pool (100 to 1000 years) and the passive pool (maybe 1000 or more years).

Because these pools do not have a physical identity or represent any distinguishable chemical fractions in the soil, an appropriate characterization of the physical and chemical nature of these pools is extremely difficult, although there have been many attempts to apply various fractionation and analytical techniques to measure these pools. The issues related to inorganic soil carbon have been addressed in the proceedings of a meeting held in Tunisia in 1997 (Lal et al., 2000). In dynamic C simulations models (e.g., Jenkinson and Rayner, 1977; van Veen and Paul, 1981; Parton et al., 1987), data on the transformation constants of the different pools are required to run these models. The Century model (Parton et al., 1987) can partition the total C by using their C:N ratios, but an understanding of the various pools would help in the usefulness of this model. These models are used over long time scales to assess long-term impacts and SOC dynamics. Data are needed to test and validate these models over shorter time periods which may be more relevant to trading of C credits. Work of Voroney and Angers (1995) has demonstrated that the Century model is robust enough to look at short-term management practices.

In spite of their uncertainties, many C transformation models have attempted to use various known physical or chemical fractions to represent the various pools, without any assurance that these fractions can truly represent the totality of these associated pools. Although the transformation rates for specific pools have been estimated or generated, the size of the pool is often less certain, making it difficult to verify that the specific fractions measured actually coincide with the specific pools. In recent years, a number of process models (Molina and Smith, 1997; Smith et al., 1997) are becoming refined enough to define the size of the various pools in a soil system. The major challenge to researchers will be to match and reconcile the size of the model-defined pools and the transformation rates of the pools in relation to soil management practices with the analytically identified SOC fractions or components.

What is learned from soil dynamics characterization is very relevant to biological simulations. As more is learned about different physical and chemical separates, the better the models become and their use in scaling data from a point to a field, to an eco-region, and even to larger areas improves. The understanding of different fractions allows us to make better management decisions and to decide which practice will give the most benefits within a given eco-region or under different cropping systems.

IV. Soil Organic Carbon Pools and Carbon Sequestration

Many studies have shown that as much as 40 to 60% of the SOC in many soils have been lost as a result of converting the soil from its natural conditions to agricultural production, especially when the soil has been subjected to intensive and repeated cultivation (Lal et al., 1998). On the other hand, because of the depleted state of SOC in these cultivated soils, there is also the potential for restoring the C content of these soils by changing cultural and tillage management practices imposed on the soil (e.g., Salter and Green, 1933; Wood et al., 1991; Franzluebbers et al., 1994; Franzluebbers and Arshad, 1996; Hunt et al., 1996; Grant, 1997; Hu et al., 1997). By applying knowledge on the C cycling processes, it should be possible to enhance the sequestration of the atmospheric CO_2 as SOC. This concept of sequestration raises many questions that will need to be addressed related to long-term or short-term gains and losses of SOC. Since the various fractions and pools of SOC do not react and transform at similar rates, it will be essential to determine which SOC fractions or pools would be representative of that soil's capability to sequester C and how this measurement can be made and verified or certified.

If the goal of sequestration is to remove C from the C cycling process, the atmospheric CO_2 should be converted into carbonates in the vadose zone of the terrestrial systems or in the depth of ocean. Unfortunately, this conversion process is extremely slow and cannot be readily adopted to meet the present need for C sequestration. For more practical purposes, C can be sequestered by converting SOC from the active pools to less reactive intermediate or passive pools. Even in the passive pools, the C sequestered may eventually become available for conversion to CO_2. Therefore, a major decision for policy makers is to agree upon at what level of stability one should strive to achieve. In active pools, C sequestered may only reside in the soil for one season to a few years, whereas in more passive pools, the C sequestered may remain for tens and hundreds of years. However, corollary to the stability of the pools, the more stable the pools, the slower they are built up; whereas the more active the pools, the faster they can change in size.

When SOC pools are selected for characterization, there is need to consider what the pools really represent. Are they the ones that are of importance to a policy maker or even to an industry that wants to trade C credits? Some of the questions which need to be addressed are related to measurement, verification, and certification of the pool and their dynamics. Can the data obtained from a point be scaled to a larger area? Alternatively, if the increase in pool size is only transient in nature, what time of the year should the pool be measured? Under the temperate conditions, the input of litter would be the highest in the fall. However, under tropical and subtropical conditions, crops and cover crops are grown year round (Sombroek et al., 1993). Timing for sampling needs to be consistent. Samples taken immediately after harvesting would give a much higher "soil" C content because of the high plant residues contents, yet much of this increase is temporary, as the residues would mostly be decomposed over the next growing season. The more stable (humidified) pools are probably the more appropriate and representative fractions for C sequestration characterization. Preferably they should be measured at or near the end of the growing season when the rate of decomposition of the residues from the preceding year is diminished and the C fractions remaining are becoming stabilized.

As illustrated in the previous section, SOC can be divided or separated into components or fractions. The C contents within each component or fraction would possess similar physical properties, or similar chemical characteristics, or similar biochemical reactivities. Some of the fractions are actively involved in the C transformation processes, whereas other fractions could be more stable and less involved in C transformations. Even though some of these fractions have been taken as equivalent to certain SOC pools, it is a difficult task to verify this connection. The transformation dynamics of these fractions as separate entities can be quite different from those with the fractions present as a component of the total SOC. Unless these fractions can be specifically labeled with different C isotope tracers to differentiate their reactivities under various natural conditions, the stability of these fractions *in situ* would be extremely difficult to determine precisely.

In the context of C sequestration in the soil, SOC within a pool should be considered to have similar reactivity. Under a given set of environmental parameters, the kinetics of transformation of C in each pool should be characterizable by a rate constant, which can be determined by a standardized incubation experiment or by a specific set of conditions characteristic to a specific location. As pointed out in the previous section, such pools have been useful for model characterization of the dynamics of soil C transformation.

In Article 3.3 of the Kyoto Protocol, trees and forest are specifically mentioned for their potential contributions to C sequestration, whereas in Article 3.4, C sequestration by soil is only implied. One of the reasons for this difference appears to be that one can see trees whereas one cannot really see soil carbon quantitatively. Many in the policy arena had made the point that SOC is very variable over the landscape. However, the same variability also occurs in any estimation of the C contents of the forest biomass. Anyone who has ever walked through a forest can readily see the difference in the size of trees and in some areas, the absence of any trees. The variability in the biomass of trees is probably not any less than that of the soils. That being the case, there is still need to account for the variability. In forestry, trees are measured and the above ground biomass is estimated using a statistical sample. In soils, C measurements of soil samples can be scaled up using landscape and computer models. By using appropriately selected point measurements and modeling, C values for a farm, a region, or even a country can be estimated.

If soil is to be considered for C sequestration, the first need is to determine which SOC pool should be measured. One can expect only a very limited C enrichment in the long-term SOC pool over short time periods, as the build-up process is likely to be slow and the size of the change is normally small on a yearly or even a 5- or 10-year basis. Even though soil cultivation has caused decline in soil C contents, the decrease is usually gradual. Similarly, in restoration of soil C contents, the increase would also be gradual.

Procedures that measure specific physical or chemical fractions (e.g., POM, microbial biomass, or other pools) have more utility in the development of models (Janzen et al., 1992). They often are more complex and time consuming and would not be run on as many samples as total C. Their use would be in calibrating and testing of models and not in the verification of C stocks relative to "C credits." They help understand overall C dynamics and the influences of different climatic conditions. However, there does not seem to be a strong case for using them to measure C stock changes over time. Some of them are the very active pools and change very rapidly in soils. The measurement of these would be included when total stocks are determined. The C making up the active pool is the one which would change the most in a 5- to 10-year period. The passive and slow release pools have mean resident times of hundreds to thousands of years. There is a slow change to these pools but no great change is expected in any accounting period.

V. Technical Considerations for Carbon Credit Assessment

If C sequestration by soil has the potential for providing C credit in reducing the atmospheric CO_2, a number of questions will have to be addressed before a procedure for providing reliable C credit can be established. The most basic question is how C credit should be established. One obvious approach is to base the credit on the amount of soil C increase. However, this is easier said than done.

Because there is already a large amount of C stored in the soil, any increase in C from year to year would be small. As the global soil may contain 1500 or more Pg of C, a change of 0.1% (i.e., 1.5 Pg) is significant when the total C is considered, as it is equivalent to the annual output of all U.S. emissions (1485 MMT-CE yr^{-1} according to DOE/EIA, 1996). Therefore, even small changes in the SOC pool can be important. However, making reliable measurement of changes in total C contents can be difficult, especially in view of the inherent variability of soil C contents across a field (Kern, 1994; Cheng and Kimble, 1999). To overcome the variability problem may require a large number of samples and analyses, which can be very costly.

Since it is hard to measure a small change in a large pool with a high level of precision, an alternative approach would be to select a specific SOC pool within the total SOC as an indicator of change in C sequestration. The problem of deciding which of the SOC pools would be the most appropriate as the indicator of increases in SOC contents was discussed in the previous section. Another approach could be based on model estimation of the influence of crop and soil management practices on C increase in the soil. This approach will require extensive information on soil properties and past management history. The ensuing discussion will not be a comprehensive list of criteria for C credit assessment decision. Only a few of the technical aspects will be discussed to illustrate the magnitude of the challenge.

A. Management Influence on C Sequestration

Whether a field has the potential for increasing C sequestration depends on the physical and chemical properties of the soil and past history of management that has been imposed on the field. The C content of a soil in its native state is usually a good indicator of the potential of this soil to sequester C. Even though by careful management the C content of a soil can be built up to a higher level than that in the native state, a more realistic goal would be to rebuild the C content of a soil to the same level as in its native state, after its C content had been depleted by excess degradation or loss. The soil C content is a reflection of the environmental conditions imposed on the soil. The native state of a soil represents a steady state condition for the soil to maintain its organic matter level in response to the surrounding environment. If the inherent C content is low (e.g., in sandy soils), the potential to increase the soil's C storage capacity would be low.

Cultivation of a soil for agricultural production purposes usually results in a decrease in soil C contents. However, certain management practices can reduce the C loss and even restore the soil C content to its native state. Such factors as crop varieties, crop rotation, soil tillage practices, water management practices (e.g., using various irrigation methods), nutrient management practices (e.g., use of mineral fertilizers vs. manure), can all influence the C status in a soil. Furthermore, the compatibility of the various management practices can influence whether C build-up is possible. Thus knowing the past management history can be useful in assessing the soil's potential for increasing its C content.

In addition, as C is sequestered under a given management practice or system, there is need to look at possible long-term changes that are management-controlled. If a management system is used for 10 years and an increase in the C level from X to X+Xi is obtained, there is also the possibility that this SOC could be lost very quickly if the production system reverts to the old practice which had caused the C loss in the first place. Therefore, when management options are considered for C crediting purposes, they need to be locked in for long periods through legally binding contracts. This does not mean that a land manager could not opt to change management down the line to meet changing demands for food and fiber. However, if changes were made, considerations must be given to its impact on C loss. There may be the need to purchase "C credits" from others to offset the net loss in the contracted credit.

Under certain conditions, the growing of some crops and the use of some cropping systems will lead to a decrease in the overall C stock, no matter what is done in a given field. For instance, certain crops may not produce enough biomass to replace the C lost through normal oxidative process in the field. Some crops produce very little biomass and, no matter what farming system is used, will have a negative impact on the SOC levels. In some areas, this impact can be overcome by the addition of manure and other organic wastes or through the growing of cover crops or the use of crop rotations. The net profit of a crop rotation system may be less than a monoculture of a high biomass producer, yet the overall effect of crop rotation can be positive. It is essential to consider the total system, accounting over a period of years and not on a year-to-year or crop-to-crop basis. With a more complete understanding of the overall SOC dynamics, the combination of other practices (cover crops, crop rotation, reduced fallow, etc.) a net negative can be changed to a positive gain. This cannot be done unless soil dynamics and the interactions of crops and management systems are known.

In regions where flood or furrow irrigation is used, it may be difficult to initiate no-till or minimum tillage practices, even though in the long run such tillage practices would be needed to enhance the soil's capacity to sequester C. By switching from flood to sprinkler irrigation, it would also improve irrigation efficiency, improve fertility management, and prevent the tendency to over-water part of field through flood irrigation which leads to the removal of added nutrients from the profile into the groundwater. Thus, increasing carbon sequestration can lead to other benefits.

The fact that not all soils will sequester the same amount of C needs to be considered. Giving a specific C credit to all soils under no-till management will not work. As the environment changes from cool humid to warm humid climates, the rate of decomposition increases and the net increase in the pool size will not be equal.

Biomass production may increase under warmer and wetter climatic conditions but so does the decomposition rates. On the other hand, in these systems there may be a faster change in the passive and slow release pools as the degree of humification is increased. The material is broken down faster but it ends up in the more stable pools through this breakdown. Thus, the climatic influence on C transformation must be understood.

B. Modeling Approach for C Credit

What a C credit should be based on is a hypothetical net increase over time that is discounted for externalities. These externalities could be climatic variations, management changes, differences in crops grown, soil differences, and even catastrophic events. The end product is a net change in a C pool from year 1 to year 1+X. The only change that needs to be verified is the difference between 1 and 1+X. The rate of change is not important, as it will be expected to vary from year to year because of differences in climatic factors. The expected credit would be X reduced for the expected risk. If this level was exceeded then a premium could be paid at the end of the measurement period. The discount should be enough so that there would not need to be a payback for advanced payments on the credit. The possibility of catastrophic events could even be insured against when the system is developed.

Measurement of specific pools and their relationship to different management can be improved by using yield data gathered at sites under specific management. Yields are measured and used to determine residue return, and a model can be used to estimate the SOC that will result. By modeling over a 5- to 10-year period, year-to-year variation can be overcome. The model can be run for each year and summed for a desired period of time. Specific pools may well need to be measured to evaluate the SOC dynamics, but in the long run what is of interest to policy makers and "C credit traders" is the change in a stock (pool) over a given period of time. It is the total change for the period (gain or loss), not the rate of change, that is important.

Policy makers and traders of C credits are interested in the relevance of the actual measurements and models to an understanding of the pool size and how it will change with different management. Someone looking for an increase in SOC may well want to model different management systems to see what system(s) or practice(s) will meet his or her need. If an energy company wants to come up with Ƅ×Ƅ MMT-CE yr^{-1} it will want to know the expected result before the system or practice is put into place. This can only be determined by modeling and modeling can only be done if there are on-site measurements at specifically designed sites to verify the model.

C. Sampling and Measurement Considerations

The time to sample soil for C credit is important. It is suggested that this be done immediately after harvest in the cropland. That is the time when the maximum amount of decomposition and humification has gone on for a given year, i.e., the system has come to a steady state. It is also a good time to measure the inputs from the current crop. Under grazing systems and in forest systems, the time of sampling may be somewhat different, but the end of a growing season may still be the best time. Under more tropical conditions, where there is year-long production and this will create more concern on the time of sampling because there is really not a fixed end of a growing season. In addition, in many areas there are now multiple crops grown in the same field. That would also affect the timing of sampling. Nevertheless, a standardized time is needed for a given "credit." It can vary from area to area but should not be changed for a given "tradable credit."

The question of what depth of soil to sample needs to be addressed. Most short-term changes are in the top 20 to 30 cm of the soil profile, as that is where the greatest amount of roots and litter inputs occur. In no-till systems, the changes may be in the top few centimeters. Therefore, it is suggested that sampling be conducted to assess pool changes in the top part of the soil profile. There is need to recognize that in many tropical forests systems there may be inputs deep within the profile. Work of Susan Trumbore (1993) has shown that at depths of even 18 m there are sequestration/oxidation process going on in tropical forest systems. Again, the depth of interest may change for the crop grown, for different land uses, and for different ecoregions.

If measurement of specific C pools is needed, the time frame for measurement is important. Measurement of yearly change may be too expensive if a measurements on every field is required. But if done by modeling, the cost could be decreased. It must be kept in mind that changes in any SOC pool may be very small or even negative on a yearly basis. A land manager can make all the correct management decisions, but if the driving climatic variables (e.g., rainfall) did not cooperate, then what was expected may not meet the perceived values for a given year. Therefore, the change in a pool size must be evaluated over longer periods (5 to 10 years). The predictions of yearly rates of change are of little use unless it is just shown as an expected average.

VI. Epilogue

The potential of managing soil to increase C sequestration is both attractive and challenging. This chapter has not considered in any detail how C can be sequestered. Many studies have shown that the SOC contents can be increased by different management practices. However, before a concerted effort is spent on these management practices for the sole purpose of increasing C sequestration (even though these practices could have value for other purposes, such as improving soil fertility, reducing erosion, enhancing soil quality, etc.), there must be agreement on how C sequestration is defined and what constitutes an acceptable level of sequestration both in terms of C accumulation over time and the eventual fate of C sequestered. The parameters for measuring C sequestration over an extended period of time must be clearly understood.

References

Balesdent, J. and A. Mariotti. 1987. Natural ^{13}C abundance as a tracer for studies of soil organic matter dynamics. *Soil Biol. Biochem.* 19:25-30.

Beck, T., R.G. Joergensen, E. Kandeler, F. Makeschin, E. Nuss, H.R. Oberholzer and S. Scheu. 1997. An inter-laboratory comparison of ten different ways of measuring soil microbial biomass C. *Soil Biol. Biochem.* 29:1023-1032.

Bremer, E., B.H. Ellert and H.H. Janzen. 1995. Total and light-fraction carbon dynamics during four decades after cropping changes. *Soil Sci. Soc. Am. J.* 59:1398-1403.

Bremner, J.M. and H. Lees. 1949. Studies on soil organic matter: II. The extraction of organic matter from soil by neutral reagents. *J. Agric. Sci.* 39:274-279.

Cambardella, C.A. and E.T. Elliott. 1992. Particulate soil organic matter changes across a grassland cultivation sequence. *Soil Sci. Soc. Am. J.* 56:777-783.

Cambardella, C.A. and E.T. Elliott. 1993. Methods of physical separation and characterization of soil organic matter fractions. *Geoderma* 56:449-457.

Cheng, H.H. 1990. Organic residues in soils: Mechanisms of retention and extractability. *Intern. J. Environ. Anal. Chem.* 39:165-171.

Cheng, H.H. and J. Kimble. 1999. Methods of analysis for soil carbon: An overview. p. 333-339. In: R. Lal, J.M. Kimble and B.A. Stewart (eds.), *Global Climate Change and Tropical Ecosystems.* Lewis Publishers, Boca Raton, FL.

Cheng, H.H. and J.A.E. Molina. 1995. In search of the bioreactive soil organic carbon: The fractionation approaches. p. 343-350. In: R. Lal, J. Kimble, E. Levine and B.A. Stewart (eds.), *Soils and Global Change.* Lewis Publishers, Boca Raton, FL.

Christensen, B.T. 1992. Physical fractionation of soil organic matter in primary particle size and density separate. *Adv. Soil Sci.* 20:1-90.

DOE/EIA. 1996. Emissions of greenhouse gases in the United States 1995. Energy Information Administration, U.S. Department of Energy, Washington, D.C.

Franzluebbers, A.J. and M.A. Arshad. 1996. Soil organic matter pools during early adoption of conservation tillage in northwestern Canada. *Soil Sci. Soc. Am. J.* 60:1422-1427.

Franzluebbers, A.J., F.M. Hons and D.A. Zuberer. 1994. Long-term changes in soil carbon and nitrogen pools in wheat management systems. *Soil Sci. Soc. Am. J.* 58:1639-1645.

Grant, F.R. 1997. Changes in soil organic matter under different tillage and rotations: Mathematical modeling in ecosystems. *Soil Sci. Soc. Am. J.* 61:1159-1175.

Hu, S., N.J. Grunwald, A.H.C. Van Bruggen, G.R. Gamble, L.E. Drinkwater, C. Shennan and M.W. Demment. 1997. Short-term effects of cover crop incorporation on soil C pools and N availability. *Soil Sci. Soc. Am. J.* 61:901-911.

Hunt, P.G., D.L. Karlen, T.A. Matheny and V.L. Quisenberry. 1996. Changes in carbon content of a Norfolk loamy sand after 14 years of conservation or conventional tillage. *J. Soil Water Conserv.* 51:255-258.

Jansson, S.L. 1958. Tracer studies on nitrogen transformations in soil with special attention to mineralisation-immobilization relationship. *Kungl. Lantbrukshogsk. Ann.* (Sweden) 24:101-361.

Janzen, H.H., C.A. Campbell, S.A. Brandt, G.P. Lafond and L. Townley-Smith. 1992. Light-fraction organic matter in soils from long-term crop rotations. *Soil Sci. Soc. Am. J.* 56:1799-1806.

Jenkinson, D.S. and J.N. Ladd. 1981. Microbial biomass in soil: Measurement and turnover. p. 415-471. In: E.A. Paul and J.N. Ladd (eds.), *Soil Biochemistry*, Vol. 5. Marcel Dekker, New York.

Jenkinson, D.S. and J.H. Rayner. 1977. The turnover of soil organic matter in some of the Rothamsted classical experiments. *Soil Sci.* 123:298-305.

Kern, J.S. 1994. Spatial patterns of soil organic carbon in the contiguous United States. *Soil Sci. Soc. Am. J.* 58:439-455.

Lal, R., J.M. Kimble, H. Eswaran and B.A. Stewart. 2000. *Global Climate Change and Pedogenic Carbonates*. Lewis Publishers, Boca Raton, FL.

Lal, R., J. Kimble and B.A. Stewart. 1995. World soils as a source or sink for radiatively-active gases. p. 1-8. In: R. Lal, et al. (eds.), *Soil Management and Greenhouse Effect*. Lewis Publishers, Boca Raton, FL.

Lal, R., J.M. Kimble, R.F. Follett and C.V. Cole. 1998. *The Potential of U.S. Cropland to Sequester Carbon and Mitigate the Greenhouse Effect*. Ann Arbor Press, Chelsea, MI.

Leavitt, S.W., R.F. Follett and E.A. Paul. 1997. Estimation of slow and fast cycling soil organic carbon pools from 6 N HCl hydrolysis. *Radiocarbon* 38:231-239.

Molina, J.A. E., and P. Smith. 1997. Modeling carbon and nitrogen processes in soils. *Adv. Agron.* 62:253-298.

Nikiforoff, C.C. 1936. Some general aspects of the Chernozem formation. *Soil Sci. Soc. Am. Proc.* 1:333-342.

Parkinson, D. and E.A. Paul. 1982. Microbial biomass. p. 821-830. In: A.L. Page, et al. (eds.), *Methods of Soil Analysis, Part 2: Chemical and Microbiological Properties*. Second edition. Agronomy No. 9, Part 2. American Society of Agronomy, Madison, WI.

Parton, W.J., D.S. Schimel, C.V. Cole and D.S. Ojima. 1987. Analysis of factors controlling soil organic matter levels in Great Plains grasslands. *Soil Sci. Soc. Am. J.* 51:1173-1179.

Paul, E.A., R.F. Follett, S.W. Leavitt, A. Halvorson, G.A. Peterson and D.J. Lyon. 1997. Radiocarbon dating for determination of soil organic matter pool sizes and dynamics. *Soil Sci. Soc. Am. J.* 61:1058-1067.

Paul, E.A., W.R. Horwath, D. Harris, R. Follett, S.W. Leavitt, B.A. Kimball and K. Pregitzer. 1995. Establishing the pool sizes and fluxes of CO_2 emissions from soil organic matter turnovers. p. 297-308. In: R. Lal et al. (eds.), *Soils and Global Change*. Lewis Publishers, Boca Raton, FL.

Salter, R.M. and T.C. Green. 1933. Factors affecting the accumulation and loss of nitrogen and organic carbon from cropped soils. *J. Am. Soc. Agron.* 25:622-630.

Smith, P., J.U. Smith, D.S. Powlson, W.B. McGill, J.R. M. Arah, O.G. Chertov, K. Coleman, U. Franko, S. Frolking, D.S. Jenkinson, L.S. Jensen, R.H. Kelly, H. Klein-Gunnewiek, A.S. Komarov, C. Li, J.A.E. Molina, T. Mueller, W.J. Parton, J.H.M. Thornley and A.P. Whitmore. 1997. A comparison of the performance of nine soil organic matter models using datasets from seven long-term experiments. *Geoderma* 81:153-225.

Sombroek, W.G., F.O. Nachtergale and A. Hebel. 1993. Amounts, dynamics, and sequestration of carbon in tropical and subtropical soils. *Ambio* 22:417-426.

Sparling, G.P. and D.L. Ross. 1993. Biochemical methods to estimate soil microbial biomass: Current developments and applications. p. 21-37. In: K. Mulongoy and R. Merckx (eds.), *Soil Organic Matter Dynamics and Sustainability of Tropical Agriculture*. John Wiley & Sons, Chichester, U.K.

Stevenson, F.J. 1965. Gross chemical fractionation of organic matter. p. 1409-1421. In: C.A. Black, (ed.), *Methods of Soil Analysis*. Agronomy 9. American Society of Agronomy, Madison, WI.

Swift, R.S. 1996. Organic matter characterization. p. 1011-1069. In: D.L. Sparks et al. (eds.), *Methods of Soil Analysis. Part 3: Chemical Methods*. SSSA Book Ser. No. 5. Soil Science Society of America, Madison, WI.

Trumbore, S.E. 1993. Comparison of carbon dynamics in tropical and temperate soils using radiocarbon measurements. *Global Biogeochem. Cycles* 7:275-290.

van Veen, J.A. and E.A. Paul. 1981. Organic C dynamics in grassland soils. 1. Background information and computer simulation. *Can. J. Soil Sci.* 61:185-201.

Voroney, R.P. and D.A. Angers. 1995. Analysis of the short-term effects of management on soil organic matter using the Century model. p. 113-120. In: R. Lal et al. (eds.), *Soil Management and Greenhouse Effect*. Lewis Publishers, Boca Raton, FL.

Wardle, D.A. 1998. Controls of temporal variability of the soil microbial biomass: A global-scale synthesis. *Soil Biol. Biochem.* 30:1627-1637.

Wood, C.W., D.G. Westfall and G.A. Peterson. 1991. Soil C and N changes on initiation of no-till cropping systems. *Soil Sci. Soc. Am. J.* 55:470-476.

Xu, J.M., H.H. Cheng, W.C. Koskinen and J.A.E. Molina. 1997. Characterization of potentially bioreactive soil organic carbon and nitrogen by acid hydrolysis. *Nutr. Cycl. Agroecosyst.* 49:267-271.

CHAPTER 10

Measuring and Comparing Soil Carbon Storage

B.H. Ellert, H.H. Janzen and B.G. McConkey

I. Introduction

The ability to measure the amount carbon (C) stored in soils is crucial to understand C cycling in terrestrial ecosystems. Globally, soils contain more than twice as much C as the atmosphere (Schimel et al., 1995). Moreover, the two pools are intricately linked, so that a loss of soil C increases atmospheric CO_2, while a gain in soil C withdraws CO_2 from the air. Recently, concern about climate change has promoted interest in mitigating atmospheric CO_2 increases by sequestering organic C in soils.

The interest in changes to soil C is not new – soil scientists have long studied the response of soil C to management practices. But much of this earlier work focused more on soil quality than on the quantity of C stored. Consequently, methods often used to measure C may not be precise enough to quantify and verify C sequestration.

The net exchange of C between land and the atmosphere often is inferred from comparisons among soil C stored in contrasting ecosystems or management regimes. Such comparisons are hampered by vertical, lateral and temporal variability in soil horizonation, C concentration and bulk density. In this chapter, we describe a possible approach to minimize these difficulties. The approach was devised to compare soil C among contrasting sampling times, ecosystems or management regimes, and is now used at several sites on the Canadian Prairies. The procedures we describe are applicable to areas without appreciable soil redistribution and without woody perennial vegetation.

II. Changes in Soil C Storage

A. Carbon Cycling

Given enough time, agricultural and native ecosystems evolve to a stable state in which the biota maximize homeostasis with the physical environment (Odum, 1969). In stable or quasi-stable ecosystems soil C inputs and outputs are equal, and soils are neither a source nor a sink of atmospheric CO_2. Perturbations strong enough that they cannot be dampened by homeostatic controls cause ecosystems to evolve or undergo succession to a new stable state. Inputs and outputs of soil C may be imbalanced and soils may function as a source or a sink of atmospheric CO_2 *only* during the evolutionary or successional stage of ecosystem development. The extent to which soils may function as a source or sink of atmospheric CO_2 during this stage depends on the magnitude of the imbalance between inputs and outputs and the duration of the imbalance before a new steady state is attained.

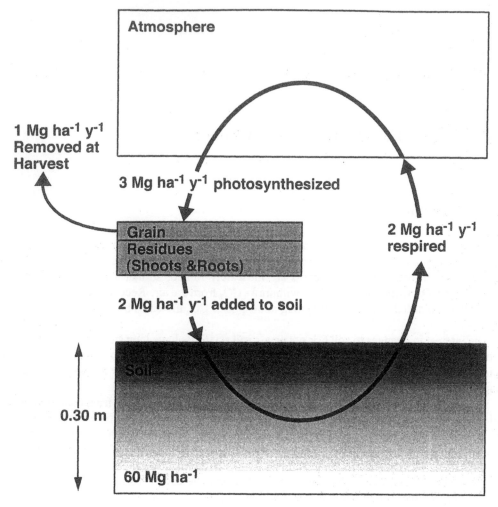

Figure 1. Hypothetical carbon cycle for spring wheat on a Dark Brown Chernozemic soil.

Land management changes soil C storage when the balance between C inputs to and outputs from the soil is altered. The amount of C stored in soil typically decreased when native prairie grasslands were cultivated to produce annual agricultural crops. This decrease was attributable both to decreases in the amount of plant residues returned to the soil and to increases in the amount of CO_2 released from soil organic matter to the atmosphere. In recent decades, the rate of C loss from agricultural soils in Canada appears to have abated (Anderson, 1995; Janzen et al., 1998). Indeed, the amount of C stored in many soils may be increased by adoption of soil-conserving practices such as extended crop rotations (especially those with perennial forages and reduced summerfallow frequency), improved fertilization, reduced tillage and organic amendments (e.g., green and livestock manures).

The balance between inputs and outputs of soil C can be measured in two ways. One approach is to measure directly all fluxes into and out of the soil (Figure 1). Although very sensitive, this method is technically demanding and estimates of net C balance are predisposed to reinforcing or compensating errors among flux measurements. Furthermore, since fluxes depend on fluctuating environmental conditions (e.g., diurnal radiation, seasonal temperature and episodic precipitation), estimates of the cumulative C balance vary widely from month to month and year to year. A second approach is to measure the difference in soil C storage between two or more sampling times. If soil C inputs and outputs are balanced, as depicted in Figure 1, then the amount of soil C will remain unchanged. This approach, though not as sensitive in the short term as the first, measures the net C imbalance over the entire period between sampling times, thereby smoothing out seasonal and inter-year fluctuations.

Measurements of temporal changes in soil C storage are less demanding than calculating C budgets from frequent and detailed estimates of short-term fluxes, but quantitative assessments of temporal changes require careful soil sampling and analysis. Temporal changes often are difficult to quantify, because the annual soil C inputs and outputs are small relative to soil C storage in many biomes like the grasslands of the North American Great Plains. Furthermore, changes in response to a new agronomic practice on existing cropland are usually much smaller than the initial change following conversion of native forest or grassland to arable agriculture. Because of the large area of Canadian cropland (on the order of 45 million ha), even a very small change in soil C (e.g., 0.2 Mg C ha^{-1} or 9 Tg C) may be an important proportion of the reduction target for CO_2 emissions (e.g., 6% of Canada's CO_2 emissions in 1990 amounts to about 8 Tg C yr^{-1}).

The time required for a detectable change in the amount of C sequestered in Chernozemic soil typically might be 4 years or more, because annual inputs and outputs are small relative to C stored in the soil. For example, suppose we adopt a new practice that results in a very large C imbalance – inputs of 2 Mg C ha^{-1}yr^{-1} vs. outputs of 1 Mg C ha^{-1}yr^{-1} – for wheat in a semi-arid region (Figure 1). This imbalance, probably much larger than is normally feasible, would have to persist for at least 3 years before an increase in soil C would be perceptible: 1 Mg C ha^{-1} yr^{-1} for 3 years would result in a 5% increase in C stored in a soil originally containing 60 Mg C ha^{-1} in the 0 to 30 cm layer. Variability of soils and plant production in time and space further complicate detection and estimation of changes in soil C.

B. Comparisons among Treatments vs. Sampling Times

The effect of a proposed practice on C storage is often evaluated by comparing the new 'C-sequestering' practice to a 'conventional' system in a randomized, replicated experiment. The increased C storage attributable to the new practice can be estimated in two ways: by measuring the change over time of C storage under the new practice or by comparing C storage under the new practice relative to that under the conventional system after an elapsed time period (Figure 2). The first approach provides an absolute measure of soil C change, but the cause of the temporal change cannot be unequivocally attributed to the C-sequestering practice. For example, the temporal change may overestimate the effect of the new practice during an interval of above normal productivity (depicted in Figure 2). The second approach may not provide a true measure of absolute C change, but it describes more clearly the soil C gain in response to the new practice. Provided initial C storage under both practices is comparable (hopefully attained by adequate randomization and replication), the difference between practices may be more informative than the temporal change. Comparisons among practices allow for estimation of both increases and avoided losses of soil C (Izaurralde et al., Chapter 37, this volume).

In large-scale projects monitoring changes in soil C storage under improved management, it may not always be practical to include a conventionally managed control. In such projects, results need to be interpreted with the recognition that factors other than the new management practice may contribute

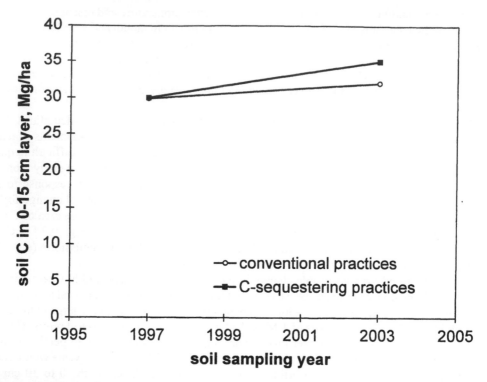

Figure 2. Estimate of temporal soil carbon change versus estimate of net soil carbon change attributable to a specific management practice.

to C change. The influence of management on soil C storage may be confounded by factors such as excessive spatial variability, weather patterns (e.g., a series of wet or dry years) that upset the balance between soil C inputs and outputs, soil deposition or erosion, which redistributes soil C without atmospheric exchanges, and environmental changes (e.g., enhanced growth from CO_2 fertilization or increased atmospheric N deposition).

III. Calculating Soil C Storage

A. Soil C Concentration vs. Mass per Soil Volume

Early studies of soil C change in response to management usually reported decreases in concentrations rather than decreases in nutrient mass per unit area. For example, early estimates of cultivation-induced 'losses' of C from Chernozemic soils usually ranged from 40 to 50% of the original organic matter present under native grassland (Rennie, 1978; Nicholaichuk, 1986). But these values did not necessarily describe quantitative C loss, because mass of C per unit volume of field soils also depends on soil bulk density and horizonation (i.e., the sequence and thickness of pedogenically distinct layers).

More recently, attempts to quantify changes in soil C storage were based on estimates of C mass per unit soil volume calculated as the product of soil bulk density, horizon thickness, and C concentration.

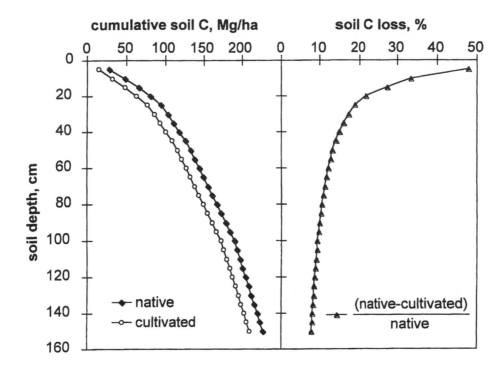

Figure 3. Proportional loss of soil organic C with increasing sampling depth.

Schlesinger (1986) calculated that the mean loss of soil C was 29% in 24 studies of temperate grasslands that contained an average of 19 kg C m^{-2} prior to cultivation. McGill et al. (1988) reported that, for the Canadian prairies, the cultivation-induced loss of soil organic matter ranged from 15 to 30%, and that the actual loss was considerably lower than earlier estimates because organic C in the A horizons was diluted by lower horizon material during cultivation.

B. Soil Thickness and Bulk Density

Proportional changes in soil C storage (e.g., percent of original C present or some other appropriate baseline) usually diminish as increasingly thick soil layers are considered because C near the soil surface is more dynamic (Figure 3). Thus, for an estimate of proportional change to be meaningful, the depth, volume or mass of soil considered must be specified. The soil depth for evaluating C change has varied widely, depending on cultivation depth, study objectives and duration, depth of root penetration, solum thickness, and researcher preference.

Concentrations of organic C in most soils decrease with depth, because C inputs (plant residues) usually are localized at the surface. To account for changes in C concentrations with soil depth, the total sampling thickness usually is subdivided into several layers, and the total soil C mass per unit area is calculated as the sum of C stored in successive soil layers to a specified depth.

For a given C concentration and soil thickness, the quantity of soil C per unit area depends on soil bulk density, and this also varies with management, moisture content, soil depth, and other properties.

To calculate the bulk density of gravelly or stoney soils, the mass of coarse rock fragments contained in the volumetric sample is subtracted from total sample mass, because coarse fragments (>2 mm) are excluded from most soil chemical analyses. Surface soils especially are prone to wide fluctuations in bulk density, such as those caused by tillage operations which loosen the soil and decrease bulk density, and by intense rainfall that compacts unprotected soil. When C concentrations are identical, the amount of C stored in a fixed soil thickness or volume varies in direct proportion to soil bulk density.

Most investigators now recognize the dependence of C storage on soil thickness and bulk density, and calculate soil organic matter and C storage as the product of concentration, soil bulk density, and soil thickness (e.g., Tiessen et al., 1982; Aguilar et al., 1988). Since C storage is expressed as mass per unit area to a fixed depth (e.g., 60 Mg C ha^{-1} to 30 cm), comparisons among soils typically are based on the same soil thickness or volume.

C. Equivalent Soil Mass Approach

Most recent evaluations have expressed soil C storage in units of mass of C per unit area to a fixed depth (e.g., Mg C ha^{-1} to 30 cm). This approach allows quantitative comparisons among treatments, but it overlooks the influence of variable soil mass among contrasting soils. Recently Boone et al. (1999) emphasized the importance of calculating the influence of coarse fragments or rock on soil nutrient status. While they recommended that soil data be expressed as mass per unit area, the soil depth, volume, and mass to be considered was unclear. Unless soil bulk densities are identical, comparisons among C stored in the same volume or depth of contrasting soils will be confounded by differences in soil mass.

The potential error from comparing C stored in unequal soil masses can be illustrated by a simple thought experiment. Suppose you extract a core of soil to a depth of 15 cm, and determine that it has a C concentration of 20 g C kg^{-1} and a bulk density of 1.6 Mg m^{-3} (Figure 4). You now pull a moldboard plow through this same soil, take a second core of soil to a depth of 15 cm, and determine that its C concentration is unchanged (20 g C kg^{-1}) but that its bulk density has now decreased to 1.2 Mg m^{-3}. Now you calculate that C content to 15 cm was 48 Mg C^{-1} before tillage but only 36 Mg C^{-1} after tillage. This implies that you have lost 25% even though you know that all of the C has remained. The 'loss', clearly, is a simple artifact arising from the difference in amount of soil sampled: in the former case, the 0 to 15 cm layer contains 2400 Mg soil ha^{-1}, in the latter, only 1800 Mg soil ha^{-1}. This simple example illustrates that, if C storage is calculated to constant depth, small differences in bulk density can introduce biases that are much larger than the management-induced differences we are looking for.

To circumvent bias introduced by differences in bulk density, comparisons of soil C storage should be expressed, we submit, in units of mass of soil C in an equivalent soil mass per unit area. This value can be easily calculated by estimating the thickness of the deepest soil layer required to attain the equivalent mass, and summing the mass of soil C contained in this and the above layers (Ellert and Bettany, 1995). The equivalent mass approach had been used by Nye and Greenland (1964), who observed that comparisons of soil organic matter "...should be based on the same soil mass, or the possible error involved in sampling to a fixed depth clearly recognized." Since then, this potential error has been overlooked by most investigators, with the exception of those at the Rothamsted Experimental Station (e.g., Jenkinson and Johnston, 1976; Powlson and Jenkinson, 1981). More recently, potential errors from comparisons among unequal soil masses have been considered by Duxbury (1995) and Campbell et al. (1998).

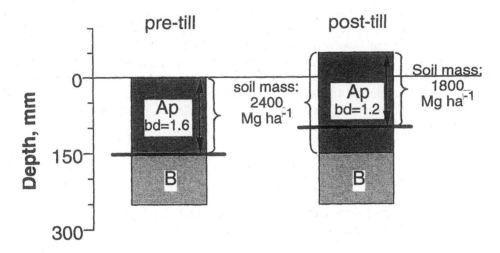

Figure 4. Hypothetical influence of tillage-induced changes in soil bulk density on comparisons between soil C stored in an equal soil volume and in an equivalent mass. (From Ellert and Bethany, *Can. J. Soil Sci.* 75:530. With permission.)

D. Equivalent Soil Mass Calculation

The mass of C stored in an equivalent soil mass can be calculated for any sampling point in the field, provided accurate data are available for thickness, bulk density and C content of successive soil layers. Although the thickness required to attain an equivalent soil mass usually is unknown at the time of sampling, it can be estimated from soil bulk density after sampling. The mass of soil, per unit area, in each soil layer can be calculated as follows:

$$M_{soil} = bd \cdot T \cdot 10,000 \ m^2 \ ha^{-1} \qquad [1]$$

where:

  M_{soil} = soil mass per unit area, Mg ha^{-1}
     bd = soil mass per sample volume, Mg m^{-3}
      T = thickness of soil layer, m

Similarly, the mass of C per unit area in each soil layer is calculated as follows:

$$M_C = (conc/1000) \cdot bd \cdot T \cdot 10,000 \ m^2 \ ha^{-1} \qquad [2]$$

where:

  M_C = C mass per unit area, Mg ha^{-1}
  con = C concentration from lab analysis, g kg^{-1} = kg Mg^{-1}
   bd = soil mass per sample volume, Mg m^{-3}
    T = thickness of soil layer, m

Conventional estimates of C storage involved summing the masses of C stored in successive soil layers to a constant depth, without any regard to the total soil mass under consideration. The 'equivalent mass method' involves a slight modification to normalize the mass of C per unit area for differences in soil mass. Since soil mass per sample volume refers to the mass of rock-free soil per unit sample volume, nutrient status is adjusted for differences in rock content. Essentially, the amount of C per unit area is calculated for an identical, standard reference or equivalent soil mass. This is done by summing the masses of C stored in successive soil layers, plus an additional layer required to attain an equivalent soil mass:

$$M_{C,\,equiv} = M_{C,\,surf} + M_{C,\,add} \tag{3}$$

Where:

$M_{C},$ equiv = C mass per unit area in an equivalent soil mass, Mg ha^{-1}
$M_{C},$ surf = C mass in surface layer(s), Mg ha^{-1}
$M_{C},$ add = C mass in additional layer required to attain equivalent soil mass, Mg ha^{-1}

Carbon masses in successive soil layers are calculated using equation [2]. The thickness of the additional soil layer (T_{add}) required to attain the equivalent soil mass is calculated as follows:

$$T_{add} = \frac{(M_{soil,\,equiv} - M_{soil,\,surf}) \cdot 0.0001\ ha\ m^{-2}}{bd_{subsurface}} \tag{4}$$

where:

T_{add} = thickness layer required to attain the equivalent soil mass, m
$M_{soil},$ equiv = equivalent soil mass = mass in heaviest layer(s), Mg ha^{-1}
$M_{soil},$ surf = sum of soil mass in surface layer(s), Mg ha^{-1}
$bd_{subsurface}$ = soil mass per sample volume of sub-surface layer, Mg m^{-3}

In this approach, the 'equivalent mass' is the minimum amount of soil that encompasses all soil in the layers most susceptible to management. For example, in a system regularly plowed to 25 cm, the equivalent mass might be the mass of soil in the most dense treatment to a depth of 30 cm. The actual value selected as the equivalent mass is less important than its consistency for all comparisons of C storage at each study site.

The calculation described earlier assumes that the bulk density and C concentration are uniform within the soil layer used to reach equivalent mass in the lighter profiles; that is, the C concentrations and bulk densities of the entire subsurface horizons are assumed to represent those in the top portions of the horizon required to attain the equivalent soil mass. Because the C concentration usually declines with increasing depth within a layer, this assumption may cause conservative increases in the C mass of light soils relative to heavy soils (i.e., soils near the equivalent mass). The magnitude of this error diminishes with progressively smaller sampling depth increments.

Another error may arise if bulk density estimates are made independently of C concentrations. Bulk density should be calculated from the same core used to determine C concentration. Although this method may not yield a true measure of field bulk density because of compression during sampling, the calculation requires the density of the soil core, *not* the actual field bulk density.

IV. Soil C Variability

Soil properties, including C concentration, bulk density and A horizon thickness, vary widely in both time and space. To estimate the exchange of C between land and the atmosphere, variations in soil C from one location to another must be distinguished from variations from one sampling time to another. Systems of soil taxonomy and land use classification recognize spatial variations associated with differences in factors such as climate, plant community, topography, parent materials and human activity. Variability exists among regions, within landscapes (e.g., Roberts et al., 1989), and even within a few grams of soil, where C exists as a heterogeneous mixture including readily decomposable fragments of plant material, refractory coal-like material, and inorganic carbonates.

A. Lateral Variability

Spatial or lateral variability in soil organic matter was recognized already in early studies. Almost a century ago, organic C content of native prairie at the Indian Head Experimental Farm in Saskatchewan was shown to vary from 4 to 11% over short distances (60 to 90 cm), making it difficult to assess the impact of cultivation on C and N storage (Alway and Vail, 1909).

Soil variability has now been assessed in many studies aiming to devise optimal soil sampling strategies for characterizing soil on a field scale (e.g., Bowman, 1991). Variations in C content along catenas or toposequences have been used to study soil processes under contrasting environmental conditions (e.g., Roberts et al., 1989). More recently, researchers have quantified spatial variability at scales of less than 1 km by collecting samples at grid points superimposed on complex landscapes (e.g., Pennock et al., 1994).

B. Biotic and Geomorphic Processes

Transformations influencing contemporary soil organic C storage may be partitioned between biotic and geomorphic processes. The biotic processes – photosynthesis and respiration – determine the balance between C inputs to soil and CO_2 outputs from decomposition. The geomorphic processes – soil redistribution by wind, water, tillage and gravity – influence soil C by removing soil and associated C from one area and depositing it elsewhere. The equivalent soil mass approach assumes implicitly that geomorphic processes are negligible. If this assumption is invalid, then comparisons among C stored in contrasting soils (or at contrasting sampling times) must be based on soil masses which differ according to the amount of redistributed soil.

The distinction between biotic and geomorphic processes helps clarify the mechanisms of soil C change, but the processes are interdependent, as soil redistribution may influence both production and decomposition, and land with low production or C retention tends to be more prone to erosion (Gregorich et al., 1998). Both biotic and geomorphic processes tend to increase the spatial variability of soil C. For example, in Chernozemic soils on hummocky terrain, plant C inputs and thus soil organic C generally are lower on the knolls, because water runs to lower slope positions. When such soils are cultivated to produce annual crops, erosion by tillage, water and wind removes the already thin topsoil from the knolls and deposits it at lower slope positions (e.g., Martz and deJong, 1987). Zero net erosion or loss of soil from the landscape (erosion from knolls ≈ deposition at lower slope positions) is typical of the Canadian Prairies where a major proportion of the area is internally drained (Lemmen et al., 1997) and drainage basins are closed.

C. Soil Redistribution

Estimating soil C change is hampered by variability in soil properties and by the combined interplay of biotic and geomorphic processes that determine soil C storage. Although earlier studies focused on the loss of soil C with eroded soil, it now is clear that soil deposition is equally important, because most eroded soil is deposited elsewhere in the terrestrial system. To estimate changes in soil C storage, it is essential to distinguish between absolute removal of soil from the watershed by long-distance transport in water or air, and mere redistribution of soil within the watershed. Whether erosion increases or decreases the decomposition of soil organic C to atmospheric CO_2 remains unclear. The effect of soil redistribution on organic C decomposition is complicated by preferential movement of C-enriched soil particles, organic C burial, and exposure of C-depleted sub-soil.

Estimates of management-induced changes in C storage at specific sites in the landscape may be simplified by selecting sites with negligible soil redistribution. If appreciable erosion does occur, the redistribution needs to be measured to account for differences in masses of soil and associated C. If temporal differences in soil mass at eroded sites are ignored, actual soil organic C losses will be partially obscured by inclusion of deeper layers during sampling. Conversely, at sites with deposition, soil C gains will be underestimated and thus obscured by omission of deeper layers.

D. Forms of Soil C

The forms included in 'soil C' must be strictly defined before changes can be properly measured. To evaluate the net exchange of C between land and the atmosphere, the important pool is ecosystem C, including plant residues on and in the soil. Soil C traditionally has been regarded as that associated with dry soil crushed to pass a sieve with 2 mm openings. Rarely are methods to dry and crush the soil standardized, and they exclude variable proportions of roots and soil-incorporated stems. Clearly this arbitrary definition of soil C biases quantitative assessments of C storage under rival management practices, especially among those which differ in the extent of residue incorporation.

A more meaningful approach to assess soil and ecosystem C is to first collect and quantify above-ground C, and include whatever remains on and in the soil as 'soil C'. Since soil C thus defined also includes plant crowns, the area and position of the soil core should represent the distribution of plants in the ecosystem; for example, with crops seeded in rows, the sampling must avoid bias between row and inter-row areas. The soil C contained in such samples includes coarse fragments of plant residues, which may be separated and quantified or uniformly mixed and ground with the soil. Total organic C storage in the ecosystem is the sum of aboveground and soil organic C.

Carbon is present in soil as both inorganic and organic materials. Inorganic C is present mainly as carbonate minerals, including primary carbonates originally present in the parent material and secondary or pedogenic carbonates formed during soil genesis. Smaller amounts of charcoal or graphite-like compounds made up of carbon atoms may also be present. Automated instruments are commercially available for estimating total soil C by dry combustion, and some instruments simultaneously determine total N using the Dumas method. Older methods based on wet oxidation to determine soil organic matter are not recommended as they are less reliable, more time consuming, and more hazardous. Since soil organic matter is a heterogeneous mixture of poorly defined compounds with variable C contents, it is more accurate and meaningful to analyze and report soil organic C rather than soil organic matter.

Studies of soil C change have usually focused on organic C. In soils with significant amounts of carbonate, soil organic C can be measured directly by combusting samples which have been acidified to eliminate inorganic C or by combusting samples under conditions to selectively oxidize organic C to CO_2 without decomposing inorganic C. Organic C may also be determined as the difference between separate analyses of total and inorganic C. Inorganic C typically is determined by measuring CO_2 released by adding acid. Analyses based on acidification to eliminate or quantify inorganic C include charcoal, coal-like materials and elemental C as soil organic C, because such materials persist after acidification and are recovered by combustion.

Weathering or formation of soil inorganic C, consisting of carbonate minerals, may generate or sequester CO_2, but in most North American agricultural soils these processes have a minor influence on CO_2 exchange relative to that of biological processes involving soil organic C (Curtin et al., 1994; Suarez, 1999). Thus the primary goal of soil inorganic C analysis is to estimate soil organic C by subtraction from total C.

V. Soil Sampling Method for Regional Monitoring

Regional soil C monitoring programs have been proposed as a means to measure temporal changes in terrestrial C storage, determine net CO_2 exchange between land and atmosphere, and to help verify models of such processes. Areal inventories of soil type, vegetative cover, land use, and land management are required to estimate soil C on a regional or national basis. Estimation of changes in soil C storage is hampered by variability in soil properties and by the interplay of biotic and geomorphic processes that determine soil C storage. Canada is fortunate to have comprehensive soil inventories (e.g., Lacelle et al., this volume), and a rich history of soil research documenting the temporal dynamics of soil C as influenced by management (e.g., Janzen et al., 1998). What is lacking in Canada and elsewhere is the regional-scale (i.e., > 10 km) data on temporal soil C changes required to link land resource inventories with detailed soils research concentrated at specific points.

A regional monitoring project has been established on the Canadian Prairies to measure the temporal change in soil C storage under agricultural cropland. The primary goal of the project is to assemble quantitative data on regional soil C storage which will be used to evaluate various approaches (e.g., models, empirical equations, 'expert' projections) to estimate soil C change on a regional basis. A soil sampling procedure was devised in an attempt to measure any change in soil C storage at specific points or 'microsites' between an initial sampling year and at a subsequent sampling time 4 to 8 years afterward. The initial samples have been collected at over 250 microsites within Saskatchewan alone.

A. Field Selection

Fields were selected to represent a broad range of soil taxa and textures over an extensive area of agricultural soils on the Canadian Prairies. Fields included fine, medium and coarse-textured soils within the Brown, Dark Brown, and Black Chernozems and the Gray Luvisols (i.e., Aridic, Typic and Udic Borolls and Boralfs). To avoid complications from subsequent geomorphic processes and to allow direct assessment of biotic C change, we selected only level landscapes where the risk of appreciable soil erosion and deposition was small. Whereas previous studies of soil variability and geomorphic processes typically have been based on a sampling interval of 1 to 100 m, the regional monitoring project is based on initial and subsequent soil C stored in fields separated by 1 to 100 km. The staff and members of conservation groups, along with other farmers, provided valuable assistance with field selection. The project was designed to assess contemporary changes in C storage, but it was recognized that previous land use could also influence current C stocks and the C sequestration potential. The project focused specifically on C-conserving benefits of reduced tillage with continuous cropping. Consequently, we selected fields with a narrow range of cropping systems based on annual cereals and oilseeds. Farms with perennial forages, organic amendments (livestock manure, sewage sludge, food processing wastes), or irrigation were excluded so that any change in soil C could be attributed more easily to changes in tillage and cropping intensity.

B. Microsite Selection and Configuration

Within each field of 30 to 65 ha, soil C storage was measured at one to six microsites, each 4 m by 7 m. Restricting the sampling locations to relatively small microsites minimizes lateral variability so that temporal changes can be determined with greater precision. The microsites represent dominant soils within the field, and consider factors such as topography, cropping history, and soil texture. Thus rather than positioning microsites at random, experienced pedologists positioned the microsites on modal soils. If the pedologist observed unexpected variability among soil morphology of the six initial cores, the position was abandoned and the microsite was re-located to a more uniform area. The number of microsites selected for each field depends on study objectives (strictly monitoring versus more detailed research) and landscape uniformity (more sites required in complex landscapes). The microsites typically are at least 100 m from field edges or uncultivated patches (e.g., depressions and rock piles) and are separated by 200 m or more to ensure adequate spatial representation of the field. Microsites usually are located to represent specific slope positions (typically midslope), and to avoid unconformities in the field, such as gullies, gravel outcroppings, saline seeps, depressions and knolls. The configuration of each microsite was a 2 by 5 m grid of six cores separated by 2 m and surrounded by a 1-m buffer (Figure 5). The precise location of each microsite was recorded. The location was measured by traditional surveying techniques, using a fixed benchmark such as a survey marker, or by the Global Positioning System (GPS). To pinpoint the exact location, electromagnetic markers (ScotchMark™ by 3-M Corp., Telecom Systems Group, Austin TX) were buried at the bottom of the two most widely separated initial core holes, below the depth of cultivation. Alternatively, the microsite was oriented toward magnetic north, and one electromagnetic marker was buried at the northeast core.

C. Sampling Procedure

Soil C change at each microsite will be determined by comparing C in soil sampled at time 0 (initial sampling) with that in subsequent sampling times. The six soil cores collected at the initial sampling are systematically positioned within the microsite to avoid re-sampling the same spot later (Figure 5). The 2 by 5 m sampling grid are marked with a template, and soil cores are removed at the grid line intersections using a hydraulic, truck-mounted soil coring machine. The recommended core size is 7 cm and minimum coring depth is 50 cm. All sampling will be performed in the spring or fall when the land is accessible and when sampling will not cause crop damage. Seasonal effects (i.e., differences

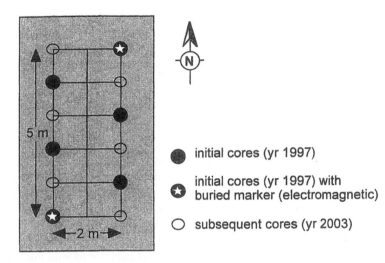

Figure 5. Microsite.

between spring and fall samples) may be avoided by re-sampling soils at the same time of year when the initial samples were collected.

After extraction, each core is carefully subdivided into 10 cm segments (0 to 10, 10 to 20, 20 to 30, 30 to 40, and 40 to 50 cm). To help account for vertical changes, the depths of recognizable soil horizons or other discontinuities within the 50-cm core also are recorded. Core descriptions, including information such as the depth of the Ap horizon, presence of carbonates, soil color, stoniness, and changes in texture, will supplement data on C storage by identifying additional sources of variability at both initial and subsequent samplings. Since land management information is collected for each microsite, valuable data is being assembled for management practices on individual farms in Saskatchewan. To prevent topsoil burial and undue disruption of the microsite, the holes remaining after soil core removal and electromagnetic marker placement are refilled with soil obtained from well outside the microsite.

The original intent was to combine, by profile depth, the six cores from each microsite to minimize analytical costs. At some sites, however, individual cores have been retained separately, because the additional analytical cost was a relatively small increment to total site costs, and data on variability within microsites was desired. By retaining individual soil cores, within-microsite variability may be compared with any temporal C change between initial and subsequent sampling. Combining the six cores, by 10-cm depth increments, at each sampling time may provide a more efficient allocation of resources by allowing for analysis of more microsites, but information on variability within the microsite is lost.

Soil bulk density, moisture content and coarse fragments are determined after the initial samples are collected. The soil samples are air-dried, crushed to pass a 2-mm sieve (including coarse fragments of plant litter), subdivided, and stored in glass jars at two separate locations to safeguard against sample destruction. To assess changes in soil C storage, initial (e.g., collected 1997) and subsequent samples (e.g., to be collected 2003) should be analyzed at the same time. This assumes that soil organic C remains unchanged during storage under appropriate conditions (dry, cool, dark). Previous experience suggests that estimates of soil organic C content are influenced more by temporal changes in analytical techniques than by true changes in C content during storage (though the same may not apply to more labile soil constituents, such as plant-available N). On the grounds that systematic analytical errors can be minimized by analyzing initial and subsequent samples at the same time, the bulk of the analyses will be delayed until after the subsequent set of samples is collected. At subsequent sampling times, the core positions within each microsite will be offset from those at previous sampling times to avoid prior disturbance (Figure 5).

V. Summary and Conclusions

Estimates of temporal changes in soil C storage are crucial to quantify the net exchange of CO_2 between land and the atmosphere. Such estimates encompass the net effect of several fluxes and transformations. Detailed measurements of individual fluxes provide insights into ecosystem physiology and help verify process-level models. But temporal variability, potential for compensating errors, and the expense of continuous monitoring make detailed flux measurements less practical for estimating net CO_2 exchange at multiple sites for a long time (> 3 years). In contrast soil C is a measurable pool which dominates terrestrial C storage in most ecosystems with non-woody biomass. The main obstacle to estimating net CO_2 exchange from temporal changes in soil C is the small magnitude of anticipated changes in soil C storage relative to the amount present. In an attempt to overcome this obstacle, a robust soil sampling procedure has been devised to measure temporal C change in agricultural soils at more than 250 sites on the Canadian Prairies. The influence of spatial variability within fields and regions is avoided by measuring soil C storage at carefully marked microsites (4 by 7 m). The three to six microsites sampled within each field provide only a crude estimate of within field variability. By measuring C storage at two or more sampling times separated by 4 to 6 years, the procedure assesses temporal soil C change within specific microsites, not within entire fields. Complexity arising from geomorphic changes, such as soil redistribution, were avoided by selecting sites where such processes are unlikely.

The procedure includes several provisions to minimize errors, including:

1. Errors from small-scale spatial variability within microsites (2 by 7 m) are reduced by collecting six cores per microsite.
2. Errors from unequal soil masses are eliminated by calculating the masses of C stored in an equivalent soil mass.
3. Errors from lateral variability of samples collected at different sampling times are avoided by marking sampling locations so that subsequent samples may be closely interspersed with initial ones.
4. Errors from systematic changes in analytical methodology are minimized by storing initial samples, and analyzing initial and subsequent samples at the same time using identical analytical procedures.

Investigations of soil C storage on a regional basis may be justified on the grounds that most previous research has been site-specific, and data collected during soil surveys was insufficient. Whether acknowledged or not, most researchers tend to focus their studies on a relatively narrow range of spatial and temporal scales. Now that humans are recognized as major contributors to global environmental change, researchers are being asked to extend or scale-up their results obtained at relatively small scales. The sampling procedure described here is intended to quantify any changes in soil C storage at various microsites distributed across the Canadian Prairies, but it will not provide explanations for the changes. Two or more management treatments would have to be imposed and maintained to determine unambiguously the influence of management on C storage. Despite this, it is hoped that the general principles discussed here will lead to the development of robust procedures to measure the exchange of CO_2 between land and the atmosphere. Eventually, these estimates at specific points will need to be linked via models, whether simple empirical equations or complex process-level models, to assess regional changes in soil C storage.

References

Aguilar, R., E.F. Kelly and R.D. Heil. 1988. Effects of cultivation on soils in northern Great Plains rangeland. *Soil Sci. Soc. Am. J.* 52:1081-1085.

Alway, F.J. and C.E. Vail. 1909. A remarkable accumulation of nitrogen, carbon and humus in a prairie soil. *J. Industrial and Engineering Chemistry* 1:74-76.

Anderson, D.W. 1995. Decomposition of organic matter and carbon emissions from soils. p. 165-175. In: R. Lal, J. Kimble, E. Levine and B.A. Stewart (eds.), *Soils and Global Change*. Lewis Publishers, Boca Raton, FL.

Boone, R.D., D.F. Grigal, P. Sollins, R.J. Ahrens and D.E. Armstrong. 1999. Soil sampling, preparation, archiving, and quality control. p 3-28. In: G.P. Robertson, D.C. Coleman, C.S. Bledsoe and P. Sollins, (eds.), *Standard Soil Methods for Long-Term Ecological Research*. Oxford University Press, New York.

Bowman, R.A. 1991. Spatial variability of selected carbon, nitrogen, and phosphorus parameters on acid and calcareous rangeland soils. *Commun. Soil Sci. Plant Anal.* 22:205-212.

Campbell, C.A., F. Selles, G.P. Lafond, B.G. McConkey and D. Hahn. 1998. Effect of crop management on C and N in long-term crop rotations after adopting no-tillage management: Comparison of soil sampling strategies. *Can. J. Soil Sci.* 78:155-162.

Curtin, D., F. Selles, C.A. Campbell and V.O. Biederbeck. 1994. Canadian prairie agriculture as a source and sink of the greenhouse gases, carbon dioxide, and nitrous oxide. Agriculture & Agri-Food Canada Publication No. 379M0082. Swift Current, Canada.

Duxbury, J.M. 1995. The significance of greenhouse gas emissions from soils of tropical agro-ecosystems. p.279-291. In: R. Lal, J. Kimble, E. Levine and B.A. Stewart (eds.), *Soil Management and the Greenhouse Effect*. Lewis Publishers, Boca Raton, FL.

Ellert, B.H. and J.R. Bethany. 1995. Calculation of organic matter and nutrients stored in soils under contrasting management regimes. *Can. J. Soil Sci.* 75:529-538.

Gregorich, E.G., K.J. Greer, D.W. Anderson and B.C. Liang. 1998. Carbon distribution and losses: erosion and deposition effects. *Soil & Tillage Research* 47:291-302.

Janzen, H.H., C.A. Campbell, R.C. Izaurralde, B.H. Ellert, N. Juma, W.B. McGill and R.P. Zentner. 1998. Management effects on soil C storage on the Canadian prairies. *Soil & Tillage Research* 47:181-195.

Jenkinson, D.S. and A.E. Johnston. 1976. Soil organic matter in the Hoosfield continuous barley experiment. p. 87-101. In: *Rothamsted Experiment Station Report for 1976, Part 2*, Harpenden, England.

Lemmen, D.S., R.E. Vance, S.A. Wolfe and W.M. Last. 1997. Impacts of future climate change on the southern Canadian Prairies: A palaeoenvironmental perspective. *Geoscience Canada* 24:121-133.

Martz, L.W. and E. de Jong. 1987. Using cesium-137 to assess the variability of net soil erosion and its association with topography in a Canadian Prairie landscape. *Catena* 14:439-451.

McGill, W.B., J.F. Dormaar and E. Reinl-Dwyer. 1988. New perspectives on soil organic matter quality, quantity, and dynamics on the Canadian prairies. p. 30-48. In: *Land Degradation and Conservation Tillage*. Partial Proceedings of the 34th annual CSSS/AIC Meeting, Calgary, Alberta.

Nicholaichuk, W. 1986. Management of prairie soils, a fragile resource. *Transactions of the Royal Society of Canada, Series V*. 1:53-66.

Nye, P.H. and Greenland, D.J. 1964. Changes in the soil after clearing tropical forest. *Plant and Soil* 21:101-112.

Odum, E.P. 1967. The strategy of ecosystem development. *Science* 164:262-270.

Pennock, D.J., D.W. Anderson and E. de Jong. 1994. Landscape-scale changes in indicators of soil quality due to cultivation in Saskatchewan, Canada. *Geoderma* 64:1-19.

Powlson, D.S. and D.S. Jenkinson. 1981. A comparison of the organic matter, biomass, adenosine triphosphate and mineralizable nitrogen contents of ploughed and direct-drilled soils. *J. Agric. Sci. Camb.* 97:713-721.

Rennie, D.A. 1977. Soils of the prairie provinces. p. 23-27. In: G.K. Rutherford, B.P Warkentin and P.J. Savage (eds.), *The Geosciences in Canada, 1977 – Annual Report and Review of Soil Science*. Geological Survey Paper 78-6. Energy Mines and Resources Canada, Ottawa.

Roberts, T.L., J.R. Bettany and J.W.B. Stewart. 1989. A hierarchical approach to the study of organic C, N, P, and S in western Canadian soils. *Can. J. Soil Sci.* 69:739-749.

Schimel, D.S. 1995. Terrestrial ecosystems and the carbon cycle. *Global Change Biology* 1:77-91.

Schlesinger, W.H. 1986. Changes in soil carbon storage and associated properties with disturbance and recovery. p. 194-220. In: J.R.Trabalka and D.E. Reichle (eds.), *The Changing Carbon Cycle – A Global Analysis*. Springer-Verlag, New York.

Suarez, D.L. 2000. Impact of agriculture on CO_2 as affected by changes in inorganic carbon. p. 257-272. In: R. Lal, J.M. Kimble, H. Eswaran and B.A. Stewart (eds.), *Global Change and Pedogenic Carbonates*. Lewis Publishers, Boca Raton, FL.

Tiessen, H., J.W.B. Stewart and J.R. Bettany. 1982. Cultivation effects on the amounts and concentration of carbon, nitrogen, and phosphorus in grassland soils. *Agron. J.* 74:831-835.

CHAPTER 11

Estimating Total System Carbon in Smallhold Farming Systems of the East African Highlands

P.L. Woomer, N.K. Karanja and E.W. Murage

I. Introduction

The Framework Convention on Climate Change, or Kyoto Protocol (McGivern, 1998), formalized the "net" approach to carbon offsets, allowing emissions from fossil fuels to be sequestered elsewhere, primarily through reforestation. It also established legally binding reduction targets thereby creating new markets for forest services (Hedger, 1998). Mitigation of atmospheric change through forestry initiatives is viewed as a viable approach to carbon offsets and many such projects are under way (Kinsman and Trexler, 1993; Dixon et al., 1993; Tipper and De Jong, 1998) including in East Africa (Verweij and Emmer, 1998). These projects assume that massive carbon stocks were lost from tropical forests to the atmosphere in the past and that it is practical and economical to reaccumulate carbon in this biome. Much evidence exists to support this view. Sanchez et al. (1994) estimate that over 600 Mha of the 1500 Mha ($=10^6$ ha) of the original tropical forest zone is now deforested and that 250 Mha of degraded lands are eligible for reclamation. Greenhouse gas mitigation through tropical forestry is sufficiently recognized that the Swiss-based firm Société Générale de Surveillance offers a verification service for project emission reduction targets (Trines, 1998).

The needs of small-scale land managers, who are often responsible for deforestation, may not be compatible with carbon mitigation through tropical reforestation. Saurbeck (1993) projected that the 787 Mha of arable land in the tropics will increase to 1286 Mha by 2025 with the largest regional increase taking place in Africa (160 Mha). Barnett (1992) expressed concerns that the establishment of tree plantations fails to meet local needs and results in diminished agricultural production, land scarcity and rural poverty. A conflict also exists between the food security needs of developing nations versus that of society as-a-whole for the stabilization of global atmospheric change, with the expectations of improved standards of living by rural poor at risk of being overlooked. Agroforestry interventions are viewed as one means of mobilizing smallhold farmers as agents of carbon sequestration (Unruh et al., 1993; Kursten and Burshel, 1993) but difficulties surround legitimacy of incorporating these systems into formal carbon offset projects, particularly when compared to large-scale plantation forestry approaches (Dixon et al., 1993).

Woomer et al. (1998a) identified organic resource availability as a constraint in smallhold farming systems in the Kenyan Highlands. Smallholds consist of different farm enterprises each with its own standing carbon stocks and potential for increase. These enterprises are connected through organic resource flows and are dynamic in response to new markets and technologies. Methods for characterizing carbon stocks and flows in smallhold farming systems are not well established and the size, distribution and opportunities for increasing farm carbon stocks are insufficiently documented.

Table 1. A profile of small-scale farming in the Central Kenyan Highlands (Kapkiyai et al., 1998) and Uganda's Lake Victoria Basin (Bekuda and Woomer, 1996)

Feature	Central Highlands	Uganda Lake Basin
Number of households	103	510
Cropping system	Maize-based	Banana-based
Farm size (ha)	3.3	1.8
Occupants (farm⁻¹)	9.5	7.0
Cattle (farm⁻¹)	6.2	2.4
Proportion of household that		
Practice intercropping	81	69
Apply some crop residues to soil (%)	100	81
Feed crop residues to livestock (%)	95	46
Apply animal manure at planting (%)	97	31
Apply mineral fertilizer (%)	81	4
Prepare compost (%)	50	16

In this chapter, we explore an approach to characterizing the carbon stocks in smallhold farming systems in East Africa and suggest targets for carbon offsets within different farm enterprises.

II. Site Selection

A. Agroecological Zones and Farming Systems

The approaches described in this chapter are based on field work and past observations in the Kenyan Highlands (Kapkiyai et al., 1998; Woomer et al., 1997) and the Kenyan and Ugandan Lake Victoria Basin (Bekunda and Woomer, 1996; Woomer et al., 1998a). These small-scale farming systems are situated close to the equator at 1100 to 2800 m elevation, allowing for efficient use of the 800 to 1500 mm annual precipitation (Hargreaves and Samani, 1986). The soils are predominantly Nitisols in the Kenyan Highlands and Acrisols and Ferralsols in the northern Lake Victoria Basin (FAO, 1977) corresponding to Alfisols, Ultisols and Oxisols, respectively (USDA, 1992). The area was naturally vegetated by Afromontane and Guino-Congolean Forests (White, 1983) most of which were cleared for cultivation. Rotational fallows were practised by natives prior to the arrival of British colonist farmers (Odingo, 1971). Following independence in the early 1960s, many lands were returned to displaced natives who today practice continuous, mixed farming with maize (*Zea mays*) and/or banana (*Musa* cvs)-legume intercrops for food closely integrated with animal raising and market cultivation of vegetables, fruits, coffee (*Coffea arabica*), pyrethrum (*Chrysanthemum cinerariaefolium*) and tea (*Camellia sinensis*) (Woomer et al., 1998a). Much of this area is now densely populated by 300 to 1200 persons km⁻², with the upper range resulting in impoverished peri-urban settlement. Other characteristics of these small-scale farming systems are profiled in Table 1. The farm sizes, number of occupants and management of crop residues are very similar between cropping systems in the Central Highlands of Kenya and Lake Victoria Basin. Greater use of fertilizers and stronger integration of livestock in Kenya reflect the better access to markets and longer continuity of agricultural development. The cropping systems themselves are not dissimilar considering that dry

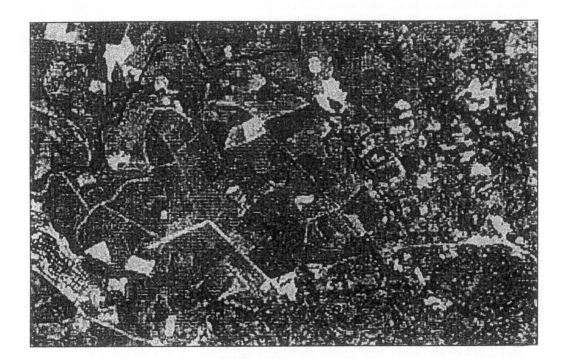

Figure 1. Landsat Thematic Mapper Image of a sugarcane plantation (left) and smallhold farms (right) in Kakamega, Western Kenya.

bean (*Phaseolus vulgaris*) is the most frequent intercrop and that maize and bananas are often grown together in both systems.

B. Selecting Farm Households

Smallhold farming in East Africa is undergoing transition from fallow-based to continuous cropping and from subsistence to market orientation, resulting in complex mosaics of land use. Many features are identifiable from remote sensing images of areas under smallhold management including cultivated and bare fields, roads and tree lots, but these are less easily interpreted than images from larger-scale, plantation systems. Figure 1 presents a Landsat Thematic Mapper Image from Kakamega, Western Kenya with pixels of 30 m x 30 m and grey-scaled from normal "false-colour" infrared. The "jumbled" pixels to the right represent smallhold farming systems and the better defined fields to the left belong to Mumias Sugar plantation. Difficulties with interpreting images from smallhold systems arise from the relatively small size of different enterprises as each pixel may contain complex mixtures of structures, trees, crops and bare land. This situation is unlikely to change until either the resolution is improved or extensive ground-truthing is conducted and linked to interpretive software that spectrally unmixes pixels, allowing each to be subdivided into component land uses (Marcus Walsh, personal communication).

Ground-truthing biomass and carbon stocks of these images prior to spacial analysis require that measurements be conducted in representative land uses within randomly selected households. Political realities and security concerns are such that researchers are unable to simply select holdings at random from remotely sensed images and visit them unannounced to farmers and authorities. An alternative is to coordinate visits through agricultural extension or local developmental non-governmental organizations who maintain lists of community members and are able to provide liaison. Researchers should be aware that local organizations are often inclined to recommend their "model" farmers who may not be representative of the larger community and must be prepared to explain their needs to farm liaison specialists. Once an individual household is selected, researchers are advised to obtain the coordinates of an easily recognized feature (e.g., a bare field adjacent to the household) using a Geographic Positioning System. The number of households examined within each agroecological zone depends upon the purpose of the study and the distribution of enterprises within the landscape but in general 5 to 10 per contrasting zone are sufficient for purposes of preliminary inventory.

C. Household Introduction, Background and Incentives

The research team is introduced to members of the farm household and explains the purpose of their visit and measurements. Even the best informed of farmers in East Africa find carbon sequestration and atmospheric change esoteric, and the objectives of the work are better explained in terms of biomass inventory, agroforestry and organic resource management. It is unlikely that the household will derive any direct benefit from carbon stock characterization and it is unwise to raise farmers' expectations. One option is to offer compensation through gifts, such as improved seed or fertilizer, or to hire members of the household to assist in field work on their farm. The household is characterized through responses to a short survey (Bekunda and Woomer, 1996). A preliminary sketch map of the farm is prepared with cartographic excellence secondary to facilitating communication with the household (Conway, 1989) and identifying various farm enterprises and land use history. Smallhold farmers are generally certain about the size of their family holding, but less so about the sizes of individual fields. More accurate estimates of field areas are obtained later in the investigation.

III. Characterizing Farm Enterprises

Smallhold farming systems operate at dual purpose, to feed the household and to market crops and livestock for income. A farm may be divided into several, interacting enterprises (Woomer et al., 1998a), each with its own purpose, management and average standing biomass. From the context of carbon stocks, these enterprises may be grouped into the following categories: (1) annual food crops; (2) perennial crops and woodlots; (3) annual markets crops; (4) fodder, forage and fallow; (5) farm and field boundaries; and (6) household livestock activities and traditional gardens (Figure 2). One approach to estimating system carbon stocks of a smallhold is to assign coverage to each of these categories, identify representative areas within each enterprise for randomized biomass and carbon measurement and then derive a total through summation.

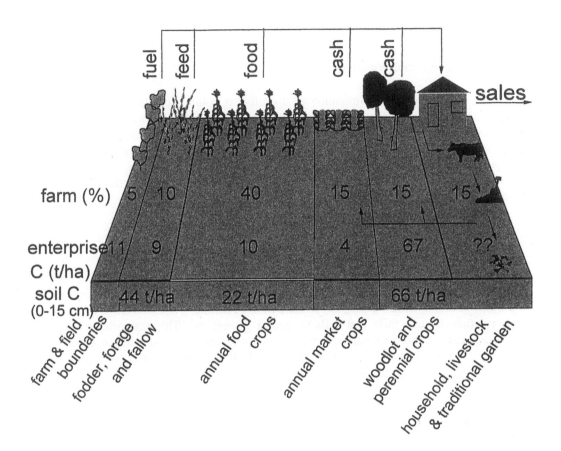

Figure 2. Smallhold farming systems consist of different enterprises connected by resource flows. The enterprises have different potentials to gain and store carbon.

A. Annual Food Crops

The most common annual food production system in Kenya is cereal-legume intercropping, with maize and beans the most widespread combination (Byerlee and Heisey, 1997). Banana replaces maize as the staple crop in Uganda's Lake Victoria Basin and is usually grown with several food and market intercrops (Bekunda and Woomer, 1996). Root crops are also important within these systems, including sweet potato (*Ipomoea batatus*), Irish potatoes (*Solanum tuberosum*), yams (*Dioscorea* spp.) and cassava (*Manihot esculenta*). Annual field crops intended primarily for market, such as cotton (*Gossypium* cvs.), sesame (*Sesamum indicum*) and sunflower (*Helianthus annuus*) are also considered part of this enterprise because these are usually grown on the same lands as intercrops or in rotation with food crops. Scattered trees often appear in this enterprise and should not be omitted from biomass estimates.

B. Perennial Crops and Woodlots

Many smallhold systems are superimposed upon former colonial plantations, particularly tea and coffee, and this enterprise remains important to household incomes. Farmers also establish orchards and woodlots. Natural and planted tree fallows are included within this category. Perennial crops and woodlands contain the largest biomass per unit land area and a likely avenue for increasing system carbon is through increasing the size and importance of this land use. Two common woodlot tree species in the highlands are eucalyptus (*Eucalyptus saligna*) and wattle (*Acacia mearnsii*). *Grevillea robusta* is a popular agroforestry species (ICRAF, 1992).

C. Annual Market Crops

Annual market crops, usually vegetables such as cabbage (*Brassica oleracea* var. *capitata*), kale (*Brassica oleracea* var. *acephala*), tomato (*Lycopersicon esculentum*), French bean (*Phaseolus vulgaris*) and onion (*Allium cepa*), are considered to be distinct from food crops for three reasons. Farmers often dedicate their best soils to income generation. This enterprise often receives a disproportionate share of farm inputs, particularly domestic composts and manures. Finally, whatever lands may be feasibly irrigated are most often placed into this land use and production occurs year-round, in contrast to seasonal annual food crop production.

D. Fodder, Forage and Fallow

As was stated previously, cropping systems in East Africa are undergoing transition from traditional, fallow-based, subsistence agriculture to continuous, mixed cropping. Livestock remain important throughout this transition and naturally occurring, grazed fallow species of the past are now cultivated for cut-and-carry fodder (Boonman, 1993). Murage (1998) reported that most farmers in Kenya's Central Highlands establish Napier grass (*Pennisetum purpureum*) for this purpose, particularly in nutrient-depleted soils. A common feature of fodder, forage and fallow plots is their domination by perennial herbs and grasses although scattered trees may occur and should not be omitted from biomass inventories.

E. Farm and Field Boundaries

Trees and dense shrubs grow along farm boundaries to delineate holdings and exclude neighboring livestock and wild animals. These boundaries may be natural and used as fuelwood, planted for fodder, fuel or fruit, or some combination of the two. Shrubs dominate internal field boundaries as trees may excessively shade annual crops. The potential of this farm enterprise is often overlooked by agricultural scientists and farmers alike, and it remains one of the few areas of the farm where land and organic resources are underutilized.

F. Household, Livestock and Traditional Gardens

The area surrounding the home is devoted to raising confined livestock and poultry, collecting manure and processing compost, cultivating traditional herbs and trees and practicing cottage industries. The complexity of the household area discourages its inventory but, considering its predominance of wooden structures and trees, accounting for this "enterprise" becomes more important as densely populated smallholder landscapes become increasingly peri-urban.

IV. Carbon Stock Measurement in Smallholder Agroecosystems

Estimating the carbon stocks in a smallhold farm requires that carbon measurements of woody and herbaceous biomass, litter, roots, tilled soil layer and subsoils be performed, and these measurements be compiled in a manner that reflects their proportion within the farming system. The carbon content of biomass and intact litter may be inferred from mass (e.g., 0.45 to 0.50) but the C contents of soils and partially decomposed litter requires chemical analysis or combustion. Carbon measurements are best made by teams of three to eight with one individual designated as leader and responsible for randomization procedures, data entry and codifying samples (Woomer and Palm, 1998). A list of field equipment useful in conducting carbon measurements is presented in Table 2. Much of this equipment is readily available in developing countries with the exception of geographical positioning systems, range finders and hand saws suitable for cutting roots in soil.

A. Woody Biomass

The presence and arrangement of woody biomass governs the approach to aboveground carbon measurement. Trees in remnant forests and woody fallows are measured in replicated, elongated transects (e.g., 25 m x 4 m) with the origin and direction established at random (Woomer and Palm, 1998). Systematically planted trees are measured along rows with the width of the transect adjusted to the inter-row spacing. Trees and large shrubs within boundaries are measured in similar fashion with the transect width also adjusted to conditions. The diameter at breast height (DBH in cm) of all stems greater than 2.5 cm is recorded using callipers or a diameter tape where circumference is expressed in units of diameter. Biomass is assigned to individual trees through allometric equations either empirically derived from local conditions or adopted from previous work in various ecological zones (Brown et al., 1989; FAO, 1997). Allometric equations based upon power functions, which intercept the origin, are recommended above quadratic approaches because of their greater accuracy for assigning biomass to smaller trees (Figure 3). For general purposes, we recommend the equations from FAO (1997) in Dry Zones (<1500 mm y^{-1}):

$$\text{Aboveground tree biomass (kg tree}^{-1}) = \exp^{(-1.996 + 2.32 \ln D)} \qquad \text{[Equation 1]}$$

and in Moist Zones (1500 to 4000 mm y^{-1}):

$$\text{Aboveground tree biomass (kg tree}^{-1}) = \exp^{(-2.134 + 2.53 \ln D)} \qquad \text{[Equation 2]}$$

Table 2. Important field tools and their uses in a field campaign to measure total system carbon in smallhold farming systems

Tool	Use
Geographic positioning system	Identify geographic coordinates of site and land use
Local or aerial map	Assist in site location, establish rapport with farmers
Random number table	Assist in randomization decisions
Compass	Assist in mapping and randomization direction
Data sheets and clip board	Enter DBH and labelling codes with sketch maps on reverse side
Range finder	Measure farm and field dimensions
Metric tape measure	Establish 25 m central linear axis of major quadrate
	Measure field dimensions
Diameter tape	Measure tree diameter
Fluorescent tape	Mark approximate location of major quadrates
Dial calliper	Measure DBH of smaller trees
1 m x 1 m wooden quadrate	Establish boundaries for understorey recovery
0.5 m x 0.5 m wooden quadrate	Establish boundaries for surface litter recovery
Hand shears	Recover understorey vegetation
Small hand rake	Recover surface litter
Hand saw	Recover small trees, woody litter and larger roots
Flat-bladed shovel	Excavate soil and roots
Bulk density cylinders	Recover soil bulk density samples for each land use
Wooden mallet	Drive bulk density cylinders into soil
Flat-bladed knife	Trim soil cylinders and others
Camera and film	Document procedures and land use
Plastic tarp	Establish sample processing area
Vegetable seeds	Establish rapport with farmers
Small domination currency	Compensate farmers for assistance in locating land uses

where Y is the aboveground tree biomass in kg and D is the measured DBH in cm. Other equations are available for drier (<900 mm y^{-1}) and wet zone (>4000 mm y^{-1}) from FAO (1997). The above equations are based upon lowland tropical moisture-temperature relationships and judgment is required when applying them to higher elevations as evapotranspiration decreases and climate becomes "wetter" at a given rainfall (FAO, 1997). Tree biomass is converted to carbon by a factor of 0.45 (Woomer and Palm, 1998). The average biomass (kg) of field-replicated 100 m^2 quadrates is adjusted to Mg ha^{-1} with a factor of 0.1 (100 quadrates ha^{-1}/1000 kg Mg^{-1}).

B. Herbaceous Vegetation

Herbaceous and woody vegetation with DBH less than 2.5 cm is harvested from randomized, replicated 1.0 m x 1.0 m frames as described by Woomer and Palm (1998). Briefly, transects are laid in each farm enterprise and 1.0 m^2 quadrate positions assigned at random intervals along them. The quadrates may be "nested" within the 100 m^2 tree quadrates, or when trees are absent or sparse, located independently of tree measurement. Plant tissues originating outside of the quadrate but falling within it are recovered and plant tissues originating within the quadrate but grown beyond it are discarded. Then all remaining vegetation is cut at groundlevel and recovered. Care is taken to

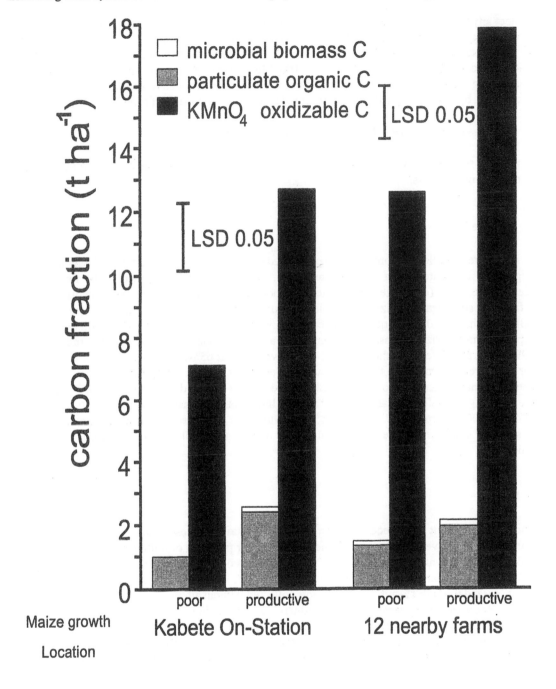

Figure 3. Labile carbon fractions in a Humic Nitisol obtained from the Kabete long-term experiment and in nearby farmers' fields in the Kenyan Highlands.

collect any fresh tissues that fall during harvest. Samples are weighed, sub-sampled, dried at 65° C to constant weight and corrected for moisture. Once dried, live vegetation is assumed to contain 0.45 C. The average biomass (kg) of field-replicated 1.0 m² herbaceous quadrates is adjusted to Mg ha⁻¹ with a factor of 10 (10000 quadrates ha⁻¹/1000 kg Mg⁻¹). Measurement of annual crop biomass is time consuming considering the size of their carbon stocks. An alternative approach is to reconstruct biomass C by adjusting yields with harvest index and the proportion of roots using the equation:

$$\text{biomass C (Mg ha}^{-1}) = 0.45 \times ((CY/HI) + RR(CY/HI)) \qquad \text{[Equation 3]}$$

where CY = reported average crop yield, HI = harvest index and RR = the ratio of below-ground to aboveground biomass.

C. Surface Litter

Surface litter is collected from centrally positioned 0.5 m x 0.5 m frames within the larger herbaceous vegetation quadrates using a small hand rake. Surface litter is assumed to be necromass of identifiable origin (e.g. leaves, fine branches) although judgment is often necessary in differentiating it from the soil organic horizon in grasslands or under trees. Woody necromass <10 cm in diameter falling within the 0.25 m² quadrate is collected with a hand saw. Logs >10 cm diameter require separate characterization based upon geometric and wood density approaches (Woomer and Palm, 1998). Surface litter is washed over a 2-mm sieve, dried at 65° C to constant weight and corrected for moisture. Alternatively, the litter is subsampled and ashed in a muffle furnace to remove mineral contaminants. Once dried or combusted, surface litter is assumed to contain 0.45 C. The average biomass (kg) of field-replicated 0.25 m² surface litter quadrates is adjusted to Mg ha⁻¹ with a factor of 40 (40000 quadrates ha⁻¹/1000 kg Mg⁻¹). Measurement of surface litter is not time consuming but this carbon stock tends to be small in enterprises other than woodlots, perennial crops, fodders and fallows.

D. Roots

Root measurement is a necessary component of detailed investigations comparing candidate management interventions but is too time consuming for purposes of routine monitoring. Roots are collected by excavating an area 0.2 m x 0.2 m to a depth of 50 cm with a narrow, flat-bladed shovel and hand saw. Coarse roots are hand sorted and washed. The remaining sample is dispersed in tap water, passed through a 2-mm sieve and roots collected without attempt to differentiate live and dead roots. Roots are then washed of gross mineral contamination, dried at 65° to constant weight, weighed and a subsample ground and ashed. Ash-corrected dry weight is assumed to contain 0.45 C. The average biomass (kg) of field-replicated 0.04 m² root quadrates is adjusted to Mg ha⁻¹ with a factor of 250 (250000 quadrates ha⁻¹/1000 kg Mg⁻¹). Other methods of sampling root biomass are described by Anderson and Ingram (1993).

F. Total Soil Organic Carbon

Total soil organic carbon is a major carbon pool within all farm enterprises. Estimation of total soil organic carbon stocks requires that soil be collected, analyzed for the total organic carbon content and then expressed on a soil volume basis by correction with bulk density. The tillage layer is usually sampled and analyzed separately from lower horizons because it is more subject to change through land management. Soil samples are collected by repeated auguring or coring from 0 to 20 and 20 to 50 cm in and around the biomass quadrates. Samples from each layer are bulked and subsampled, air dried, ground and total organic carbon determined through complete oxidation in sulfuric acid (Nelson and Sommers, 1975). Soil bulk density measurements are obtained by driving thin-walled metal cylinders of known volume into the vertical face of the root excavations with a short-handled wooden mallet at depths central to sampled soil horizons (in the above case at 10 and 35 cm depth). The cylinders are withdrawn, soil protrusions trimmed with a knife, the sample removed and dried for moisture and bulk density determination. Total soil organic carbon (Mg ha^{-1}) for each depth (D) may be calculated from soil organic carbon (%) and soil bulk density as:

Soil organic C (Mg ha^{-1}) = 100 cm m^{-1} x 100 cm m^{-1} x D cm x BD (g cm^{-3}) x SOC
x 10000 m^2 ha^{-1} x 0.000001 Mg g^{-1} [Equation 4]

where D = the depth of each soil layer, BD is soil bulk density (g cm^{-3}) and SOC is soil organic carbon content expressed in decimals (SOC (%)/100). Soil C in rocky soils and horizons may be adjusted for each soil depth as:

[D x (1 - proportion of coarse fragments)]   [Equation 5]

G. Labile Soil Carbon Fractions

Labile carbon refers to those fractions most subject to management changes and includes microbial biomass (Beare et al., 1997), particulate organic matter (POM) (Cambardella and Elliott, 1992) and POM density fractions (Meijeboom et al., 1995). Labile carbon is closely associated with nutrient mineralization and reflects organic input management over the past several years (Barrios et al., 1996; Kapkiyai et al., 1998). This fraction is the first to be depleted by continuous cultivation (Woomer et al., 1994) and the first to reform during fallow (Barrios et al., 1996; Kotto Same et al., 1997) or other management practices favoring soil C increase such as minimum tillage, crop residue retention or addition of manures (Lal et al., 1998).

Several methods of carbon fractionation were undertaken in studies of a long-term experiment and farmers' fields of the Central Kenyan Highlands (Kapkiyai et al., 1998; Murage, 1998) including microbial biomass by chloroform fumigation/extraction (Vance et al., 1987), physical separation of POM (Cambardella and Elliott, 1992; Okalebo et al., 1993), density separation of POM with Ludox, an alkaline polysilicate soluble in water (Meijboom et al., 1995) and KMnO$_4$ oxidizable soil organic matter (Blair et al., 1995). The sizes of several organic matter fractions resulting from the long-term soil management experiment at Kabete, Kenya (Kapkiyai et al., 1998) and in 12 nearby farmers' fields (Murage, 1998) are presented in Figure 3. All soils belong to the Humic Nitisols (FAO, 1977), corresponding to Paludalfs by USDA Soil Taxonomy (USDA, 1992). Poor maize growth results from 18 years of stover removal without external inputs at the Kabete experiment and from a paucity of external inputs at the nearby farms. Microbial biomass and particulate organic carbon are combined as one measure of labile soil C and compared to KMnO$_4$ oxidizable soil organic carbon. Significant differences exist between pool sizes and field types. Based upon the comparative size of the carbon fractions in Figure 3, KMnO$_4$ oxidizable C detects materials other than POM and microbial biomass. One shortcoming of the oxidation method is that it is a destructive measurement, rather than one resulting in recovery of a characterizable labile fraction. However, the procedure was sensitive to land and organic resource management and yields environmentally significant results (e.g., in Mg ha^{-1}). Researchers are advised to include a measure of labile soil organic carbon when examining management options designed to protect or sequester soil organic matter and are unsure if changes in total soil carbon will be sufficient to detect differences (Barrios et al., 1996).

H. Data Compilation and Sampling Considerations

To calculate total system carbon stocks, the individual pools within each farm enterprise, woody biomass, herbaceous biomass, litter, roots and soil, are totalled and expressed as Mg carbon ha^{-1}. Next, the total area of each enterprise, expressed in hectares (ha), is multiplied by their respective stocks and then farm enterprises are totalled. Different farms, not field replicates within farms or individual quadrates, must be regarded as statistical replicates when estimates of error are produced.

Approaches to estimating biomass and wood volumes within different farm enterprises (Figure 4) are presented in Table 3. Annual crops should be measured at peak biomass or biomass data is better reconstructed from yield records. Banana presents a problem in biomass measurement. It is a giant herb with a large underground storage organ (Purseglove, 1972). The application of allometric equations for woody biomass is inappropriate and destructively sampling bananas as herbaceous biomass is extremely difficult. Modelling approaches may assist in reconstructing banana biomass C from yield records when detailed soil data is available and land history known (Woomer et al., 1998b). Younger fallows and woodlots are undergoing successional changes and researchers must exercise judgement in randomization procedures, particularly for woody biomass measurement (Woomer and Palm, 1998). Fields of annual crops and fodder often contain scattered trees, the DBH of each should be measured as these may contain greater biomass than the crop. It may be necessary to separate farm and field boundaries because the former usually contains more and larger trees, requiring a set of total length, average width and unit biomass measurements for each. Guidelines for estimating wood volume of farm structures are not well established but information is available on the density of wood from many tree species (FAO, 1997).

V. Confluence of Interest in Farm Production and Carbon Sequestration

Crop production is an overriding objective in agriculture, particularly so in Africa with its growing food insecurity (World Bank, 1996). Kenya and Uganda have received over 3 million tonnes of food aid over the past two decades (FAO, 1994), a situation that requires innovative approaches to agricultural development (Sanchez et al., 1997). Many principles of sustainable agriculture address more efficient use of organic inputs and biological processes (Harwood, 1990). Elements of more sustainable agriculture in Africa include stronger integration of crop and animal production systems, increased reliance upon agroforestry systems and protection of soil organic matter through recycling and conservation (Okigbo, 1990). Unruh et al. (1993) report that 15,500,000 km^2 in Africa are suitable for agroforestry and conversion to this system could offset carbon emissions by the continent for many decades. At the same time, it must be remembered that agriculture in Africa results from cleared natural vegetation, often tropical forests (Gaston et al., 1998) and that the carbon stocks within smallholder landscapes may only be increased to a limited extent before food production, particularly by annual crops, becomes compromised. The contradiction between global carbon imbalances and more immediate economic concerns leads to scepticism over the practicality of the carbon offset concept among policy-makers in developing countries (Trexler, 1993). Yet, the objectives of farmers and environmentalists may coincide over a modest increase in smallhold farm

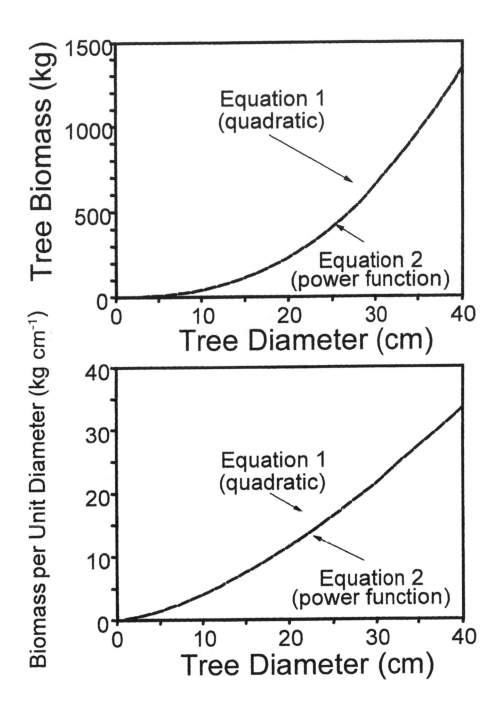

Figure 4. Relationship between tree diameter at breast height and tree aboveground biomass through allometric equations based upon power and quadratic equations.

Table 3. Approaches to estimating biomass and wood volumes in different enterprises of smallhold farming systems

Smallhold enterprise	Approach
Annual food crops	Measure random sample of herbaceous vegetation by destructive sampling of small quadrates with plot size dependent upon row spacing. Include scattered woody biomass by measuring DBH >2.5 cm. Crop biomass may be reconstructed from yield through harvest index. Adjust for cropping pattern.
Woodlots and perennial crops	Calculate total row length of woody biomass. Measure DBH of a random sample of trees >2.5 cm in 25-m long quadrates with rows adjusted for row spacing. Adjust randomization procedures in woodlots resembling natural forest.
Annual market crops	Destructively sample herbaceous vegetation in randomized, replicated 1.0 m^2 quadrates or reconstruct biomass based on yields.
Fodder, forage and fallow	Destructively sample herbaceous vegetation in randomized, replicated 1.0 m^2 quadrates. Measure DBH of scattered trees. Tree fallows are considered woodlots.
Farm and field boundaries	Establish total length and average width of farm boundaries. Measure DBH of woody vegetation >2.5 cm along randomized, replicated 25-m sections. Nest quadrates for destructively sampled herbaceous vegetation within woody biomass sample if necessary.
Household area	Estimate the wood volume of structures and adjust for wood density. Measure DBH of woody biomass >2.5 cm. Estimate mass of manure piles and compost, sample and analyze for total C.

carbon stocks and, given the importance of smallhold agriculture in developing countries and the unlikelihood that these lands will return to forest, these lands may warrant attention for future carbon mitigation.

Reports of carbon sequestration resulting from tropical agroforestry interventions range between 23 Mg ha^{-1} to 60 Mg ha^{-1} (Unruh et al., 1993; Kursten and Burshel, 1993) with the higher estimate assuming shade tolerance of associated crops and three decades of tree growth. Woomer et al. (1997) described a series of farm improvements to smallhold farming systems of the East African Highlands over 20 years that could result in the sequestration of 66 Mg ha^{-1}, improve yields of food crops and diversify farm enterprises with much of the increase resulting from tree biomass and soil conservation. However, it is unrealistic to expect all farm enterprises in the East African Highlands (Figure 2) to spontaneously convert to agroforestry. Rather, the proportion of land within various enterprises will change as new market opportunities appear (Woomer et al., 1998a) and land conservation intensifies (Tiffen et al., 1994) which may in turn lead to greater incentive for tree cultivation (Young, 1989).

Likely ranges of carbon stocks contained within the biomass of farm enterprises (Figure 2) and in Nitisols of the Central Kenyan Highlands (Woomer et al., 1997) are presented in Table 4. Note the disproportionate amount of carbon in woody biomass and soil compared to herbaceous biomass. Clearly, greater opportunity exists to store carbon through increasing woody biomass within the farm

Table 4. Lower and upper carbon stock estimates for different smallhold enterprises and surface soil in the East African Highlands

Component[1]	Carbon stock estimate	
	Lower	Upper
	——————Mg C ha[-1]——————	
Farm and field boundary[2]	1	11
Fodder and fallow (Napier grass)[3]	5	9
Annual food crop (maize)[4]	3	10
Annual market crop (cabbage)[5]	0.5	4
Woodlot and perennial crops[6]	20	114
Soil (0-15 cm)[7]	22	66

[1] All estimates account for root and aboveground biomass at 0.45% C. [2] The lower estimate for farm boundaries assumes 300 trees of 5 cm DBH ha[-1], the upper estimate increases tree populations to 600, DBH to 10. [3] Estimates are based upon Napier Grass (Boonman, 1993) standing biomass C of 4.5 and 9 Mg C ha[-1] for the lower and upper levels, respectively. [4] Estimates for annual food cropland are based upon yield increase from 1.4 to 5.4 Mg ha[-1], 33% harvest index and 15% root biomass, two crops per year (Kapkiyai et al., 1998) and are adjusted to 0.5 for stover harvest and periodic bare fallow. [5] The annual market crop estimate is based on cabbage yield increasing from 11 to 45 Mg ha[-1] season[-1] over two seasons y[-1] (Lekasi et al., 2000), 90% moisture, 15% roots. [6] The woodlot/perennial crop estimates are based upon 400 trees ha[-1] (5 m x 5 m) of 15 cm DBH increasing to 625 trees ha[-1] (4 m x 4 m) of 25 cm DBH. [7] Soil organic C in the 0-15 layer assumes bulk density of 1.1 g cm[-1] and an increase from 1.0% to 3.0% SOC.

than in improving the yields of annual crops, yet the latter approach is often in the farmer's immediate interest. Methods of sequestering soil organic carbon are not examined in detail in this chapter and readers with interest in this topic are referred to Lal et al., 1998. Based on results from the Kabete (Kenya) long-term experiment, increasing total soil organic matter and its particulate fraction may offer immediate benefit in terms of crop yield (Figure 5). Note that crop yield significantly responded to the amount of POC with each ton resulting in yield increases of 1.76 and 0.35 Mg ha[-1] of maize grain and dry beans, respectively. When commodity prices and exchange rates are considered, soil improvement associated with POC increased crop value by US $582 Mg[-1]ha[-1].

An innovative attempt to increase woody biomass within a smallholding is occurring at Wagacha Farm (Woomer et al., 1998a) where 320 *Grevillea robusta* were planted on 2.8 ha of sloped land about 30 km west of Nairobi, Kenya. The trees were planted to provide fuelwood for domestic use, poles for sale and improve soil erosion control. The tree spacing (78 m[2] tree[-1], rows 20 m apart) is not expected to greatly reduce yields of maize and Napier grass. After eight years the trees are expected to reach approximately 20 cm DBH and contain 13 Mg C ha[-1]. This approach stands in stark contrast to the eroded, poor yielding fields of neighbours and serves as an example of the confluence between farmer's and society's interests in tree planting. Agroforestry is emerging as a powerful force in rural development and it is not the purpose of this chapter to further elaborate other than to comment that the carbon sequestration potential of various agroforestry interventions are too seldom considered, particularly by the field researchers in the tropics responsible for evaluating the advantages of various agroforestry systems.

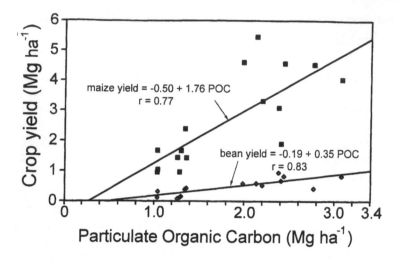

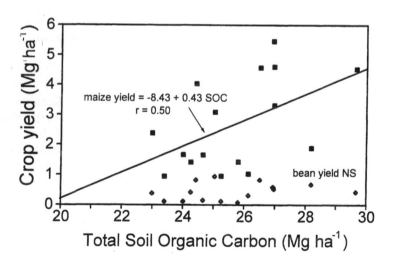

Figure 5. The relationship between soil organic matter fractions and field crop yields in an unfertilized Humic Nitisol of the Central Kenyan Highlands.

VI. Conclusions

Densely populated, intensively cultivated, smallhold agricultural landscapes in the East African Highlands operate at a fraction of their original carbon stocks and may be viewed as organic resource depleted. Many resource management strategies formulated to improve farm livelihoods seek to increase the availability of organic materials and to utilize them more efficiently, resulting in parallel opportunities for increased agricultural productivity and carbon sequestration. However, changes in total system carbon are seldom inventoried within smallhold farming systems for three reasons: (1) threats to food security occupy greater importance within research and development agendas; (2) the linkages between farm productivity and system carbon dynamics are not widely appreciated; and (3) methods for rapid estimation of carbon stocks within smallholder systems are not well established. A farm may be divided into six carbon pools: woody vegetation, herbaceous vegetation and annual crops, litter, roots, soils and wooden farm structures, each of which is sampled, measured or monitored in a different manner. First, representative farm enterprises must be identified and their areas measured. The biomass of woody perennials is estimated through allometric relationships. Annuals and herbaceous perennials are destructively sampled with adjustments necessary for biomass fluctuation throughout the year. Surface soils are more subject to disturbance and are more intensively sampled than subsoils. Carbon sequestered in wooden physical structures are often abundant in peri-urban settings and their composition may be estimated on the basis of cubic meters per hectare accompanied by a short formal survey to ascertain their longevity. Assuming that worst-case farming systems represent a minima in carbon stocks and that several interventions, particularly agroforestry, aimed at improving food supply also sequester carbon, potentials exists for credible inclusion of these landscapes within carbon offset projects after concerns of reliability and cost-effectiveness are addressed. While the basic criteria for comparing carbon offsets within smallholder and plantation forestry systems are the same (carbon accumulation/costs per unit area and time), approaches within the smallhold sector are much more complicated in part because of the more complex interrelation of environmental and social goals.

Acknowledgments

The authors thank Jane J. Kapkiyai of the Ministry of Agriculture and Livestock Development for sharing data from the Kabete long-term experiment and Marcus Walsh of the International Centre for Research in Agroforestry for generating the Landsat image used in this chapter. The data reported in this paper were gathered during projects funded by The Rockefeller Foundation Forum for Agricultural Resource Husbandry, whose assistance is gratefully acknowledged.

References

Anderson, J.M. and J.S.I. Ingram (eds.). 1993. *Tropical Soil Biology and Fertility — A Handbook of Methods.* CAB International, Wallingford, U.K.

Barnett, A. 1992. *Desert of Trees: The Environmental and Social Impacts of Large-Scale Tropical Reforestation in Response to Global Climate Change.* Friends of the Earth.

Barrios, E., R.J. Buresh and J.I. Sprent. 1996. Organic matter in soil particle size and density fractions from maize and legume cropping systems. *Soil Biol. Biochem.* 28:185-193.

Beare, M.H., M.V. Reddy, G. Tian and S.C. Srivastava. 1997. Agricultural intensification, soil biodiversity and ecosystem function in the tropics: the role of decomposer biota. *Applied Soil Ecology* 6:87-108.

Bekunda, M.A. and P.L. Woomer. 1996. Organic resource management in banana-based cropping systems of the Lake Victoria Basin, Uganda. *Agriculture Ecosystems and Environment* 59:171-180.

Blair, G.J., R.D. Lefroy and L. Lisle. 1995. Soil carbon fractions based on their degree of oxidation and the development of a Carbon Management Index for agricultural systems. *Australian Journal of Agricultural Resources* 46:1459-1466.

Boonman, J.G. 1993. *East Africa's Grasses and Fodders: Their Ecology and Husbandry.* Kluwer Academic Publishers, Dordrecht. 343 pp.

Brown, S., A.J.R. Gillespie and A.E. Lugo. 1989. Biomass estimation methods for tropical forests with applications to forest inventory data. *Forest Science* 35:881-902.

Byerlee, D. and P.W. Heisey. 1997. Evolution of the African maize economy. p. 9-22. In: D. Byerlee and C.K. Eicher (eds.), *Africa's Emerging Maize Revolution.* Lynn Rienner Publishers, Boulder, CO.

Cambardella, C.A. and E.T. Elliott. 1992. Particulate soil organic matter changes across a grassland cultivation sequence. *Soil Science Society of America J.* 56:777-783.

Conway, G.R. 1989. Diagrams for farmers. p. 77-86. In: R. Chambers, A. Pacey and L.A. Thrupp (eds.), *Farmer First.* Intermediate Technology Publications, London.

Dixon, R.K., K.J. Andrasko, F.G. Sussman, M.A. Lavinson, M.C. Trexler and T.S. Vinson. 1993. Forest sector carbon offset projects: Near-term opportunities to mitigate greenhouse gas emissions. *Water Air and Soil Pollution* 70:561-577.

Food and Agriculture Organization of the United Nations (FAO). 1977. *Soil Map of the World: Volume VI, Africa.* FAO, Rome. 299 pp.

Food and Agriculture Organization of the United Nations (FAO). 1994. *Food Aid in Figures* 12:52-55. FAO, Rome.

Food and Agriculture Organization of the United Nations (FAO). 1997. *Estimating Biomass and Biomass Change of Tropical Forests: A Primer.* FAO Forestry Paper 134. FAO, Rome. 55 pp.

Gaston, G., S. Brown, M. Lorenzini and K.D. Singh. 1998. State and change in carbon pools in forests of tropical Africa. *Global Change Biology* 4:97-114.

Hargreaves, G.H. and Z.A. Samani. 1986. *World Water for Agriculture.* International Irrigation Center, Utah State University, Logan. 617 pp.

Harwood, R.R. 1990. A history of sustainable agriculture. p. 3-19. In: C.A. Edwards, R. Lal, P. Madden, R. Miller and G. House (eds.), *Sustainable Agricultural Systems.* Soil and Water Conservation Society, Ankeny, IA.

Hedger, M.M. 1998. Making Kyoto work — the complex agenda on forestry. *Commonwealth Forestry Review* 77:172-180.

International Centre for Research in Agroforestry (ICRAF). 1992. *A Selection of Useful Trees and Shrubs for Kenya.* ICRAF, Nairobi. 225 pp.

Kapkiyai, J.J., N.K. Karanja., P.L. Woomer and J.N. Qureshi. 1998. Soil organic carbon fractions in a long-term experiment and the potential for their use as a diagnostic assay in highland farming systems of Central Kenya. *African Crop Science Journal* 6:19-28.

Kinsman, J.D. and M.C. Trexler. 1993. Terrestrial carbon management and electric utilities. *Water Air and Soil Pollution* 70:545-560.

Kotto-Same, J., P.L. Woomer, A. Moukam and L. Zapfack. 1997. Carbon dynamics in slash-and-burn agriculture and land use alternatives of the humid forest zone in Cameroon. *Agriculture Ecosystems and Environment* 65:245-256.

Kursten, E. and P. Burshel. 1993. CO_2-mitigation by agroforestry. *Water Air and Soil Pollution* 70:533-544.

Lal, R., J.M. Kimble, R.F. Follett and C.V. Cole. 1998. *The Potential of U.S. Cropland to Sequester Carbon and Mitigate the Greenhouse Effect.* Sleeping Bear Press, Ann Arbor, MI. 128 pp.

Lekasi, J.K., P.L. Woomer and M. Bekunda. 2000. Mulching cabbage with banana residues affects yields, nutrient supply, soil biota, weed expression and soil moisture. *African Crop Science Journal* (in press).

McGivern, B.P. 1998. Conference of the parties to the Framework Convention on Climate Change: Kyoto Protocol. *International Legal Matters* 37:22-43.

Meijboom, F.W., J. Hassink and M. van Noordwijk. 1995. Density fractionation of soil macroorganic matter using silica suspensions. *Soil Biology and Biochemistry* 27:1109-1111.

Murage, E.W. 1998. *Soil Carbon Pools Based on Physical, Chemical and Biological Fractionation and their Relationship to Fertility of Humic Nitisols in the Central Kenyan Highlands.* MSc. Thesis, Department of Soil Science, University of Nairobi, Kenya. 142 pp.

Nelson, D.W. and L.E. Sommers. 1975. A rapid and accurate method for estimating organic carbon in soil. *Proc. of the Indiana Academy of Science* 84:456-462.

Odingo, R.S. 1971. *The Kenya Highlands: Land Use and Agricultural Development.* East Africa Publishing House, Nairobi, Kenya. 229 pp.

Okalebo, J.R., K.W. Gathua and P.L. Woomer. 1993. *Laboratory Methods of Soil and Plant Analysis: A Working Manual.* TSBF, Nairobi, Kenya. 88 pp.

Okigbo, B.N. 1990. Sustainable agricultural systems in tropical Africa. pp. 323-352. In: C.A. Edwards, R. Lal, P. Madden, R. Miller and G. House (eds.), *Sustainable Agricultural Systems.* Soil and Water Conservation Society, Ankeny, IA.

Purseglove, J.W. 1972. *Tropical Crops: Monocotyledons.* John Wiley & Sons, New York. 607 pp.

Sanchez, P.A., K.D. Shepherd, M.J. Soule, F.M. Place, A.U. Mukwunye, R.J. Bursch, F.R. Kwesiga, A-M.N. Izac, C.G. Ndiritu and P.L. Woomer. 1997. Soil fertility replenishment in Africa: An investment in natural resource capital. p 1-46. In: R.J. Bursch, P.A. Sanchez and F. Calhoon (eds.), *Replenishing Soil Fertility in Africa.* Soil Science Society of America Special Publication No. 51. Soil Science Society of America, Madison, WI.

Sanchez, P.A., Woomer, P. and C.A. Palm. 1994. Agroforestry approaches for rehabilitating degraded lands after tropical deforestation. *JIRCAS International Symposium Series* (Japan) 1:108-119.

Saurbeck, D.R. 1993. CO_2 emissions from agriculture: Sources and mitigation potentials. *Water Air and Soil Pollution* 70:381-388.

Tiffen, M., M. Mortimore and F. Gichuki. 1994. *More People, Less Erosion: Environmental Recovery in Kenya.* John Wiley & Sons, Chichester, U.K. 311 pp.

Tipper, R. and B. De Jong. 1998. Quantification and regulation of carbon offsets from forestry: comparison of alternative methodologies, with special reference to Chipatas, Mexico. *Commonwealth Forestry Review* 77:219-228.

Trexler, M.C. 1993. Manipulating biotic carbon sources and sinks for climate change mitigation: can science keep up with practice? *Water Air and Soil Pollution* 70:579-593.

Trines, E.P. 1998. SGS' carbon offset verification service. *Commonwealth Forestry Review* 77:209-213.

United States Department of Agriculture, Soil Conservation Service (USDA). 1992. *Keys to Soil Taxonomy, Fifth Edition.* Soil Management Support Service Monograph No. 19. Washington, D.C. 541 pp.

Unruh, J.D., R.A. Houghton and P.A. Lefebvre. 1993. Carbon storage in agroforestry: an estimate for sub-Saharan Africa. *Climate Research* 3:39-52.

Vance, E.D., P.C. Brookes and D.S. Jenkenson. 1987. An extraction method for measuring soil microbial biomass C. *Soil Biology and Biochemistry* 19:703-707.

Verweij, H. and Emmer, I.M. 1998. Implementing carbon sequestration projects in two contrasting areas: the Czech Republic and Uganda. *Commonwealth Forestry Review* 77:203-208.

White, F., 1983. *The Vegetation of Africa.* United Nations Educational Scientific and Cultural Organization, Paris. 356 pp.

Woomer, P.L. and Palm, C.A. 1998. An approach to estimating system carbon stocks in tropical forests and associated land uses. *Commonwealth Forestry Review* 77:181-190.

Woomer, P.L., M.A. Bekunda, N.K. Karanja, T. Moorehouse and J.R. Okalebo. 1998a. Agricultural resource management by smallhold farmers in East Africa. *Nature and Resources* 34:22-33.

Woomer, P.L., M.A. Bekunda and D. Bwamiki. 1998b. Modelling banana growth and organic matter dynamics with the Century model. *African Crop Science Journal* 6:205-214.

Woomer, P.L., C.A. Palm, J.N. Qureshi and J. Kotto-Same. 1997. Carbon sequestration and organic resource management in African smallholder agriculture. p. 58-78. In: R. Lal, J.M. Kimble, R.F. Follett and B.A. Stewart (eds.), *Management of Carbon Sequestration in Soils.* Advances in Soil Science, CRC Press, Boca Raton, FL.

Woomer, P.L., Martin, A., Albrecht, A., Resck, D.V.S. and Scharpenseel, H.W. 1994. The importance and management of soil organic matter in the tropics. pp 47-80. In: P.L. Woomer and M.J. Swift (eds.), *The Biological Management of Tropical Soil Fertility.* John Wiley & Sons, Chichester, U.K.

World Bank. 1996. *African Development Indicators 1996.* The World Bank, Washington, D.C.

Young, A. 1989. *Agroforestry for Soil Conservation.* International Council for Research in Agroforestry, Nairobi and CAB International, Wallingford, U.K.

CHAPTER 12

Assessment and Significance of Labile Organic C Pools in Forest Soils

P.K. Khanna, B. Ludwig, J. Bauhus and C. O'Hara

I. Introduction

The main source of soil organic C (SOC) is the input of organic matter via belowground and aboveground return of plant residues. Part of this input of organic C is transformed by soil biota into organic compounds of varying composition, a small part is leached as dissolved organic carbon, and the rest is returned to the atmosphere as CO_2. Transformations of SOC depend upon the decomposability of the litter and the biotic activity. Organic compounds in the soil undergo successive transformation by biotic action until they are protected either physically by becoming inaccessible to microbial activity (occluded fraction) or chemically by forming strong bonds with mineral surfaces (organo-mineral complexes). Transformation of SOC into lesser or least labile forms is part of the C sequestration process.

Conceptually, SOC has been divided into fractions of different decomposability. Three functional fractions, active, slow and recalcitrant (passive) have been widely used to describe decomposition dynamics (Paul and Juma, 1981; Parton et al., 1988). These fractions do not correspond to any defined part of soil organic matter (SOM) that can be readily determined. The transient phases of soil C which can be readily mobilized by biological processes are grouped as labile soil organic C (LSC) pool. The LSC contains microbial biomass and other non-microbial labile C and it is expected to respond most rapidly to physical, chemical and microclimatic changes in the soil environment. Time is thus an essential component of the definition of LSC expressed in terms of its turnover rates.

The contribution of LSC to nutrient fluxes is proportionally much higher than those of passive SOM pools. Consequently, LSC pools play a central role in many forest ecosystem processes by supplying energy (as available C) for microbial activity and by affecting mineralization and immobilization of nutrients. The quasi-equilibrium between the LSC and the refractory (protected, passive) SOM fractions is dependent upon SOM quality (usually assessed by the C/N ratio, lignin content, microbial C-to-total SOC ratio, and other fractions of its components) and the nature and activity of the microbial populations. The latter is largely driven by climatic variables. Litter composition and factors determining litter decomposition will therefore determine the potential of soils to sequester C and to reduce its outputs as atmospheric CO_2.

Forest soils are less frequently disturbed, are less exposed to climatic extremes, and receive litter inputs that are mostly of lower quality than in agricultural soils. In addition, in many regions forest soils are of lower inherent site fertility than agricultural soils, where clearing of forest land for agriculture concentrated on the fertile sites. Also forest soils generally receive less fertilizer and lime than most agricultural soils. It therefore follows that the content and dynamics of SOC in forest soils differ significantly from those in agricultural soils. Soil analytical methods developed for agricultural soils have often failed in forest soils. For that reason we will concentrate in this review on the performance of methods for the determination of the LSC in forest soils.

A number of techniques is used to describe the labile C fractions in forest soils. As microorganisms themselves constitute a significant fraction of LSC, biological techniques of measuring microbial biomass C and microbial respiration (indicating mineralizable C) are frequently employed to study the dynamics of LSC pools. However, LSC pools may include non-microbial fractions. Therefore, methods which employ physical fractionation or chemical extraction of soil to assess LSC may provide complementary information to that attained via biological techniques (Hu et al., 1997). This chapter reviews some of these techniques, namely physical fractionations, chemical extractions and biological procedures. These techniques are reviewed in terms of their attributes, and their usefulness and limitations for defining practical measures of assessing LSC. Additionally, several measures of labile organic C pools were compared for forest soils, which had been subjected to different management treatments. This comparison was used to assess the sensitivity of LSC pools to forest management practices.

II. Measurement of Labile Organic C Pools

A. Physical Fractionation Techniques (Density and Particle Size Fractionations)

The composition of SOM follows a continuum of particle sizes from fresh plant residues of large particle size to decomposed residues in the finest fractions (Baldock et al., 1992). The usefulness of physical fractionation techniques for the determination of LSC was shown by applying ^{13}C mass spectrometry investigations to different SOM size fractions, and it was concluded that the lifetime of organic C may be determined more by the physical position and protection than by the chemical nature of the SOM (Balesdent, 1996).

Commonly used physical fractionation techniques include (a) density fractionations and (b) particle size fractionations. Fractionation techniques employ either a single method suitable for each set of fractions or a combination of methods separating particle sizes and density fractions (Magid et al., 1996).

1. Density Fractionation Techniques

Density separation of forest soil organic matter (SOM) yields a light fraction (LF) and a heavy fraction (HF). These two fractions may represent SOM at different stages of decomposition. The LF is not firmly associated with the mineral part of the soil (Christensen, 1992) and is primarily composed of younger plant debris, which is only partially decomposed and still resembling its original form. Gregorich et al. (1996) demonstrated that the chemistry and isotopic composition of the LF was similar to the plant residue from which it was derived. However, they reported a reduction in carbohydrate C in the LF when compared with the plant material. This reduction was attributed to early stages of decomposition because carbohydrate C is an easily mineralized source of C. The HF is composed of amorphous material in an advanced stage of humification adsorbed to mineral particles, forming organo-mineral complexes.

Separation of light and heavy fractions is achieved by suspending soil in a dense liquid. However, numerous variations on the basic method have been employed. These include: (a) stirring of the suspension; (b) centrifugation of the suspension (in particular, for different duration and centrifugal force); (c) removal of the LF from the suspension surface by suction; (d) dispersion of soil by low or high power sonication, or by using hexametaphosphate; (e) use of surfactants; and (f) the standing time of the suspension prior to removing the LF. Details of such variations are described in studies referenced in Table 1. Young and Spycher (1979) tested a range of densities using tetrabromoethane-ethanol for several soils under different vegetation types. They suggested that a density of approximately 1.6 g cm^{-3} was the optimum for separating LF organic matter from HF organo-mineral complexes. Many such organic liquid media are toxic and hazardous (Strickland and Sollins, 1987; Meijboom et al., 1995). Therefore, organic liquid media have been largely replaced by aqueous solutions of inorganic salts, such as sodium polytungstate and sodium iodide, and occasionally by other media such as colloidal silica and water. Although mostly less toxic than organic liquids, the effect of inorganic salt solutions on the nature of SOM fractions remains a matter of concern if these fractions are to undergo further analysis based on biological processes. For example, Sollins et al. (1984) found that N mineralization was reduced by 70% when soil was washed in sodium iodide. Magid et al. (1996) reported that C mineralization was retarded by sodium polytungstate.

For a number of forest soils listed in Table 1, carbon in the LF (LFC) was about 28% of the total soil C (Table 2). The remaining 72% of soil C were associated with HF. However, the mean C content of the LF was 30%, which was well below the C content in the undecomposed plant tissue (approximately 50%), indicating that LF is highly oxidized. The high C/N ratio in LF when compared with HF indicates that as decomposition advances, more of the C associated with LF is mineralized by microorganisms, whereas N is partially immobilized. This results in the incorporation of organic matter of a lower C:N ratio into the HF (Ladd et al., 1993). In agreement with this, Sollins at al. (1984) and Boone (1994) have reported that the HF is the principal source of N in forest soils.

Studies by Spycher et al. (1983), Swanston and Myrold (1997) and Entry and Emmingham (1998) considered the effects of soil profile depth and forest age on the LF of organic matter. With increasing soil depth or decreasing forest age, the percent of total soil organic C (SOC) and LFC in the fine soil fraction decreased. However, as a proportion of SOC, the contribution from the LF diminished with increasing forest age. Chemical analysis indicated that the LF organic matter from the old growth forest was of poorer quality (higher lignin; lower sugar, cellulose and hemicellulose content) than that from the young and secondary regrowth forests. It was suggested that this would result in slower rates of decomposition and greater C storage in the soil under older compared with younger growth forest.

Golchin et al. (1994) used a combination of suspension, decantation and sonication to distinguish the LF further into a 'free' LF and an 'occluded' LF. The 'free' LF was comparable to the LF reported in other studies (i.e., younger and only partially decomposed plant debris), whereas the 'occluded' LF belonged to the pool of 'old' C.

Some limitations of the density fractionation have been reported. LF may contain a considerable amount of charcoal and charred material that may have no or very little decomposability. Similarly LF found in peaty soils may represent inert SOM (Oades, 1972). In tropical soils, Theng et al. (1989) did not find that density fractionation was a useful technique to define labile organic matter, because the colloidal nature of oxide particles and their association with SOM may result in similar effective densities for soil colloids.

2. Particle Size Fractionation

Particle size fractionations showed that the C/N ratio declines with decreasing particle size and that SOM in the sand or coarser fractions consists of relatively unaltered plant material (Schulten et al., 1993). This indicates that decomposability of SOM decreases with decreasing particle size owing to physical protection (Christensen, 1992; Balesdent, 1996). Guggenberger et al. (1994) studied the effects of land use on the composition of organic matter in various particle size fractions. They found that the SOM composition of the sand fraction was largely affected by land-use changes, whereas SOM bound by silt- and clay-sized minerals was more influenced by the chemical and physical soil

Table 1. Total C in the soil (C_{tot}), light fraction carbon (LFC), and C/N ratios of various fractions (soil, light fraction (LF), and heavy fraction (HF)) in forest soils of different depths and clay contents collected from various locations and under different vegetation cover. nd: not determined

Location	Vegetation	Soil	Depth (cm)	Clay (%)	C_{tot} g C kg⁻¹ (soil)	LFC g C kg⁻¹ (LF)	LFC g C kg⁻¹ (soil)	LFC % of total soil C	C/N Soil	C/N LF	C/N HF
[1]Wisconsin, USA	Oak (control)	Alfisol	0-10	10	30.4	350	1.28	4.2	nd	nd	nd
"	Oak (no litter)	"	"	10	18.4	351	0.46	2.5	nd	nd.	nd
"	Oak (2x litter)	"	"	10	45.4	351	2.36	5.2	nd	nd	nd
"	Oak (control)	"	"	10	27.2	350	1.71	6.3	nd	nd	nd
"	Oak (no litter)	"	"	10	12.7	351	1.09	8.6	nd	nd	nd
"	Oak (2x litter)	"	"	10	33.9	351	1.46	4.3	nd	nd	nd
[2]Bavaria, Germany	Beech	Inceptisol	0-4	15	70	370	39.20	56	18	20	18
"	Spruce	Entisol	"	10	30	460	23.70	79	21	30	20
[3]Oregon, USA	Douglas fir	Inceptisol	0-3	10	117	340	62.01	53	25.8	30.8	21.9
"	"	"	3-13	10	53	396	19.08	36	22.7	56.8	17.3
"	"	"	13-23	10	30	403	10.2	34	17.1	40.8	13.2
[4]South Australia	Mixed forest	Alfisol	0-10	12	35.3	283	6.81	19.3	15.4	22.5	14.3
[4]Queensland, Australia	Poplar box	"	"	23	25	258	3.68	14.7	11.4	13.4	11.2
[4]Queensland, Australia	Brigalow forest	Vertisol	"	52	41.7	301	13.05	31.3	11.6	16.3	10.2
[5]Washington State, USA	Douglas fir	Inceptisol	0-12	35	36	313	20.05	35.8	37.2	55.4	22.1
"	Red alder/ douglas fir	"	"	35	97	303	41.23	42.5	25.6	28.2	16
[5]Oregon, USA	Conifer	"	0-10	nd	125	327	30.38	24.3	21.3	38.6	23.6
"	Red alder	"	"	nd	152	324	47.58	31.3	17.7	21.4	17.5
"	Conifer	"	0-15	nd	43	164	11.95	27.8	26	44	23.2
"	Mountain hemlock	Spodosol	"	nd	24	102	11.33	47.2	27.5	52.2	23.5
[5]Costa Rica	Tropical wet forest	Inceptisol	0-15	nd	159	336	40.39	25.4	11.9	12.6	11.8
"	"	"	"	70	46	313	2.48	5.4	11.8	20.7	11.5
"	Successional vegetation	"	"	nd	51	233	4.39	8.6	10.6	18.4	10.5
"	Tropical wet forest	"	"	40	73	295	6.28	8.6	12.2	17.2	11.5
[6]Florida, USA	Slash pine	Spodosol	0-15	3.3	31	404	4.37	14.1	36.2	68.6	32.8
[6]Mass., USA	Oak/pine	Entisol	"	4	19	346	1.71	9	4	86.1	42.3

Table 1. (continued)

Location	Vegetation	Soil	Depth (cm)	Clay (%)	C_{tot} g C kg⁻¹ (soil)	LFC g C kg⁻¹ (LF)	g C kg⁻¹ (soil)	% of total soil C	C/N Soil	LF	HF
[6]Oregon, USA	Burnt and clear cut	Inceptisol	0-15	22.2	50	293	18.15	36.3	28	49.1	23.2
"	Conifer/red alder	"	"	13.7	152	293	77.22	50.8	20.1	23.9	17.2
[7]Oregon, USA	Douglas fir	"	0-10	45	115	nd	52.90	46	nd	nd	nd
"	"	"	"	50	47	nd	27.26	58	nd	nd	nd
"	"	Ultisol	"	45	25	nd	16.00	64	nd	nd	nd
[8]Oregon, USA	Red alder (clear cut)	Inceptisol	0-5	25	109	332	24.42	22.4	34.7	59.7	23.5
"	"	"	5-15	30	71	249	12.78	18	29	83	21.4
[9]Queensland, Australia	Mixed eucalypt forest	Vertisol	0-10	40	17	163	3.52	20.7	13.1	nd	nd
"	"	"	"	50	8	73	1.17	14.6	12.1	nd	nd
"	"	Alfisol	"	15	13	156	3.90	30	14.8	nd	nd

All studies used separation densities of 1.6 to 1.7 g cm⁻³, except Dalal and Mayer (1986) who used a density of 2.0 g cm⁻³, and Jollins (1984) who used 1.4 g cm⁻³.
[1]Boone (1994); [2]Kögel-Knabner and Ziegler (1993); [3]Spycher et al. (1983); [4]Golchin et al. (1994); [5]Sollins et al. (1984); [6]Strickland and Sollins (1987); [7]Entry and Emmingham (1998); [8]Swanston and Myrold (1997); and [9]Dalal and Mayer (1986).

Table 2. Some statistical parameters of soil C in forest soils (summarized from data given in Table 1); C fractions are LF (light fraction), HF (heavy fraction) and C_{tot} (total soil organic C) (n = number of cases)

	C_{tot} g (C) kg⁻¹ (soil)	LFC g (C) kg⁻¹ (LF)	g (C) kg⁻¹ (soil)	% of total soil C	C/N Soil	LF	HF
Mean	57	300	17.9	28	21.4	37.9	19.1
Minimum	8	73	0.5	3	10.6	12.6	10.2
Maximum	159	460	77.2	79	44	86.1	42.3
n	36	33	36	36	27	24	24

environment. Carbohydrates in the sand size fraction were mainly of plant origin, while those in the clay size fraction were enriched in carbohydrates of microbial origin. The silt size fraction was depleted of carbohydrates and it was suggested that microorganisms effectively use carbohydrates in the vicinity of silt. By using ^{13}C-NMR spectra after photo-oxidation of soil fractions Skjemstad et al. (1996) measured the amount and nature of protected C (charcoal and charred materials) in soils. They found that finely divided charcoal was a major constituent of many Australian soils, and in some soils more than 88% of the charcoal present occurred in the <53 μm fractions.

For particle size fractionations, there is no general agreement about the sample treatment, such as whether the samples should be dried or not and whether high or low sonification should be used for sample dispersion (Christensen, 1992; Balesdent, 1996).

B. Chemical Methods (Oxidations or Extractions) for the Determination of LSC

By assuming that LSC includes components of SOM which are readily oxidizable or easily hydrolyzable, chemical methods that use mild oxidizing agents (Conteh et al., 1997) or acid solvents are employed to measure labile C in soils. Other chemical methods are milder and include cold or hot water extractions followed by chemical analysis for different SOM constituents.

1. Oxidation Methods

Commonly used agents for a mild oxidation of SOM are: MnO_4^-, $Cr_2O_7^{2-}$, H_2O_2 or O_2. Unfortunately, the oxidized SOM fraction cannot be ascribed to any defined chemical compounds or active SOM pools (Balesdent, 1996). However, changes in the readily oxidizable C fraction on addition of plant residues, which was measured by using MnO_4^-, were related to polysaccharides (Conteh et al., 1998).

We tested the usefulness of the $KMnO_4$ oxidation method to determine labile SOC on a set of six forest soils from different sites and depths. The soils were incubated for 180 days under laboratory conditions (20°C, field moisture). The difference in $KMnO_4$ labile C measured before and after incubation was not consistently related to the amount of CO_2 respired during incubation (Figure 1).

Chemical oxidation methods for the determination of LSC have a number of limitations. Different soil samples may have variable amounts of readily oxidizable fractions which will make standardization of any method a difficult task. The amount of C in the sample, MnO_4^- concentration and contact time will influence the results (Blair et al., 1995). Moreover, Shang and Tiessen (1997) found that the measurement of oxidizable C by using spectrophotometry of the MnO_4^- solution was much lower than that measured separately in the soil residue after oxidation. Chemical oxidation using H_2O_2 or thermic oxidation was not found to be useful for the determination of labile carbon (Balesdent, 1996).

2. Acid Extractions

Polglase et al. (1992) used a sequential extraction of organic matter in cold and hot trichloroacetic (TCA) acid, modified from the method described by Kedrowski (1983). The sum of C in cold (4°C, 0.30 M TCA) and hot (90°C, 0.15 M TCA) extracts was considered LSC. The labile C content (g/kg) of the organic (L and F layers) and mineral soil (0 to 5 cm) horizons was determined for loblolly and slash pine plantations, both subject to a weed control plus fertilizer treatment. Under both species the LSC content decreased in the order of L layer > F layer > mineral soil.

Hu et al. (1997) used 12 M (at 22°C) and 1 M (at 100°C) H_2SO_4 to obtain soil carbohydrates (C_{carb}), and the choloroform fumigation-extraction method to obtain microbial C (C_{mic}), of subtropical forest and agricultural soils. They found that the C_{carb}/SOC and C_{mic}/SOC ratios provided complementary information on substrate quality and availability, respectively. It was observed that the C_{carb}/SOC and C_{mic}/SOC ratios of the subtropical forest soils studied were low indicating that substrate quality of SOM was low, and thus its availability to microorganisms was limited.

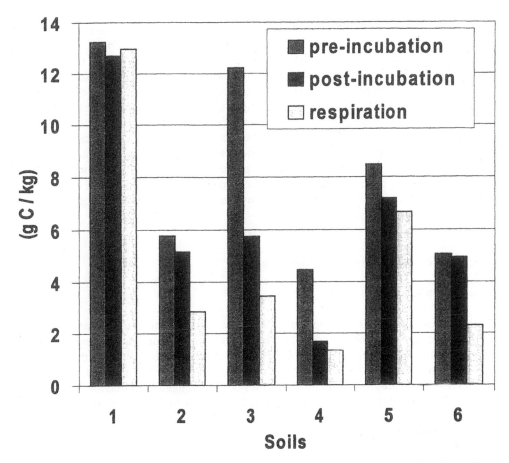

Figure 1. Amount of labile C measured by the KMnO$_4$ extraction method in soils (pre-incubation) is compared with the amount measured by the KMnO$_4$ extraction in post-incubated samples. Incubation period was 180 days. Amount of CO$_2$ lost through respiration is included for comparison.

Cheshire et al. (1992) suggested that soil carbohydrates, which can be obtained by hydrolysis, are a readily mineralizable pool of soil C, and are an important source of energy for microorganisms. Soil carbohydrates may comprise 5 to 25% of soil organic matter (Hu et al., 1997), and exist in a range of forms from simple sugars to polysaccharides including cellulose and hemicellulose (Hu et al., 1997; Kögel-Knabner et al., 1988). Soil carbohydrate content and composition is classically determined through a combination of acid hydrolysis (extraction phase) and chromatography (determination phase) (Cheshire et al., 1992; Guggenberger et al., 1994; Gregorich et al., 1996).

Some of the carbohydrates and proteins, which may be protected from biodegradation by association with humic material or inorganic colloids (Oades, 1995) can be extracted by acid extraction methods, thereby limiting the use of these methods to assess LSC.

3. Hot and Cold Water Extractions

Extractions with water have been used to obtain a LSC fraction as an alternative to acid extractions. Davidson et al. (1987) included separate cold and hot water extractions as part of a study comparing several techniques used to assess forest soil available C. Hot water extractable C was significantly correlated with the denitrification potential, a bioassay of C availability, whereas cold water extractable C was not. Both measures strongly correlated with mineralizable-C. Similarly, Sparling et al. (1998) found that hot water soluble C strongly correlated with both total C and microbial biomass C across a range of soil types under a variety of land uses. It was concluded from this study that while hot water soluble C may represent a measure of labile C, it was not necessarily a substitute for microbial C, and that the use of hot water has good potential for wide scale soil quality monitoring but would require further validation. However, the potential of using water extractions as a measure of LSC was questioned by Balesdent (1996) who reported that SOM extracted by hot water or autoclaving contained some of the metabolic components and the microbial biomass, but it also contained a substantial amount of 'old' SOC.

C. Biological Methods of Measuring Labile Soil C

1. Microbial Biomass

Microbial biomass is an essential component of labile SOC. It is sometimes used as a surrogate for labile C pool, because it can be readily determined through physiological and biochemical methods. In a review on microbial biomass C (C_{mic}) and N (N_{mic}) in forest soils, Bauhus and Khanna (1999) compiled literature data from 517 soil samples from tropical, temperate and boreal forests, and the summarized results are presented here. In the forest floor C_{mic} ranged from 0.3% to 8.1% of the total organic C and N_{mic} from 3 to 17% of the total organic N (Table 3). In the mineral soil C_{mic} ranged from 0.3% to 9.9% and N_{mic} from 2.5% to 27.5%. The C_{mic}/SOC ratio has been suggested as a measure of C availability to decomposers or the quality of SOM (Bosatta and Ågren, 1994; Sparling et al., 1994). Theng et al. (1989) suggested that C_{mic}/SOC ratios greater than 3% and smaller than 1% were due to soil factors other than SOM quality. Some C_{mic}/SOC values reported in the literature (Table 3) exceeded 3%, however, many values were below 0.5%, suggesting that soil factors (e.g., acidity) may cause low microbial biomass in many forest soils. Some high values of soil C_{mic}/SOC may result from fresh litter additions. A decrease in C_{mic}/SOC with profile depth was observed in the forest floor (Bauhus and Khanna, 1999). However, in many cases this trend did not continue in the mineral soil. The decline of these ratios in the forest floor suggests a decrease in the availability of the substrate to microorganisms as decomposition proceeds.

Bosatta and Ågren (1994) demonstrated that the C_{mic}/SOC ratio is a function of the percent organic carbon remaining for a cohort of litter undergoing decomposition. Despite the expected use of microbial biomass C for estimation of LSC, two limitations remain. First, this ratio is influenced by microbial mortality, which is partially dependent on soil physical properties such as clay content (Bosatta and Ågren, 1994). High clay content may offer greater protection to microbes against predation or desiccation (Hassink, 1994). Second, microbial biomass C forms only a part of LSC pool.

Table 3. Some descriptive statistics for microbial biomass C (C_{mic}) and ratios between C_{mic} and total organic C in soils (SOC), and microbial biomass N (N_{mic}) in a number of forest soils (n = number of cases)

	C_{mic} (mg kg^{-1})	C_{mic}/SOC (%)	C_{mic}/N_{mic} (%)
Forest floor			
Mean	7700	1.8	9.2
Maximum	40,860	8.1	17.3
Minimum	590	0.3	3.0
n	101	78	39
Mineral soil			
Mean	608	1.6	8.2
Maximum	5506	9.9	27.5
Minimum	19	0.3	2.5
n	313	292	119

Table 4. Potentially mineralizable soil C (C_o), microbial biomass C (C_{mic}), and metabolic quotient qCO_2 (mg C respired / kg C_{mic} per day) in a forest soil (0-5 cm, total soil C = 37.5 g kg^{-1}) with following treatments: control (C), P = fertilized with phosphorus, L = leached, I = inoculum from a garden soil (for experimental details refer to Bauhus et al., 1993)

Treatment	C_o		C_{mic}		qCO_2
	g C kg^{-1}	% of total C	mg kg^{-1}	% of C_o	
C	2.42	6.46	198	8.2	43.7
P	2.34	6.25	201	8.6	47.7
L	2.58	6.89	217	8.4	47.4
I	1.74	4.64	217	12.5	38.8
PI	1.57	4.19	223	14.2	34.7
LI	1.85	4.94	228	12.3	40.4

2. Potentially Mineralizable C

Long-term aerobic incubations (Stanford and Smith, 1972) have been gainfully employed to determine the potentially mineralizable C and N in forest soils based on the kinetics of C and N mineralization (Zak et al., 1993).

Figure 2 shows a typical cumulative curve for soil respiration when a soil is incubated under laboratory conditions of constant temperature and moisture. Typically a first order kinetics model, $C_t = C_o (1-e^{-kt})$, can be used to describe the cumulative curve where C_t refers to the amount of soil C respired at time t (days), C_o is the potentially mineralizable C pool, and k is the decay rate constant. In a forest soil (0 to 5 cm) with a number of treatments studied by Bauhus et al. (1993) and outlined in Table 4, the goodness of fit for the above model was $r^2 = 0.95$ or greater in most cases. The experiment included treatments of fertilizer application (phosphorus), soil leaching and inoculation

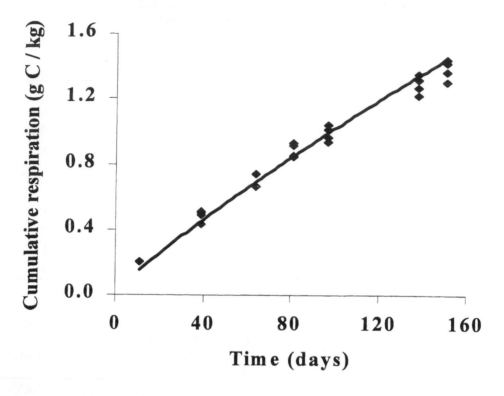

Figure 2. Cumulative respiration from a forest soil (0-5cm) incubated under laboratory conditions (Bauhus et al., 1993). The line gives the fitted values using the equation: $C_t = C_o$ $(1-e^{-kt})$ which accounted for 98.5% of variance (details in the text).

with a garden soil to induce nitrification through a change in microbial populations (details in Bauhus et al., 1993). The C_o values were obtained by fitting the data for CO_2 respired at each time interval (Table 4).

Inoculation with different microbial populations led to a decrease in C_o values, an increase in microbial biomass C and a decrease in the metabolic quotient (Table 4). The mean C_o value was 6.5% of SOC in non-inoculated soils, which was reduced to 4.6% in inoculated soils. The reduction of C_o in inoculated soils may be related to autotrophic nitrifiers; these organisms occurred only in the inoculated soils. It is speculated that in soils with a high population of autotrophic nitrifiers, a fraction of the C respired from the soil may be assimilated by autotrophic microbes, thus decreasing the value of C_o. This can be considered as a limitation of the above method, which may have a significant influence when comparing soils where changes in microbial population have occurred.

Carbon in the microbial biomass (C_{mic}) accounted for 8.2 to 14.2% of C_o in this soil. This was within the range of 3.5 to 37% found in the literature (mean = 12 for n = 26). Thus, a significant fraction of C_o is derived from organic forms other than soil microbial biomass. The above results also suggest that C_o values can represent the LSC fraction and any management practice that influences the microbial population will affect the labile C in soils in a significant way. However, the results represent short-term changes in LSC fraction. Appropriate studies are needed in forest soils to consider long-term changes in C_0 due to management practices.

Of the 56 forest soils for which C_0 values were available in the literature, 69% had C_0 values lower than 10% of the total SOC, indicating again that only a small fraction of the total C in soils is potentially mineralizable and will be involved in the fast turnover processes. For forest soils where the turnover factor, k, was available, it was related to C_{mic}/SOC ratio in soils (Figure 3a), but a closer linear relationship was observed when it was related to C_{mic}/C_0 (Figure 3b). These relationships suggest that soils with a high proportion of microbial biomass, and in particular with a high proportion of microbial biomass relative to LSC pool (potentially mineralizable), have a higher turnover of organic C.

A first order kinetics equation was used to determine C_0 values. However, some studies have differentiated the total C_0 values into two C pools – fast and slow pools by using a two-component model to describe the respiration data (e.g., Motavalli et al., 1994). Not only is the data requirement for fitting a two-component model high, but the uncertainty in assessing four unknown parameters may make them of limited use.

The estimation of LSC using the aerobic incubation technique to measure potentially mineralizable C has one major drawback: long-term incubations for many months are required with continuous measurements of CO_2 evolved. Moreover, the incubation conditions are not standardized, and the consequence of excluding microfauna and of fine roots (variable amounts present) during incubation remains unknown.

III. Factors Influencing C Sequestration in Forest Soils

Fractions of labile SOC and sequestered soil C are related, and it seems that this relationship involves a number of feedbacks depending upon tree and understorey species, site productivity, site factors determining nutrient supply, nutrient uptake, litter decomposition, and SOM protection mechanisms. Total C inputs to soils are largely based on the plant productivity at a site which in turn is related to climate and nutrient supply – the higher the nutrient supply in soils, the higher is the productivity and hence the carbon input to soils. Labile pools of soil organic matter are intimately associated with ecosystem productivity in the short term. The fertility or nutrient content of a forest soil also has a large influence on C sequestration. Higher soil nutrient levels will produce higher litter quality in the vegetation (which may mean lower phenolic compounds and lower lignin content). Conversely the lower the nutrient levels, the lower the litter quality and thus the decomposition. Hims (1998) calculated that the amount of N, S and P needed for the net sequestration of 104 kg C into the slowly decomposed organic matter (humus) will be 833, 200 and 143 kg, respectively. The nutrients needed may come from a number of sources available for forest soils – inorganic fertilizers, nutrient-rich organic substances (sewage sludge, manures etc), atmospheric precipitation, N-fixing plants and nutrients released from mineral weathering. Johnson and Henderson (1995) noted that an increase in soil C is possible only if N-fixers are introduced in the rotation or are present as part of the system. The reasons for the greater C accumulation under nitrogen-fixers are not explicitly known, but they hypothesized that increased N inputs could cause greater soil organic matter stabilization. A significant amount of other nutrients is locked into freshly formed soil organic matter.

Sequestration of C in forest soils is also influenced by tree species. The role of seven tree species in SOM formation was studied by Bernhard-Reversat (1987) in sahelian landscapes of the dry tropics in Senegal. He proposed two models of C sequestration based on the quality of litter produced by tree species. The first case represented litter from native *Acacia* species that had low levels of readily soluble compounds in fresh and partly decomposed residues. This litter resulted in high microbial

P.K. Khanna, B. Ludwig, J. Bauhus and C. O'Hara

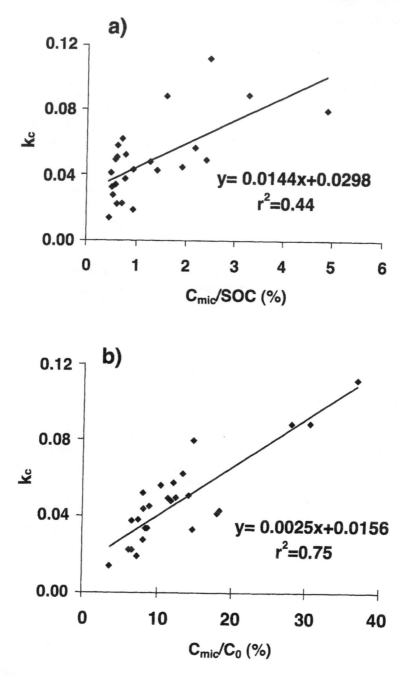

Figure 3. Relationship between the turnover factor for soil carbon (k_c) and (a, top) the fraction of microbial biomass C (C_{mic}) of the soil organic C (SOC), and (b, bottom) the fraction of microbial biomass C (C_{mic}) of the labile soil C (C_o).

respiration rates, and thus had a large component of organic light fractions that entered the surface soil to form organo-mineral complexes. The second case was represented by *Eucalyptus* litter, which was characterized by high levels of soluble components. This litter resulted in moderate microbial respiratory outputs, and had low amounts of organic-light fraction to form SOM. Consequently, planting *Eucalyptus*, especially in sandy soils, did not increase SOM (Bernhard-Reversat, 1993). Similar results were reported by Wang et al. (1991) for a *Eucalyptus* plantation in Puerto Rico.

Clearcutting of a primary forest will influence C sequestration in soils depending upon the fate of the harvested slash and availability of the 'so far protected' SOM. In most cases a decrease in SOC has been observed during the first year (Khanna et al., 1998). Soil C remains highly dynamic in the surface layers for a number of years after clearfelling and slash burning depending upon the microsites (Khanna et al., 1998). Johnson and Henderson (1995) observed that little change in soil C is expected if minimum soil disturbance occurs during the establishment of a new forest stand on a forest land.

Decomposition is a multi-phased process, with at least three phases: leaching, initial fast decomposition (high C/N) and slow decomposition (narrowing C/N). Thus, for assessing the role of litter for C sequestration, one has to consider the fate and form of C remaining in soil after each of the above phases. For example, the higher the leaching losses (litter quality, textural classes – sandy soils) the lower are the amounts left for formation of SOM. Furthermore, both physical and chemical protection mechanisms and the factors affecting them would determine the amount of C sequestered during decomposition processes. The exact nature of these mechanisms and the relationships between the amount of C sequestered and the factors affecting them have not been worked out properly.

IV. Conclusions

The labile organic C pool does not correspond to any single physical or chemical fraction of soil organic matter, but refers to a readily metabolized SOM component. This pool, which may contain soil microbial biomass and other non-microbial labile components of SOM, is expected to respond most rapidly to physical, chemical, and microclimatic changes in the soil environment. A number of methods are used to assess LSC pools which include physical fractionation techniques, chemical extractions, and procedures based on soil biological activity. Separation of SOM into physical fractions of light and heavy density and of different sizes can provide useful information on pools of labile soil C, but there is still no agreement on the size and density for separating LSC. Among the chemical methods used for measuring LSC, oxidation, acid hydrolysis and water extraction are common techniques. Most of these methods including the one using MnO_4^- oxidation may not always be useful to assess LSC fraction. Data from the literature and our own studies were used to show that the potentially mineralizable C in forest soils obtained from long-term laboratory incubations can assess LSC fraction and the turnover parameter of LSC, which may prove sensitive to management impacts. The LSC pools and turnover parameters can be used to compare sites, to assess C sequestration in soils and to provide information and parameterization of SOC dynamic models. However, there are additional merits in considering the suggestion by Paul and Morris (1999) that the best estimates of the SOC pools can be achieved through a combination of chemical, biological and physical fractionation. According to them, the resistant soil C pools, which may be associated with chemical recalcitrance and clay protection, are best measured by acid hydrolysis. The active SOC pool and its mean residence time (MRT) can be obtained by extended soil incubations to measure respiration. The slow SOC pool and its MRT are estimated from non-linear curve analysis.

Acknowledgments

B. Ludwig wishes to thank the German Research Council (DFG) for the research grant. We wish to thank Dr. Keryn Paul and Dr. Heather Keith, CSIRO, Forestry and Forest Products, for making useful comments on an earlier draft of the chapter.

References

Baldock, J.A., J.M. Oades, A.G. Walters, X. Peng, A.M. Vassallo and M.A. Wilson. 1992. Aspects of chemical structure of soil organic materials as revealed by solid-state ^{13}C NMR spectroscopy. *Biogeochem.* 16:1-42.

Balesdent, J. 1996. The significance of organic separates to carbon dynamics and its modelling in some cultivated soils. *Europ. J. Soil Sci.* 47:485-493.

Bauhus, J. and P.K. Khanna. 1999. The significance of microbial biomass in forest soils. p. 77-110. In: Rastin, N. and Bauhus, J. (eds.), *Going Underground — Ecological Studies in Forest Soils.* Research Signpost, Trivandrum, India.

Bauhus, J., P.K. Khanna and J.R. Raison. 1993. The effect of fire on carbon and nitrogen mineralization and nitrification in an Australian forest soil. *Austr. J. Soil Res.* 31:621-639.

Bernhard-Reversat, F. 1987. Litter incorporation to soil organic matter in natural and planted tree stands in Senegal. *Pedobiologia* 30:401-417.

Bernhard-Reversat, F. 1993. Dynamics of litter and organic matter at the soil-litter interface in fast-growing tree plantations on sandy ferrallitic soil (Congo). *Acta Oecologia* 14:179-195.

Blair, G.J., R.D.B. Lefroy and L. Lisle. 1995. Soil carbon fractions based on their degree of oxidation and the development of a carbon management index for agricultural systems. *Aust. J. Agric. Res.* 46:1459-1466.

Boone, R.D. 1994. Light-fraction soil organic matter: origin and net contribution to net nitrogen mineralization. *Soil Biol. Biochem.* 26:1459-1468.

Bosatta, E. and G.I. Ågren. 1994. Theoretical analysis of microbial biomass dynamics in soils. *Soil Biol. Biochem.* 26:143-148.

Cheshire, M.V., J.D. Russell, A.R. Fraser, J.M. Bracewell, G.W. Robertson, L.M. Benzing-Purdie, C.I. Ratcliffe, J.A. Ripmeester and B.A. Goodman. 1992. Nature of soil carbohydrate and its association with soil humic substances. *J. Soil Sci.* 43:359-373.

Christensen, B.T. 1992. Physical fractionation of soil and organic matter in primary particle size and density separates. *Adv. Soil Sci.* 20:1-89.

Conteh, A., R.D.B. Lefroy and G.J. Blair. 1997. Dynamics of organic matter in soil as determined by variations in $^{13}C/^{12}C$ isotopic ratios and fractionation by ease of oxidation. *Aust. J. Soil Res.* 35:881-890.

Conteh, A., G.J. Blair and I.J. Rochester. 1998. Soil organic carbon fraction in a Vertisol under irrigation cotton production as affected by burning and incorporating cotton stubble. *Aust. J. Soil Res.* 36:655-667.

Dalal, R.C. and R.J. Mayer. 1986. Long-term trends in fertility of soils under continuous cultivation and cereal cropping in southern Queensland. IV. Loss of organic carbon from different density functions. *Aust. J. Soil Res.* 24:301-309.

Davidson, E.A., L.F. Galloway and M.K. Strand. 1987. Assessing available carbon: Comparison of techniques across selected forest soils. *Comm. Soil Sci. Plant Anal.* 18:45-64.

Entry, J.A. and W.H. Emmingham. 1998. Influence of forest age on forms of carbon in Douglas-fir soils in the Oregon Coast Range. *Can. J. For. Res.* 28:390-395.

Golchin, A., J.M. Oades, J.O. Skjemstad and P. Clarke. 1994. Study of free and occluded particulate organic matter in soils by solid state ^{13}C CP/MAS NMR spectroscopy and scanning electron microscopy. *Aust. J. Soil Res.* 32:285-309.

Gregorich, E.G., C.M. Monreal, M. Schnitzer and H.-R. Schulten. 1996. Transformation of plant residues into soil organic matter: Chemical characterization of plant tissue, isolated soil fractions, and whole soils. *Soil Sci.* 161:680-693.

Guggenberger, G., B.T. Christensen and W. Zech. 1994. Land-use effects on the composition of organic matter in particle-size separates of soil: I. Lignin and carbohydrate signature. *Europ. J. Soil Sci.* 45:449-458.

Hassink, J. 1994. Effects of soil texture on the size of the microbial biomass and the amount of C mineralization per unit of microbial biomass in Dutch grassland soils. *Soil Biol. Biochem.* 26:1573-1581.

Hassink, J. 1995. Density fractions of soil macroorganic matter and microbial biomass as predictors of C and N mineralization. *Soil Biol. Biochem.* 27:1099-1108.

Hims, F.L. 1998. Nitrogen, sulfur, and phosphorus and the sequestering of carbon. p. 315-319. In: R. Lal, J.M. Kimble, R.F. Follett and B.A. Stewart (eds.), *Soil Processes and Carbon Cycle*. Adv. Soil Sci. Lewis Publishers, Boca Raton, FL.

Hu, S., D.C. Coleman, C.R. Carroll, P.F. Hendrix and M.H. Beare. 1997. Labile soil carbon pools in subtropical forest and agricultural ecosystems as influenced by management practices and vegetation types. *Agric. Ecosys. Environ.* 65:69-78.

Johnson, D.W. and P. Henderson. 1995. Effects of forest management and elevated carbon dioxide on soil carbon storage. p 137-145. In: R. Lal, J. Kimble, E. Levine and B.A. Stewart (eds.), *Soil Management and Greenhouse Effect*. Adv. Soil Sci. Lewis Publishers, Boca Raton, FL.

Kedrowski, R.A. 1983. Extraction and analysis of nitrogen, phosphorus and carbon fractions in plant material. *J. Plant Nutr.* 6:989-1011.

Khanna, P.K., B. Ludwig, J. Bauhus, I. Serrasolsas and J. Raison. 1998. Changes due to management practices in chemical and biological parameters of soils under eucalypt stands. Proc. Congress Intern. Soc. Soil Sci., Montpellier, France.

Kögel-Knabner, I., W. Zech and P.G. Hatcher. 1988. Chemical composition of the organic matter in forest soils: The humus layer. *Z. Pflanzenernähr. Bodenk.* 151:331-340.

Kögel-Knabner, I. and F. Ziegler. 1993. Carbon distribution in different compartments of forest soils. *Geoderma* 56:515-525.

Ladd, J.N., R.C. Foster and J.O. Skjemstad. 1993. Soil structure – Carbon and nitrogen metabolism. *Geoderma* 56:401-434.

Magid, J., A. Gorissen and K.E. Giller. 1996. In search of the elusive fraction of soil organic matter: three size-density fractionation methods for tracing the fate of homogeneously ^{14}C-labelled plant materials. *Soil Biol. Biochem.* 28:89-99.

Meijboom, F.W., J. Hassink and M. Van Noordwijk. 1995. Density fractionation of soil macroorganic matter using silica suspensions. *Soil Biol. Biochem.* 27:1109-1111.

Motavalli, C., C.A. Palm, W.J. Parton, E.T. Elliot and S.D. Frey. 1994. Comparison of laboratory and modeling simulation methods for estimating soil carbon pools in tropical forest soils. *Soil Biol. Biochem.* 26:935-944.

Oades, J.M. 1972. Studies on soil polysaccharides. III. Composition of polysaccharides in some Australian soils. *Aust. J. Soil Res.* 10:113-126.

Oades, J.M. 1995. An overview of processes affecting the cycling of organic carbon in soils. p. 293-303. In: R.G. Zepp and C. Sonntag (eds.), *The Role of Nonliving Organic Matter in the Earth's Carbon Cycling.* John Wiley & Sons, New York.

Parton, W.J., J.W.B. Stewart and C.V. Cole. 1988. Dynamics of C, N, P and S in grassland soils: a model. *Biogeochem.* 5:109-131.

Paul, E.A. and N.G. Juma. 1981. Mineralization and immobilization of soil nitrogen by microorganisms. p 179-195. In: F.E. Clark and T. Rosswall (eds.), *Terrestrial Nitrogen Cycles.* Ecological Bulletin, Stockholm.

Paul, E.A. and S.J. Morris. 2001. The determination of soil C pool sizes and turnover rates: tracers and biophysical fractionation. In: R. Lal, J.M. Kimble, R.F. Follett and B.A. Stewart (eds.), *Assessment Methods for Soil Carbon.* Lewis Publishers, Boca Raton, FL.

Polglase, P.J., E.J. Jokela and N.B. Comerford. 1992. Phosphorus, nitrogen, and carbon fractions in litter and soil of southern pine plantations. *Soil Sci. Soc. Am. J.* 56:566-572.

Schulten, H.R., P. Leinweber and C. Sorge. 1993. Composition of organic matter in particle-size fractions of an agricultural soil. *J. Soil Sci.* 44:677-691.

Shang, C. and H. Tiessen. 1997. Organic matter lability in tropical Oxisol: Evidence from shifting cultivation, chemical oxidation, particle size, and magnetic fractionations. *Soil Sci.* 162:795-807.

Skjemstad, J.O., P. Clarke, J.A. Taylor, J.M. Oades and S.G. McClure. 1996. The chemistry and nature of protected carbon in soil. *Aust. J. Soil Res.* 34:251-271.

Sollins, P., G. Spycher and C.A. Glassman. 1984. Net nitrogen mineralization from light- and heavy-fraction forest soil organic matter. *Soil Biol. Biochem.* 16:31-37.

Sparling, G.P., P.B.S. Hart, J.A. August and D.M. Leslie. 1994. A comparison of soil and microb-ial carbon, nitrogen, and phosphorus contents, and macro-aggregate stability of a soil under native forest and after clearance for pastures and plantation forest. *Biol. Fertil. Soils* 17: 91-100.

Sparling, G., M. Vojvodic-Vukovic and L.A. Schipper. 1998. Hot-water soluble C as a simple measure of labile soil organic matter: the relationship with microbial biomass C. *Soil Biol. Biochem.* 30:1469-1472.

Spycher, G., P. Sollins and S. Rose. 1983. Carbon and nitrogen in the light fraction of a forest soil: vertical distribution and seasonal patterns. *Soil Sci.* 135:79-87.

Stanford, G. and S.J. Smith. 1972. Nitrogen mineralization potentials of soils. *Soil Sci. Soc. Am. Proc.* 36:465-472.

Strickland, T.C. and P. Sollins. 1987. Improved method for separating light- and heavy-fraction organic material from soil. *Soil Sci. Soc. Am. J.* 51:1390-1393.

Swanston, C.W. and D.D. Myrold. 1997. Incorporation of nitrogen from decomposing red alder leaves into plants and soil of a recent clearcut in Oregon. *Can. J. For. Res.* 27:1496-1502.

Theng, B.K.G., K.R. Tate, P. Sollins, N. Moris, N. Nadkarni and R.L. Tate. 1989. Constituents of organic matter in temperate and tropical soils. p. 5-32. In: D.C. Coleman, J.M. Oades and G. Uehara (eds.), Dynamics of Soil Organic Matter in Tropical Ecosystems. NifTAL Project, University of Hawaii at Manoa.

Wang, D., F.H. Bormann, A.E. Lugo and R.D. Bowden. 1991. Comparison of nutrient use efficiency and biomass production in five tropical tree taxa. *For. Ecol. Manage.* 46:1-21.

Young, J.L. and G. Spycher. 1979. Water-dispersible soil organic mineral particles: I. Carbon and nitrogen distribution. *Soil Sci. Soc. Am. J.* 43:324-328.

Zak, D.R., D.F. Grigal and L.F. Ohmann. 1993. Kinetics of microbial respiration and nitrogen mineralization in Great Lakes forests. *Soil Sci. Soc. Am. J.* 57:1100-1106.

Section IV

Assessment and Analytical Techniques

CHAPTER 13

Interlaboratory Carbon Isotope Measurements on Five Soils

R.F. Follett and E.G. Pruessner

I. Introduction

No "reference soil materials" (RSMs) are available for use by research laboratories when they are determining the natural abundance of the ^{13}C isotope. Consequently, laboratories have resorted to using other reference materials as either reference or working standards. These laboratories also obtain RSMs of their own or from other laboratories, but these soils have not necessarily been adequately characterized through interlaboratory comparisons or against suitable reference materials (RMs). The ^{13}C isotope concentration for soil and plant material is expressed as a ratio to the ^{12}C isotope concentration and that ratio is, by convention, expressed relative to the same ratio of a $CaCO_3$ standard known as PDB from the Cretaceous Pee Dee formation in South Carolina (Boutton, 1991). The sign of the $\delta^{13}C$ value indicates whether the sample has a higher or lower $^{13}C/^{12}C$ isotope ratio than PDB.

$$\delta\ ^{13}C = [\frac{(^{13}C/^{12}C(\text{sample}))}{(^{13}C/^{12}C(\text{reference}))} - 1] * 1000 \qquad \text{Eq. (1)}$$

The ^{13}C isotope abundance, calculated from equation 1, is expressed as $\delta^{13}C$ and is in "per mil" (‰) units. Important is that original PDB material is no longer available. Consequently, in December 1995 during an Advisory Group Meeting held in Vienna, Austria, guidelines were set out by the International Union of Pure and Applied Chemists, the National Institute of Science and Technology (NIST), and the International Atomic Energy Administration that the relative $\delta^{13}C$ should be expressed as VPDB (Eos, 1996). The modification of the PDB scale is realized by assigning a $\delta^{13}C$ value of +1.95 to NBS 19 calcite. The authors of this chapter have not attempted to recast the data presented in terms of VPDB.

Because $\delta^{13}C$ of soil organic carbon (SOC) results from plant derived sources, then the measured $\delta^{13}C$ of soil reflects the plant types responsible for the deposition of the SOC into the soil. Thus, if the photosynthetic pathway of the dominant plants whose signature is measured in the SOC is of C_3 species, then the measured $\delta^{13}C$ will be closer to -26‰. Plants such as wheat (*Triticum aestivum* L.), sugar beets (*Beta vulgarus* L), and alfalfa (*Medicago sativa* L.) are C_3 species. If the photosynthetic pathway of the dominant plants whose signature is measured in the SOC is of C_4 species, such as blue grama (*Boutaloua gracilis* Willd. ex Kunth) or corn (*Zea mays* L.), then the measured $\delta^{13}C$ will be nearer to $-12‰$ (Follett, et al. 1997). C_3 species selectively exclude the uptake of the ^{13}C isotope from atmospheric CO_2 relative to their uptake of the ^{12}C isotope during photosynthesis more than do C_4 species. This is reflected by the more negative $\delta^{13}C$ measured in either plant material or SOC derived from C_3 species.

Besides having a characteristic $\delta^{13}C$ range of between -26 and $-12‰$, mineral soils also have a range of SOC concentration of between ≤ 1 to perhaps $\sim 6\%$ SOC. Although often used for $\delta^{13}C$ measurements of soil, RMs available from NIST or other sources have either or both a $\delta^{13}C$ value and carbon (C) concentration that is far outside the range of that of soil. The NIST has no soil as RMs for $\delta^{13}C$.[1] The NIST does have RMs available, as gaseous CO_2,[2] for use by atmospheric scientists and possibly others. Even if such RMs have an appropriate $\delta^{13}C$ range, they are not in the same form as soil nor do they have similar C concentration. Thus, all available RMs are less suitable and/or more difficult to use than if they were soil and an effort is still greatly needed to work with NIST to have soils as RMs with $\delta^{13}C$ values that are on the same scale as are the gaseous CO_2 RMs.

The purpose of this study was to compare the suitability of five soils having wide, but suitable, ranges of $\delta^{13}C$ and SOC concentrations as possible RSMs. The study was conducted with the participation of recognized laboratories and individuals (see Acknowledgments section) who conduct C-isotope ratio work within the scientific community, are equipped with automated carbon analysis-isotope-ratio mass spectrometry equipment, routinely run soil samples, were willing to participate in an interlaboratory comparison study, would use soil material provided by our laboratory, and would conduct the analyses as if the soil sent to them were routine samples. Laboratories involved were asked to continue to use their own RMs and normal operating procedures.

II. Material and Methods

Five soils were selected for use in this interlaboratory study. The selected soils had $\delta^{13}C$ values ranging from about -13 to about $-26‰$ and SOC concentrations ranging between about 0.5 to 6.3%. Soils used for the test were obtained from sites near Dalhart, TX; Akron, CO; Mandan, ND; Swift Current, SASK; and Melfort, SASK. Soils from these respective sites were a Dallam sandy loam (*fine-loamy, mixed, thermic* Aridic Haplustalf); a Weld silt loam (*fine-loamy, mixed, mesic* Calcidic Argiustalff); a Temvik silt loam (*fine-silty, mixed* Typic Haploborall); a Swinton loam (*fine-silty, mixed, frigid* Typic Haploborall); and a Melfort clay loam (*fine-loamy, mixed, frigid* Virtic Cryoborall).

All soils were pre-treated within our laboratory before being sent to the other laboratories for analyses. The purpose of the pre-treatment was to remove inorganic soil carbonates that may have been present, provide uniform samples, and eliminate the need for sample pre-preparation by the participating laboratories. Soil carbonate was removed by the addition of $0.3\ M\ H_3PO_4$ (20:1 solution: soil ratio) and shaking for 1 h, then filtering though glass microfiber filters. The procedure was repeated until the pH of the soil solution remained within 0.2 pH unit of that of the original acid solution (Follett et al., 1997). However, Chesire (1979) has reported that polysaccharides, primarily neutral sugars (hexoses, pentoses, and deoxyhexoses) are removed by dilute acid extraction. The difference we measured between non-acidified and acidified samples of the Dalhart soil, which showed

---

[1] Personal communication with Mr. Bruce MacDonald and Dr. Robert Vocke, National Institute of Standards and Technology, Standard Reference Materials Program, Gaithersburg, MD.

[2] Reference materials numbered 8562, 8563, and 8564 are as CO_2 with $\delta^{13}C$ values of -3.72, -41.51, and -10.40, respectively.

detectable amounts of soil carbonate when 0.1 M HCl was dripped on the soil, was 0.09% total C and amounted to a 14.7% decrease in total soil C following acidification compared to the non-acidified soil sample. The value for $\delta^{13}C$ was 0.82‰ less negative for the acidified than for the non-acidified Dalhart soil, and the observed change likely resulted from removal of the $CaCO_3$. The other four soils used in this study show no detectable carbonates when 0.1 M HCl was dripped on them. The difference between the non-acidified and acidified samples across these four soils averaged 0.54 ± 1.17 (AVG. ± STD) %, as a fraction of the total soil C in the non-acidified samples. The $\delta^{13}C$ was 0.04 ± 0.12 (AVG ± STD) ‰ more negative in the acidified than in the non-acidified soils, or essentially no change with acidification. From the above data, our conclusion is that for the soils used in this study, acidification removed soil inorganic C (carbonates), but little or no organic C. Irrespective, it should be recognized that for some soils, acidification may remove neutral sugars and possibly other soluble organic compounds. The importance of these compounds and their amounts and vulnerability to extraction and/or mild acids require further investigation.

Following pre-treatment and oven drying (55 °C), all soils were ground to pass through a "80 Tyler screen equivalent" (0.18-mm) screen. Vials of each of the uniformly prepared soils (10 g each) were distributed to all participating laboratories. Based upon our experience, we requested that each laboratory do at least three different runs and include at least five duplicate samples of each soil in each run. However, some laboratories used a different number of runs and number of duplicate samples per run than was requested.

Each of the seven laboratories provided information about the mass-spectrometer system and reference material(s) they used. However, it was not an objective of this study to make comparisons between either the laboratories or the equipment they were using. Therefore, and only for informational purposes, we report the various types of equipment[3] and range of reference samples used. Among the participating laboratories the mass spectrometers used included three Europa-Tracermass, one Europa 20-20, two Fisons Optima, and one VG Isochrome machines. Both Carlo-Erba and Roboprep carbon/nitrogen analyzers were used to combust and introduce the resulting gaseous samples into the mass spectrometers. Reference materials used by individual laboratories included ANU sucrose ($\delta^{13}C$ = -10.4‰, C = 42%), beet sugar ($\delta^{13}C$ = -26.45‰, C = 42.2%), Urea ($\delta^{13}C$ = -18.2‰, C = 20%), and soil RMs ($\delta^{13}C$ = -16.55, -17.64 and -25.27‰, SOC = 1.61, 2.24, and 3.75%), NBS 22 oil ($\delta^{13}C$ = -29.7‰), and vacuum pump oil ($\delta^{13}C$ = -27‰). The two oils likely are composed of about 80% C (CRC, 1987).

III. Results and Discussion

If means were calculated for all observations across all laboratories, the resulting values would be unevenly weighted because the different laboratories made different numbers of observations. For example, laboratory 1 reported up to 12 observations across each of 8 separate runs for a total of between 90 to 95 individual analyses for each soil (Table 1). Meanwhile laboratories 4 and 5 made five observations across each of three separate runs for a total of 15 individual analyses for each soil. Thus, to obtain the means of $\delta^{13}C$ and SOC for the five soils using data from all seven laboratories, we used the data provided by each laboratory within only their first to fifth observations from each of their first to third runs (Table 1, 2a, and 3a). Laboratories 6 and 7 did not report data for SOC (Table 1) and thus mean and standard deviations for SOC were calculated using data from within the first five observations from each of the first to third runs by Laboratories 1 through 5. Besides the "across laboratory" means and standard deviations, the average values for $\delta^{13}C$ and SOC obtained by each

---

[3] Trade and company names are included for the benefit of the reader and do not imply endorsement or preferential treatment of the product by the authors or the USDA.

Table 1. Comparison by laboratory for measures of $\delta^{13}C$ and % soil organic carbon

| Soil | | Laboratory number | | | | | | | n | Runs 1 to 3, Samples 1 to 5; across laboratories | |
		1	2	3	4	5	6	7		Mean	Std. Dev.
Dalhart, Texas	$\delta^{13}C$	-13.27	-13.12	-12.28	-13.23	-12.74	-12.59	-12.63	80	-12.89	0.834
	% SOC	0.51	0.44	0.52	0.59	0.55	—	—	59	0.53	0.055
Akron, Colorado	$\delta^{13}C$	-16.55	-16.05	-15.90	-16.45	-16.05	-15.72	-16.06	79	-16.15	0.426
	% SOC	1.61	1.45	1.54	1.61	1.67	—	—	60	1.58	0.165
Mandan, North Dakota	$\delta^{13}C$	-21.34	-21.23	-20.69	-20.53	-20.97	-20.65	-20.77	78	-20.88	0.412
	% SOC	2.73	2.39	2.80	2.86	2.87	—	—	58	2.75	0.179
Swift Current, Saskatchewan	$\delta^{13}C$	-24.66	-23.92	-23.74	-23.87	-23.82	-23.58	-23.71	78	-23.95	0.586
	% SOC	1.89	1.69	1.85	1.89	1.93	—	—	58	1.86	0.098
Melfort, Saskatchewan	$\delta^{13}C$	-26.20	-25.76	-25.49	-24.86	-25.49	-25.09	-25.31	80	-25.46	0.518
	% SOC	6.37	6.01	6.31	5.82	6.39	—	—	59	6.16	0.309

Table 2a. Mean, standard deviation, and coefficient of variation of δ ^{13}C for five soils across seven laboratories for runs 1 to 3, sample numbers 1 to 5

Soil	Dalhart, TX	Akron, CO	Mandan, ND	Swift Current, SASK	Melfort, SASK
Mean	−12.890	−16.154	−20.879	−23.951	−25.464
Standard deviation	0.834	0.426	0.412	0.586	0.518
Coefficient of Variation (%)	6.47	2.64	1.97	2.45	2.03

Table 2b. Maximum likelihood variance component estimates for δ ^{13}C for five soils

σ^2 Laboratory	0.0715	0.0623	0.0814	0.0568	0.1877
σ^2 Run	0.0571	0.0567	0.0161	0.2064	0.0198
σ^2 Number	0.5758	0.0756	0.0755	0.0917	0.0548
Total σ^2	0.7071	0.1946	0.1730	0.3549	0.2623

individual laboratory across their individual data runs are also shown in Table 1. The individual laboratory averages for δ ^{13}C shown in Table 1 ranged from 0.30‰ less negative to 0.74‰ more negative than the means across laboratories. Individual laboratory averages for SOC shown in Table 1 ranged from 0.09% lower to 0.23% higher than the means across all laboratories.

Means for the five soils used in this "round robin" study had δ ^{13}C values that ranged from −12.89 to −25.46‰. The standard deviation for δ ^{13}C was lowest for the soils from Mandan, ND and Akron, CO. The highest standard deviations were for the soils from Dalhart, TX, Swift Current, SASK, and Melfort, SASK. Thus, there were differences between soils for the consistency with which their δ ^{13}C was measured. The Dalhart soil had the least negative and the Melfort soil had the most negative δ ^{13}C of the soils tested. Means for SOC ranged from 0.53 to 6.16%. The standard deviations when measuring SOC were lowest for soils from Dalhart, TX and Swift Current, SASK and highest for the soils from Melfort, SASK and Mandan, ND.

Data in Table 2a repeat the means and standard deviations by soil for δ ^{13}C that are in Table 1 and include the coefficients of variation (c.v.). Table 2b shows the results of maximum likelihood variance component estimate (MLVCE) analysis (SAS, 1990) of the variance (σ^2) components of the total σ^2 that resulted from laboratory, run number and sample number and allows an evaluation of the different soils for possible use as δ ^{13}C reference samples. The total σ^2 shown at the bottom of Table 2b is the sum of laboratory σ^2, run σ^2, and sample σ^2. As with the standard deviation (Table 2a), total σ^2 for δ ^{13}C was smallest for the Mandan and Akron soils. Also the c.v. for the Mandan soil was smaller than for any of the other soils. The c.v.'s for the Dalhart and Akron soils were higher than for any of the other soils. However, because of its smaller standard deviation and total σ^2, the Akron soil is considered as a possible choice for a RSM. The total σ^2 for δ ^{13}C was largest for the Dalhart and the Swift Current soils.

It was not our purpose to compare one laboratory to another. The use of MLVCE analyses, however, does allow the σ^2 for laboratory, the σ^2 for run, and the σ^2 for sample number to be considered individually and also summed separately from the laboratory σ^2. Such an evaluation is important for determining whether to use one particular soil versus another as a reference sample.

Table 3a. Mean, standard deviation, and coefficient of variation of % SOC for five soils across seven laboratories for runs 1 to 3, sample numbers 1 to 5

Soil	Dalhart, TX	Akron, CO	Mandan, ND	Swift Current, SASK	Melfort, SASK
Mean	0.53	1.58	2.75	1.86	6.16
Standard deviation	0.055	0.165	0.179	0.098	0.309
Coefficient of Variation (%)	10.46	10.42	6.50	5.27	5.01

Table 3b. Maximum likelihood variance component estimates for $\delta\ ^{13}C$ for five soils

σ^2 Laboratory	0.0026	0.0000	0.0305	0.0072	0.0513
σ^2 Run	0.0003	0.0158	0.0055	0.0015	0.0251
σ^2 Number	0.0006	0.0124	0.0057	0.0032	0.0260
Total σ^2	0.0035	0.0282	0.0417	0.0119	0.1025

Thus, because of the low laboratory fraction of total σ^2 for the Dalhart soil, it might be chosen when comparing measurements of $\delta\ ^{13}C$ between laboratories. However, because of its high σ^2 for sample number and total σ^2, the Dalhart soil should likely not be chosen. Conversely, the low combined sum of σ^2 for run and sample number suggest that the Melfort soil might be superior for use by an individual laboratory but because of its high laboratory σ^2, it is likely not as suitable for interlaboratory comparisons as are the other soils. Based upon their MLVCE analyses, the Mandan and Akron soils are likely the most suitable soils evaluated in this study to use as RSMs, followed by the Swift Current soil. However, the σ^2 of the Swift Current soil for run number was somewhat high, which could make it less suitable overall than the Mandan and Akron soils.

A similar evaluation of MLVCE to that made for $\delta\ ^{13}C$ was also made for SOC. The means, standard deviations, and c.v.'s for SOC are shown in Table 3a. The standard deviations and the total σ^2 (Table 3b) for the SOC measurements were lowest for the Dalhart and Swift Current soils while those for the Melfort and Mandan soils were highest. The c.v.'s were lowest for the Melfort and Swift Current soils.

Whereas there were seven laboratories that reported data for the $\delta\ ^{13}C$ analyses, there were only five laboratories that provided data for the SOC analyses. With fewer laboratories reporting SOC data than there were for the $\delta\ ^{13}C$ data, a note of caution is needed that the use of the MLVCE analysis becomes less sensitive as the number of laboratories decreases. The laboratory σ^2 as a fraction of the total σ^2 was quite high for the Dalhart, Mandan, and Melfort soils and accounted for from half to nearly three-fourths of the total σ^2, thus causing some concern in the use of these soils as well as the Swift Current soil for interlaboratory comparisons. Conversely, the laboratory σ^2 for the Akron soil was very small indicating that it was the best soil for these interlaboratory tests with run and sample number being the largest contributors to the total σ^2. The run σ^2 and sample number σ^2 were smallest for the Dalhart and the Swift Current soils followed by the Mandan soil, indicating that these soils might be the most suitable for use within a single laboratory to obtain reproducible results.

IV. Summary

Available RMs have either or both their $\delta^{13}C$ values or SOC concentrations outside the general range of those for mineral soils. During combustion analyses these non-soil RMs often do not combust the same as does soil and their use can be more difficult than if they were soils. Individual laboratories generally obtain one or more soil materials on their own to use as RSMs, but such material may not be adequately characterized and may not provide the desired reproducibility.

The five mineral soils used in this study had a wide range of $\delta^{13}C$ values and SOC concentrations. In measuring $\delta^{13}C$, the standard deviation of the mean values averaged across data provided by seven laboratories for these five soils ranged from 0.412‰ for the Mandan soil to 0.834‰ for the Dalhart soil. Standard deviations of the mean values across data provided by five laboratories for SOC for these five soils ranged from 0.055% for the Dalhart soil to 0.309% for the Melfort soil.

We evaluated the suitability of these five soils as possible RSMs for $\delta^{13}C$ measurements and that could also be shared among different laboratories or possibly used solely within a single laboratory. The results show that the soils were different, but that certain soils were likely more suitable to exchange between laboratories than were others. For example, the MLVCE analyses indicated that 32% of the total σ^2 for the Akron soil was from the laboratory component and the remaining 68% from the sum of the run and observation number components. For the Melfort soil 71% of the total σ^2 was from the laboratory component and the remaining 29% from the sum of the run and observation number components. Thus, for interlaboratory comparisons the Akron soil is a better choice. Conversely, and if properly standardized for a single laboratory, the Melfort soil would provide much more consistent data across runs and observations.

We also evaluated the suitability of these same five soils as possible SOC RSMs that again might be shared among different laboratories or used solely within a single laboratory. The results show that the soils were different, but again that certain soils were likely more suitable to exchange between laboratories than were others. The MLVCE analyses indicated that for SOC measurements the Akron, Dalhart, and Swift Current soils would be the most suitable soils for interlaboratory exchanges. For within laboratory use the Dalhart and Swift Current soils followed by the Mandan soil appear to be the most suitable because of their low total σ^2 and low run σ^2 and sample number σ^2. The authors have reservations concerning the use of the Dalhart soil because its SOC is at the lower end of the general range of soil SOC for mineral soils, thus requiring an instrument that is sensitive at a low SOC concentration. The soil is also quite sandy, which makes its preparation and grinding more difficult than for other soils. Therefore, for SOC the Akron and Swift Current soils may be the most suitable, while for use in a single laboratory the Swift Current and Mandan soils may be the most suitable.

For soil $\delta^{13}C$, the Mandan and Akron soils are likely the most suitable soils evaluated in this study to use as RSMs followed by the Swift Current soil. For SOC, the Swift Current and Mandan soils followed by the Akron soil may be best to use as RSMs. Mandan, Akron, and Swift Current soils provide average soil RSM values for $\delta^{13}C$ and SOC of about -20.9, -16.2, and -24.0‰ and of 2.75, 1.58, and 1.86%, respectively. These ranges should be quite suitable for most RSM needs for either $\delta^{13}C$ or SOC.

In summary, the preliminary value assignments for the $\delta^{13}C$ and SOC in the soils tested with this "round robin" study is an appropriate first step in characterizing these soil materials. However, a much more rigorous study design that adequately considers between-laboratory and within-laboratory biases is eventually needed.

Acknowledgments

The authors gratefully acknowledge the participation and willingness of the following individuals to conduct the necessary analyses and to provide the data they obtained to the authors: Dr. Ted Elliott and Mr. Dan Reuss, Colorado State University, Fort Collins, CO; Dr. John Duxbury and Ms. Safia Naqi, Cornell University, Ithaca, NY; Dr. Sean Denby and Ms. Marie-Claire Olivant, EUROPA Scientific Ltd, Crewe, U.K.; Dr. Roger Burke, USEPA/NREL, Athens, GA; Dr. Eldor Paul, Michigan State University, East Lansing, MI; Dr. David Harris, University of California, Davis, CA; and Dr. Ed Clapp, USDA/ARS and Ms. Meg Layese, University of Minnesota, St. Paul, MN. Special appreciation is expressed to Dr. Al Frank for providing soil from Mandan, ND; to Dr. Con Campbell for providing soil from Swift Current and Melfort, SASK; and to Dr. Gary Richardson, USDA/ARS statistician, Fort Collins, CO, for his assistance with statistical analyses and interpretation.

References

Boutton, T.W. 1991. Stable carbon isotope ratios of natural materials: I. Sample preparation and mass spectrometric analysis. p. 155-171. In: D.C. Coleman and B. Fry (eds.), *Carbon Isotope Techniques*. Academic Press, New York.

Chesire, M.V. 1979. *Nature and Origin of Carbohydrates in Soils*. Academic Press. London. 216 pp.

CRC. 1987. *Handbook of Chemistry and Physics*. 67th ed. CRC Press, Boca Raton, FL.

Eos. 1996. Guidelines for reporting stable H, C, and O isotope-ratio data. *Eos, Transactions American Geophysical Union* 77:255.

Follett, R.F., E.A. Paul, S.W. Leavitt, A.D. Halvorson, D. Lyon and G.A. Peterson. 1997. Carbon isotope ratios of Great Plains soils and in wheat-fallow systems. *Soil Sci. Soc. Amer. J.* 61:1068-1077.

SAS Institute. 1990. SAS/STAT Users Guide. Vol. 2. Version 6 ed. Cary, NC.

CHAPTER 14

The Determination of Soil C Pool Sizes and Turnover Rates: Biophysical Fractionation and Tracers

E.A. Paul, S.J. Morris and S. Böhm

I. Introduction

The interpretation of soil organic-C (SOC) dynamics in ecosystem functioning, sustainable agriculture, forestry management and global change requires knowledge of the size and fluxes of the SOC pools involved. Accurate assessment of these pools requires analysis of well-taken, representative soil samples of known bulk density to the depth of soil formation. Samples must reflect an understanding of microscale patterns that control SOC dynamics and associated reactions in the field (Robertson and Gross, 1994). Much of the information for global change calculations is interpreted on a regional or global basis. The sites studied must be representative of all areas in the extrapolation. Some soils such as sands have only a limited SOC accumulation capacity; others such as clays have a much higher sequestration capacity.

Soil organic matter is a continuum of a complex of related humic materials. No one fractionation, technique or pool can be expected to adequately characterize the turnover rates of the whole soil. The description of SOC dynamics and modeling has shown that first order kinetics and a three-pool concept can reasonably effectively describe SOC dynamics (Paustian et al., 1992). The ability to estimate soil C is dependent on the accuracy with which C contents of individual layers and horizons can be measured. Total C and N are usually determined with an accuracy of ±2% of the total amount present; this means that management effects cannot usually be determined until after at least 10 years have elapsed. Techniques that measure the more dynamic fractions such as microbial biomass (Horwath et al., 1996), the light fraction (LF) (Janzen et al., 1992), particulate organic matter (POM) (Cambardella and Elliott, 1993) and laboratory incubation (Paul et al., 1999) are more sensitive. Laboratory incubation, especially when combined with C_3-C_4 plant switches, can give turnover estimates for a fast and slow pool after only a few years of management change.

We have used a combination of biological and chemical approaches to analytically determine the pool sizes and their turnover rates (Figure 1). Not all of the approaches in Figure 1 will necessarily be used on a particular sample. The routine determination of pool sizes and fluxes involves acid hydrolysis and incubation. The residue of acid hydrolysis is used to determine the size of the resistant pool (C_r). Carbon dating measures its mean residence time (MRT) (Leavitt et al., 1997). Acid hydrolysis dissolves the polysaccharides and most of the nitrogenous constituents and is known to leave behind the aromatic humics (Martel and Paul, 1974; Scharpenseel and Schiffman, 1977). It will not dissolve modern lignin residues nor does it effectively, as one would hope, separate the effects of treatment such as extended fallow that should result in the loss of all but the resistant fraction. The

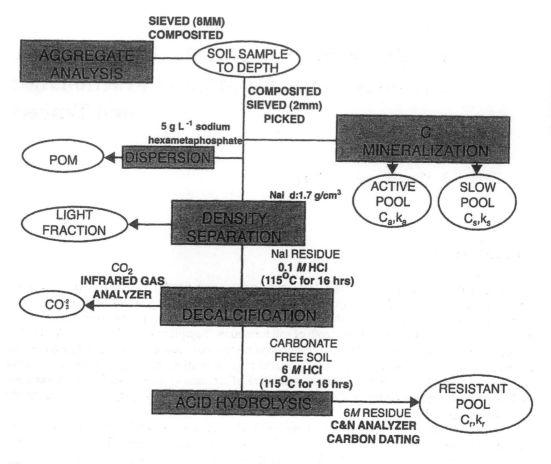

Figure 1. Methods utilized in analytical determination of pool sizes and fluxes.

residue of hydrolysis, however, ranges from 30 to 70% of the SOC and is on average 1400 years older than the total SOC (Leavitt et al., 1996; Paul et al., 1997). The proportion of non-hydrolyzable C is reproducible between samples and is dependent on soil type and parent materials (Collins et al., 1999). Carbon dating is expensive and not readily available. Our calculations show that an MRT of greater than approximately 500 years will not affect the calculations of the pool dynamics of the more bioavailable C_a and C_s pools. The size of the Cr pool however affects the size of the other two and must be known.

Extended incubation utilizes the degradative capacity of the soil biota's enzymes to release the active and slow SOC pools. The CO_2 evolution curves, based on a unit time basis, are used to calculate the dynamics of the active and slow pools. The other approaches shown in Figure 1 provide additional information to that which is determined by acid hydrolysis and incubation. Particulate organic matter and the light fraction are good indicators of turnover (Carter and Stewart, 1996). Soil carbonates can store large amounts of C and result in misinterpretation if not removed or measured separately.

There is a great deal of interest in obtaining some rapid chemical index of the bioavailable fraction. Results from hot water extraction, weak acid hydrolysis, etc., can be correlated to the results from incubation and can be useful in some interpretations (Baker et al., 1998). The results of aggregate analysis are also useful in measuring the effect of physical protection that controls the dynamics of the C_a and C_s pools. These are described elsewhere (Carter and Stewart, 1996; Smucker et al., 1998).

The ^{13}C present in many agricultural soils, which have recently grown C_4 corn or sorghum on original C_3 forest sites or C_3 wheat on previously C_4 grasslands, provides a useful signal. It is not used in our SOC dynamics calculations directly and therefore can be utilized to validate modeling using the analytically derived data. The ^{13}C signal also provides data on the percent of the SOC derived from recent residues and the proportion of recent residues remaining (Follett et al., 1997). Measurement of the $^{13}CO_2$ produced during incubation can be used to measure the pool sizes and decomposition rate constants of the individual C_3 and C_4 SOC components (Collins et al., 2000).

II. Site Selection and Sampling

Agronomic samples often can be obtained from well-replicated, long-term plots with known histories of climate, management and plant residue inputs (Gregorich et al., 1995; Follett et al., 1997; Paul et al., 1997). We routinely utilize six, 5.4-cm diameter cores taken at 0 to 20, 20 to 50 and 50 to 100 cm depths or from the appropriate soil horizons from each replicate of each treatment. The effects of zero till are best measured by taking additional, near-surface soil samples (0 to 5 cm) and litter samples.

Soil type, history prior to afforestation and present forestry practices affect C storage in forests (Johnson, 1992). Present and former owners and Forest Service and Department of Natural Resources officers can provide management records and oral histories. This information together with topographic maps, county soil surveys, satellite imagery, etc. are utilized for locating afforested, native undisturbed sites and associated current agricultural sites with well-documented histories on equivalent soil type, slope, aspect and vegetation. Accurate measurements of SOC pools in forested sites or among landscapes are especially difficult to achieve. The factors that control SOC turnover rates on these sites are patterned at a number of scales that include differences in regional climate, soil type topography, and vegetation (Morris and Boerner, 1999). Soil characteristics at the base of individual trees have been found to differ from those found 2 m from the base of those trees (Zinke, 1962). The extent of the difference varies with tree species (Finzi et al., 1998 a, b). Measurements of soil C can be impacted by these differences. Soil C was significantly greater downslope than upslope of red oaks in four southern Ohio mixed oak forests (Morris and Boerner, 1999) and drip line estimates would have underestimated downslope organic C by 7% and overestimated upslope organic C by 20%.

High spatial variability, at even smaller scales, is associated with components of the C and N pools (Starr et al., 1992) and the microorganisms (Morris, 1999) that regulate turnover rates in both agricultural and natural systems. Research on 2 m x 1m plots on soils under a current wheat field and under poplar after 10 years regrowth on previous agricultural land revealed a 5- to 10-fold variation in C and N in both systems (Stoyan et al., 2000). Phosphorus had a 100-fold variation in the wheat and 30-fold in the poplar plot soils. High heterogeneity at the microscale can be minimized through compositing of multiple samples. Hauck et al. (1994) present a concise discussion of field variability and sample collection. Compositing samples does not eliminate heterogeneity at larger scales. Research on site-specific agriculture and the interpretation of processes at the niche level at which they actually occur in ecosystem analysis will require new sampling protocols.

III. Soil Analysis Methodologies

The determination of pool sizes and fluxes involves the partitioning of a composited soil sample by utilizing quartering techniques to ensure adequate subsampling. Soils to be used for C and N analysis and incubation are transported from the field and stored at 4°C. If aggregate analysis (Six et al., 1998; Six et al., 1999) is to be performed, collection and transportation must ensure minimum disturbance and compaction. Samples are gently separated by hand, sieved through an 8-mm sieve and a subsample for aggregate analysis is dried. Following removal of the subsample for aggregate analysis, the soil is sieved through a 2-mm sieve. Since plant residues are modeled separately in our program, we remove identifiable plant fragments prior to analysis. The subsample to be used in determination of C mineralization rates by the long-term incubation method is then refrigerated. The remaining soil is air-dried.

Inorganic C is a significant component of arid soils and occurs at depth in many humid soils. It can be present in recently limed agricultural soils in sufficient quantities to affect C storage calculations. The turnover of pedogenic carbonates is sufficiently rapid to affect global change calculations. We eliminate inorganic C in agricultural soils for SOC analysis and carbon dating but measure its content in soils used for global change calculations. The soil is treated with 100 ml of 0.25 M HCl added to 20 g soil. Care must be taken to remove excess HCl by repeated evaporation. Washing with deionized H_2O after acid treatment can remove soluble C and should be avoided. Alternatively, 100 ml of $0.03M$ H_3PO_4, which does not break down in the C analyzer or mass spectrometers can be used (Follett et al., 1997). The weak phosphoric acid treatment is repeated until the solution remains within 0.2 pH units of the original acid solution. Others have added dilute HCl directly to soil already weighed into silver sampling cups without apparent damage to their instrumentation. The soil inorganic C is measured by difference or by analysis of the CO_2 evolved during acid treatment.

Dried soils, to be utilized for C and N analyses, are ground to pass a 180-μm screen in a roller mill that can handle multiple samples in glass storage containers containing steel bars (Harris and Paul, 1989). The quantity of soil used is corrected for moisture by drying a subsample to 105°C. Samples for C and N and mass spectrometry are weighed to six decimal places on a Cahn balance. Where a Cahn balance is not available, a five-place analytical balance will provide reasonable accuracy. Normal protocols for a Carlo Erba type CHN analyzer are followed (Sollins et al., 1999). Mass spectrometry to measure ^{13}C follows the procedures of Harris and Paul (1989). Appropriate working standards for both SOC and ^{13}C should be inserted in the sampling tray every 12 samples. Working standards for ^{13}C can be corn (–10.75‰) and beet (–25.68‰) sugars standardized against NIST 22 (–29.74‰) or AIEA-C-6, ANU sucrose at –10.43‰. Multiple analysis of the reference samples yields a standard deviation of 0.05‰.

A. Soil Physical and Chemical Analysis

Extensive information is required for interpretation of data gathered on the pool sizes and fluxes of SOC. This includes available P content, pH, cation exchange capacity, soil texture and moisture retention curves. These are conducted according to standardized methods (Page et al., 1982; Carter, 1993). Bulk density can be determined by a number of different methods, but none is without problems (Brady and Weil, 1999). We weigh a number of 5.4-cm-diameter cores obtained either with truck mounted hydraulics or by hand using a hammer driven sampling tube and removal with a portable tripod and associated jack. Alterations of horizon depth as a consequence of management are noted. Root and rock content determinations are made for each profile sampled so that the profile C inventory reflects non-soil particles.

Across-treatment differences in bulk density can impact C content differences among treatments. Corrections for differences in bulk density between sites or treatments involve calculations to equivalent weights to depth among sites (Gregorich et al., 1995). Bulk density differs with horizon, horizon depth differs with treatment and corrections for both must be made to compare equal weights of soil across treatments. A comparison of the effects of bulk density per horizon and horizon depth between a forest site and agricultural site demonstrates how these components differ among soils (Figure 2). Bulk density, weight of soil per horizon and horizon depth differ among all horizons meas-

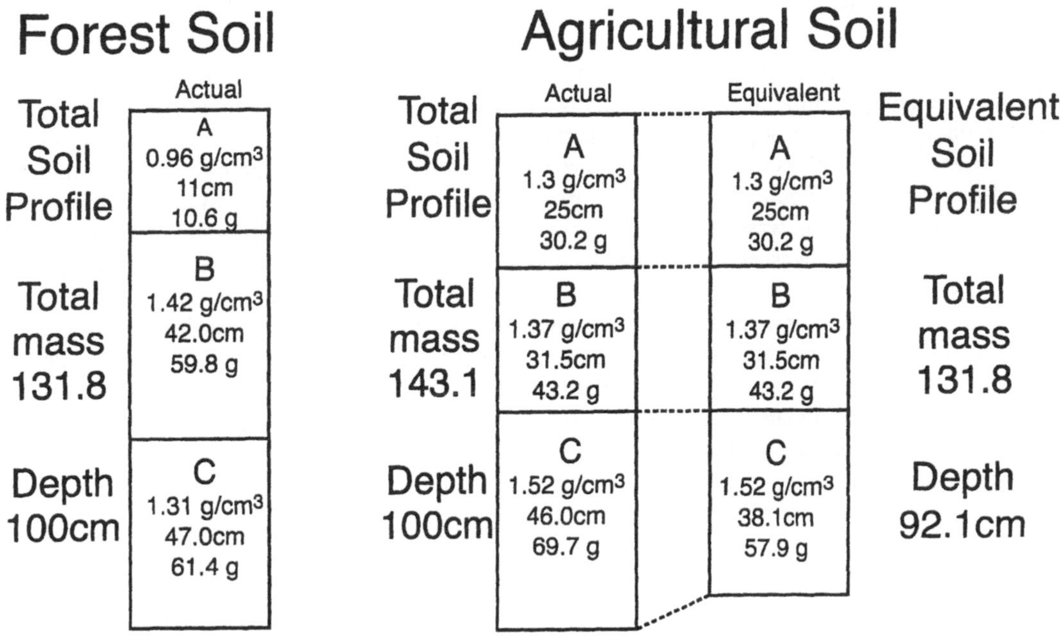

Forest Soil

Total Soil Profile	Actual
	A 0.96 g/cm³ 11cm 10.6 g
Total mass 131.8	**B** 1.42 g/cm³ 42.0cm 59.8 g
Depth 100cm	**C** 1.31 g/cm³ 47.0cm 61.4 g

Agricultural Soil

Total Soil Profile	Actual	Equivalent	Equivalent Soil Profile
	A 1.3 g/cm³ 25cm 30.2 g	**A** 1.3 g/cm³ 25cm 30.2 g	
Total mass 143.1	**B** 1.37 g/cm³ 31.5cm 43.2 g	**B** 1.37 g/cm³ 31.5cm 43.2 g	Total mass 131.8
Depth 100cm	**C** 1.52 g/cm³ 46.0cm 69.7 g	**C** 1.52 g/cm³ 38.1cm 57.9 g	Depth 92.1cm

Figure 2. The correction of soil carbon pools on an equivalent weight basis.

ured. The forest soil has a shorter A horizon (11 cm), a lower bulk density (0.96 g cm^{-3}), and less mass of soil (10.6 g) than the A horizon of the agricultural soil (25 cm, 1.3 g cm^{-3}, 30.2 g). A 1 cm x 1cm x 100 cm profile of the forest soil weighs 131.8 g whereas that of the agricultural soil weighs 143.1 g. To compare C contents of these soils equivalent weights must be compared so the agricultural horizon must be shortened to 92.1 cm so that 131.8 g of each soil is compared.

B. Soil C Pools and Fluxes

1. Physical Fractionation

The light fraction (LF), if measured, is separated from the air-dried soil (usually only surface subsamples) by floating 10 g soil in 40 ml NaI solution (**SG** ~ 1.70 g cm^{-3}) in a 100-ml beaker (Janzen et al., 1992). We use NaI rather than molybdenum polytungstate to lower costs in our multisample analysis. The floating material is washed, finely ground (<180 µm) and analyzed for total C and δ^{13}C. The residual soil is washed three times with 50 ml distilled water to remove NaI.

The measurement of particulate organic matter (POM) involves dispersion of 10 g soil in 30 ml of 5 g L^{-1} sodium hexametaphosphate by shaking for 15 h on a reciprocal shaker (Cambaradella and Elliot, 1992). The slurry is passed through a 53-µm sieve and rinsed thoroughly with water to remove silt and clay. The POM plus sand retained on the sieve is oven dried at 50°C, ground to pass a 180-µm screen and analyzed for C, ^{13}C and N (Harris and Paul, 1989). The POM in this analysis often contains most of the LF unless it is removed prior to analysis. We previously remove all identifiable plant residues.

2. Acid Hydrolysis

The resistant fraction is determined by refluxing 1 g soil in 6 M HCl at 115 °C for 16 h. This is most easily done using a temperature-controlled digestion block. The use of reflux condensers fitted with fritted glass joints makes it possible to retain the acid-water mixture. The tubes can be used without condensers if small, glass filter holders or large glass marbles are used to partially enclose the tops. The digestion liquid should condense no more than one half way to the top and the tubes should be shaken occasionally to wash down soil materials that accumulate at the point of condensation. Refluxed samples are washed three times with deionized water, dried at 55°C and ground to pass a 180-μm screen.

Soils to be utilized for C dating are very prone to contamination by the low levels of anthropogenic ^{14}C that are found in many laboratories. We take special precautions when sampling and sieve and dry our samples in an environment, such as a home garage, known not to have been exposed to tracer C and send it to a C dating laboratory (Leavitt et al., 1997) for further processing. The accelerator mass spectrometric analysis allows the use of milligram sized samples that must, however, be representative of the area sampled.

3. C Mineralization

Mineralizable soil C is measured during extended laboratory incubation at 25°C. Moist, sieved, surface samples (approximately 80 g soil) are placed in tared, 6-oz specimen containers and adjusted to optimum moisture content by adding calculated quantities of H_2O. The optimum moisture content for mineralization can be estimated by placing a known mass of soil into a funnel with a filter paper. The soil, filter paper, and funnel are then weighed together. The funnel, with soil, is then placed into a beaker of water and the water is allowed to move into the soil until the surface of the soil appears shiny. The funnel is then removed from the beaker of water, placed into an airtight container to eliminate evaporation and is allowed to drain. After 24 hours, the funnel containing the drained soil is weighed. Gravimetric water content is determined. If this method is used, the moisture content of the soil to be incubated is raised to 50 to 70% of this value depending on the soil texture. Sandy soils require the lower values so they do not lose their structure. We currently use a method in which 80 g soil, sieved to 2 mm and air dried, is placed into a 5-cm-diameter hollow glass cylinder capped with nylon mesh that is held in place by a rubber band. The cylinder is inverted in 2 mm of water until the water has moved 80% of the way up the soil column (water must not become limiting). The lower, saturated soil is then removed and gravimetric water content is determined on the soil above the saturated soil to just below the layer of dry soil. The samples to be incubated are then raised to this gravimetric water content: 70% of this water content is used when the soil used in the incubation is sandy, highly heterogeneous or comprised of more than one pedogenic horizon to avoid saturating the soil.

The specimen container, with soil, is placed into a quart-size, wide-mouth canning jar containing 20 ml of water to maintain humidity throughout the incubation. When the amount of soil available for testing is limited, smaller jars (i.e. wide-mouth pints or half-pints) may be used in place of the quart jars. The lids of the canning jars are boiled, fitted with septa, placed on the canning jar and sealed with rings. Controls with known CO_2 concentrations (1%) are prepared by injecting CO_2 from sources such as a calibration gas into empty containers. These detect leakage and serve as analytical controls. The samples should be flushed with CO_2 free air to establish that the initial CO_2 concentration is zero. Alternatively if the jar contains ambient CO_2 concentrations at time zero, a baseline CO_2 concentration may be determined with an infrared analyzer or gas chromatography immediately after sealing the jar. Samples are incubated at 25°C in the dark.

Incubations are best conducted with moist field soils as described above. Occasionally it is necessary to use dried, stored soil. Dried soils produce different initial CO_2 evolution curves because of initial recycling of dead biomass and recovery of the biota once water is added. We have found that dried soils produce essentially the same C_s and k_s values as soils utilized immediately in the moist condition (Collins et al., 2000). The decomposition rate constant k_a, however, is higher after air drying. This can be partly corrected for by preincubating dried samples for seven days after wetting.

During incubation, we allow CO_2 to accumulate to some extent between samplings to reduce errors due to improper flushing. Jars should be flushed after four to five samplings or when the CO_2 concentration is expected to exceed 6% before the next sampling period. Two air hoses are inserted into the septa (in from the air source and out to the lab) using small gauge needles. Samples may need to be flushed 20 to 30 min if the concentration of CO_2 is high (>2%) and should be allowed to return to atmospheric pressure following flushing. This is best achieved by placing a water filled syringe with the plunger removed on the out needle. The water in the syringe will bubble if the jar is properly sealed when the jar is being flushed and allow for the pressure to equilibrate with environmental pressure without laboratory CO_2 flowing into the jar when the air source is removed. If the water does not bubble or water flows into the jar rather than maintaining equal pressure with the interior of the jar when the flush air source is removed there is a leak present.

Rather than flushing, samples may be opened and allowed to equilibrate with laboratory air; this compromises seals on jars and laboratory air is very variable in its CO_2 content. Compressed air, from cylinders, may be used with a soda lime scrubber but there may be a 2 to 3°C chilling effect as the compressed air comes in contact with the soil. Decreases in respiration over short measurement increments may result. In-house compressed air may be used with a soda lime scrubber but an oil filter or trap is needed to keep oil from reaching the soil samples. Aquarium pumps may be used in conjunction with appropriate traps but several may need to be used in tandem to achieve pressures great enough to flush several samples at once. Finally, CO_2 concentration should be checked following sample flushing to establish that previous CO_2 has been cleared and to establish a new sample baseline. The new baseline should be established at least 24 h after flushing to allow for a new CO_2 equilibrium to be established within the jar.

The amount of CO_2 evolved is determined, every few days initially, by infrared gas analysis, gas chromatography or by using base traps such as NaOH. If a base trap is used high surface area traps should be replaced and titrated periodically. The time period between CO_2 determination is lengthened at appropriate periods during the incubation. Curve fitting to determine the pool sizes and decomposition rate constants of the active and slow pools requires a minimum of ten values in incubation periods that run from 200 to 800 days. The amount of evolved CO_2 is converted to a rate function by determining headspace CO_2 (usually in ppm or ml CO_2 min^{-1}), converting to moles then grams CO_2-C and dividing by soil weight adjusted for bulk density and gravimetric moisture content and time (see also Robertson et al., 1999).

Where mass spectrometers with gas handling capability are available for ^{13}C analysis, CO_2 samples can be directly injected (Barrie and Prosser, 1996). If gas-handling facilities are not available or if the samples need to be transported or stored, the CO_2 is trapped in excess NaOH; the amount can be calculated or determined ahead of time. We use 5ml of 2 M NaOH for 25 g of soil in the 160-ml jars. Blank jars contain NaOH but no soil. The evolved CO_2 is precipitated as $SrCO_3$ (Harris et al., 1997) using 4 M SrCl$_2$: residual NaOH is measured by back titration with 0.3 M HCl to pH 7 using an automatic titrator. The $SrCO_3$ is analyzed for ^{13}C content. The calculations involving ^{13}C are shown by Balesdent et al. (1988) or Boutton (1996). The control jars described above also supply useful information on accuracy and leaks.

4. Curve Analysis

The size and turnover rates of the active (C_a) and the slow (C_s) pool are determined by curve analysis of the CO_2 evolved per unit time. The three-pool first order model utilized is

$$C_{(t)} = C_a\, e^{-k_a t} + C_s\, e^{-k_s t} + C_r\, e^{-k_r t}$$

where $C_{(t)}$ is total carbon pool in soil at time t, C_a and k_a represent the active pool and C_s and k_s represent the slow pool. C_r has been previously estimated by acid hydrolysis. Carbon dating that measures the MRT is used to calculate k_r where $k_r = 1/MRT$ as in a first order reaction at steady state. For ease of interpretation, we prefer to show our decomposition kinetics as MRT rather than as the decomposition rate constant k. The laboratory-derived values are scaled to field, mean-annual temperatures (MAT) by assuming a Q_{10} of 2, ($2^{(25-MAT)/10}$). Where carbon dates are unavailable, MRT is assumed to be 1000 years. Studies of grassland and forest sites (Leavitt et al., 1996; Paul et al., 1997) have shown the non-hydrolyzable, resistant pool to be at least this old in the majority of samples analyzed. The size of the slow pool is defined as $C_s = C_{soc} - C_a - C_r$ with C_{soc} representing the soil C at time 0 (the time of sampling).

The statistical evaluation of the CO_2 evolution data to determine pool sizes and rates using nonlinear regression is relatively straightforward. The parameters C_a, k_a and k_s are calculated using nonlinear regression (NonLIN; Systat, Inc, Evanston IL or PROC NLIN; SAS 1995). Differences in pool sizes or kinetics among several treatments can be determined using a t-test on the asymptotic SE estimates, a sum of squares reduction test or a repeated measures ANOVA (Willson et al., 2000). We have included an example from a Corn Belt soil to highlight each step in the overall analysis from the mineralization data (Table 1). The rate of CO_2 evolution per day (Figure 3) was utilized in the regression analysis to evaluate C_a, k_a, and C_s for each of the depths sampled. We had 18 sampling dates over 500 days for the top 20 cm, 8 sampling dates over 8 dates over 260 days for the 25 to 50 cm depth and 8 sampling dates over 212 days for the 50 to 100 cm depth. SAS PROC NLIN METHOD=MARQUARDT (SAS, 1995) was used in the analysis because it was more robust with our data set than some of the other methods, such as DUD and GAUSSIAN. The choice of specific method is dependent on the data set.

The most important consideration when evaluating pool sizes and rates in this manner is the type of model to fit. Both two- and three-pool models either unconstrained or constrained are available. Our basic model, if carbon dates are available, consists of three pools as follows:

$$dC/dt = C_a * k_a e^{(-k_a * days)} + (C_{SOC} - C_r - C_a) * k_s e^{(-k_s * days)} + C_r * k_r e^{(-k_r * days)}$$

This model utilizes sample SOC, the results of the acid hydrolysis and the age of C_r. Radiocarbon dating of the residue of the acid hydrolysis determines the MRT of C_r. Carbon dating is relatively expensive and therefore C dates are often unavailable. The MRT of C_r is usually so long that the turnover of this pool does not affect the overall calculations of k_a and k_s. This is demonstrated in Table 1 where the measured field MRT was 530 years for the surface soil, 895 years for the 25 to 50 cm depth and 4406 years for the 50 to 100 cm depth. The values for the two more rapid pools are determined originally in laboratory incubation. Use of the Q_{10} equation to establish the equivalent MRT's for incubation under laboratory temperatures results in values of 175, 295 and 1453 years for the three depths. We tested the three pool model by using both the actual measured C dates and by using assumed values of 1000 years as the MRT for the nonhydrolyzable C of these soils (Table 1). The use of 530 years rather than the assumed value of 1000yr resulted in an increase in the MRT of C_s by 1.3 years. The parameters for the dynamics of the fast and slow pools of the lower depths were the same when using either assumed or actual values for the MRT of C_r. These results indicate that

Table 1. Calculation of pool size and rate of turnover for active (C_a, k_a) and slow pools (C_s, k_s) to a depth of 100 cm in an agricultural field at the Kellogg Biological Station (KBS), MI using a three-pool constrained model[a]

Depth (cm)	C_{SOC} mg kg^{-1}	C_r Mg kg^{-1}	k_r day^{-1}	MRT yr laboratory	MRT yr field	C_a mg kg^{-1}	k_a day^{-1}	MRT day laboratory	MRT day field	C_s mg kg^{-1}	k_s day^{-1}	MRT yr laboratory	MRT yr field
Using Measured Values for k_r													
0-20	10700	4800	1.6E-05	175	530	570	5.01E-02	20	60	5330	2.26E-04	12	37
25-50	2600	800	9.3E-06	295	895	132	5.65E-02	18	54	1670	4.55E-04	6	18
50-100	1300	400	6.2E-07	1453	4406	84	3.27E-01	3	9	816	1.24E-03	2	7
Using Assumed Values for k_r													
0-20	10700	4800	8.3E-06	330	1000	569	5.01E-02	20	60	5332	2.34E-04	12	36
25-50	2600	800	8.3E-06	330	1000	132	5.65E-02	18	54	1670	4.55E-04	6	18
50-100	1300	400	8.3E-06	330	1000	84	3.26E-01	3	9	816	1.24E-03	2	7

[a] $C_{(t)} = C_a*k_a e^{(-k_a*days)} + (C_{soc} - C_r - C_a)*k_s e^{(-k_s*days)} + C_r*k_r e^{(-k_r*days)}$, $C_{(t)}$ = rate of C evolution per unit time (d(CO_2) dt^{-1}), C_{soc} = SOC measured at time 0, C_r = resistant C (non-hydrolyzable C), k_r = 1/MRT when using carbon dating or 1/1000 year when using an assumed MRT. C_a = active C, C_s = slow C ($C_{soc} - C_r - C_a$).

[b] Mean annual temperature for KBS is 9.0°C, incubation temperature is 25°C and the Q_{10} correction is $2^{(25-9)/10}$.

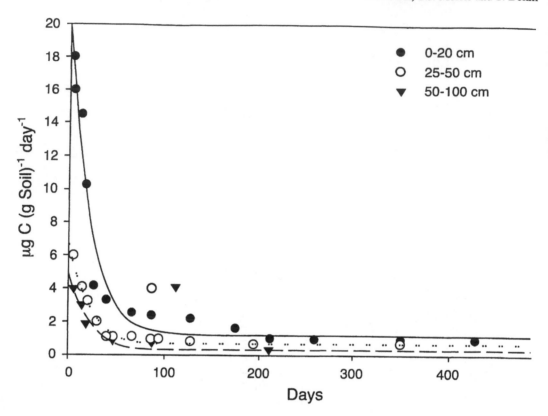

Figure 3. Evolution of CO_2 from three depths of a Michigan field soil expressed on a per unit time basis. Data presented are from Collins et al. (2000).

the three-pool constrained model is robust using either measured or assumed MRT for the C_r pool as long as the size of the C_r pool is known.

An alternative to running the three-pool constrained model is the two-pool constrained model:

$$dC/dt = C_a * k_a e^{(-k_a)} + (C_{soc} - C_r - C_a) * k_s e^{(-k_s * days)}$$

In this model the size of the C_r pool is used in determining the size of the Cs pool but is not modeled separately. Results from this model differ little from the three-pool model, again suggesting that the size of the C_r pool is important but the age of this pool adds little to the analysis.

A two-pool unconstrained model has also been tested to determine pool sizes and turnover rates. In this model only CO_2 evolution is known and all pools and rates are modeled. The main differences between the results produced by the constrained and unconstrained models are in the size and MRT of the slow pool. In our Corn Belt soil, the unconstrained model underestimated the size of C_s up to 75% and decreased the MRT by 29 years, whereas C_a and k_a were equivalent to the results of the constrained model. Examples of the commands used to determine C_a, k_a, and k_s with the two- and three-pool constrained models, using PROC NLIN (SAS, 1995), are located in Table 2.

The turnover rates produced by the regression analysis can be extrapolated to field conditions by the use of Q_{10} corrections. These adequately describe the temperature dependence of decomposition for soil temperatures of 5 to 35°C (Katterer et al., 1998). Soil moisture has strong effects in arid en-

Table 2. Commands[a] for determining the size and turnover rate of the active (C_a, k_a) soil C pool and turnover rate of the slow (k_s) pool[b] with two- and three-pool constrained models using SAS PROC NLIN (SAS, 1995) for a single soil depth or horizon at a single site

Model	SAS commands

Two pool constrained model	Data;
	Infile 'c:\data.txt;
	Input csoc cr co2 day;
	run;
	proc nlin method=marquardt;
	parameters ka=0.001 ks=0.003 ca=100 to 1000 by 100;
	model co2 = ca*ka*exp(-ka*day) + (csoc-cr-ca)*ks*
	exp(-ks*day);
	der.ca = ka*exp(-ka*day)-ks*exp(-ks*day);
	der.ka=ca*exp(-ka*day)-ca*exp(-ka*day)*ka*day;
	der.ks=(csoc-cr-ca)*exp(-ks*day)-(csoc-cr-ca)*
	exp(-ks*day)*ks*day;
	run;

Three pool constrained model	data;
	infile 'c:\data.txt';
	input csoc cr kr co2 day;
	run;
	proc nlin method=marquardt;
	parameters ka=0.01 ks=0.001 ca=100 to 1000 by 100;
	model co2 = ca*ka*exp(-ka*day) + (csoc-cr-ca)*ks*exp(-ks*day) +
	(cr)*kr*exp(-kr*day);
	der.ca = ka*exp(-ka*day)-ks*exp(-ks*day);
	der.ka=ca*exp(-ka*day)-ca*exp(-ka*day)*ka*day;
	der.ks=(csoc-cr-ca)*exp(-ks*day)-(csoc-cr-ca)*exp(-ks*day)*ks*day;
	run;

[a]Data necessary for this analysis include total soil C (Csoc) determined by combustion, resistant C (Cr) determined with acid hydrolysis, CO_2 evolution rate (CO_2) determined using long-term incubations, day of incubation on which the rate was determined (day) and age of the resistant pool (kr) determined with carbon dating corrected for laboratory incubation temperatures. Derivatives are indicated by der and must be included for the Marquart model used here.
[b]The size of the slow (Cs) pool is constrained to be Csoc-Cr-Ca.

vironments but has been shown to have limited effects on field CO_2 decomposition rates in the Corn Belt and associated forests (Buyanovsky and Wagner, 1998; Paul et al., 1998). Moisture effects are not incorporated into our direct determinations of SOC dynamics of Corn Belt and forest soils. They are used in our general modeling of SOC dynamics (Paustian et al., 1992).

The analytically derived values for the pools sizes and fluxes make it possible to compare management and ecosystem processes on specific soil types. The pool sizes and fluxes vary enough on specific soils that measurements on representative soils need to be made for modeling purposes (Collins et al., 1999, 2000). Our analysis does not include the effects of pool to pool transfers or of the recycling of C during microbial growth. This, as well as calculations on a landscape basis, are best done by using one of the available models. These should be parameterized by the use of analytically derived data such as that described in this chapter.

Acknowledgments

These approaches were developed under the auspices of U.S. Department of Energy grant: Soil Organic Matter Dynamics and Management Decision Making in an Enriched CO_2 Environment (DE-FG02-94ER61924). The soils were obtained from the NSF-KBS-LTER project.

References

Baker, A.D., D.A. Zuberer and L.P. Wilding. 1998. Inconsistencies in methods for measuring bioavailable carbon and microbial biomass in soils from the NRCS wet soils monitoring project. In: *Soil C Pools*, An International Workshop, November 2–5, 1998.

Balesdent, J., G.H. Wagner and A. Mariotti. 1988. Soil organic matter turnover in long-term field experiments as revealed by carbon-13 natural abundance. *Soil Sci. Soc. Am. J.* 52:118-124.

Boutton, T.W. 1996. Stable carbon isotope ratios of soil organic matter and their use as indicators of vegetation and climate change. p. 47-82. In: T.W. Boutton and S. Yamasaki (eds.), *Mass Spectrometry of Soils*. Marcel Dekker, New York.

Barrie, A. and S. J. Prosser. 1996. Automated analysis of light-element stable isotopes by isotope ratio mass spectrometry. p. 1-46. In: T.W. Boutton and S. Yamasaki (eds.), *Mass Spectrometry of Soils*. Marcel Dekker, New York.

Brady, N. C., and R. R. Weil. 1999. *The Nature and Properties of Soils*. 12th ed. Prentice-Hall, Upper Saddle River, NJ.

Buyanovsky, G.A. and G.H. Wagner. 1998. Carbon cycling in cultivated land and its global significance. *Global Change Biology*. 4:131-141.

Cambardella, C.A. and E.T. Elliott. 1992. Particulate soil organic matter changes across a grassland cultivation sequence. *Soil Sci. Soc. Am. J.* 56:777-783.

Cambardella, C.A. and E.T. Elliott. 1993. Carbon and nitrogen distribution in aggregates from cultivated and native grassland soils. *Soil Sci. Soc. Am. J.* 57:1071-1076.

Carter, M.R. and B.A. Stewart (eds.). 1996. *Structure and Organic Matter Storage in Agricultural Soils*. Lewis Publishers, Boca Raton, FL.

Carter, M. R. (ed). 1993. *Soil Sampling and Methods of Analysis*. Canadian Society of Soil Science. Lewis Publishers, Boca Raton, FL.

Collins, H.P., R.L. Blevins, L.G. Bundy, D.R. Christenson, W.A. Dick, D.R. Huggins and E.A. Paul. 1999. Soil carbon dynamics in corn-based agroecosystems: Results from ^{13}C natural abundance. *Soil Sci. Soc. Amer. J.* 63:584-591.

Collins, H.P., E.T. Elliott, K. Paustian, L.G. Bundy, W.A. Dick, D.R. Huggins, A.J.M. Smucker and E.A. Paul. 2000. Soil carbon pools and fluxes in long-term Corn Belt agroecosystems. *Soil Biology and Biochemistry* 32:157-168.

Finzi, A.C., N. Van Breemen and C.D. Canham. 1998a. Canopy tree-soil interactions within temperate forests: species effects on soil carbon and nitrogen. *Ecological Applications*. 8(2):440-446.

Finzi, A.C., C.D. Canham and N. Van Breemen. 1998b. Canopy tree-soil interactions within temperate forests: species effects on pH and cations. *Ecological Applications*. 8(2):447-454.

Follett, R.F., E.A. Paul, S.W. Leavitt, A.D. Halvorson, D. Lyon and G.A. Peterson. 1997. Carbon isotope ratios of Great Plains soils in wheat-fallow systems. *Soil Sci. Soc. Amer. J.* 61:1068-1077.

Gregorich, E.G., B.H. Ellert and C.M. Monreal. 1995. Turnover of soil organic matter and storage of corn residue carbon estimated from natural ^{13}C abundance. *Can. J. Soil Sci.* 75:161-167.

Harris, D. and E.A. Paul. 1989. Automated analysis of ^{15}N and ^{14}C in biological samples. *Comm. Plant and Soil Anal.* 20:935-947.

Harris, D., L.K. Porter and E.A. Paul. 1997. Continuous flow isotope ratio mass spectrometry of $^{13}CO_2$ trapped as strontium carbonate. *Comm. Plant and Soil Anal.* 28:747-757.

Hauck, R.D., J.J. Meisinger and R.L. Mulvaney. 1994. Methods of soil analysis, Part 2. p. 907-950. In: *Microbiological and Chemical Properties.* SSSA Book Series 5, Soil Science Society of America, Madison, WI.

Horwath, W.R., E.A. Paul, D. Harris, J. Norton, L. Jagger and K.A. Horton. 1996. Defining a realistic control for the chloroform-fumigation incubation method using microscopic counting and ^{14}C-substrates. *Can. J. Soil Sci.* 96:459-467.

Janzen, H.H., C.A. Campbell, S.A. Brandt, G.P. Lafond and L. Townley-Smith. 1992. Light-fraction organic matter in soils from long-term crop rotations. *Soil Sci. Soc. Am. J.* 56:1799-1806.

Johnson, D. W. 1992. Effects of forest management on soil carbon storage. p. 83-120. In: J. Wisneiewski and A. E. Lugo (eds.), *Natural Sinks of CO_2.* Kluwer Academic Publ., Dordrecht, Netherlands.

Katterer, T., M. Reichstein, O. Andren and A. Lomander. 1998. Temperature dependence of organic matter decomposition: a critical review using literature data analyzed with different models. *Biol. Fertil. Soils* 27:258-262.

Leavitt, S.W., E.A. Paul, E. Pendall, J.R. Pinter and B.A. Kimble. 1997. Field variability of carbon isotopes in soil organic carbon. *Nuclear Instruments and Methods in Physics Research B.* 123:451-454.

Martel, Y.A. and E.A. Paul. 1974. Effects of cultivation on the organic matter of grassland soils as determined by the fractionation and radiocarbon dating. *Can. J. Soil Sci.* 54:419-426.

Morris, S.J. 1999. Spatial distribution of fungal and bacterial biomass in hardwood forest soils in southern Ohio: fine scale variability and microscale patterns. *Soil Biology and Biochemistry* 31:1375-1386.

Morris, S. J. and R.E.J. Boerner. 1999. Spatial distribution of fungal and bacterial biomass in southern Ohio hardwood forest soils: scale dependency and landscape patterns. *Soil Biology and Biochemistry* 31:887-902.

Page, A. L., R.H. Miller and D.R. Keeney (eds.). 1982. *Methods of Soil Analysis, Part 2, Chemical and Microbiological Properties.* 2nd edition. American Society of Agronomy, Madison, WI.

Paul, E.A., R.F. Follett, S.W. Leavitt, A. Halvorson, G.A. Peterson and D.J. Lyon. 1997. Radiocarbon dating for determination of soil organic matter pool sizes and dynamics. *Soil Sci. Soc. Am. J.* 61:1058-1067.

Paul, E.A., H.P. Collins and S. Haile-Mariam. 1998. Analytical determination of soil C dynamics. *Trans.* 16 Int. Congr. Soil Sci., Montpellier, France.

Paul, E.A., D. Harris, H.P. Collins, U. Schulthess and G.P. Robertson. 1999. Evolution of CO_2 and soil carbon dynamics in biologically managed, row-crop agroecosystems. *Applied Soil Ecology,* 11:53-65.

Paustian, K., W.J. Parton and J. Persson. 1992. Modeling soil organic matter in organic-amended and nitrogen-fertilized long-term plots. *Soil Sci. Soc. Am. J.* 56:476-488.

Robertson, G. P. and K.L. Gross. 1994. Assessing the heterogeneity of below ground resources: quantifying pattern and scale. p. 237-253. In: M.M. Caldwell and R.W. Pearcy (eds.), *Exploitation of Environmental Heterogeneity by Plants,* Academic Press, San Diego, CA.

Robertson, G. P., D. Wedin, P.M. Groffman, J. M. Blair, E. Holland, K. J. Nadelhoffer and D. Harris. 1999. Soil Carbon and Nitrogen Availability: Nitrogen Mineralization, Nitrification, and Soil Respiration Potentials. p. 258-271. In: G. P. Robertson, C. S. Bledsoe, D. C. Coleman and P. Sollins (eds.), *Standard Soil Methods for Long-Term Ecological Research.* Oxford University Press, New York.

SAS. 1995. *Statistical Analysis System User's Guide: Statistics*. Version 6.2. SAS Institute, Cary, NC.

Scharpenseel, H.W. and H. Schiffman. 1977. Radiocarbon dating of soils, a review. *Z. Pflanzenernahr. Bodenkd.* 140:159-174.

Six, J., E.T. Elliott, K. Paustian and J.W. Doran. 1998. Aggregation and soil organic matter accumulation in cultivated and native grassland soils. *Soil Sci. Soc. Am. J.* 62:1367-1377.

Six, J., E.T. Elliott and K. Paustian. 1999. Aggregate turnover and SOM dynamics in conventional and no-tillage soils. *Soil Sci. Soc. Am. J.* 63:1350-1358.

Smucker, A.J.M., D. Santos, Y. Kavdir and E.A. Paul. 1998. Concentric gradients within stable soil aggregates. Trans. 16 Int. Congr. Soil Sci., Montpellier, France.

Sollins, P., C. Gassman, E.A. Paul, C. Swanston, K. Lajtha, J.W. Heil and E.T. Elliott. 1999. Soil carbon and nitrogen: pools and fractions. p. 89-105. In: G.P. Robertson, C.S. Bledsoe, D.C. Coleman and P. Sollins (eds.), *Standard Soil Methods for Long-Term Ecological Research.* Oxford University Press, New York.

Starr, J. L., T.B. Parkin and J.J. Meisinger. 1992. Sample size consideration in the determination of soil nitrate. *Soil Sci. Soc. Am. J.* 56:1824-1830.

Stoyan, H., H. De-Polli, S. Bohm, G.P. Robertson and E.A. Paul. 2000. Spatial variability of soil respiration and related soil properties at the plant scale. *Plant and Soil* 219:1-12.

Willson, T.C., O. Schabenberger, E.A. Paul and R.R. Harwood. 2000. Seasonal changes in nitrogen mineralization potential in soils under agricultural management. *Soil Sci. Soc. Am. J.* (in press.)

Zinke, P.J. 1962. The pattern of influence of individual forest trees on soil properties. *Ecology* 43:130-133.

CHAPTER 15

Ecozone and Soil Profile Screening for C-Residence Time, Rejuvenation, Bomb ^{14}C Photosynthetic δ^{13}C Changes

H.W. Scharpenseel, E.M. Pfeiffer and P. Becker-Heidmann

I. Introduction

Several major parameters are decisive in assessing the formation of humic substance and the size of the SOM pool:

- Living biomass and annual biomass input of the ecozones;
- Rate of biomass turnover into SOM;
- SOM residence time;
- Amount/concentration of SOM in the ecozones.

Biomass formation varies, depending on climate factors (mainly temperature and rainfall), atmospheric CO_2 concentration, nutrient availability, type of vegetation and its protein -N content, soil pH, and soil texture, including HAC or LAC clay mineral dominance. Table 1 shows rough estimates of biomass for major ecozones. The rate of turnover of biomass into SOM in many cases can be derived from decomposition studies with uniformly ^{14}C-labelled plant substance (Scharpenseel and Pfeiffer, 1997). Commonly used standard methods (Table 2) can analyze the SOM concentration in an ecozone rather exactly. According to the Century Model (Parton et al., 1987), SOM fractions vary considerably in different landscapes (Woomer et al., 1994). Post et al. (1982) and Degens (1989) indicate that 24.5% of the total Soil-C is in wetlands, 13.7% in tundra, 12.0% in croplands and agricultural areas, 9.5% in wet boreal forests, 9.6% in tropical woodland and savanna, and 8.6% in cool temperate steppes. The 375 million ha of peats contain an additional C-pool of ca 1500 PG C.

^{14}C-dating can detect the mean residence time (MRT) of SOM-C. However, some factors such as rejuvenation by leaching of modern C and bomb-^{14}C entering the soil profile from atomic testing complicate these measurements.

d^{13}C measurement reveals changes in vegetation involving different photosynthetic mechanisms –C3, C4, and CAM (d^{13}C of C3 = ca –25 to 27‰, d^{13}C of C4 = ca – 12 to 14‰ d^{13}C of CAM = ca – 17‰) in the course of the dated period and earlier.

Undoubtedly, if one can afford it, a (thin) layer sampling and ^{14}C dating and d^{13}C MS measurement of the soil profile is superior to single or few sample tests. Mainly respiratory biotic C and O contribute to the development of secondary pedogenic carbonates δ^{13}C and eventually also δ^{18}C.

Table 1. Biomass in ecozones

Ecozone	Ref.	Million ha	Pg C pool	Pg C y^{-1}	10^6 g C ha^{-1} y^{-1}
Land surface	7	12,800	560-600	ca 60	
Tundra	6,9	800	2		1.0-2.5
Woodland (total)	3	4,200	359+		
Temperate and northern forests	3,6,8,9	2,400	276	7.4	12
Tropical forests	3,6,9,10	2,500	306-450	17.7	19
Grassland (total)	6,9	3,200	ca 50		
Temp. grassland	6,9	900	13	5.4	6
Bush savanna	6	2,300		18.4	8
Savanna	9	1,500	27	5.4	
Dryland	4	4,500	1-5	0.5-1.0 pot. C sink	
Tundra	6,9	800	2	0.6	1
Bush dryland	6	1,800			1
Mountain desert	6	2,400			0.03
Cropland	6	1,450		4.2	3
Wetlands	7	1,020		20	
Boreal, temperate wetlands	1,2	7,900			
Swamps and marshes	6	200		6	
Riceland	7,9	136		0.52	3.8

(1) Armentano, 1980
(2) Aselmann and Crutzen, 1990
(3) Dixon et al., 1994
(4) Glenn and Squires, 1993
(5) Neue et al., 1994
(6) Paul and Clark, 1989
(7) Scharpenseel et al., 1995
(8) Sombroek et al., 1993
(9) Whittaker and Likens, 1973
(10) Wisniewski and Lugo, 1992

The SOM-C pool directly relates to biomass input, plus rate of turnover to SOM and SOM stability as expressed by C residence time. A typical example is the frequently observed lower biomass input but higher SOM-C pool of higher C residence time in grassland, compared with higher biomass input but lower SOM-C pool due to lower SOM-C residence time in woodland (both located under similar climate and geomorphology) (Scharpenseel and Becker-Heidmann, 1997).

II. Residence Time of SOM-C

As early as the 1950s and 1960s, scientists were cautiously using the ^{14}C dating method by Arnold and Libby (1949) and Libby (1952) to measure for C residence time in soils.

$$t = - \frac{\ln 2}{T1/2} \ln \frac{(a)}{a_o}$$

Table 2. Carbon in soil organic matter (SOM)

C compartment	Reference	Area 10⁶ ha	C stocks g C m⁻²	C stocks Pg	C stocks % total C	C input, pot. source, sink Pg C y⁻¹
Ice free continuous surface	7	12,800		1550		
Woodland, total	2,10	ca 4200		787		2.5 si
Temperate + N. forests	4,12,17,18	2500	8-11,700			
Tropical forests	4,17,18	1700	13-14,500			
Bush savanna	12,17,18	2300	5.4-12,400			0.2 si
Savanna	1,18	1500		81		
Tundra	9,12,13,18	ca 800	22,000	ca 185	ca 13	0.5-1.0 si
Grassland, total	7,14	3200		300-350		0.3-0.5 si
Temperate grassland	12,14,18	900	23,000	170		0.5-1.0 si
Dryland	8,16	4500		or 250 carb 1200		
Bush dryland	12	1800		110		
Mountain deserts	12	2400		20	ca 12	
Cropland	12,13,14	1450		150-180	14.5	ca 1.0
Wetlands	13,14	1020		650		ca 0.1
Temperate + Boreal wetlands				455		
Swamps and marshes	12,13	200		140		
Riceland	11,15	136		12		C-sink <28°C CH₄ 60 Tg per year, 12-15 stay ca 55 Pg total
SOM-C loss by plowing	3					1.5-2.0 (CO₂ and mineral fertilizer
SOM-C sink by modern agric.	2,6					18-24
Leaf droppings, veget. rel.	2					ca 24 (so)
C from wood clearing	2					ca 0.25 (si)
Lacustrice C-sedimentation	2					ca 1 (si)
C aquatic transpiration to ocean	5,10					

(1) Alexandrov and Oikawa, 1995
(2) Aselmann and Lieth, 1983
(3) Cole, IPCC W. Gr. II, 1994
(4) Dixon et al., 1994
(5) Downing et al., 1993
(6) Esser, 1990
(7) Eswaran et al., 1993
(8) Glenn and Squires, 1993
(9) Tans et al., 1990
(10) Meybeck, 1981
(11) Neue et al., 1989
(12) Paul and Clark, 1989
(13) Post et al., 1982
(14) Scharpenseel, 1995
(15) Scharpenseel et al., 1996
(16) Schlesinger, 1982
(17) Sombroek et al., 1993
(18) Whittaker and Likens, 1973

Deriving an expression of soil age from natural ¹⁴C measurements of SOM-C caused concern mainly because of SOM-C leaching and resultant rejuvenation. Induced ¹⁴C from thermonuclear testing caused new fears. Classic conventional stratigraphic, palynologic (Averdieck, 1978; Behre, 1978; Faegri and Iverssen, 1993), and dendrochronologic (Becker, 1993) dating methods were used for comparison with D¹⁴C measurements.

Methodological improvements of ¹⁴C gas and benzene liquid scintillation counting or accelerator mass spectrometry, with its miniscule sampling technique, led to a widely accepted methodological layout and assessment of induced/involved sources of error (Geyh et al., 1971; Scharpenseel et al., 1992; Trumbore, 1993, 1994).

Obviously a single ^{14}C-date, representing SOM-C residence time, cannot be considered sufficient to calculate an annual rate of biomass converted into SOM-C, since a >0.1% modem C addition per annum would limit the ^{14}C-date of existing SOM-C to less than 1,000 yrs (Becker-Heidmann, 1989). In addition, the limited soil substrate homogeneity requires sampling repetitions. In general, ^{14}C-dating of soils by a set of a few sample measurements permits at best a qualified estimate of apparent mean residence time (AMRT).

A 2 to 5 cm (thin) layer soil profile scan for % C, as well as for D^{14}C and also for δ^{13}C, reveals for the individual (thin) layers the level of C, the corresponding ^{14}C-age D^{14}C and based on the d^{13}C levels any vegetation changes of different photosynthetic mechanism (C3, C4, CAM), their depth.

Relevance, and roughly the residence time of the different C3- or C4- or CAM vegetation regimes, is as follows:

$$\delta^{13}C = \frac{\dfrac{^{13}C}{^{12}C}(sample) \ - \ \dfrac{^{13}C}{^{12}C}(PDB)}{\dfrac{^{13}C}{^{12}C}(PDB)} 1000 \text{ ‰}$$

(PDB = marine carbonate standard of Belemnitella americana of Cretaceous Peed De Formation according to Craig (1953)).

Table 3. ^{14}C-dates of (aquic) Hapludolls (loessic belt, Lower Saxonia, Germany)

Depth (cm)	Soellingen 1 S of Brunswig	Soellingen 2 S of Brunswig	Soellingen/ Jerxheim 1	Soellingen/ Jerxheim 2	Adlum near Hildesheim
10 - 20	1210 ± 70	1340 ± 80	2270 ± 110	1740 ± 70	
20 - 30	2070 ± 80	1920 ± 80		2040 ± 70	1040 ± 60
30 - 40	2560 ± 90	1760 ± 60	2450 ± 80		1690 ± 70
40 - 50	2310 ± 90	1780 ± 80	3120 ± 80	3010 ± 70	1920 ± 70
50 - 60	2830 ± 80	2470 ± 90	3470 ± 80	3790 ± 80	2260 ± 70
60 - 70	3020 ± 80	2680 ± 70	4060 ± 80	4720 ± 80	2770 ± 70
70 - 80	4800 ± 100	3310 ± 70	4320 ± 80	5290 ± 80	3010 ± 70
80 - 90			4060 ± 80	5550 ± 80	4000 ± 80
90 - 100		5300 ± 80			

(*Radiocarbon*, 38, 2, 1996, p. 277, B 26-31; B 98-104; B 33-40; B 106-113, B 121-128)

After the introduction of the ^{14}C dating method for assessing SOM residence time, the risk of ignoring an effect of leaching and, consequently, rejuvenation, quickly led to detailed control measurements in soils of fairly well-known geologic/pedologic age. Table 3 shows a set of loessic Mollisols, which formed during the end of the Boreal/Atlantic phases of the Holocene. Their age gradient to depth and maximum age of 4000 to 5000 years BP, measured already in 1966 at low bomb-C level, complies quite well with age estimates (Scharpenseel et al., 1967; Scharpenseel et al., 1996).

In soils with inherent SOM/clay migration, such as Alfisols, the slightly higher, δ^{13}C level in the argillic horizon, which often is observed, indicates a small rise in level of the heavier ^{13}C isotope

(isotope discrimination) in course of the many times repeated sorption and release processes of leaching SOM-C at the higher number, of clay surfaces of the argillic horizon. The slight enrichment of the heavier isotope, in this case of the ^{14}C, would reduce the carbon's apparent mean residence time (AMRT) somewhat in the argillic horizon and would add some age to the profile zone of slightly lower ^{14}C below the argillic horizon. Additional bomb-^{14}C infiltration, could produce a fairly drastic rejuvenation of the AMRT-measurement, especially in the near surface horizons in sticky gleyey soils, also in deeper parts of the soil profile in the case of courser sandy or volcanic ash soils (see bomb-^{14}C curve over time, Scharpenseel and Becker-Heidmann, 1994). Thus, measurements in various soil profiles and soil substrates indicate a considerable and gradual difference in bomb ^{14}C leaching and effect on AMRT, depending on climate factors, especially level of precipitation, and on soil permeability, especially clay content and also on pH. Table 4 reveals the depth range of bomb ^{14}C penetration and the extent in increase of ^{14}C ecoregions.

Figure 1, which concerns 6 widespread soil orders very important for land use, indicates the ^{14}C age of dated soil samples (dated by Bonn and Hamburg University Labs) vs. soil depth (plus correlation factor and number of dated soil samples). Only soils of (thin) layer scanned and dated profiles are used, to avoid statistically non-existent frequencies of soil samples which might show certain age and depth levels due merely to sampling depth habits.

Table 5 shows the residence time of 129 samples from five soil orders at six depth levels (10, 20, 50, 100, 150, 200 cm) for the whole soil sample pool. Alfisols, Mollisols, and Vertisols most of the time seem to have the highest residence times below the depth of an eventual plowing zone. As a reflection of regional diversity impact two series of Mollisols and Vertisols are plotted separately for their countries of origin in Figures 2 and 3, with identification of the numbers of the regression lines in Tables 6 and 7.

Figures 1-3 and Tables 5-7 clearly indicate, at different soil profile depth levels, signs of an inverse relationship between SOM-C residence time and sustained biomass delivery to the soil-C pool. In many observed cases, this typically leads to higher SOM-C residence time in grassland soils with but moderate annual biomass input, compared with woodland soils having considerably higher rates of annual biomass delivery and SOM-C replenishment (Figure 4) (Scharpenseel and Becker-Heidmann, 1997). The ages vs. depth diagrams and tables show that the C residence time measurement over the whole soil profile is a strong indicator for the soil's/landscape's SOM stability or restoring capacity.

Stuiver and Braiunas (1988), Kerr (1996), and Crowley and Kim (1996) discuss sun climate links with solar components and forcings contributing directly to climate change. They also discuss variations of the "C" record, which could occur on a global scale due to cosmic impacts, changing the specific activity of the naturally formed ^{14}C (^{14}N (n,p) ^{14}C +â radiation).

III. Soil Profile Scan for SOM-C Residence Time ($D^{14}C$) plus SOM-C History ($\delta^{13}C$)

The proceedings of an IAEA Symposium (IAEA, 1994) collected key information on stable isotope use in the soil plant system. Boutton and his coworkers contributed much about the orbit of use of ^{13}C (Boutton et al., 1993), its existence in atmospheric, terrestrial, marine, and freshwater environments (Boutton, 1991), and its use in identifying vegetation and climate change events in regions like Utah (Dzurez et al., 1985), Arizona (McPherson et al., 1993), and Central Texas (North et al., 1994). Martin et al. (1990) reported on landscape changes in the Ivory Coast, reflected in the δ^{13}C measurements.

Table 4. Bomb-^{14}C in thin layer wise ^{14}C dated and δ^{13}C scanned soil profiles

Lab code HAM	Soil max. cm depth	Bomb-C cm depth	Mod.+bomb ^{14}C-max	Max. SOM- ^{14}C-age	Max. age cm depth	Bomb-C- range-δ^{13}C	AMRT range δ^{13}C
Temperate climate and subtropics							
Germany							
Ohlendorf A 1652-1690	78	6	113	3170 ± 90	68 - 70	-28.0 to -26.8	-27.1 to -25.5
Ohlendorf B 1910-1973	146	7	107	2730 ± 80	111 - 114	-26.7 to -26.1	-26.6 to -24.0
Wohldorf 1750-1806	110	11	145	2340 ± 90	82 - 84	-30.4 to -27.1	-28.8 to -25.0
Timmendorf 1860-1901	84	10	119	2730 ± 60	74 - 76	-26.8 to -26.5	-26.7 to -24.7
Trittau forest 1704-1734	60	10	101	510 ± 60	48 - 50	-27.6 to -26.7	-26.8 to -24.2
Israel							
Akko 2501-2507	187	15	111	7080 ± 100	183 - 187	-26.5 to -24.7	-25.2 to -21.8
Qedma 2552-2614	250	30	110	14,390 ± 150	148 - 152	-25.0 to -18.9	-22.7 to -12.0
Tropics, especially riceland							
India							
Patencheru 1975-2053	158	12	115	6100 ± 90	156 - 158 also 148-150	-17.3 to -14.9	-18.1 to -10.8
Philippines							
Los Banos 2147-2187	82	82	125	150 ± 50	70 - 72	-23.0 to -15.7	-16.9 to -16.6
Paugil 2365-2399	70	18	124	2180 ± 80 5470 ± 60	66 - 68	-22.5 to -21.4	-22.9 to -19.7
Pao 2330-2364	88	16	117	1230 ± 70	86 - 88	-18.7 to -14.9	-15.7 to -13.5
Bugallon 2315-2329	30	16	117	3630 ± 90	28 - 30	-22.0 to -24.5	-13.9 to -13.3
Tiaong L 2188-2220	66	24	118	1020 ± 50	36 - 38	-22.2 to -16.6	-25.2 to -18.9
Tiaong H 2222-2248	54	54	121	360 ± 60	44 - 46	-20.2 to -15.8	-17.7 to -15.8
San Dionisio 2249-2301	510	20	122	28,600 ± 1040	390 - 410	-19.4 to -17.4	-23.0 to -12.2
China							
Namtou Hsien 2661-2688	80	18	108	1800 ± 100	78 - 80	-26.3 to -21.7	-20.6 to -14.7
Pingtung 2700-2718	100	16	112	5700 ± 100	90 - 100	-23.1 to -22.1	-21.9 to -17.2
Chum Pae 2401-2467	92	14	119	1460 ± 70	90 - 92	-20.7 to -19.4	-21.4 to -18.1
Thailand							
Klong Luang 2468-2499	64	16	106	3450 ± 90	60 - 62	-22.6 to -21.2	-22.2 to -20.2
Tachiat 1 3070-3100	62	24	107	1440 ± 70	54 - 56	-25.9 to -23.9	-25.9 to -24.0
Tachiat 2 3101-3129	58	30	110	1060 ± 70	54 - 56	-26.1 to -24.8	-27.7 to -26.1
Tonsang 2837-2915	42	28	122	490 ± 60	36 - 38	-22.0 to -19.3	-21.1 to -20.0

(From Becker-Heidmann et al., *Radiocarbon*, 1996.)

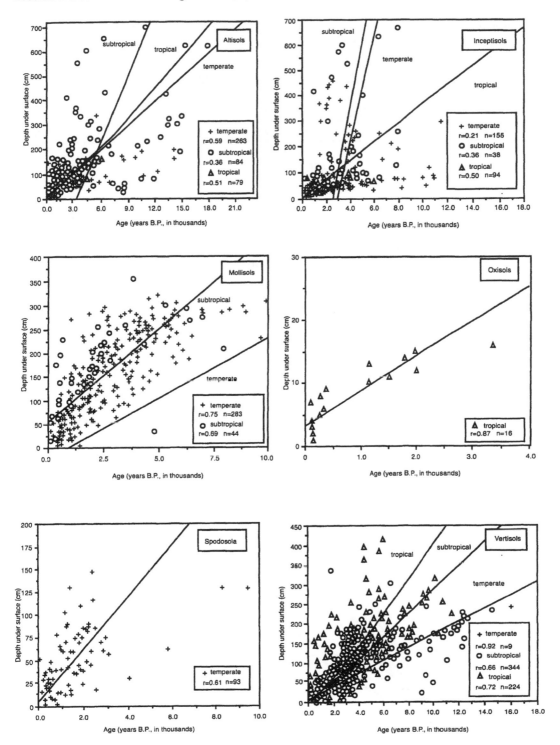

Figure 1. C residence time of (thin) layer ^{14}C-dated soil profiles of 6 major soil orders (regression indicated separately for temperate, subtropical and tropical origin.

Table 5. Average regression; correlation factor and corresponding ^{14}C-age (AMRT = apparent mean residence time) for different depth levels of layerwise dated soil profiles (13 Alfisols, 16 Inceptisols, 47 Mollisols, 9 Spodosols, 44 Vertisols). Paleosols were not included; soil profiles sampled in different W-, E-, S- European countries, in Argentinia, Australia, Israel, Sudan and Tunisia

Soil order	Ascend Regression line Corr. F.	AMRT of regression line (years B.P.)					
		10 cm	20 cm	50 cm	100 cm	150 cm	200 cm
Alfisols	0.4651						
	0.739	480	960	2400	4800	7200	9600
Inceptisols	0.0225						
(Plaggepts)	0.209	870	920	1000	1160	1350	1490
Mollisols	0.4695						
	0.888	750	1240	2700	5150	8050	10000
Spodosols	0.0747						
	0.332	1350	1430	1680	2100	2520	2930
Vertisols	0.4014						
	0.772	0	410	1620	3650	5670	7700
All soils ALF+EPT+OD+ OLL+ERT	0.4415 0.772	460	920	2300	4600	6900	9200

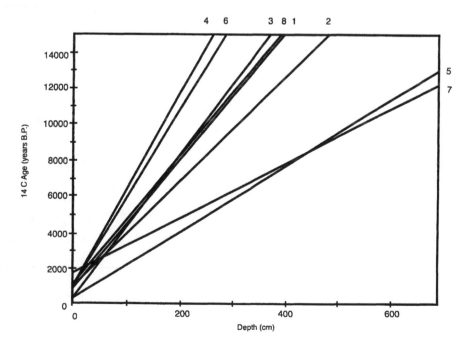

Figure 2. Mollisol dates; regression lines represent countries of origin (see Table 6, 1–8).

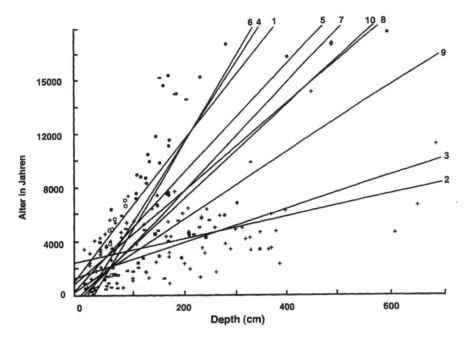

Figure 3. Vertisol dates; regression lines represent countries of origin (see Table 7, 1–10).

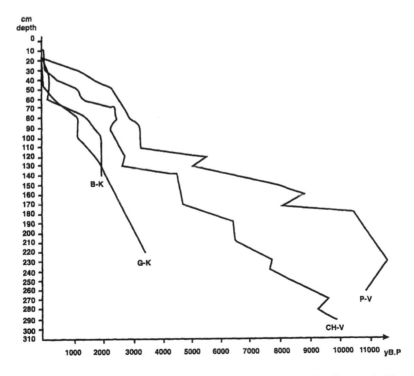

Figure 4. Australian Vertisols and Krasnozems under forest or grassland, sampled by G.D. Hubble, CSIRO, Queensland; P-V, Paget Vertisol under grassland; CH-V, Chinchilla Vertisol under Acacia harpophylla forest; G-K, Gabbinbar Krasnozem, wooded grassland; B-K, Beechmont Krasnozem under subtropical rainforest.

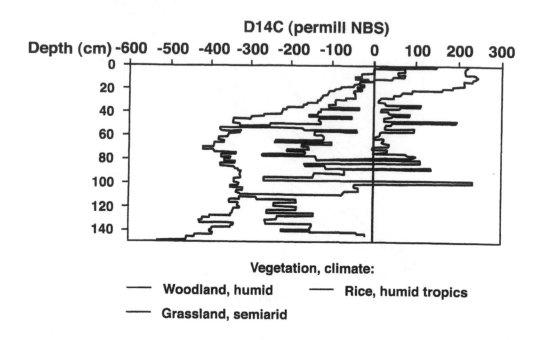

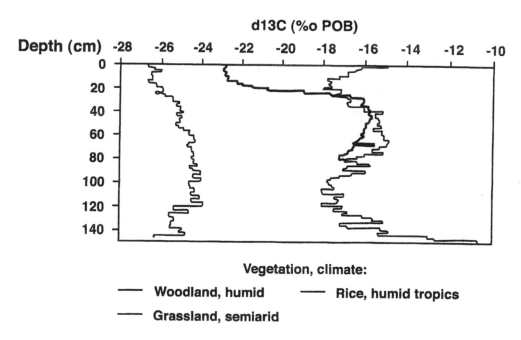

Figure 5. $D^{14}C$ and $\delta^{13}C$ 2 cm thin layer soil profile scan of typical woodland, grassland, and riceland soils. Features: bomb-^{14}C entrance in soil profiles, riceland deepest, grassland the least $\delta^{13}C$, high rise in riceland's deeper layer with methanogenesis small rise in woodland, especially in argillic horizon by isotope discrimination in course of SOM migration (repeated sorption and release) cropland sorghum, (C4, -12 to -14 ‰) originally woodland (C3, -25 ‰), phases of grassland (mix of -25 and -12 to -14 ‰), at greatest depth begin of free carbonates with $\delta^{13}C$ tending to ± 0.

Table 6. Regression equations for age vs. depth of ^{14}C-dated Mollisol samples from different countries of origin (*=0.95, **=0.99, ***=0.999 as levels of significance)

Code	Country of origin	N	Regression equation and significance of correlation for age versus depth, N = number of samples	
1	Germany	115	Y = 35.5 X + 810	R = 0.69***
2	Tschechien	38	Y = 28.4 X + 1175	R = 0.89***
3	Russia	15	Y = 39.5 X + 284	R = 0.93***
4	Hungary	17	Y = 49.0 X + 893	R = 0.69**
5	Australia	24	Y = 18.0 X + 364	R = 0.56**
6	Bulgaria	10	Y = 54.3 X + 818	R = 0.97***
7	Tunisia (recent)	56	Y = 14.9 X + 1719	R = 0.70***
8	Tunisia (covered)	16	Y = 35.3 X + 1107	R = 0.74**

Table 7. Regression equations for age vs. depth of ^{14}C-dated Vertisol samples from different countries of origin (*=0.95, **=0.99, ***=0.999 as levels of significance)

Code	Country of origin	N	Regression equation and significance of correlation for age versus depth, N = number of samples	
1	Germany	18	Y = 48.07 X + 1469.6	R = 0.71***
2	Sudan	97	Y = 8.22 X + 2387.3	R = 0.47
3	Tunisia	26	Y = 12.21 X + 1341.3	R = 0.66***
4	Argentina	11	Y = 58.32 X - 755.4	R = 0.67*
5	Israel	31	Y = 39.79 X + 801.5	R = 0.85***
6	Bulgaria	12	Y = 63.01 X - 1735.5	R = 0.89***
7	Italy	65	Y = 39.42 X - 436.9	R = 0.69***
8	Spain	29	Y = 32.23 X + 1026.9	R = 0.79***
9	Portugal	30	Y = 24.53 X + 541.9	R = 0.54**
10	Australia	59	Y = 34.55 X - 80.3	R = 0.76***

Our lab in Hamburg conducted a number of studies based on (thin) layer δ^{13}C (plus in some cases δ^{18}O) scans of soil profiles (SOM δ^{13}C as well as secondary soil carbonate crust's δ^{13}C) (Becker-Heidmann et al., 1996; Scharpenseel and Becker-Heidmann, 1997, 1994).

D^{14}C plus δ13C soil profile combinations derived from woodland, grassland, and riceland (Figure5) clearly reflect rigid vegetation changes in the grassland profile (C3-woodland vs. C4-grassland vs. C3-cropland vs. C4-sorghum dominated cropland), which could be consequences of climate changes or anthropogenic land use impacts. Moderate δ^{13}C isotope discrimination, mainly in the argillic horizon of the woodland profile, as well as prominent ^{13}C enrichment in the methanogenic zone of the riceland soil profile, are typical phenomena. Figure 6a reflects the latter, showing tile δ^{13}C scan of a river Elbe marsh soil (at Assel) with maximum methanogenesis and highest δ^{13}C level occurring at the 70 to 110 cm depth. Figure 6b shows an analog maximum of cellular activity by methanogenic bacteria.

References

Alexandrov, G.A. and T. Oikawa. 1995. Net ecosystem production resulted from CO_2 enrichment: Evaluation of potential response of a savannah ecosystem to global changes in atmospheric composition. p.117-119. In: *Proceedings of the Tsukuba Global Carbon Cycle Workshop*. Tsukuba, Japan.

Armentano, T.V. (ed.). 1980. The role of organic soils in the world carbon cycle. CONF-7905135. United States Department of Energy, Washington, D.C.

Arnold, J.R. and W.F. Libby. 1949. Age determinations by radiocarbon content: Checks with samples of known age. *Science* 110:678-680.

Aselmann, I. and H. Lieth. 1983. Implementation of agricultural productivity into existing global models of primary productivity. In: E.T.Degens, S. Kempe and H. Soliman (eds.), *Transport of Carbon and Minerals in Major World Rivers. Vol. II*, Mitt. Geol. Paläontol Inst. Univ. Hamburg, Special Publication of SCOPE/UNEP.

Aselmann, J. and P.J. Crutzen. 1990. A global inventory of wetland distribution and seasonality, net primary productivity and estimated methane emissions. p.441-455. In: A.F. Bouwman (ed.), *Soils and the Greenhouse Effect*. John Wiley & Sons, New York.

Averdieck, F.R. 1978. Palynologischer Beitrag zur Entwicklungsgeschichte des Großen Plöner Sees und der Vegetation seiner Umgebung. *Archiv f. Hydrobiologie* 83:1-46.

Becker-Heidmann, P. 1989. Die Tiefenfunktionen der natürlichen Kohlenstoff-Isotopengehalte von vollständig dünnschichtweise beprobten Parabraunwerden (Hapludalf) und ihre Relation zur Dynamik der organischen Substanz in diesen Böden. (Dissertat.) *Hamburger Bodenkundliche Arbeiten* 13:1-228.

Becker-Heidmann, P., H.W. Scharpenseel and H. Wiechmann. 1996. Hamburg radiocarbon thin layer soils database. *Radiocarbon* 38:295-345.

Becker, B. 1993. An 11,000 year German oak and pine dendrochronology for radiocarbon calibration. *Radiocarbon* 35:201-213.

Behre, K.E. (ed.). 1978. Beiträge zur Paläo-Ethnobotanik von Europa. Fischer, Stuttgart.

Boutton, T.W. 1991. Stable carbon isotope ratios of natural materials: II. Atmospheric, terrestrial, marine, and freshwater environments. In: *Carbon Isotope Techniques*, Academic Press, San Diego.

Boutton, T.W., L.C. Nordt, S.R. Archer, A.C. Midwood and I. Casar. 1993. Stable carbon isotope ratios of soil organic matter and their potential use as indicators of paleoclimate. p. 445-459. In: Proceedings International Symposium of IAEA, International Atomic Energy Agency, Vienna. p. 445-459.

Cole, C.F. 1994/1995. IPCC W.Gr. II. Soils Chapter in IPCC-Report.

Craig, H. 1953. The geochemistry of stable carbon isotopes. *Geochim. Cosmochim. Acta* 3:53-92.

Crowley, T.J. and K.Y. Kim. 1996. Comparison of proxi records of climate change and solar forcing. *Geophysical Research Letters* 23:359.

Degens, E.T. 1989. *Perspectives on Biogeochemistry*. Springer-Verlag, Berlin, 423 pp.

Dixon, R.K., S. Brown, R.A. Houghton, A.M. Solomon, M.C. Trexler and J. Wisniewski. 1994. Carbon pools and flux of global forest ecosystems. *Science* 263:185-190.

Downing, J.P., M. Meybeck, J.R. Orr, R.R. Twilley and H.W. Scharpenseel. 1993. Land and water interface zones. *Water, Air and Soil Pollution* 70:123-137.

Dzurec, R.S., T.W. Boutton, M.M. Caldwell, and B.N. Smith. 1985. Carbon isotope ratios of soil organic matter and their use assessing community composition changes in Curlew Valley, Utah. *Oecologia* (Berlin) 66:17-24.

Esser, G. 1990. Modelling global terrestrial sources and sinks of CO_2 with special reference to soil organic matter. p. 247-263. In: A.F. Houwman (ed.), *Soils and the Greenhouse Effect*. John Wiley & Sons.

Eswaran, H., E. Van Den Berg and P.F. Reich. 1993. Organic carbon in soils of the world. *Soil Sci. Soc. Am. J.* 57:192-194.

Faegri, V. and J. Iversen. 1993. Bestimmungsschlüssel für die nordwesteuropäische Pollenflora. G. Fischer, Jena, Germany.

Geyh, M.A., J.H. Benzler and G. Röschmann. 1971. Problems of dating Pleistocene and Holocene soils by radiometric methods. p. 63-75. In: D.H. Yaalou (ed.), *Paleopedology – Origin, Nature and Dating of Paleosols.* Jerusalem Press, Jerusalem.

Glenn, E., V. Squires, M. Olsen and R. Freye. 1993. Potential for carbon sequestration in the drylands. *Water, Air and Soil Pollution* 70:341-355.

IAEA and FAO. 1994. Nuclear techniques in soil-plant studies for sustainable agriculture and environmental preservation. Symposium Proceedings. Vienna, IAEA Printing.

Keff, R.A. 1996. New dawn for sun-climate links? *Science* 271:1360-1361.

Martin, A., A. Mariotti, J. Balesdent, P. Lavelle and R. Vuattoux. 1990. Estimate of organic matter turnover rate in a savanna soil by ^{13}C natural abundance measurements. *Soil Biol. Biochem.* 22:517-523.

McPherson, G.R., T.W. Boutton and A.J. Midwood. 1993. Stable carbon isotope analysis of soil organic matter illustrates vegetation change at the grassland/woodland boundary in southeastern Arizona, USA. *Oecologia* 93:95-101.

Meybeck, M. 1981. Flux of organic carbon by rivers to the oceans. p. 219-269. In: National Technical Information Service, Springfield, VA.

Neue, H.U., P. Becker-Heidmann and H.W. Scharpenseel. 1989. Organic matter dynamics, soil properties and cultural practices in ricelands and their relationship to methane production. p. 457-466. In: A.F. Bouwman (ed.), *Soils and the Greenhouse Effect.* John Wiley & Sons, New York.

Neue, H.U., J.L. Gaunt, Z.P. Wang, P. Becker-Heidmann and C. Quijano. 1994. Carbon in tropical wetlands. p. 201-220. In: Transactions 15th World Congress Soil Science, Acapulco, Vol. 9, Supplement.

Nordt, L.C., T.W. Boutton, C.T. Hallmark and M.R. Waters. 1994. Late quarternary vegetation and climate changes in Central Texas based on the isotopic composition of organic carbon. *Quarternary Research* 41:109-120.

Parton, W.J., D.S. Schimel, C.V. Cole and D.S. Ojima. 1987. Analysis of factors controlling soil organic matter levels in Great Plains grasslands. *Soil Sci. Soc. Amer. J.* 51:1173-1179.

Paul, E.A. and E.E. Clark. 1989. *Soil Microbiology and Biochemistry,* Academic Press, San Diego. 273 pp.

Post, W.M., W.R. Emanuel, P.J. Zinke and A.G. Stangenberger. 1982. Soil carbon pools and world life zones. *Nature* 298:156-159.

Scharpenseel, H.W., M.A. Tamers and F. Pietig. 1967/68. Altersbestimmungen von Böden durch die Radiokohlenstoffdatierungsmethode. II. Eigene Datierungen. *Pflanzenernährung und Bodenkunde* 119:44-52.

Scharpenseel, H.W. and P. Becker-Heidmann. 1992. Twenty-five years of radiocarbon dating soils: Paradigm of erring and learning. *Radiocarbon* 34:541-549.

Scharpenseel, H.W. and P. Becker-Heidmann. 1994. ^{14}C-dates and ^{13}C-measures of different soil species. In: R. Lal, J. Kimble and E. Levine (eds.), *Soil Processes and Greenhouse Effect.* USDA, Soil Conservation Service, National Soil Survey Center, Lincoln, NE.

Scharpenseel, H.W., J. Freytag and E.M. Pfeiffer. 1995. The carbon budgets in drylands: assessments based on carbon residence time and stable isotope formation. In: V.R. Squires, E.P. Glenn and A.T. Ayoub (eds.), *Proceedings UNEP Workshop on Combatting Global Climate Change by Combatting Land Degradation.* UNEP, Nairobi.

Scharpenseel, H.W. and E.M. Neiffer. 1995. Carbon cycle in the pedosphere. In: *Proceedings of the Tsukuba Global Carbon Cycle Workshop.* CGER-Report ISSN 1341-4356 1018-'95.

Scharpenseel, H.W., F. Pietig, H. Schiffmann and P. Becker-Heidmann. 1996. Radiocarbon dating of soils: database contribution by Bonn and Hamburg radiocarbon labs. *Radiocarbon* 39:277-293.

Scharpenseel, H.W., L.M. Pfeiffer and P. Becker-Heidmann. 1996. Organic carbon storage in tropical hydromorphic soils. p. 361-392. In: M.R. Carter and B.A. Stewart (eds.), *Structure and Organic Matter Storage in Agricultural Soils*. Advances in Soil Science. Lewis Publishers, Boca Raton, FL.

Scharpenseel, H.W. and E.M. Pfeiffer. 1997. Carbon turnover in different climates and environments. p. 577-590. In: R. Lal, J. Kimble, R.F. Follett and B.A. Stewart (eds.), *Soil Processes and the Carbon Cycle*, Advances in Soil Science. Lewis Publishers, Boca Raton, FL.

Scharpenseel, H.W. and P. Becker-Heidmann. 1997. Carbon sequestration by grassland and woodland soils of different climate zones as revealed by (thin)layer wise carbon-14 dating. p. 9-3 to 9-5. In: *Proceedings XVIII. Intern. Grassland Congress*, Winnipeg-Saskatoon, Canada. ID No. 52.

Scharpenseel, H.W., A. Mtimet and J. Freytag. 1998. Soil inorganic carbon and global change. p. 27-42. In: R. Lal, J. Kimble, H. Eswaran and B.A. Stewart (eds.), *Global Climate Change and Pedogenic Carbonates*. Lewis Publishers, Boca Raton, FL.

Schlesinger, W.H. 1982. Carbon storage in the caliche and arid soils, a case study from Arizona. *Science* 133:247-255.

Sombroek, W.G., F.O. Nachtergaele and A. Hebel. 1993. Amounts, dynamics and sequestering of carbon in tropical and subtropical soils. *AMBIO* 22:417-425.

Stuiver, M. and T.F. Brayiunas. 1988. The solar component of the atmospheric ^{14}C record. p. 245-266. In: F.R. Stephenson and A.W. Wolfendale (eds.), *Secular and Geomagnetic Variations in the Last 10,000 Years*. Kluwer Academic, Dordrecht.

Tans, P.P., L.Y. Fung and T. Takahashi. 1990. Observational constraints on the global atmospheric budget. *Science* 247:1431-1438.

Trumbore, S.E. 1993. Comparison of carbon dynamics in tropical and temperate soils using radiocarbon measurements. *Global Biogeochemical Cycles* 7:275-290.

Trumbore, S.E. 1994. Applications of accelerator mass spectrometry to soil science. AMS Laboratory, Department of Geosciences, University of California, Irvine.

Whittaker, R.H. and G.E. Likens. 1973. The primary production of the biosphere. *Human Ecology* 1:299-369.

Woomer, P.L., A. Martin, A. Albrecht, D.V.S. Resck and H.W. Scharpenseel. 1994. The importance and management of soil organic matter in the tropics. p. 47-80. In: P.L. Woomer and M.J. Swift (eds.). *The Biological Management of Tropical Soil Fertility*. John Wiley & Sons, New York.

CHAPTER 16

Use of [13]C Isotopes to Determine Net Carbon Sequestration in Soil under Ambient and Elevated CO$_2$

W.R. Horwath, C. van Kessel, U. Hartwig and D. Harris

I. Introduction

The limitation in determining soil C sequestration under elevated CO$_2$ conditions is the accuracy of the methods used to assess changes in total soil C. Methodologies to determine changes in soil C range from mass balance approaches to C isotope dilution studies. An accurate measure of soil bulk density is required by these methods to assess changes in soil C. Both approaches are direct but yield different types of information. The mass balance approach yields information on changes in the total soil C pool. However, small changes in total soil C are difficult to detect because C entering the soil through decomposition processes is small compared to the background soil C (Hungate et al., 1996). The ^{13}C isotope dilution approach can overcome this problem by yielding additional kinetic information that describes the rate of C accumulation, turnover of recently added C, and pools of C within the total soil C pool (Balesdent and Mariotti, 1996). In combination, the mass balance and ^{13}C isotope dilution methods provide information that can accurately determine the size and kinetics of standing soil C pools that are influenced by such factors as cultural practices or elevated CO$_2$. However, both methods are limited by factors that affect soil variability, such as soil heterogeneity.

The power of the stable C isotope approach is in its ability to follow the gain or loss of new C entering the soil (Balesdent et al., 1988). Depending on its source, the CO$_2$ used for elevated CO$_2$ experiments, such as in open top chambers or FACE (Free Atmospheric Carbon Enrichment), has a $\delta^{13}C_{PDB}$ value between -35 to -50 ‰ (Leavitt et al., 1996; Nitschelm et al., 1997). Upon a 1:1 mixing with atmospheric CO$_2$, the $\delta^{13}C_{PDB}$ of the elevated CO$_2$ obtains a ^{13}C signature of about -22 to -29 ‰. Plants grown under the mixed gas elevated CO$_2$ are depleted in ^{13}C compared to plants exposed to ambient CO$_2$ conditions. A simple isotopic mixing model can then be used to calculate the percentage of new C entering the soil from the plant material labeled under ^{13}C depleted elevated CO$_2$ (Balesdent et al., 1988). For example, a cotton field soil exposed to elevated and depleted [13]CO$_2$ for 3 years was found to contain 10% new C (Leavitt et al., 1994).

The purpose of this study was to evaluate the amount of new C entering the soil in a fertilized pasture system. In this study we used a FACE system to maintain an elevated CO$_2$ regime depleted in ^{13}C to assess both the amount and rate at which new C entered the soil.

II. Materials and Methods

A. Site and Treatment Description

The FACE experiment is located at 550 m above sea level at the Swiss Federal Institute of Technology (ETH) field station at Eschikon, Switzerland. Three ambient CO_2 rings and three elevated CO_2 rings were established in a random block design on a clay loam soil classified as an eutric cambisol. The soil contained 31% sand, 38% silt and 31% clay. Soil organic C content averaged 2.5% and the pH was 7.0.

The control (i.e., ambient) CO_2 rings were exposed to ambient (35 Pa pCO_2) air having a $^{13}CO_2$ signature of $-8‰$ (Figure 1). The three CO_2-fumigated rings were maintained at 60 Pa pCO_2 during daylight hours and for the duration of the growing season. The source of CO_2 in the fumigated rings was highly depleted in ^{13}C ($-45‰$) and when mixed with ambient air resulted in a $^{13}CO_2$ signature of approximately $-18‰$.

Monoculture plots (2.8 by 1.9 m), each of perennial ryegrass (*Lolium perenne*) and red clover (*Trifolium repens)*, were established within each ring. Both species are C3 plants. The plots received either a low (14 g N m^{-2} yr^{-1}) or high (56 g N m^{-2} yr^{-1}) level of NH_4NO_3 fertilizer and were cut four times a year. Fertilization was equally split into four applications usually following the cutting of the sward. Plants fixing the ambient CO_2 had a ^{13}C value of $-28‰$, while the plant in the fumigated rings had a ^{13}C value of $-38‰$ (data not shown).

B. Soil and Plant Sampling and Preparation

Plants were sampled at each harvest and soil samples were taken in the fall. Soils were sampled to 10 cm depth using a 5-cm diameter corer. Soil samples were air-dried, crushed by hand, sieved through a 2-mm sieve, ball milled to < 53μm and leached with 3 M HCl to remove carbonates. Plant and root samples were washed free of adhering soil and subsequently oven-dried at 60°C. Litter and root samples were milled to <53μm.

C. ^{13}C Analysis

The isotopic composition ($\delta^{13}C$) of plant and soil samples was determined in an automated single-inlet isotope ratio mass spectrometry system. Results of the C isotope analysis were determined in relation to Pee Dee Belemnite (PDB). The fraction of new carbon, f, was calculated as follows:

$$f = \frac{\delta_e - \delta_0}{\delta_1 - \delta_0}$$

where δ_0 and δ_e refer to soil organic C at the beginning and at the end of the experiment, and δ_1 is the ^{13}C signal of red clover and ryegrass plant material. The average above-ground $\delta^{13}C$ values of the plants within CO_2 treatment collected throughout the growing seasons were used to obtain the δ_1 value (Balesdent et al., 1988).

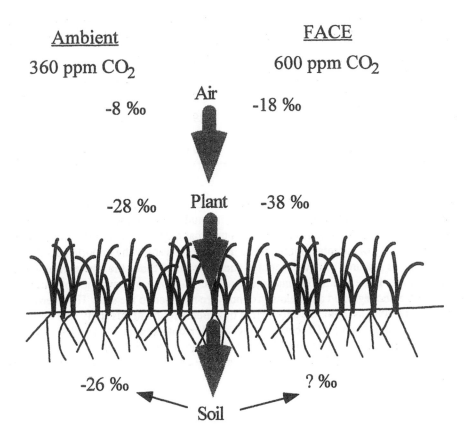

Figure 1. The change in the plant and soil ^{13}C signature under ambient and the more depleted ^{13}C signature found under FACE conditions.

D. Data Analysis

Consistent differences in total soil C among ryegrass and red clover were not found so their results were pooled. New C input was calculated using the following exponential function based on a single pool substitution model:

$$C_{new} = C_{max} * \left(1 - e^{-k*year}\right)$$

where C_{max} is the maximum amount of new C in the soil and k is the decomposition rate constant.

III. Results and Discussion

A. Fraction of New Carbon in the Soil

By the third year of elevated CO$_2$ treatment, the average fraction of new C in the soil in the 0 to 10 cm depths for all treatments reached 0.24 after which there was no further increase (Table 1). The effect

Table 1. The fraction of new C in the soil (f-value) calculated for each year for the low and high fertilizer N treatments in the fumigated rings; standard error of the mean shown in parentheses (n = 6)

	Year of elevated CO_2				
	94	95	96	97	98
			f		
High N	0.15 (0.02)	0.24 (0.02)	0.29 (0.04)	0.24 (0.03)	0.22 (0.02)
Low N	0.15 (0.02)	0.24 (0.02)	0.25 (0.02)	0.22 (0.02)	0.21 (0.02)

of fertilization on the fraction of new C in the soil was non-significant. The main source of new C in the soil in the 0 to 10 cm soil depth will be roots and exudates since the aboveground plant material was removed for forage. Hebeisen et al. (1997) found that the top 10 cm of the soil in these systems contained the majority of the total root biomass. These results show the importance of root exudates and root residues as a source of new C input in these systems. Since the aboveground biomass was removed for forage, these intensively managed pasture systems may have the potential to sequester more C if both the aboveground and below ground were returned to the soil. The average fraction of new C in the soil (24%) for both low and high N treatments found in this study was two to four times higher than those found in a FACE cotton project in Arizona (Leavitt et al., 1994). Differences in climate and crop species most likely explain these differences. In Arizona, the warmer climate, tillage versus no tillage in the Swiss FACE, annual versus a perennial crop and a shorter growing season for cotton could lead to less C sequestration in soil.

B. Quality of New Soil C

New C input to soil and its stabilization will depend on the ability of the decomposer community to cope with a change in residue quality and amount of C entering the soil. An increase in atmospheric CO_2 can cause an increase in plant C to N ratio through limitation in N availability (Ball, 1997). The importance of litter quality in decomposition processes under elevated CO_2 has been contradictory. Bazzaz (1990) summarized studies on litter quality or increased C input under elevated CO_2 and concluded that C storage in soil should increase. Similarly, other studies have concluded that increases in the C to N ratios of plant residues, lignin to N and lignin to P ratios decrease decomposition rates and should lead to increased C input to soil (O'Neill, 1994). However, other studies have found minimal effect of residue quality on decomposition rates under elevated CO_2 (Torbert et al., 1996). Most of these studies of residue quality have been conducted on natural ecosystems where nutrient limitation might be expected. The intensively managed Swiss FACE experiment is unlikely to face nutrient limitation.

C. Calculating New Soil C

Models that simulate the decomposition of C substrates in soil rely on first order reaction kinetics using single or double pool substitution functions (Jenkinson and Rayner, 1977). Figure 2 depicts an increase of soil C from an increase in plant C input of 125% seen in this study under elevated CO_2 conditions (data not shown). The model is based on two values consisting of the present total soil C

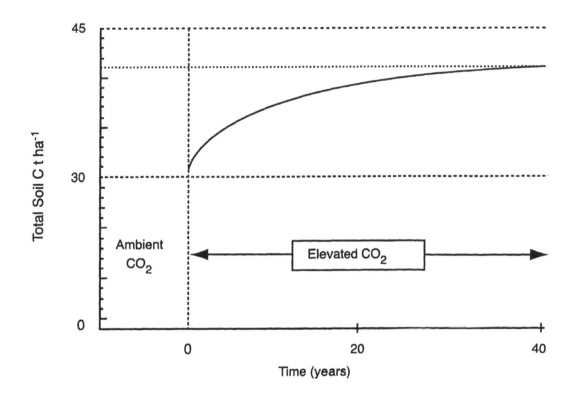

Figure 2. Predicted increase in soil C under elevated CO_2 conditions at the Swiss FACE plots. The model assumes an increase of 25% in plant production and no change in the decomposition rate constant.

(30 t ha[-1], Table 2) and an annual C input under ambient CO_2 conditions of 3.5 t ha[-1] at the Swiss FACE site. Using a single pool substitution function, soil C should increase from 30 to 41 t ha[-1] within 40 years of doubling atmospheric CO_2 levels using a decomposition rate constant of k = 0.117. The model assumes that the decomposition rate constant remains unchanged following the increase in C input from increased production following an increase in CO_2 levels (Jenkinson and Rayner, 1977). The mean residence time (MRT) of the total C pool would be 8 years indicating that it should be possible to detect changes in soil C within 4 to 10 years of doubling the atmospheric CO_2 level.

In our study, no statistically significant (p = 0.17) change in total soil C was detected after 6 years of elevated CO_2 (Table 2), though there were indications that that this might be possible in the near future. There was no effect of nitrogen on soil C accrual. Accurate estimates of soil C sequestration in the field under elevated CO_2 let alone ambient conditions is a complex task (Hungate et al., 1996). Reliable measurements of soil bulk density are required to differentiate small changes soil C in comparison to large standing C pools in soils. Leaching the soil with HCl to remove carbonates may have removed new labile C leading to an underestimation of new C sequestration. In addition, soil het-

Table 2. Average total soil (0-10 cm) C under ryegrass and red clover after 6 y of elevated CO_2 and fertilization; standard error of the mean shown in parentheses (n = 6)

Treatment	Fertilizer	t C ha^{-1}
Ambient		
	Low N	28.6 (3.2)
	High N	28.7 (3.8)
Elevated CO_2		
	Low N	29.3 (3.2)
	High N	31.0 (3.7)

erogeneity introduces a large variability component, which often tests the resilience of statistical techniques to detect small differences in total soil C on a spatial or temporal scale. For these reasons, detecting small changes in soil C remains a formidable task.

The accumulation of new C in the soils exposed to elevated CO_2 was determined using the value for the fraction of new C in the soil described above and then applied to a single exponential substitution function. In both the low and high fertilizer treatments, the amount of new C entering the soil plateaus after 4 years of elevated CO_2 (Figures 3 and 4). The size of the new C pool in the low N treatment is 7.1 t C ha^{-1} and it has a mean residence time (MRT) of 1.4 years. In the high fertilizer treatment, the size of the new C pool is 8.2 t C ha^{-1} and it has an MRT of 1.5 years. These results indicate that the new C entering the soil represents approximately 25% of the total C pool. The short MRT indicates that the new C cycles rapidly and represents a very labile pool of C.

Conceptual models divide SOM into fractions with different turnover rates to simulate soil C and nutrient dynamics (van Veen et al., 1984). Jansson (1958) postulated that two decomposable SOM fractions, called active and passive, could be used to describe the availability of soil N. Jenkinson and Rayner (1977) applied the same concept to soil C pools and considered these pools had distinct turnover rates. The active fraction contains easily mineralizable organic matter consisting of recently deposited organic material, including fine roots, fungal hyphae, and microbial products (Tisdall and Oades, 1982). The labile or active C pool is considered to have a turnover rate of less than a few months, while the passive soil organic pool contains older or protected C with turnover rates of greater than 1 year (Paul and Clark, 1996). Paul et al. (1997) suggest that a third pool of protected C, representing half the soil C, with an age of greater than 1000 years. The third pool is considered stable and does not contribute to short-term (50 to 100 years) soil processes.

The rapidly cycling C pool described by our research is characterized by an MRT of 1.5 years or less. The rapid turnover rate indicates that the pool is comprised of labile C and possibly small components of more resistant C. The labile pool of C was replaced with new C within 4 years of starting elevated CO_2 conditions (Figures 3 and 4). The replacement of this rapidly cycling pool of C occurred without a significant increase in total soil C (Table 2). The lack of increase in soil C is perplexing since above ground production increased on average 20% for red clover and 7% for ryegrass under the first 2 years of elevated CO_2 (Hebeisen et al., 1997). Hebeisen et al. (1997) observed an increase in the root to shoot ratio of ryegrass but not in the red clover. Overall below-ground production would have most likely increased through increased root exudation under elevated CO_2 (Darra, 1996). The results indicate that the larger amount of C entering the soil under elevated CO_2 conditions may have been cycling at a faster rate explaining the observed non-significant increase in total soil C. Further changes in the overall storage of soil C would manifest itself through the accumulation of additional new C presumably in the more protective or passive soil C pool, which has a mean residence time much longer than 2 years (Jenkinson et al., 1992).

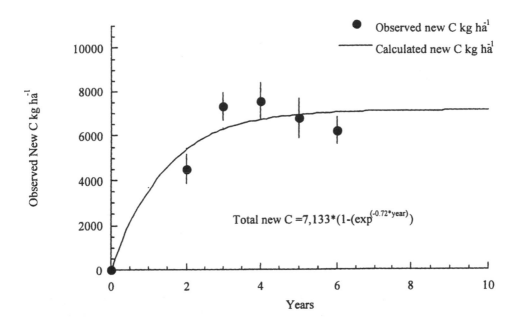

Figure 3. The accumulation of new C in soil from the low fertilizer elevated CO_2 at the Swiss FACE Plots. Actual soil C values are indicated as closed circles and the simulated new C in soil is estimated using a single pool substitution exponential model.

D. Influence of Decomposition on Soil C Sequestration

In Figure 5, scenarios are presented depicting the influence of the decomposition rate constant on the accumulation of soil C. Under ambient conditions, the level of soil C is expected to remained unchanged unless their are changes in crop productivity or SOM from the implementation of alternative agronomic practices, such as reduced tillage or use of legume cover crops (Drinkwater et al., 1998). Under scenario A, additional C input to soil results in the accumulation of total soil C. Initially this accumulation is in the labile pool which is soon completely replaced with new C and reaches a new equilibrium pool size. The assumption in scenario A is that the decomposition rate constant is unchanged and that the kinetics are first order regardless of the amount of C entering the soil (Jenkinson and Rayner, 1977). The unchanged specific decomposition rate implies that the activity of the decomposer community is limited only by substrate availability. In scenario B, the amount of soil C remains unchanged regardless of the additional C input under elevated CO_2. This condition could occur if the specific decomposition rate increases along with substrate availability leading to higher order kinetics.

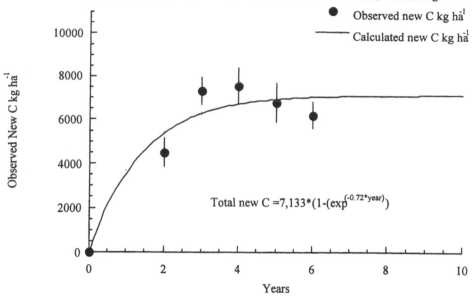

Figure 4. The accumulation of new C in soil from the high fertilizer elevated CO_2 at the Swiss FACE Plots. Actual soil C values are indicated as closed circles and the simulated new C in soil is estimated using a single pool substitution exponential model.

In our study, 6 years of elevated CO_2 lead to a small increase in total soil C that was not statistically significant. This indicates that scenario B in Figure 5 best describes the results of our experiment. The increase in the decomposition rate could occur as a result of increased metabolic reactions, changes in microbial community size or the ability of soil to protect C. It is highly unlikely that the rate of metabolic reactions would change. Metabolic reactions are governed by temperature and moisture limitations of which there were no differences between the ambient and elevated CO_2 treatments (data not shown).

Changes in microbial communities under elevated CO_2 could be expected through changes in the amount and quality of substrate C entering the soil. Studies on elevated CO_2 affects on soil microorganisms have been inconsistent. Whipps (1985) showed no consistent pattern on the total heterotrophic bacterial population in the rhizoplane, rhizosphere, and bulk soil of *Zea mays* under elevated CO_2. Similarly, Schortemeyer et al. (1996), using the same field-level FACE system as in this study, found non-significant CO_2 effects on microbial numbers in the rhizosphere of ryegrass and red clover. Other studies have found inconsistent effects of elevated CO_2 on microbial biomass and diversity probably because of the short-term nature of these types of studies (Zak et al., 1993; Rice et al., 1994).

Hassink (1997) hypothesized that the amount of C that can be associated with clay and silt particles is limited. He found that the amount of C associated with < 20 µm fractions in a series of soils across temperate and tropical regions had a linear relationship (r = 0.89 for C). Protection of C can also occur within soil aggregates (Camberdella and Elliott, 1993). This may explain the lack of C sequestration in our experimental pasture system. The intensive fertilizer management and cutting used in this study may have also limited the system's ability to sequester C.

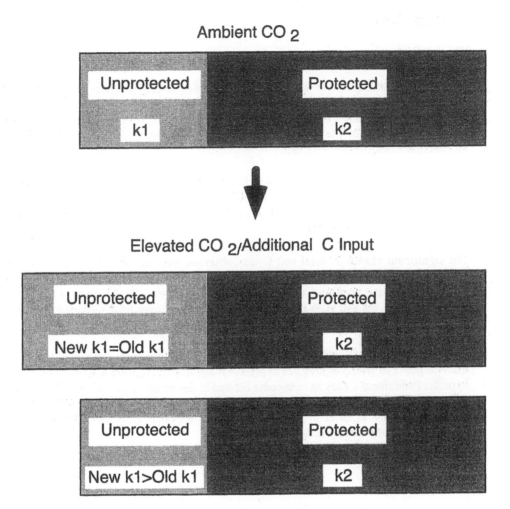

Figure 5. Fate of soil C under conditions of elevated CO_2 and increase plant C input to soil. Scenario A depicts no change and B shows an increase in the decomposition rate constant.

The ^{13}C approach as presented has provided additional information to demonstrate C sequestration in soil compared to the mass balance approach. The mass balance approach showed only a non-significant trend in C accumulation under elevated CO_2 after 6 years. The ^{13}C isotope approach showed that significant new C was entering the soil under elevated CO_2. The results of the ^{13}C approach also provided useful insight into determining the processes that control the sequestration of C in soil. However, ^{13}C studies on elevated CO_2 and on the ability of agricultural systems to sequester soil C has been limited. In addition, the effect of ample nutrient supply and cutting of perennial plants under elevated CO_2 on soil C sequestration has received little attention. Additional studies on the effects of elevated CO_2 will be required to assess the potential of highly fertilized and managed agroecosystems to sequester C.

IV. Conclusions

The data in this study support the following conclusions:

1. Using first order kinetics to describe C accumulation in soil, we estimated that it would take 40 years to increase soil C in our system from 30 to 41 t C ha[-1] if the decomposition rate constant remained constant and a 25% increase in C input to soil occurred. A significant accumulation of soil C should occur by 4 to 10 years making it feasible to test models relying on first order kinetics to describe C sequestration in soil.

2. The maximum accumulation of new C in the soil occurred by 4 years regardless of fertilizer treatment. The new C had an MRT of approximately 1.5 years showing that the new C represented a labile pool of soil C. The new labile C pool was 25% of the total soil C pool and represents the most actively cycling C in the soil.

3. No significant change in total soil C was observed even though C inputs to soil most likely increased through increased below-ground production. This indicates that the decomposition rate constant may have increased and could explain the non-significant increase of soil C. Alternatively, leaching the soil with HCl to remove carbonates may have removed newly sequestered labile C leading to an underestimation of C sequestration.

4. To avoid being mineralized, the C entering the soil needs to become intimately associated with the mineral phase or within aggregates to protect it against decomposer activity. The capacity of soil to protect additional C may have become limited in this study. Alternatively, the high availability of nutrients and frequent cutting of the sward could have promoted decomposition processes. Thus, the ability of soils to sequester C under elevated CO_2 will depend on soil texture characteristics and management practices.

Acknowledgments

Financial support for this project was provided by the Natural Sciences and Engineering Research Council of Canada, the University of California-Davis, and the Swiss Federal Institute of Technology, Zürich, Switzerland.

References

Balesdent, J. and A. Mariotti. 1996. Measurement of soil organic matter turnover using ^{13}C natural abundance. In: T.W. Boutton and S. Yamaski (eds.), *Mass Spectrometry of Soils.* Marcel Dekker, New York.

Balesdent, J., Z.G.H. Wagner and A. Mariotti. 1988. Soil organic matter turnover in long-term field experiments as revealed by carbon-13 natural abundance. *Soil Sci. Soc. Am. J.* 52:118-124.

Ball, A.S. 1997. Microbial decomposition at elevated CO_2 levels: effect of litter quality. *Global Change Biol.* 3:379-386.

Bazzaz, F.A. 1990. The response of natural ecosystems to the rising global CO_2 levels. *Ann. Rev. Ecol. Sys.* 21:167-196.

Camberdella, C.A. and E.T. Elliott. 1993. Carbon and nitrogen distribution in aggregates from cultivated and native grassland soils. *Soil Sci. Soc. Am. J.* 57:1071-1076.

Darra, P.R. 1996. Rhizodeposition under ambient and elevated CO_2 levels. *Plant and Soil* 187:265-275.

Drinkwater, L.E., P. Wagnoner and M. Sarrentonio. 1998. Legume based cropping systems have reduced carbon and nitrogen losses. *Nature* 396:262-265.

Hassink, J. 1997. The capacity of soils to preserve organic C and N by their association with clay and silt particles. *Plant and Soil* 191:77-87.

Hebeisen, T., A. Luscher, S. Zanetti, B.U. Fischer, U.A Hartwig., M. Frehner, G.R. Hendrey, H. Blum and J. Nosberger. 1997. Growth response of *Trifolium repens* L. and *Lolium perenne* L. as monoculture and bi-species mixture to free air CO_2 enrichment and management. *Global Change Biol.* 3:149-160.

Hungate, B.A., R.B. Jackson, C.B. Field and F.S. Chapin II. 1996. Detecting changes in soil carbon in CO_2 enrichment experiments. *Plant and Soil* 187:135-145.

Jansson, S.L. 1958. Tracer studies on nitrogen transformations in soil. *Ann. Roy. Agric. Coll. Sweden.* 24:101-361.

Jenkinson, D.S. and J.H. Rayner. 1977. The turnover of soil organic matter in some of the Rothamsted classical experiments. *Soil Sci.* 123:298-305.

Jenkinson, D.S., D.D. Harkness, E.D. Vance, D.E. Adams and A.F. Harrison. 1992. Calculating net primary production and annual input of organic matter to soil from the amount and radiocarbon content of soil organic matter. *Soil Biol. Biochem.* 24:295-308.

Leavitt, S.W., E.A. Paul, A. Galadima, F.S. Nakayama, A.R. Danzer, H. Johnson and B.A. Kimball. 1996. Carbon isotopes and carbon turnover in cotton and wheat FACE experiments. *Plant and Soil* 187:147-155.

Leavitt, S.W., E.A. Paul, B.A. Kimball, G.R. Hendrey, J.R. Mauney, R. Rauschkolb, H.H. Rogers, K.F. Lewin, J. Nagy, P.J. Pinter and H.B. Johnson. 1994. Carbon isotope dynamics of free-air CO_2-enriched cotton and soils. *Agricul. For. Met.* 70:87-101.

Nitschelm, J.J., A. Luscher, U.A. Hartwig and C. van Kessel. 1997. Using stable isotopes to determine soil carbon input differences under ambient and elevated atmospheric CO_2 conditions. *Global Change Biol.* 3:411-416.

O'Neill, E.G. 1994. Responses of soil biota to elevated atmospheric carbon dioxide. *Plant and Soil* 165:55-65.

Paul, E.A. and F.E. Clark. 1996. *Soil Microbiology and Biochemistry*. 2nd ed. Academic Press, New York.

Paul, E.A., R. Follett, S.W. Leavitt, A. Halvorson, G.A. Peterson and D.J. Lyon. 1997. Radiocarbon dating for determination of soil organic matter pool sizes and dynamics. *Soil Sci. Soc. Am. J.* 61:1058-1067.

Rice C.W., F.O. Garcia, C.O. Hampton and C.E. Owensby. 1994. Soil microbial response in tallgrass prairie to elevated CO2. *Plant and Soil* 165:67-74.

Schortemeyer, M., U.A. Hartwig, G.R. Hendrey and M.J. Sadowsky. 1996. Microbial community changes in the rhizospheres of white clover and perennial ryegrass exposed to free carbon dioxide enrichment (FACE). *Soil Biol. Biochem.* 28:1717-1724.

Tisdall, J.M. and J.M. Oades. 1982. Organic matter and water-stable aggregates in soils. *J. Soil Sci.* 33:141-163.

Torbert, H.A., H.H. Rogers, S.A. Prior, W.H. Schlesinger and G.B. Runion. 1996. Composition and decomposition of soybean and sorghum tissues grown under elevated atmospheric carbon dioxide. *J. Environ. Qual.* 25:822-827.

van Veen, J. A., J. N. Ladd and M. J. Frissel. 1984. Modeling C and N turnover through the microbial biomass in soil. *Plant and Soil* 76:257-274.

Whipps, J.M. 1985. Effect of CO_2 concentration on growth, carbon distribution and loss of carbon from the roots of maize. *J. Exp. Bot.* 36:644-651.

Zak, D.R., K.S. Pregitzer, P.S. Curtis, J.A. Teeri, R. Fogel and D.L. Randlett. 1993. Elevated atmospheric CO_2 and feedback between carbon and nitrogen cycles. *Plant and Soil* 51:105-117.

CHAPTER 17

Methods Using Amino Sugars as Markers for Microbial Residues in Soil

W. Amelung

I. Introduction

Microbial transformations of soil organic N are crucial for plant nutrition. During such transformations, microbes buildup and decompose cell-wall constituents. In the short term, soil organic matter (SOM) transformations are thus linked to fluctuations in dead and living microbial biomass (van Veen et al., 1984). Several methods can be used to assess living soil biota (Alef and Nannipieri, 1995; Paul and Clark, 1996), but very few are suitable to determine microbial C and N residues in soil. Among them, analyses of amino sugars provide a clue to the assessment of both dead and living biomass in soil, because plants do not synthesize amino sugars at significant scale (Stevenson, 1982).

The amino sugar pattern can be used as an indicator for the origin of different microbial C and N residues in soil (review: Parsons, 1981). Muramic acid is almost uniquely produced by bacteria (Parsons, 1981; Kenne and Lindburg, 1983), whereas fungal chitin consists of glucosamine only (Parsons, 1981). Since galactosamine is absent in many fungi but commonly produced by bacteria, increasing glucosamine to galactosamine ratios have been applied to indicate accumulation of fungal residues in acid forest soils (Sowden, 1959; Kögel and Bochter, 1985).

The quantitative results obtained by different authors, however, are difficult to compare, because of the lack of standardization among the different procedures used for sample preparation, hydrolysis, purification, and determination of amino sugars in soil. Amino sugar extraction has been performed on either air-dried (e.g., Joergensen and Meyer, 1990), field-moist (e.g., Chantigny et al., 1997) or frozen soil (e.g., Nannipieri et al., 1979), either at high temperatures in closed vessels (e.g., Benzing-Purdie, 1981), under reflux (e.g., Zelles, 1988), or in combination with cold extraction (e.g., Stevenson, 1983). Usually acid hydrolysis with 6 N HCl has been applied to release amino sugar monomers from both cell structures and soil (Bremner, 1965), but also other HCl concentrations or organic acids have been used (e.g., Stevenson, 1982, 1983). Hydrolysis times ranged from 3 (Zelles, 1988) to 66 hours (Stevenson, 1965). Reviewing the common hydrolytic procedures should facilitate future standardization in amino sugar extraction from soil. The same applies to the methods of amino sugar isolation and determination. Sum parameters are easily obtained with colorimetric and distillation techniques (Bremner, 1965), whereas the determination of individual carbohydrate N requires chromatographic procedures, such as paper chromatography (Grant and West, 1986) or ion-exchange columns which are utilized by amino acid analyzers (Sowden, 1967, Nannipieri et al., 1979; Joergensen and Meyer, 1990), high performance liquid chromatography (HPLC; e.g., Zelles, 1988; Ekblad and Näsholm, 1996), or capillary gas chromatography (= gas liquid chromatography, GLC; e.g., Benzing-Purdie, 1981; Hicks and Newell, 1983; Zhang and Amelung, 1996). The different methods have never been compared together by using the same samples, thus rendering their evalua-

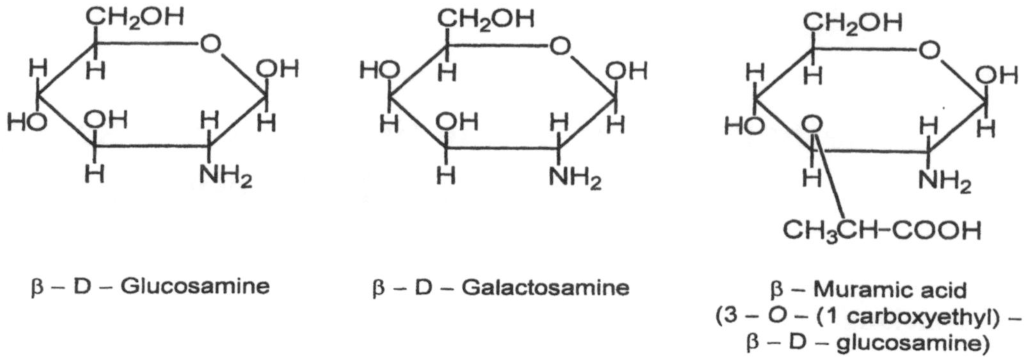

CH₂OH ... (chemical structures)

β – D – Glucosamine   β – D – Galactosamine   β – Muramic acid
                                                        (3 – O – (1 carboxyethyl) –
                                                        β – D – glucosamine)

CH₃CH–COOH

Figure 1. Chemical structure of glucosamine, galactosamine, and muramic acid (3-carboxyethyl-glucosamine).

tion difficult. Nevertheless, evaluation of sample preparation should be possible when considering that amino-sugar contents do not reflect living microbial biomass in soil (e.g., Parsons, 1981). The effect of different hydrolyses has been tested by independent experiments (e.g., Bremner and Shaw, 1954; Zelles, 1988), and determination procedures can be evaluated with respect to their selectivity, sensitivity, precision, and potential interference with other compounds.

It is my objective to review and evaluate common extraction, isolation and determination procedures that have been used to assess amino sugars in terrestrial soils. As determination of amino sugars from soil generally requires methods to remove interfering compounds, procedures developed only for biological materials are not considered here. Some comments will be addressed to the question, how reliably amino sugars can be used for assessing the origin of different microbial residues in soil.

II. Identification of Amino Sugars in Soil

At least 26 amino sugars have been identified in microorganisms or products synthesized by them (review: Sharon, 1965), but only a few of them have been found in soil. As reviewed by Bremner (1965), the first evidence for the presence of amino sugars in soil hydrolysates were positive tests for 2-amino sugars (hexosamines; Figure 1), later, glucosamine was detected in paper chromatograms of soil hydrolysates. Chromatographic analyses of soil hydrolysates gave additional evidence for the presence of galactosamine (Stevenson, 1954, 1956). Up to now, there are indications for 11 additional amino sugars in soil (Stevenson, 1983). Wang and Cheng (1964) reported the occurrence of talosamine in a German Peat. Benzing-Purdie (1980) positively identified fucosamine by mass spectrometry, and Stevenson (1983) suggested further identification of gulosamine, but only muramic acid (Miller and Casida, 1970; Casagrande and Park, 1978; Zelles, 1988; Figure 1)), and mannosamine (Zhang and Amelung, 1996) have been quantified. One essay proved existence of the N-acetylated derivative of glucosamine in soil (Skujins and Pukite, 1970), which is more common in living organisms than the non-acetylated form (Parsons, 1981). Other peaks for amino sugars isolated by chromatographic analyses of soil hydrolysates were not identified with certainty because of lack of available amino sugar standards (Stevenson, 1983). There appears to be three major problems in identification of amino sugars, namely (i) isolation of additional amino sugars, (ii) sensitivity of methods to detect minor amino sugars, and (iii) destruction of labile amino sugars during hydrolysis. Identification of amino sugars in soil was usually performed using paper chromatography (Bremner, 1958; Crumpton, 1959; Stevenson, 1983), by co-chromatography of standard additions to soil hydrolysates (Stevenson, 1983), or comparing mass spectra of soil amino sugars with those of standards (Benzing-Purdie, 1980; Mielniczuk et al., 1995; Zhang and Amelung, 1996). Unless further amino sugar standards become available, future improvements in the identification of amino sugars in soil will be difficult.

Glucosamine and galactosamine constitute 5 to 10% of soil N (Stevenson, 1982), whereas muramic acid concentrations are usually 10 to 30 times, mannosamine levels up to 50 times lower than those of glucosamine (Zelles et al., 1990; Chantigny et al., 1997; Amelung et al., 1999). Future quantification of additional amino sugars may thus be dependent on the development of more sensitive detectors. For analysis, amino sugars have to be liberated from their polymers using strong acids at high temperatures. Under these conditions N-acetylated amino sugars will be de-acetylated, and sialic acid and arabinosamine might not survive the treatment at all since they decompose already in weak acids (Parsons, 1981). In search for other amino sugars in soil, other kinds of hydrolysis procedures should be considered, such as acetic acid for extraction of N-acetylglucosamine (Skujins and Pukite, 1970).

III. Determination of Amino Sugars in Soil Hydrolysates

Quantitative determination of amino sugars from soil usually involved the three steps (i) hydrolysis, (ii) isolation, and (iii) detection. Different methods are subject to various interfering substances. Consequently, methods for amino sugar release and isolation will be described in conjunction with methods of the detection in the following pages. An evaluation of the different hydrolysis procedures used in soil science is carried out in Section IV.

The methods for determination of amino sugars in soils have been modified from those used in the medical and biological sciences. The total of amino sugars was commonly estimated by colorimetric procedures adapted and modified from Elson and Morgan (1933), and by alkaline decomposition according to Tracey (1952). The use of ion-exchange columns according to Boas (1953) or in modifications (Eastoe, 1954; Stevenson, 1983) did not only help to eliminate interfering substances from the soil hydrolysate, but it also allowed separation of different amino sugar monomers in various effluents (Eastoe, 1954; Marumoto et al., 1972; Stevenson, 1983). Specific colorimetric procedures have been applied to quantify N-acetylglucosamine (Skujins and Pukite, 1970) or lactic acid released from muramic acid in soil hydrolysates (Millar and Casida, 1970). The development of amino acid analyzers, HPLC, and GLC gave access to routine determination of individual amino sugars.

Bremner (1965) and Stevenson (1965) reviewed in detail the application of the methods of Elson and Morgan (1933) and Tracey (1952) to soil hydrolysates. I will therefore put my main emphasis on those procedures that were introduced to soil science after 1965.

A. Alkaline Decomposition

The original method of Tracey (1952) was based on the observation that heating of glucosamine and galactosamine for a few minutes with alkali at 100°C will release practically all of their nitrogen in form of ammonia. Bremner and Shaw (1954) modified this procedure for application to soil hydrolysates by distilling the total NH_3 released with a borate buffer at pH 8.8. Preformed NH_3 was

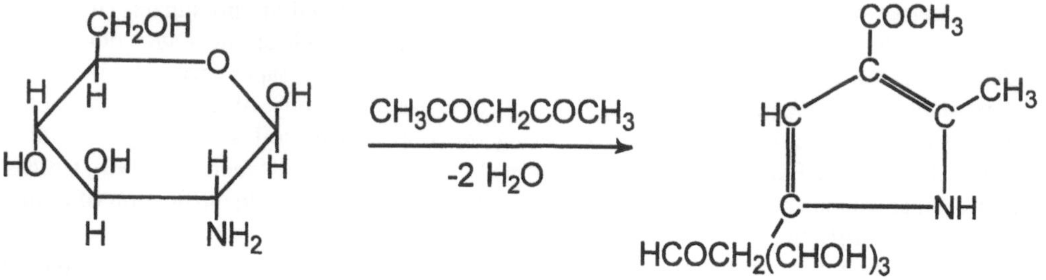

Figure 2. Formation of pyrrole by reaction of glucosamine with acetylacetone.

estimated by distillation with MgO in a microdiffusion unit, a procedure which gave complete recovery of NH_3 without deaminating amino sugars. The detailed outline of the procedure as well as comments on its technical use in soil science is given by Stevenson (1965).

The method gained particular importance when combined with other procedures for N-fractionation of soil (e.g., Stevenson, 1982, González-Prieto and Carballas, 1988). In this context also the chemical transformations of ^{15}N to Amino-sugar ^{15}N could be studied (Azam et al., 1989a,b).

B. Specific Colorimetric Procedures

The most common colorimetric method for the estimation of amino sugars in acid (HCl) soil hydrolysates is that of Elson and Morgan (1933) which requires that both the amino group and the OH on C-1 are free. After reaction in alkaline solution with acetylacetone (pentane-2,4-dione) to form pyrrole derivatives (Figure 2; Parsons, 1981), a stable red color is obtained by addition of an acid, alcoholic solution of p-dimethylaminobenzaldehyde (Ehrlich's reagent), which may be detected at 512 nm (Belcher et al., 1954), or 530 nm (Stevenson, 1965). Neutral sugars or amino acids alone do not interfere, but since mixtures of the two produce a red color when heated with dimethylaminobenzalde-hyde in alkaline solution, neutral sugars should be removed. This was most commonly done using the method of Boas (1953). On percolating the solution through a Dowex 50 exchange resin (usually after removal of HCl in the hydrolysate by evaporation) only the amino but not the neutral sugars were retained on the column. The amino sugars could be further eluted with 2 N HCl. A more serious problem in the determination might be caused by the formation of red-colored complexes between iron and acetylacetone. This could be prevented by precipitation of iron on neutralizing the hydrolysate with an anion exchange resin which also adsorbs dark colored, humic materials (Stevenson, 1957).

As a detailed outline of the methodological steps as well as technical comments have been given by Stevenson (1965), they will not be repeated here. Several improvements have been suggested. Apart from hydrolyses (see Section IV.B.) they refer either to purification (Beckwith and Parsons, 1980; Stevenson, 1983) or colorimetric reaction (Belcher et al., 1954; Allison and Smith, 1965).

Scheidt and Zech (1990) presented a simplified modification of the color reaction in the method of Elson and Morgan (1933), using samples from a forest soil profile. Based on studies of Burtseva et al. (1986) the chromophore typical of amino sugars after color reaction was separated from others by transferring it into an organic phase after basification. It should be noted, however, that apart from hydrolysis, the concentrations of reagents also differed from those outlined by Stevenson (1965). The method of Scheidt and Zech (1990) included the following steps of sample processing: After acid hydrolysis with 6 N HCl, and evaporation of HCl, ultrasound was applied to suspend the residue completely in 10 ml of distilled water. Two 2-ml aliquots were transferred into test tubes. Then, without further purification, 1 ml of acetylacetone reagent was added (1.5 ml of acetylacetone dissolved in 50 ml of 1 N Na$_2$CO$_3$), and the mixture was heated in a water bath at 85 to 90°C for 40 min. After adding 1 ml of ethanol and 1 ml of Ehrlich's reagent (400 mg of p-dimethylamino-benzaldehyde dissolved in a mixture of 15 ml of ethanol and 15 ml of concentrated HCl), the mixture was heated again at 60 to 70°C for 10 min to remove carbon dioxide. To each tube 1.5 ml of 5 N NaOH was added, and the whole mixture was gently shaken to allow complete mixing of the solvents. The addition of NaOH changed the color from red to yellowish. Finally, 2 ml of diethyl ether was added. Then photometric detection was carried out at 467 nm using the ether phase and pure diethylether as a reference.

There are other methods for colorimetric determination of amino sugars in soil, such as the indole-HCl-method of Dische and Borenfreund (1950), or that of Tsuji et al. (1969) who also used the deamination step but produced a color with 3-methyl-2-benzothiazolone hydrochloride and ferric chloride. These methods were less selective than that of Elson and Morgan (1933). After sugar removal by resin purification, however, the indole method gave similar results to those obtained by the Elson and Morgan (1933) method (Stevenson, 1983). The method of Tsuji et al. (1969) was claimed not to be affected by mixtures of neutral sugars and amino acids; however, iron may interfere with the deamination reaction and should be removed (Parsons, 1981). Coloration with ninhydrin is the basis for simultaneous determination of amino acids and amino sugars after ion-exchange chromatography using amino acid analyzers (see Section III.D).

N-acetylated amino sugars

According to Elson and Morgan (1933), N-acetylglucosamine and N-acetylgalactosamine also yield a reddish-purple coloration with p-dimethylaminobenzaldehyde, but only after heating in alkaline solution. The color reaction is not due to formation of pyrrole derivatives, but to the elimination of water, followed by ring-closure with the formation of a disubstituted oxazole, which condenses with p-dimethylaminobenzaldehyde. On performing the color reaction with a concentrated borate buffer, Reissig et al. (1955) were able to estimate N-acetylamino sugars. Skujins and Pukite (1970) later succeeded in extracting them from air-dried soil by shaking it three times with 50 ml of 2.5 M acetic acid for 10 min. The hydrolysates were purified with a combination of an anion and a cation exchange resin, evaporated and dissolved in 0.05 M Na phosphate buffer, pH 7.0.

Muramic acid

Since muramic acid is produced by bacteria in concentrations close to those of glucosamine, it had been earlier suggested that it should occur in soil (e.g., Bremner, 1965). Nevertheless, attempts to quantify it in soil with the method of Elson and Morgan (1933) failed, probably due to the low concentration of muramic acid in soil, the low color intensity of the reaction products with p-dimethylaminobenzaldehyde, but also the sorption of muramic acid on the anion exchange resin column (Millar and Casida, 1970). Only Stevenson (1983) succeeded in identifying muramic acid with the method of Elson and Morgan (1933) after ion-exchange chromatography. The first, although indirect, experimental evidence for the occurrence of muramic acid in soil was provided by Millar and Casida (1970). It was based upon the release of lactic acid from muramic acid in purified soil hydrolysates at alkaline conditions, separation of lactic acid from excess lysine and methionine, and subsequent colorimetric analysis using the method of Markus (1950). Durska and Kaszubiak (1983) later applied the method of Millar and Casida (1970) for screening muramic acid in different mineral soils and humic acids.

Comments

There were controversial findings whether different amino sugars yield different optical densities when coloration was performed according to the method of Elson and Morgan (1933). Crumpton (1959) found similar color intensities for glucosamine and galactosamine whereas those for mannosamine and muramic acid were much lower. In contrast, Swann and Balasz (1966) reported that glucosamine and galactosamine were also detected at different intensities. To overcome this problem they suggested simultaneous recording of the absorbance at 530 and 570 nm. When only the sum of amino sugars is to be estimated, errors resulting from low absorbance of mannosamine and muramic acid are negligible, due to the much lower concentration of these two amino sugars in soil (Bremner, 1965; Zelles et al., 1990; Amelung et al., 1999).

Amino sugar contents in soil obtained by alkaline distillation were generally similar or slightly higher than those obtained by colorimetric methods (Bremner, 1965; Stevenson, 1965). According to Stevenson (1965) it was not possible to say which of these methods were more accurate. Nevertheless, in my opinion the determination of amino sugars by alkaline decomposition might be more accurate than the simple colorimetric determination adapted from Elson and Morgan (1933), unless laborious sample purification is applied (Stevenson, 1965, 1983). On the one hand, decomposition of amino sugars during hydrolysis is simultaneously allowed for, when only amino sugar-derived ammonia is detected. On the other hand, Tracey (1952) reported that a mixture of 2.5 M lysine and 2.5 M glucose increased apparent glucosamine levels only by 5%, whereas the same compounds increased apparent glucosamine levels by 18% when determination was performed with the method of Elson and Morgan (1933). It seems, therefore, as if amino-sugar N determination by alkaline decomposition might be less affected by interference with mixtures of amino acids and other sugars.

The method of Scheidt and Zech (1990) avoids all purification steps. The measured amino sugar concentrations of samples from forest litter and mineral surface soil (4 h hydrolysis) were in the same range as independently determined by Kögel (1987) using a gas chromatographic procedure with a nitrogen-selective detector (Kögel and Bochter, 1985), but after 6 h hydrolysis time. Since the recovery of glucosamine spiked to the hydrolysates ranged between 96 and 107%, the method of Scheidt and Zech (1990) seems reliable for rapid screening of amino-sugar N in soil hydrolysates. Unfortunately, there has as yet been no direct comparison of this method with the procedures reviewed by Stevenson (1965). Moreover, there was no spiking of soil hydrolysates with compounds known to interfere with the other methods. It is my personal impression that the method of Scheidt and Zech (1990) offers a valuable, much more rapid alternative to the other colorimetric procedures. However, its accuracy remains to be verified by more data.

The colorimetric determination of N-acetylamino sugars is presently the only method for detecting these compounds in soil. According to Reissig et al. (1955) there is hardly any interference from mixtures of sugars and amino acids or from Mg salts. Skujins and Pukite (1970) stated that there was no interference from added glucosamine, i.e., the method seems to be quite specific. How far hydrolysis products of soil nucleotides may interfere with coloration after all purification steps remains to be investigated (Skujins and Pukite, 1970). The analytical range for pure N-acetylglucosamine was 1 to 40 µg ml^{-1}.

The indirect estimation of muramic acid according to Millar and Casida (1970) is laborious and still based on the assumption that there are no other compounds in soil that behave like muramic acid and would liberate lactic acid under the experimental conditions. This method, while certainly important for the very first proof of muramic acid in soil, has in the meantime been replaced by more sophisticated analyses of muramic acid using HPLC (Moriarty, 1983; Zelles, 1988) or GLC (Casagrande and Park, 1977; Zhang and Amelung, 1996).

C. Chromatographic Separation followed by Colorimetric Determination

As already indicated, reactions of the amino group with ninhydrin can be utilized for detecting both amino-acid and amino-sugar N (Stevenson and Cheng, 1970; Parsons; 1981), provided that they can be separated. This was laboriously achieved by sampling different effluent volumes in simple cation-exchange chromatography (Eastoe, 1954; Stevenson, 1983), and sometimes long resin columns were used (1 m, Wagner and Mutatkar, 1968). These problems were easily overcome using paper chromatography or suitable columns and buffer solutions with automated amino acid analyzers.

Determination of Glucosamine Using Quantitative Paper Chromatography

Stevenson (1983) used paper chromatography for identification of amino sugars from soil, but Grant and West (1986) introduced a procedure to determine glucosamine quantitatively. Briefly, after hydrolysis of 2.5 g of fresh soil with 12 M HCl for 48 h (25°C) followed by a 6-h hydrolysis under N_2 (100°C), the solutions were filtered and HCl was removed by evaporation at 45 to 50°C. Duplicate aliquots of the hydrolysate and of an authentic glucosamine solution were placed on Whatman No. 1 paper as 2 cm bands and chromatographed in butan-1-ol/pyridine/glacial acetic acid/water (60/40/3/30, v/v/v/v) for 48 h. The papers were dried in air, then dipped in ninhydrin reagent (1 g of ninhydrin, 100 mg of Cd acetate, 5 ml of glacial acetic acid, 100 ml of acetone, 10 ml of water). Glucosamine spots developed over 4 to 5 days over concentrated H_2SO_4 in darkness. They were cut out, eluted with 4 ml of methanol and absorbance was measured spectrophotometrically at 500 nm (Grant and West, 1986). For homogeneity tests, chromatography in the second dimension was performed with pyridine/water(4/1, v/v), similar to Stevenson (1983).

Determination of Glucosamine and Galactosamine Using Amino Acid Analyzers

After separating amino sugars and amino acids via exchange chromatography with amino acid analyzers, detection was usually performed by post-column derivatization, ending in a coloration after reaction of ninhydrin with amino acid and amino sugar-derived ammonia (e.g., Spackman et al., 1958; Sowden, 1969; Wagner and Mutatkar, 1968). The direct detection of amino-sugar chelates at 254 nm might be a promising alternative (Navratil et al., 1975; Parsons, 1981). The following procedure was taken from Joergensen (1987) as slightly modified by Joergensen and Meyer (1990), because it was the most recent one based on the ninhydrin reaction.

REAGENTS
- Buffer A: 0.06 M Na$_3$-citrate, pH 3.25
- Buffer B: 0.10 M Na$_3$-citrate, pH 4.35
- Buffer C: 0.10 M Na$_3$-citrate + 1.3 M NaCl, pH 4.75
- Buffer F: 0.5 M NaOH with 0.1% Na$_2$-EDTA for regeneration
- Ninhydrin reagent: 40 g of ninhydrin is dissolved in 1500 ml of methoxy-ethanol and mixed with 500 ml of Na-acetate buffer (4M, pH 5.3) and 14 ml of 15% titanium-III-chloride solution.

PROCEDURE:
The HCl hydrolysate (6 M HCl for 24 h at room temperature followed by 6 h under reflux) was adjusted to pH 1.55 with NaOH. Then, amino acids and amino sugars were determined with ninhydrin after cation exchange chromatography (Biotronic LC 2000). Chromatographic conditions were:
- Pre-separation column: Durrum DC-3 resin (120 × 12 mm)
- Separation column: Durrum DC-6A resin (210 × 6 mm)
- Flow rate for buffers: 32 ml h^{-1}
- Flow rate for ninhydrin reagent: 29 ml h^{-1}

- Post-column reaction coil: PTFE (30 m × 0.3 mm ID)
- Reaction temperature: 100°C
- Temperature program:

min	°C	buffer	ninhydrin
5	45	A	no
20	45	A	yes
4	45	B	yes
36	61	B	yes
41	61	C	yes
20	45	F	yes
35	45	A	no
5 (break)			

Recording of absorbance was performed at 440 and 570 nm. Later Scholle et al. (1993) suggested the following modifications: After hydrolysis with 6 N HCl, membrane filtration (PTFE, 0.2 μm), and removal of HCl by evaporation, the dried residue was taken up in 1 ml of a 0.06 M Na$_3$-citrate buffer solution containing 16.65 g L^{-1} and 80 ml of thiodiglycol L^{-1} as an antioxidant.

Determination of Muramic Acid Using Amino Acid Analyzers

Cheshire (1979) pointed out that it should also be possible to detect muramic acid with amino acid analyzers after ninhydrin reaction. However, to my knowledge, this has not yet been achieved in soil research. Nevertheless, Joergensen et al. (1995) succeeded in determining muramic acid after reaction with Na$_2$-bicinchoninic acid (BCA) and subsequent photometric detection at 570 nm; separation was performed with borate complex ion-exchange chromatography (Sinner and Puls, 1978, modified). The BCA reagent consisted of a mixture of three solutions in a ratio of A:B:C = 100:5:2 (v/v/v), (A) = 8% Na$_2$CO$_3$ (anhydrous) + 0.2% aspartic acid, (B) = 4% bicinchoninate, (C) = 4% CuSO$_4$×5H$_2$O. Pre-separation column was again Durrum DC-3 resin (110 × 12 mm), the separation column was a Durrum DC-6A resin (280 × 6 mm). The column temperature was set to 62°C, that of the reaction bath remained at 100°C. Compound separation was achieved by gradient programming with buffer A (0.1 M sodium-citrate dihydrate and 5.5% methanol, adjusted with HCl to pH 3.25) for 15 min and with buffer B (0.40 M sodium-citrate dihydrate, pH 4.35) for 45 min. The flow rate for buffers amounted to 28.05 ml h^{-1} and that of the reagent to 21.09 ml h^{-1}, with a buffer-to-reagent ratio of 1.28:1. Using this experimental set-up, muramic acid was eluted at 24 min, glucosamine at 48 min, and galactosamine at 50 min. Colorimetric detection was conducted at 570 nm.

Comments

The quantitative paper chromatography of Grant and West (1986) certainly offers the advantage that it can be rapidly conducted without sophisticated instrumentation. Whilst paper developing times are long, the actual working time is rather short. The method has been tested only for glucosamine, ergosterol and diamino-pimelic acid. It seems therefore suitable for a more or less rapid screening of different biomarkers, whereas its use for amino sugar determination is still limited to glucosamine. The analytical range has not been reported, the coefficient of variation ranged from 6 to 23% of the means.

Using amino acid analyzers, both glucosamine and galactosamine can be conveniently determined together with amino acids (e.g., Nannipieri et al., 1979; Yoshida and Kumada, 1979; Joergensen and Meyer, 1990). For a better understanding of amino sugar dynamics, the simultaneous assessment of muramic acid was of great advantage in this context (Joergensen et al., 1995). Determination of muramic acid was achieved down to 1 μg g^{-1} soil, but incomplete separation from presumably neutral sugars in some samples may eventually interfere with quantification (Joergensen, 1998, personal communication). Neutral sugars, however, might be easily removed by a prior purification step using cation-exchange resins (Boas, 1953).

The main problem in quantifying amino sugars simultaneously with amino acids was extraction, because amino sugars decompose at hydrolytic conditions required for exhaustive release of amino acids (usually 6 N HCl for 16 to 24 hours under reflux; Stevenson, 1982). Sowden (1959) suggested using correction factors for the decomposition of amino sugars when simultaneously quantifying them with amino acids in a single run. In my opinion, this may not be reliable, because optimum hydrolytic conditions for the liberation of amino sugars from soil material may vary between different types of samples (e.g., Zelles, 1988; Zhang and Amelung, 1996). Before applying correction factors, I therefore suggest determining the degree of decomposition of amino sugar during hydrolysis for the specific samples under investigation. Alternatively, it might be even less time-consuming but more accurate to apply successive hydrolyses, i.e., to release amino sugars under optimum conditions (e.g., 6 h) with 6 N HCl, and then to hydrolyze the residue again until complete release of amino acids. Both hydrolysates might be later combined. A similar approach had been proposed by Gallali et al. (1975) for exhaustive extraction of amino sugars from soil.

Goh and Edmeades (1979) desalted hydrolysates prior to determination of amino sugars with an amino acid analyzer. As pointed out by Sowden (1969), high concentrations of Fe and Al may affect the recovery of individual amino acids, but not that of amino sugars. Desalting of hydrolysates can therefore be omitted when only amino sugars are investigated, but it should be kept in mind that increasing Al concentrations in the hydrolysate have been shown to shift the position of peaks (Sowden, 1969). The author observed that at higher Al concentrations the signal for valine may overlay that of galactosamine. In such cases, the pH of the buffer solution had to be adapted Sowden (1967; 1969).

D. High-Performance-Liquid Chromatography (HPLC)

After conversion of amino sugars into fluorescent products, reversed-phase HPLC can be utilized for amino sugar determination at ng level (Caroll and Nelson, 1979; Mimura and Delmas, 1983). For environmental samples, derivatization was usually performed with o-phthaldialdehyde (OPA; Moriarty, 1983; Zelles, 1988). More recently, phenylisothiocyanate (Schmitz et al., 1991) and 9-fluorenylmethylchloroformate (Ekblad and Näsholm, 1996) have been suggested. With the OPA reaction it was also possible to quantify muramic acid in sediments and soil (Moriarty, 1983; Zelles, 1988; Chantigny et al., 1997). The OPA method had the additional advantage that glucosamine formed only one signal compared to three in the method of Ekblad and Näsholm (1996), for instance. The following outline was adapted from Zelles (1988); the methods of Schmitz et al. (1991) and Ekblad and Näsholm (1996) have not yet been explicitly tested for soil samples, the method of Moriarty (1983) was applied only to sediments but not yet to terrestrial soil.

REAGENTS

- Borate buffer: 0.04 M boric acid adjusted to pH 9.5 with 1 M NaOH
- OPA reagent: 270 mg o-phthaldialdehyde dissolved in 5 ml of methanol and 200 µl 2-mercaptoethanol. The solution was adjusted to 50 ml with the borate buffer, and "aged" for > 12 h. The strength was maintained by adding 20 µl 2-mercaptoethanol every 2 days.
- Eluent: 0.05 M sodium citrate, 0.05 M sodium acetate (pH 5.3), and (50:50)-methanol tetrahydrofuran, all three at a solvent mixture ratio of 90:8.5:1.5 (v/v/v) (Mimura and Delmas, 1983, modified)

All reagents should be of analytical or chromatography grade.

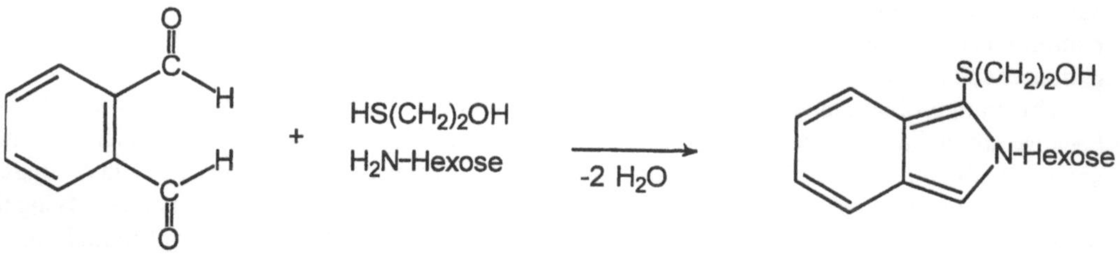

Figure 3. Reaction of o-phthaldialdehyde with amino-sugar N.

Prior to hydrolysis of 0.5 to 1 g soil, samples were flushed with N_2 for > 30 s in screw-cap test tubes (25 × 100 mm). Anaerobic hydrolysis was performed under reflux with 6 M HCl (3 h; soil:solution ratio = 1:20, w/v) under N_2 atmosphere. After cooling to room temperature, the hydrolysate was centrifuged at 1500 g for 10 min, and stored in a refrigerator until analyses. An aliquot (0.2 to 1 ml) was evaporated to dryness at 45 to 50°C under reduced pressure in a rotary evaporator. The dried hydrolysate was redissolved in 500 µl OPA reagent, and the precipitate was removed by centrifugation (1 min). The reagent mixture was left for 2 min. Under the alkaline pH conditions both aldehyde groups react with the amino group in the presence of the reducing agent 2-mercaptoethanol, to form a substituted 1,2-isoindole under elimination of two molecules of water (Figure 3).

After reaction, a full loop (20 µl) was injected onto the column. Conditions for reversed-phase chromatography were:
- Column: ODS Hypersil 5 µm, 125 × 4.6 mm I.D.
- Flow rate of eluent: 1.5 ml min^{-1}
- Fluorescence detection: Excitation wavelength, 340 nm; emission wavelength, 445 nm

Modifications (Chantigny et al., 1997; Chantigny, 1998, personal communication)

Since peak separation was partly incomplete, some modifications were introduced to the procedure of Zelles (1988), working with an OPA reagent of 5 mg o-phthaldialdehyde ml^{-1} 0.5 M potassium borate buffer (pH 10). The dried hydrolysate was taken up in 1 ml of OPA reagent and centrifuged for 3 min (21000 g). The reaction time was set to 5 min ± 15 s. Following injection of 25 µl to a 5-µm C18 column (8 × 100 mm), complete separation of peaks was obtained with a gradient program using the eluent A of Lindroth and Mopper (1979) and Zelles (1988, see above) in combination with an eluent B, being 65% methanol in water (Chantigny, 1998, personal communication). At initial conditions the gradient was 85% eluent A and 15% eluent B, after 17.5 min the mixture ratio was 70% eluent A and 30% eluent B, respectively. Chromatographic separation was accomplished after 22 min (Figure 4).

Comments

The modifications introduced by Chantigny et al. (1997; Chantigny, 1998, personal communication) have three major advantages: (1) the peaks were separated with the gradient program at baseline resolution (Figure 4), (2) resolution was achieved within a shorter period (14 compared to 25 min), and (3) addition of eluent B partly cleaned up the column. The better resolution of glucosamine probably reduced its error of determination. Shorter retention times allowed to increase the reaction time from 2 to 5 min, and centrifugation time from 1 to 3 min, thus making the method more feasible

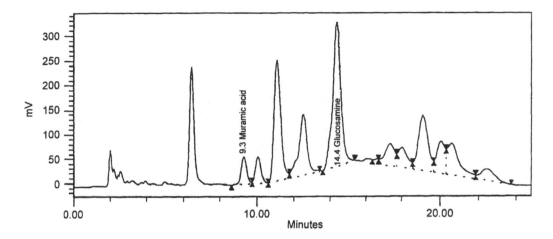

Figure 4. HPLC chromatogram of p-phtaldialdehyde amino sugar derivatives (Chantigny, 1998, unpublished data).

for routine analyses. There was no real time-consuming step, the longest being separation on the HPLC column. The column must be cleaned, however, once a day.

Reaction of OPA with amino-sugar N creates derivatives that are still anomers, e.g., Jahnel and Frimmel (1996) detected α- and β-glucosamine in water humic acids. The chromatograms of Zelles (1988) and Chantigny et al. (1997; Chantigny, 1998, personal communication), however, showed only one peak per compound, i.e., the different anomers were not separated. In contrast, Ekblad and Näsholm (1996) found three peaks for glucosamine. Since the ratio of the peaks was rather constant, the first peak was used for quantification. Losses in sensitivity were of minor importance in this respect, since the method itself was still sensitive enough to detect glucosamine in the ng range. Thus, multiple peak formation is not necessarily a disadvantage but may also facilitate identification of glucosamine when other compounds are co-eluted.

The o-phthaldialdehyde does not exclusively react with amino sugars but also with amino acid and degradation products containing free NH_2 functional groups. Peaks of unknown origin appeared thus in the chromatograms of Zelles (1988) and Chantigny et al. (1997; Chantigny, 1998, personal communication; Figure 4). It remains to be verified, therefore, that amino sugar signals in the chromatograms are not underlain by other compounds. Moriarty (1983) proved identity for muramic acid by co-chromatography of standard additions, and by destroying muramic acid in alkali, which resulted in the absence of the respective peak. Whether glucosamine is underlain by another compound might perhaps be investigated with mass spectrometry. Nevertheless, since glucosamine-to-muramic acid ratios at sites under grass in the study of Chantigny et al. (1997) were of similar magnitude to those of the study of Amelung et al. (1999) for bulk soils from similar climate, it seems reasonable to assume that there were only small if any contributions of unknown signals to the glucosamine (or muramic acid) peak.

Ekblad and Näsholm (1996) suggested to pre-extract amino acids from proteinaceous material two times with 0.2 N NaOH (6 h, 20°C, followed by 17.5 h, 100°C). This rather simple approach removed almost all amino acid material whereas glucosamine was quantitatively recovered. I doubt, however, that this purification step might be useful for soil samples. Schnitzer (1982) recommended to use cold 0.1 or 0.5 N NaOH for extraction of humic acids from soil, and it was found that humic acids contain amino sugars (Gallali et al., 1975; Calderoni and Schnitzer, 1984; Coelho et al., 1997). In addition, it remains unknown whether a pretreatment with 0.2 N NaOH does not include the risk of muramic acid degradation to glucosamine, recalling that Millar and Casida (1970) destroyed muramic acid after short heating in alkali at pH 12.5. Thus, similar purification for the OPA method is not recommended.

Fluorescence yields of the OPA reaction products change with incubation time and solvent volume, i.e., both parameters should be kept constant to sustain reproducibility of analysis (Zelles, 1988). Failure to stay within the narrow interval of derivatization time in the modified procedure (5 min ± 15 s), however, required to repeat amino sugar determination (Chantigny, 1998, personal communication). The main disadvantage of the OPA method is its sensitivity against variations of pH.

High fluorescence yields were obtained only when the pH of the solution was between 8.5 and 10, which was not the case for one of the soil samples investigated by Zelles (1988). In my opinion there are two reasons that might have accounted for the low pH value in the mineral soil sample of the study of Zelles, namely incomplete removal of HCl during evaporation and contribution of metallic cations to the low pH in solution or both. As this specific sample had the highest clay content of the samples investigated by Zelles (1988), it seems reasonable to speculate that formation of $Fe(III)Cl_4^-$ and $Fe(III)Cl_4(H_2O)_2^-$ complexes hindered complete removal of HCl upon evaporation. If true, the method was sensitive to Fe in solution, i.e., unless pH was controlled using an aliquot from the reaction mixture, the method would include the risk of creating artifacts when applied to soils with high Fe content (e.g., Oxisols) or contrasting textures. According to Chantigny (1998, personal communication), however, the texture effect is not very likely, since he hardly observed losses in fluorescence yields when applying their method to soils with 20 to 60% clay. If the pH of the reaction mixture was found to be < 8.5, it might be re-increased with additional OPA reagent (Chantigny, 1998, personal communication). Other procedures, such as repeated HCl removal by evaporation after re-dissolving the dried hydrolysate in water (Benzing-Purdie, 1981), or Fe removal from the hydrolysate, e.g., with acetylacetone (Sowden, 1969), would be more time-consuming.

Schmitz et al. (1991) reported that high salt contents interfered with the HPLC separation of glucosamine-phenylisothiocyanate derivatives. Ekblad and Näsholm (1996) removed salts during their NaOH washing step. Zelles (1988) indicated that high ionic strength decreased fluorescence intensity. The extent to which high salt concentrations in soil hydrolysates (e.g., after dissolution of lime in soils) affect fluorescence yields remains to be tested.

In summary, the details of the method of Zelles (1988) have to be followed with care, and pH should be controlled when the procedure is applied to different soil samples. To the method's credit, the details have been thoroughly evaluated, and even with additional control of pH the method remains simple and rapid. The coefficient of variation ranged from 2 to 37% for glucosamine, and from < 1 to 31% for muramic acid (n = 3; calculated from Zelles et al., 1990), but there is good reasoning to assume that with the modifications introduced by Chantigny et al. (1997), these errors can be significantly reduced. Thus, when modifications of Chantigny et al. (1997) are included into the method of Zelles (1988), the procedure seems to be suitable for routine determination of glucosamine and muramic acid in trace amounts. Moreover, the procedure can likely be expanded to the simultaneous determination of amino acids (Moriarty, 1983; Jahnel and Frimmel, 1996), as well as to that of galactosamine and mannosamine, since their peaks have already been identified (Zelles, 1988).

E. Gas Chromatography (GC)

Since amino sugars are not volatile, they have to be converted to volatile derivatives for gas chromatographic analyses. This was achieved by various procedures. Alditol acetates have been used

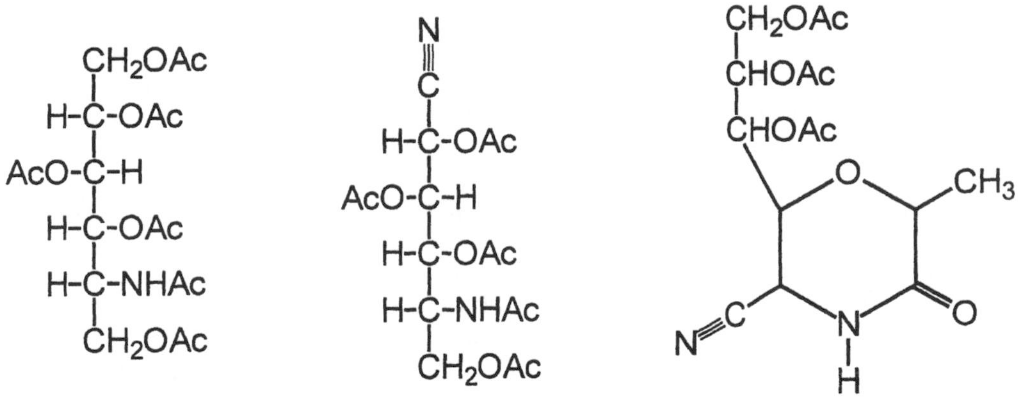

Figure 5. Glucosaminitol acetate (left), glucosaminonitrile acetate (middle), and muramic acid tetra acetate after nitrile and lactam formation (right structure).

for analyzing glucosamine and galactosamine in soil (e.g., Spiteller, 1980; Benzing-Purdie 1981, 1984), the procedure being also suitable for determination of mannosamine (Coelho et al., 1997). Converting amino sugars to aldononitrile acetate derivatives (Figure 5) allowed next to the measurement of the three hexosamines the additional detection of muramic acid in soil hydrolysates (Zhang and Amelung, 1996;). Other procedures included the formation of trimethylsilyl derivatives (for soil matrixes solely applied to muramic acid; Casagrande and Park, 1977, 1978; Mielniczuk et al., 1995), trifluoroacetyl-*n*-butyl ester derivatives (suitable for determination of galactosamine, Casagrande, 1970), trifluoroacetyl derivatives (used for determination of glucosamine and galactosamine; Kögel and Bochter, 1985), and O-methyloxime acetyl derivatives (used for determination of glucosamine and muramic acid; Hicks and Newell, 1983).

In principle, the conversion of amino sugars to trimethylsilyl derivatives is a simple and rapid step, but it yielded two anomeric peaks for muramic acid (Casagrande and Park, 1977; Mielniczuk et al., 1995). This caused losses in sensitivity by a factor of 2 compared to methods where signal intensity is concentrated in one peak. Thus, silylation was suitable for rapid screening of muramic acid in peat (Casagrande and Park, 1978) and organic dust (Mielniczuk et al., 1995). The procedure has to my knowledge not yet been introduced for analyzing other amino sugars in impure soil hydrolysates, possibly due to resolution problems.

Multiple peak formation can be overcome by using other derivatization procedures. Trifluoroacetyl-n-butyl esters were suggested by Casagrande (1970). Because the derivatives of amino sugars coeluted with those of amino acids, glucosamine could not be quantified. This was achieved by Kögel and Bochter (1985), who used a SE 54 capillary column in combination with a nitrogen-selective, flameless, thermionic detector. Nevertheless, the chromatogram obtained from the hydrolysate of an Oe horizon still contained signals from interfering compounds, causing a slight overestimation of galactosamine. Moreover, as N-selective detectors are not commonly used in laboratories, the method has not been applied again after development. Converting amino sugars to alditol acetates was first tested with soil hydrolysates by Benzing-Purdie (1981, 1984). The chromatograms were almost free from interfering signals and did not depend on the use of a N-selective detector. Using aldononitrile acetates as amino sugar derivatives for soil hydrolysates, Zhang and Amelung (1996) obtained chromatograms of similar quality. Other derivatization procedures, such as the formation of methyloxime-acetate derivatives for measuring glucosamine and muramic acid (mannosamine and galactosamine could not be separated; Hicks and Newell, 1983) are presently of limited use for soil hydrolysates, because they have been applied only to biological materials so far, i.e., the effect of interfering compounds remains to be investigated. The following discussion is therefore restricted to methods using alditol acetates or aldononitrile acetates for amino sugar determination in soil hydrolysates.

Determination of Amino Sugars as Alditol Acetate Derivatives (Benzing-Purdie, 1981, 1984; Coelho et al., 1997)

After hydrolysis of freeze-dried soil (100 mg) for 18 h with 6 N HCl at 105°C, myo-inositol solution was added as first internal (= *surrogate*) standard. Then the mixture was centrifuged, and supernatant was decanted and evaporated in vacuo. Water was added twice during evaporation. The residue was dried in a desiccator over KOH for 18 h. Water (2 ml) was added, followed by 60 mg sodium borohydride. The solution was left 2 h at room temperature and centrifuged, and the supernatant was acidified with acetic acid and > 5 times evaporated with methanol in vacuo. After the residue was visibly dry, 2 ml of acetic anhydride were added, and the reaction mixture was heated for 2 h at 120°C and evaporated again. The acetylated product was extracted with small amounts of $CHCl_3$, and 1 µl was used for high resolution GC analysis with a flame ionization detector (FID). The chromatographic conditions were (Benzing-Purdie, 1984):

- Capillary GC column: SE-54 (15 m × 0.2 mm ID, fused silica)
- Injector temperature: 250°C Column temperature: 210°C
- Detector temperature: 250°C
- Split ratio: 80:1
- Column flow (He): 0.67 ml min^{-1}

Separation of peaks was achieved within 11 min (Benzing-Purdie, 1984).

Coelho et al. (1997) replaced the 15 m SE-54 column with a 20 m SE-54 column in GLC analyses. They worked also under isothermal conditions (190°C column temperature), using H_2 as carrier gas (flow rate = 1 ml min^{-1}). Injector and detector temperatures were set at 210 and 240°C, respectively. OV-101 and OV-225 capillary columns failed to separate the peaks at baseline resolution (Benzing-Purdie, 1981, 1984; Coelho et al., 1997).

Determination of Amino Sugars as Aldononitrile Derivatives (Zhang and Amelung, 1996)

REAGENTS

- 1 M and 6 M HCl, 0.4 M KOH, methanol, dichloromathene, ethyl acetate-hexane (1:1), myo-inositol, standard solutions of myo-inositol, 3-methylglucamine, N-acetylglucosamine, N-acetylgalactosamine, N-acetylmuramic acid, and N-acetylmannosamine.
- Derivatization reagent 1: 32 mg ml^{-1} of hydroxylamine hydrochloride and 40 mg ml^{-1} of 4-(dimethylamino)pyridine in pyridine-methanol (4:1 v/v).
- Derivatization reagent 2: acetic anhydride.

Soil samples (containing about 0.3 mg N) were heated under N_2 in closed hydrolysis flasks with 10 ml of 6 N HCl (105°C, 8 h), containing 100 µg of myo-inositol as *surrogate* standard. After cooling, the hydrolysate was filtered through glass fibre filters GF6 (Schleicher and Schuell, FRG), and the filtrate was dried completely by rotary evaporation at 40 to 45°C in vacuum. Then, the residue was dissolved with 20 ml of water, and pH adjusted to 6.6 to 6.8 with the 0.4 N KOH solution. The precipitate was removed by centrifugation (2000 g for 10 min) and the clear supernatant was freeze-dried or evaporated (40°C) to dryness again. The residue was dissolved in 3 ml of dry methanol and centrifuged (2000 g for 10 min). The methanolic amino sugar fraction was transferred to a 5 ml reactivial, and dried again by using dry air or nitrogen gas at 45°C. Finally, the residue was dried completely in a vacuum. Amino sugars were transformed into aldononitrile derivatives as described by Guerrant and Moss (1984). The derivatization reagent 1 (0.3 ml) was added to the reactivial containing a dry sample or a mixture of standards (no more than 3 mg of amino sugars in total). Prior to derivatization 50 µg of methyl glucamine was added as second internal standard. After capping the vial and shaking for a few seconds, the solution was heated for 30 min at 75 to 80°C. Then, the vial was cooled to room temperature, and 1 ml of acetic anhydride was added. The vial was closed, shaken again, and reheated at 75 to 80°C for 20 min. After cooling, 2 ml of dichloromethane were added. Excess derivatization reagents were removed by two washing steps. First, after 1 ml of 1 N HCl was added and strongly shaken for 30 s, the upper aqueous phase was removed. Second, in the same manner, the organic phase was extracted three times with distilled water (1 ml each). In the last washing step, the water was removed as completely as possible. The final organic phase was dried with dry air at 45°C, and finally, dissolved in 300 µl of ethyl acetate-hexane (1:1). One µl was used for separation of aldononitrile derivatives by GLC (HP 5890 or 6890) equipped with a flame ionization detector and N_2 as carrier gas. The chromatographic conditions were (Zhang and Amelung, 1996, modified):

- Capillary GC column: HP Ultra-2 (25 m × 0.2 mm × 0.33 µm, fused silica)
- Injector temperature: 250°C
- Column temperatures: programmed
- Detector temperature: 300°C
- Split ratio: 30:1
- Column head pressure: 60 kPa

The temperature programme was set as follows: The initial column temperature of 120°C was held for 1 min and then temperature was increased at 10°C min^{-1} to 250°C for 2.5 min. Thereafter, the temperature was increased again at 20°C min^{-1} to 270°C, held for 2 min. For column cleaning, it was heated again at 60°C min^{-1} to 300°C, held for 7 min. Separation of peaks was achieved in 14 min (Figure 6). Recent experience showed that the Ultra-2 column can be easily replaced by the cheaper HP-5 fused silica column (30 m × 0.25 mm × 0.33 µm) when holding initial temperature for 4 instead of 1 min.

Comments

GLC analysis of soil has the advantage, that the accuracy of sample processing can be easily controlled within a single run using *surrogate standards* (Hübschmann, 1996). The surrogate standard is an internal standard that is added to the sample prior to processing. Using another internal standard at the end of sample processing, it is possible to determine the recovery of the surrogate standard, and thus to correct for the losses of amino sugars during sample work-up. This implies that the surrogate standard behaves similarly to the amino sugars. This is normally the case for myo-inositol (Benzing-Purdie et al., 1984; Coelho et al., 1997; Zhang and Amelung, 1996). Recovery of myo-inositol after hydrolysis and sample clean-up usually ranged from 70% to 90% with the aldononitrile acetate method (Zhang and Amelung, 1996), and exceeded 80% with the alditol acetate method as modified by Coelho et al. (1997; Coelho, 1998, personal communication).

Both derivatization procedures, the alditol and aldononitrile acetate method, produce a single peak for the amino sugar derivatives that is chemically stable for several hours to a few days. Both methods

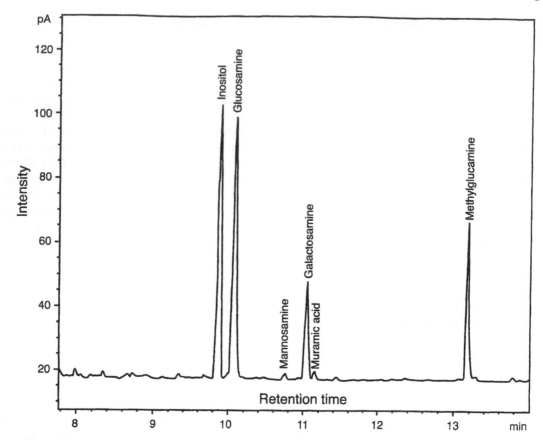

Figure 6. GLC chromatogram of aldononitrile acetate devivatives (mineral soil sample).

are sensitive and precise, with an average error of determination < 10% for glucosamine and galactosamine, and < 15% for muramic acid (Coelho et al., 1997; Zhang and Amelung, 1996). The principal disadvantage of the alditol acetate method includes a large number of manual preparation steps, especially, for the removal of excess borate generated in the reduction step, which make the procedure time-consuming and tedious to perform. According to Coelho (1998, personal communication), 4 samples can be processed in a week, compared to about 16 to 24 samples with the aldononitrile method. In addition, the alditol acetate method failed to quantify muramic acid in soils, although this should be possible (Whiton et al., 1985).

Any method using acetic anhydride for acetylation includes the risk that excessive acetic anhydride will destroy the capillary column. In this regard, the final drying step in the method of Benzing-Purdie (1984) should be conducted with care. To be safe from losing the capillary column, excessive acetic anhydride can be destroyed using 1 M HCl and water while extracting the amino sugar derivatives with chloroform (Hicks and Newell, 1983). Zhang and Amelung (1996) applied this procedure for aldononitrile derivatives (see above), but it should work for alditol acetates as well (Whiton et al., 1985). During the liquid-liquid extraction steps with chloroform and HCl/water, amino sugar derivatives are additionally liberated from water-soluble impurities. This may eventually result in additional decolorization of the chloroform phase (personal observation).

When impurities in the final derivatization mixtures are not removed, they may deteriorate the column. Coelho (1998, personal communication) used the alditol acetate method to determine amino sugars in humic acids and fungal melanins; she had to purify the GC column after each run by heating it at maximum temperature for 30 min. In contrast, the capillary column used for determination of aldononitriles can be easily restored by injecting 1 µL of hexane after 10 or more analytical runs (Zhang, 1996, unpublished).

Metals present in soil hydrolysates may interfere with the production of volatile derivatives (Parsons, 1981). Benzing-Purdie (1981) controlled the presence of iron and aluminium by atomic absorption analyses and found that almost all metallic cations had precipitated during complete reduction. It seems, therefore, that removal of Fe and Al by additional purification steps using ion exchange (Stevenson, 1983) or solvent extraction (Sowden, 1969) is not absolutely necessary for GLC analyses of amino sugars in 6 N HCl soil hydrolysates, as long as iron contents are not extremely high. Thus, the alditol acetate methods worked well for samples with low to medium iron contents, such as Mollisols, Inceptisols, Alfisols or organic materials. The situation may be different when analyses are performed on Fe rich soils, such as Oxisols.

Coelho et al. (1997) applied the alditol acetate method to humic acids extracted from Oxisols rich in Fe and Al. These elements were found in the humic acid extracted, and interfered with the reduction step; as a result, amino sugar levels were very low (Coelho, 1998, personal communication). The problem was overcome by using additional borohydride, and additional time for clean-up. With this modification the values obtained for the amino sugars became higher, and recovery of amino sugars and internal standard added prior to hydrolysis exceeded 80% (Coelho, 1998, personal communication). From these results it seems reasonable to speculate that high Fe(III) contents may consume too much of the borohydride, thus rendering sugar reduction incomplete. Also Benzing-Purdie (1981) observed that amino sugar yields decreased by 40% if too much borohydride was consumed for Fe reduction. It is suggested that the alditol acetate method in its present form has to be applied with care on samples with high iron contents, such as Oxisols, and perhaps Ultisols or spodic B horizons. If done, more reduction reagent than presently recommended will be needed for sample processing.

In contrast, determining amino sugars from Oxisols as aldononitrile derivatives (Amelung and Zech, 1998) did not create any analytical problems. In the method of Zhang and Amelung (1996) most iron and aluminum is probably precipitated during the neutralization step, although this effect has not yet been explicitly tested. Nevertheless, high recoveries of amino sugars and myo-inositol added to the soil hydrolysate suggested that inhibition of derivatization by iron and co-precipitation of amino sugars was negligibly small for the samples under investigation (Zhang and Amelung, 1996).

Extremely high cation concentrations may occur in soil hydrolysates from calcareous soils, due to the release of Ca^{2+} cations from lime dissolution after addition of 6 N HCl. When applying the method of Zhang and Amelung (1996) to samples containing > 50% of lime, I observed that the derivatization mixture may become solid during aldononitrile formation (unpublished experiments). The problem was easily overcome by removing excessive Ca^{2+} prior to hydrolysis. Calcareous soil samples have to be pre-treated, therefore, with either 0.1 N HCl or, better, with 0.5 N trifluoroacetic acid at room temperature, until the lime is destroyed and pH remains low for > 2 h. After centrifugation, the weak acid and Ca^{2+} is removed by decanting before hot hydrolysis with 6 N HCl is applied to liberate amino sugars from soil material. The procedure did not alter amino sugar contents or patterns in mineral soils that were free of lime, and recovery of the surrogate standard added to lime-containing samples was sustained at > 80% (Amelung, 1998, unpublished data). The alditol acetate method has to my knowledge not yet been applied to soil samples with high lime content, i.e., it is uncertain whether excessive Ca^{2+} may create problems during derivatization.

F. Enzymatic Procedures

Five years after Millar and Casida (1970) had introduced the first method for indirectly assessing muramic acid in soil by colorimetric determination of lactic acid after alkaline hydrolysis, Moriarty (1975) suggested an enzymatic measurement of D-lactic acid derived from muramic acid hydrolysis. The method was designed for sediments ingested by animals, but it has to my knowledge not yet been tested for terrestrial soil.

IV. Comparative Evaluation of the Different Procedures Involved in the Determination of Amino Sugars from Soil and Tentative Suggestions for Future Standardization

As already indicated, different procedures have been used for (A) sample preparation, (B) hydrolysis, (C) purification, and (D) determination. As in the former sections procedures have been discussed concerning their technical feasibility, it is the objective of this section to evaluate these procedures.

A. Sample Preparation

There is as yet little consensus about sample pretreatment prior to hydrolysis. Extraction is performed on either air-dried, field-moist or frozen soil, but there are only few studies on the effect of sample pretreatment and storage on amino sugar yields. Moreover, the optimum sample storage may be different for different types of samples (West et al., 1987; Zelles et al., 1991). Errors from improper sample pretreatment may result from both (i) poor representativity of the sample aliquot used for amino sugar determination and (ii) changes of amino sugar contents with storage time.

In order to make measurements more reproducible, a representative aliquot has to be taken from the sample. This is most easily achieved when the samples have been ground. Especially for heterogeneous samples, such as those rich in particulate organic matter, it is recommended, therefore, to use air or freeze drying followed by milling. Chantigny (1998, personal communication) observed that compared to the analyses of moist aggregates, previous drying and grinding reduced the error of determination by about 5%. Amino sugars not only reflect living but also dead biomass and stabilized microbial residues in soil (Parsons, 1981; Chantigny et al., 1997). Thus, a decrease of microbial activity due to the removal of water (Sparling et al., 1986; Stott et al., 1986) should not affect the result significantly as long as storage time is not too long.

Storing air-dried soils for 2 weeks did not systematically change the glucosamine contents on the average of three samples. Amino sugar levels were less stable if samples were kept under moist conditions (West et al., 1987). Thus, for amino sugar determination and short-term storage of soil (2 weeks), air-drying is preferred to storage under moist conditions. This is different from the determination of metabolic related parameters such as microbial biomass, ATP content, and enzyme activities which have to be assessed as soon as possible after sampling (Zelles et al., 1991).

A detailed study on the effect of long-term sample storage on amino sugars in different organic and mineral soils was conducted by Zelles et al. (1991). As individual samples behaved differently, I used their data and conducted a (4 × 3) multivariate analysis of variance (StatSoft, Inc. 1996) for statistical evaluation, the effect of different sample storage and storage times on amino sugar levels. It should be noted, however, that Zelles et al. (1991) used a 5-day reactivation at 4°C before determining glucosamine and muramic acid in thawed, formerly frozen samples or re-moistened, formerly air-dried samples.

The statistical evaluation of the data of Zelles et al. (1991) revealed that different kinds of sample storage did not affect muramic acid, and little affected glucosamine levels, except that storage at 4°C and –140°C caused significant glucosamine losses after the first month of storage time (Table 1). The amino sugar pattern was also not much affected by different kinds of sample storage. Only after 2 months the increase in glucosamine exceeded the proportional increase of muramic acid in air-dried and –140°C frozen samples. This effect was no longer present, however, after 20 months of storage time (Table 1).

None of the storage practices prevented an increase in amino sugar levels after 20 months (Table 1). Obviously, storage time had a more important effect on amino sugar levels than storage practice. The increase in amino sugar concentrations after storage of 20 months is difficult to explain, and according to Zelles et al. (1991) it cannot be excluded that it is a methodological artifact. Nevertheless, as amino sugars should not be used as an indicator for living cells (see above), I do not recommend reactivating frozen, air-dried or any other long-term stored soil samples for amino sugar determination. Whether amino sugar levels might be better sustained for longer storage periods when microbial biomass is killed prior to storage, e.g., by fumigation or, perhaps, by freeze-drying or milling, remains to be tested.

Among the different procedures used, freezing is preferred, because it sustained best other biological parameters, and also because it was the only storage practice that kept amino sugar levels constant for the first 2 months (Table 1). Recalling the requirements for making the analyses more reproducible (see above), I recommend air-drying the frozen samples and to mill them prior to hydrolysis, or, to keep air-dried and milled samples frozen until processed. This should guarantee save storage of samples for at least the first 2 months after sampling.

B. Amino Sugar Extraction

As alkaline conditions may deaminate amino sugars, acid hydrolysis has to be used to liberate them from soil. The commonly used procedures for assessing total amounts of the respective amino sugars in soil, however, differ with respect to type of acid, acid strength, hydrolysis time, and hydrolytic conditions (Table 2).

Type of Acid and Acid Strength

Hydrolysis with 2.5 N acetic acid has been suggested for the determination of N-acetylated amino sugars in soil (Skujins and Pukite, 1970). N-acetylated amino sugars might also be recovered in 4 M trifluoroacetic acid hydrolysates (1 h, 125°C; Neeser and Schweizer, 1984). However, this has not yet been explored for soil. Total amino sugar hydrolyses with 4 M trifluoroacetic acid (4 h, 105°C) was incomplete (Zhang and Amelung, unpublished data). Kraus et al. (1990) suggested using sulfuric acid for amino sugars from biological material, but Spiteller (1980) found that sulfuric acid (26 N, 20°C, 16 h, followed by 1 N, 5 h, reflux) was not suitable for the amino sugar release from soil.

A noticeable proportion of amino sugars was already extracted with 1 N HCl (3 h, reflux), being the first step in procedures suggested for sequential amino sugar extraction (González-Prieto and Carballas, 1988, 1992; González-Prieto et al., 1997). This acid concentration, however, is too low for complete recovery of chitin-derived glucosamine. Smithies (1952) reported that after presoaking with 12 N HCl for 2 to 3 days, 6 N HCl (6h, 100°C) was suitable for hydrolyzing chitin, the hydrolysis being incomplete with lower HCl concentration. Thus, on assuming that chitin was a major source of

Table 1. Changes of muramic acid and glucosamine during sample storage (averages from a sandy, loamy, forest A, peat, and forest Oi horizon; data calculated from Zelles et al., 1991)

	Storage time (months)					
	Muramic acid			Glucosamine		
Storage temperature	1	2	20	1	2	20
	changes compared to fresh soil samples (%)					
4 °C	−21a	03a	57b*	−25a*	17c	39d*
−18 °C	−28a	−08a	74b*	14b	05bc	45d*
−140 °C	−20a	−07a	69b*	−21a*	21c*,#	42d*
+21 °C	−08a	04a	60b*	03b	30c*,#	58d*

Identical letters (a–d) within a column or within a row of muramic acid or within a row of glucosamine indicate that the changes were not significantly different at the $P < 0.05$ probability level. The asterisks (*) indicate that the amino sugar contents were significantly different from those of freshly analyzed soil ($P < 0.05$). The mark (#) indicates that the changes of glucosamine were significantly different from those of muramic acid at similar storage time and conditions.

amino sugars in soil, this hydrolysis procedure was frequently adopted (Table 2). Later experiments confirmed that 6 N HCl generally yielded more amino sugars than lower (3 N, Benzing-Purdie, 1981; 4.5 N, Zelles, 1988) or higher HCl concentrations (9 N and 12 N; Zelles, 1988). Parsons and Tinsley (1961) suggested that in order to suppress amino sugar decomposition, a mixture of 6 N HCl and 90% formic acid (1 h, 100°C, closed vessel) is superior to 6 N HCl alone, but these experiments need verification for different soils and prolonged hydrolysis times.

It is unclear to me whether pretreatment of soil with 6 N (Joergensen and Meyer, 1990; Scholle et al., 1993) or concentrated HCl solutions (Stevenson, 1965) at room temperature is really necessary. According to Stevenson (1957), however, the presoaking step with 12 N HCl facilitated hydrolysis, probably because better wetting of particulate SOM rendered hydrolyses more reproducible.

In summary, 6 N HCl gave the best results for complete amino sugar release from soil and has been most frequently recommended (Table 2). Nevertheless, it should be recalled that hot 6 N HCl may destroy rare amino sugars, such as sialic acid, arabinosamine (Parsons, 1981), or fucosamine (Benzing-Purdie, 1980), and that N-acetylated amino sugars are deacetylated (Parsons, 1981).

Hydrolysis Time

Smithies (1952) found that > 86% of chitin-N was released as ammonia upon the distillation method of Tracey (1952) during a 3 to 15 h incubation period with 6 N HCl at 100°C (samples were presoaked in 12 N HCl for 2 to 3 days), without showing distinct maxima. Ekblad and Näsholm (1996) hydrolyzed chitin at 1, 4, 7, 10, 13, 16, 19, and 26 h (6 N HCl, 100°C, without presoaking in 12 N HCl) and found that chitin was recovered best after a 7-h hydrolysis. Experiments with standard solutions confirmed that about 87% of glucosamine was recovered after presoaking the soil for 2 to 3 days in 12 N HCl followed by 6-h hydrolysis with 6 N HCl (100°C; Bremner and Shaw, 1954; Stevenson, 1957). As acceptable losses were also reported for 9-h hydrolysis (19%; Stevenson, 1957), 6 to 9 h of hydrolysis time has been recommended to release amino sugars from soil (Mehta et al., 1961; Bremner, 1965). With the introduction of modern chromatographic techniques, the effect of different hydrolysis times on yields of different amino sugar monomers could be reexamined. Zhang and Amelung (1996) confirmed that maximum amino sugar yields were obtained after 6 h for an organic soil and after 8 h for a mineral soil sample (Figure 7). Zelles (1988) observed substantial glucosamine degradation after 9 h, and even after 6 h in one of his samples (Figure 7), whereas Chantigny (1998, personal communication) did not find significantly different amino sugar yields with 3 and 6 h hydrolysis time. In agreement with the previous work on degradation of chitin, I suggest that minimum hydrolysis time should be 6 h, unless specifically tested for a given sample (e.g., Zelles, 1988). Maximum hydrolysis time must not exceed 8 h (Sowden, 1959; Zhang and Amelung, 1996), because longer times may cause substantial amino sugar degradation (Sowden, 1969; Petruzzelli et al., 1974; Zelles, 1988). After prolonged hydrolysis (24 h, reflux) 6 N HCl may even yield less amino sugars than 3 N HCl (Schnitzer and Hindle, 1981). Gallali et al. (1975) suggested applying a second hydrolysis (24 h, 6 N HCl, 105°C) after the first extraction of amino sugars with 6 N HCl.

The recovery of glucosamine upon HCl hydrolysis was not significantly affected by the presence of soil and various authors recommended the use of a correction factor to allow for decomposition of hexosamines during the hydrolysis (Bremner and Shaw, 1954; Stevenson, 1957). Proposed correction factors for glucosamine are 1.14-1.25 for 6-h hydrolysis (Bremner and Shaw, 1954; Stevenson, 1957), 1.25 for 8-h hydrolyses (Sowden, 1959) and 1.19 for 9-h hydrolysis (Stevenson, 1957). Nevertheless, when correcting for amino sugar decomposition, it must be considered that decomposition may not only occur during hydrolysis but also during sample processing. In addition, Zelles (1988) showed that optimum hydrolysis time depended on sample type, and that decomposition of muramic acid was different than that of glucosamine. Consequently, a correction factor that is observed for a specific soil and a specific method is not necessarily the same for another soil or another method. Therefore, the use of correction factors is not recommended, unless specifically tested.

Hydrolysis Conditions

Hydrolysis has been either performed continuously or successively, under reflux or in closed vessels (Table 2), and with different ratios of HCl:soil (see Section III). In all cases it was performed under anaerobic conditions. Oxygen in solution may decompose amino sugars (Zelles, 1988; Chantigny, 1998, personal communication).

Janel et al. (1978) reported that successive hydrolyses yielded better estimates on stability and concentration of nitrogenous compounds than did continuous hydrolysis, but these experiments were conducted on beech litter and not on mineral soil, and they were conducted with 3 N HCl and not with 6 N HCl. Higher yields by successive hydrolysis were not confirmed by Gonzáles-Prieto and Carballas (1988). Schnitzer and Hindle (1981) found even more amino-sugar N by continuous hydrolysis.

In many cases amino sugars were extracted in closed vessels under N_2 at 100 to 110°C (Table 2). According to Benzing-Purdie (1981) this yielded similar amino sugar concentrations as under reflux conditions (18 h, mineral soil). In contrast, Zelles (1988) reported for an organic soil a more effective hydrolysis of glucosamine under reflux than in a closed vessel under nitrogen (6 h), while muramic acid was not affected. An explanation was not given. I am not able to say whether lower amino sugar yields in experiments in closed vessels might be due to insufficient N_2 supply from a 30-s gassing with N_2. However, Chantigny et al. (1997; Chantigny, 1998, personal communication) had to extend the gassing period to 2 min. Moreover it seems reasonable to speculate that hydrolyses in closed vessels may increase the risk of insufficient contact of particulate SOM and HCl, especially when swimming light material is attached to the glass walls (personal observation, and Cheshire, 1996, personal communication). The latter effect is certainly a bigger problem for nonmilled material as used by Zelles (1988), but it can be minimized when using N_2 to "blow" these particles back into the solution (Zhang, 1997, personal communication). I conclude that reflux may facilitate obtaining a better amino sugar yield from soil. Nevertheless, hydrolysis in closed vessels is simpler and allows the simultaneous hydrolysis of more samples. Longer gassing with N_2 (e.g., 2 min; Chantigny, 1998, personal communi-

Table 2. Examples for sample pretreatment and recommended hydrolyses used in common procedures for amino sugar determination in soil samples

Method of determination	Authors	Acid	Hydrolysis time	Hydrolysis temperature
Reviews				
	Bremner (1965) Mehta et al. (1961)	6 N HCl	6 - 9 h	100°C
	Stevenson (1965)o	12 + 6 N HCl	60 h + 6 h	25°C + reflux
Alkaline Decomposition				
	Azam et al. (1989a,b)a	6 N HCl	9 h	reflux
	Gallali et al. (1975)a	6 N HCl	8 h + 24 h (successive)	105°C
	Gonzáles-Prieto and Carballas (1992)a	1 + 3 N HCl	3 h each (successive)	110°C
	Smithies (1952)[1]	6 N HCl	6 h	100°C
Colorimetry				
DMAB	Stevenson (1983)a	12 + 3 N HCl	48 h + 6 h	25°C + reflux
Ninhydrin	Grant and West (1986)	4 N HCl	4 h	100°C
DMAB, modified	Parsons and Tinsley (1961)a	6 N HCl + 90 % HCOOH (1:1, v/v)	1 h	100°C
DMAB, modified	Scheidt and Zech (1990)m	6 N HCl	4 h	100°C
DMAB, modified	Skujins and Pukite (1970)	2.5 N acetic acid[2]	3 x 10 min.	not mentioned
Lactic acid detection	Millar and Casida (1970)o	6 N HCl	4 h	reflux
Amino acid analyzers				
	Calderoni and Schnitzer (1984)x	6 N HCl	24 h	reflux
	Joergensen and Meyer (1990)o	6 N HCl	24 h + 6 h	25°C + reflux
	Nannipieri et al. (1979)x	6 N HCl	24 h	110°C
	Schnitzer and Hindle (1981)x	3 N HCl	≥ 4 h (successive or continuous)	reflux
	Sowden and Schnitzer (1967)	6 N HCl	20 h	reflux
	Yoshida and Kumada (1979)x	6 N HCl	6 h	reflux

HPLC

FMOC-Cl	Ekblad and Naesholm (1996)[1]	6 *N* HCL	7 h	100°C
OPA-fluorescence	Zelles (1988)o,	6 *N* HCL	3 h	reflux
	Chantigny et al. (1997)m			

GLC

Alditol acetates	Benzing-Purdie (1981; 1984)o	6 *N* HCL	18 h	105°C
	Coelho et al. (1997)m	6 *N* HCL	6 h	105°C
Aldononitrile acetates	Zhang and Amelung (1996)o	6 *N* HCL	6-8 h	105°C
Trifluoracetates	Kögel and Bochter (1985)o	6 *N* HCL	6 h	105°C
Trimethylsilyation	Casagrande and Park (1978)a	4 *N* HCL	4 h	reflux ?[3]
Trimethylsilyation	Mielniczuk et al. (1995)o	4 *N* HCL	overnight	100°C
Timethylsilyl oximes	Hicks and Newell (1983)o[1]	6 *N* HCL	4.5 h	100°C
	Hicks and Newell (1984)a[1]	6 *N* HCL	11 h	100°C

a = application of the original method; o = original method (as described); m = modified method; x = own analytical chromatographic conditions; DMAB = dimethylaminobenzaldehyde; HPLC = high performance liquid chromatography; FMOC-Cl = 9-fluorenmethylchloroformate; OPA = o-phthaldialdehyde; GLC = gas-liquid chromatography
[1] not tested for soil; [2] for N-acetylated glucosamine. [3] temperature not mentioned.

Table 3. Comparison of methods used for amino sugar determination with regard to detection of individual monomers, sensitivity, precision, and speed

Method of determination	Sum	GlcN	GalN	Mur	ManN	CV (%)	LLQ (ng)	Sample wt. (g)	Speed (samples week^{-1})
Alkaline decomposition	x	—	—	—	—	15	5000	5-10	<10[†]
Colorimetry									
Dimethylaminobenzaldehyde (DMAB)	x	(x)[a]	(x)	—	—	15	5000[b]	2-10	<10[†]
DMAB, modified to make purification redundant	x	—	—	—	—	10	n.d.	200	>40[‡]
DMAB, modified for N-Ac	N-Ac	—	—	—	—	n.d.	1000[c]	2	>40[‡]
Paper chromatography	—	x	x	—	—	15	n.d.	2.5	10-20[‡]
Amino acid analyzers	—	x	x	—	—	15	500	2-4	40-50
HPLC									
0-phthaldialdehyde (OPA)	—	x	—	x	—	<15[d]	10[c]	0.5-1	40-60
9-fluorenmethylchloroformate (FMOC-Cl)	—	x	—	—	—	<10	35	n.d.;0.1[#]	40-60[§]
GLC (+Detector)									
Alditol acetylation (flame ionization)	—	x	x	—	—	10	10[¶]	0.1	4
Aldononitrile acetylation (flame ionization)	—	x	x	x	x	10	10[f]	0.1-0.3	16-24
Trifluoroacetylation (thermionic N,P selective)	—	x	x	—	—	15[g]	50	n.d.; 1[#]	4[§]
Trimethylsilylation (mass spectrometry)	—	—	—	x	—	<10	5	0.02-0.05[#]	20-30[§]

GlcN: glucosamine; GalN: galactosamine; Mur: muramic acid; ManN: mannosamine; CV: coefficient of variation; LLQ: routine limit of detection; N-Ac: N-acetylated derivatives; n.d. = not determined, [#]: Sample weight given for organic samples only; multiplied by a factor of 10 for samples from mineral soils; [§]: time for sample processing estimated from similarities to the other HPLC (OPA) and GLC (alditol acetate) method, respectively. [†]: ion-exchange procedure needed for purification is said to be suitable for processing only a very limited number of samples (Benzing-Purdie (1981); [‡]: estimated from publication; [¶]: assuming a 1 μL injection volume; a) only with modifications; b): Allison and Smith (1965); c): 100 ng according to the original method of Reissig et al. (1955); d): after improving peak resolution by Chantigny et al. (1997); e): muramic acid, 100 ng for glucosamine (Chantigny et al., 1997), can probably be extended to 25 pg used for linearity tests (Zelles, 1988); f): muramic acid, 20 ng for glucosamine; g): recorded for spiked samples.

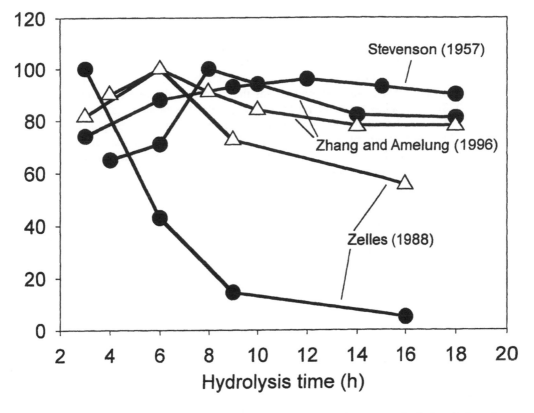

Figure 7. The effect of hot 6 *N* HCl on glucosamine release from organic (triangles) and mineral soil samples (dotted symbols) using sample averages from different authors.

cation), the use of air-dried, milled material, and presoaking with 12 *N* HCl (Stevenson, 1957) might perhaps improve amino sugar yields obtained at 100 to 110°C under N_2.

Next to hydrolytic conditions, Zelles (1988) investigated the effect of different HCl volumes (10, 20, 30 ml) on the release of muramic acid and glucosamine from 1-g samples of different soils. He found that increasing amounts of HCl per g sample released different amounts of muramic acid and glucosamine upon hydrolysis, and recommended to use 20 ml g^{-1} soil. The results, however, were difficult to standardize, because (i) the effect was different for different samples, (ii) the effect was different for glucosamine and muramic acid, (iii) 1 g of different samples contained different amounts of amino sugars, and (iv) the solid volume varied due to more space needed by 1 g of organic compared to 1 g of mineral soil. Thus, it is simply recommended here not to work at a ratio of soil:HCl below 1:20 (w/v).

I could not give any explanation how a higher HCl volume (at constant molarity) might negatively affect the amounts of amino sugars released upon hydrolysis. Nevertheless, amino sugar may also be destroyed upon evaporation. In this context it seems possible that an increasing HCl volume may lead to decreasing amino sugar recoveries, because longer evaporation times are needed for HCl removal.

C. Amino Sugar Isolation

Isolation procedures for amino sugars involved removal of HCl and other interfering compounds. The HCl removal was usually accomplished by rotary evaporation under reduced pressure (40 to 45°C). The temperature should not exceed 45°C (Allison and Smith, 1965). HCl removal could as well be achieved by freeze-drying, but evaporation was the milder of the two procedures (Dawson and Mopper, 1978). Even at low temperatures (< 40°C) considerable amino sugar losses may occur as a result of wall-induced condensation reactions with other co-extracted components (Dawson and Mopper, 1978). Among them, phenolic groups in particular are known to form condensation products with amino sugars (Bondietti et al., 1972). As this effect depends on quantity and quality of the co-extracted compounds, concentration of pure standard mixtures to dryness does not necessarily reveal these losses. To overcome these drawbacks, the walls of the vessels should be either silylated using trimethylchlorosilane, 1% (v/v) in toluene (Grant and West, 1986) or coated with glycerol (50% in ethanol (Dawson and Mopper, 1978). As silylation of glassware is expensive and requires long reaction times, addition of 0.6 ml of the glycerin solution seem advisable to prevent amino sugar losses during evaporation (Hicks and Newell, 1983).

Interference of other compounds with amino sugar determination has been mainly reported for methods based on alkaline decomposition or coloration reactions (including amino acid analyzers). For alkaline decomposition and coloration detection of amino sugars with Ehrlich's reagent, both organic and inorganic impurities have to be removed with resins (e.g., Stevenson, 1965). For the removal of dark-colored humic material, anion exchange resins (Stevenson, 1957; 1983) or neutralization to pH 6.6 to 6.8 (Zhang and Amelung, 1996) have been recommended. The latter probably promoted additional removal of Fe. However, the introduction of new salts changes ionic strength and may thus not be advisable for the colorimetric determination of amino sugars. If only iron is to be removed, the use of solvent extraction with acetylacetone (pH 4) was superior to alkaline elution of amino compounds after resin extraction (Sowden, 1969). Alternatively, solvent extraction of iron with methyl isobutyl ketone (4-methyl-2-pentanone; Beckwith and Parsons, 1980) has been recommended. Parsons and Tinsley (1961) washed the DOWEX 50 cation exchange columns with 5 ml of 0.005 N EDTA in 0.1 N NaCH$_3$COO buffer (pH 5.8) for Fe removal, thus allowing for the final elution of amino sugars with 2 N HCl as outlined by Boas (1953).

When Fe was extracted with acetylacetone chloroform, care had to be taken to watch the pH value. Sowden (1969) observed considerable losses of both glucosamine and galactosamine at pH 6. At pH 4 Sowden (1969) recovered 95% of both amino sugars after removing iron with acetylacetone from 6 N HCl hydrolysates obtained from clayey soil, whereas recoveries decreased to 78 and 89% of glucosamine and galactosamine, respectively, when this method was applied to very impure hydrolysates of a Spodosol. In such cases, repeated Fe removal using acetylacetone might be advisable.

As amounts and effects of interfering compounds vary with sample type and method, it must be concluded that there is no uniform purification procedure for all methods and all soils, i.e., recoveries of spiked amino sugar levels should be tested when a method is newly applied to sample types that had not been included into the methodological tests, such as Oxisols and spodic horizons rich in Fe.

D. Amino Sugar Measurement

There has not yet been a comparative study on the determination of amino sugars from soil using all common methods described in Section III. Consequently, it is not possible to say which of the methods best approached the real amino sugar contents in soil. Nevertheless, comparison of methods is possible with respect to their (i) precision and accuracy, (ii) sensitivity, (iii) simplicity, (iv) selectivity, (v) potential interference with other compounds, and yet proven applicability to various types of soil samples.

Precision

The overall accuracy of any method is usually split into random errors (= precision) and systematic errors (= bias). Precision actually refers to repeatability and reproducibility in interlaboratory analyses; errors of this type are caused by apparent random factors as they are not under the control of the analyst (Egli, 1995). As the real amino sugar content in soils remains unknown, and as interlaboratory comparisons are lacking, the "precision" of a given method in soil science was estimated by the repeated measurement with known amounts of spiked amino sugar contents (Stevenson, 1957). In doing so, the real precision is overestimated due to the control of errors by the analyst.

The recovery of standard additions of amino sugars generally exceeded 85% of initial spike level during sample processing (data not shown). Only Kögel and Bochter (1985) recovered less glucosamine (62%) and mannosamine (45%) when adding these compounds to an Oi horizon (it was rich in lignin) prior to hydrolysis. It was assumed that these amino sugar losses might be due to condensation reaction with phenolic groups upon hydrolysis (Bondietti et al., 1972; Kögel and Bochter, 1985). All other methods succeeded to reproduce high recovery of spiked amino sugars at ≤ 5% analytical error, i.e., they appear to have similar "precision" in terms of Stevenson (1957). More or less similar results from amino sugar determination with either alkaline decomposition and colorimetric techniques (Bremner, 1965) or with GLC based upon alditol acetate derivatization and measurements with amino acid analyzers (Benzing-Purdie, 1981) support this finding.

The reproducibility of a measurement usually decreases with decreasing concentration of the analyte, i.e., for a more comprehensive discussion of precision its reproducibility at routine analyses also should be considered. The reproducibility of a determination was frequently assessed by the coefficient of variation (CV), and the average CV commonly ranged from 10 to 15% (Table 3). Individual errors may well amount to 23% (Grant and West, 1986) or 37% deviation of the means (Zelles et al., 1991). Higher CV reported for amino sugar determination with amino acid analyzers (25% on average; Joergensen and Meyer, 1990, Table 3) resulted from repeated analyses of field replicates, and did not represent only analytical errors which range about 15% as well (Table 3; Joergensen, 1998, personal communication). Obviously, none of the methods available is substantially more reproducible than the other ones.

Sensitivity and Lower Limits of Quantification

In a strict sense, the *sensitivity* of a method corresponds to the slope of the calibration curve (Beyermann, 1982). As far as this curve was shown, its slope usually approached the value one, i.e., all methods seemed to have similar sensitivity. In a broader respect, however, sensitivity also should be discussed with respect to the *lower limit of quantification* (LLQ).

The *limit of quantification*, i.e., the lowest signal at which a given measurement precision (such as with 10% relative standard deviation) can be assured for the quantification of a particular *detected* compound (Currie, 1968; Beyermann, 1982), has not been assessed for the methods used for amino sugar determination from soil. Hence, the LLQ has been suggested for practical purposes, corresponding to the lowest concentration of a compound used for calibration (Egli, 1995). In this context, the HPLC and GLC methods were superior to alkaline decomposition and colorimetric procedures by two orders of magnitude (Table 3).

The LLQ is assessed for pure standard mixtures, whereas consideration of average sample weight might highlight the sensitivity of the method for soil hydrolysates. The lowest sample weights have been used by Mielniczuk et al. (1995). Slightly higher sample weights have been used f or HPLC com-

Table 4. Comparison of methods used for amino sugar determination with regard to applicability tested for hydrolysates from different terrestrial soil samples

Method of determination	Applicability proved for					
	Oxisol	Andisol	>50% lime	Organic soil	Other mineral soil
Alkaline decomposition					
Colorimetry					
Dimethylaminobenzaldehyde (DMAB)	x	—	—	x	x
DMAB, modified to make purification redundant	x	—	—	x	x
DMAB, modified for N-Ac	—	—	—	x	x
Paper chromatography	—	—	—	—	soil not described
Amino acid analyzers	x	x	x	x	x
HPLC					
O-phthaldialdehyde (OPA)	—	—	—	x	x
9-fluorenmethylchloroformate (FMOC-Cl)	—	—	—	x	—
GLC (+Detector)					
Alditol acetylation (flame ionization)	x	—	—	x	x
Aldonitrile acetylation (flame ionization)	x	x[1]	x	x	x
Trifluoroactylation (thermionic N,P selective)	—	—	x	x	—
Trimethylsilylation (mass spectrometry)	—	—	—	—	x (dust)

x: application tested; —: applicability not proved; [1]unpublished data.

pared to GLC, but for HPLC analyses only an aliquot (usually 1 of 10 mL of the hydrolysate) is finally used for determination, and its LLQ can probably be extended to pg level (Zelles, 1988). Much higher amounts of sample were needed for alkaline decomposition and colorimetric procedures including amino acid analyzers (Table 3). Only the modified procedure of Scheidt and Zech (1990) seemed to be as sensitive as GLC and HPLC procedures.

Simplicity

Colorimetric methods including paper chromatography as well as distillation procedures have the advantage that they may be performed in laboratories with simple technical equipment. Nevertheless, the purification procedures needed to remove interfering iron and neutral sugars make these procedures time-consuming. In this respect, amino acid analyzers or the method of Scheidt and Zech (1990) provide more convenient tools for amino sugar determination. HPLC procedures seemed to be the fastest alternative for amino sugar determination, having the additional advantage that muramic acid can be detected after reaction with o-phthaldialdehyde and mercaptoethanol (Zelles, 1988; Chantigny et al., 1997). The GLC procedures of Zhang and Amelung (1996) or Mielniczuk et al. (1995) have been of intermediate speed, whereas only few samples may be processed in a week with the alditol acetate method (Table 3).

Selectivity

As indicated earlier, different classes of soil microorganisms produce different amino sugars. Identification of individual amino sugars is not possible with colorimetric and distillation techniques, because they only allow for detection of bulk amino-sugar N. Separation of glucosamine and galactosamine can be achieved with different ion-exchange chromatography designs (Stevenson, 1983; Joergensen and Meyer, 1990), but only GLC and HPLC techniques gave access to the simultaneous quantification of glucosamine and muramic acid (Table 3). The separation efficiency of capillary GLC is superior to the other methods, so much so that different anomers are commonly separated. This problem can be overcome by suitable derivatization techniques. In this respect, the aldononitrile acetate method of Zhang and Amelung (1996) is presently the only one that can be used for the simultaneous determination of glucosamine, galactosamine, muramic acid, and mannosamine from soil, i.e., it is most selective for the determination of individual monomers.

Interfering Compounds: Specificity, and Applicability

Interfering compounds have two effects on the amino sugar determination. First, they may show up in chromatograms as an additional signal during detection, i.e., they may reduce specificity. Second, they may interfere with the derivatization reaction, i.e., they may reduce sensitivity and cause erratic results, due to incomplete derivatization.

The *specificity* corresponds to the ability of a given method, to determine only the compound the method has been designed for (Beyermann, 1982). It can be easily estimated from peak numbers in chromatographic techniques. In this context, the GLC methods of Benzing-Purdie (1981, 1984) and Zhang and Amelung (1996) provide a higher specificity than the HPLC procedure of Zelles (1988), even after the improvements by Chantigny et al. (1997), due to less additional peaks in the GLC chromatogram. In contrast, the method of Kögel and Bochter (1985) showed more additional peaks, i.e., lower specificity. The specificity of amino sugar determination by amino acid analyzers cannot be high in principle, because these methods were not explicitly designed for amino sugar determination.

The low selectivity and specificity of colorimetric and distillation procedures was additionally complicated by the fact that they showed interference from iron and amino acid plus neutral sugar mixtures, rendering these procedures time-consuming due to the additional steps required for sample purification. The method of Scheidt and Zech (1990) was claimed to be free from interfering compounds. Amino acid analyzers can be easily used for amino sugar determination without purification steps, but as iron and aluminum may alter retention times of amino acids, longer columns and alterations of the buffer programs may be required for simultaneous assessment of amino sugars with amino acids (Sowden, 1967).

The HPLC method of Zelles (1988) does not include a purification step; however, as discussed in Section III.D, additional control of pH after the derivatization step might be advisable, especially when the procedure is applied to samples other than those tested by the authors. How far primary amines and amino acids interfered in this method remains to be examined, e.g., by mass spectrometry. According to Chantigny (1998, personal communication), interference from amino acids is not very likely, because elution conditions used to determine amino sugars on HPLC are unlikely to be the same hydrolytic for amino acids. Moreover, hydrolytic conditions used to extract amino sugars are not optimal for the release of amino acids (see Section IV.B). Iron may also interfere with the formation of alditol acetate derivatives (Benzing-Purdie, 1981), especially in Oxisol samples (Coelho, 1998, personal communication), due to consumption of borohydride needed for the reduction step. The problem can be overcome by increasing the amount of borohydride for the reduction step. In contrast, there seemed to be no important influence of iron on aldononitrile acetate formation. I observed, however, that > 50% lime in the samples may release too much Ca^{2+} upon hydrolysis, rendering further sample processing impossible due to the formation of gel-like structures during the derivatization step. The extent to which high Ca^{2+} concentrations may cause problems in the HPLC and other GLC procedures remains to be tested. A prior dissolution of lime with cold 0.5 M trifluoroacetic acid is recommended.

New analytical problems usually occur when a method is newly applied to samples it has not been tested for. In my opinion, special problems generally may occur when soils contain considerable amounts of Fe (i.e., Oxisols), Al (i.e., Andisols), Ca (i.e., lime bedrock or lime in special soil fractions), or humic-like substances (i.e., in Oa horizons, peat or humic acid extracts). In this context, least surprise might be expected when using amino acid analyzers for amino sugar determination, whereas the HPLC methods or the modified colorimetric procedures (Skujins and Pukite, 1970; Scheidt and Zech, 1990) have been applied only to a small range of soils; the range of soils analyzed by the aldononitrile acetate method was intermediate (Table 4).

V. Amino Sugars as Indicators for Microbial Residues in Soil: Problems of Interpretation

In the earlier sections discussion was restricted to technical details of the different methods and to efforts of quantifying individual amino sugar monomers, but little has been said about how reliably amino sugar data may be used as markers for different microbial residues in soil. In order to highlight this question, and because new information has become available after Parsons' review (1981), a short explanation is given about the origin of amino sugars in soil.

Plants do not synthesize significant amounts of amino sugars. Thus, amino sugars in soil are mainly of microbial origin (Parsons, 1981; Stevenson, 1982).

Glucosamine (chitosamine) is widely distributed among soil organisms. Higher fungi and invertebrates, especially arthropods, contain chitin that consists of β-(1 ⟶ 4)-linked N-acetylglucos-amine. Non-acetylated glucosamine polymers form chitosan, which can be also an important constituent of fungal cell walls (Parsons, 1981). Moreover, glucosamine has been found in the gut lines of earthworms and nematode egg shells (Bird and McClure, 1976), in polysaccharides of mollusks, and in the gelatin of snails (Sharon, 1965). In bacteria glucosamine is most common in the peptidoglycan cell wall, where its N-acetylated derivative is β-(1 $\mapsto$ 4)-linked to N-acetyl muramic acid. The peptidoglycan is an important constituent of gram positive bacteria and has minor importance in gram negative ones. Additional N-acetyl-glucosamine is also found in teichoic acids of the gram positive bacteria, in lipopolysaccharides of the gram negative ones, and in cell walls and capsular polysaccharides.

Due to the occurrence of glucosamine in bacterial products other than peptidoglycan, the glucosamine to muramic acid ratio of bacteria is > 1. Joergensen et al. (1995) found that the ratio of glucosamine to muramic acid ranged from 2 to 8 (3.7 ± 2.5, on average) for four different bacteria. In soil samples this ratio varied between 4 (Ap horizon under wheat, Chantigny et al., 1997) and 65 (forest Oe horizon; Zelles et al., 1990), with values between 8 and 20 common for mineral soil (Zelles et al., 1991; Joergensen et al., 1995; Amelung et al., 1999). It seemed thus reasonable to assume that the additional glucosamine was mainly produced by fungi. Glucosamine produced by invertebrates is of minor importance in this respect, due to the lower biomass of amino sugar-containing invertebrates compared to that of microorganisms. Grassland soils, for instance, usually have 0.1 to 10 g m^{-2} arthropods (which contain chitin-derived glucosamine), whereas fungal biomass (chitin is a cell wall component of higher fungi) amounts to 100 to 1000 g m^{-2} in temperate climates (Dunger, 1983; Curry and Good, 1992).

Muramic acid levels have been used for comparative assessment of bacterial biomass in soil (Millar and Casida, 1970). It should be noted, however, that the occurrence of muramic acid is not limited to bacteria (including actinomycetes). Also the cell walls of blue-green algae contain muramic acid in concentrations up to 50% of the dry weight (Sharon, 1965; Drews, 1973). Using muramic acid as a marker for bacterial residues is thus not possible in sediments and estuarine soils (Parsons, 1981). If algae thrive at the surface of terrestrial soils, it may be advisable to remove the first 2 to 3 mm of top soil when using muramic acid as a marker for bacterial residues.

Another important amino sugar is galactosamine (chondrosamine). It frequently occurs in capsular and extracellular polysaccharides of bacteria but it may also be part of the cell walls of bacteria, especially in actinomycetes (Sharon, 1965; Parsons, 1981), attached to lipopolysaccharides (Parsons, 1981) or teichoic acids (Ladd and Jackson 1982). Certain taxonomic classes of fungi, such as trichonomycetes, and myxomycetes, may produce small amounts of galactosamine (Sharon, 1965); the role of myxomycetes for saccharide production is still in question (Herrera, 1992). The contribution of galactosamine from polysaccharides of mollusks (Sharon, 1965) to total galactosamine content should be rather small. Although the source of galactosamine is less clearly defined, it was assumed that most of the galactosamine in soil originates from bacteria (Parsons, 1981). Increasing ratios of glucosamine to galactosamine have thus been employed to indicate a shift from bacterial to fungal derived amino-sugar N in forest soils (Sowden, 1959; Kögel and Bochter, 1985). For a more comprehensive study on fate and origin of galactosamine, however, it should be analyzed simultaneously with the other bacterial sugar, muramic acid.

Less is known about the origin of mannosamine in soils. While it was commonly found in bacterial products (Sharon, 1965; Kenne and Lindburg, 1983), Coelho et al. (1997) only recently found it in fungal melanin. Parsons (1981) suggested that mannosamine might be a decay product of sialic acid that decomposes in weak acid.

Although different amino sugars may indicate different origin, it is not possible to deduce actual numbers of bacteria and fungi from amino sugar concentrations, for several reasons: (i) As the glucosamine content of fungi varies among different species and among different age classes of a certain fungus (Hicks and Newell, 1984), average glucosamine concentrations do not quantitatively indicate fungal biomass. (ii) As gram-positive bacteria have higher amino sugar contents than gram-negative ones, bacterial biomass may not be exactly deduced from muramic acid concentrations, and changes in muramic acid concentration in soil might as well result from changes in composition of the bacterial community. (iii) Amino sugars generally survive their producers, resulting in an accumulation of amino sugars during N immobilization and initial SOM decomposition (Marumoto et al., 1972; Sowden and Ivarson, 1974), i.e., amino sugars represent both living and dead microbial residues in soil (Parsons, 1981; Durska and Kaszubiak, 1983; Chantigny et al., 1997). (iv) The turnover time of different amino sugars in dead microbial biomass may not be the same. Low concentrations of muramic acid in soil, for instance, may reflect a faster turnover, because muramic acid standard solutions do not survive for long in water (Zhang, 1998, personal communication) and have to be stored below 0°C compared with those of the other amino sugars. In addition, bacterial cell-wall constituents may be more accessible to microbial decay than fungal ones, because fungal chitin may be protected from decay by a layer of glucan or by melanin (Parsons, 1981). (v) Glucosamine and chitin are rapidly degraded when freshly added to soil, but glucosamine contents of soil are not low, thus indicating that glucosamine in soil may be stabilized against microbial attack (Bondietti et al., 1972; Parsons, 1981), i.e., it might have survived from turnover of other dead microbial residues. (vi) The stabilization of individual amino sugars in soil might be linked to soil properties; Amelung (1997) indicated by principal component analyses that the ratio of glucosamine to muramic acid was closely related to physical soil properties while the ratio of glucosamine to galactosamine did not correlate with soil parameters, although both galactosamine and muramic acid should indicate bacterial residues. However, muramic acid is larger and has a carboxylic group. Carboxylic groups favor sorption of SOM compounds to mineral particles (Kaiser et al., 1997). This might be the reason why a larger proportion of the total muramic acid concentration was attached to clay particles compared with that of other hexosamines (Zhang et al., 1998). (vii) The efficacy of amino sugar stabilization by soil minerals might be different for bacterial and fungal derived products. Clay minerals may arrange around the bacterial cells to form microaggregates, the bacterial core of it being protected from cell lysis when treated with chloroform (Foster and Martin, 1981, Foster, 1988; Ladd et al., 1993). In Mollisols, microaggregates may be held together by particulate organic matter including fungal hyphae to form macroaggregates (Tisdall and Oades, 1982; Waters and Oades, 1991; Tisdall, 1996), the organic matter of which being more susceptible to microbial decay than that of microaggregates (Elliott, 1986; Gupta and Germida, 1988). As Guggenberger et al. (1999) found a wider glucosamine-to-muramic acid ratio in macro- than in microaggregates, it seems reasonable to speculate that despite the protection by melanin or glucan (see above), fungal chitin may be more accessible to microbial degradation than bacterial peptidogycan that is encrusted by clay particles in microaggregates (for more details concerning SOM stabilization, also in other soil types, see, e.g., reviews of Tiessen et al., 1984; Martin and Haider, 1986; Oades, 1988; Ladd et al., 1993; Christensen, 1996; Feller and Beare, 1997; Guggenberger et al., 1998, and individual contributions in Lal et al., 1998).

In summary, bulk amino sugar concentrations are a quantitative indicator of microbial residues in soil. As they do not vary with weather conditions at significant scale (Chantigny et al., 1997), their investigation might be even more useful than detection of living biomass for comparative studies of long-term dynamics such as land-use effects on microbe-derived SOM (e.g., Zhang et al., 1997; 1999; Guggenberger et al., 1999), or for determining the impact of microorganisms on soil functioning such as aggregation (Chantingy et al., 1997). Different amino sugars are indicative of different microbial origin. However, quantitative estimates about the proportions of fungi and bacteria in the living biomass seem impossible. Different amino sugar ratios thus are only qualitative indicators for changes in the composition of the decomposer community. Using muramic acid as a marker for bacterial residues has the advantage that its origin is well known. However, due to a different structure its residence time in soil possibly is different from that of glucosamine. Moreover, it cannot yet be absolutely excluded that muramic acid may be decomposed to glucosamine. This reaction seems to be limited to high pH values (Millar and Casida, 1970). High pH values may occur in the gut of soil animals, such as in the hindgut of termites (pH > 12, Brune, 1998). In contrast to glucosamine, the origin of galactosamine is less clear. Its use as a marker for bacterial residues in soil has the advantage that, because of its structural similarities to glucosamine, the turnover time of the galactosamine monomer is more likely similar to that of glucosamine. Hence, differences in residence time are more likely due to differences in the turnover of the corresponding polymers. To assure that changes in amino sugar ratios really indicate a shift from fungi to bacteria or vice versa, I recommend investigating whether changes in glucosamine-to-galactosamine ratios are consistent with changes in the glucosamine-to-muramic acid ratios. This seemed to be the case for comparisons between soils (Amelung et al., 1999), but not within a given soil and its fractions (Zhang et al., 1998, 1999). Future work might succeed in identifying amino sugars in living and dead microorganisms, i.e., to utilize amino sugar data with simultaneous measurements of living biomass for an estimation of the turnover time of the microbial population in soil. This might perhaps be achieved using GC-IR-MS (gas chromatography-isotope ratio-mass spectrometry).

VI. Conclusions

Amino sugars have been used to trace microbial residues in soil. Indications on different microorganisms may be obtained by analyses of individual amino sugar monomers. This was quantitatively achieved for glucosamine, muramic acid, galactosamine, and mannosamine. At present, amino sugars can be routinely determined at ng level with about 10 to 15% accuracy. The main source of error probably results from incomplete hydrolysis.

Different procedures have been used for sample preparation, hydrolysis, purification, and determination. For future work it is suggested to air-dry and mill the samples, and to presoak them in 12 N HCl for 1 to 3 days prior to hydrolysis to improve the reproducibility of extraction. Final continuous hydrolyses should be performed with 6 N HCl (6 to 8 hours, unless specifically tested) either under reflux (if the samples are rich in floatable organic matter) or under an N_2 atmosphere in closed vessels at 105°C (which is less time-consuming). Amino sugar losses upon following evaporation of HCl may be minimized by adding 0.6 ml of glycerol to the hydrolysate (50% in ethanol). Further purification depends on the analytical procedure. For final determination of amino sugars, the use of HPLC or GLC is superior to other methods with respect to sensitivity.

If only glucosamine and muramic acid are of interest, their determination can be most rapidly and most simply achieved by an HPLC procedure based on the reaction of amino sugars with o-phthaldialdehyde and subsequent fluorescence detection. Disadvantages of this method include uncertainties whether the derivatization reaction is performed under optimum conditions. It is recommended that pH should be controlled for HPLC analyses of glucosamine and muramic acid, despite prolonging the analysis times. Possible interferences by other amino-containing compounds are presumably small but remain to be tested. In contrast to the HPLC procedure, GLC separation of amino sugars as aldononitrile derivatives is ca. three times more time-consuming, but provides advantages compared with the HPLC procedure with regard to selectivity (galactosamine and mannosamine could be additionally quantified), stability of the derivatives, and robustness of quantification against pH variations upon HCl evaporation. Peak identity has been verified for at least a range of samples (mainly Mollisols). Thus, I prefer the aldononitrile method when fundamental questions such as origin and fate of amino sugars need to be considered in a limited number of samples, but the HPLC method detecting OPA derivatives offers an interesting alternative for rapid screening of relative differences between glucosamine and muramic acid concentrations in large sample sets.

It should be pointed out that different soils may exhibit different interferences with any method of amino sugar determination and, up to now, only few soil types and diagnostic horizons have been analyzed. Before applying a method to soils it has not yet been tested for, precision and peak origin should be re-examined, e.g., by spiking experiments and mass spectrometry. For final standardization and comparative evaluation of existing data, we need to compare and calibrate our different methods on a large, diverse sample set, including samples from different soil orders and different diagnostic horizons. This has never been done for the determination of amino sugar monomers and is an urgent research demand.

Acknowledgments

I am indebted to M. Chantigny, R. Coelho, R. Joergensen, and L. Zelles for fruitful discussions of their methods, and acknowledge additional comments from K.W. Flach, L. Haumaier, and X. Zhang.

References

Alef, K. and P. Nannipieri. 1995. *Methods in Applied Soil Microbiology and Biochemistry*. Academic Press, San Diego.

Allison, D.J. and Q.T. Smith. 1965. Application of a rapid and reliable method for determination of hexosamines in the skin of estrogen-treated rats. *Anal. Biochem.* 13:510-517.

Amelung, W. 1997. Zum Klimaeinfluß auf die organische Substanz nordamerikanischer Prärieböden. Bayreuther Bodenkundliche Berichte 53, 140 pp.

Amelung, W. and W. Zech. 1998. Organic compounds of ultrasonically dispersible and non-dispersible clay fractions in an Oxisol, Brazil. CD-ROM of the Proceedings of the 16th World Congress of Soil Science, August 20–26 in Montpellier, France. International Soil Science Society.

Amelung, W., X. Zhang, K.-W. Flach and W. Zech. 1999. Amino sugars in native grassland soils along a climosequence in North America. *Soil Sci. Soc. Am. J.* 63:86-92.

Azam, F., R.L. Mulvaney and F.J. Stevenson. 1989a. Chemical distribution and transformations of non-symbiotically fixed ^{15}N in three soils. *Soil Biol. Biochem.* 21:849-855.

Azam, F., R.L. Mulvaney and F.J. Stevenson. 1989b. Transformation of ^{15}N-labeled leguminous plant material in three contrasting soils. *Biol. Fertil. Soils.* 8:54-60.

Beckwith, C.P. and J.W. Parsons. 1980. The influence of mineral amendments on the changes of organic nitrogen components of composts. *Plant Soil* 54:259-270.

Belcher R., A.J. Nutten and C.M. Sambrook. 1954. The determination of glucosamine. *Analyst* 79:201-208.

Benzing-Purdie, L. 1980. Identification of 2-amino-2,6-dideoxygalactose hydrochloride in a soil hydrolysate. *J. Agric. Food Chem.* 28:1315-1317.

Benzing-Purdie, L. 1981. Glucosamine and galactosamine distribution in a soil as determined by gas liquid chromatography in soil hydrolysates: Effect of acid strength and cations. *Soil Sci. Soc. Am. J.* 45:66-70.

Benzing-Purdie, L. 1984. Amino sugar distribution in four soils determined by high resolution gas liquid chromatography. *Soil Sci. Soc. Am. J.* 48:219-222.

Beyermann, K. 1982. Organische Spurenanalyse. In: Hulpke et al. (eds.), *Analytische Chemie für die Praxis*. Thieme-Verlag, Stuttgart.

Bird, A.F. and M.A. McClure. 1976. The tylenchid (nematode) egg shell: Structure, composition and permeability. *Parisitology* 72:19-28.

Boas, N. 1953. Method for the determination of hexosamines in tissues. *J. Biol. Chem.* 204:553-563.

Bondietti E., J.P. Martin and K. Haider. 1972. Stabilization of amino sugar units in humic type polymers. *Soil Sci. Soc. Am. J.* 36:597-602.

Bremner, J.M and K. Shaw. 1954. Studies on the estimation and decomposition of amino sugars in soil. *J. Agric. Sci.* 44:152-144.

Bremner, J.M. 1958. Amino sugars in soil. *J. Sci. Food Agric.* 9:528-532.

Bremner, J.M. 1965. Organic nitrogen in soils. p. 93-149 In: W.V. Bartholomew and F.E. Clark (ed.), *Soil Nitrogen*. American Society of Agronomy, Madison, WI.

Brune, A. 1998. Termite guts: the world's smallest bioreactors. *Trends Biotechnol.* 16:16-21.

Burtseva, T.I., S.A. Cherkasova and Yu. S. Ovodov. 1985. Quantitative determination of amino sugars in bacterial lipopolysaccharides. *Khimiya Prirodnykh Soedinenii* 6:739-743.

Calderoni, G. and M. Schnitzer. 1984. Nitrogen distribution as a function of radiocarbon age in Paleosol humic acids. *Org. Geochem.* 5:203-209.

Caroll, S.F. and D.R. Nelson. 1979. Fluorometric quantitation of amino sugars in the picomole range. *Anal. Biochem.* 98:190-197.

Casagrande, D.J. 1970. Gas-liquid chromatography of thirty-five amino acids and two amino sugars. *J. Chromatogr.* 49:537-540.

Casagrande, D.J. and K. Park. 1977. Simple gas-liquid chromatographic technique for the analyses of muramic acid. *J. Chromatogr.* 135:208-211.

Casagrande, D.J. and K. Park. 1978. Muramic acid levels in bog soils from the Okefenokee swamp. *Soil Sci.* 125, 181-183.

Chantigny, M.H., D.A. Angers, D. Prévost, L.-P. Vézina and F.-P. Chalifour. 1997. Soil aggregation and fungal and bacterial biomass under annual and perennial cropping systems. *Soil Sci. Soc. Am. J.* 61:262-267.

Cheshire, M.V. 1979. *Nature and Origin of Carbohydrates in Soils*. p. 23-67. Academic Press, London.

Christensen, B. T. 1996. Carbon in primary and secondary organomineral complexes. p. 97-165 In: M.R. Carter and B.A. Stewart (eds.), *Structure and Organic Matter Storage in Agricultural Soils*. Advances in Soil Science, CRC Press, Boca Raton, FL.

Coelho, R.R.R., D.R. Sacramento and L.F. Linhares. 1997. Amino sugars in fungal melanins and soil humic acids. *Eur. J. Soil Sci.* 48:425-529.

Crumpton, M.J. 1959. Identification of amino sugars. *Biochem. J.* 72:479-486.

Currie, L.A. 1968. Limits for qualitative detection and quantitative determination, application to radiochemistry. *Anal. Chem.* 40:586-593.

Curry, J.P. and J.A. Good. 1992. Soil faunal degradation and restoration. *Adv. Soil Sci.* 17:171-215.

Dawson, R. and K. Mopper. 1978. A note on the losses of monosaccharides, amino sugars, and amino acids from extracts during concentration procedures. *Anal. Biochem.* 84:186-190.

Dische, Z. and E. Borenfreund. 1950. A spectrophotometric method for the microdetermination of hexosamines. *J. Biol. Chem.* 185:517-522.

Drews, G. 1973. Fine structure and chemical composition of the cell envelopes. p. 99-116. In: N.G. Carr and B.A. Whitton (eds.), *The Biology of Blue Green Algae*. Bot. Monograph No. 9. Blackwell, Oxford.

Dunger, W. 1983. Tiere im Boden. A. Ziemsen-Verlag, Wittenberg.

Durska, G. and H. Kaszubiak. 1983. Occurrence of bound muramic acid and α, ε-diaminopimelic acid in soil and comparison of their contents with bacterial biomass. *Acta Microbiol. Pol.* 12:257-263.

Eastoe, J.E. 1954. Separation and estimation of chitosamine and chondrosamine in complex hydrolysates. *Nature* 20:540-541.

Egli, H. 1995. Assurance of data quality. p. 235-255. In: H.-J. Stan (ed.), *Analyses of Pesticides in Ground and Surface Water I: Progress in Basic Multi-residue Methods*. Springer-Verlag, Berlin.

Ekblad, A. and T. Näsholm 1996. Determination of chitin in fungi and mycorrhizal roots by an improved HPLC analysis of glucosamine. *Plant Soil* 178:29-35.

Elliott, E.T. 1986. Aggregate structure and carbon, nitrogen, and phosphorus in native and cultivated soils. *Soil Sci. Soc. Am. J.* 50:627-633.

Elson, L.A. and W.T. Morgan. 1933. A colorimetric method for the determination of glucosamine and chondrosamine. *Biochem. J.* 27:1824-1828.

Feller, C. and M.H. Beare. 1997. Physical control of soil organic matter dynamics in the tropics. *Geoderma* 79:69-116.

Foster, R.C. and J.K. Martin. 1981. In situ analyses of soil components of biological origin. p. 75-110. In: E.A. Paul and J.N. Ladd (eds.), *Soil Biochemistry*, Volume 5. Marcel Dekker, New York.

Foster, R.C. 1988. Microenvironments of soil microorganisms. *Biol. Fert. Soils* 61:189-203.

Gallali, T. A. Guckert and F. Jaquin. 1975. Étude de la distribution des sucres aminés dans la matiére organique des sols. Bull. Ec. Natl. Sper. *Agron. Ind. Aliment.* 17:53-59.

Goh, K.M. and D.C. Edmeades. 1979. Distribution and partial characterization of acid hydrolysable organic nitrogen in six New Zealand soils. *Soil Biol. Biochem.* 11:127-132.

González-Prieto, S.J. and T. Carballas. 1988. Modified method for the fractionation of soil organic nitrogen by successive hydrolyses. *Soil Biol. Biochem.* 20:1-6.

González-Prieto, S.J. and T. Carballas. 1992. Simple step-wise acid hydrolyses method for the fractionation of soil organic nitrogen. *Soil Biol. Biochem.* 24:925-926.

González-Prieto, S.J., L. Joncteur-Monrozier, J.N. Hétier and T. Carballas. 1997. Changes in the soil organic N fractions of a tropical Alfisol fertilized with ^{15}N-urea and cropped to maize or pasture. *Plant Soil* 195:151-160.

Grant, W.D. and A.W. West. 1986. Measurement of ergosterol, diaminopimelic acid and glucosamine in soil: evaluation as indicators of microbial biomass. *J. Microbiolog. Meth.* 6:47-53.

Guerrant, G.O. and C.W. Moss. 1984. Determination of monosaccharides as aldononitrile, O-methyloxime, alditol, and cyclitol acetate derivatives by gas chromatography. *Anal. Chem.* 56:633-638.

Guggenberger, G., S.D. Frey, J. Six, K. Paustian and E.T. Elliot. 1999. Bacterial and fungal cell-wall residues in conventional and no-tillage agroecosystems. *Soil Sci. Soc. Am. J.* 63:1188-1198.

Guggenberger, G., K. Kaiser and W. Zech. 1998. SOM pools and transformation determined by physical fractionation. *Mittlgn. Dtsch. Bodenkundl. Gesellsch.* 87:175-190.

Gupta, V.V.S.R. and J.J. Germida. 1988. Distribution of microbial biomass and its activity in different soil aggregate size classes as affected by cultivation. *Soil Biol. Biochem.* 20:777-786.

Herrera, J.R. 1992. *Fungal Cell Wall: Structure, Synthesis and Assembly*. CRC Press, Boca Raton, FL.

Hicks, R. and S.Y. Newell. 1983. An improved gas chromatographic method for measuring glucosamine and muramic acid concentrations. *Anal. Biochem.* 128:438-445.

Hicks, R. and S.Y. Newell. 1984. A comparison of glucosamine and biovolume conversion factors for estimating fungal biomass. *Oikos* 42:355-360.

Hübschmann, H.-J. 1996. *Handbuch der GC/MS, Grundlagen und Anwendung.* VCH-Verlagsgesellschaft mbH, Weinheim.

Jahnel, J.B. and F.H. Frimmel. 1996. Detection of glucosamine in acid hydrolysis solution of humic substances. *Fresenius J. Anal. Chem.* 354:886-888.

Janel, Ph., J. Monrozier and F. Toutain. 1978. Caracterization de l'azote des litieres et des sols par hydrolyse acide. *Soil Biol. Biochem.* 11:141-146.

Jörgensen, R.G. 1987. *Flüsse, Umsatz und Haushalt der Postmortalen Organischen Substanz und Ihrer Stoffgruppen in Streudecke und Bodenkörper eines Buchenwald-Ökosystems auf Kalkgestein.* Göttinger Bodenkundliche Berichte 91, 409 pp.

Joergensen, R.G. and B. Meyer. 1990. Chemical change in organic matter decomposing in and on a forest Rendzina under beech. *J. Soil Sci.* 41:17-27.

Joergensen, R.G., G. Scholle and V. Wolters. 1995. Die Bestimmung von Muraminsäure als Biomarker von Bakterien. *Mittl. Dtsch. Bodenkundl. Ges.* 76:627-630.

Kaiser, K., G. Guggenberger, L. Haumaier and W. Zech. 1997. Dissolved organic matter sorption on subsoils and minerals studied by ^{13}C-NMR and DRIFT-spectroscopy. *Eur. J. Soil Sci.* 48:301-310.

Kenne, L.K. and B. Lindburg. 1983. Bacterial polysaccharides. p. 287-353. In: G.O. Aspinall (ed.), *The Polysaccharides.* Academic Press, New York.

Kögel, I. 1987. *Organische Stoffgruppen in Waldhumusformen und ihr Verhalten während der Streuzersetzung und Humifizierung.* Bayreuther Bodenkundliche Berichte, Band 1.

Kögel, I. and R. Bochter. 1985. Amino sugar determination in organic soils by capillary gas chromatography using a nitrogen-selective detector. *Z. Pflanzenernaehr. Bodenkd.* 148, 260-267.

Kraus, R.J., F.L. Shinnik and J.A. Marlett. 1990. Simultaneous determination of neutral and amino sugars in biological materials. *J. Chromatogr.* 513:71-81.

Ladd, J.N. and R.B. Jackson. 1982. p. 173-228. In: F.J. Stevenson (ed.), *Nitrogen in Agricultural Soils.* American Society of Agronomy, Madison, WI.

Ladd, J.N., R.C. Foster and J.O. Skjemstad. 1993. Soil structure: Carbon and nitrogen metabolism. *Geoderma* 56:401-434.

Lal, R., J. Kimble, R.F. Follett and B.A. Stewart. 1998. *Soil Processes and the Carbon Cycle.* Adv. Soil Sci., CRC Press, Boca Raton, FL.

Lindroth, P. and K. Mopper. 1979. High performance liquid chromatographic determination of subpicomole amounts of amino acids by precolumn fluorescence derivatization with o-phthalaldehyde. *Anal. Chem.* 51:1667-1674.

Markus, R.L. 1950. Colorimetric determination of lactic acid in body fluids utilizing cation exchange for deproteinization. *Arch. Biochem.* 29:159-165.

Martin, J.P. and K. Haider. 1986. Influence of mineral colloids on turnover rates of soil organic carbon. p. 283-304. In: P.M. Huang and M. Schnitzer (eds.), *Interactions of Soil Minerals with Natural Organics and Microbes.* Soil Science Society of America, Madison, WI.

Marumoto, T., K. Furukawa, T. Yoshida, H. Kai and T. Harada. 1972. Effect of the application of rye-grass on the contents of individual amino acids and amino sugars contained in the organic nitrogen in soil. *J. Fac. Kyushu Univ.* 17:37-47.

Mehta, N.C., P. Dubach and H. Deuel. 1961. Carbohydrates in soil. p. 335-355. In: M.L. Wolfrum and R.S. Tipson (eds.), *Advances in Carbohydrate Chemistry.* Academic Press, New York.

Mielniczuk, Z., E. Mielniczuk and L. Larsson. 1995. Determination of muramic acid in organic dust by gas chromatography-mass spectrometry. *J. Chromatogr.* 670:167-172.

Millar, W.N. and L.E. Casida. 1970. Evidence for muramic acid in soil. *Can. J. Microbiol.* 16:299-304.

Mimura, T. and D. Delmas. 1983. Rapid and sensitive method for muramic acid determination by high-performance liquid chromatography with precolumn fluorescence derivatization. *J. Chromatogr.* 208:91-98.

Moriarty, D.J.W. 1975. A method for estimating the biomass of bacteria in aquic sediments and its application to trophic studies. *Oecologia* 20:219-229.

Moriarty, D.J.W. 1983. Measurement of muramic acid in marine sediments by high performance liquid chromatography. *J. Microbiol. Methods* 1:111-117.

Nannipieri, P., F. Pedrazzini, P.G. Arcara and C. Piovanelli. 1979. Changes in amino acids, enzyme activities, and biomasses during soil microbial growth. *Soil Sci.* 127:26-34.

Navratil, J.D., E. Murgia and H.F. Walton. 1975. Ligand exchange chromatography of amino sugars. *Anal. Chem* 47:122-125.

Neeser, J.-R. and T.F. Schweizer. 1984. A quantitative determination by gas-liquid chromatography of neutral and amino sugars (as O-methyloxime acetates), and a study on hydrolytic conditions for glycoproteins and polysaccharides in order to increase sugar recoveries. *Anal. Biochem.* 142:58-67.

Oades, J. M. 1988. The retention of organic matter in soils. *Biogeochem.* 5:35-70.

Parsons, J.W. 1981. Chemistry and distribution of amino sugars in soils and soil organisms. p. 197-227. In: E.A. Paul and J.N. Ladd (eds.), *Soil Biochemistry*, Vol. 5. Marcel Dekker, New York.

Parsons, J.W. and J. Tinsley. 1961. Chemical studies of polysaccharide material in soils and composts based on extraction with anhydrous formic acid. *Soil Sci.* 92:46-53.

Paul, E.A. and F.E. Clark. 1996. *Soil Microbiology and Biochemistry*. Academic Press, San Diego.

Petruzzelli, G.G., G. Guidi and M. LaMarca. 1974. Determinazione degli aminoacidi nella substanza organica del terreno. *La Nuova Chimica* 50:82-88.

Reissig, J.L., J.L. Strominger and L.F. Leloir. 1955. A modified colorimetric method for the estimation of N-acetylamino sugars. *J. Biol. Chem.* 217:959-966.

Scheidt, M. and W. Zech. 1990. A simplified procedure for the photometric determination of amino sugars in soil. *Z. Pflanzenernähr. Bodenkd.* 153:207-208.

Scholle, G., R.G. Joergensen, M. Schaefer and V. Wolters. 1993. Hexosamines in the organic layer of two beech forest soils: effects of mesofauna exclusion. *Biol. Fertil. Soils* 15:301-307.

Schmitz, O., G. Dannenberg, B. Hundshagen, A. Klinger and H. Bothe. 1991. Quantification of vescilar-arbuscular mycorrhiza by biochemical parameters. *J. Plant Physiol.* 139:106-114.

Schnitzer, M. 1982. Organic matter characterization. p. 581-594. In: A.L. Page, R.H. Miller and D.R. Keeney (eds.), *Methods of Soil Analysis. Part 2.* 2nd ed. Agron. Monogr. 9. American Society of Agronomy and Soil Science Society of America, Madison, WI.

Schnitzer, M. and D.A. Hindle. 1981. Effects of different methods of acid hydrolysis on the nitrogen distribution in two soils. *Plant Soil* 60:237-243.

Sharon, N. 1965. Distribution of amino sugars in microorganisms, plants and invertebrates. p. 1-45 In: E.A. Balasz and R.W. Jeanlanx (eds.), *The Amino Sugars, Part 2A. Distribution and Biological Role.* Academic Press, New York.

Sinner, M. and J. Puls. 1978. Non-corrosive dye reagent for detection of reducing sugars in borate complex ion-exchange chromatography. *J. Chromatogr.* 156:197-203.

Skujins, J. and A. Pukite. 1970. Extraction and determination of N-acetylglucosamine from soil. *Soil Biol. Biochem.* 2:141-143.

Smithies, W.R. 1952. Chemical composition of a sample of mycelium of *penicillium griseofulvum* Dierckx. *Biochem. J.* 51:259-264.

Sowden, F.J. 1959. Investigations of the amounts of hexosamines found in various soils and methods for their determination. *Soil Sci.* 88:138-143.

Sowden, F.J. 1967. Determination of amino acids and amino sugars in soil hydrolysates with the autoanalyzer using long and short columns. Technicon Symp.: Automation in Analytical Chemistry, New York, Oct. 1966, Vol. 1, 129-132.

Sowden, F.J. 1969. Effect of hydrolysis time and iron and aluminium removal on the determination of amino compounds in soil. *Soil Sci.* 107:264-371.

Sowden, F.J. and K.C. Ivarson. 1974. Effects of temperature on changes in the nitrogenous constituents of mixed forest litters during decomposition after inoculation with various microbial cultures. *Can. J. Soil Sci.* 54:387-394.

Spackman, D.H., H.W. Stein and S. Moore. 1958. Automatic recording for use in the chromatography of amino acids. *Anal. Chem* 30:1190-1206.

Sparling, G.P., T. Speir and K.N. Whale. 1986. Changes in microbial biomass C, ATP content, soil phospho-monoesterase and phospho-diesterase activity following air-drying of soils. *Soil Biol. Biochem.* 18:363-370.

Spiteller, M. 1980. Kapillargaschromatographische Bestimmung von Zuckern unterschiedlicher Böden. *Z. Pflanzenernähr. Bodenkd.* 143:720-729.

StatSoft, Inc. 1996. Statistica für Windows 5.1 [Computerprogramm - Handbuch]. Tulsa, OK.

Stevenson, F.J. 1954. Ion exchange chromatography of the amino acids in soil hydrolysates. *Soil Sci. Soc. Am. Proc.* 18:373-377.

Stevenson, F.J. 1956. Isolation and identification of some amino compounds in soils. *Soil Sci. Soc. Am. Proc.* 20:201-204.

Stevenson, F.J. 1957. Investigations of aminopolysaccharides in soils: I. Colorimetric determination of hexosamines in soil hydrolysates. *Soil Sci* 83:113-122.

Stevenson, F.J. 1965. Amino sugars. p. 1429-1436. In: C.A. Black (ed.), *Methods of Soil Analysis, Part 2. Chemical and Microbial Properties.* American Society of Agronomy, Madison, WI.

Stevenson, F.J. 1982. Organic forms of soil nitrogen. p. 101-104. In: F.J. Stevenson (ed.), *Nitrogen in Agricultural Soils.* American Society of Agronomy, Madison, WI.

Stevenson, F.J. 1983. Isolation and identification of amino sugars in soil. *Soil Sci. Soc. Am. J.* 47:61-65.

Stevenson, F.J. and C.-N. Cheng. 1970. Amino acids in sediments: recovery by acid hydrolyses and quantitative estimation by a colorimetric procedure. *Geochim. Cosmochim. Acta* 34:77-88.

Stott, D.E., L.F. Elliot, R.I. Papendick and G.S. Campbell. 1986. Low temperature or low water potential effects on the microbial decomposition of wheat residue. *Soil Biol. Biochem.* 18:577-582.

Swann, D.A. and E.A. Balasz. 1966. Determination of the hexosamine content of macromolecules with manual and automated techniques using the p-dimethylaminobenzaldeyde reaction. *Biochim. Biophys. Acta* 130:112-129.

Tiessen, H., J.W.B. Stewart and H.W. Hunt. 1984. Concepts of soil organic matter transformations in relation to organo-mineral particle size fractions. *Plant Soil* 76:287-295.

Tisdall, J.M. and J. M. Oades, 1982. Organic matter and water-stable aggregates in soils. *J. Soil Sci.* 33:141-163.

Tisdall, J.M. 1996. Formation of soil aggregates and accumulation of soil organic matter. p. 57-96. In: M.R. Carter and B.A. Stewart (eds.), *Structure and Organic Matter Storage in Agricultural Soils.* Adv. Soil Sci., CRC Press, Boca Raton, FL.

Tracey, M.V. 1952. The determination of glucosamine by alkaline decomposition. *Biochem. J.* 52:265-267.

Tsuji, A., T. Kinoshita and M. Hishino. 1969. Analytical chemical studies on amino sugars, II: Determination of hexosamines using 3-methyl-2-benzothiazolone hydrochloride. *Chem. Pharm. Bull.* 17:1505-1510.

Van Veen, J.A., J.N. Ladd and M.J. Frissel. 1984. Modelling C and N turnover through the microbial biomass. *Plant Soil* 76:257-274.

Wagner, G.H. and V.K. Mutatkar. 1968. Amino components of soil organic matter formed during humification of ^{14}C glucose. *Soil Sci. Soc. Am. Proc.* 32:683.686.

Wang, T.S.C. and S.C. Cheng. 1964. Amino sugars in soil hydrolysates. *Rep. Taiwan Sugar Expt. Sta.* 34:86. *Chem Abstr.* 62:10247.

Waters, A. G. and J.M. Oades. 1991. Organic matter in waterstable aggregates. p. 163-174. In: W. Wilson (ed.), Advances in Soil Organic Matter Research: Proceedings of a Symposium, Colchester, U.K. 3–4 September, 1990. Royal Society of Chemistry, Cambridge, U.K.

West, A.W., W.D. Grant and G.P. Sparling. 1987. Use of ergosterol, diaminopimelic acid and glucosamine contents of soils to monitor changes in microbial populations. *Soil Biol. Biochem.* 19:607-612.

Whiton, R.S., P. Lau, S. Morgan, J. Gilbart and A. Fox. 1985. Modifications of the alditol acetate method for analysis of muramic acid and other neutral and amino sugars by capillary gas chromatography-mass spectrometry with selected ion monitoring. *J. Chromatogr.* 347:109-120.

Yoshida, M. and K. Kumada. 1979. Studies on the properties of organic matter in buried humic horizon derived from volcanic ash. III. Sugars in hydrolysates of buried humic horizon. *Soil Sci. Plant Nutr.* 25:209-216.

Zelles, L. 1988. The simultaneous determination of muramic acid and glucosamine in soil by high-performance liquid chromatography with precolumn fluorescence derivatisation. *Biol. Fertil. Soils* 6:125-130.

Zelles, L., K. Stepper and A. Zsolnay. 1990. The effect of lime on microbial activity in spruce (*Picea abies* L.) forests. *Biol. Fertil. Soils* 9:78-82.

Zelles, L., P. Adrian, Q.Y. Bai, K. Stepper, M.V. Adrian, K. Fischer, A. Maier and A. Ziegler. 1991. Microbial activity measured in soils stored under different temperature and humidity conditions. *Soil Biol. Biochem.* 13:955-962.

Zhang, X., W. Amelung, Y. Yuan and W. Zech. 1997. Amino sugars in soils of the North American cultivated prairie. *Z. Pflanzenernähr. Bodenkd.* 160:533-538.

Zhang, X. and W. Amelung. 1996. Gas chromatographic determination of muramic acid, glucosamine, galactosamine, and mannosamine in soils. *Soil Biol. Biochem.* 28:1201-1206.

Zhang, X., W. Amelung, Y. Yuan and W. Zech. 1998. Amino sugar signature of particle-size fractions in soils of the native prairie as effected by climate. *Soil Sci.* 163:220-229.

Zhang, X., W. Amelung, Y. Yuan, S. Samson-Liebig, L. Brown and W. Zech. 1999. Land-use effects on amino sugars in particle-size fractions of an Argiudoll. *Appl. Soil Ecol.* 11:271-275.

CHAPTER 18

Characterization of Soil Organic Matter

C.L. Ping, G.L. Michaelson, X.Y. Dai and R.J. Candler

I. Introduction

Soil organic matter (SOM) can be characterized in different ways, depending on the purposes. Total organic carbon (TOC) and carbon to nitrogen ratio (C:N) are commonly used to quantify and qualify SOM. Many different extraction and fractionation schemes have been developed in order to study the composition of SOM. However, the most commonly used extractant is a dilute alkali solution (NaOH), which was introduced by Russian scientists in the 18th century (Kononova, 1961). The dilute alkali solution remains the most effective and most widely used extractant for SOM.

Dilute alkali solution (NaOH) has its limitations. It has been criticized as being a harsh extractant because it may induce hydrolysis, degradation or esterization of SOM components and thus create artifacts. The more mild extractants, such as organic solvents and water, have very low recovery rates leading to the question of how representative the extracted organic fraction is of the "whole" SOM (Malcolm, 1990). Schnitzer and Schuppli (1989) tried a sequential extraction with satisfactory results, using different organic solvents followed by NaOH and pyrophosphate solutions. In spite of its drawbacks, NaOH remains the preferred extractant due to its high percentage of extracted organic fraction. However, dilute NaOH solution extracts not only humic substances mainly humic and fulvic acids, but also many nonhumic substances such as low molecular weight acids, and saccharides along with their derivatives including simple sugars, oligosacchrides, polysaccharides, amino sugars and sugar acids (Thurman and Malcolm, 1989). Thus, the commonly referred to humic and fulvic acids are actually humic and fulvic acid fractions (MacCarthy et al., 1990).

Many methods have been developed for evaluation of SOM. However, most methods are designed for specific soils for a specific purpose or for the evaluation of specific organic fractions. Methods in general have been established to evaluate two separate groups of soils: the organic (peat soils and forest litter) and the mineral soils (often agricultural soils). SOM evaluation methods for the organic soils have concentrated on the analysis of the fibrous fractions of SOM, which, in the absence of a significant mineral component, dominate these soils. Methods for organic soils are commonly adapted from wood fiber analytical methods (Rayan et al., 1990) or from forage and feed fiber analytical methods for application to forest litter or organic surface soil (Goering and Van Soest, 1975; Van Cleve, 1974). In contrast, the SOM of mineral soils is often evaluated as discussed above, with alkali extract for the humic and fulvic acid fractions with the residual SOM regarded ash humin. More sophisticated, complex and comprehensive techniques, such as those of Beyer et al. (1993), have been developed to address SOM as a whole and recognize that soils, in reality, exist with a continuum of organic matter contents from essentially zero to 100%. The method outlined here attempts to utilize, simplify and combine aspects of many of the above-mentioned procedures, while adding some more definition to the soluble-extractable fractions present which are so important to the biochemical activity of all soils. Specifically, the method joins the XAD resin technique of Malcolm and MacCarthy (1992) to the basic fiber analysis of Goering and Van Soest (1975). The humin technique of Beyer (1993) is integrated into the technique to separate humin (non-alkali extractable SOM) from the more resistant lignin of fibrous organic matter. As with most SOM fractionation techniques the resulting fractions are operationally defined and must be recognized as such.

II. Method of Soil Organic Matter Characterization by Fractionation

The fractionation method outlined below (see flow chart in Figure 1) adapts and combines aspects of several methodologies established in the literature into a single procedure. This method allows the soil organic matter (SOM) to be assessed as a whole or in parts and can be performed in its entirety as a sequential procedure. The resulting fractions are, as in other methodologies, operationally defined, but fractions obtained are related to commonly cited literature methods of similar technique from which the various segments are adapted. This method more clearly separates the humic substances (humic and fulvic acids and humin) which are so significant in soils but are overlooked or overestimated by other techniques. The advantage of this approach is in the assessment of the SOM as a whole through the combination of techniques developed for organic soils useful for fibrous SOM, with techniques developed for mineral soils useful for assessing humified SOM. The combined techniques allow for a more complete characterization of SOM by which all soils can be compared. This becomes useful when studying soils with a continuum of variation in SOM content such as exists for soils from within a single soil profile and among soils from different ecosystems.

A. Extractable SOM

Fresh field moist soils of known water content are extracted repeatedly with $0.1 M$ NaOH (Ping et al., 1995; Michaelson and Ping, 1997), extracts are combined, pressure filtered through a 0.45-μm polysulfanone membrane, and the filter cake returned to the non-extractable soil. This separation creates two parts: first, the extractable portion (extractable SOM) containing the soluble humified and soluble non-humic SOM components and second, the portion containing lipids, fibrous, and organo-mineral SOM components along with soil minerals (non-extractable SOM and soil).

The extractable SOM is fractionated into its component humic acid (HA), fulvic acid (FA), hydrophobic neutral (HON), low molecular weight acids (LMA), low molecular weight neutrals (LMN), and hydrophilic (HI) fractions. This separation is accomplished by passing the extractable SOM through tandem columns of first XAD-8 and then XAD-4 resins (Malcolm and MacCarthy, 1992) according to the procedure outlined by Malcolm et al. (1995). The resins selectively adsorb organic compounds based on hydrophobicity with respect to the resin resulting from the functional group to carbon ratio of SOM molecules. This ratio has been found to vary inversely with the molecular weight trend, which goes from high to low in the order of HA, HON, FA, LMA, HI, and non-humic compounds. The HA, FA and HON fractions are retained on the XAD-8 resin. The HA and FA fractions are then removed with NaOH and separated by solubility at pH 1 (HA precipitates and FA remains in solution). The HON is removed with ethanol or acetonitrile and determined colorimetrically at 400 nm by comparison to known solutions. The SOM exiting the XAD-8 resin is passed through the XAD-4 resin, which retains the LMA and LMN fractions. These fractions are determined by calculation of the amount of carbon retained on the column and the amount removed from the column by NaOH (LMA fraction) with the difference between the former and latter being the LMN fraction. The LMN fraction can be removed for study, using an ethanol or acetonitrile eluent.

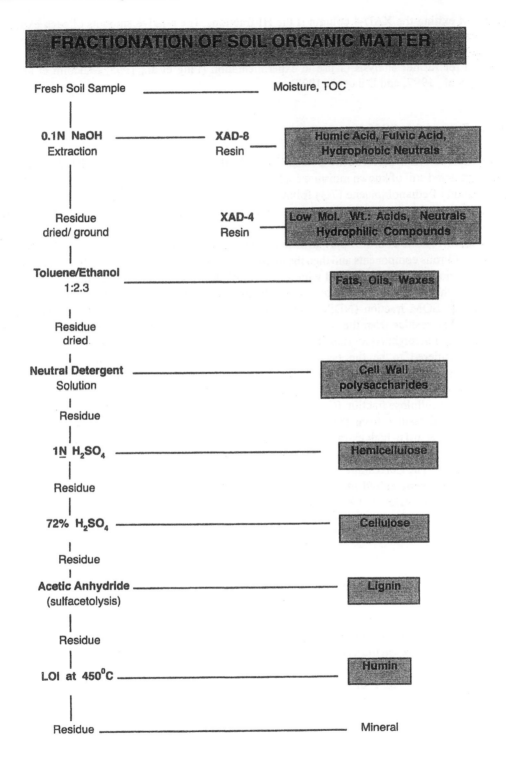

Figure 1. General flow of the sequential SOM characterization procedure.

The SOM exiting the XAD-4 column is the HI fraction. The relative amounts of each extractable SOM fraction are determined by dissolved organic C analysis of extracts and column inputs and efluents as they enter and exit the resins. The various fractions isolated by the procedure may be collected for further analysis or use in experimentation (Ping et al., 1995; Malcolm et al., 1995; Candler et al., 1997; and Dai et al., 1998).

B. Non-Extractable SOM

The residual soil from part A above is washed with distilled water until free of NaOH salts. A portion of this extracted soil of known moisture content is then weighed into an alundum thimble and boiled in a solution of ethanol/toluene (7:3) followed by a reflux wash (adapted from Beyer et al., 1993). The lipid SOM fraction is determined gravimetrically by evaporation of the ethanol/toluene reflux solution.

Dried soil residue from the lipid extraction above is used for a sequential dissolution determination of various fibrous components and then the humin (recalcitrant organo-mineral) SOM fractions. The lipid extraction residue is refluxed with a neutral pH detergent solution (Goering and Van Soest, 1975) and the weight loss due to this treatment is considered the neutral detergent soluble carbohydrate SOM fraction (NDSC) consisting of easily soluble cell wall components of plant materials. The residue from the NDSC dissolution is then treated with a 1 N sulfuric acid solution also containing detergent (Goering and Van Soest, 1975). The weight loss by treatment with the 1 N acid is considered to be the hemicellulose SOM fraction (Hc) containing the less soluble hemicellulose component of cell walls with some attached proteins. Residue from the Hc fraction determination is then treated with 72% sulfuric acid. The weight loss by dissolution in the strong acid is taken as the cellulose fraction (Cel) of SOM (Goering and Van Soest, 1975).

The washed residue from the Cel determination above contains the most recalcitrant SOM fractions with regard to both extraction and microbial degradation. These fractions are largely lignin (polyphenolics) and humin (organo-mineral and physically or chemically protected humic material) SOM fractions. The Cel residue is dried by washing with acetone and then treated with a mixture of acetic acid and acetic anhydride in the presence of sulfuric acid catalyst to solublize the phenolic lignin components (Beyer, 1993). The difference in SOM content before treatment and after treatment is taken as lignin (Lg) SOM fraction, and the SOM in the treatment residue as the humin (Hu) SOM fraction (Beyer et al., 1993).

III. Utility of SOM Characterization by Fractionation

The SOM fractionation method described above has been useful in the elucidation of various aspects of the quality of carbon in soils and their role in soil processes. Studies briefly summarized in the following sections present the utility of the method in three areas of soil carbon quality studies. Areas reviewed are, first, application of SOM fractionation in pedogenesis; second, chemical characterization of SOM resultant from soil C processes; and third, biological production of CO_2 in soils and soil waters in relation to SOM quality.

A. Application of SOM Fractionation in Pedogenesis

Soil organic matter is one of the most important controlling factors in soil genesis and thus used as criteria in soil classification. Fractions of SOM are useful in assessment of soil characteristics as displayed in soil color and development through quantification of relative amounts of stable humic acids (HA) versus the mobile fulvic acids (FA). Soil color has good correlation with SOM composition, except in cases where soils have parent materials of characteristic dark color. The humic acids are stabilized through formation of Ca-humates that are biologically stable, as indicated by the C-14 dating of Tsutsuki et al. (1988), which indicated that humic acids have existed in Mollisols for more than 7000 years. With increased leaching, Ca in the metal humates is replaced with H^+ and Al^{+++}, and the Mollic horizons will turn into Umbric horizons while retaining their black color (Soil Survey Staff, 1998). In the Andisol classification, soils in the melanic great group have a dark surface horizon and humus fractions dominated by humic acids (Otowa et al., 1988).

Soils classified in the fulvic great group have light brown or light reddish brown surface horizons, and their humus fractions are dominated by fulvic acids (Ping et. el., 1988). Orlov (1996) related the humic/fulvic acid fraction ratio (HA/FA) to biological activity period (BAP) in days. He calculated the ratio of about 0.5 for Tundra soils, 0.5 to 0.8 for Spodosols, 0.8 to 1.8 for soils of Aquic suborders (Gleysols), and 1.8 to 2.8 for Mollisols. Ping et al. (1997) calculated the HA/FF ratio for Alaska tundra soils and found that the HA/FA ratio for the wet tundra soils on the coastal plains to be 1.7 and 1.3 for the moist tundra soils on the Arctic foothills. Such a discrepancy is due to the different soil classification systems. According to the Russian soil classification system (Mazhotova et al., 1994) most of the Alaska arctic tundra soils are classified as Gleysols because of the poor drainage, and the HA/FA ratios are comparable. The relationship between HA/FA ratio is clearly related to precipitation and thus leaching in some Alaska soils (Ping et al., 1995; 1998). The HA/FA ratio increases with increased leaching. Drainage also plays an important part in the humic substance composition. Humic acid dominates the extractable fraction in the poorly drained arctic Alaska tundra soils whereas fulvic acid dominates the extractable fraction the better drained soils of the region (Ping et al., 1988). Such differences are also reflected in soil color: the poorly drained soils are a dark color due to organic matter accumulation or grayish brown due to iron removal with reduced conditions. The well drained soils have light brown or reddish brown color caused by Fe and Al complexes with organic compounds dominated by fulvic acids.

B. Chemical Characterization of SOM

The SOM fractionation method as outlined is extremely useful for the isolation, concentration and purification of component classes comprising SOM for subsequent chemical characterization of decomposition products and substrates. Several studies have employed this methodology and demonstrated some distinct as well as similar chemical characteristics for SOM fractions. Character of SOM has been compared within soil horizons of soil profiles as well as for those between soil types (Malcolm et al., 1995; Ping et al., 1995; Candler et al., 1997). Investigations utilizing ^{13}C NMR spectroscopy revealed common functional groups among all fractions but each fraction exhibited distinct peak intensities and widths. The spectra demonstrated clear distinctions for HA, FA, HON, LMA, LMN and HI fractions among soil types as well as between soil horizons. These results suggest disparate biological, temperature or moisture regimes dictating different decomposition products and substrates.

Carboxylic acid , ester and phenolic group analysis determined by titration of SOM fractions suggested clear distinctions for each fraction within as well as between sites (Candler et al., 1997). Fulvic acid fractions generally contain greater carboxylic acid content compared to humic acids or low molecular weight acids. However, the low molecular weight acids of a Spodisol were found to contain higher carboxylic content than fulvic acids from the same soil, further illustrating distinctions between soils (Ping et al., 1995). Phenolic content of all fractions varies within a narrow range, less than 1meq/g, whereas significantly greater ester contents have been observed for LMA fractions. These results further support the concept of distinct biological/chemical processes occurring between sites and horizons.

C. Biological Production of CO_2 in Soils and SOM Quality

The isolation, concentration and purification of humic and non-humic soluble fractions obtained by the above method have proven useful in studies of the bioreactivity of soluble organic matter in soils and soil waters. Thus far, four types of bioactivity studies have utilized the SOM fractionation method described. First, fractions from the extractable SOM were used to evaluate the relative activity between fractions. Second, the CO_2 respiration of whole soils was studied in relation to the fraction distribution of whole soil SOM. Third, the pedological assessment of arctic soil C stocks was coupled with SOM fractionation analysis to provide data on active fraction stocks throughout various soil profiles in the active layer and upper permafrost of arctic soils. And fourth, the fraction distribution soil water dissolved organic carbon was studied in conjunction with runoff waters and stream and lake waters in the arctic system to assess quality of spring runoff from tundra.

In order to access the bioreactivity of different organic fractions, Dai et al. (1999) incubated the organic fractions at 25°C for 12 months. The low molecular weight neutral (LMN) and hydrophilic (HI) fractions were found to generate the largest amount of CO_2 followed by the hydrophobic neutrals (HON), low molecular weight acids (LMA), fulvic acids (FA) and humic acids (HA). Decreasing order of bioactivity corresponded to an increasing degree of aromaticity. Dai et al. (1999) further studied the relationship between CO_2 evolution and the chemical composition as revealed by ^{13}C NMR of the fractions. They found a significant negative correlation between both the carboxyl and carbonyl-C, and CO_2 evolution. With a decrease in bioavailability, there was an increase in carboxyl-C and carbonyl-C for the fractions in the order of LMN, HON, LMA, and HA and FA. This might be explained that from neutrals to LMA to HA and FA, humification increases, and the more humified, the more carboxyl-C. Therefore, carboxyl-C can be a good indicator for degree of humification.

Utilization of the above method for a series of whole soil profiles can provide information of the stocks of various fractions of SOM at all depth in the soil. These data exist for a series of carbon flux study sites in arctic Alaska (C.L. Ping, unpublished data). When these data are taken with the findings on bioactivity of SOM fractions (Dai et al., 1999 and Michaelson et al., 1998) it may be possible with more study to apply these SOM fraction stock data to scenario-based greenhouse gas emission modeling efforts. As an example, one of the most recent studies of arctic winter carbon dioxide emission activity was published by Oechel et al. (1997). In this study, results of monthly carbon dioxide emissions were presented for the 1993-94 cold-season months of October, November, March and May, along with soil temperature profile data. Late 1993 carbon dioxide emissions were at a peak for all sites in October. Soil temperature profiles were presented for two sites, U-pad and Happy Valley (coastal plain and foothills sites, respectively). Carbon storage and quality data for the soil profiles of these sites are available from the soils investigations group of the NSF-LAII cooperative study (C.L. Ping, unpublished data). Combining these data makes it possible to look at three features of this system that logically should be related: first, the carbon dioxide emissions for a peak month; second, the carbon stocks of the zone of the soil profile, which is at maximum temperature for biological activity; and third, the stocks of the most biologically active organic substrates present for each study site. Figure 2 presents the October 1993 carbon dioxide flux data (top) from Oechel et al. (1997) with the total carbon stores (middle) and stores of the most labile-soluble carbon fraction (bottom) from C.L. Ping (unpublished data). For the high flux month of October, zones of the active layer that were at maximum temperature contained equal amounts of carbon stocks. But these zones contained amounts of both soluble carbon and the most labile fraction (hydrophilic components) of carbon, which were proportional to the flux activities observed by Oechel et al. (1997).

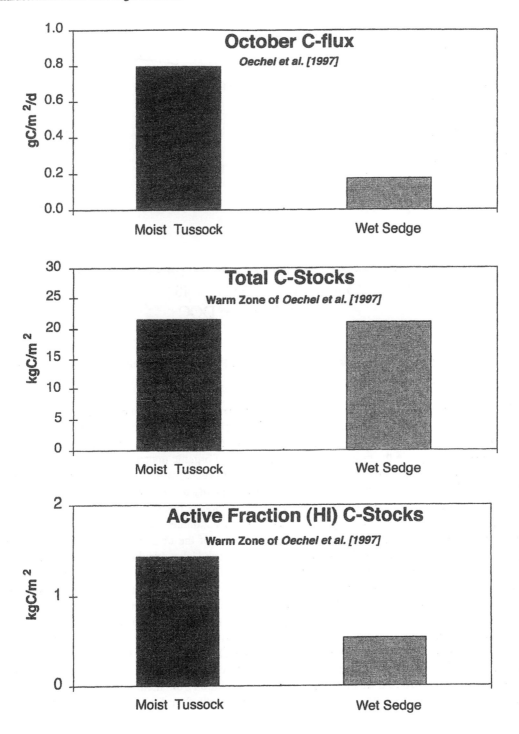

Figure 2. October cold-season flux of CO_2 (from Oechel et al., 1997) in relation to total and extractable carbon fractions in cold-season subsurface active layer warm zones (C.L. Ping, unpublished data).

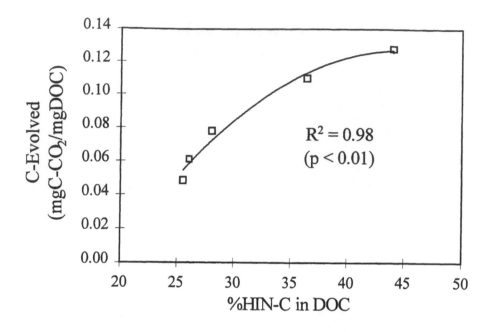

Figure 3. Hydrophylic fraction in soil water DOC relative to CO_2-C evolution in a 4°C incubation.

The XAD resin section of the above method was originally developed as a method for characterization, isolation and concentration of the dissolved organic matter (DOM) of surface waters (Malcolm and MacCarthy, 1992). Its adaption for use on the alkali soluble portion of SOM makes it well suited for use in the more comprehensive study of organic matter in watersheds by allowing for the use of the same methods for soil, soil water and runoff waters to surface stream and lake waters. Such a study was carried out to evaluate the quality and character of DOM at thaw in the arctic ecosystem (Michaelson et al., 1998). The DOM of waters from the frozen soil, thawed soil and soil runoff water to the early spring stream flows were fractionated by the XAD resin section of the above method. The results allowed a coordinated view of the character of DOM in the various component stages of water transfer from tundra to stream and lake. Incubation techniques combined which DOM character evaluation identified the relatively high activity of the same fractions identified for SOM extracts by Dia et al. (1998), discussed above. Reduction in the content of the active fraction (HI) was highly correlated to CO_2 emissions from incubating DOM. Reductions in DOM concentration as well as quality (HI content) were observed for waters as they moved from soil at thaw to thawed soil to soil runoff to stream waters. Soil DOC averaged initially 116 mgC l⁻¹ with 71% HI fraction for soils upon thaw but reduced to a DOC of 12 mgC l⁻¹ with only 9 to 20% HI fraction upon reaching the streams. Figure 3 from Michaelson et al. (1998) illustrates the strong relationship between the HI fraction present and CO_2 evolution from incubating soil DOM. Fractions present in the nonextractable portion of SOM have also been found to relate to bioactivity of soils.

IV. Summary

Characterization of SOM is necessary in the study of soil carbon sequestration and dynamics, change in storage and in the understanding of transformations occurring in soils affecting the atmosphere.

Prediction of soil carbon transformations, for example, the respiration of greenhouse gases, is of special interest in SOM studies. Although many procedures exist for fractionation of the SOM pool, one must choose a method suited for the soils studied, considering that soils contain a complete range of both quality and quantity of SOM. The adaption of certain aspects of the available methods and combining them as presented here offers several advantages. The primary advantage of this procedure is that it is an approach that recognizes SOM as a whole and includes all classes of SOM that may be present in soils ranging from organic to mineral soils and from humified SOM to fibrous non-decomposed and organo-mineral complexes. A second advantage of this method is that it is relatively simple and is well suited to study soils in relation to the aquatic environment and allows for humified fractions to be isolated and concentrated for further chemical characterization as with C^{13}NMR. The general procedure has been outlined above and results highlighted from several different studies that have successfully utilized the method for various objectives. The method initially separates alkali extractable SOM and separates these soluble fractions by XAD resin technology in parallel to methods developed in the study of surface water DOM. The nonextractable portion of the soil is then sequentially treated to remove the lipid, cell wall components, fibers including cellulose and lignin and, finally, the resistant humified SOM known as humin is quantified along with the mineral soil component.

References

Beyer, L. 1993. Estimation of soil organic matter composition according to a chemical separation of litter and humic compounds. *Commun. Soil Sci. Plant Anal.* 24:211-226.

Beyer, L., C. Wachendorf and C. Koebbemann. 1993. A simple wet chemical extraction procedure to characterize soil organic matter (SOM). I. Application and recovery rate. *Commun. Soil Soil Plant Anal.* 24:1645-1663.

Candler, R., C.L. Ping, G.J. Michaelson and R.L. Malcolm. 1997. Characterization of fractionated soil organic matter from two Alaskan arctic soils by ^{13}C NMR and titration analysis. p. 301-307. In: J. Drozd, S.S. Gonet, N. Senesi and J. Weber (eds.), *The Role of Humic Substances in the Ecosystem and in Environment.* Polish Society of Humic Substances, Wroclaw, Poland.

Dai, X.Y., C.L. Ping and G.J. Michaelson. 2000. Bioactivity of soil organic matter in tundra soils. p. 29-38. In: R. Lal, J. Kimble and B.A. Stewart (eds.), *Global Climate Change and Cold Regions Ecosystems.* Advances in Soil Science, Lewis Publishers, Boca Raton, FL.

Dai, X.Y., C.L. Ping and W. Zech. 1998. Bioavailability and chemical structure of organic matter fractions in tundra soils. p. 216. 1998 Agronomy Abstracts. American Society of Agronomy, Madison, WI.

Goering, H.K. and P.J. Van Soest. 1975. Forage fiber analyses, apparatus, reagents, procedures, and some applications. Agriculture Handbook No. 379. USDA-Agricultural Research Service, U.S. Government Printing Office, Washington, D.C.

Kononova, M.M. 1961. *Soil Organic Matter, Its Nature, Its Role in Soil Formation and Soil Fertility.* Pergamon Press, New York.

MacCarthy, P., R.L. Malcolm, C.E. Clapp and P.R Bloom. 1990. An introduction to humic substances. p.1-12. In: P. MacCarthy, C.E. Clapp, R.L. Malcolm and P.R. Bloom (eds.), *Humic Substances in Soil and Crop Sciences: Selected Readings.* American Society of Agronomy, Madison, WI.

Malcolm, R.L. 1990. Evaluation of humic substances from Spodosols. p. 200-210. In: J.M. Kimble and R.D. Yeck (eds.), *Proceedings of the Fifth International Soil Correction Meetings (ISCOM IV) Characterization, Classification, and Utilization of Spodosols.* USDA, Soil Conservation Service, Lincoln, NE.

Malcolm, R. L. and P. MacCarthy. 1992. Quantitative evaluation of XAD-8 and XAD-4 resins used in tandem for removing organic solutes from water. *Environ. Internat.* 18:597-607.

Malcolm, R.L., K. Kennedy, C.L. Ping and G.J. Michaelson. 1995. Fractionation, characterization, and comparison of bulk soil organic constituents in selected cryosols of Alaska. p. 315-327. In: R. Lal, J. Kimble, E. Levine and B.A. Stewart (eds.), *Soils and Global Change*. Advances in Soil Science, Lewis Publishers, Boca Raton, FL.

Mazhitova, G.G., C.L. Ping, S.V. Gubin, J.P. Moore and C.A.S. Smith. 1994. Comparison of Russian, American, and Canadian classification of selected tundra and taiga soils in N.E. Russia. *Pochvovedeniye* 12:26-33 (in Russian).

Michaelson, G.J. and C.L. Ping. 1997. Comparison of $0.1N$ sodium hydroxide with $0.1M$ sodium pyrophosphate in the extraction of soil organic matter from various soil horizons. *Commun. Soil Sci. Plant Anal.* 28:1141-1150.

Michaelson, G.J., C.L. Ping, G.W. Kling and J.E. Hobbie. 1998. The character and bioactivity of dissolved organic matter at thaw and in the spring runoff waters of the arctic tundra north slope. *Alaska. J. Geophys. Res.* 103 (D22) 28,939-28,946.

Oechel, W.C., G. Vourlitis and S.J. Hastings. 1997. Cold-season CO_2 emission from arctic soils. *Global Biogeochem. Cycles* 11:163-172.

Orlov, D.S., O.N. Biryukova and N.I. Sukhanova. 1996. *Soil Organic Matter of Russia*. Nauka, Moscow (in Russian).

Otowa, M., S. Shoji and M. Saigusa. 1988. Allic, melanic and fulvic attributes of Andisols. p. 192-202. In: R.I. Kinlock et al. (eds.), Proceedings of the IX International Soil Classification Workshop, Kanto, Japan. 20 July – 1 August, 1987. Japanese Committee for the 9th International Soil Classification Workshop, for the Soil Management Support Service, Washington, D.C.

Ping, C.L., G.J. Michaelson, X.Y. Dai and R. J. Candler. 1998. The geochemistry of organic matter in selected Cryosols of Alaska. I Proceedings, 16th World Congress of Soil Science, Symposium no. 39, Paper 1478. 20-26, August, 1998. Montpellier, France. 4 pp. (CD-ROM).

Ping, C.L., G.J. Michaelson, W.M. Loya, R.J. Candler and R.L. Malcolm. 1997. Characteristics of soil organic matter in Arctic ecosystems of Alaska. p. 157-167. In: R. Lal, J.M. Kimble and B.A. Stewart (eds.), *Carbon Sequestration in Soils*. Advances in Soil Science, Lewis Publishers, Boca Raton, FL.

Ping, C.L., G.J. Michaelson and R.L. Malcolm. 1995. Fractionation and carbon balance of soil organic matter in selected cryic soils in Alaska. p. 302-314. In: R. Lal, J. Kimble, E. Levine, and B.A. Stewart (eds.), *Soils and Global Change*. Advances in Soil Science, Lewis Publishers, Boca Raton, FL.

Ping, C.L., S. Shoji, T. Ito and J.P. Moore. 1988. The classification of cold Andisols and associated Spodosols in eastern Hokkaido and northern Honshu, Japan and southern Alaska, U.S.A. p. 178-191. In R.I. Kinlock et al. (eds.), Proceedings of the IX International Soil Classification Workshop, Kanto, Japan. 20 July – 1 August, 1987. Japanese Committee for the 9th International Soil Classification Workshop, for the Soil Management Support Service, Washington, D.C.

Ryan, M.G., J.M. Melillo and A. Ricca. 1990. A comparison of methods for determining proximate carbon fractions in forest litter. *Can. J. For. Res.* 20:166-171.

Schnitzer, M. and P. Schuppli. 1989. Method for the sequential extraction of organic matter from soils and soil fractions. *Soil Sci. Soc. Am. J.* 53:1418-1424.

Soil Survey Staff. 1998. *Keys to Soil Taxonomy*, Eighth Edition, Natural Resources Conservation Service, USDA, Washington, D.C.

Thurman, E.M. and R.L. Malcolm. 1989. Nitrogen and amino acids in fulvic and humic acids from the Suwannee River. p. 99-118. In A.C. Averett, J.A. Leeheer, D.M. McKnight and K.A. Thorn (eds.), Humic substances in the Suwannee River, Georgia: Interaction, properties, and proposed structures. U.S. Geological Survey Open-File Report 87-557.

Tsutsuki, K., C. Suzuki, S. Duwutsuka, P. Becker-Heidmann and H.W. Scharpenseel. 1988. Investigation on the stabilization of the humus in Mollisols. *Z. Pflanzenernahr. Bodenk.* 151:87-90.

Van Cleve, K. 1974. Organic matter quality in relation to decomposition. p. 311-324. In: A.J. Holding, O.W. Heal, S.F. MacLean, Jr. and P.W. Flanagan (eds.), *Soil Organisms and Decomposition in Tundra.* Tundra Biome Steering Committee, Stockholm, Sweden.

CHAPTER 19

Fractionating Soil in Stable Aggregates Using a Rainfall Simulator

G.C. Starr, R. Lal and J.M. Kimble

I. Introduction

Methods of fractionating soil into aggregate size fractions usually involve some variation of the wet sieve method (Yoder, 1936). This method is cumbersome and laborious (Kemper and Rosenau, 1986), and unable to accurately simulate the physical energies imparted to aggregates in rainfall and runoff events (Bruce-Okine and Lal, 1975; De Vleeschauwer et al., 1978; Mbagwu, 1986). Hence, the wet sieving alone cannot supply a very accurate measure of the size pools of stable soil aggregates and associated soil organic carbon (SOC) and nutrients available for transport by water erosion. A raindrop impact method for testing aggregate stability (Bruce-Okine and Lal, 1975) represents a better simulation of erosion-induced aggregate breakdown. The purpose of this chapter is to report on testing of a realistic method to fractionate raindrop-stable soil aggregates into size pools using a rainfall simulator.

The traditional wet sieve method (Yoder, 1936) involves gently raising and lowering a nest of sieves in water to assess the stability of aggregates to rapid wetting. Typically, intact macro-aggregates in some known size range are placed on the top sieve. The action of the water infiltration into the aggregates (slaking) and the movement of water around the aggregates cause disintegration and the water stable aggregates (WSA) settle to the appropriately sized sieve. Variations of the method (Kemper and Rosenau, 1986) include: type of pretreatment (e.g. air dry, capillary wetting, vacuum wetting, equilibration at 100% relative humidity, etc.), selected sieve sizes (usually >0.1 mm), the initial size range of macro-aggregates (commonly 5 to 8 mm), and the time of wet sieving.

In addition to establishing the wet sieve method of fractionation, Yoder (1936) showed that the majority of soil transported from runoff plots in Ohio was in the form of WSA and hypothesized that size distributions of aggregates would be closely related to vulnerability to soil erosion. However, the wet sieve approach to fractionation does not involve raindrop impact to disintegrate the soil structure as occurs in natural rainstorms. Also, the wet sieve method is laborious (Kemper and Rosenau, 1986), particularly when analyzing the finest material remaining in a dilute mixture of water, micro-aggregates, and clay domain sizes. The amount of water used (about 9 L) for the Yoder method makes filtering, drying, and centrifuging the smallest WSA material very cumbersome. The quantity of fine material can be estimated by subtracting the amount retained on the sieves from the total, but the fine material itself is difficult to analyze.

Soil organic carbon (SOC), a stabilizing agent for aggregates (Tisdall and Oades, 1982; Feller and Beare, 1997; Kay, 1998), reduces soil susceptibility to erosion (Le Bissonnias et al., 1997; Piccolo et al., 1997; Singer and Le Bissonnias, 1997). The soil and SOC movement from the landscape in erosion events is probably much greater for the fine fraction because of preferential transport by runoff into inland waterways and aquatic ecosystems (Lal, 1995; Starr et al., 2000). It was expected that the transport and fate of soil moved by erosion would depend on the WSA pool size. Therefore, the objective of this study was to develop and describe a realistic technique for assessing the WSA size fraction pools available for transport by water erosion.

II. Materials and Methods

The approach and methodology described in this chapter involved the use of a rainfall simulator to rain on aggregates placed on a nest of sieves. When raindrops fall on soil at the top of a nest of sieves much of the energy is absorbed in an initial impact on the top sieve or largest aggregates. Water droplets and WSA are accelerated between sieves by gravity and strike the smaller sieve sizes with a somewhat lower energy. As the WSA pass through the nest of sieves, some impact and disintegration take place and the soil is fractionated into WSA that can withstand raindrop impact.

A schematic of the essential components of the system (Figure 1) shows the nest of five 12.6-cm inside diameter sieves (0.25-, 0.50-, 1.0-, 2.0-, and 5.0-mm mesh sizes), a cylindrical splash-guard, and a cylindrical collection trough (12 cm diameter × 10 cm height). The splash-guard (12 cm diameter × 15 cm height) effectively contained all WSA within the system. The collection trough holds about 8 cm of rain before overflowing through the spout and could withstand over 110 °C if oven drying were desired. The splash-guard and collection trough were built using galvanized steel. In these fractionations, the rainfall simulator (Choudhary et al., 1997) produced 50 mm/hr of rain for 30 min with a sprayer raining at 340 KPa pressure from 2 m above the top sieve with about 2 mm mean drop diameter.

The soil (Rayne series – 2 to 6% slopes, fine loamy mixed mesic typic Hapludult) for this study was taken from plateaus of the USDA – North Appalachian Experimental Watersheds (Kelley et al., 1975). Samples were taken under moist conditions from the near surface (0 to 5 cm) of four management systems: hardwood forest (>100 yr old), conventional tillage continuous corn (*Zea mays*), no-till continuous corn, and pasture. Coarse (>8 mm) litter, plants, and roots were removed, samples from several random locations were composited, and the soil was air-dried. Soil samples were then sieved to obtain 5 to 8 mm macro-aggregates and equilibrated 3 days at 100% relative humidity under a slight vacuum in a chamber containing 1 N sulfuric acid (Collis-George and Lal, 1971).

Fifty grams of prepared macro-aggregates were placed on the top (5-mm) sieve and treated with the rainfall simulator with the settings described above. Six sieve sets were placed directly beneath the rainfall simulator in a circular pattern (radius of about 30 cm) and rotated 120° around the circle every 10 min for a total of 30 min and 360° of rotation. The rotations were done so that all sieves would be treated about equally by raindrop impact. The splash guards were then removed and WSA on them were rinsed onto the top sieve, taking care to use the minimum amount of de-ionized water from a water bottle. To complete the fractionation, two or more minutes of settling allowed the coarser micro-aggregates to settle to the bottom of the collection trough and about 250 ml of supernatant was decanted through the spout into a beaker. The supernatant was subsequently poured back through the nest of sieves to break up any conglomerations of aggregates. The sieves were then separated, starting with the top (5 mm) and working down through the nest, while using the de-ionized water and supernatant sparingly to break up conglomerations and rinse the sides.

Material on each sieve was rinsed into a pan where the macro-aggregates were allowed to settle to the bottom quickly. The floatable organic matter (FOM) and water were decanted through a 0.25-mm sieve. The FOM from all sieve sizes were combined in a single sieve and transferred to a drying container. The WSA were transferred to drying containers and all pools fractionated in this fashion

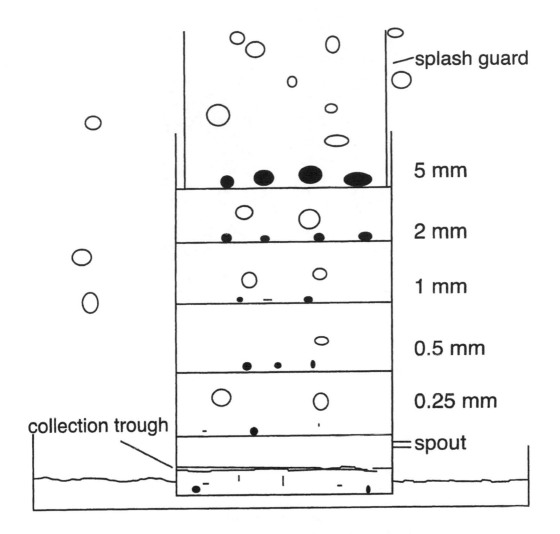

Figure 1. Schematic of the rainfall fractionator showing aggregates (solid) and water.

were allowed to air dry on a warm surface atop of an oven (~35 °C surface temperature). The pools were then weighed, a sub-sample was taken for gravimetric water content analysis, and another sub-sample was analyzed for total organic C (USDA-NRCS, 1995).

Three 50-g sub-samples were prepared from the soil of each management system. Each sub-sample repetition was fractionated into seven pools (>5, 2 to 5, 1 to 2, 0.5 to 1, 0.25 to 0.5, <0.25 mm WSA, and > 0.25 mm FOM). The same sieve sets were used to obtain the mass distribution of primary particles (Gee and Bauder, 1986) and water stable aggregates (Kemper and Rosenau, 1986). The relevant statistical means represented a best estimate of the WSA size distributions. Standard deviations and coefficient of variability were used to assess the repeatability of the method. Grapher software (Golden Software Inc., Golden, CO) was used to calculate and display the linear regressions between SOC and MWD for the comparison of various wet sieve fractionation methods.

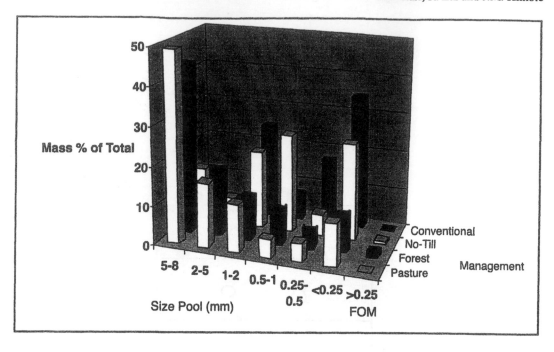

Figure 2. Mass of aggregate pools obtained using the rain fractionator.

III. Results and Discussion

Initial feasibility studies showed efficient use of water with only 400 ml of concentrated micro-aggregates and good repeatability so the system capabilities were expanded to fractionate six sieve sets simultaneously. The system was further tested on the same soil under four management systems that were expected to have very different aggregate stability. The relatively stable size distributions and lower micro-aggregate pool (<0.25 mm) of forest and pasture soils (Figure 2) were assessed using the rainfall simulator and compared with the no-till and conventional-till continuous corn. The management system with the least stable aggregates and most transportable micro-aggregates (35%) was conventional till continuous corn showing more than three times as much transportable micro-aggregates as pasture and forest. No-till was intermediate in its stability under raindrop impact and had a surprisingly high transportable micro-aggregate pool (25%). This would suggest that low erosion from no-till agriculture is more dependent on factors such as high infiltration rates, low runoff, and the protection of surface aggregates from raindrop impact by litter, rather than inherent stability of large aggregates to raindrop impact. The FOM was a small fraction (ranging from 0.2 to 2%) of total mass.

Expressed as a percent of total mass (50 g) there was an average standard error of 2.5% (1.3 g) for a single sample estimation of a pool using the rainfall fractionator. This standard error compares favorably with 5.3% for the Yoder method (Table 1) and is comparable to but slightly more than 0.94% for primary particles. The repeatability of the calculated MWD was good with a coefficient of variability averaging 6.8% compared with 14% for primary particles and 19% for the Yoder method. The percent mass recovery, calculated for each trial then averaged, was 98.5% for the rainfall fractionator. Mass recovery was not attempted on the microaggregates (Yoder method) and primary particles (0.25 mm).

Table 1. Four measures for assessing repeatability of the various fractionation methods

Aggregate fractionation method	Standard deviation (% of total soil mass) for individual pool assessment	Average coefficient of variability for individual pool assessment (%)	Standard deviation (mm) for calculation of mean weight diameter (MWD)	Average coefficient of variability for MWD (%)
Rain fractionator	2.5	24	0.15	6.8
Yoder fractionator	5.3	48	0.56	19
Primary particles	0.94	23	0.14	14

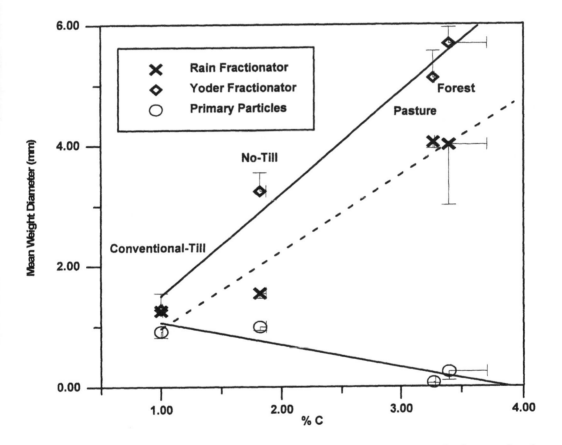

Figure 3. Relationship between aggregation and C percent; bars are standard error for three replications.

Four measures of repeatability (shown in the four data columns of Table 1) were considered: standard deviation (% of total soil mass) for individual pool assessment, average coefficient of variability for individual pool assessment (%), standard deviation (mm) for calculation of MWD, and average coefficient of variability for MWD (%). The Yoder method had the lowest repeatability in all four categories. Although the rainfall simulator had more than two times the repeatability of the

Table 2. Regression parameters expressing the correlation between mean weight diameter (MWD) and soil organic carbon content (SOC)

Aggregate fractionation method	Regression equation	Number of points used	Average mean weight diameter (MWD, mm)	Average soil organic carbon (SOC, %)	Coefficient of determination
Rain fractionator	MWD = 1.28×SOC −0.31	4	2.72	2.37	0.95
Yoder fractionator	MWD = 1.70×SOC −0.20	4	3.83	2.37	0.95
Primary particles	MWD = −0.37×SOC +1.43	4	0.56	2.37	0.84

Yoder method in all four categories, it was comparable to or slightly less repeatable than primary particle fractionation in three of four categories.

A positive, approximately linear relationship between aggregation (expressed as mean weight diameter) and C% (Figure 3) was observed for both the aggregate fractionation methods; however, the correlation was negative for primary particles. Aggregation was consistently greater when assessed using the Yoder method, intermediate for the rain fractionator, and least for primary particles, as would be expected considering the increased physical energies imparted to the large aggregates by raindrop impact. The statistics of the regression analysis (Table 2) showed coefficients of determination of 0.84, 0.95, and 0.98 for primary particles, rain fractionator, and Yoder method, respectively.

IV. Conclusions

Testing of the rain fractionator system of aggregate fractionation and comparison with traditional wet sieving was conducted and results presented in this study suggest the following inferences may be conclusively drawn:

1. The rainfall fractionation procedure is easy to use.
2. The variability of the new method is good when compared with the Yoder method and is comparable to the variability obtained in primary particle fractionation.
3. The rainfall frationator is consistently more destructive than the Yoder method because of the energies imparted by rainfall impact and this simulates natural rainfall conditions.
4. The method is sensitive enough to detect treatment-induced differences in aggregation and structural attributes.
5. Differences in aggregation (expressed as mean weight diameter) are mostly explained by treatment-induced differences in SOC content. This is evidenced by the linear correlations between MWD and SOC (coefficients of determination ranged from 0.84 to 0.98).

References

Bruce-Okine, E. and R. Lal. 1975. Soil erodibility as determined by a raindrop technique. *Soil Sci.* 119:149-157.

Chaodhary, M.A., R. Lal and W.A. Dick. 1997. Long-term tillage effects on runoff and soil erosion under simulated rainfall for a central Ohio soil. *Soil Till. Res.* 42:175-184.

Collis-George, N. and R. Lal. 1971. Infiltration and structural changes as influenced by initial moisture content. *Aust. J. Soil Res.* 11: 93-105.

De Vleeschauwer, D., R. Lal and M. De Boodt. 1978. Comparison of detachability indices in relation to soil erodibility for some important Nigerian soils. *Pedologie* 28:5-20.

Feller, C. and M.H. Beare. 1997. Physical control of soil organic matter dynamics in the tropics. *Geoderma* 79:69-116.

Gee, G.W. and J.W. Bauder. 1986. *Methods of Soil Analysis. Part I. Physical and Mineralogical Methods,* Second Edition, American Society of Agronomy, Madison, WI.

Kay, B.D. 1998. Soil structure and organic carbon: a review. p. 169-197. In: R. Lal, J.M. Kimble, R.F. Follett and B.A. Stewart (eds.), *Soil Processes and the Carbon Cycle,* Lewis Publishers, Boca Raton, FL.

Kemper, W.D. and R.C. Rosenau. 1986. *Methods of Soil Analysis. Part I. Physical and Mineralogical Methods,* Second Edition, American Society of Agronomy, Madison, WI.

Kelley, G.E., W.M. Edwards, L.L. Harrold and J.L. McGuinness. 1975. Soils of the North Appalachian experimental watershed. USDA-ARS Miscellaneous Publication No. 1296.

Le Bissonais, Y. and D. Arrouays. 1997. Aggregate stability and assessment of soil crustability and erodibility: II. Application to humic loamy soils with various organic carbon contents. *Eur. J. Soil Sci.* 48:39-48.

Mbagwu, J.S.C. 1986. Erodibility of soils formed on a catenary toposequence in southeastern Nigeria as evaluated by different indices. *East Afr. Agric. For. J.* 52:74-80.

Piccolo, A., G. Pietramellara and J.S.C. Mbagwu. 1997. Reduction in soil loss from erosion-susceptible soils amended with humic substances from oxidized coal. *Soil Technology* 10:235-245.

Singer, M.J. and Y. Le Bissionas. 1998. Importance of surface sealing in the erosion of some soils from a Mediterranean climate. *Geomorphology* 24:79-85.

Starr, G.C., R. Lal, R. Malone, L. Owens, D. Hothem and J. Kimble. 2000. Modeling soil carbon transported by water erosion processes. *Land Degrad. Develop.* 11:83-91.

Tisdall, J. M. and J.M. Oades. 1982. Organic matter and water-stable aggregates in soils. *J. Soil Sci.* 33:141-163.

USDA-NRCS. 1995. *Soil Survey Investigations Report No. 45 Version 1.0.* National Soil Survey Center Soil Survey Laboratory, Lincoln, NE.

Yoder, R. E. 1936. A direct method of aggregate analysis and a study of the physical nature of erosion losses. *J. Am. Soc. Agron.* 28:337-351.

CHAPTER 20

Toward an Efficient Method for Measuring Soil Organic Carbon Stocks in Forests

G.R. Smith

I. Need for Efficient Carbon Measurement Methods

A. Climate Change Mitigation

In November 1998, the United States signed the Kyoto Protocol, agreeing to reduce its greenhouse gas emissions to 7% below 1990 levels by about 2010. While the U.S. Congress has not ratified the treaty and it is not yet in effect, many people and organizations are seeking ways to meet emission reduction commitments at minimal cost. One focus of attention has been on using forests as sinks for atmospheric carbon. Forests offer opportunities because as they grow and develop, they take substantial amounts of the greenhouse gas carbon dioxide (CO_2) from the atmosphere and, through photosynthesis, bind the carbon into organic matter. In order to count sequestered carbon as mitigation for other emissions, the quantity and timing of sequestration must be reliably quantified.

This chapter discusses methods for efficiently making on-site measurements and calculating the quantity of carbon in a forest. These methods address issues relevant to implementing greenhouse gas emission mitigation projects. Making multiple measurements over time allows measurement of net sequestration or release of carbon. The strategy divides a forest into several carbon pools—such as trees and mineral soil—statistically samples each pool, and combines the calculations into a measurement of total forest carbon stock. The chapter addresses major assumptions underlying the proposed approach, general strategies for reducing per ton measurement costs, stratified sampling designs, some of the practical and logistical issues of implementing field measurements, and data analysis. The chapter closes with suggestions for future work. To limit length, this chapter focuses on methods for measuring and calculating soil organic carbon in mineral soils, and provides minimal information about measuring other carbon pools in forests.

B. Soil Carbon Stocks Change over Time

Many efforts quantifying forest carbon assume that organic carbon stocks in mineral soils are constant (e.g., Birdsey, 1996 a, b; U.S. Environmental Protection Agency, 1995; Krankina et al., 1996). Soil organic carbon stocks are substantial, typically ranging from 50 to 150 metric tons per hectare (t ha^{-1}) (Brady and Weil, 1996). They develop primarily through biological activity in a mineral substrate, interacting with climate (Kern, 1994). Stocks are not necessarily in equilibrium, however, and rates of carbon accretion or loss can be significant. Even in hot, dry climates, soil carbon accretion may be significant. In an undeveloped soil on a dry site in southern California, carbon accretion has been measured at 0.9 to 3.8 t ha^{-1} in 40 years, depending on the type of vegetative cover (Quideau et al., 1998). On well-developed soils, management may significantly change soil carbon stocks. Agriculture in the central U.S. corn belt is believed to have reduced soil organic carbon by roughly 30 t ha^{-1} in the top 20 cm of soil during the first half of the 20th century (Lal et al., 1998). Significant opportunities for carbon accretion in temperate forest and agricultural soils exist where anthropogenic disturbance has increased soil respiration and/or decreased organic matter inputs, reducing soil carbon stocks below what they otherwise would have been (Lal et al., 1998; Harrison et al., 1995; Black and Harden, 1995; Johnson, 1992).

C. Models and Tables Not Reliable Predictors at Site Scale

Regional averages and models of modest complexity can be used to estimate total national terrestrial carbon sequestration. However, sites of similar forest type and forest stand age often vary widely in the amount of carbon they are sequestering or change in the amount of carbon sequestered. As a result, regional or national averages are not adequate for predicting the quantity of carbon sequestered in individual projects covering a few tens of hectares to a few thousands of hectares.

Similarly, tables and equations exist for predicting the change in forest stand carbon through time for major tree species in the United States (e.g., Birdsey, 1996 a, b; U.S. Environmental Protection Agency 1995; Brown et al., 1997; Brown and Schroeder, 1999), but individual sites vary widely from published average values. Published tables are typically keyed off stand age or the size of dominant trees. Two major difficulties exist when trying to use simple indicators of site-level carbon stock with few or no field measurements of site conditions. First, on individual sites within a classification, the physical conditions that guide site productivity and forest type vary widely from average. Variation in soil parent material, temperature, and precipitation leads to differences in potential carbon stocks. Second, sites with different disturbance histories often have very different stocks of carbon in pools other than live trees, even when the live tree pools are similar. Some tables seek to address the largest of these differences by differentiating stands of different origins, such as regeneration after logging versus regeneration on former pasture (e.g., Birdsey, 1996a, b).

Uncertainties and complex interactions of soil, plants, and disturbance history make it difficult to model changes in carbon stocks as a function of management activities. In some cases, even predicting the direction of change is problematic. For example, coarse woody debris tends to increase in pulses after logging, fire or wind disturbance. Rates of decay (and carbon emission resulting from the decay of wood) vary as a function of soil disturbance, ground temperature, and moisture. Disturbance may open the forest canopy, warming the ground and increasing decay rates. Interactions between factors result in total carbon stock changes which are not monotonic with tree age (Spies et al., 1988; Krankina and Harmon, 1994).

D. Carbon Measurement Presently Too Expensive

Achieving the potential efficiencies of a market for emissions credits requires that credits reflect actual carbon storage and be well documented. Few buyers would purchase credits if they fear that those individual credits might become invalid because they are not derived from verified mitigation activity. However, if the cost of carbon measurement is greater than the market price for mitigation by another means, few will invest in terrestrial carbon sequestration. Our challenge is to develop forest carbon sequestration methods that cost less than potential market price of the quantity of carbon sequestration they would document.

Methods suitable for supporting carbon sequestration crediting should detect terrestrial carbon stock changes occurring over 5- to 10-year periods, at the scale of a land ownership or group of land ownerships. To satisfy this criterion, measurements must detect carbon stock changes that are small relative to the spatial variability in the stock. For example, in productive, temperate forests, changes in carbon stocks on order of 0.5 to 4 tonnes per hectare per year (t ha^{-1} yr^{-1}) are common on sites with stocks of 50 to 600 t ha^{-1} (Sampson and Hair, 1996). Many stands sequestering carbon do so at a rate of 2 t ha^{-1} yr^{-1} or less. A significant factor contributing to the difficulty and cost of measuring forest carbon is the relatively fine scale heterogeneity of forests. Measurements taken as little as a few meters apart might yield numbers that vary by a factor of two. On a site with an average stock of 200 t ha^{-1} the change over 10 years is often less than 20 tons, or 10% of stock. If combined statistical and measurement errors are 10%, change is not likely to be detected. If the coefficient of variation of a population is large, such as 35% or more, it will take several dozen to a few hundred samples to estimate the mean with a confidence of 5 to 10% (Reed and Mroz, 1997). Also, a forest is composed of several pools of carbon. Each pool requires different methods of measurement, and multiple measurements drive up costs.

While techniques exist for measuring the quantity of carbon in components of a given parcel of forest land, current methods require making detailed, site-specific measurements involving substantial labor by skilled technicians and analysts, and may involve use of expensive equipment and sophisticated software programs. Making measurements that are accurate enough to detect the amount of carbon sequestered by a decade of forest growth could easily cost thousands of dollars. Even with optimization of sampling, current soil carbon measurement techniques are too expensive to use on all but very large projects.

II. Assumptions

The most fundamental assumption in this chapter is that increases in the carbon stock on a forest site are composed of carbon that is withdrawn from the atmosphere. It is assumed that plants take carbon dioxide from the atmosphere, sequester some of that carbon in their tissues and products, and portions of that carbon move to the woody debris, forest floor, and soil carbon pools, and that carbon does not enter a site from any other source. This chapter addresses only organic carbon and does not address soil inorganic carbon. On some arid and semiarid sites, soil inorganic carbon sequestration may be significant (Lal et al., 1998).

Methods described here use the convention of including fine roots, less than 2 mm in diameter, with soil. Coarse roots are to be measured separately. The amount of organic carbon adhered to and incorporated in rocks is not measured because the quantity is assumed to be small or unchanging. While typically ignored, quantities of organic carbon adsorbed by rocks may be significant, precipitating from organic acids originating from plants (Gurung, 1998).

The default assumption about the fate of carbon leaving the site is that the carbon is returned to the atmosphere as carbon dioxide, as a product of oxidation or respiration. However, carbon may be transported off the site and remain sequestered. Mechanisms for transport and continued sequestration include erosion transport to a location where carbon is buried and stored (such as some sediment eroded from a site and buried on a lake bottom) and continued sequestration of carbon in wood products made from wood removed from the site. Also, reductions in the quantity of carbon measured on a site could result from movement of carbon to forms or locations beyond reach of measurement techniques. Specifically, if soil carbon is measured only to a depth of 1 m, any carbon which is transported to depths greater than 1 m will not be measured and will be assumed to be released.

This chapter assumes that released carbon is emitted as carbon dioxide. However, if woody material is deposited in wet environments, some of that material may decompose anaerobically, with part of the carbon released in the form of methane (Lal et al., 1998). Because methane has a global warming potential 21 times that of carbon dioxide (Energy Information Administration, 1996) the climate effect is not a simple linear function of the amount of carbon sequestered or released.

Measuring carbon stocks avoids difficulties of measuring carbon fluxes. Accurate measurement of fluxes requires continuous measurement of flux at representative points and is expensive. Continuous measurement would be required because flux may vary substantially from hour to hour, as a function of air temperature, soil temperature, humidity, soil moisture, wind speed, and light level. Also, some weather conditions may make measured values vary from actual flux. Patterns of difference between flux measurements and actual flux are not well understood.

III. Outline of a Method for Measuring Terrestrial Carbon

Measuring soil carbon stock is a component of measuring total forest carbon stock. Dividing the forest into different carbon pools allows utilization of existing methods for measuring carbon stock in each of those pools. Convenient pools are:

- Live and standing dead tree trunks
- Tree parts other than trunks (including coarse roots > 2 mm diameter)
- Vegetation other than trees
- Woody debris and the forest floor
- Mineral soil (including fine roots < 2 mm diameter)

While this chapter focuses on measuring soil carbon, many of the strategies described here apply to making cost-effective measurements of the total carbon stock in a forest.

Central to a strategy for cost-effective measurement is using established techniques. Often others have estimated or tested measurement errors of the techniques. If methods are already accepted within the scientific community, then measurements made using those methods will be more easily accepted.

Standard techniques for sampling soil for carbon content analysis are described by National Soil Survey Center (1995; 1996). Carbon calculations require measurement of bulk density and rock content; an accurate but labor intensive method suitable for rocky soil is described by Vincent and Chadwick (1994). Bulk density varies by soil water content (National Soil Survey Center, 1995) so it is preferable to measure bulk density at a standard water content. Sampling the full depth of the soil profile would provide a more accurate estimate of total soil carbon. However, complete sampling of deep soil profiles requires a lot of labor or equipment and is expensive. Methods suggested here consider sampling to a depth of 1 m (or to bedrock if bedrock is shallower than 1 m) to be a reasonable compromise between thoroughness and expediency. It has been asserted that in most soils, sampling to a depth of 1 m will measure much of the carbon (Brady and Weil, 1996). However, substantial carbon is present at depths greater than 1 m (Jobbagy and Tachsch, in press). Even if substantial carbon exists in depths greater than 1 m, studies in agricultural soils have shown most short to medium term changes in soil carbon stock resulting from changed management practices occurring in the top 10 to 20 cm of the soil profile (e.g., Dick et al., 1986). However, under some conditions significant carbon stock increases may be induced at greater depths (Fisher et al., 1994; Jackson et al., in press).

In soils with few coarse fragments larger than one half the diameter of the corer, core sampling is relatively easy and often can be done to a depth of 1 m. If soils are too dense for easy coring to

1 m, coring to 50 cm could be considered. In soils with more coarse fragments, augers may work. If augers fail, a common post hole digger may work faster than a shovel.

Many soil carbon studies stratify by soil depth or profile and calculate carbon stock separately at different depths. Since the purpose of the methods described here is to get an estimate of total site carbon, to save cost, the entire soil sample core may be processed as one sample. However, homogenizing entire cores for laboratory analysis risks failing to detect changes occurring over a few years, as these changes become smaller relative to the total stock measured and become masked by variability of soil carbon. If the goal is to detect changes occurring over a few years, rather than assess total carbon stock on site, measuring soil carbon only in the top 10 cm of soil is more likely to produce a useful outcome than measuring soil carbon to 1 m depth and homogenizing cores to the full depth for laboratory analysis of carbon content. Methods for processing soil samples are described by National Soil Survey Center (1996). Soil is sieved through a 2-mm screen to remove coarse material, then the fine material is homogenized and a sample is retained for laboratory measurement of organic carbon content. Laboratory methods for soil organic carbon measurement are described in National Soil Survey Center (1996) and Nelson and Sommers (1982). Measurement of bulk density and amount of fine soil are necessary for calculating carbon stock per unit land area. Standard technique for measuring bulk density from cores is described in National Soil Survey Center (1996). This method requires digging a hole and taking cores from the walls. When soil sampling is done by coring, a less time-consuming practice may be adequate. It is possible to dry the entire core, remove coarse fragments and weigh the fines, and divide by the total volume of the core sample to get a measurement of fine soil density

$$F/V = D_f$$

where F is the dry weight of fines in grams, V is the volume of the entire core sample in cubic centimeters, and D_f is the density of fine material in grams per cubic centimeter. This method avoids having to weigh or measure the volume of wood and coarse mineral fragments but note that the calculated value for fine soil density is not the bulk density that would be found by measuring a clod containing no coarse fragments. Instead, this density figure is the mass of fines expanded to fill the entire volume sampled, as if coarse fragments were removed and the remaining fines fitted to the original volume within the corer. This measurement method may be biased toward low values because as the corer is pressed into the soil, coarse fragments struck by the corer may push some fine soil out of the sample. This bias is expected to be significant if there is a significant amount of buried wood or coarse mineral fragments.

The percent organic carbon in fine soil is multiplied by the area, depth, and density of fine soil to get total carbon stock. Mass of organic carbon per hectare is

$$(\%C/100) * D_f * Depth * 100 = Carbon\ t\ ha^{-1}$$

where %C is the percentage of carbon by weight in fine soil determined by laboratory testing, D_f is the density of fine material in grams per cubic centimeter, and Depth is the soil depth over which carbon is measured and calculated in centimeters. Multiplying by 100 converts from grams per square centimeter to tons per hectare. When density is calculated for each core sample, estimates of mass of carbon per hectare may be calculated from each sample. Then descriptive statistics can be applied to the population of estimates to calculate the mean estimate of carbon stock, the standard error of estimate of the mean, and standard deviation of samples around the mean. If a few bulk density measurements are made using methods described in National Soil Survey Center (1996), then for each soil sample where carbon is measured adjustments must be made for coarse fragments and wood.

Then a mean bulk density estimate must be applied to each laboratory test result to calculate carbon for the area sampled.

IV. General Strategies for Cost-Effective Carbon Stock Measurement

A. Choice of Pools to Measure

The first strategy for cost-effective carbon stock change measurement is focusing attention on carbon pools most likely to have large changes in absolute terms. Total change may be large because it is large per unit area, or may be large because it occurs across a very wide area. Use literature and the professional judgement of experts who know about local conditions to decide what pools are likely to have significant changes over the measurement period, and focus effort on these pools. Note that if a pool has very large absolute stock, then a small proportional change in stock may be large in absolute terms. Soil carbon, as a large pool relative to the amount of carbon in vegetation, need only change a small percentage to have a large effect on the net carbon sequestration of a forest.

In forests, the strategy of focusing on carbon pools that are likely to have large absolute changes means focusing on lands with high rates of vegetation growth, and paying more attention to trees and woody debris and litter than understory shrubs. It may be cost-effective not to measure understory carbon stock, or make only very inexpensive measurements of the understory, even if carbon stock in the understory is expected to be increasing. Literature shows that the understory is often about 2% of biomass of mature forest (Birdsey et al., 1993; U.S. Environmental Protection Agency, 1995). If conditions suggest that understory biomass is constant or increasing, it may be acceptable to assume the biomass is constant and not measure it. If the change in biomass is highly unlikely to be more than the statistical error of measurements, then it may be acceptable to treat the biomass as constant. When understory carbon stock is expected to decline, measurement is probably merited. For example, when converting brush land to forest, understory biomass is likely to decrease when canopy closure occurs. Instead of spending substantial amounts on measuring pools which are small but likely to decline, it may be acceptable and cost effective to assume that the pool declines at the greatest plausible rate, and not spend money measuring that pool.

Where techniques are compatible, combining field surveys of different pools can save money, while allowing later separate analyses of the different pools. Specifically, live and dead trees, and coarse woody debris and the organic layers of a forest floor could be surveyed together.

B. Measuring Larger Areas

Because carbon in terrestrial ecosystems is spatially and temporally variable, many samples are needed to get accurate estimates of mean amounts of carbon on a project area. If it is possible to apply the results of a given set of measurements across a larger area, the cost of measurement is spread over a larger total number of tons of change in carbon stock, reducing the per ton cost of measurement. However, to extrapolate measurements across larger areas requires that the additional lands be similar to lands where sampling was done. Unfortunately, determining what is "similar" may not be possible prior to sampling.

C. Measuring Infrequently

Measuring less frequently as a strategy for reducing per ton measurement costs has two benefits. First, if a rate of sequestration persists relatively monotonically for a number of years, then the total amount sequestered on a given area is greater over a longer time period. Given that it generally does not cost more to measure a larger stock than a smaller stock on a given area assessed (unless some pool is absent, and thus no measurement is needed), per ton cost decreases as it is spread across a greater number of tons. Second, if rate of sequestration is monotonic, then less frequent measurements will seek to detect a proportionally larger change in stock, relative to total stock, which is a larger amount relative to a given coefficient of variation. As a result, it may be possible to accept a less precise measurement, which is much cheaper because it uses fewer samples or techniques that are less precise but easier to carry out. For example, change in carbon stock in a growing forest might be on the order of 1% per year. This is less than the error (sampling or measurement) of most measurement techniques. Thus, over 1 year, it is unlikely that sampling will detect change, even if change is occurring, even with large sample sizes and use of the most accurate techniques available. However, if measurements are made at 10-year intervals, there may be a 10% change in stock, which might be documented by normal techniques with feasible sample sizes.

D. Choice of Techniques

When seeking to increase cost-effectiveness of measurements, one should compare the costs of different techniques. Sometimes more accurate techniques are much more expensive than somewhat less accurate techniques. It may be possible to use techniques where each individual sample value is less accurate, but achieve a needed level of accuracy by taking many samples. Alternatively, it may be possible to calibrate less expensive techniques against expensive techniques and adjust the less expensive measurements by a correction factor determined for the given field crew or site (e.g., Harmon and Sexton, 1996).

Maintaining a high level of field crew training and experience can minimize the time needed to take a given set of measurements and can reduce measurement errors. With a skilled crew, it may be possible to make ocular estimates of piece size or percent cover. Two-phase sampling is key to limiting costs where delineation of different strata is possible prior to field sampling. With experience, designers of field sampling layouts may become skilled in determining boundaries of homogenous strata that capture much of the variance across the project area, allowing smaller sample sizes within each stratum and still obtain desired precision. Where it is not possible to delineate strata prior to sampling, post-stratification may be useful (Reed and Mroz, 1997). Where both expensive but accurate and inexpensive but inaccurate measurement techniques exist, ranked set sampling or other double-sampling techniques may allow efficient combination of many inexpensive measurements and a few expensive measurements.

V. Stratified Sampling Designs

A. Reasons to Use Stratified Sampling

Sampling is a standard technique for limiting the cost of finding out characteristics of a population that involves measuring a subset of a population and extrapolating to the whole population. Sampling is so common that we often do not think of it as a strategy for efficiency. Two main reasons for sampling are (1) limiting cost when a population is large and (2) limiting disturbance when measurement is destructive.

Much of inferential statistics deals with making extrapolations and predicting the likelihood of difference between population characteristics estimated from a sample and the actual population characteristics. Sampling introduces error in the form of probable difference between a sample mean and population mean, often expressed in terms of standard deviations or confidence intervals. Statistics can tell us the likely size of sampling error. Sampling should be designed to balance the size of likely sampling errors against cost of taking additional samples that would reduce size of sampling errors.

Stratification is a common technique for blocking population variance into different strata, to achieve lower variance across samples taken within each stratum. Statistics for calculating the variance and error of the estimate within a stratum are outlined below. Statistics for combining measurements across different strata within a pool are also addressed. When combining measurements from different pools, care must be taken because different pools typically will have different variances.

B. Difficulties with Stratified Sampling

The key issue in stratification is choosing strata to minimize the variation within each stratum and maximize the variation between different strata (Reed and Mroz, 1997). When variance within a pool is blocked in different strata, then a relatively small number of sample measurements within each stratum may give an estimate with a smaller statistical error than a larger number of sample points taken without stratification. In addition to choosing boundaries of strata prior to sampling, one must estimate the degree of differences between strata to be able to calculate the optimal number of strata and number of samples needed within each stratum to shrink standard errors to desired levels.

Defining strata and their boundaries requires a readily visible indicator of likely change in the value of the variable of primary interest. For example, soil typing maps or landforms may suggest boundaries for soil sampling strata within a project area. If an area appears relatively homogeneous it may be most efficient to not stratify sampling. Because each stratum must be sampled with sufficient intensity to limit the sample variance within the stratum, unless the area is very large (such as tens of thousands of hectares or larger), it probably it will not be efficient to have more than three or four strata.

Systematic sampling is a sampling method where a starting point is chosen (randomly or not), then subsequent sampling points are determined by a system. The advantage of systematic sampling is that time to locate and travel to sample points may be reduced, with significant labor costs savings. However, systematic sampling is not random sampling because at most the first sampling point is randomly chosen. Thus, it violates the requirements of many statistics for random samples. In theory, the variance of a set of measurements collected from points on a systematic grid is not calculated the same as for a random sample. In practice, however, calculating the variance of a systematically gathered set of measurements using equations for simple random sampling typically gives a result that is quite accurate (Reed and Mroz, 1997). However, some experts recommend using systematic sampling only where the focus is on estimating the population mean, and using random sampling when the focus is on determining the precision of the estimate (Avery and Burkhart, 1994). Because the emphasis in detecting change in carbon stocks depends on accurate estimates of means at two different times, and finding the difference between these two means, systematic sampling may be appropriate. If one uses systematic sampling, one must be careful that the distance between systematically located measurement points does not correspond to some regular pattern in the landscape, such as the distance between ridges or the distance between regularly spaced trees.

C. Predicting Number of Samples Needed

Achieving the goal of reliably measuring change in terrestrial carbon stocks requires measurements having precision such that combined measurement and sampling errors are less than likely changes in carbon stocks. Measurement error is the difference between an individual measurement of a sample and the actual value of that sample. Sampling error is statistically likely difference between characteristics of a measured subset of a population and the actual characteristics of the entire population. If one is willing to accept less confidence that measurements of carbon stocks are very close to actual amounts of stocks, then one accepts larger sampling errors and may take fewer samples. However, a risk of accepting large sampling errors is that calculated amounts of sequestration may be substantially different from actual sequestration.

Efficient sampling requires taking enough measurements to reduce the error of the estimated quantity of carbon to an acceptable level. Basic statistics says that increasing the sample size reduces the error of the estimate of a population mean. However, taking and processing samples costs money. To optimize, one wishes to take the minimum number of samples necessary to obtain the desired level of accuracy. Formulas for calculating the number of samples needed to get a desired degree of statistical accuracy can be obtained by transforming formulas describing confidence limits around estimated means. An equation for calculating optimal sample size is

$$n = \left(\frac{ts}{E}\right)^2$$

where n is the number of samples required, t is a value taken from a table of the Student's t distribution for a given number of degrees of freedom and desired confidence interval, s is the estimated standard deviation of the measured values, and E is the desired half width of the confidence interval (Avery and Burkhart, 1994). The half width of the confidence interval is the desired level of statistical error. Note that s and E must be expressed in the same units. Often statistical error is referred to with a probability such as "p of 0.05." In this equation E and s would typically be expressed in units of tons of carbon per hectare.

A difficulty with this equation is that it requires an estimation of values that have not yet been measured, s and t. Looking up a t value in a table of the distribution of t requires knowing the number of degrees of freedom for t. It is reasonable to choose t from a table of distribution of t using the desired confidence probability and a relatively high number of degrees of freedom such as 60. Alternatively, estimates of t can be refined by iterative calculation. However, because of lack of precision in the estimate of the size of the standard deviation, iterative calculation is not merited (Avery and Burkhart, 1994). Estimating s is a bit trickier. It can be estimated from a small pilot study or from previous studies done in locations believed to be similar to the area to be sampled. For example, say one wishes to know, with 95% confidence, the stock of carbon on a project area to within 4 t ha^{-1}. Thus E equals 4. Suppose there is a large study in the literature having a standard deviation of soil carbon measurements of 50 t ha^{-1} making s equal 50. Making the assumption that the population of area to be measured is roughly similar to sample in the published study, and assuming 60 degrees of freedom, a table of values of t says that $t = 2$ for a 95% confidence interval. Inserting these values into the equation $n = (2*50/4)^2 = 625$ samples. Taking and processing 625 samples is expensive and one may wish to see if one can tolerate a decrease in the expected accuracy of the study.

This equation ignores the correction for sampling a finite population. While there is a finite number of soil cores or trees on a parcel being measured, the number of items that might be sampled is large enough that the correction can be ignored for the sake of simplicity. If one is sampling less than 5% of the population, leaving out the finite population correction factor does not significantly change the estimate of the sample size needed to get a given level of precision (Reed and Mroz, 1997).

Inspection of this equation reveals that because the right side of the equation is squared and the left side is not. When the desired confidence interval is decreased by half (the precision is doubled), which is to say the acceptable standard error of the estimate of the mean is halved, the sampling intensity increases four times. Because a substantial portion of sampling costs are field labor costs that vary essentially linearly with the number of samples taken, this relationship makes it difficult to measure heterogeneous conditions both accurately and inexpensively.

Because small plots often exhibit more variability than larger plots, the effect of plot size on the observed standard deviation should be considered when designing a field sampling system. However, if sampling costs vary more as a function of total area sampled rather than by the number of plots, it may be cheaper to measure a greater number of smaller plots than fewer larger plots.

When designing a sampling plan, always collect some extra samples. Some samples may turn out to be improperly collected, or may be lost or damaged before laboratory analysis. Variance may be higher than expected and the extra samples may be needed to achieve the desired level of precision. By the time of remeasurement, some sample points are likely to be impossible to locate, destroyed, or on sold lands. Also, in light of knowledge developed through the initial measurement, strata boundaries may be moved after the initial sampling, thus reducing the number of sample points in a given stratum.

D. Predicting Number of Samples Needed with Stratification

Stratified sampling is a strategy for dealing with high variance, when variance is believed to correlate to some other variable that may be used to defined strata. For example, suppose one is attempting to get an accurate estimate of the carbon stock on a property on the east side of the Cascade Mountains in Washington or Oregon. Suppose the property encompasses steep, dissected landforms. Knowledge of the climate and vegetation suggests that the steep, south facing slopes, with a harsh environment, are likely to have substantially lower soil carbon that the more moist, moderate temperature north slopes. Measuring the entire ownership with a simple random sample is likely to yield a very high variance. However, because soil carbon content is believed to be somewhat correlated to aspect, the study designer may choose to divide the property into two strata, north facing and south facing ground. Each stratum would be sampled and estimates of carbon stock generated separately. Then the estimates of the different strata would be combined.

When stratifying sampling, one must decide how to allocate sampling effort across strata. Often designers of inventories choose to sample each stratum in proportion to the size of the population in that stratum compared to the size of the population as a whole. This proportional allocation of sampling is usually more efficient than taking an equal number of samples from each stratum. But proportional allocation often is not the most efficient means of allocating sampling effort. In proportional sampling, effort is allocated among strata in proportion to the variance of each stratum: More samples are taken from strata having larger samples. If strata have relatively similar variances, proportional sampling is efficient. However, if some strata have variances much greater than others, it may be efficient to take more samples within strata with large variances, and fewer samples within strata with smaller variances. This steering of sampling effort toward strata with large variances tends to reduce the overall error of the estimate, for a given amount of sampling effort. An equation for allocating samples among strata is

$$n_h = \frac{N_h s_h}{\displaystyle\sum_{h=1}^{k} N_h s_h}$$

where n_h is the number of sample units in stratum h, N is the number of sample units in the population (proportions may be used), and s_h is the sample standard deviation in stratum h (Reed and Mroz,

1997). This equation requires estimating the sample variance within each stratum. Note that if the variance is constant across strata, this equation reduces to distributing samples in proportion to the proportion of the population in each stratum.

When increasing the sampling effort in strata having greater variances, if one is sampling a small proportion of the total population, the total sample size needed to achieve some specified error bound is

$$n = \frac{t_{\alpha/2,df}^2 \left[\sum_{h=1}^{k} N_h S_h \right]^2}{N^2 E^2}$$

where $t_{\alpha/2,df}$ is the αth percentage point of a t-distribution having df degrees of freedom, and other variables are as described above (Reed and Mroz, 1997). The value of n must be approximated because the number of degrees of freedom is not known in advance. An adequate approximation of $t_{\alpha/2,df}$ is 2, if the confidence interval is set at 95% (with a two-tailed test). Note that this equation requires reasonable estimates of the sample variances of individual strata, which may make it—in practice—impossible to use this optimization.

When the cost of taking a sample varies across strata it may be worth adjusting the allocation of samples to account for cost differences. Sampling cost may vary because one stratum is hard to travel through (because it is steep or has dense vegetation), is remote, or because the item being measured occurs in a form which is more time consuming to measure. An equation for allocating sampling effort with respect to cost is

$$n_h = \frac{N_h S_h / \sqrt{c_h}}{\sum_{h=1}^{k} N_h S_h / \sqrt{c_h}} (n)$$

where C_h is the actual or relative cost of measuring a sample unit in stratum h and the other variables are as previously defined (Reed and Mroz, 1997). Note that to make this optimization, before sampling one must have information about the cost of sampling in each stratum. If calculated quantities of carbon are discounted by some error term, as the value of carbon increases, in general, it becomes economically efficient to sample more intensively, to reduce error terms. Under this scheme, reducing the error term would allow crediting a larger proportion of the sequestered carbon.

VI. Implementing a Sampling Design in the Field

One would expect to have limited control over defining the area to be measured. If the goal is to document the carbon on an ownership, then one is stuck with the ownership boundaries. However, measurement might be much less expensive if it is possible to not measure a small portion of an ownership that is very different from the rest of the ownership and would require substantial effort to sample. If an ownership is small, and other owners in the vicinity with similar lands wish to measure their carbon stocks, and if the management regimes are similar across ownerships, it may be possible to measure multiple ownerships as one unit and spread the cost of measurement over a larger area.

Once the area is defined, one must decide which carbon pools to measure. Some investigators make the assumption that soil carbon stocks under forest management are constant and choose not to measure them. If stocks are indeed constant or are increasing, this approach may be acceptable, depending on regulatory requirements and the goals of the landowner. Given public and regulatory concerns about the amount of greenhouse gases in the atmosphere, under-counting of sequestration is much more acceptable than over-counting.

While defining strata within the tree pool is relatively simple (one may readily observe trees and roughly identify major differences in tree volume), defining soil strata is not so simple. U.S. Department of Agriculture Natural Resources Conservation Service (NRCS) county soil survey reports may provide a useful starting point for stratifying soils. Another approach is to divide soils by landform, such as measuring slopes separately from riverbottom lands. An experienced soils expert may be able to reconnoiter the area, dig some shovel pits, evaluate vegetation and landform, and define soil strata. On projects of few hundred to few thousand hectares, it may be most efficient to stratify by landform. Regardless of the methods used for choosing strata, for most projects it is likely to be efficient to make the first cut at defining strata in the office, using maps or aerial photos. Field work can adjust and verify choices made in the office.

A key aspect of defining strata is accurately determining the area of each stratum. The least expensive way to determine areas may be to calculate them from orthophoto maps or other geocorrected photos or maps, using GIS or manual area estimation techniques. If technology is available and it is easy to traverse boundaries of units, it may be efficient to determine areas by locating corners or traversing boundaries with a global positioning system (GPS) receiver, especially if a differential correction signal is available. However, it is time consuming to traverse boundaries and on small parcels the errors of uncorrected GPS readings are often significant relative to total parcel area.

Because soil bulk density and rock content are spatially heterogeneous it is desirable to measure as many points as possible, but the cost of measurement usually permits measurement at only a few points. At minimum, bulk density and rock content should be assessed in each soil stratum. Consider two-phase sampling where crews make an ocular rating of the density and rockiness of soil, and relate ocular estimates to measurements made using Soil Survey Laboratory methods (e.g., National Soil Survey Center, 1995; National Soil Survey Center, 1996).

To control for fine-scale heterogeneity of carbon distributions, use permanent sample points. The goal is to measure the same points, or sampling elements, at each time when the carbon stock is measured. When remeasuring permanently located points, one is measuring change in individual elements. This requires less sampling intensity than is needed to make equally accurate estimates of change in a heterogeneous population made using different sampling points at each time of measurement. Remeasuring permanently located points is desirable if the primary goal is accurate knowledge of change in stock, and there is less concern about having an accurate estimate of absolute stock. Permanent plots are particularly appropriate for soil because soil carbon content may vary significantly across locations a few meters apart, with no apparent change in vegetation (Pritchard, 1996).

When establishing plots, permanently mark the plot center with a means that will survive major disturbance. A half-meter long piece of rebar pounded completely into the ground is inexpensive and will survive fire and most logging disturbance, and may be relocated with a metal detector. Actual samples may be taken a given distance and direction from the plot center to minimize trampling effects and interference between samplings. Consider how one sampling might interfere with subsequent samplings both within and across pools. Establish procedures to prevent problems. For example, trampling and scuffing during soil sampling might affect subsequent understory vegetation measurements.

When recording plot locations, make sure to use information that can be followed by a different person, years later, to find each plot. Unless one is using a system with real time error correction, coordinates read from a GPS receiver will get one close to a plot but will not be adequate for relocating plots without additional information or marking. Plots can be located with reference to land survey markers or boundaries, or with respect to geographic features unlikely to move. Over time, "permanent" geographic features such as road intersections and survey monuments often are destroyed or moved. As a result, for long-term studies, redundant plot location information often ends up being useful.

There are many reasons why data archives must be designed to facilitate easy retrieval many years after initial measurements. Most importantly, to be able to pair remeasurement data with original measurements, subsequent analysts must be able to recover raw measurement data by plot and understand how data were initially collected and analyzed. Also, it is likely that subsequent analysts will wish to reanalyze data using different statistics or constants. At minimum, people making subsequent remeasurements must be able to recover directions for locating plots. With time, measurements of soil bulk density and rock content may improve, and subsequent analysts must be able to recalculate stocks using new factors.

VII. Data Analysis

A. Estimating Population Mean and Error of Estimate

If a stratified sampling design is used, estimating the overall population mean involves computing the average of sample observations within each stratum, then weighting these averages and combining them. Following Avery and Burkhart (1994), an equation for pooling samples is

$$\overline{X}st = \frac{\sum_{h=1}^{L} N_h \overline{x}_h}{N}$$

where $\overline{X}st$ is the overall population mean, L is the number of strata, N_h is the total number of units in statum h $(h = 1, ..., L)$, and N is the total number of units in all strata.

Calculating the confidence intervals around an estimated population mean requires calculation of the standard error of the mean, which is in turn calculated from the variances of each stratum. The variance of each stratum is calculated in the same way as calculating the variance of a simple random sample. For simplicity, the standard error of the estimated population mean can be calculated as if there were sampling with replacement, using the equation

$$S\overline{x}st = \sqrt{\frac{1}{N_2} \sum_{h=1}^{L} \frac{N_h^2 S_2^h}{n_h}}$$

where $S\overline{x}_{st}$ is the standard error of the population mean, s^2_h is the variance for stratum h, and other symbols are as previously described. The confidence interval of the mean can be described as

$$\overline{x}st \pm ts\overline{x}_{st}$$

where the degrees of freedom of t can be estimated as

$$\sum_{h=1}^{L} \left(n_h - 1 \right)$$

B. Estimating the Accuracy of Measurements of Change in Stock

Measuring change in stock typically requires a measurement of stock at the start of an accounting period and remeasurement of the stock at the end of the accounting period. Sequestration (or emission) would be the difference between the mean estimates of the stock at the end of the accounting period and the beginning of the period. Each time the stock is measured using sampling, the measured value is a sample mean. Basic statistics tell how to calculated the confidence interval around a difference of means. Calculating statistical error in the measurement of stock is calculating the confidence interval around a difference of means.

For modest accounting periods like 5 or 10 years, large areas of land that are sequestering carbon typically do not change dramatically in their land cover and total carbon stock. It is reasonable to assume that the "population" of carbon sampled at the end of the accounting period is very similar to the population sampled at the beginning of the accounting period. Thus it is reasonable to assume that the variance at the end of the period is essentially the same as the variance at the beginning. Finding a confidence interval around the difference of sample means having a common variance is

$$\left(\mu_2 - \mu_1 \right) = \left(\overline{X}_2 - \overline{X}_1 \right) \pm t_x s_p \sqrt{\frac{1}{n_2} + \frac{1}{n_1}}$$

where μ_2 is the actual carbon stock at the end of the accounting period, μ_1 is the actual carbon stock at the beginning of the accounting period, $\overline{X}_2$ is the measured carbon stock at the end of the accounting period, $\overline{X}_1$ is the measured carbon stock at the beginning of the accounting period, t_x is the appropriate Student's t value for the chosen level of confidence x (in one tail of the distribution) and degrees of freedom, s_p is standard deviation of the samples, n_2 is the number of samples taken at the end of the accounting period, and n_1 is the number of samples taken at the beginning of the accounting period (Wonnacott and Wonnacott, 1984). Note that this equation is for independent samples. Often achieving desired precision will require measurement of permanently located sampling points. Remeasurement of permanent plots gives greater precision because the analyst may find the differences between earlier and later measurements on individual plots, and analyze the differences as a single sample.

C. Measurement Error

The above equations estimate the statistical error resulting from sampling rather than counting an entire population. In addition to sampling error, measurement error may exist. Measurement error is the accuracy of a measurement, with respect to the actual attribute of an entity being measured. For example, if an entity is actually 100 cm long but is measured to be 102 cm, the measurement error is 2%. Among other things, measurement error arises from crudeness of measuring instruments, difficulty in observing whatever is being measured, and operator error. Measurement error is reduced by training data collectors and checking some data against independent remeasurements. Sampling protocols should include quantified limits for measurement error, and procedures for checking achieved measurement accuracy. If initial data collection does not meet predetermined accuracy standards, the data must be discarded and recollected.

Measurement error typically increases the variance of measured numbers. If measurement errors are random and not biased in one direction, then measurement error does not change the estimate of a population mean, though it increases calculated variance.

D. Effect of Errors in Bulk Density and Rock Content Measurements

Measuring soil bulk density, especially in soils with significant rock content, is expensive and often produces relatively unreliable measurements (Vincent and Chadwick, 1994). These difficulties are mitigated somewhat when focusing on measurement of change in soil carbon stock, rather than measurement of absolute carbon stock. Because of the role of density and rock content in calculating carbon stocks as multipliers and the fact that they change relatively slowly over time, they may be used as constants by which other measurements are multiplied. Specifically, percentages of fine soil organic carbon content determined through laboratory testing of soil samples are multiplied by soil depth, area, and density and reduced by volume of rock content to produce a calculation of total soil carbon in a given area. Treating bulk density and rock content as constants avoids introducing the costs and errors of making independent measurements of density and rock content each time the carbon stock is counted. If bulk density and rock content were measured independently each time carbon stocks were measured, measurement errors would be compounded and measurement costs would be increased.

E. Discounting Measured Changes

Some have proposed "discounting" the amount of credit given for calculated amounts of sequestered carbon, to compensate for errors in measurement. However, discounting calculated carbon quantities has two negative effects. First, mean estimates of carbon stock are, statistically, the best estimate of the quantity of carbon sequestered. If errors in measurements are unbiased, then when aggregating measurements across units and projects, overestimates and underestimates pretty much balance out, giving a good estimate of what is happening to the atmosphere. Discounting for sampling uncertainty would bias measurements and reduce the accuracy of aggregate numbers compiled from project reports. Second, decreasing credits for sequestration by a proportion of uncertainty in measurements of carbon stocks decreases the number of tons of credits authenticated by a given amount of measurement effort, making measurement more costly on a per ton basis. As a result, fewer carbon sequestration projects will provide an acceptable return on investment, which is likely to decrease total terrestrial carbon sequestration effort.

VIII. Topics for Future Work

Work on a variety of issues could make carbon stock measurement more economical and useful. One potentially fruitful avenue for improving terrestrial carbon stock measurement is examining technologies and techniques developed for other uses, to see if they may be adapted to measuring carbon stocks. A special need is for cheaper and more accurate measures of soil bulk density and rock content. For example, it may be possible to adapt technologies developed to locate land mines using variation in soil density, and use these instruments to measure soil bulk density. A potentially promising technique that uses neutron probes for measuring bulk density has, unfortunately, not worked well enough for its use to become widespread. Also, inexpensive techniques for sampling soil at depths greater than 1 m are needed because carbon may be transported to depths greater than one meter and stored for long periods. Additional capacity for measuring deep soil carbon should be complemented by enhanced understanding of the persistence and dynamics of deep soil carbon.

As protocols and technologies are developed and applied to a wide variety of sites, broadly applicable and cost-effective measurement methods may become apparent. If methods can be standardized, design costs will be substantially reduced. As knowledge increases, models calibrated to sites with inexpensive measurements of site conditions may produce highly reliable estimates of carbon stock change at the site scale. Increased knowledge about ecological processes will increase the robustness of application of carbon stock change relationships to new locations. Also, modeling may help to identify specific management actions and locations likely to yield large carbon and ecological benefits. Adaptive management, where effects of management activities are measured and compared to expected effects, is essential to increasing the cost-effectiveness of practices and verifying the degree of goal achievement (Gunderson, Holling and Light, 1995; Walters and Holling, 1990; Walters, 1986).

Future work should make measuring changes in terrestrial organic carbon stocks both more reliable and less expensive. Improvements would make terrestrial carbon sequestration more available as a verifiable basis for tradable carbon credits, allowing sequestration to make a larger contribution to efforts to mitigate climate change.

Acknowledgments

Financial support for this work was provided by the Environmental Defense Fund and Environmental Resources Trust.

References

Avery, T.E. and H.E. Burkhart. 1994. *Forest Measurements*, 4th ed. McGraw-Hill, New York.

Birdsey, R.A., A.J. Plantinga and L.S. Heath. 1993. Past and prospective carbon storage in United States forests. *Forest Ecology and Management* 58:33-40.

Birdsey, R.A. 1996a. Carbon storage for major forest types and regions in the conterminous United States. p. 1-26. In: R.N. Sampson and D. Hair (eds.), *Forests and Global Change, Volume 2: Forest Management Opportunities for Mitigating Carbon Emissions*. American Forests, Washington, D.C.

Birdsey, R.A. 1996b. Appendix 2-4: Regional estimates of timber volume and forest carbon. p. 261-372. In: R.N. Sampson and D. Hair (eds.), *Forests and Global Change, Volume 2: Forest Management Opportunities for Mitigating Carbon Emissions*. American Forests, Washington, D.C.

Black, T.A. and J.W. Harden. 1995. Effect of timber harvest on soil carbon storage at Blodgett Experimental Forest, California. *Can. J. For. Res.* 25:1385-1396.

Brady, N.C. and R.R. Weil. 1996. *The Nature and Properties of Soils*, 11th ed. Prentice-Hall, Upper Saddle River, NJ.

Brown, S.L. and P.E. Schroeder. 1999. Spatial distribution of aboveground production and mortality of woody biomass forests based on inventory data for eastern U.S. *Ecological Applications* 9:968-980.

Brown, S.L., P. Schroeder and R. Birdsey. 1997. Aboveground biomass distribution of U.S. eastern hardwood forests and the use of large trees as an indicator of forest development. *Forest Ecology and Management* 96:37-47.

Dick, W.A., D.M. Van Doren, Jr., G.B. Triplett, Jr. and J.E. Henry. 1986. Influence of long-term tillage and rotation combinations on crop yields and selected soil parameters: results obtained from a Mollic Ochraqualf soil. OARDC Res. Bull. 1181, Wooster, OH.

Energy Information Administration. 1996. Emissions of greenhouse gases in the United States. DOE/EIA-0573(95) U.S. Dept. of Energy, Washington, D.C.

Fisher, M.J., I.M. Rao, M.A. Ayarza, C.E. Lascano, J.I. Sanz, R.J. Thomas and R.R. Vera. 1994. Carbon storage by introduced deep-rooted grasses in the South American savannas. *Nature* 371:236-238.

Gunderson, L.H., C.S. Holling and S.S. Light. 1995. *Barriers and Bridges to the Renewal of Ecosystems and Institutions.* Columbia University Press, New York.

Gurung, J. 1998. Organic Matter Characterization of a Douglas-fir Forest Soil and the Effects of Biosolids Application. Master's Thesis, College of Forest Resources, University of Washington, Seattle.

Harmon, M.E. and J. Sexton. 1996. Guidelines for Measurements of Woody Debris in Forest Ecosystems. Publication No. 20. U.S. LTER Network Office, University of Washington, Seattle.

Harrison, A.F., P.J.A. Howard, D.M. Howard, D.C. Howard and M. Hornung. 1995. Carbon storage in forest soils. *Forestry* 68:335-348.

Jackson, R.B., H.J. Shenk, EG. Jobbagy, J. Danadell, G.D. Colello, R.E. Dickinson, C.B. Field, P. Friedlingstein, M. Heimann, K. Hibbard, D.W. Kicklighter, A. Kleidon, R.P. Neilson, W.J. Parton, O.E. Sala and M.T. Sykes. Belowground consequences of vegetation change and their treatment in models. *Ecological Applications* (in press).

Jobbagy, E.G. and R.B. Jackson. The vertical distribution of soil organic carbon and its relation to climate and vegetation. *Ecological Applications* (in press).

Johnson, D.W. 1992. Effects of forest management on soil carbon storage. *Water Air Soil Pollut.* 64:83-120.

Kern, J.S. 1994. Spatial Patterns of Soil Organic Carbon in the Contiguous United States. *Soil Sci. Soc. Am. J.* 58:439-455.

Krankina, O.N. and M.E. Harmon. 1994. The impact of intensive forest management on carbon stores in forest ecosystems. *World Resource Review* 6:161-177.

Krankina, O.N., M.E. Harmon and J.K. Winjum. 1996. Carbon storage and sequestration in the Russian forest sector. *Ambio* 25:284-288.

Lal, R., J.M. Kimble, R.F. Follett and C.V. Cole. 1998. *The Potential of U.S. Cropland to Sequester Carbon and Mitigate the Greenhouse Effect.* Sleeping Bear Press, Chelsea, MI.

National Soil Survey Center. 1995. Soil Survey Laboratory Information Manual: Soil Survey Investigations Report No. 45, Version 1.0. Natural Resources Conservation Service, United States Department of Agriculture. U.S. Government Printing Office: 1995-658-181.

National Soil Survey Center. 1996. Soil Survey Laboratory Methods Manual: Soil Survey Investigations Report No. 42, Version 3.0. Natural Resources Conservation Service, United States Department of Agriculture. U.S. Government Printing Office: 1996-756-515.

Nelson, D.W. and L.E. Sommers. 1982. Total carbon, organic carbon, and organic matter. p. 539-579. In A.L. Page et al. (eds.), *Methods of Soil Analysis.* Part 2. 2nd ed. Agron. Monogr. 9. American Society of America and Soil Science Society of America, Madison, WI.

Pritchard, S. 1996. Carbon Storage and Soil Properties of Subalpine Forests and Meadows of the Olympic Mountains, Washington. Master's Thesis, College of Forest Resources, University of Washington, Seattle.

Quideau, S.A., R.C. Graham, O.A. Chadwick and H.B. Wood. 1998. Organic carbon sequestratin under chaparral and pine after four decades of soil development. *Geoderma* 83:227-242.

Reed, D.D. and G.D. Mroz. 1997. *Resource Assessment in Forested Landscapes*. John Wiley & Sons, New York.

Sampson, R.N. and D. Hair. 1996. *Forests and Global Change Volume 2: Forest Management Opportunities for Mitigating Carbon Emissions*. American Forests, Washington, D.C.

Spies, T.A., J.F. Franklin and T.B. Thomas. 1988. Coarse woody debris in Douglas-fir forests of western Oregon and Washington. *Ecology* 69:1689-1702.

U.S. Environmental Protection Agency. 1995. Climate Change and Mitigation Strategies in the Forest and Agriculture Sectors. EPA 230-R-95-002. Office of Policy, Planning and Evaluation, Climate Change Division.

Vincent, K.R. and O.A. Chadwick. 1994. Synthesizing bulk density for soils with abundant rock fragments. *Soil Sci. Soc. Am. J.* 58:455-464.

Walters, C. 1986. *Adaptive Management of Natural Resources*. MacMillan, New York.

Walters, C.J. and C.S. Holling. 1990. Large-scale management experiments and learning by doing. *Ecology* 71:2060-2068.

Wonnacott, T.H. and R.J. Wonnacott. 1984. *Introductory Statistics for Business and Economics*, 3rd ed. John Wiley & Sons, New York.

CHAPTER 21

Soil Organic Matter Evaluation

R.A. Rosell, J.C. Gasparoni and J.A. Galantini

I. Introduction

Soil organic matter (SOM) is a complex mixture of animal and plant residues, fresh and at all stages of decomposition, living and decaying microbial tissue and heterotrophic biomass, and the relatively resistant humic substances. It has a series of beneficial properties for soils, even at low percentage levels. Organic matter (OM) is one of the most complex, dynamic and reactive soil component. The various organic fractions have different properties which affect the soil behavior and fertility. Unfortunately, it is difficult to separate and quantify those fractions in order to be able to correlate soil constituents and plant productivity.

Carbon is the main element present in SOM, comprising from 48 to 60% of the total weight. Organic carbon determination is often used as the basis for SOM estimation by multiplying the organic carbon (OC) value by a conversion factor. Many researchers use the so-called Van Bemmelen factor of 1.724 (see below). Actual evidence indicates that the factors most appropriate for converting OC to OM are 1.9 to 2.5 for surface and subsoil, respectively (Broadbent, 1953; Nelson and Sommers, 1982).

On the other hand, Rasmussen and Collins (1991) have indicated that the conversion factor to transform % OC to % SOM varies between 1.4 to 3.3. For that reason it is suggested to use the "% of OC," which is determined, instead of "% of SOM," which must be estimated.

The wet oxidation procedure is based on the oxidation of OC in soil with a dichromate-sulfuric acid mixture, either without or with external heat in order to accelerate and complete the oxidation. It is known that these methods do not produce complete oxidation of the OC, requiring therefore a correction (oxidation) factor for calculating the results. In both groups of procedures the excess dichromate ($Cr_2O_7^{2-}$) is back titrated with ferrous ammonium sulfate by using one redox indicator (diphenylamine, o-phenanthroline, or N-phenylanthralinic acid). The use of a correction (incomplete oxidation) factor may introduce an error because the OM in different soils, or even in different horizons of a soil profile, is not always oxidized to the same degree. The correction (oxidation) factor may vary markedly among surface soils. Reported values range from 1.19 to 1.40 (Nelson and Summer, 1982) and are greater for subsurface than for surface soils.

Non-cultivated, native, virgin soils decrease their carbon content and negatively change their physical, chemical and biological properties upon cultivation. Cultivated, pasture and prairie soils also lose SOM rapidly after they are cultivated (Miglierina et al., 1988). The loss is usually exponential, declining sharply during the first 10 to 20 years. Later losses continue declining until an equilibrium is reached after 50 to 60 years (Jenny, 1941).

Dalal and Mayer (1986) have found that, after cultivation, OC bound to sand-sized particles declined rapidly; those associated with silt-sized particles increased from 48 to 61% (Sikora et al.,

1996). The OC differential behavior when it is associated with or is present in several soil particle size fractions is a clear indication of the importance of physically fractionating SOC.

It has been recognized that the interaction of SOM fractions with clay minerals is an essential factor in the behavior of the organic fractions (Christensen, 1992). Soil aggregate fractionation is based on the principle that the fractions of SOM are associated and/or bound to inorganic particles of different sizes, which differ in structure and function, playing important and differential roles in SOM turnover. Size fractionation is mainly applied to whole soil samples. Soil samples are usually dispersed in water and then separated by wet sieving. The procedure is considered to be very efficient because of the solution and mechanical effects of water. In sandy soils dry sieving is commonly used.

In general, most of the humified SOM in agricultural soils is found associated with the silt- and clay-sized particle fractions, which are called the fine fraction. The clay fraction can contain more than 50% of SOM (Tiessen and Stewart, 1983; Bonde et al., 1992). Other authors (Whatson and Parsons, 1974; Elustondo et al., 1990) have reported that the silt-size fraction appeared to have the highest proportion of SOM.

Recently, Galantini and Rosell (1997) studied the evolution of SOC, N, P, and S in two particle size fractions of an Entic Haplustoll under several crop sequences after 10 years of cultivation. They used wet sieving and obtained the denominated fine fraction (FF, < 100 μm particle size) and coarse fraction (CF, 100 to 2000 μm particle size). They determined that the level of some SOC fractions associated with the FF was little affected after cultivation. On the other hand, mineralization of OM was more intensive in the coarse fraction, thus showing its effects on the changes and dynamics of SOC and plant nutrients. Knowledge of the SOC distribution, mainly the composition of the CF, may therefore predict the OC and nutrient turnover and the present crop-soil fertility status.

The concept behind physical fractionation of soils emphasizes the role of soil minerals in SOM stabilization and turnover. The physical fractionation techniques are considered chemically less destructive, and the results obtained from physical soil fractions are considered to be related more directly to the structure and function of SOM *in situ*.

The different SOM fractions and their transformation rates require a knowledge of their composition and functions. In this chapter, some SOM definitions, fractionation (physical, biological, and chemical), and determination methodologies as well as their agronomic implications are discussed.

II. Definitions

Soil OM has been defined as "the organic fraction of the soil exclusive of undecayed plant and animal residues" (SSSA, 1997) and has been used synonymously with "humus." However, for laboratory analysis, the SOM generally includes only those organic materials that accompany soil particles through a 2-mm sieve (Nelson and Sommers, 1982). Organic residues larger than 2 mm are denominated plant debris.

Other SOM definitions have been used by numerous authors throughout the years. Historically, chemical concepts and procedures have primarily influenced the methodology employed for SOM research. Years of research efforts have been expended to determine the chemical structure and properties of SOM, such as elemental composition, cation exchange capacity (CEC), total and/or partial acidity, nutrient availability and/or sequestration, etc. in the humified fractions. In general, chemical characterization of OM has been unsuccessful in predicting agronomic responses.

A present concept of SOM components or composition can be defined as follows:

•plant debris: organic residues larger than 2000 μm.

•organic residues (> 53 μm or 100 to 2000 μm): relatively fresh residues (Magdoff, 1996), decaying plant and animal tissues or debris and their partial decomposition products, including compounds of different biochemical origin (amino acids, carbohydrates, fats, waxes, organic acids, etc.) present in decaying tissues. This compartment is also defined as the organic matter of the coarse fraction (CF) of the soil (Andriulo et al., 1990). Cambardella and Elliott (1992) have defined this fraction as particulate organic matter (POM). Other authors have defined this fraction as young SOM (Janssen, 1984; Andriulo et al., 1990) or light fraction (Greenland and Ford, 1964; Richter et al., 1975) of SOM.

•soil biomass (0.1 to 2 µm): the SOM present as live and dead microbial tissues, which carry out important functions related to residues decomposition, nutrient cycling, pest controls, etc., and also promote porosity channels (Magdoff, 1996).

- the OM compartment of the soil biomass and the humic substances associated to the less than 53 μm (or 100 μm) size is also called the fine fraction (FF) of the SOM.

- humic substances (< 53 µm): "a series of relatively high-molecular-weight, brown to black colored substances formed by secondary synthesis reactions " (Stevenson, 1982). According to their molecular size and their solubility in alkali and acid aqueous solutions, well-decomposed humic substances (HS) include:

humin, the alkali insoluble fraction of the HS, formed possibly by the condensation of humic and fulvic acids and residual lignin components;

humic acids, the dark-colored organic material insoluble in acid media but soluble in alkali solutions and specific organic solvents;

fulvic acid fraction, the yellow-reddish (depending on pH), water-soluble, colored material which remains in solution after precipitation of the humic acids by acidification.

The OM compartments of the soil biomass and the HS associated with < 53 μm size particles is also called the fine fraction (FF) of the SOM. The agronomic or functional aspects of SOM and its fractions have been important in situations related to soil fertility and plant productivity. For over a century, SOM has been separated by chemical procedures (water or alkali-acidic extractions) to be used more in pedogenic studies than in agricultural applications. However, Feller (1993) considers that:

- humic substances have a low turnover rate (Anderson and Paul, 1984; Duxbury et al., 1989) and are, therefore, not necessarily implicated in the short-term processes which occur in agronomic situations; and

- the behavior of these chemical compartments in relation to major soil processes such as aggregation, mineralization and surface properties are not yet fully understood.

In the last 30 years, SOM characterization based on physical (particle size or density) separation (Feller, 1979; Tiessen and Stewart, 1983; Elliott and Cambardella, 1991; Christensen, 1992) has intensified. Much evidence has accumulated to demonstrate that fractionation of soils according to particle size provides a useful tool for the study of SOM dynamics and distribution. A key point in these separations is to achieve adequate soil dispersion to prevent inclusion of microaggregates of smaller size particles in the silt and sand size separates. Chemical dispersants used regularly in the past may be too drastic, causing denaturalization of part of the SOM.

III. Functional Organic Matter Fractions

Normally, SOM is used as an indicator of soil quality. The most labile compartments of SOM are thought to participate in supplying nutrients (N, P, S and minor elements) for plant growth. The most labile compartments are rapidly depleted as a result of cultivation or soil movement. The intermediate labile compartments, which have a turnover rate of 10 to 20 years, must be periodically replenished to maintain an adequate level of SOM, fertility and sustainability.

Soil physical or granulometric separation is considered to provide fractions related to the structure and function *in situ* of the SOM. Rosell et al. (1996) studied the interactive effects between a 9-year-old wheat (*Triticum aestivum* L.) - sunflower (*Helianthus annuus* L.) rotation and the physically separated organic fractions of two texturally different Entic Haplustolls. The most important SOC changes were observed in the so-called young SOC or POC fraction (100 to 2000 μm) in both soils.

Three different compartments of SOM have been suggested for tropical soils (Feller et al., 1991):

plant debris (> 20 μm), with C/N >15;

organo-silt (2 to 20 μm), with properties intermediate between the former two and a C/N ratio ranging from 10 to 15;

organo-clay (< 2 μm), amorphous organic material closely associated to clays' bacterial debris and metabolites with a C/N ratio lower than 10.

A. Particulate SOM

Actual research indicates that the POM properties are similar to the characteristics of the intermediate labile SOM compartments. In native grassland soils, POM can account for up to 48% of the total SOC and 32% of the total soil N (Greenland and Ford, 1964).

Numerous authors have found that POM (or POC) shows promise as a short-term or early warning indicator of long-term changes in soil quality. The POC has been shown to be much more sensitive than SOC to changes in agricultural management and, as such, has been proposed as an indicator of soil quality (Gregorich and Ellert, 1993).

Others authors (Sikora et al., 1996) have indicated that the POC pool consists of partially decomposed fractions of plant residue and has a density of < 1.85 g cm^{-3} and a C/N ratio > 20. Light fraction (LF) SOM, as defined originally by Greenland and Ford (1964), has a similar density (< 2.0 g cm^{-3}) and an approximate C/N ratio of about 25. In this chapter, LF SOM is considered to be analogous to POM or young SOM.

B. Soil Microbial Biomass

Live soil biomass is responsive for the mineralization of plant residues and/or POM and the genesis of more stable humic substances as well as the production of plant nutrients. For that reason soil biomass is essential to the dynamics of SOM quality. Besides microbial biomass, soil respiration, enzyme activity and active OC component (carbohydrates, peptides, etc.) are factors associated with total OC which may function as indicators of soil quality. However, Campbell et al. (1991) suggested that an increase in biomass N does not necessarily imply an improvement of soil quality. On the other hand, Chien et al. (1964) reported that respiratory capacity increased with improvement of soil fertility. Sikora et al. (1996) consider that "data suggest that microbial biomass, CO_2 evolution, and specific respiration changes are much larger than SOM changes and that these compartments may allow more rapid evaluation of treatment effects" on soil fertility and plant production.

Presently, the more widely used biomass indicators are: Biomass C (kg ha^{-1}); Biomass C / total OC ratio (%); CO_2-C evolution (kg ha^{-1} d^{-1}); potentially mineralizable N (kg ha^{-1} y^{-1}); and CO_2-C / Biomass C (d^{-1}). The biomass indicators must have a common feature: they require a good, fresh sample, with no excess of humidity and temperature. Enzyme production and evolution have been studied, but results are not conclusive yet (Doran et al., 1996).

C. Physical Separation and Characterization of SOM

Many authors (Turchenek and Oades, 1979; Anderson et al., 1981; Tiessen and Stewart, 1983; Christensen, 1985; Balesdent et al., 1988; Christensen, 1992; Cambardella and Elliott, 1993) have employed soil physical fractionation according to particle size or density to study the behavior of SOM in these fractions.

The procedures have been useful to determine differences in the structural and dynamics properties of SOM. Ultrasonic energy disruption (sonication) is commonly used to obtain a good level of dispersion of soil structure, though different ultrasonic levels may generate changes in the distribution of the SOM among particle size fractions. Increasing the intensity of sonication may result in the distribution of proportionately more organic material into finer soil fractions (Elliott and Cambardella, 1991).

Cambardella and Elliott (1993) developed a dry and wet sieving procedure to isolate and characterize SOM fractions that were originally occluded in the soil aggregates and evaluated their effect on the turnover of the OM in several cultivation systems.

Nutrient contents were determined for discrete size/density OM fractions isolated from within the macroaggregate structure. Eighteen percent of the total SOC and 25% of the total N in a no-till soil was associated with fine-silt size isolated from inside macroaggregates (enriched labile fraction, ELF) and particles having a density of 2.07 to 2.21 Mg m^{-3}. Sodium metatungstate was used to obtain different solution/suspension densities (1.85, 2.07 and 2.22 Mg m^{-3}). The amount of OC and N sequestered in the ELF decreased as the intensity of tillage increased, when comparing no-till and conventional tillage.

The procedure proposed by Cambardella and Elliott (1993) has improved the quantitative determination of SOM fractions, which in turn increases understanding of SOM dynamics in cultivated grassland systems.

Aggregate disruption was done by suspending 10 g dried aggregate soil in 60 mL of water at room temperature and immersing an ultrasonic probe to a depth of 3 to 5 mm into the soil suspension. The disrupted soil suspension was passed through a series of sieves in order to obtain four size fractions: (1) > 250 μm; (2) 53 to 250 μm; (3) 20 to 53 μm; and (4) < 20 μm, which were dried overnight at 50°C.

The specific rate of mineralization (μg net mineral N / μg total N in the fraction) from microaggregate-derived ELF was not different for the various tillage treatments (bare fallow, stubble mulch, and no-till) but was greater than for intact macroaggregates.

IV. Estimation of SOM

It is difficult to quantitatively estimate the amount of SOM. Procedures used in the past involve determination of change of weight of the soil sample resulting after destruction or decomposition of organic compounds by hydrogen peroxide (H_2O_2) treatment or by ignition at high temperature. Both techniques are subject to errors.

Numerous methods are available for the determination of OC and the estimation of OM, all of them with inherent associated problems. For that reason, the method more applicable to the soils to be analyzed and the required accuracy of the results should be taken into account. All current methods are included in one of the following principles (see Table 1 for a classification and comments on the different methodologies):

Table 1. Determination of total organic carbon in soils

Method	Principles	CO_2 Determination	Comments
Wet combustion			
Combustion train	Sample is heated with $K_2Cr_2O_7$ – H_2SO_4 – H_3PO_4 mixture in a CO_2-free air stream to convert OC in CO_2.	Gravimetric Titrimetric	Available equipment Good precision. Adapted to solution analysis. Time-consuming. Titrimetric determination of CO_2 less precise.
Van-Slyke-Neil apparatus	Sample is heated with $K_2Cr_2O_7$ – H_2SO_4 – H_3PO_4 mixture in a combustion tube attached to a Van-Slyke-Neil apparatus to convert OC in CO_2.	Manometric	Expensive and easily damaged apparatus.
Titrimetric dicromate redox procedure	Sample is treated with a $K_2Cr_2O_7$ – H_2SO_4 mixture. Excess dichromate is back titrated with ferrous ammonium sulfate.		Requires simple volumetric equipment. It is easy to handle large number of samples and it is not costly. Oxidation factor is needed.
Dry combustion			
Resistance furnace	Sample mixed with CuO and heated to 1500°C in a O_2 stream to convert all C in CO_2.	Gravimetric Titrimetric	Reference method but time-consuming.
Induction furnace	Sample mixed with Fe, Cu, and Sn is rapidly heated to ~ 1650°C in a O_2 stream to convert all C in CO_2.	Gravimetric Titrimetric	Rapid combustion. Induction furnace is expensive.
Automated	Sample is mixed with catalysts and heated with resistance or induction furnace in a O_2 stream to convert all C in CO_2.	Gas chromatography Gravimetric Conductimetric IR absorption spectometry	Rapid, simple and precise. Expensive. Slow release of contaminant CO_2 from alkaline earth carbonates with resistance furnace.
Weight-loss-on-ignition	Sample is heated to 430° C in a muffle furnace during 24 hours.	Gravimetric	Weight losses are due to water, CO_2 and other volatile compounds. Gives too high levels of soil organic matter.

A. Wet Oxidation of OC

1. Using Acid Dichromate Solution

a. Titrimetric Dichromate Redox Methods

The potassium dichromate procedures are widely used in soil OC determination because of their rapidity and simplicity compared with others wet or dry combustion methods. Advantage of acid oxidation is the removal of carbonate C.

Dichromate redox methods potentially suffer from a number of interferences and low recoveries. In the Walkley-Black (1934) method, OC is oxidized solely by the action of dichromate, heated as by the exothermic mixing of aqueous dichromate and concentrated sulfuric acid ($\approx 120°C$). As indicated in the Introduction, a correction factor must be used due to the incomplete oxidation reached with this temperature (Walkley and Black, 1934). This factor varies for different soils (Gillman et al., 1986), different horizons of a soil profile or organic fractions (Galantini et al., 1994).

The main chemical reactions are:

$$2 \, Cr_2O_7^{2-} + 3 \, C^° + 16 \, H^+ \;\Rightarrow\; 4 \, Cr^{3-} + 3 \, CO_2 + 8 \, H_2O$$
$$Cr_2O_7^{2-} + 6 \, Fe^{2+} + 14 \, H^+ \;\Rightarrow\; 2 \, Cr^{3-} + 6 \, Fe^{3-} + 7 \, H_2O$$

The most common correction oxidation factor is based on the supposition that 76% of OC is oxidized in this procedure. Therefore, a correction factor which takes into account that only the easily oxidizable carbon (approximately the 76% of total OC) is measured has to be included and calculated as follows:

$$\text{Oxidation factor} \;=\; 1 \, / \, 0.76 \;=\; 1.33$$

The $K_2Cr_2O_7$ tritimetric determination of organic carbon is then obtained by applying the following equation:

$$OC \, (\%) \;=\; \frac{(A - B) \, 0.003 \times 1.33}{w}$$

where:
A, milliequivalents of $K_2Cr_2O_7$ added to the soil sample
B, milliequivalents of $FeSO_4$ needed to back titrate the excess $K_2Cr_2O_7$
0.003, milliequivalent weight of OC (g)
1.33, oxidation factor
w, oven-dried soil sample weight (g).

Another factor (1.724) converts OC to OM. This conversion factor of 1.724 is the result of attributing 58% to the content of OC in the OM (Tabatabai, 1996). Other authors, reported by Shulte and Hopkins (1996), recovered 49 to 96% (average 78%) of the OC in 96 Wisconsin soils by this method.

Some authors have proposed supplementary heating on a hot plate or block digestor (Mebius procedure) to improve OC oxidation (Mebius, 1960; Schollemberger, 1945). In some cases two blanks are recommended due to the thermal instability of dichromate acid.

Several errors can affect the quantification of OC by this procedure:
• a source of positive errors is the presence of Cl⁻ (producing CrO_2Cl_2) and Fe^{2+} or Mn^{2+} (producing Cr^{3+} and Fe^{3+} or Mn^{3+})

• the higher oxides of Mn (largely MnO_2) compete with dichromate for oxidizable substances when heated in an acid medium, resulting in a negative error (Tabatabai, 1996)

• the acid dichromate digestion solution decompose at temperature above 150°C, limiting the heating to relatively low temperature (Nelson and Sommers, 1975)

Despite the difficulties and inaccuracies, the dichromate redox method is widely used because it requires a minimum of equipment, can be adaptable to handle large number of samples, is suitable for comparative work on similar soils and it is not costly.

b. Wet Oxidation of OC with Collection and Determination of Evolved CO_2

The collection and determination of evolved CO_2 can eliminate many of the interferences of the titrimetric dichromate redox procedure and can be employed at a higher digestion temperature (Allison, 1960). The evolved CO_2 may be trapped in solid absorbents (followed by measurement of weight changes) or in an alkali solution (followed by back titration). An adaptation described by Snyder and Trafymov (1984) allows the processing of a large number of samples at a time.

2. Wet Oxidation with Hydrogen Peroxide

The H_2O_2 method proposed by Robinson (1927) produces the incomplete oxidation of OM, the extent of which varies among soils (Gallardo et al., 1987). Additional errors are introduced both in filtering and drying the oxidized residue and the filtrate at 110 °C. The method is, therefore, unsatisfactory as a means of determining total SOM.

B. Dry Oxidation of OC in a Furnace

1. Dry Combustion Followed by Measurement of Weight Changes

The dry combustion is followed by the measurement of changes in weight loss-on-ignition (WLOI). This is an alternative method (avoiding Cr waste) for the dichromic acid digestion, which is widely used in soil testing laboratories. The method assumes that SOC is oxidized completely within a narrow temperature range at which losses (water of hydration, for example) from minerals are negligible. Unfortunately, this is not the case, so temperature selection is somewhat arbitrary but, at the same time, critical to minimize errors (Schulte and Hopkins, 1996). Temperatures higher than 500°C can result in errors from loss of CO_2 from carbonates, structural water from clay minerals, oxidation of Fe^{2+}, and decomposition of hydrated salts (Schulte and Hopkins, 1996) and oxides (e.g., gibbsite). Temperatures below 500°C should eliminate many of these errors but may result in incomplete SOM oxidation (Gallardo et al., 1987).

Some authors have reported a complete SOM oxidation at temperatures between 430°C and 500°C (Davies, 1974; Giovanni et al., 1975), with no loss of carbonates. Other studies showed that part of the soil humic substances resists oxidation even at 600°C (Gallardo et al., 1987).

Differential thermal analysis showed an endothermic peak around 250°C (due to gibbsite interference), a first exothermic peak around 350°C (oxidation of humic acid lateral chains) and a second peak around 450°C (oxidation of humic acid central nucleus).

Dupuis (1971) found that the presence of sodium salts thermally stabilized the SOM, while aluminum slows down (humic-aluminum complexes) or accelerates (excess of aluminum) combustion.

Gallardo et al. (1987) concluded that part of the SOM may be thermostable in presence of certain inorganic compounds such as the transitional elements.

A review of papers comparing WLOI method was presented by Schulte and Hopkins (1996). Significant correlation equations were found in all cases and r^2 higher than 0.90 in most of them. Differences in slopes for regressions may be attributed to differences in heating time and temperature, and/or differences in the nature of the clay and SOM fractions.

WLOI may be reasonably accurate and economical for estimating SOM if precautions such as hygroscopicity and hydrated salt content are taken into account.

2. Dry Oxidation of OC with Collection and Determination of Evolved CO_2

The use of a stream of O_2 in the tube furnace at temperature over $900\,^{\circ}C$ provides for near complete combustion of OC. The addition of MnO_2 and CuO as catalysts convert any evolved CO to CO_2 (Tiessen et al., 1981).

The most convenient combustion methods use automated instruments, which heat the sample by induction and determine the evolved CO_2 by infrared absorption. Many of the commercially available units are equipped for simultaneous C, H, N, and/or S analysis.

In soils with significant levels of carbonates, however, organic carbon can be overestimated if carbonates are not removed before analysis or accounted for by analysis.

V. Comments

SOC determination and SOM estimation can be obtained by using wet and dry combustion methods. The wet combustion titrimetric method is rapid, simple and rather accurate. The WLOI oxidation determination is also rapid and accurate, but both procedures have several errors to be evaluated and corrected by using correction (oxidation and conversion) factors.

The most convenient, but expensive, determinations are obtained with automated instruments.

The main sources of errors in the SOM estimation are erratic field sampling; the use of correction factors (oxidation of OC ($1/0.76 = 1.33$) and conversion of OC to OM (1.724)) coefficients; inherent titration and volumetric analytical procedures; presence of inorganic C (carbonates); and the need to express results on a volumetric basis by using the soil bulk density taken at sampling.

Acknowledgments

Thanks to Prof. Dr. John W. Doran of the University of Nebraska, U.S.A., and Prof. Drs. Pedro Sánchez and Poul Smithson of the International Center for Research in Agroforestry (ICRAF), Nairobi, Kenya, for their academic and technical suggestions during the manuscript review. Financial support from the National Research Council (CONICET) and the Comisión de Investigaciones Científicas (CIC), 1900 LA PLATA is appreciated.

References

Allison, L.E. 1960. Wet-combustion apparatus and procedure for organic and inorganic carbon in soil. *Soil Sci. Soc. Am. Proc.* 24:36-40.

Allison, L.E. 1965. Organic carbon. p. 1367-1378. In: C.A. Black et al. (eds.), *Methods of Soil Analysis, Part 2, Chemical and Microbiological Properties*. American Society of Agronomy, Madison, WI.

Anderson, D.W. and E.A. Paul. 1984. Organo-mineral particles and their study by radiocarbon dating. *Soil Sci. Soc. Am. J.* 42:298-301.

Anderson, D.W., S. Saggar, J.R. Bettany and J.W.B. Stewart. 1981. Particle size fractions and their use in studies of soil organic matter: I. The nature and distribution of forms of carbon, nitrogen and sulfur. *Soil Sci. Soc. Amer. J.* 45:767-772.

Andriulo, A., J. Galantini, C. Pecorari and E. Torioni. 1990. Materia orgánica del suelo en la región pampeana. I. Un método de fraccionamiento por tamizado. *Agrochimica XXXIV* (5-6) 475-489.

Balesdent, J., G.H. Wagner and A. Mariotti. 1988. Soil organic matter turnover in long-term field experiments as revealed by carbon-13 natural abundance. *Soil Sci. Soc. Am. J.* 51:1200-1207.

Bonde, T.A., B.T. Christensen and C.C. Cerri. 1992. Dynamics of soil organic matter as reflected by ^{13}C natural abundance in particle size fractions of forested and cultivated oxisols organic matter. *Soil Biol. Biochem.* 23:275-277.

Broadbent, F.E. 1953. The soil organic fraction. *Adv. Agron.* 5:153-183.

Cambardella, C.A. and E.T. Elliott. 1992. Particulate soil organic matter changes across a grassland cultivation sequence. *Soil Sci. Soc. Am. J.* 56:777-783.

Cambardella, C.A. and E.T. Elliott. 1993. Methods for physical separation and characterization of soil organic matter fractions. *Geoderma* 56:449-457.

Campbell, C.A., V.O. Biederbeck, R.P. Zentner and G.P. Lafond. 1991. Effect of crop rotations and cultural practices on soil organic matter, microbial biomass and respiration in thin black chernozem. *Can. J. Soil Sci.* 71:363-376.

Chien, T., F. Ho, H. Feng, S. Liu and P. Chen. 1964. The microbial characteristic of red soils. *Acta Pedol. Sin.* 12:399-400.

Christensen, B.T. 1985. Carbon and nitrogen in particle size fractions isolated from Danish arable soils by ultrasonic dispersion and gravity sedimentation. *Acta Agric. Scand.* 35:175-187.

Christensen, B.T. 1992. Physical fractionation of soil and organic matter in primary particle size and density separates. *Adv. Soil Sci.* 20:1-90.

Dalal, R.C. and R.J. Mayer. 1986. Long-term trends in fertility of soils under continuous cultivation and cereal cropping in Southern Queensland: II. Total organic carbon and its rate of loss from the soil profile. *Aust. J. Soil Res.* 24:281-292.

Davies, B.E. 1974. Loss-on-ignition as an estimate of soil organic matter. *Soil Sci. Soc. Am. Proc.* 38:150-151.

Doran, J.W., M. Sarrantonio and M.A. Liebig. 1996. Soil health and sustainability. *Adv. Agron.* 56:1-54.

Dupuis, T. 1971. Characterization par analyse thermique differentielle des complexes de l'aluminium avec les acides fulviques et humiques. *J. Thermal Anal.* 3:281-288

Duxbury, J.M., M.S. Smith and J.W. Doran. 1989. Soil organic matter as a source and a sink of plant nutrients. In: D.C. Coleman, J.M. Oades and M. Uehara (eds.), *Dynamics of Soil Organic Matter in Tropical Ecosystems*, NIFTAL project, University of Hawaii, Hawaii.

Elustondo, J., D.A. Angers, M.R. Laverdiere and A. N'Dayegamiye. 1990. Étude comparative de l'ágrégation et de la matiére organique associée aux fractions granulométriques de sep sols sous culture de maïs ou en prairie. *Can. J. Soil Sci.* 70:395-402.

Elliott, E.T. and C.A. Cambardella. 1991. Physical separation of soil organic matter. *Agric. Ecosyst. Environ.* 34:407-419.

Feller, C. 1979. Une méthode de fractionnement granulométrique de la matiere organique des sols: application aux soils tropécaux, à textures grossières, tres pauvres en humus. *Edh. ORSTOM, Ser. Pedol.* 17:339-346.

Feller, C. 1993. Organic inputs, soil organic matter, and functional soil organic compartments in low-activity clay soils in tropical zones. p. 77-78. In: K. Mulongov and R. Mercks (eds.), *Soil Organic Matter Dynamics and Sustainability of Tropical Agriculture.* John Wiley-Sayce, Chichester.

Feller, C., E. Fritsch, R. Ross and C. Valentin. 1991. Effects of the texture on the storage and dynamics of organic matter in some low activity clay soils (West Africa, particularly). *Cah. ORSTOM, Ser. Pedol.* 26:25-36.

Galantini, J.A. and R.A. Rosell. 1997. Organic fractions, N, P, and S changes in a semiarid Haplustoll of Argentina under different crop sequences. *Soil and Tillage Research* 42:221-228.

Galantini, J.A., R.A. Rosell and J.O. Iglesias. 1994. Determinación de materia orgànica en fracciones granulométricas de suelos de la región semárida bonaerense. *Ciencia del Suelo* 12:81-83.

Gallardo, J.F., J. Saavedra, T. Martin-Patino and A. Millan. 1987. Soil organic matter determination. *Commun. Soil Sci. Plant Anal.* 18:699-707.

Gilman, G.P., D.F. Sinclair and T.A. Beach. 1986. Recovery of organic carbon by Walkley and Black procedure in highly weathered soils. *Commun. Soil Sci. Plant Anal.* 17:885-892.

Giovanni, G., Poggio and P. Sequi. 1975. Use of an automatic CHN analyzer to determine organic and inorganic carbon in soils. *Commun. Soil Sci. Plant Anal.* 6:39-49.

Greenland, D.J. and G.W. Ford. 1964. Separation of partially humified organic materials from soils by ultrasonic dispersion. In Trans. 8th Intern. Congr. Soil Science, 137-148. Budapest, Hungary, 31 Aug. - 9 Sept.

Gregorich, E.G. and B.H. Ellert. 1993. Light fraction and macroorganic matter in mineral soils: p. 397-407. In: M.R. Carter (ed.), *Soil Sampling and Methods of Analysis.* Lewis Publishers, Boca Raton, FL.

Janssen, B.H. 1984. A simple method for calculating decomposition and accumulation of "young" soil organic matter. *Plant Soil* 76:297-304.

Jenny, H. 1941. *Factors of Soil Formation.* McGraw-Hill, New York. 281 pp.

Magdoff, F. 1996. Soil organic matter fractions and implications for interpreting organic matter tests. p. 11-19. In: F. Magdoff et al. (eds.), *Soil Organic Matter: Analysis and Interpretation.* SSSA Spec. Pub. 46. American Society of Agronomy, Madison, WI.

Mebius, L.J. 1960. A rapid method for the determination of organic carbon in soil. *Anal. Chim. Acta* 22:120-124.

Miglierina, A.M., R.A. Rosell and A.E. Glave. 1988. Change of chemical properties of Haplustoll soil under cultivation in semiarid Argentina. p. 421-425. Proc. Intern. Conf. on Dryland Farming, Amarillo/ Bushland, TX.

Nelson, D.W. and L.E. Sommers. 1975. A rapid and accurate procedure for estimation of organic carbon in soil. *Proc. Indiana Acad. Sci.* 84:456-462.

Nelson, D.W. and L.E. Sommers. 1982. Total carbon, organic carbon, and organic matter. p. 539-579. In: A.L. Page et al. (eds.), *Methods of Soil Analysis, Part 2, Chemical and Microbiological Properties.* American Society of Agronomy, Madison, WI.

Rasmussen, P.E. and H.P. Collins. 1991. Long term impact of tillage, fertilizer and soil organic matter in temperate semiarid regions. *Adv. Agr.* 45:93-134.

Richter, M., G. Massen and I. Mizuno. 1975. Total organic carbon and "oxidable" organic carbon by Walkley-Black procedure in some soils of the Argentine Pampa. *Agrochimica* 17:462-472.

Robinson, W.O. 1927. The determination of organic matter in soils by means of hydrogen peroxide. *J. Agric. Res.* 34:339-356.

Rosell, R.A., J.O. Iglesias and J.A. Galantini. 1996. Organic carbon changes in soil fractions of two texturally different Haplustolls under cultivation. p. 161-167. In: C.E. Clapp et al. (eds.), *Humic Substances and Organic Matter in Soil and Water Environments: Characterization, Transformations and Interactions.* International Humic Substances Society, St. Paul, MN.

Schollenberger, C.J. 1945. Determination of soil organic matter. *Soil Sci.* 59:53-56.

Schulte, E.E. and B.G. Hopkins. 1996. Estimation of soil organic matter by weight loss-on-ignition. p. 21-31. In: F. Magdoff et al. (eds.), *Soil Organic Matter: Analysis and Interpretation.* SSSA Spec. Pub. 46. American Society of Agronomy, Madison, WI.

Sikora, L., V. Yakovchenko, C. Cambardella and J.W. Doran. 1996. Assessing soil quality by testing organic matter. In: F. Magdoff et al. (eds.). *Soil Organic Matter: Analysis and Interpretation.* SSSA Spec. Pub. 46. American Society of Agronomy, Madison, WI.

Snyder, J.D. and J.A. Trofymow. 1984. A rapid accurate wet oxidation diffusion procedure for determining organic and inorganic carbon in plant and soil samples. *Commun. Soil Sci. Plant Anal.* 15:587-597.

Soil Science Society of America (SSSA). 1997. *Glossary of Soil Science Terms.* American Society of Agronomy, Madison, WI. 138 pp.

Stevenson, F. 1982. *Humus Chemistry.* Wiley-Interscience. 443 pp.

Tabatabai, M.A. 1996. Soil organic matter testing: An overview. p. 1-10. In: F. Magdoff et al. (eds.), *Soil Organic Matter: Analysis and Interpretation.* SSSA Spec. Pub. 46. American Society of Agronomy, Madison, WI.

Tiessen, H. and J.W.B. Stewart. 1983. Particle size fractions and their use in studies of soil organic matter: II. Cultivation effects on organic matter composition in size fractions. *Soil Sci. Soc. Am. J.* 47:509-514.

Tiessen, H., J.R. Bettany and J.W.B. Stewart. 1981. An improved method for the determination of carbon in soils and soil extracts by dry combustion. *Commun. Soil Sci. Plant Anal.* 12:211-218.

Turchenek, L.W. and J.M. Oades. 1979. Fractionation of organo-mineral complexes by sedimentation and density techniques. *Geoderma* 21:311-343.

Walkley, A. and I.A. Black. 1934. An examination of Degtjareff method for determining soil organic matter and a proposed modification of the chromic acid titration method. *Soil Sci.* 37:29-38.

Whatson, J.R. and J.W. Parsons. 1974. Studies of soil organo-mineral fractions. I. Isolation by ultrasonic dispersion. *J. Soil Sci.* 25:1-8.

CHAPTER 22

The Development of the KMnO$_4$ Oxidation Technique to Determine Labile Carbon in Soil and Its Use in a Carbon Management Index

G. Blair, R. Lefroy, A. Whitbread, N. Blair and A. Conteh

I. Introduction

Soil carbon plays a central role in both the cycling of nutrients and in soil structure (Swift and Woomer, 1993; Syers and Craswell, 1995). Plant biomass produced in agricultural or natural systems may be either removed from the system in products or residues, or remain on the soil surface. The fate of this biomass has a profound effect on C cycling.

When plant debris or animal manures enter the soil system they are used by microbes as an energy and nutrient source. This degradation process results in a range of other carbon compounds. Such compounds may continue to be broken down to simpler compounds or, in some soils, may become intimately mixed with clay particles such that they are protected from further microbial attack. In systems where burning is, or has been, practiced, the soil may contain charcoal that is microbially inert but which may possess surface charge properties that can bind organic molecules.

Such a "cascading" effect presents real difficulties in both the development of analytical techniques to determine the size of the various C pools in the soil and for computer modellers.

Soil carbon also plays a key role in binding soil into aggregates, with a range of carbon-rich materials involved in this process (Allison, 1973). Transient binding of soil aggregates is brought about by microbial and plant-derived polysaccharides. Temporary binding is provided by roots and fungal hyphae, whereas persistent bonding is brought about by aromatic humic materials, with amorphous iron and aluminium compounds and polyvalent metal cations involved.

There are marked differences between agroecosystems and natural ecosystems in terms of the carbon pools and turnover rates (Campbell, 1978). Undisturbed forests and grassland systems are characterized by a large pool of total carbon, comprised of many sub-pools of carbon turning over at different rates. This results in an almost continuous supply of carbon, of differing quality, being returned to the system. Such a system provides a relatively continuous supply of energy for microbes and aggregate binding materials; hence, soil fertility and stability are generally favorable under such systems (Sanchez and Logan, 1992). When these natural systems are put into agricultural use, major changes occur in both the carbon pool size and turnover rates. The most significant change results from the incorporation of standing biomass and surface litter into the soil, which breaks down the size of the organic matter particles and brings about intimate contact between the organic matter and soil organisms. This results in a rapid decline in soil carbon when soil moisture and temperature are favorable (Capriel, 1991). Such a decline results in reduced nutrient supply through lower microbial activity and lower physical fertility through a lower supply of "microbial glues" and contributes to rising global atmospheric CO_2 levels. These changes, in turn, lead to increasing dependence on inorganic fertilizers, increased erosion risk, lower crop and pasture yields, and global warming. In systems where a green manure crop is incorporated into the soil, major changes in carbon and nutrient cycling occur. In these situations a large amount of residue, with a low carbon to nutrient ratio and easily accessible carbohydrates and polysaccharides, are incorporated in the soil. This results in a rapid breakdown of these materials and a subsequent rapid release of nutrients from them.

The loss of carbon that results from agricultural practices has serious consequences for both soil chemical and physical fertility. The loss of carbon from a small pool is more serious than the loss of carbon from a large pool. That is, any practice that reduces soil carbon in an already depleted system is likely to have more serious effects on the productivity and sustainability of that system compared with the loss of the same amount of carbon from a forest or natural system. The loss of carbon from a dynamic, or labile pool, i.e., one turning over rapidly and providing a ready supply of nutrients and carbon to the system, is more serious than the loss of carbon from a recalcitrant, non-labile pool, which is turning over in terms of decades.

The major difficulty encountered when attempting to sequester C in soils is to know, over a short time frame, whether the altered practice is having any effect. The techniques described in this chapter achieve this aim.

II. Measurement of Soil C and Effects of Management on C Dynamics

A. Measurement of Carbon

A large number of techniques have been used to fractionate soil carbon. Whitbread (1996) summarized these procedures as follows: "Physical fractionation based on size and density has long been employed to study SOM. Such procedures are usually preceded by sonication, shaking or chemical pre-treatments, using various chemical pre-treatments which solubilize or oxidize organic matter. Density fractionation is generally carried out in solutions ranging in density from 2 to 1.6 g cm^{-3}. One of the early attempts at fractionation was based on chemical separation of components into fractions such as humic and fulvic acids and polysaccharides. Many studies have employed the use of fumigation to estimate microbial biomass as a measure of active carbon. The problem with this technique is that it only represents the organisms present at the time of sampling which may be affected as much by environmental conditions as substrate availability. A range of modern spectroscopic techniques have been employed to understand the components of SOM. These range from nuclear magnetic resonance (NMR) through to pyrolysis and field ionization mass spectrometry.

Eriksen et al. (1995) extracted organic compounds from soil with aqueous acetyl acetone following different amounts of ultrasonic dispersion. This method combines both chemical and physical fractionation. Jenkinson (1966) used the evolution of carbon dioxide in incubation studies to determine organic matter decomposition. A development of this procedure is the UV photo-oxidation technique (Skjemstadt et al., 1993), which involves the measurement of CO_2 when organic matter is oxidized in a mixture of air and irradiation from high energy UV sources. Oxidizing agents have also been used to assess the relative importance of different forms of SOM. Solutions of potassium permanganate ($KMnO_4$) have been used in many studies for the oxidation of organic compounds.

The rates and extent of oxidation of the different substrates is governed by their chemical composition (Hayes and Swift, 1978). An extensive literature review dealing with the oxidation of organic compounds by permanganate has been presented by Stewart (1965). It was Piret et al. (1957), as cited by Hayes and Swift (1978), who first brought to the attention of soil chemists the usefulness of $KMnO_4$ solutions in oxidizing soil organic matter when they showed that oxidation of peat by $KMnO_4$ yielded 12 to 22% of aromatic polycyclic acids. More recently, extensive work in understanding the chemistry of soil organic matter had utilized the oxidizing power of $KMnO_4$ (Schnitzer and Desjardin, 1964). Most of the work, however, was concerned with identification of the oxidation products of humic and fulvic acids.

Loginow et al. (1987) developed a method of fractionation based on differing strengths of $KMnO_4$, with strengths ranging from 33 to 333mM. The method is based on the supposition that the oxidative action of potassium permanganate, under neutral conditions, is comparable to that of enzymes from micro-organisms and other enzymes present in the soil. The degree of oxidation is analyzed by measuring the release of CO_2 or the consumption of oxidizing agents. This method has been standardized and modified by Blair et al. (1995) and the procedure streamlined to use only 333mM $KMnO_4$ as the oxidizing agent. This allows the calculation of the labile carbon (C_L) and non-labile (C_{NL}) components, the latter being calculated as the difference between total organic carbon (C_T), measured by combustion and the C_L. The relative amounts of these two fractions and the total carbon in a cropped and reference soil have been used by Blair et al. (1995) to calculate a Carbon Management Index (CMI). This index compares the changes that occur in total and labile carbon as a result of agricultural practice, with an emphasis on the changes in C_L, as opposed to C_{NL} in SOM. The CMI is calculated as follows:

(a) Change in total C pool size: The loss of C from a soil with a large carbon pool is of less consequence than the loss of the same amount of C from a soil already depleted of C or which started with a smaller total C pool. Similarly, the more a soil has been depleted of carbon, the more difficult it is to rehabilitate. To account for this a Carbon Pool Index is calculated as follows:

$$\text{Carbon Pool Index (CPI)} = \frac{\text{sample total C (mg g}^{-1})}{\text{reference total C (mg g}^{-1})} = \frac{C_T \text{ sample}}{C_T \text{ reference}}$$

(b) The loss of labile C is of greater consequence than the loss of non-labile C. To account for this, since it is the turnover of labile carbon which releases nutrients and the labile carbon component of SOM appears to be of particular importance in affecting soil physical factors (Whitbread, 1996), a Carbon Lability Index is calculated as follows:

$$\text{Lability of C(L)} = \frac{\text{C in fraction oxidized by KMnO}_4 \text{ (mg labile C g}^{-1} \text{ soil)}}{\text{C remaining unoxidized by KMnO}_4 \text{ (mg labile C g}^{-1} \text{ soil)}} = \frac{C_L}{C_{NL}}$$

$$\text{Lability Index (LI)} = \frac{\text{Lability of C in sample soil}}{\text{Lability of C in reference soil}}$$

c) The Carbon Management Index (CMI) is then calculated as follows:
Carbon Management Index (CMI) = C Pool Index × Lability Index × 100 = CPI × LI × 100

B. Management Effects on Carbon Pools and CMI

Twelve examples of the use of C_L measurements and CMI are presented below to demonstrate the utility of the oxidation procedure and its subsequent use in calculating the CMI.

The Glen Innes long-term experiment in New South Wales, Australia was commenced as a Maize/Spring Oats/Red Clover (M/SO/RC) rotation experiment in 1921. This site is located on the Northern Tablelands of NSW on a Vertisol soil. In the cropping system, residue management prior

Table 1. Effect of crop rotations on C dynamics in the Glen Innes crop rotation experiment

	Reference	M/SO	M/M/SO/RC	M/SO/RC
C_T (mg g^{-1})	47.2	13.7b[1]	19.7a	20.7a
CPI	1.00	0.29b	0.42a	0.44a
C_L (mg g^{-1})	7.9	2.8c	3.8b	6.3a
C_{NL} (mg g^{-1})	39.9	10.9c	15.9a	14.4b
L	0.20	0.26	0.24	0.44
LI	1.00	1.30b	1.21b	2.19a
CMI	100	38c	50b	96a
% yield reduction 1922/33-70/81		54	17	7

[1]Means within a component followed by a common letter are not significantly different at 5% level by the DMRT.
(Adapted from Blair et al., 1998.)

to 1970 was burning of maize stalks, removal of red clover hay and removal of oat straw after the grain was harvested. Post 1970, maize residues and red clover have been incorporated into the soil. Cultivation takes place before each crop.

The results of carbon measurements made in 1994 are presented in Table 1. All cropping systems have resulted in a major decline in total carbon such that the CPI had been reduced to 0.29 in the M/SO rotation. Labile carbon (C_L) measurements showed a major decline with cropping in the M/SO rotation and only a minor decline in the M/SO/RC rotation. Although there was no significant difference in C_T between the two rotations containing red clover (/RC), the one which had more red clover and less cultivation (M/SO/RC) had a significantly higher concentration of C_L and consequently a higher LI. These changes in carbon pools resulted in significant differences between the three rotations in CMI. The greatest decline in CMI occurred in the exploitative M/SO rotation with a significantly higher CMI in the M/SO/RC rotation, which can be related to the amount of cultivation, fallow and red clover in the rotation. These changes in CMI are closely reflected in the maize yield reductions which occurred between the first and fifth 12-year rotations (Table 1).

Data from a second long-term experiment being conducted at Woburn, U.K. is shown in Table 2. This experiment was established on a sandy loam soil (Cottenham series) in 1965 on the Stackyard Field (cropped since 1846) at Woburn Experimental Farm, Husborne Crawley, Bedfordshire, England, to investigate the effects of different organic manures on crop yield, organic matter content and nutrient uptake.

The experiment consists of four treatments with a build up phase for 6 years followed by a run-down phase of 4 years. The four treatments in the build-up phase consisted of the growing of a grass/legume pasture (LC6), the annual application of 9-11 t ha^{-1} farmyard manure (DG), annual applications of straw (straw) applied at 7 to 8 t ha^{-1} and annual additions of mineral fertilizer, P, K and Mg equivalent to the amount added in the farmyard manure (FD). During the build-up phase, crops were grown in the DG, straw and FD treatments and the pasture was mown two to four times a year and the cuttings removed from the plots. Following the build up phase, the pasture was ploughed in and crops were grown in all four treatments during the 4-year run-down phase. During this time no organic manures or fertilizers were applied. The phases were then repeated in a 10-year cycle. Soil samples collected following harvest after the third run-down phase were used in this study.

Table 2. Different carbon fractions and L, LI, CPI and CMI for the Woburn organic manuring experiment

	Fertilizer	Manure	Ley	Straw
C_T (mg g^{-1})	6.90b[1]	10.50a	7.90b	8.63ab
CPI	1.00b	1.60a	1.20b	1.32ab
C_L (mg g^{-1})	1.36c	2.34a	1.66bc	1.80b
C_{NL} (mg g^{-1})	5.54b	8.19a	6.24b	6.83ab
L	0.26	0.29	0.26	0.27
LI	1.00	1.07	1.05	1.05
CMI	100c	166a	122bc	134b

[1]Means within a component followed by a common letter are not significantly different at 5% level by the DMRT.

The addition of farmyard manure increased labile carbon (C_L) by 64% compared to the fertilizer-only treatment. The incorporation of straw and the 6-year ley treatment also increased C_L by 29% and 21%, respectively (Table 2). In all cases the increase in total carbon is less than labile carbon with the straw incorporation treatment (25% increase) showing the least difference (Table 2). There has been a significant increase in the Carbon Management Index (CMI) as a result of the incorporation of farmyard manure when compared to all other treatments (Table 2). The incorporation of straw has also increased the CMI when compared to the fertilizer-only treatment. The CMI was calculated using the fertilizer (FD) treatment as the reference.

The greater increase in C_L compared to C_T following organic inputs is similar to the findings of Whitbread et al. (1996) who found only slight increases in C_T following 3 years of wheat residue return at Warialda, NSW, Australia, but that there was a significant increase in C_L. The greater increase in these more easily oxidizable carbon fractions in the manure treatment compared to the fertilizer-only treatment may be attributed to the greater crop yields and resulting increased root mass as a consequence of the N released from the manure (Karlen and Camberdella, 1996). The incorporation of straw and the 6-year ley treatment have increased these more labile carbon fractions similarly, compared to the fertilizer-only treatment. Hossain et al. (1996) reported a significant increase in organic carbon following 4 years of a grass/legume ley compared to wheat crops with all residues returned on a clay soil at Warra, Queensland, Australia. However, the residues returned would have been much less than the 7 to 8 t ha^{-1} applied in this experiment. Although the higher quality residues of the pasture may have been expected to contribute more labile C to the soil than the straw treatment (Paustian et al., 1992), plowing and cropping of a pasture soil results in a rapid decrease in carbon, particularly the more labile fractions (Kandelier and Murer, 1993; Blair et al., 1995). After four years with no residues returned, the slower breakdown rate of the straw treatment compared to the 6-year ley treatment has resulted in a similar C_L concentration.

The addition of farmyard manure and straw has resulted in improved carbon dynamics as indicated by the significantly higher CMI values in these treatments compared to the fertilizer-only treatment. The greater increase in C_L shown in the manure treatment has increased the CMI to a greater extent than the incorporation of straw.

Soils from a series of cotton experiments and from commercial cotton farms in Eastern Australia have been studied. In one such experiment conducted at Warren in New South Wales on a Vertisol soil (Entic Chromustert), a series of seven rotations containing cotton are included in a farming systems

Table 3. Changes in C under 2 years of different cotton rotations at Warren, NSW; original C concentration 1993 $C_T = 6.15$ mg g^{-1}, $C_L = 0.84$ mg g^{-1}

Rotation	ΔC_T	ΔC_L	ΔCMI
Continuous cotton (CC)	0.11	0.02	3
Cotton - long fallow (CLF)	−0.45	−0.07	−9
Cotton - field peas (CFP)	0.33	0.14*	18*
Cotton - low input wheat[1] (CWlo)	1.61**	0.20*	23**
Cotton - high input wheat[2] (CWhi)	0.84*	0.14*	17*
Cotton - wheat - lablab (CWLL)	1.12**	0.20*	24**
Cotton - wheat -lablab - fertilizer (CWLLF)	1.00*	0.07	8

[1] 40 kg seed ha^{-1}, 17 kg N ha^{-1}; [2] 106 kg seed ha^{-1}, 120 kg N ha^{-1}
*Change significant at <0.05%; **Change significant at $p < 0.01$%

experiment. The experiment commenced in 1993 and the results of soil analyses on samples collected after 2 years are shown in Table 3.

A significant increase in C_L was recorded in the CFP rotation, a significant increase in C_T in the CWLLF rotation and significant increases in both C_T and C_L in the 3 rotations which included wheat (Table 3). These increases in C_L resulted in significant increases in CMI.

The effects of cotton stalk management were examined in an experiment conducted over 3 years at the Australian Cotton Research Institute, Narrabri. Samples were collected in 1991 and again in 1994 these were analyzed for C_T, C_L $^{13}\delta C$, N_T, light fraction carbon and polysaccharides. The experiment consisted of a split-plot with residue management as the main plot and nitrogen as the sub-plot. Results are shown in Table 4.

There was no significant change in N_T over the 3-year period in either the trash-retained or trash-burnt treatment. Both C_T and C_L increased significantly through time in the trash-retained treatment but not in the trash-burnt treatment. In addition to these changes in carbon a significant increase in total and labile polysaccharides was measured in the trash-retained treatment and again no significant change occurred in the trash-burnt treatment. The changes in both C_T and C_L resulted in significant changes in the CPI and LI resulting in marked differences in the CMI at the end of the experiment. In the trash-retained treatment the CMI had increased to 141 and by contrast it had declined to 94 in the burnt treatment.

The labile carbon determination has been used extensively in sugarcane cropping areas in Australia and Brazil.

In a trash management experiment at Tully, Queensland green trash management has resulted in substantial differences in C pools compared to burning (Table 5). Sugarcane cropping has resulted in a decline in C_T in the 0 to 1 cm soil layer but an increase in the 1 to 25 cm layer compared to the uncultivated reference soil (Table 5). The method of trash management has markedly affected C_L in the 0 to 1cm zone with the concentration in the green trash treatment being over four times that in the burnt treatment. There were no differences between treatments in the 1 to 25 cm soil layer. The lability of C (L) in the surface soil of the green trash treatment was some 1.7 times that in the reference soil and the burnt cane treatment. These changes in C_T and C_L are reflected in LI, CPI and hence CMI with significantly higher values in the green trash than the burnt treatment in the 0 to 1 cm soil layer and no significant differences between treatments in the 1 to 25 cm layer.

Table 4. Trash management effects on N, C and polysaccharides in a cotton experiment conducted at Narrabri, NSW, Australia

	N_T	C_T	C_L	Polysaccharides (mg C g^{-1})	
	(%)	——— mg g^{-1} ———		Total	Labile
			Trash retained		
1991	1.09a[1]	10.05b	1.13b	1.28b	0.11b
1994	1.11a	11.11a	1.67a	1.66a	0.14a
Change	+0.02m[2]	+1.06m	+0.54m	+0.38m	+0.03m
			Trash burnt		
1991	1.07a	9.67b	1.07b	1.24b	0.10b
1994	1.09a	10.08b	1.01b	1.20b	0.10b
Change	+0.02m	+0.41n	−0.06n	−0.04n	0.00n

[1]Numbers within a column for each year followed by the same letter are not significantly different at p <0.05 according to DMRT.
[2]Numbers within a column for change data followed by the same letter are not significantly different at p < 0.05 according to DMRT.

Table 5. Dynamics in a sugarcane cropping soil at Tully, Queensland as affected by cane trash management

Sample	C_T	C_L	C_{NL}	L	LI	CPI	CMI
	——— mg g^{-1} ———						
				0 to 1 cm			
Reference	45.4	5.6	39.8	0.14	1.00	1.00	100
Green trash	37.4a	7.1a	30.3a	0.24a	1.70a	0.82a	137a
Burnt	13.6b	1.6b	12.1b	0.14b	0.97b	0.30c	28bc
				1 to 25 cm			
Reference	9.0	0.7	8.3	0.09	1.00	1.00	100
Green trash	12.1b	0.9b	11.2b	0.08c	0.92b	0.74b	65b
Burnt	10.9b	1.0b	9.9b	0.10c	1.17b	0.66b	80b

In a trash management experiment being conducted at the Bureau of Sugar Experiment Stations at Mackay, Queensland the no-till treatment had significantly higher C_T and C_L than the cultivated treatment in the 0 to 10 cm soil layer. These changes resulted in a significantly higher CMI in the no-till compared to the cultivated treatment (Table 6).

In an experiment conducted by Ball Coelho et al. (1993) in Brazil, mulch management or burning did not affect C_T over a 12-month period. However, C_L was significantly increased in the mulch treatment resulting in a substantial increase in the LI which greatly influenced the CMI (Table 7).

Soil samples provided by Dr. Carlos Cerri, CENA Piriaba, Brazil have also been analyzed for C_T, C_L and $^{13}\delta C$. Burning cane for 22 years resulted in a substantial reduction in C_T and this is reflected in the low CPI shown in (Table 8). The lability (L) of the soil C was not reduced to the same extent.

Table 6. Effect of cultivation on C fractions in surgarcane at Mackay, Queensland

	No-till	Cultivated
C_T (mg g^{-1})	12.46a[1]	10.22b
CPI	0.37a	0.31b
C_L (mg g^{-1})	1.99a	1.68b
C_{NL} (mg g^{-1})	10.47a	8.54b
L	0.2a	0.2a
LI	0.98a	0.99a
CMI	35.4a	29.9b

[1]Numbers within a parameter followed by the same letter are not significantly different at p<0.05 according to DMRT. (Adapted from Blair, 1998.)

Table 7. Soil C pools immediately following and 12 months after harvest in a sugarcane trash management experiment in Brazil

	Immediately after harvest		After 12 months	
	Burnt	Mulch	Burnt	Mulch
C_T (mg g^{-1})	7.31	7.34	7.41	7.97
CPI	1.00	1.00	1.01	1.09
C_L (mg g^{-1})	1.46a[1]	1.21c	1.46b	1.69a
C_{NL} (mg g^{-1})	5.84	6.13	5.94	6.28
L	0.25	0.20	0.25	0.27
LI	1.00	1.00	1.00	1.35
CMI	100b	100b	102b	147a

[1]Means within a component followed by a common letter are not significantly different at <0.05 according to DMRT. (Adapted from Blair et al., 1998.)

Samples taken after 60 years from the same area following the introduction of trash return measured a slight increase in C_T and this was reflected in the CPI. Trash return did not increase the lability of the carbon in this system as can be seen from the lower LI value in the later sampling. The $^{13}\delta C$ value moved towards a C4 signal through time (Table 8). This allowed a calculation of the proportion of forest carbon present in the cropping system. Whereas 42% of the carbon present after 22 years of farming was derived from the forest, this had declined to only 3% after 60 years.

In an experiment examining the impact of legume leys, straw management and fertilizer management on wheat yields and soil properties being undertaken by Whitbread (1996) changes in carbon pools have been measured. The experiment was established on a red earth soil (Alfisol) in 1992 and consisted of a 1-year legume phase followed by 3 years of wheat cropping. Using an uncropped reference site adjacent to the experiment, Whitbread found marked changes in C_T and C_L fllowing wheat straw removal. When wheat straw was removed from the area C_T declined between 1995 and 1996 and C_L increased. By contrast, when wheat straw was retained in the system substantial

Table 8. Effect of sugarcane trash management on soil C in NE Brazil

	Forest	22 yr burnt	60 yr burnt
C$_T$ (mg g^{-1})	43.4	13.2	16.1
CPI	1.00	0.30	0.37
LI	1.00	0.66	0.59
CMI	100	20	22
$^{13}\delta$C ‰	−22.6	−16.5	−12.3
% forest C	100	42	3

Table 9. Effect of wheat residue return on C pools and CMI in an experiment conducted on a Red Earth soil at Wariaida, NSW, Australia

Straw management	Year	C$_T$	C$_L$	L	CMI
		——— mg g^{-1} ———			
Uncropped		25.22	5.31	0.21	100
Removed	1995	6.41a[1]	0.89c	0.16	16c
	1996	5.93a	1.04b	0.22	19b
Returned	1995	6.87a	1.01b	0.17	18b
	1996	6.51a	1.38a	0.27	27a

[1]Numbers within a year followed by the same letter are not significantly different at p<0.05 according to DMRT.

increases in C$_T$ and C$_L$ were measured. These were reflected in the CMI of the system which had recovered from 16 to 19 in the residue-removed treatment and from 18 to 29 where residues were returned (Table 9).

In another experiment in a tropical environment in Ubon, Thailand the impact of residues with slow to medium breakdown rates is being examined. After four annual additions of 1.5 tonnes per hectare both the total carbon and the lability of the carbon had increased significantly (Table 10). The increase in soil carbon was greatest in the treatments receiving *S. saman*, *P. taxodifolius* and *A. auriculiformis*.

Measurements of leaf breakdown rate by in vitro perfusion (Lefroy et al., 1995) has shown that the leaf litter of *S. saman*, *P. taxodifolius* and *A. auriculiformis* breaks down at a slower rate than the leaves of *C. cajan*. This is reflected in the carbon dynamics in the system after 4 years as shown in Table 10. This data demonstrates the importance of crop residue or organic amendment quality in sequestering C. If significant sequestration of C is to be achieved in agricultural soils there is a need to re-evaluate the quality of crop residues. Generally, present-day residues have been selected for use by animals and as a result they are of high quality and break down rapidly in soil. The opportunity exists to select residues with higher tannin or polyphenol concentrations and/or tougher cuticles, all of which will slow down the rate of breakdown in the soil.

In addition to soils from non-flooded systems soils from the long term flooded rice experiment at IRRI (Cassman et al., 1996) have been analyzed (Table 11). The experiment commenced in 1982 and has grown two flooded rice crops each year. The samples analyzed were collected in 1996. Inorganic

Table 10. Effect of four annual additions of legume residues on rehabilitation of soil carbon at Ubon, Thailand

Legume residue	C_T (mg g^{-1})	LI	CPI	CMI
No residue	3.9	1.00	1.00	100
Samanea saman	4.6	1.44	1.32	191
Phylanthus taxodifolius	4.6	1.45	1.32	192
Acacia auriculiformis	4.7	1.46	1.36	198
Cajanus cajan	4.4	1.27	1.25	158

Table 11. Soil C fractions from IRRI nitrogen/organic matter experiment

Treatment	C_T (mg g^{-1})	C_L (mg g^{-1})	CPI	L	LI	CMI
Control	2.03	0.33	1.00	0.19	1.00	100
Prilled urea	2.16	0.43	1.06	0.25	1.32	140
Azolla	2.34	0.45	1.15	0.23	1.21	139
Rice straw/urea	2.27	0.49	1.19	0.28	1.47	175
Sesbania	2.36	0.50	1.16	0.27	1.42	186

N is applied at 116 kg ha^{-1} for the dry season, and 58 kg ha^{-1} for the wet season crop. Organic inputs are adjusted to supply the same amounts of N as in the inorganic source.

Whereas the maximum increase in C_T was 16.2% where Sesbania had been applied the increase in C_L was 51.5%. These changes resulted in small differences between treatments in CPI but large differences between treatments in L and consequently LI. This meant that the treatments such as Azolla and Sesbania, which had the same C_T concentration, had vastly different CMI values (139 and 186, respectively). The C_L and L values suggest that this flooded system has an active C cycle and that yield declines measured in this experiment are unlikely to result from unfavorable C dynamics.

Loss of labile carbon is one of the most serious problems facing word agriculture. Measurements in a range of agricultural and natural systems have shown that labile carbon as measured by 333 mM KMnO$_4$ is able to detect small changes in carbon pool. Such changes in C_T and C_L in a soil, relative to a reference, can be used to calculate a carbon management index. While this index has no absolute value, it is extremely useful in that it allows a rapid assessment of the rate of change of cropping systems. Comparison of the KMnO$_4$ oxidation method as opposed to more traditional methods of soil carbon fractionation indicate that the KMnO$_4$ is oxidizing low molecular weight and non-humidified fractions of the soil.

The results of this experiment are supported by many other studies in a range of cropping and pasture systems which show that the rate of change of C_L is generally greater than C_T in both the degradation and rehabilitation cycle.

C. Relationship between Measured Carbon Fractions and Other Measures of Soil C

Conteh et al. (1998) has examined the relationship between carbon measured by KMnO$_4$ oxidation and more traditional measures of soil carbon (Table 12). Neither C_L or C_{NL} was found to be related to humic

Table 12. Relationship (r^2) between C$_L$C$_{NL}$ and other measures of soil C

Component	C$_L$	C$_{NL}$
Humic acid (HA)	0.19	0.02
Fulvic acid (FA)	0.91**	0.56
Humin	0.54	0.96**
Microbial biomass C	0.59*	0.45
Total polysaccharides	0.71*	0.63*
Labile polysaccharides	0.84**	0.40
Non-labile polysaccharides	0.60*	0.64*

(Modified from Conteh et al., 1998.)

Table 13. Size of a range of C pools in the 0 to 20 cm horizon of a gray clay soil from the Gwydir Valley in NSW, Australia

Component	Uncropped	Cropped
C$_T$ (mg g^{-1})	22.4	9.4
C$_L$ (mg g^{-1})	3.6	1.3
Humic acid (mg g^{-1})	2.7	1.9
Fulvic acid (mg C g^{-1})	8.4	4.0
Total polysaccharides (mg C g^{-1})	3.0	1.2
Labile polysaccharides (mg C g^{-1})	0.25	0.10
Light fraction (mg g^{-1})	15.1	6.9
Light fraction-C (mg C g^{-1})	6.9	1.1
Light fraction-C$_L$ (mg C g^{-1})	2.5	0.2
Microbial biomass (mg g^{-1})	0.23	0.12

(Adapted from Conteh, 1998.)

acid. By contrast, C$_L$ correlated stongly to fulvic acid whereas C$_{NL}$ was more related to the humin fraction. Both C$_L$ and C$_{NL}$ were related to total polysaccharides but only C$_L$ was related to the labile polysaccharides. C$_L$ was also related to microbial carbon whereas C$_{NL}$ was not.

An indication of the size of various C pools in a grey clay soil from the Gwidir Valley in NW NSW, Australia is shown in Table 13. The concentration of C$_L$ is some 10 times that of the microbial biomass in this soil. While the concentration of C$_L$ and total polysaccharides are similar in this soil, this is not often the case.

D. Relationship between Carbon Pools and Aggregate Stability

Blair et al. (1997) examined the relationship between various carbon pools and aggregate stability and in a survey of soils from the cotton growing areas of Eastern Australia they found that C$_L$ was more closely related to mean weight diameter of aggregates following tension wetting than was C$_T$ (Table

Table 14. Linear relationships (Y = a + bx) between aggregate stability (Y), expressed as mean weight diameter (MWD), and C fractions (X) in 20 soils collected throughout the cotton growing areas of eastern Australia

X	r^2
C_L	0.61**
C_T	0.46*
C_{NL}	0.41ns
C_{LL}	0.43ns
C_I	0.05ns

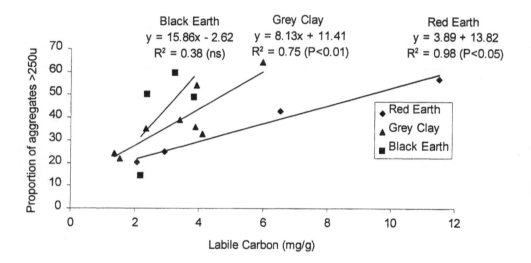

Figure 1. Relationship between aggregation, labile C and soil type. (Adapted from Whitehead et al., 1995.)

14) in soils with <52% clay. No significant relationships were found between aggregate stability and C_{NL}, less labile C ($C_{LL} = C_{Walkley-Black} - C_L$) or C_I ($C_T - C_{LL}$). When soils with >52% clay were included in the relationship no correlations were found, presumably because that in high clay soils, materials such as calcium are more important in binding aggregates than organic compounds.

In a survey of cropping soils of North Western New South Wales, Whitbread found similar relationships between labile carbon and aggregate stability (Figure 1).The relationship was strongest in the red earth soils (Alfisols) which have the lowest clay content and the weakest structure. By contrast, no significant relationship existed in the black earth soils (Vertisols) which have a high clay content and an exchange complex which is highly calcium dominated. In these soils Ca acts as a cementing agent and organic binding agents are of little or no significance.

III. Conclusions

The data presented in this chapter show that labile C (C$_L$) plays an important role in the maintenance of both chemical and physical fertility of soils and, hence, their sustainability.

Modern agricultural systems, in which crop residues are removed or burnt, are not sustainable in the long term because of the decline in C and, particularly, labile C that results from such practices. These soils offer significant scope to sequester C. The development of the KMnO$_4$ oxidation procedure to determine labile C (C$_L$) has resulted in the development of a Carbon Management Index (CMI) which takes into account changes in both labile and total C in agricultural soils, relative to a reference site. Generally, C$_L$ falls more rapidly than C$_{NL}$ when soils are exploited, and rises more rapidly during soil restoration. Measurement of C$_L$ and calculation of CMI can be used to monitor short-term changes in C pools, which is important to determine if the management practices implemented to sequester C are having a positive effect.

Matching residue decay rates to crop demand (synchrony) to maximize the recycling of nutrients, and building the total soil C pool size to sequester C should be the objective of sustainable cropping systems. The latter may require a redesign of the quality of crop residues to make them more resistant to microbial attack.

References

Allison, F.E. 1973. *Soil Organic Matter and Its Role in Crop Production.* Elsevier Scientific Publishing Co., Amsterdam. 673 pp.

Ball-Coelho, B., H. Tiessen, J.W.B. Stewart, I.H. Salcedo and E.V.S.B. Sampaio. 1993. Residue management effects on sugarcane yield and soil properties in Northeastern Brazil. *Agron. J.* 85:1004-1008.

Blair, G.J., R.D.B. Lefroy and L. Lisle. 1995. Soil carbon fractions, based on their degree of oxidation, and the development of a carbon management index for agricultural systems. *Australian J. Agricultural Res.* 46:1459-1466.

Blair, G.J., N. Blair, R.D.B. Lefroy, A. Conteh and H. Daniel. 1997. Relationships between KMnO$_4$ oxidizable C and soil aggregate stability and the derivation of a carbon management index. p. 227-232. In: J. Drozd, S.S. Gonet, N. Senesi and J. Weber (eds.), *The Role of Humic Substances in the Ecosystem and in Environmental Protection.* Proceedings 8th Meeting of the International Humic Substances Society Polish Society of Humic Substances, Grunwaldzka, Wroclaw, Poland .

Blair, G.J., L. Chapman, A.M. Whitbread, B. Ball-Coelho, P. Larsen and H. Tiessen. 1998. Soil carbon and nitrogen changes resulting from sugarcane trash management at two locations in Queensland, Australia and in North-East Brazil. *Australian J. Soil Res.* 36:873-881.

Blair, G. J., R.D.B. Lefroy, A. Whitbread, N. Blair and H. Daniel. 1998. The use of a Carbon Management Index (CMI) to monitor changes in soil carbon and to determine the sustainability of agricultural systems. Environmental Benefits of Soil Management. p. 345-348. Proceedings of the ASSSI National Soils Conference 1998. Brisbane, Qld., 27-29 April, 1998. Australian Society of Soil Science, Inc., North Sydney, N.S.W.

Blair, N. 1998. The Impact of Farming Practices on Soil Carbon and Physical Fertility. B.Sc. (Hons.)/thesis, The University of New England, Armidale, N.S.W.

Conteh, A. 1998. Soil Carbon Fractions as Indicators of Sustainability of Cotton Cropping Systems. Ph.D. dissertation, The University of New England, Armidale, N.S.W.

Conteh, A., Blair, G., and Macleod, D. 1998. $KMnO_4$-oxidizable carbon and its relationship to other soil organic carbon measurements. Proceedings of 16th World Congress of Soil Science. Symposium 7, Montpellier, France, 20-26 August 1998. 5 pp.

Campbell, C.A. 1978. Soil organic carbon, nitrogen, and fertility. p. 173-271. In: M. Schnitzer and S.U. Khan (eds.), *Soil Organic Matter*. Agriculture Canada, Ottawa.

Capriel, P. 1991. Influence of graduated management intensity on humus chemistry. *Bayerisches Landwirtschaftliches Jahrbuch* 68:367-369.

Cassman, K.G., S.K. De Datta, S.T. Amarante, S.P. Liboon, M.I. Samson and M.A. Dizon. 1996. Long-term comparison of the agronomic efficiency and residual benefits of organic and inroganic nitrogen sources for tropical lowland rice. *Experimental Agriculture* 32:427-444.

Eriksen, J., R.D.B. Lefroy and G.J. Blair. 1995. Physical protection of soil organic S studied using acetylacetone extraction at various levels of ultrasonic dispersion. *Soil Biology and Biochemistry* 27:1005-1010.

Hayes, M.H.B. and M.H.B. Swift. 1978. The chemistry of soil organic colloids. p. 179-320. In: D.J. Greenland and M.H.B. Hayes (eds.), *The Chemistry of Soil Constituents*. John Wiley & Sons, Chichester, U.K.

Hossain, S.A., R.C.Dalal, S.A.Waring, W.M. Strong and E.J.Weston. 1996. Comparison of legume-based cropping systems at Warra, Queensland. 1. Soil nitrogen and organic carbon accretion and potentially mineralisable nitrogen. *Australian Journal of Soil Research* 34:273-287.

Jenkinson, D. S. 1966. Studies on the decomposition of plant material in soil II. Partial sterilization of soil and the soil biomass. *J. Soil Sci.* 17:280-302.

Kandelier, E. and E. Murer. 1993. Aggregate stability and soil microbial processes in a soil with different cultivation. *Geoderma* 56:503-513.

Karlen, D.L. and C.A. Cambardella. 1996. Conservation strategies for improving soil quality and organic matter storage. p. 395-420. In: M.R. Carter and B.A. Stewart. (eds.), *Structure and Organic Matter Storage in Agricultural Soils*. Advances in Soil Science. Lewis Publishers, Boca Raton, FL.

Lefroy, R.D.B., Y. Konboon and G.J. Blair. 1995. An in vitro perfusion method to estimate plant residue breakdown and associated nutrient release. *Australian J. Agric. Res.* 46:1467-1476.

Loginow, W., W. Wisniewski, S.S. Gonet and B. Ciescinska. 1987. Fractionation of organic carbon based on susceptibility to oxidation. *Polish J. Soil Sci.* 20:47-52.

Paustian, K., W.J. Parton and J. Persson. 1992. Modelling soil organic matter in organic-amended and nitrogen-fertilized long-term plots. *Soil Sci. Soc. Am. J.* 56:1173-1179.

Sanchez, P.A. and T.J. Logan. 1992. Myths and science about the chemistry and fertility of soils in the tropics. p. 35-46. In: R. Lal and P.A. Sanchez (eds.), *Myths and Science of Soils in the Tropics*. SSSA Special Publication. Soil Science Society of America, Madison, WI.

Schnitzer, M. and J.G. Desjardins. 1964. Further investigations on the alkaline permanganate oxidation of organic matter extracted from a podzol Bh horizon. *Canadian J. Soil Sci.* 44:272-299.

Skjemstad, J.O., L.J. Janik, M.J. Head and S.G. McClure. 1993. High energy ultraviolet photo-oxidation: a novel technique for studying physically protected organic matter in clay- and silt-sized aggregates. *J. Soil Sci.* 44:485-499.

Stewart, R. 1965. Oxidation by permanganate. p. 1-68. K.B. Wiberg (ed.), *Oxidation in Organic Chemistry*, Part A. Academic Press, New York.

Swift, M. J. and P. L. Woomer. 1993. Organic matter and the sustainability of agricultural systems: definition and measurement. p. 3-18. K. Mulongoy and R. Merck (eds.), *Soil Organic Matter Dynamics and Sustainability of Tropical Agriculture*. John Wiley, New York.

Syers, J.K. and E.T. Craswell. 1995. Role of soil organic matter in sustainable agriculture systems. p. 7-14. In: R.D.B. Lefroy, G.J. Blair and E. T. Craswell (eds.), *Soil Organic Matter Management for Sustainable Agriculture.* Ubon, Thailand, 24–26 August, 1995. ACIAR, Canberra.

Whitbread, A. M. 1996. The Effects of Cropping System and Management on Soil Organic Matter and Nutrient Dynamics, Soil Structure and the Productivity of Wheat. Ph.D. dissertation, The University of New England, Armidale, N.S.W.

Whitbread, A.M., G.J. Blair and R.D.B. Lefroy. 1996. The impact of cropping history on soil physical properties and soil carbon. p. 311-312 In: *Proceedings of Australian and New Zealand National Soils Conference.* 1–4 July, 1996, Melbourne, Australia. ASSSI, NZSSS.

CHAPTER 23

Effects of Soil Morphological and Physical Properties on Estimation of Carbon Storage in Arctic Soils

G.J. Michaelson, C.L. Ping and J.M. Kimble

I. Introduction

Climate models have predicted warming climate with relatively large temperature increases for the northern latitudes. Over the past decade, a warming has been observed over parts of the arctic region (Oechel et al., 1993). Efforts are under way to develop models predicting the function and response of the terrestrial arctic ecosystem as it will feed back to the atmosphere under changing conditions. Developing models requires more accurate and detailed knowledge of terrestrial C-stocks in the Arctic. It is estimated that arctic ecosystems contain 13% of all terrestrial carbon (Post et al., 1982). These estimates are based on the carbon in the active layer of arctic soils. More recent detailed examination of region soils have recognized that stock estimates for many arctic soils are doubled when considering the whole pedon including the upper permafrost (Michaelson et al., 1996 and Bockheim et al., 1998).

It has become apparent that a more detailed and complete inventory of soil carbon is necessary in the study of terrestrial feedback under changing arctic conditions. Soils of the Arctic by the nature of their morphology, present a unique challenge to the easy assessment of C-stocks. The presence of permafrost and continual freeze-thaw processes in soil active layers serve to redistribute surface accumulated carbon throughout the profile of arctic soils. In temperate regions, the horizons within a soil profile are generally horizontal and often continuous across large areas. Freeze-thaw processes in the Arctic result in distortion and interruption of soil genetic horizons which can complicate soil sampling and description.

To accomplish the task of soil C-stock assessment, representative soil profiles must be sampled and described in a manner that allows for the calculation of soil carbon contents on a landscape basis (weight/area). The objective of this chapter is to examine the range and variation of key soil morphological (horizon thickness and type) and physiochemical properties (bulk density and carbon content) that are necessary to measure for the estimation of profile C-stores and to examine these properties using data collected for arctic soils.

II. Materials and Methods

A. C-Store Calculation

Soils are described and sampled to 1 m depth. Each soil genetic horizon identified in the profile, is sampled in triplicate by carefully cutting dimensional blocks (average 400 cm^3) at random from undisturbed clods which represent the horizon. These blocks are measured (length × width × height), dried, weighed and analyzed for organic carbon. The key components of the soil C-store calculation are:

(1) From the soil profile description:

Soil horizon thickness,

Field: by estimation in the soil sampling pit across the exposed surface (average measurement in exposed face of sampling pit) (Schoeneberger et al., 1998).

-or-

Sketch: by analysis of the composition of a to-scale sketch by sketching to scale the horizons as they occur in a representative 1 m^2 exposed face of a sampling pit and determining the composition of the horizons based on the sketch (Kimble et al., 1993).

(2) From block samples cut from genetic horizons:

Soil dry bulk density: BD = block dry wt./block volume (g cm^{-3})

Carbon Content: %OC on dry soil by LECO CHN Analyzer

Carbon Density$_{horizon}$ = B.D.$_{hor.}$ × %OC$_{hor}$ × 10 (kgC m^{-2} cm^{-1})

C-store calculations are performed using the above quantities as follows:

Horizon C-stores = B.D × %OC × horizon thickness by sketch × (100–v)

v = % volume of ice veins (>2 mm) and or coarse fragments

Soil C-stores = Σ(horizon C-stores to 1 m)

Results presented here are calculated using data from a series of 21 soil profiles (Table 1) examined for a study of C-stores in arctic Alaska and presented by Michaelson et al. (1996). The soils were from sites representing the major vegetation land cover classes of two major geomorphic units in northern Alaska: coastal plain and foothills (Figure 1).

III. Results and Discussion

A. Soil Horizon Thickness Estimates

Horizon thickness is a primary component of the equation (above) for calculation of soil profile C-stores and errors in its estimation will directly affect C-store estimates. Soils of many regions are comprised of a series of genetic horizons that are formed and remain relatively horizontal with only slight variation in thickness across the scale of 1 m in the soil profile. Soils of the cold regions, however, may be formed on horizontal landscapes but are subject to processes which mix and deform layers after formation. Mixing and deformation of soil horizons resulting from cold region soil processes is referred to as cryoturbation. This cryoturbation is evident in soil profiles such as those presented in Figure 2. Horizon thickness is very difficult to measure accurately in the field. The complex nature of soil profile horizons in the cold regions has necessitated the use of different methods (Kimble et al., 1993) other than the commonly used standard methods (Schoeneberger et al., 1998) of field measuring of horizon thickness.

Table 1. Study sites from Michaelson et al., 1996

Site I.D.[b]	Soil classification[c]	Land cover class Tundra type[d]	Carbon storage[a]		
			Active layer	Perma-frost	Total 1 m
			— kg C m^{-3} —		
Coastal Plain					
1	Typic Historthel	Moist Nonacidic	31	30	61
2	Typic Historthel	Wet Nonacidic	19	41	60
3	Typic Mollorthel	Wet Nonacidic	24	18	42
4	Glacic Hemistel	Wet Nonacidic	27	55	82
5	Typic Aquiturbel	Moist Nonacidic	31	16	47
6	Terric Sapristel	Moist Acidic	36	25	61
7	Typic Historthel	Wet Acidic	33	26	59
8	Typic Aquiturbel	Moist Nonacidic	42	5	47
9	Typic Historthel	Wet Nonacidic	21	15	36
Foothills					
10	Typic Aquiturbel	Moist Nonacidic	20	20	40
11	Ruptic-Histic Aquiturbel	Moist Acidic	42	46	88
12	Lithic Eutrocryept	Barren	3	0	3
13	Ruptic-Histic Aquiturbel	Moist Acidic	16	17	33
14	Ruptic-Histic Aquiturbel	Moist Acidic	11	9	20
15	Ruptic-Histic Aquorthel	Wet Acidic	27	7	34
16	Glacic Sapristel	Wet Acidic	38	32	70
17	Ruptic-Histic Aquiturbel	Moist Acidic	18	31	49
18	Typic Aquiturbel	Moist Acidic	16	0	16
19	Pergelic Cryorthent	Riparian Shrubland	4	0	4
20	Histic Aquiturbel	Wet Acidic	17	13	30
21	Aquic Dystrocryept	Alpine	10	0	10

[a] From Michaelson et al. (1996).
[b] See Figure 1 for site location map.
[c] From Soil Survey Staff (1998).
[d] From Auerbach and Walker (1995).

In this study, comparisons were made between the horizon thickness estimates of the two available methods of estimation. Estimates by pit measurements referred to here as 'field,' were compared to estimates at the same sites from soil pit sketch components referred to as 'sketch.' These two methods were compared for 13 of the study sites. Data by the two techniques were grouped by genetic horizon for comparison (Figure 3). In general, correlation of thickness estimates decreases as horizons occur at greater depth in the soil profile until the permafrost layer (Cf) is reached. The greatest discrepancies between methods occurred in the horizons of the subsurface active layer. The two techniques were highly correlated in the surface or near surface organic horizons which are most often continuous or more easily measured across a profile (Oi: $r = 0.94$, and Oe: $r = 0.96$). Method correlation for the Oa horizon was low ($r = 0.62$). As compared to the other O horizons, the Oa is genetically older, more decomposed and has been subject to more intensive frost action mixing processes (cryoturbation). For the mineral horizons, measurement techniques were highly correlated for the humus enriched relatively thin, A horizons ($r = 0.96$) with poorer correlation of techniques in the B ($r = 0.85$) and combination: O/A, O/B and A/B ($r = 0.56$) horizons. The thickness of the upper permafrost Cf horizon was measured

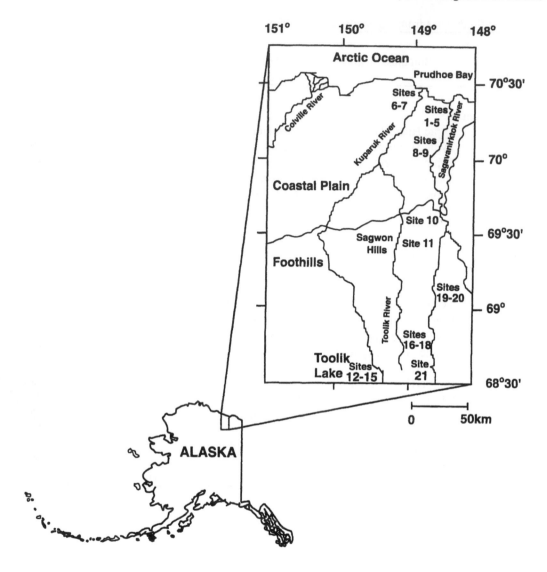

Figure 1. Location of study sites.

similarly by either method (r = 0.99) with a nearly 1:1 relationship. These results confirm that horizon thickness is most difficult to determine in the field for the horizons which are often discontinuous (Oa) or those most affected by cryoturbation, and occur in zones along with other materials (B and combination horizons). Careful and skillful use of the sketch method has the best potential to give accurate estimations of horizon thickness when combination and other horizons affected by cryoturbation are present. Thickness of the combination horizons was most often over estimated relative to the sketch method, while other horizons either had good correlation or data were more evenly skewed to both sides of a 1:1 relationship. This could be important to account for as these subsurface active layer horizons can hold relatively large amounts of the profile C-stores (data presented in Figure 4).

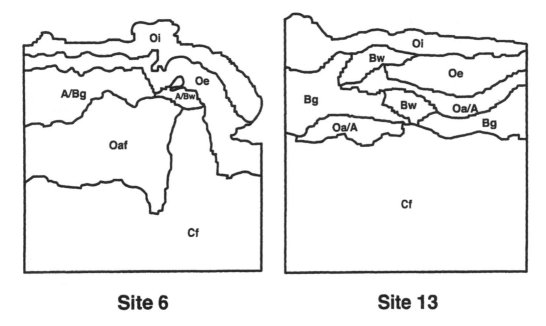

Site 6   **Site 13**

Figure 2. Examples of site profile sketches, 1 × 1 m scale.

B. Bulk Density and Organic Carbon Measurements

In Table 2, bulk density (B.D.) and organic carbon content (%OC) of commonly occurring soil genetic horizons are summarized by geomorphic region. The range of values for both soil properties varied widely within genetic horizon groups. Variability was higher in the foothills, as indicated by wider ranges and higher coefficients of variation (CV) for both B.D. and %OC, compared to the coastal plain. The span in the range of values of B.D. for coastal plain soils averaged 2.3× compared to 3.6× for the foothills soils. Excluding the most highly variable Cf for the coastal plain and A for the foothills, ranges for %OC in the horizons averaged 2.1× for the coastal plain and 2.8× for the foothill soils. The Cf is highly variable for both properties on the coastal plain due to varying amounts of small ice veins and organic inclusions.

Coefficients of variation (CV) are given in Table 2 for B.D. and %OC values grouped by genetic horizon on each geomorphic area. Also, the average CV values are given for replicate samples taken within individual sampling profiles for comparison. The CV values exhibit the same general trend as for the ranges of B.D. and %OC, with the foothills averaging higher than the coastal plain. Over all soil profiles, the CV values for both B.D. and %OC are about 2 times greater than for individual samplings of horizons. Excluding the highly variable Cf, the average CV values for coastal plain B.D and %OC were 0.26 and 0.31 over all sites compared to 0.12 and 0.13 for individual sampling sites. The CV values for B.D. and %OC over all foothill sites similarly were 0.40 and 0.44 compared to 0.18 and 0.15 at individual sites. Subsurface combination and permafrost Cf horizons have the highest variability.

The B.D. of the upper O horizons (Oi and Oe) of foothills profiles vary considerably due in part, to the distribution patterns of vegetation on the surface in tussock versus inter-tussock microsites which are difficult to sample as one unit. As an example, an eriophorum (sedge) tussock has a B.D. of 0.16 g cm^{-3} compared to the inter-tussock B.D. of 0.06 g cm^{-3}.

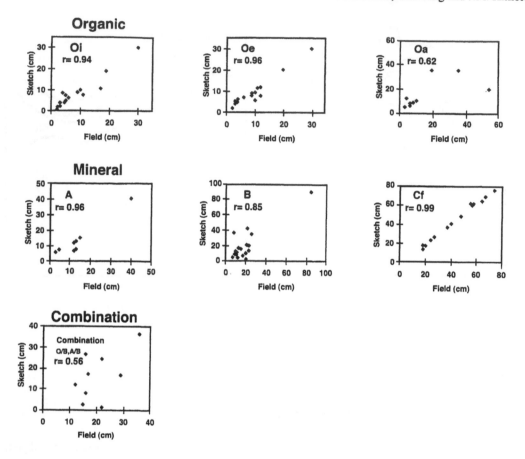

Figure 3. Relationship between field and sketch estimates of horizon thickness with correlation coefficients (r).

C. Horizon Composition and Contribution to C-Stores

The morphology of a soil profile includes the relative occurrence of genetic horizons. The contribution of each horizon to C-stores can be calculated from the B.D. and %OC contents discussed above. What can be referred to as average C-densities can be calculated for the commonly occurring horizons, in kg C m^{-2} cm^{-1} of horizon (Figure 4, top). When the average thickness of each horizon is multiplied by the C-density of the horizon, an average C-store contribution can be calculated for each horizon (Figure 4, bottom). Average data for the two geomorphic areas indicate that in general those horizons with the largest C-densities are also the ones that contribute the largest amounts of C to profile stocks. For coastal plain soils the C-density is the highest in the Oa and combination horizons. When considering thickness occurring in profiles, the Oa and Cf are the most significant contributors to profile C-stocks. In foothills profiles, the Oa, Oe and combination horizons contain the largest C-densities and the Oa, combination, and Cf horizons contribute the most to C-stocks. In the foothills region, there are slope movement processes and active layers are deeper as a result of warmer temperatures. Both of these factors likely serve to increase the thickness and %OC contents in these horizons and thus the importance of the combination and Cf horizons to C-storage in this region relative to the coastal plain.

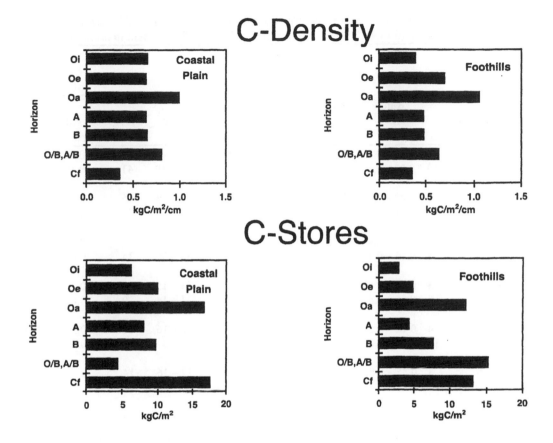

Figure 4. Comparison of soil average carbon density (kg C m^{-2} cm^{-1}) and carbon stores (kg C m^{-1}) in genetic horizons of the 21 study soils; stores are to 1 m depth.

IV. Summary and Conclusions

Recent studies have indicated that more accurate estimations of C-storage in arctic soils will depend on accounting for the subsurface mineral and organic horizon stores as well as those of surface organic layer. Some studies have indicated that this accounting could as much as double estimates of soil C-stocks in some arctic regions. Variations in soil component horizon properties that are key to the calculation of C-stores, such as B.D. and %OC, vary considerably across and within large landscape or geographic areas. Careful estimation of these properties and the relative occurrence of component soil horizons in a soil profile are essential to good accurate C-store calculations. The special character of arctic soils, especially the Gelisols, necessitate the employment of different methods that aid in description and sampling of cryoturbated soils.

The irregular and broken nature of soil horizons due to cryoturbation in arctic soils, necessitates the use of different methods in description and sampling of these soils. Particularly problematic is the determination of relative horizon thickness and composition with the commonly occurring combination or mixed horizons. In addition, these horizons vary considerably in B.D. and %OC content both within a sampling site and between sites within a geomorphic area. As it happens, the most problematic soil horizons to sample are indeed the ones contributing the larger amounts of C to

Table 2. Variation in the bulk density and organic carbon content of the soil horizons in pedons of the coastal plain (CP) and foothills (FH) regions of the Kuparuk River basin in arctic Alaska

Area	Genetic horizon	n	Range B.D. g cm^{-3}	Avg. B.D. g cm^{-3}	Average coefficient of variation		Range % O.C.	Avg. % O.C.	Average coefficient of variation	
					Avg. all pedons	Indiv. pedon			All pedons	Indiv. pedon
CP	Oi	6	0.16 – 0.34	0.28	0.21	0.19	24 – 40	30	0.26	0.12
FH		13	0.02 – 0.15	0.09	0.33	0.24	25 – 49	41	0.18	0.08
CP	Oe	6	0.20 – 0.46	0.33	0.30	0.12	16 – 43	26	0.39	0.08
FH		10	0.11 – 0.37	0.23	0.35	0.24	21 – 44	32	0.27	0.15
CP	Oa	5	0.25 – 0.66	0.45	0.36	0.14	21 – 44	27	0.37	0.20
FH		5	0.15 – 0.47	0.31	0.39	0.10	33 – 36	34	0.04	0.06
CP	A	3	0.57 – 0.69	0.62	0.10	0.08	11 – 15	14	0.15	0.07
FH		3	0.56 – 1.82	1.02	0.69	0.12	1 – 14	9	0.82	0.08
CP	B	4	0.72 – 1.53	1.13	0.38	0.09	5 – 16	11	0.52	0.02
FH		13	1.04 – 2.00	1.44	0.21	0.11	1 – 5	3	0.50	0.15
CP	Comb.	3	0.55 – 0.80	0.66	0.20	0.07	12 – 16	14	0.14	0.31
FH	(B/O,A/B)	5	0.17 – 0.74	0.49	0.43	0.29	8 – 43	18	0.80	0.35
CP	Cf	10	0.28 – 1.28	0.58	0.72	—	1 – 17	6	0.97	—
FH		4	0.44 – 0.87	0.70	0.29	0.27	6 – 9	6	0.45	0.12

profile stocks (Oa, combination, and Cf horizons). Using special methods for estimation of horizon thickness (sketch composition) and the use of multiple subsamples will increase the accuracy of C-store estimates of arctic soils. With the variation in soils evident from the range in component horizon properties, the identification of the relative occurrence of various soils across the landscape will also be very important and must be address for an accurate estimation of C-stores for a large-scale geomorphic area.

Acknowledgments

Funding support was received through the NSF ARCUS-LAII Office of Polar Programs, USDA Hatch Funds and USDA National Soil Survey Center Global Change Initiative.

References

Auerbach, N.A. and D.A. Walker. 1995. Preliminary vegetation map, Kuparuk Basin, Alaska: a Landsat derived classification. Institute of Arctic and Alpine Research, University of Colorado, Boulder.

Bockheim, J.G., D.A. Walker and L.R. Everett. 1998. Soil carbon distribution in nonacidic and acidic tundra of arctic Alaska. p. 143-155. In: R. Lal, J. Kimble, E. Levine, R.F. Follet, and B.A. Stewart (eds.), *Soil Processes and the Carbon Cycle*. Adv. Soil Sci., Lewis Publishers, Boca Raton, FL.

Kimble, J.M., C. Tarnocai, C.L. Ping, R. Ahrens, C.A.S. Smith, J. Moore and W. Lynn. 1993. Determination of the amount of carbon in highly cryoturbated soils. p. 277-291. In D.A. Gilichinsky (ed.), *Proceedings of the Joint Russian-American Seminar on Cryopedology and Global Change*, Russian Academy of Sciences, Institute of Soil Science and Photosynthesis, Pushchino, Russia.

Michaelson, G.J., C.L. Ping and J.M. Kimble. 1996. Carbon storage and distribution in tundra soils of arctic Alaska, U.S.A. *Arct. and Alp. Res.* 28:414-424.

Oechel, W.C., S.J. Hastings, G. Vourlitis, M. Jenkins, G. Riechers and N. Grulke. 1993. Recent change of Arctic tundra ecosystems from a net carbon dioxide sink to a source. *Nature* 361:520-523.

Post, W.M., W.R. Emanuel and P.J. Zinke. 1982. Soil carbon pools and world life zones. Nature 298:156-159.

Schoeneberger, P.J., D.A. Wysocki, E.C. Benham and W.D. Broderson. 1998. *Field Book for Describing and Sampling Soils*. Natural Resources Conservation Service, USDA, National Soil Survey Center, Lincoln, NE.

Soil Survey Staff. 1998. *Keys to Soil Taxonomy*, Eighth edition. Natural Resources Conservation Service, USDA, Washington, D.C.

CHAPTER 24

Estimation of Particulate and Total Organic Matter by Weight Loss-on-Ignition

C.A. Cambardella, A.M. Gajda, J.W. Doran, B.J. Wienhold and T.A. Kettler

I. Introduction

Soil organic matter (SOM) is considered one of the most important components of soil quality (Larson and Pierce, 1991) and has long been recognized as a key component of terrestrial ecosystems. It is the primary source of energy for the flora and fauna found in the soil and an important source of inorganic nutrients for plant growth. In addition, SOM is responsible for rendering the physical environment of soil suitable for plant growth, primarily through its effects on soil aggregation, which, in turn, affects water infiltration and storage, drainage, and root penetration (Allison, 1973).

Soil organic matter can be defined as a series of fractions that comprise a continuum based on decomposition rate (Schimel et al., 1985; Christensen, 1996). Conceptually, particulate organic matter (POM) is a transitory pool of organic matter that is intermediate in the decay continuum between fresh plant litter and humified, stable organic matter. Analytically, POM is a physically isolated, size-defined fraction (0.053 to 2.0 mm) that is retained on a sieve with 53-μm openings after dispersion of the soil. Several physically isolated, size-defined organic matter fractions that are similar to POM have been identified in the literature, including macro-organic matter (MOM; 0.05 to 2.0 mm) (Bolinder et al., 1999) and coarse organic matter (COM; 0.50 to 2.0 mm) (Allmaras et al., 1988). A number of studies suggest that POM is a dynamic soil property and an important pool of available C (Cambardella and Elliott, 1992; Gregorich and Ellert, 1993; Golchin et al., 1994). Particulate organic matter has been shown to be much more sensitive to change than total SOM (Elliott et al., 1994), responding selectively and rapidly to changes in land use and soil management (Greenland and Ford, 1964; Tiessen and Stewart, 1983; Dalal and Mayer, 1986; Cambardella and Elliott, 1992; Janzsen et al., 1992; Angers et al., 1993; Beare et al., 1994; Biederbeck et al., 1994; Bremer et al., 1994; Gregorich et al., 1995; Jastrow et al., 1996; Wander and Traina, 1996; Yavchenko et al., 1996).

The concept of using POM as an indicator of soil quality has been widely adopted, both nationally and internationally, in the past 10 years. The method for isolating POM from soil is simple, accurate, and rapid, and the required equipment and supplies are inexpensive and readily accessible to most laboratories. Quantification of POM C is generally done using automated dry combustion technology. These automated methods are accurate and precise but require a large capital investment and extensive technical expertise to run and maintain the instruments. The weight loss-on-ignition (WLOI) method (Schulte, 1988) is a promising alternative for quantification of SOM and only requires an accurate balance and a high temperature oven. Our objective was to develop and test the WLOI method for POM and SOM as an alternative to the more costly and technically sophisticated dry combustion methods.

II. Methods

A. Field Sampling

Two USDA-NRCS benchmark soils, of contrasting textures, were used for calibration/evaluation of the WLOI methods for POM and SOM. The Sharpsburg silty clay loam (Typic Argiudoll) soil was collected from the A horizon of a benchmark site located in Cass Co., NE (Long. 97°10'W, Lat. 40°51'N) and contained 3% sand, 63% silt, 34% clay; 17.3 g organic C (OC) kg^{-1}, 1:1 pH = 6.9, 1:1 EC = 0.63 dS m^{-1}. The Valentine sand (Typic Ustipsamment) was collected from the A horizon of a benchmark site in Brown Co., NE (Long. 100°0'W, Lat.42°20') and contained 90% sand, 7% silt, 3% clay; 4.2 g OC kg^{-1}, 1:1 pH = 6.05, 1:1 EC = 0.12 dS m^{-1}.

Surface soil (0 to 7.5 cm) cores were collected from five field sites in the central and northern Great Plains to evaluate the influence of tillage and cropping management practices on POM and total SOM for varying locations and climates. At each location the soil sampled from each management treatment and field replicate represented a composite of 15 to 30 cores depending on the diameter of the sampling probe that was used (1.75 to 3.1 cm diameter). Cores were collected randomly from each field replicate but were stratified to represent an appropriate balance of within row and between row areas for each management system. Field soil weight was recorded for each composite and subsamples were dried for 24 h at 105°C to determine soil water content. The experimental design at each field site included one treatment that was managed conventionally and one that was under an alternative management option, such as reduced tillage or a multi-year crop rotation. Field site, soil and management descriptions are detailed in Table 1.

B. Laboratory Analysis

1. Evaluation of WLOI Methods

The Sharpsburg and Valentine benchmark soils were used to evaluate the effect of ignition temperature and time on POM and SOM recovery. We tested two temperature/time treatment combinations: 360°C for 2 h and 450°C for 4 h. The minimum soil mass required for accurate and reproducible results for POM by WLOI was determined by igniting 30, 60, and 100 g subsamples from each of the two benchmark soils.

2. Soil Sample Preparation and Isolation of POM Fractions

The soils from the five field sites and the two benchmark sites were gently broken apart by hand, pushed through a 2-mm sieve, and dried overnight at 50 to 55° C in a forced air oven. Plant fragments (> 2 mm) were discarded since they are not defined as POM or SOM. The dry, 2-mm sieved soil was systematically mixed and 30-g and 10-g samples were removed from each field replicate composite for POM and SOM analysis, respectively. We dispersed the soil for POM analysis with 100 ml of 5-g L^{-1} sodium hexametaphosphate and shook the samples for 18 h on a reciprocal shaker.

The dispersed soil samples were first passed through a 500-μm sieve and then through a 53-μm sieve and rinsed thoroughly with water until the rinsate was clear. The 500-μm sieve removed larger sand grains and coarse organic matter fragments which simplified passing the soil slurry through the smaller sieve. The material retained on the sieves was back-washed into small aluminum pans and dried at 50 to 55° C for 24 h, which was sufficient to achieve a constant weight. The (silt+clay) material which passed through the 53-μm sieve was dried at 70°C for 24 h. The higher temperature was required to dry the large volume of (silt+clay) slurry to a constant weight in 24 h. The exact weight

Table 1. Soil and management descriptions for five experimental sites in the central and northern Great Plains

Site	Soil	Alternative management	Conventional management	Study time before 1998
Akron, CO	Weld silt loam	WW-corn-millet no-till	WW-fallow sweep plow	9 years
Brookings, SD	Barnes clay loam	Corn-soybean-wheat-alfalfa chisel plow	continuous corn chisel plow	8 years
Mandan, ND	Temvik-Wilton silt loam	SpW-WW-sunflower no-till	SpW-fallow Sweep plow and disk	14 years
Mead, NE	Sharpsburg silty clay loam	Corn-soybean-oat-clover reduced till	Continuous corn disk tillage	16 years
Sidney, NE	Duron silt loam	WW-fallow no-till	WW-fallow moldboard plow	27 years

of each dried sample was recorded to four decimal places, although three decimal places would suffice. The two dried POM fractions ((>500μm but < 2000 μm fraction) and (> 53 μm but < 500 μm fraction)) were stored at room temperature in small coin envelopes. The dried whole soil and (silt + clay) fraction samples were ground on a roller mill to pass through a 250-μm sieve and then stored at room temperature in glass scintillation vials.

3. POM and SOM by WLOI

The dried POM fractions and subsamples of dried/ground whole soil (for SOM) were placed in a muffle furnace and heated to 450° C. The samples were left in the oven for 4 h after the oven temperature reached 450° C. After ignition, the samples were cooled in a dessicator containing dried ascarite. After cooling, the ignited sample weight was recorded to four decimal places (although three decimal places would suffice) and the amount of total SOM or POM in each POM subfraction calculated where:

SOM (mg g^{-1}) = ([Wt. at 105°C – Wt. after 4 h at 450°C]/ Wt. at 105°C) × 1000
POM (mg g^{-1}) = ([Wt. at 55°C – Wt. after 4 h at 450°C]/ Wt. at 55°C) × 1000.

Particulate organic matter masses for the two POM subfractions were added together to obtain total POM mass.

4. POC and SOC by Dry Combustion

Total organic C in the dried/ground whole soil (SOC) and (silt + clay) fraction was determined by dry combustion in a Carlo-Erba NA 1500 NCS elemental analyzer (Haake Buchler Instruments, Paterson,

C.A. Cambardella, A.M. Gajda, J.W. Doran, B.J. Wienhold and T.A. Kettler

NJ[1]) after removal of carbonates with 1.5 N hydrochloric acid. Particulate organic matter C (POC) was calculated as the difference between SOC and (silt + clay) – associated organic C (Cambardella and Elliott, 1992).

5. Statistical Analysis

The experimental model for all field sites was a completely randomized block design with two treatments and three replicates. We tested for differences among field sites and cropping systems using one-way analysis of variance. Linear regression analysis was used to develop relationships between SOM and SOC or POM and POC.

III. Results and Discussion

A. Evaluation of WLOI Methods

One of the most common problems associated with using WLOI is the ignition method tends to overestimate organic matter in soil (Schulte and Hopkins, 1996). This occurs because inorganic constituents, primarily hydrated clays, lose weight during heating along with the organic materials. High temperature heating (>500°C) can result in loss of CO_2 from carbonates, structural water from clay minerals, oxidation of Fe^{+2}, and decomposition of hydrated salts (Ball, 1964; Ben-Dor and Banin, 1989). The presence of gypsum ($CaSO_4 \cdot 2H_2O$) can result in significant overestimates because gypsum is 21% water. These problems may be exacerbated using WLOI for POM because carbonate crystals, iron concretions leave it, and siliconized charcoal can separate out from soil with the POM fraction. Heating at temperatures below 500°C should eliminate much of the error, although careful preheating is required to completely remove water from gypsum prior to ignition

We tried two combinations of heating temperature and heating time in our study, 360°C for 2 h and 450°C for 4 h. These times and temperatures are typical of those reported in the literature in the past 20 years (Schulte and Hopkins, 1996).

We found significant but inconsistent differences in WLOI estimates for POM and SOM for the two heating treatments (data not shown). The higher temperature and longer heating time generally resulted in larger WLOI estimates for POM and SOM. The relative difference between the two methods ranged from 0.0% to 50.5% and the relative difference between the two methods was generally greater for POM (average of 22.7%) than for SOM (average of 13.6%). The two methods were most similar for the Akron site (relative difference of 2.6% for POM and 0.0% for SOM), which had the lowest amount of SOM and POM compared with the other soils.

In testing three different masses of soil (30, 60, and 100g) to determine the minimum soil sample size required for reproducible estimates of POM, we found that POM could be accurately measured using a 30-g soil sample for both the Sharpsburg and Valentine benchmark soils (Table 2).

Based on results of preliminary evaluations to determine optimal ignition temperature and sample size for WLOI, we recommend an ignition temperature of 450° C for 4 h, and a 30-g and 10-g subsample for quantification of POM and SOM, respectively.

---

[1]Reference to trade names and companies is made for information purposes only and does not imply endorsement by the USDA or the University of Nebraska.

Table 2. Values for POM and SOM obtained by WLOI for the two benchmark soils

Soil mass dispersed	Sharpsburg soil		Valentine soil	
	POM[a]	SOM	POM	SOM
	mg g^{-1}			
30 g	3.91	39.9	4.38	9.7
	(0.3)[b]	(1.1)	(0.5)	(0.5)
60 g	4.01	—	4.33	—
	(0.3)	—	(0.5)	—
100 g	3.54	—	4.14	—
	(0.2)	—	(1.0)	—

[a]POM and SOM in g kg^{-1}; [b]mean of 3 replicates, standard deviation in parantheses.

Table 3. Mass of SOM and POM quantified by loss-on-ignition; amount of C in SOC and POC quantified by dry combustion

Site/soil	System	SOM[a]	SOC	POM	POC	POM/SOM	POC/SOC
Akron	Altern	34.1 (4.2)[b]	9.3 (1.9)	6.5 (1.6)	2.5 (0.5)	0.191	0.269
Akron	Conven	28.6 (3.4)	7.2 (0.8)	4.3 (0.6)	1.5 (0.4)	0.150	0.208
Brookings	Altern	52.8 (0.5)	22.4 (1.8)	6.3 (2.1)	6.4 (1.2)	0.119	0:286
Brookings	Conven	53.5 (9.7)	22.3 (2.7)	4.9 (1.0)	6.1 (0.4)	0.092	0.274
Mandan	Altern	63.3 (9.0)	25.1 (4.4)	11.6 (1.2)	5.8 (1.3)	0.183	0.231
Mandan	Conven	49.8 (5.0)	20.4 (3.2)	7.1 (0.8)	3.5 (1.0)	0.143	0.172
Mead	Altern	51.7 (0.5)	21.4 (2.5)	9.5 (1.3)	6.3 (1.5)	0.184	0.294
Mead	Conven	47.9 (3.3)	17.6 (2.4)	7.4 (0.5)	3.3 (0.4)	0.154	0.188
Sidney	Altern	54.7 (3.9)	22.3 (1.9)	11.2 (1.0)	4.1 (0.4)	0.205	0.184
Sidney	Conven	35.7 (2.1)	13.8 (0.8)	5.3 (0.7)	2.0 (0.2)	0.148	0.145
Sharpsburg	—	39.9 (1.1)	17.3 (0.1)	3.9 (0.3)	2.1 (0.08)	0.098	0.121
Valentine	—	9.8 (0.5)	4.2 (0.2)	4.4 (0.5)	3.6 (0.6)	0.449	0.857

[a]SOM and POM in g kg^{-1} soil, SOC and POC in g C kg^{-1} soil; [b]mean of 3 field replicates, standard deviation in parentheses.

Estimates of POM by WLOI for the benchmark Sharpsburg silty clay loam were much less variable than for the Valentine sand benchmark soil (Table 3), primarily because it is much more difficult to obtain reproducibly homogeneous soil subsamples for POM analysis from a sandy soil. The coefficients of variation for the mean of total SOM by WLOI for the two benchmark soils were less than 5% (calculated from data in Table 3).

The conversion factor for estimating SOC from SOM can vary from 1.72 to 2.00 for surface soils (Broadbent, 1965). If you assume SOM is 58% C and calculate SOC from the WLOI data for SOM, the calculated values for SOC are 1.4 to 2.3 times greater than the measured values for SOC. The error is greater for the Akron soils than for soils from the other field sites. In order to avoid introducing potential error into our calculations, we report simple WLOI mass estimates for SOM rather than converting SOM to SOC. There are no reported conversion factors in the literature for calculating POC from WLOI POM. Data from a Monona silt loam in SW Iowa indicate the percentage of C in POM can range from 44.7 to 50.8 (Gale and Cambardella, in press). This range

C.A. Cambardella, A.M. Gajda, J.W. Doran, B.J. Wienhold and T.A. Kettler

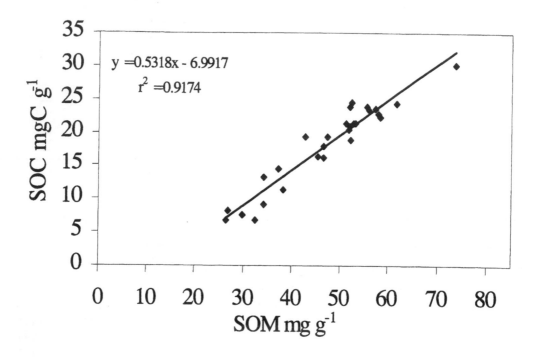

Figure 1. Relationship of SOM quantified by WLOI and SOC quantified by dry combustion.

will likely be greater for data taken from a wider variety of soils and field sites. As a result, we have reported POM mass from WLOI without attempting to convert the data to POC.

B. Evaluation of POM and SOM from WLOI as Soil Quality Indicator Variables

Total SOC quantified by dry combustion was highly correlated (r^2 = 0.92) with total SOM from WLOI for the five field sites (Figure 1). Values for SOM ranged from 28.6 to 63.3 g kg^{-1} and values for SOC ranged from 7.2 to 25.1 g C kg^{-1} (Figure 2 and Table 3). Using SOM by WLOI as a soil quality indicator variable, we were able to detect differences between field sites and management systems as effectively as using SOC by dry combustion. The Weld silt loam soil from Akron, CO cropped to winter wheat every other year and managed conventionally had the least amount of SOC and SOM. In contrast, the Temvik-Wilton silt loam from Mandan, ND planted to a 3-year rotation of spring wheat, winter wheat, and sunflower, and managed with no-tillage had the greatest amount of SOC and SOM. Total SOC and SOM averaged across cropping systems at Akron were significantly less than the other four field sites averaged across cropping systems (Table 3 and 4). Total SOC and SOM for the alternative cropping system averaged across all field sites were significantly greater than for the conventional cropping system averaged across all field sites (Table 3 and 4).

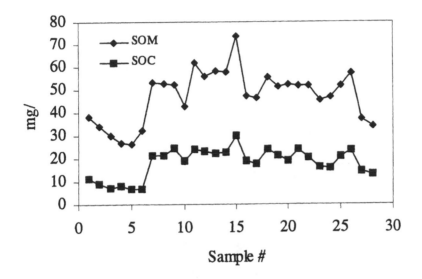

Figure 2. Data pattern of SOM quantified by WLOI and SOC quantified by dry combustion.

Table 4. ANOVA for site and cropping system main effects and the site by system interactive effect

	SOM	SOC	POM	POC
Site				
Df	4	4	4	4
F	19.63	34.96	13.59	22.45
Pr>F	0.0001	0.0001	0.0001	0.0001
System				
Df	1	1	1	1
F	7.29	14.38	46.44	30.41
Pr>F	0.0158	0.0016	0.0001	0.0001
Site*Syst				
Df	4	4	4	4
F	1.43	2.01	2.99	2.55
Pr>F	0.2687	0.1419	0.0508	0.0796

Estimates of POM mass obtained using WLOI are weakly correlated ($r^2 = 0.21$) with POC data obtained using dry combustion (Figure 3). Dry combustion estimates for POC ranged from 1.5 to 6.4 g C kg^{-1}soil and estimates of POM using WLOI ranged from 4.3 to 11.6 g kg^{-1}soil (Table 3). In general, estimations of POM by WLOI were as reproducible as our dry combustion estimates for POC (Table 3) and POM quantified by WLOI generally followed the same pattern as POC from dry combustion (Table 3 and Figure 4). Estimates of POM by WLOI were as sensitive to changes in

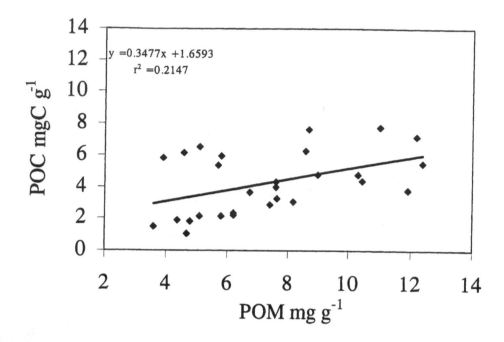

Figure 3. Relationship of POM quantified by WLOI and POC quantified by dry combustion.

management and location as POC by dry combustion, although relative differences between sites were slightly different for the two methods. The conventional soils from Akron had relatively low amounts of POC and POM, similar to our observations for SOC and SOM, but the Duroc silt loam from Sidney, NE cropped to winter wheat every other year and managed conventionally also had less POC and POM than the other soils. There was significantly more POC in the Brookings soils and significantly more POM in the Mandan soils compared to the other four sites averaged across cropping systems (Tables 3 and 4). Particulate organic matter C and POM for the alternative cropping system averaged across all field sites were significantly greater than POC and POM for the conventional cropping system averaged across all field sites (Tables 3 and 4).

IV. Conclusions and Recommendations

Based on results of preliminary evaluations to determine optimal ignition temperature and sample size for POM and SOM by WLOI, we recommend an ignition temperature of 450° C for 4 h, and a 30-g and 10-g subsample for quantification of POM and SOM, respectively. We also recommend that soil bulk density be determined for each soil sample at every sampling time when using WLOI for POM and SOM. Bulk density is needed in order to express the data on a volumetric basis, permitting comparisons of standing stocks of POM and SOM across ecosystems. The data from the five field

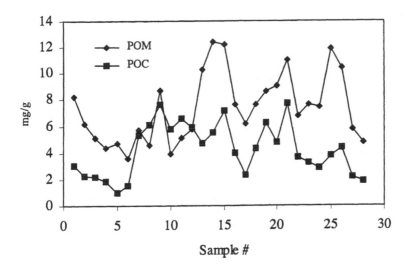

Figure 4. Data pattern of POM quantified by WLOI and POC quantified by dry combustion.

sites in this study are reported on a volumetric basis in a companion paper (Gajda et al., this volume). Conversion of POM and SOM values from WLOI to estimates of POC and SOC is not recommended because percent C of POM and SOM varies widely.

Estimates of SOM by WLOI are well correlated with total SOC by dry combustion across a number of soil types, locations and management practices. Soil organic matter determined by WLOI is equally sensitive to changes in agricultural management and location as SOC determined by dry combustion. Measurements of POM by WLOI are as sensitive to changes in tillage and cropping management as POC by dry combustion, despite the relatively weak correlation between POM and POC. Data trends for SOM, SOC, POM, and POC in the 7.5 to 30 cm depth increment (Gajda et al., Chapter 27, this volume) were similar to the surface soil (0 to 7.5 cm) estimates reported in this chapter.

Relative comparisons of POM and SOM by WLOI for one field at several points in time or between one or more fields at one point in time appears to be feasible based on our data. However, we recommend caution in interpretation of POM data obtained by WLOI and suggest researchers use dry combustion to confirm their data if the technology is available and affordable. In laboratories that are equipped with dry combustion instrumentation, WLOI for POM or SOM can be a useful tool for screening large numbers of samples. If large differences between samples are identified, dry combustion for POC and SOC can be used to verify the results.

Our data indicate that POM and SOM by WLOI qualify as sensitive indicators of soil quality change. Researchers, land managers, or producers wishing to use POM and SOM for soil quality estimates where dry combustion technology is not readily available will derive the greatest benefit from this research.

References

Allison, F.E. 1973. Soil organic matter and its role in crop production. Elsevier Publishing Co., New York.

Allmaras, R.R., J.L. Pikul Jr., J.M. Kraft and D.E. Wilkins. 1988. A method for measuring incorporated crop residue and associated soil properties. *Soil Sci. Soc. Am. J.* 52:1128-1133.

Angers, D.A., A.N. Dayegamiye and D. Cote. 1993. Tillage-induced differences in organic matter of particle size fractions and microbial biomass. *Soil Sci. Soc. Am. J.* 57:512-516.

Ball, D.F. 1964. Loss-on-ignition as an estimate of organic matter and organic carbon in noncalcareous soils. *J. Soil Sci.* 15:84-92.

Beare, M.H., M.L. Cabrera, P.F. Hendrix and D.C. Coleman. 1994. Aggregate-protected and unprotected organic matter pools in conventional- and no-tillage soils. *Soil Sci. Soc. Am. J.* 58:777-786.

Ben-Dor, E. and A. Banin. 1989. Determination of organic matter content in arid-zone soils using a simple "loss-on-ignition" method. *Commun. Soil Sci. Plant Anal.* 20:1675-1695.

Biederbeck, V.O., H.H. Janzsen, C.A. Campbell and R.P. Zettner. 1994. Labile soil organic matter as influenced by cropping practices in an arid environment. *Soil Biol. Biochem.* 26:1647-1656.

Bolinder, M.A., D.A. Angers, E.G. Gregorich and M.R. Carter. 1999. The response of soil quality indicators to conservation management. *Can. J. Soil Sci.* 79:37-45.

Bremer, E., H.H. Janzsen and A.M. Johnston. 1994. Sensitivity of total, light fraction and mineralizable organic matter to management practices in a Lethbridge soil. *Can. J. Soil Sci.* 74:131-138.

Broadbent, F.E. 1965. Organic matter. p. 1397-1400. In: C.A. Black (ed.), *Methods of Soil Analysis.* Agron. Monogr. 9. American Society of Agronomy, Crop Science Society of America, Soil Science Society of America, Madison, WI.

Cambardella, C.A. and E.T. Elliott. 1992. Particulate organic matter changes across a grassland cultivation sequence. *Soil Sci. Soc. Am. J.* 56:777-783.

Christensen, B.T. 1996. Carbon in primary and secondary organomineral complexes. p. 97-165. In: M.R. Carter and B.A. Stewart (eds.), *Structure and Organic Matter Storage in Agricultural Soils.* Advances in Soil Science. Lewis Publishers, Boca Raton, FL.

Dalal, R.C. and R.J. Mayer. 1986. Long-term trends in fertility of soils under continuous cultivation and cereal cropping in Southern Queensland. IV. Loss of organic carbon from different density fractions. *Aust. J. Soil Res.* 24:301-309.

Elliott, E.T., I.C. Burke, C.A. Monz, S.D. Frey, K.H. Paustian, H.P. Collins, E.A. Paul, C.V. Cole, R.L. Blevins, W.W. Frye, D.J. Lyon, A.D. Halvorsen, D.R. Huggins, R.F. Turco and M.V. Hickman. 1994. Terrestrial carbon pools: preliminary data from the Corn Belt and Great Plains regions. p. 179-191. In: J.W. Doran et al. (eds.), *Defining Soil Quality for a Sustainable Environment.* SSSA Spec. Pub. 35. American Society of Agronomy and Soil Science Society of America, Madison, WI.

Gajda, A.M., J.W. Doran, T.A. Kettler, B.J. Wienhold, J.L. Pikul, Jr. and C.A. Cambardella. 2001. Soil quality evaluations of alternative and conventional management systems in the Great Plains. p. 381-400. In: R. Lal, J.M. Kimble, R.F. Follett and B.A. Stewart (eds.), *Assessment Methods for Soil Carbon.* Lewis Publishers, Boca Raton, FL.

Gale, W.J. and C.A. Cambardella. 2000. Carbon dynamics of surface residue- and root-derived organic matter under simulated no-till. *Soil Sci. Soc. Am. J.* 64.

Golchin, A., J.M. Oades, J.O. Skjemstad and P. Clarke. 1994. Study of free and occluded particulate organic matter in soils by solid state ^{13}C CP/MAS NMR spectroscopy and scanning electron microscopy. *Aust. J. Soil Res.* 32:285-309.

Greenland, D.J. and G.W. Ford. 1964. Separation of partially humified organic materials from soils by ultrasonic dispersion. *Trans. 8th Int. Cong. Soil Sci.* II:137-147.

Gregorich, E.G. and B.H. Ellert. 1993. Light fraction and macroorganic matter in mineral soils. pp. 397-407. In: M. R. Carter (ed.), *Soil Sampling and Methods of Analysis*. Canadian Society of Soil Science. Lewis Publishers, Boca Raton, FL.

Gregorich, E.G., B.H. Ellert and C.M Monreal. 1995. Turnover of soil organic matter and storage of corn residue carbon estimated from natural ^{13}C abundance. *Can. J. Soil Sci.* 75:161-167.

Larson, W.E. and F.J. Pierce. 1991. Conservation and enhancement of soil quality. pp. 175-203. In: Evaluation for sustainable land management in the developing world. Vol. 2. IBSRAUM Proc. 12 (2). Thailand. International Board for Soil Research and Management, Bangkok.

Janzsen, H.H., C.A. Campbell, S.A. Brandt, G.P. Lafond and L. Townley-Smith. 1992. Light-fraction organic matter in soils from long-term crop rotations. *Soil Sci. Soc. Am. J.* 56:1799-1806.

Jastrow, J.D., T.W. Boutton and R.M. Miller. 1996. Carbon dynamics of aggregate-associated organic matter estimated by carbon-13 natural abundance. *Soil Sci. Soc. Am. J.* 60:801-807.

Schimel, D.S., D.C. Coleman and K.A. Horton. 1985. Soil organic matter dynamics in paired rangeland and cropland toposequences in North Dakota. *Geoderma* 36:201-214.

Schulte, E.E. 1988. Recommended soil organic matter tests. p. 29-31. In: W.C. Dahnke (ed.), Recommended chemical soil test procedures for the North Central Region. NCR Publ. No. 221 (revised). Cooperative Extension Service, North Dakota State University, Fargo.

Schulte, E.E. and B.G. Hopkins. 1996. Estimation of soil organic matter by weight loss-on-ignition. p. 21-31. In: *Soil Organic Matter: Analysis and Interpretation*. SSSA Special Publication no. 46. Soil Science Society of America, Madison, WI.

Tiessen, H. and J. W. B. Stewart. 1983. Particle-size fractions and their use in studies of soil organic matter: II. Cultivation effects on organic matter composition in size fractions. *Soil Sci. Soc. Am. J.* 47:509-514.

Wander, M.M. and S.J. Traina. 1996. Organic matter fractions from organically and conventionally managed soils: I. Carbon and nitrogen distribution. *Soil Sci. Soc. Am. J.* 60:1081-1086.

Yavchenko, V., L.J. Sikora and D.D. Kaufman. 1996. A biologically-based indicator of soil quality. *Biol. Fert. Soils* 21:245-251.

CHAPTER 25

Use of Near Infrared Spectroscopy to Determine Inorganic and Organic Carbon Fractions in Soil and Litter

B. Ludwig and P.K. Khanna

I. Introduction

Near infrared (NIR, 750 to 2500 nm wavelength) spectroscopy is a well-known analytical technique and has been adopted by many agricultural and manufacturing industries. In agriculture it is commonly used to measure the nutritional components of forage and animal feeds (Norris et al., 1976; Shenk and Westerhaus, 1994; Fahey and Hussein, 1999), and to measure secondary metabolites in plants (Clark et al., 1987; Windham et al., 1988). This method has obtained certified status (AOAC, 1990) to measure moisture, crude protein and acid detergent fiber in forages. NIR spectroscopy is known for its rapidity, convenience, accuracy and ability to analyze many constituents at the same time (Stark et al., 1986). It is now believed that this technique has the capacity to be used to study complex functional attributes of natural systems, for example, the susceptibility of plants to insect attack (Rutherford and van Staden, 1996), the yield of pulp from trees (Wright et al., 1990) and the biological degradability of sawlogs (Hoffmeyer and Pedersen, 1995; Schimleck et al., 1996). It is still being tested for its usefulness to analyze soils, other biological materials and ecological parameters (Ben-Dor and Banin, 1995; Foley et al., 1998).

Owing to the complex chemical nature of soil and litter, assigning a defined part of NIR spectra to chemical characteristics may not always be possible. However, by relating the results of chemical analysis to the NIR spectra of a large set of samples, the composition of soil and litter constituents can potentially be determined indirectly (McLellan et al., 1991; Morra et al., 1991). The increased use of NIR spectroscopy (NIRS) for soil analysis during the last decade may partly be attributed to improved computer hardware and to improved and user-friendly statistical NIRS software (Shenk and Westerhaus, 1991a). Some researchers have applied NIRS enthusiastically to determine a wide range of plant, litter and soil constituents (Ben-Dor and Banin, 1995; Foley et al., 1998), and to study litter decomposition (Gillon et al., 1999), whereas others have reported discrepancies and problems when using NIRS for these types of analysis (Krishnan et al., 1980; Henderson et al., 1992; Rossel and McBratney, 1998).

The objective of this chapter is to critically review the usefulness and limitations of NIRS to determine different carbon fractions (inorganic, microbial, non-microbial, readily decomposable, charcoal and other charred materials and plant residues) in the litter and soil.

II. Principle of the Method, Particle Size and Moisture

Near infrared radiation is absorbed by different chemical bonds (e.g., C-H, O-H, N-H, C=O, S-H, CH_2, C-C) and results in bending, twisting, stretching or scissoring of the bonds (Osborne and Fearn, 1986). An NIR spectrum does not contain distinct or sharp peaks, because the peaks may consist of overtones and combinations resulting from primary absorption of radiation. Furthermore, some of the radiated light may be scattered, causing further problems in the evaluation of NIR spectra (Shenk and Westerhaus, 1993).

The spectral absorbance in the NIR region greatly depends on the particle size of the sample. Increasing particle size results in an increased apparent path length for the incident light, which increases the absorbance (Casler and Shenk, 1985).

The moisture content of the sample also influences the spectra (O-H vibrations). Thus, most studies use dried samples to avoid problems arising from different water contents. In addition, in some applications, parts of the spectrum corresponding to water may be excluded from any calibrations.

III. Calibration and Validation

Calibration of a set of soil values is carried out by developing a regression equation between the absorbance spectra and the laboratory analyses. The calibration data set should be homogenous with respect to the properties being measured. Spectra that differ greatly from the average spectra can be identified by the CENTER algorithm of Shenk and Westerhaus (1991b) and should be removed from the set. The regression equations commonly used for calibration include multiple linear regression (MLR), principal component regression (PCR), partial least square (PLS) and a modified PLS. MLR uses only a few wavelengths, whereas the other methods use the entire spectrum or most of it. Prior to calibration, the spectra should be corrected by a detrending method (Barnes et al., 1989). For calibrations, different mathematical treatments to the spectrum can be performed, e.g., taking derivatives of 1st or 2nd or 3rd order, defining the gap over which the derivative is to be calculated and smoothing of the spectra. The calibration equation, which gives the best results in terms of standard error of validation, should be selected. No single mathematical treatment gives the best prediction for all variables and all products. Thus, the only way to obtain the lowest error of prediction is by following a trial-and-error procedure (Couteaux et al., 1998).

There is no general agreement on sample numbers required for the calibration. However, the sample set should be reasonably diverse and should cover the expected variation. For all applications, validation is the only proof to check whether the calibration was successful. After a successful validation, routine NIR predictions should be tested continuously on a subsample (e.g., 10% of all samples). Predictions for samples from outside the calibrated range may be inaccurate. The accuracy of predictions may change with time through the expansion of the calibration data set or other reasons.

IV. Usefulness and Limitations

In soil and litter C occurs in various forms and undergoes turnover processes at varying rates. The analysis of some of these C components is easy, and reliable conventional methods are available, but for others specialized methods are required, which are not only time consuming but also need special equipment. Moreover, there is a scarcity of available methods that can be used under field conditions to measure C of the soil and litter. The NIR technique has been gainfully applied with varying success to assess some of the C forms, as described below.

A. Carbonate

Carbonate contents in soils were predicted in arid soils after preheating them to 600°C for 8 h (Ben-Dor and Banin, 1990). The preheating resulted in the removal of the strong absorption features of the OH groups from organic matter and clay minerals. It was concluded that carbonate concentrations of 10 to 75% can be satisfactorily estimated (confidence limits of ± 5 to 7% $CaCO_3$) by NIRS (Ben-Dor and Banin, 1995). However, prediction of carbonate contents using unheated soil samples gave only poor correlation coefficients for the validation set.

Heating of soils was not necessary, if the mid-infrared range (2500 to 25000 nm) was used for the prediction of carbonate contents using a locally tested linear PLS model (Janik and Skjemstad, 1995). Carbonate contents were predicted in the range of 0 to 12% with a correlation coefficient of 0.97 (n = 48). The major peaks used for prediction occurred in a window between 3000 and 1500 cm^{-1} which is relatively free from peaks resulting from other soil components.

B. Organic Matter (OM)

Ben-Dor and Banin (1995) assessed NIRS to estimate the OM content for 91 soil samples (calibration: n = 39, validation: n = 52) from Israel with a range of 0.1 to 13.2% OM. The considered wavelengths (nm) were 1042.9, 1412.2, 1584.9, 1941.0, 2016.5 and 2388.3. The validation resulted in a correlation coefficient of 0.74 and in a relative standard error of prediction (SEP) of 1.3%. In the range of 0 to 4% the prediction of soil OM content was quite useful with some large exceptions, whereas at contents above 4% (n = 4), the agreement between measured and predicted values was poor. Ben-Dor and Banin (1995) suggested that the soil population in their study represented two different groups: soils with highly decomposed OM (0 to 4% OM), and those with less decomposed OM (4 to 13% OM). They assumed that functional properties of OM in the first group were more similar to humic substances, whereas the functional groups in the second group were more similar to fresh litter and vegetation (Ben-Dor and Banin, 1995). However, the quality of calibrations for samples with OM contents above 4% was limited by the small number of samples.

Soil organic C was successfully estimated (r = 0.93, SEP = 0.2%, n = 72) in soils of low C content (0 to 2.5% C) for three major soil series of the Darling Downs, Queensland, Australia, by considering the wavelengths 1744, 1870 and 2052 nm (Dalal and Henry, 1986). However, two limitations were noted: NIR predictions for organic C were poor for soils low in C (below 0.3%) and for red earths (Rhodic Paleustalfs) due to color interference. Thus, Dalal and Henry (1986) concluded that NIRS may be used in routine soil testing within a moderate range of OM and a narrow range of soil colors. Morra et al. (1991) estimated total C contents in soils ranging from 0 to 8% and in different particle size fractions of 12 soil series. The considered wavelengths were 2426, 1736, 2210, 2032, 2250 and 2170 nm for the silt plus coarse clay fractions and 2206, 2226, 1766, 2346 and 2246 nm for the silt fractions. The C contents of the silt and silt plus coarse clay fractions were estimated quite well with SEPs being 0.6% and 0.3%, respectively. Morra et al. (1991) reported two main limitations to the technique, first, the necessity of a large sample set for calibrations, and second, the necessity of using a defined or closed sample population.

In our own study (Figure 1), the soil C content was predicted quite well by considering the entire NIR spectra for different Australian soils (red earths) collected from various depth intervals to 30 cm. The standard error of calibration was 0.3% and SEP was 1.3%. However, there were a few exceptions where the prediction was not good (Figure 1).

In contrast to the above findings, Krishnan et al. (1980) did not find an absorption peak related to soil organic matter for wavelengths between 800 and 2400 nm. They suggested that the visible wavelength region (optimal wavelengths: 623.6 and 564.4 nm) provided better information than the NIR region for determining soil organic matter. Janik et al. (1998) found the mid-infrared range (2500

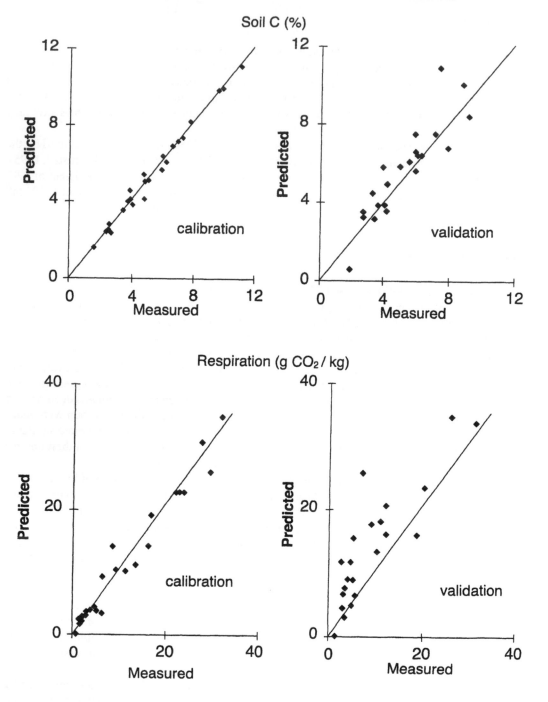

Figure 1. Measured and predicted values of C and respiration during 180 days incubation under laboratory conditions in a range of forest soils. The lines indicate 1:1 values. The standard errors of prediction are given in the text.

to 25000 nm) superior over the NIR range for the determination of soil carbon in the range from 1.1 to 8.0%. A correlation coefficient of 0.96 (n = 188) was obtained for their validation. Rossel and McBratney (1998) could not predict OM contents in some Australian soils at various additions of freeze-dried liquid compost. This lack of agreement may have occurred because the freeze-dried compost may not have accurately represented the complex composition of soil organic matter and hence its reflectance (Rossel and McBratney, 1998). Henderson et al. (1992) fractionated OM of four agricultural soils from Indiana into seven components (crude organic matter, crude humic acid (HA) and crude fulvic acid (FA), purified HA and FA, oxidized mineral soil and oxidized mineral soil without Fe and Mn oxides). Samples of all organic fractions were freeze-dried. They obtained spectra in the visible and NIR range of all fractions and found the spectra useful to provide information about soil organic C content, but not organic matter composition.

C. Charcoal and Charred Material

It is becoming increasingly important to assess the amount of carbon present in charcoal or charred materials or black C in soils and sediments (Skjemstad et al., 1999; Masiello and Druffel, 1998), as it is considered to be a form of long-term sequestered C. However, the methods employed to assess these materials are highly sophisticated and require specialized and expensive equipment. Skjemstad et al. (1996) have successfully measured charred materials by using solid state ^{13}C NMR spectroscopy on various fractions of four surface soils that were repeatedly subjected to high energy ultraviolet photo-oxidation. Kuhlbusch (1995) used an extraction procedure followed by thermal treatment to differentiate black carbon from organic carbon. Thermal treatment was used to measure H and C in the ash and plant material. The molar H/C ratio was lower (0.19 ± 0.05) in the ash than in the plant material. However, this technique has not been tested on soil. There has been no attempt to use NIRS for assessing charred materials in soils.

D. Fine Roots

A significant amount of organic C in many soils may be present in the form of fine roots of varying degree of decomposition. The usual method of soil preparation (drying and sieving) does not allow the complete separation of this fraction. Use of NIRS to quantify this C fraction may be possible, although no work has been reported so far. However, an indication is available from a palaeoecological study of peat by McTiernan et al. (1998), who noted a strong relationship (regression coefficient of 0.85) between the conventional analysis and NIR predicted values for *Ericales* roots.

E. Litter Decomposition

Chemical composition of the biological substances in soils is an essential determinate of their decomposition rates especially under conditions where the climatic and microbial factors are either optimum or non-limiting. Forest foliage of 13 different species at varying stages of decomposition was studied using laboratory measurements of total N, lignin and cellulose contents and NIRS (McLellan et al., 1991). The contents varied from 0.4 to 3.2% for total N, from 16 to 44% for lignin and from 36 to 57% for cellulose. For total N, the wavelengths considered were 2174 (protein), 2378 (oil), 2090 (OH) and 1230 (CH) nm. For lignin, they were 1438 (CH_2), 2154 ($CONH_2$), 1708 (CH_3) and 2320 (starch) nm. The wavelengths for cellulose were 2252 (cellulose), 1898 (CO_2H), 2076 (oil) and 1754 (CH_2). The NIR estimates were good for total N (n = 148, r = 0.94, SEP = 0.2%) and satisfactory for lignin (n = 57, r = 0.87, SEP = 3%) and cellulose (n = 55, r = 0.84, SEP = 3%). Similarly, Joffre et al. (1992) successfully predicted C, N, ash content, acid-detergent fiber and acid-detergent lignin for the litter of eight species of evergreen and deciduous broad-leafed trees, conifers, and shrubs using NIRS. They tested MLR and PLS and found that both regression methods provided acceptable validation statistics for C, N and ash. However, PLS had a better prediction accuracy for acid-detergent fiber and acid-detergent lignin.

Decomposition (expressed as percentage of ash-free litter mass remaining) was studied for leaf litter of 10 different species with variable initial chemical compositions and at different decomposition stages (Gillon et al., 1993). The C/N ratios varied from 34 to 71, and the lignin/N ratios from 9 to 29 for the leaf litter used for decomposition experiments. Decomposition experiments were carried out in microcosms at 22°C and moisture conditions of 80% of field capacity, and also in the field using litterbags. For the calibration, the entire spectra were considered with the exception that part of the spectra corresponding to water (1948 to 1968 nm) was excluded. The litter mass remaining in microcosms decreased from 100 to 22% (n = 216) and was predicted by NIRS with an r^2 value of 0.98 and a SEP of 3.5%. Application to litterbag experiments gave slightly less agreement between NIR estimations and measured results (n = 29, r^2 = 0.93, SEP = 4.5%). From the results obtained by additional studies, Gillon et al. (1999) observed that the litter mass remaining and the decay constants were more closely related with the NIR spectra of the initial litters than with their other initial characteristics. Therefore, NIRS provided the opportunities of studying spatial and temporal variability under field conditions, and of characterizing the gradual changes in litter quality during decomposition.

Changes of absorption in the visible and NIR range were studied in decomposition experiments with grape marc and cattle manure (Ben-Dor et al., 1997). The slope values in the visible range were found useful for monitoring the compost maturity with (slope between 638 and 450 nm) and without (slope between 800 and 680 nm) relying on the organic matter species.

F. Microbial Activity and Biomass in Soils

Basal respiration (for intact sample cores and sieved samples) and substrate-induced respiration (SIR) (for sieved samples) were successfully described for samples from a mor layer of a Haplic podsol in Sweden using NIRS on freeze-dried and milled samples (Palmborg and Nordgren, 1993). Before sieving, 82% of the variation in basal respiration could be explained by NIR. After sieving, 93 to 98% of the variation in basal respiration and 88% of the variation in SIR was explained (Palmborg and Nordgren, 1993). However, the small sample number (n = 30) leaves some doubts whether the findings may be generalized.

The following example shows that a successful calibration is not sufficient to guarantee a good prediction. Soil respiration after 180 days was calibrated successfully for different Australian soils (red earths) (Figure 1) with SEC being 2.1%. However, the validation gave less agreement between measured and predicted values (Figure 1) and some large deviations were noted. SEP was 7.7% and a considerable bias was evident, indicating that additional parameters may be required to predict soil respiration for the considered soils.

The effects of clearcutting and clearcutting followed by prescribed burning on humus quality from a Norway spruce dominated stand in Finland were studied using microbiological data and NIRS (Pietikäinen and Fritze, 1995). Total microbial carbon (C_{mic}) was measured by substrate-induced respiration (SIR) and fumigation-extraction (FE). The NIR data could explain 75 to 82% of the variation in C_{mic}-FE, C_{mic}-SIR and ergosterol concentration, but NIR explained only 62% of the variation in respiration. A major shortcoming of this study, however, was that the sample number (n = 12) was too small to test the usefulness of NIRS for such applications.

Additional work is required before microbial parameters can be obtained from NIR spectra. This would include the use of field moist soils without any treatment (sieving or grinding) and developing additional relationships to include environmental parameters essential to describe microbial activity.

G. Determination of $^{12}C/^{13}C$ Isotope Ratios

Okano et al. (1983) developed a simple calibration method for the determination of ^{13}C in plant material using IR absorption spectrometry. The principle is that plant material is oxidized to CO_2 and the absorption intensity of infrared radiation by $^{12}CO_2$ and $^{13}CO_2$ is measured at two different IR regions (2390 to 2370 cm^{-1} for $^{12}CO_2$ and 2280 to 2260 cm^{-1} for $^{13}CO_2$) separated by a diffraction grating equipment that is capable of a resolution of 20 cm^{-1}. The method could predict ^{13}C abundance in a range of C3 and C4 plants within 97% of measured values. Clark et al. (1995) successfully calibrated NIR spectra of ground plant materials (without oxidizing them to CO_2) to their ^{13}C contents. Plant materials consisted of a range of grasses and legumes. Coefficients of variation for all validation sample sets used for the prediction of the ^{13}C discrimination (Δ) by NIR were less than 3% and 77 to 82% of the samples with the lowest Δ were correctly identified by NIRS. Clark et al. (1995) and Foley et al. (1998) pointed out that NIRS is not expected to replace isotope ratio mass spectrometry, but NIRS may be useful to rapidly screen a large number of samples so that the more expensive mass spectrometer measurements can be targeted on those samples which are of interest.

H. Field Measurements with Portable NIR Spectrometers

Sudduth and Hummel (1993a) developed a portable NIR spectrophotometer, which included a circular variable filter monochromator, a fibre optic coupling for sensing of remote samples, and a software algorithm which corrected the instrument readings to a zero baseline. The instrument had a sensing range of 1650 nm to 2650 nm, an optical bandwidth of under 55 nm and could acquire a spectrum every 200 ms. Laboratory calibrations and validations to determine soil organic matter content were satisfactory (SEP = 0.40%) for 30 Illinois soils at 1.5 MPa and 0.033 MPa moisture tension levels. However, in-furrow field operation produced unsatisfactory results, probably mainly due to the movement of soil past the sensor during data acquisition (Sudduth and Hummel, 1993b).

V. Summary and Conclusions

NIR spectroscopy appears to be a very promising tool to determine the amount and composition of plant and litter components, and also for determining litter decomposition. Additionally, NIRS may prove useful to rapidly screen a large number of samples for their $^{12}C/^{13}C$ isotope ratios so that the more expensive mass spectrometer measurements can be targeted on those samples which are of interest.

For the determination of soil carbon fractions, the accuracy of the NIR technique seems to be lower than for litter samples. A reason for the low accuracy may be that soil carbon consists of components of varying ages and composition. Most researchers would hesitate to replace the classical methods for the determination of soil C in the laboratory (e.g., Walkley-Black, total element analyzers) with NIRS because such methods are also fast and cheap to run. However, the use of portable NIR spectrometers for the determination of soil C in the field would be of great interest. Useful applications of the NIR technique may lie in estimating those soil C parameters, which require expensive and time-consuming analytical methods (litter decomposition rates, microbial activity, microbial biomass, C-isotope ratios, and the amount of charred material and fine roots). It has been proposed that NIRS was useful to determine microbial biomass and activity, but only a limited number of studies have been conducted to support this. Soil respiration could be described by NIRS only qualitatively. However, the ability of NIRS to determine multiple parameters from a single non-destructive analysis of soils is a real potential that should be further investigated.

References

AOAC. 1990. Official Methods of Analysis. Association of Official Analytical Chemists, Washington, D.C.

Barnes, R.J., M.S. Dhanoa and S.J. Lister. 1989. Standard normal variate transformation and detrending of NIR spectra. *Appl. Spec.* 43:772-777.

Ben-Dor, E. and A. Banin. 1990. Near-infrared reflectance analysis of carbonate concentration in soils. *Appl. Spec.* 44:1064-1069.

Ben-Dor, E. and A. Banin. 1995. Near-infrared analysis as a rapid method to simultaneously evaluate several soil properties. *Soil Sci. Soc. Am. J.* 59:364-372.

Ben-Dor, E., Y. Inbar and Y. Chen. 1997. The reflectance spectra of organic matter in the visible near-infrared and short wave infrared region (400-2500 nm) during a controlled decomposition process. *Remote Sens. Environ.* 61:1-15.

Casler, M.D. and J.S. Shenk. 1985. Effect of sample grinding size on forage quality estimates of smooth bromegrass clones. *Crop Sci.* 25:167-170.

Clark, D.H., M.H. Ralphs and R.C. Lamb. 1987. Total alkaloid determinations in larkspur and lupin with near infrared reflectance spectroscopy. *Agron. J.* 79:485-490.

Clark, D.H., D.A. Johnson, K.D. Kephard and N.A. Jackson. 1995. Near infrared reflectance spectroscopy estimation of ^{13}C discrimination in forages. *J. Range Manage.* 48:132-136.

Couteaux, M.M., K.B. McTiernan, B. Berg, D. Szuberla, P. Dardenne and P. Bottner. 1998. Chemical composition and carbon mineralisation potential of Scots pine needles at different stages of decomposition. *Soil Biol. Biochem.* 30:583-595.

Dalal, R.C. and R.J. Henry. 1986. Simultaneous determination of moisture, organic carbon and total nitrogen by near infrared reflectance spectroscopy. *Soil Sci. Soc. Am. J.* 50:120-123.

Fahey, G.C. and H.S Hussein. 1999. Forty years of forage quality research: Accomplishments and impact from an animal nutrition perspective. *Crop Sci.* 39:4-12.

Foley, W.J., A. McIlwee, I. Lawler, L. Aragones, A.P. Woolnough and N. Berding. 1998. Ecological applications of near infrared reflectance spectroscopy — a tool for rapid, cost-effective prediction of the composition of plant and animal tissues and aspects of animal performance. *Oecologia* 116:293-305.

Gillon, D., R. Joffre and P. Dardenne. 1993. Predicting the stage of decay of decomposing leaves by near infrared reflectance spectroscopy. *Can. J. For. Res.* 23:2552-2559.

Gillon, D., R. Joffre and A. Ibrahima. 1999. Can litter decomposability be predicted by near infrared reflectance spectroscopy? *Ecology* 80:175-186.

Henderson, T.L., M.F. Baumgardner, D.P. Franzmeier, D.E. Stott and D.C. Coster. 1992. High dimensional reflectance analysis of soil organic matter. *Soil Sci. Soc. Am. J.* 56:865-872.

Hoffmeyer, P. and J.G. Pedersen. 1995. Evaluation of density and strength of Norway spruce by near infrared reflectance spectroscopy. *Holz als Roh- und Werkstoff* 53:165-170.

Janik, L.J. and J.O. Skjemstad. 1995. Characterization and analysis of soils using mid-infrared partial least-squares. II. Correlations with some laboratory data. *Aust. J. Soil Res.* 33:637-650.

Janik, L.J., R.H. Merry and J.O. Skjemstad. 1998. Can mid infrared diffuse reflectance analysis replace soil extractions? *Aust. J. Exp. Agric.* 38:681-696.

Joffre, R., D. Gillon, P. Dardenne, R. Agneessens and R. Biston. 1992. The use of near-infrared reflectance spectroscopy in litter decomposition studies. *Ann. Sci. For.* 49:481-488.

Krishnan, P., J.D., Alexander, B.J. Butler and J.W. Hummel. 1980. Reflectance technique for predicting soil organic matter. *Soil Sci. Soc. Am. J.* 44:1282-1285.

Kuhlbusch, T.A.J. 1995. Method for determining black carbon in residues of vegetation fires. *Environ. Sci. Techn.* 29:2695-2702.

Masiello, C.A. and E.R.M. Druffel. 1998. Black carbon in deep-sea sediments. *Science* 280:1911-1913.

McLellan, T.M., J.D. Aber, M.E. Martin, J.M. Melillo and K.J. Nadelhoffer. 1991. Determination of nitrogen, lignin, and cellulose content of decomposing leaf material by near infrared reflectance spectroscopy. *Can. J. For. Res.* 21:1684-1688.

McTiernan, K.B., M.H. Garnett, D. Mauquoy, P. Ineson and M.M. Couteaux. 1998. Use of near-infrared reflectance spectroscopy (NIRS) in palaeoecological studies of peat. *Holocene* 8:729-740.

Morra, M.J., M.H. Hall and L.L. Freeborn. 1991. Carbon and nitrogen analysis of soil fractions using near-infrared reflectance spectroscopy. *Soil Sci. Soc. Am. J.* 55:288-291.

Norris, K.H., R.F. Barnes, J.E. Moore and J.S. Shenk. 1976. Predicting forage quality by near-infrared reflectance spectroscopy. *J. Anim. Sci.* 43:889-897.

Okano, K., O. Ito, N. Kokubun and T. Totsuka. 1983. Determination of ^{13}C in plant materials by infrared absorption spectrometry using a simple calibration method. *Soil Sci. Plant Nutr.* 29:369-374.

Osborne, B.G. and T. Fearn. 1986. *Near Infrared Spectroscopy in Food Analysis.* Longman Scientific and Technical, Essex.

Palmborg, C. and A. Nordgren. 1993. Modelling microbial activity and biomass in forest soil with substrate quality measured using near infrared reflectance spectroscopy. *Soil Biol. Biochem.* 25:1713-1718.

Pietikäinen, J. and H. Fritze. 1995. Clear-cutting and prescribed burning in coniferous forest: comparison of effects on soil fungal and total microbial biomass, respiration activity and nitrification. *Soil Biol. Biochem.* 27:101-109.

Rossel, R.A.V. and A.B. McBratney. 1998. Laboratory evaluation of a proximal sensing technique for simultaneous measurement of soil clay and water content. *Geoderma* 85:19-39.

Rutherford, R.S. and J. van Staden. 1996. Towards a rapid near-infrared technique for prediction of resistance to surgarcane borer *Eldana saccharina* Walker (Lepidoptera Pyralidae) using stalk surface wax. *J. Chem. Ecol.* 22:681-694.

Schimleck, L.R., A.J. Mitchell and P. Vinden. 1996. Eucalypt wood classification by NIR spectroscopy and principal component analysis. *Appita J.* 49:319-324.

Shenk, J.S. and M.O. Westerhaus. 1991a. ISI NIRS-2. Software for Near-infrared Instruments. Infrasoft International, Silver Spring, MD.

Shenk, J.S. and M.O. Westerhaus. 1991b. Population definition, sample selection and calibration procedures for near infrared reflectance spectroscopy. *Crop Sci.* 31:469-474.

Shenk, J.S. and M.O. Westerhaus. 1993. Analysis of Agriculture and Food Products by Near-infrared Reflectance Spectroscopy. Infrasoft, Port Matilda.

Shenk, J.S. and M.O. Westerhaus. 1994. The application of near infrared spectroscopy (NIRS) to forage analysis. p. 406-449. In: G.C. Fahey, L.E. Mosser, D.R. Mertens and M. Collins (eds.), National conference on forage quality evaluation and utilization. American Society of Agronomy. Madison, WI.

Skjemstad, J.O., P. Clarke, J.A. Taylor, J.M. Oades and S.G. McClure. 1996. The chemistry and nature of protected carbon in soil. *Aust. J. Soil Res.* 34:251-271.

Skjemstad, J.O., J.A. Taylor, L.J. Janik and S.P. Marvanek. 1999. Soil organic carbon dynamics under long-term sugarcane monoculture. *Aust. J. Soil Res.* 37:151-167.

Stark, E., K. Luchter and M. Margoshes. 1986. Near-infrared analysis (NIRA): A technology for quantitative and qualitative analysis. *Appl. Spec. Rev.* 22:335-399.

Sudduth, K.A. and J.W. Hummel. 1993a. Portable, near-infrared spectrophotometer for rapid soil analysis. *Transactions of the ASAE* 36: 85-193.

Sudduth, K.A. and J.W. Hummel. 1993b. Soil organic matter, CEC, and moisture sensing with a portable NIR spectrophotometer. *Transactions of the ASAE* 36: 1571-1582.

Windham, W.R., S.L. Fales and C.S. Hoveland. 1988. Analysis for tannin concentration in *Sericea lespedeza* by near infrared reflectance spectroscopy. *Crop Sci.* 28:705-708.

Wright, J.A., M.D. Birkett and M.J.T. Gambino. 1990. Predictions of pulp yield and cellulose content from wood samples using near infrared reflectance spectroscopy. *Tappi J.* 73:164-166.

CHAPTER 26

Development of Rapid Instrumental Methods for Measuring Soil Organic Carbon

G.W. McCarty and J.B. Reeves III

I. Introduction

Evidence for global climate change has intensified international efforts toward reducing anthropogenic CO_2 sources and increasing sink capacities of the terrestrial biosphere. With agricultural ecosystems comprising an important part of the terrestrial biosphere, assessment of sink/source relationships within these ecosystems is essential for assessing sink capacity of the global biosphere. Agricultural ecosystems are thought to have potential as substantial CO_2 sinks when management practices are optimized for carbon (C) sequestration (Lal et al., 1998). The U.S. has proposed regulating CO_2 emissions by setting up a "carbon economy" whereby producers and consumers of CO_2 can buy and sell C credits. A similar economy setup for sulfur (S) emissions, has been credited with saving 90% of the initial estimated cost for S emission reduction, thereby demonstrating the efficiency of market-based systems for regulation (Kerr, 1998). Within a carbon economy, agricultural production systems could earn credits and make substantial contributions towards meeting net emission goals. It is unclear how verifiable credits would be applied to agricultural ecosystems and many issues and uncertainties need to be resolved concerning the proper system for total accounting of sinks and sources in agriculture and assigning C sequestration credits to production systems (Schlesinger, 1999). It is clear that development of a system for assigning credits requires greater understanding of the ability of agricultural systems to store C and metrics for verifying C storage in the landscape. The difficulty of this task is increased by the typically large spatial structure of organic C within agricultural landscapes.

The CENTURY model for soil organic matter (Metherell et al., 1993) provides an important framework for predicting changes in soil organic C within agricultural ecosystems and provides a set of parameters used in assessing the influence of production management on soil C storage. For example, the model includes the lignin content of plant residue as an important soil parameter for predicting C sequestration within ecosystems, but limited information is known about fate and transformation of plant residue constituents such as lignin with the formation of stable soil organic matter in agricultural ecosystems. Assessing the status and fate of residue lignin can provide important information about the tendency of the ecosystem to store organic C. Classical chemical methods for assessing composition of soil organic matter are cumbersome and largely arbitrary in their type of characterization. For example, it is difficult to directly relate the classical measurements of humic and fulvic acid to ability of soil to sequester C. New techniques are needed to provide more relevant assessments of compositional changes in soil organic matter in different ecosystems. Analytical pyrolysis has been shown to be a powerful tool for assessing composition of complex organic materials such as forages and wood including information on lignin content (Caballero et al., 1997; Reeves,

371

Table 1. Statistics on organic C content of soil samples collected from plots under plow and no-tillage management

Source	Mean	Range	Std
All samples (n = 179)	13.3	6.13 - 33.9	4.6
No tillage (n = 90)	14.9	6.13 - 33.9	5.9
Plow tillage (n = 89)	11.8	6.67 - 15.2	1.6

1990; Rodrigues et al., 1999). With the ability to detect lignin, pyrolysis may provide the means to measure important changes in soil organic matter quality such as lignin composition and transformation during formation of stable soil C. Evaluation of such parameters will likely be relevant in assessments of the potential for different soil ecosystems to sequester C.

We present evaluations of near infrared reflectance spectroscopy (NIRS) and pyrolysis-GC/MS as methods for rapid measurement of soil organic C, and discuss their potential use for assessing storage and sequestration of C in ecosystems.

II. Materials and Methods

The soil samples studied were obtained from long-term field experiments located in the Piedmont region near the town of Clarksville in central Maryland, and in the coastal plain region near the town of Queenstown on the Delmarva Peninsula of Maryland. The field plots in the Piedmont region were established in 1976 on well-drained Delanco silt loam (Aquic hapludult) and those in the coastal plain region were established in 1973 on a somewhat poorly drained Bertie silt loam (Aquic Hapludult). The plots studied at each location have been under annual treatments of plow and no tillage and different amounts of N fertilizer and maintained in a continuous culture of corn [*Zea mays* (L.)]. Soil cores from each plot were segmented into depth intervals of 0 to 2.5, 2.5 to 5.0, 5.0 to 7.5, 7.5 to 12.5, and 12.5 to 20 cm. Paired samples of soil from central Iowa (Webster series) were also collected from land under long-term plow tillage (34 mg C g^{-1} soil) and adjacent land under tall grass prairie vegetation (60 mg C g^{-1} soil). Before analysis, soil samples were ground with an IKA 10A Mill fitted with a tungsten blade (Tekmar-Dohrmann, Cincinnati, OH). Samples were analyzed for organic C by combustion using a Leco CNS-2000 Elemental Analyzer (Leco Corp., St. Joseph, MI).

Near infrared reflectance spectroscopy (NIRS) analysis of soil samples was performed by scanning from 1100 to 2498 nm on a FOSS-NIRSystems model 6250 scanning monochromator (FOSS-NIRSystems, Silver Spring, MD) equipped with a rotating sample cup. All analysis of NIRS data was performed by partial least squares regression (PLS) using Grams/386 PLSPlus V2.1G (Galactic Industries Corp., Salem, NH). Preliminary efforts using a variety of data subsets, spectral data point averaging, derivatives and the data pre-treatments (mean centering, variance scaling, multiplicative scatter correction and baseline correction) showed that the use of a first derivative pre-treatment with a gap of 16 data points produced the best calibration. Table 1 provides summary statistics for organic C contents of soils used for NIRS evaluation.

Pyrolysis gas chromatography/ mass spectral (Py-GC/MS) analysis was performed using a Chemical Data System Pyroprobe model 2000 equipped with an AS-2500 Pyrolysis Autosampler (Chemical Data System Inc, Oxford, PA). Pyrolysis was carried out in quartz tubes containing samples with 250 μg of organic C using a temperature ramp from 200 to 1000°C (40°C s^{-1}) and a 20 s hold at 1000°C. The pyrolyzate was swept directly into a Finnigan MAT GCQ gas chromatograph coupled

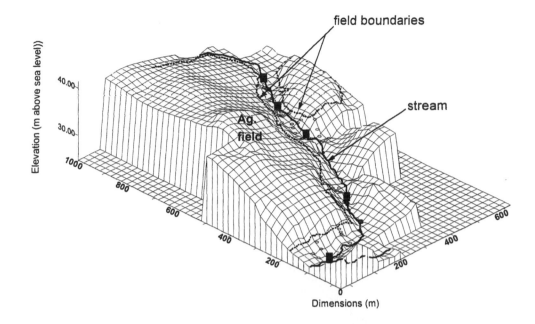

Figure 1. Agricultural ecosystem containing production fields and riparian system.

to a Finnigan MAT GCQ ion trap mass spectrometer (Finnigan MAT, San Jose, CA). The gas chromatographic column was DB-5ms (J&W Scientific, Inc., Folsom, CA) (30 m × 0.25 mm i.d., 0.25 μm film thickness) operated from 50 to 300°C at 5°C min⁻¹ holding the initial temperature for 10 min. The injector temperature was 300°C. The Py GC interface temperature was 250°C. Carrier gas (He) was held at a constant velocity of 40 cm sec⁻¹ under splitless conditions. The GC/MS transfer line temperature was 300°C and the ion source temperature was 200°C. Mass spectra were obtained by electron impact at 70 eV from 40 to 650 h/z (1 scan sec⁻¹). Peak identification was based on mass spectral interpretation and published libraries of mass spectra of lignocellulose pyrolyzates (9 to 13).

III. Results and Discussion

A typical agricultural ecosystem may consist of production fields in close contact with a riparian system as shown by the watershed in Figure 1. Management of this watershed for C sequestration should include assessment of the entire ecosystem including the riparian/wetland systems because of the major impact agricultural activities can have on wetland function and the potential for redistribution of C resources within the watershed through erosion. The spatial structure of soil organic C in agricultural ecosystems can be immense. In this small watershed, for example, soil organic C contents range from 150 g C kg⁻¹ soil in the wooded wetland soil to 10 g kg⁻¹ in upland slopes.

Current technologies have very limited ability to accurately assess the amount of C stored in organic matter within the landscape. With conventional methods of analysis, a substantial amount of resource is invested in making relatively few measurements of soil organic C. This leads to a tendency of under-sampling the landscape for accurate measurement of spatial structure. The other approach is to make many more less accurate measures of soil organic C thereby providing better assessment of spatial structure. It seems clear that for typical landscapes the greatest potential error

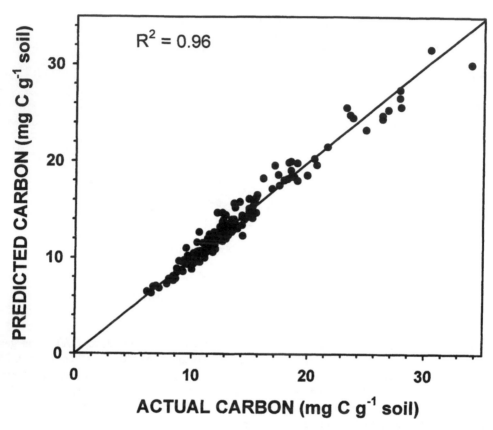

Figure 2. NIRS calibration for organic C in soil (n = 179). (Adapted from Reeves et al., 1999a.)

would be associated with the spatial assessment. Near-infrared spectroscopic analysis of soils has been shown to provide rapid and largely accurate measurement of soil organic C (Reeves and McCarty, 1999) and to provide a better means for assessing spatial structure of soil C.

A. Near Infrared Analysis for Determining Amount of Soil Organic C

First-derivative spectral data for soil in the range 1100 to 2500 nm were used to develop NIRS calibrations of soil organic C. Comparison of the derivative data for two soils with large relative differences in organic C provide indication of the different regions of the spectrum with sensitivity to soil organic C content (Figure 2). An overall calibration for soil organic C (Figure 3) was developed using soil samples collected from the long-term tillage experiments across factors of geographic location, tillage, N fertilization, and depth of sampling. The use of samples across these factors afforded the ability to determine the influence of these different factors on robustness of calibrations.

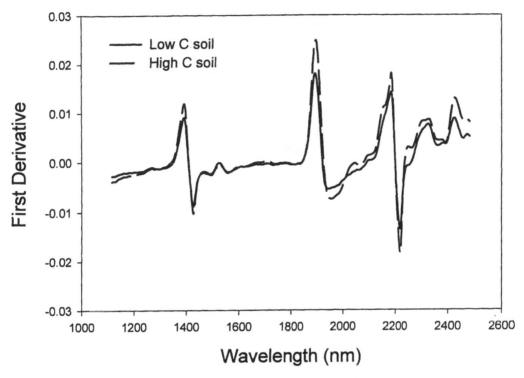

Figure 3. First derivative of near infrared spectra for soils with the greatest difference in organic C content within the set of samples studied. (Adapted from Reeves et al., 1999a.)

These results demonstrate that good within-field and within-region calibrations can be developed for estimating organic C of soil spanning across soil management factors.

The field sites under study were located in the Coastal Plain and Piedmont region of Maryland separated by approximately 150 km. Although the soils sampled from both geographic areas had silt loam textures, mineralogy of their parent material was significantly different, with the Piedmont soils containing visible forms of mica while absent in the other. Apparently, the spectral analysis could discern differences in parent material and discriminant analysis of data provided complete segregation of soils from the two field sites (data not shown). Use of a calibration based solely on the 90 soil samples taken from the Coastal Plain site produced a notably biased estimate of organic C in the Piedmont soils (Figure 4a). Addition of eight samples from the Piedmont region to the original calibration set containing the 90 Coastal Plain soils largely eliminated the bias (Figure 4b). This illustrates the considerable ability in NIRS to adjust calibration based on new conditions. In this case only a few samples from the second field would have to be analyzed by chemical analysis. Various strategies are available for development and maintenance of calibrations including use of spectral diversity assessments as the basis for selecting samples in developing the calibration set. Additional work is needed to evaluate the utility of more universal calibrations for soil C based on regional or national collections of soil to provide soil diversity. Short of a universal calibration, however, localized calibrations appear to have good utility. The use of NIRS in developing spatial maps of soil organic C based on within-field calibrations is being explored using the watershed shown in Figure 1 (Reeves and McCarty, 1999). The concept is that a small subset of the total soil samples can be

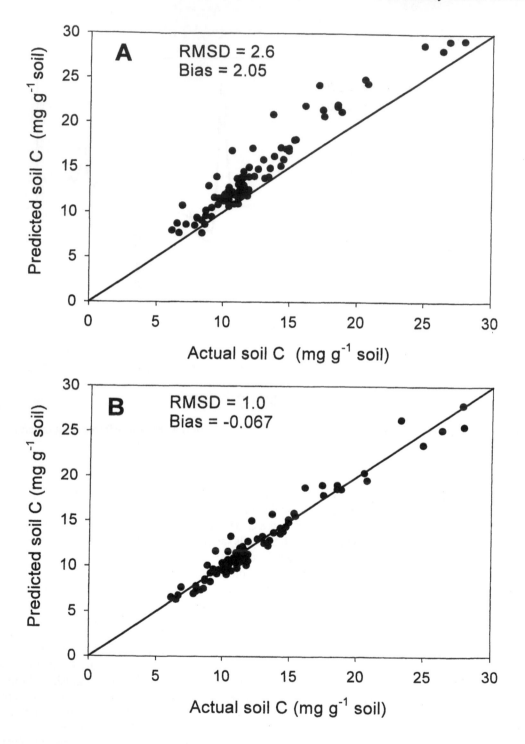

Figure 4. Assessment of robustness for NIRS calibrations for soil organic C which was developed within one geographic region and applied to another region. (A) Estimation of Piedmont soil samples based on a calibration containing the 90 samples of Coastal Plain soil; (B) Estimation of Piedmont soil samples based on calibration containing the 90 Coastal Plain soils and 8 samples from the Piedmont.

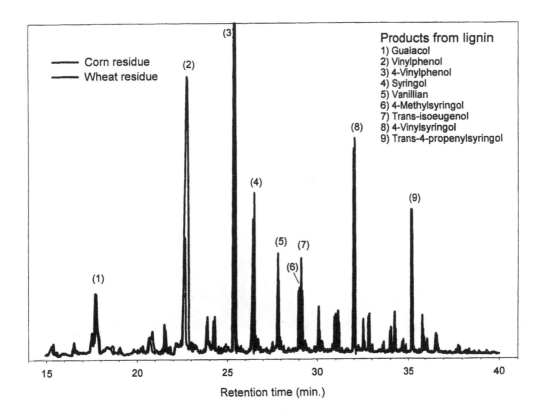

Figure 5. Identification of the pyrolytic products formed from lignin in corn and wheat residue.

chemically analyzed to generate the calibration for soils of the field to be mapped. This could include the use of spectral diversity assessments by NIRS to select the set of local calibration soils. As indicated by work with the Coastal Plain and Piedmont sets of soil, the majority of a calibration set may be comprised of regional soil samples, and bias, due to local conditions, may be eliminated by including a small subset of local soils in the calibration.

B. Analytical Pyrolysis for Assessing Composition of Soil Organic C

Modern pyrolysis-GC/MS instruments provide sufficient resolution and sensitivity for quantifying and identifying hundreds of products formed with pyrolysis of the complex organic C structures in soil. With proper sample preparation and analytical protocols, pyrograms of soil are highly reproducible in terms of retention time, peak areas and patterns (Reeves et al., 1999b). Comparing the pyrograms for surface samples (0 to 2.5 cm) of soil under no-tillage and plow-tillage management (Figure 5) illustrates the differences in organic matter composition detected by pyrolysis. By performing the analysis on an equivalent C basis, the pyrograms provide direct assessment of the compositional differences in soil organic matter independent of differences in total C content. In this example, the C content of the soil samples ranged from 34 mg C g^{-1} soil (no-till) to 13 mg C g^{-1} soil (plow-till) and the analysis was performed using 250 µg of soil C.

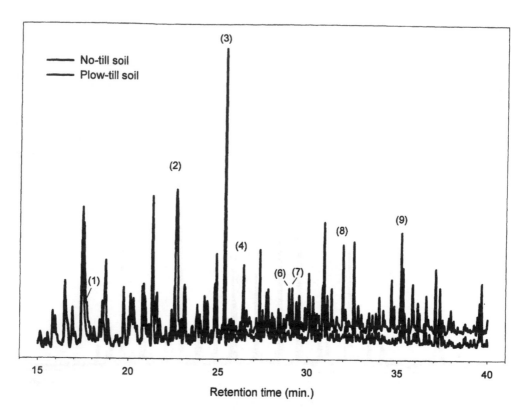

Figure 6. Pyrograms for surface samples (0 to 2.5 cm) of soil under plow and no-tillage management.

It is well established that pyrolytic analysis gives a clear "fingerprint" for lignins in organic materials with the production of highly characteristic signature products (Caballero et al., 1997; Dijkstra et al., 1998; Reeves, 1990; Rodrigues et al., 1999). Pyrograms for wheat and corn residues (Figure 6) provide identification of the abundant signature products formed with lignin pyrolysis. Many of these signature products are present in soil under no tillage, but are absent or reduced under plow tillage (Figure 5). The lignin signature observed in the top 2.5 cm of no-till soil attenuates with depth and is absent within the 12.5 to 20 cm depth interval (not shown). This likely reflects differences in both amount and degree of decomposition of plant residues associated with soil organic matter. Previous work provided evidence for compositional differences in soil organic matter within the profile of no-till soil as crudely reflected by stratification of the C:N ratio within profile (McCarty and Meisinger, 1997; McCarty et al., 1998). Pyrolytic analysis can provide a more refined assessment of compositional stratification. Comparison of a tall grass prairie soil and nearby cultivated soils from Iowa (Figure 7) also demonstrated compositional differences in the soil organic matter associated with these paired ecosystems. The strong lignin signature associated with the prairie grassland soil is greatly attenuated with cultivation.

The pyrolytic products from no-till and grassland soils are indicative of significant lignin pools in these soils. Given evidence that soils under natural prairie vegetation and under no-tillage management have strong capacity to sequester C and the general conclusion that lignin is an important precursor for stabilized soil organic matter, it is likely that pyrolytic analysis provides a characterization of soil organic matter that is directly relevant to the capacity of ecosystems to sequester C. Analytical

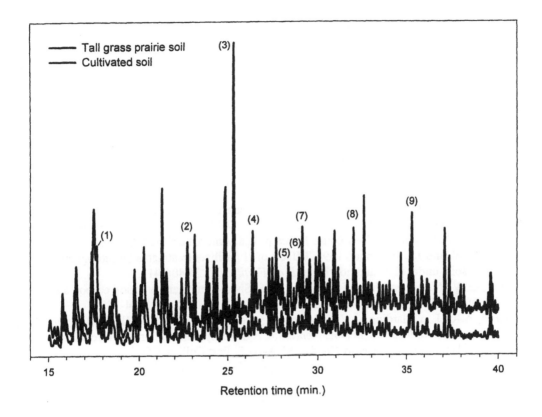

Figure 7. Pyrograms for a tall grass prairie soil and nearby cultivated soil from Iowa.

pyrolysis is also well suited for plant residue decomposition studies and can provide important insight to the processes involved with humification and soil organic matter stabilization and to the extent by which such processes differ in various ecosystems and management systems.

IV. Conclusions

The difficulties in accurately assessing C storage in agricultural ecosystems has provided increased impetus for new approaches and methodologies for measurement of C in soil. The extensive spatial structure of organic C in natural and agricultural ecosystems usually constrains the accuracy of C storage assessments for landscapes. A complex spatial structure will often lead to higher error associated with under-sampling the landscape compared with the analytical error associated with measurements. Tradeoffs may be necessary between the analytical accuracy and the speed and ease of analysis in order to increase sample handling capacity. Near infrared reflectance spectroscopy (NIRS) offers a simple and rapid procedure for measurement of soil organic C with some possible loss in accuracy for the individual analyses when compared to standard chemical methods. The feasibility of using NIRS for assessing spatial structure of soil C on a landscape is being evaluated by using within-field calibrations for organic C. The results have indicated that NIRS has good utility for assessing C storage in agricultural ecosystems. Pyrolysis-GC/MS is another instrumental method with good utility for rapid assessment of composition of complex organic matrices such as soil organic matter. This method provides a clear signature for lignin which is an important precursor for stabilized soil organic C. Initial studies indicate that some soil ecosystems that have been characterized with high capacity to store C also carry a strong lignin signature detected by pyrolysis. Characterization of soil ecosystems by these techniques may permit better assessment of C sequestration in ecosystems.

References

Caballero, J.A., R. Font and A. Marcilla. 1997. Pyrolysis of Kraft lignin: yields and correlations. *J. Anal. Appl. Pyrolysis* 39:161-183.

Dijkstra E.F., J.J. Boon and J.M. van Mourik. 1998. Analytical pyrolysis of a soil profile under Scots pine. *Eur. J. Soil Sci.* 49:295-304.

Kerr, R.A. 1998. Acid rain control: Success on the cheap. *Science* 282:1024-1027.

Lal, R., J.M. Kimble, R.F. Follett and C.V. Cole. 1998. *The Potential of U.S. Cropland to Sequester Carbon and Mitigate the Greenhouse Effect.* p. 128. Ann Arbor Press, Chelesa, MI.

McCarty, G.W. and J.J. Meisinger. 1997. Effects of N fertilizer treatments on biologically active N pools in soils under plow and no tillage. *Biol. Fertil. Soils* 24:406-412.

McCarty, G.W., N.N. Lyssenko and J.L. Starr. 1998. Short-term changes in soil carbon and nitrogen pools during tillage management transition. *Soil Sci. Soc. Am. J.* 62:1564-1571.

Metherell, A.K., L.A. Harding, C.V. Cole and W.J. Parton. 1993. *CENTURY Soil Organic Matter Model Environment Technical Documentation.* Agroecosystem version 4.0. Great Plains System Research Unit Technical Report No. 4. USDA-ARS, Fort Collins, CO.

Reeves III, J.B. 1990. Pyrolysis gas chromatography of five ruminant feedstuffs at various stages of growth. *J. Dairy Sci.* 73:2394-2403.

Reeves III, J.B., G.W. McCarty and J.J. Meisinger. 1999a. Near-infrared reflectance spectroscopy for analysis of organic matter in soils. *J. Near Infrared Spectrosc.* 7:179-193.

Reeves III, J.B., B.A. Francis and G.W. McCarty. 1999b. Effect of pyrolysis conditions on PY/GC/MS results of low organic matter soils and forages: Temperature, program, sample size, split ratio, and fingerprints. 50th Pittsburgh Conference, Orlando, FL, Abstr. 1291.

Reeves III, J.B. and G.W. McCarty. The potential of NIRS as a tool for spatial mapping of soil composition for use in precision agriculture. Proceedings of the 9th International Conference on Near-Infrared Spectroscopy, Verona, Italy (in press).

Rodrigues, J., D. Meier, O. Faix and H. Pereira. 1999. Determination of tree to tree variation in syringyl/guaiacyl ratio of Eucalyptus globulus wood lignin by analytical pyrolysis. *J. Anal. Appl. Pyrolysis* 48:121-128.

Schlesinger, W.H. 1999. Carbon sequestration in soils. *Science* 284:2095.

CHAPTER 27

Soil Quality Evaluations of Alternative and Conventional Management Systems in the Great Plains

A.M. Gajda, J.W. Doran, T.A. Kettler, B.J. Wienhold, J.L. Pikul, Jr. and C.A. Cambardella

I. Introduction

Concerns over environmental effects of intensive agriculture, such as soil organic matter loss and degradation of soil-water-air resources, have promoted consideration of change from the prevailing high-input agricultural systems to crop management systems that are more diverse, require less tillage and purchased inputs, and enhance biological diversity (Doran et al., 1996; Lupwayi et al., 1998). Soil health, productivity, and sustainability are affected by numerous complex interactions between physicochemical and biological factors. New monitoring approaches to determine simultaneous changes in soil physical, chemical and biological properties are needed to evaluate system function and sustainability of the soil resource. Soil organic matter (SOM) is a major determinant of soil condition and an important source of inorganic nutrients for plant production. Soil organic matter components that are responsive to management practices such as particulate organic matter (POM), mineralizable N, and microbial biomass need to be evaluated as soil quality attributes for assessing soil function and system sustainability (Jenkinson and Ladd, 1981; Doran, 1987; Jenkinson, 1988; Cambardella and Elliott, 1992; Sparling, 1992; Kaiser, 1994; Doran and Parkin, 1996). Particulate organic matter, that part of SOM greater than 0.050 mm in size, has been identified as the organic matter fraction most sensitive to changes in soil management practices (Cambardella and Elliott, 1992). Increases in surface SOM are closely associated with increases in microbial biomass in the surface 5 to 7.5 cm of soil for no-tillage management as compared to previously tilled systems (Doran, 1987; Gupta et al., 1994). Soil microbial biomass C and N contents, mineralizable C and N levels, and respiration may be useful measures and sensitive indicators of change in SOM dynamics (Anderson and Domsch, 1989; Sparling, 1992; Parkin et al., 1996).

The objective of this study was to measure a critical set of soil quality indicators, as contained in the minimum data set given by Doran and Parkin (1996), to determine if consistent relationships exist between SOM, POM, and C and N mineralization in long-term conventional and alternative crop management systems. Our hypotheses were: (i) soil under alternative management systems, where either tillage is reduced and/or crop rotation intensified, will have greater SOM, POM, and soil microbial biomass contents; and (ii) mineralizable N will be greater in soil under alternative rather than under conventional management systems.

II. Materials and Methods

Soil samples were collected from five sites located in four states in the Western Corn Belt and Great Plains of the U.S.A. The three experimental fields in North Dakota (ND), South Dakota (SD), and Colorado (CO) are managed by the USDA-ARS Experimental Research Stations located in those areas, whereas the long-term experimental fields at Sidney and Mead, Nebraska (NE) are collaborative projects with the University of Nebraska. Soil type and experimental description for tillage and crop rotation are given in Table 1. The soils ranged in texture from loam to clay loam.

Composite soil samples were taken in April or May of 1998 at depths of 0 to 7.5 cm (depth A) and 7.5 to 30 cm (depth B) from field replicates of long-term experimental trials. A sufficient number of core samples were taken to result in a minimum of 800 and 1400 g of soil from each treatment replicate for the A and B depths, respectively. The number of cores sampled per composite varied with probe tip diameter; 1.75 cm or 1.90 cm (A-30 cores; B-15-20 cores) and 3.1 cm (A-15 cores; B-15 cores). Sampling across each replicate was done in a random manner, but was stratified for between and within row locations according to their relative proportions for the respective tillage, cropping management, and vehicular traffic patterns at each location. More detailed discussion of stratified composite sampling for soil quality analyses is given by Dick et al. (1996) and Sarrantonio et al. (1996). At Brookings, SD all samples were taken from the in-row planting area. At other locations, between crop row areas were sampled from traffic and non-traffic areas such that the number of cores from each area reflected the proportion of between row areas which were comprised of wheel tracks. Immediately after field sampling, samples were placed into double plastic bags and stored at 4°C for analyses within one week.

A. Laboratory Analyses

Composite samples from each site were first weighed, then thoroughly mixed, and 13 to 15 g sub-samples were dried for 24 h at 105°C to determine soil water content. The bulk density of each sample was calculated by dividing its total oven-dry mass by the volume of soil collected in each composite sample. All soil analysis results were converted to a volumetric basis (Doran and Parkin, 1996), to reflect the actual condition sampled in the field, as follows:

Volumetric results (kg ha^{-1}-depth) =
Gravimetric results (μg g^{-1}) × soil bulk density (g cm^{-3}) × (cm depth of sample × 10^{-1}).

Prior to laboratory analyses each soil sample was sieved through a 4.75-mm sieve. The samples were then split into two parts: one part was sieved < 2.00 mm in field moist condition and was used for biologically sensitive analyses (described later), the other part was dried in a forced air oven at 55°C for 24 hours and then sieved < 2.00 mm. These sample splits are subsequently referred to as fresh and air-dried soil, respectively.

1. Physical and Chemical Analyses

Soil textural analyses were done on air-dried soil using the hydrometer method as described by Gee and Bauder (1986). Soil electrical conductivity (EC) and pH were done using fresh soil on a 1:1 (soil:water) mixture as described by Dahnke and Whitney (1988) and Eckert (1988). EC measurements were corrected for water content to an exact 1:1 soil:water basis, and were taken before pH to avoid erroneously high conductivity, which can result from salt leakage from pH electrodes. Soil NO_3- and NH_4-N contents were determined by extracting 10 g (fresh soil) of each soil sample with 100 ml

Table 1. Site and experimental descriptions for tillage and crop rotation for five experimental sites in the Great Plains

Site Soil series Classification	Location Lat. °N	Location Long. °W	Climate[a] MAT °C	Climate[a] MAP mm	Climate[a] PET(GS) mm	Cropping and tillage management comparison[b] Alternative	Cropping and tillage management comparison[b] Conventional	Fertilizer N rate Alternative kg ha^{-1}	Fertilizer N rate Conventional kg ha^{-1}	Study time to 1998 Years
Mandan, ND Temvik-Wilton silt loam (fine-loamy, mixed Typic Argiboroll)	47	101	4	410	1194(644)	SpW/WW/Sun No-till	SpW/fallow Conv. till	67	22	14
Brookings, SD Barnes clay loam (fine-loamy, mixed Udic Haploboroll)	44	97	6	581	1180(567)	Corn/Soy/SpW/Alf Conventional till	Continuous corn Conv. till	0	113	8
Mead, NE Sharpsburg silty clay loam (fine, montmorillonitic, mesic, Typic Argiudoll)	41	97	10	745	1275(578)	Corn/Soy/Sorg/ Oat-cl Disk-reduced till	Continuous corn Disk	90/34	180	16
Akron, CO Weld silt loam (fine, smectite, mesic Aridic Agriustolls)	40	103	9	412	1983(733)	WW/corn/millet No-till	WW/fallow Sweep plow	67	67	9
Sidney, NE Duroc loam (fine-silty, mixed, mesic Pachic Haplustolls)	41	103	9	446	1891(650)	WW/fallow No-till	WW/fallow Moldboard plow	0	0	27

[a]MAT, mean annual temperature; MAP, mean annual precipitation; PET, annual potential evapotranspiration (average growing season ET)
[b]Cropping designations; SpW, spring wheat; WW, winter wheat; Sun, sunflower; Soy, soybean; Srg, sorghum; Alf, alfalfa; Oat-cl, oats-clover.

of $2M$ KCl on a reciprocal shaker for 30 min and filtering (Keeney and Nelson, 1982). Nitrate in the filtered soil extracts was reduced to nitrite with copperized cadmium, and quantified using the Griess-Ilosvay method (Keeney and Nelson, 1982), while ammonium was quantified with the indophenol blue procedure (Keeney and Nelson, 1982). Both nitrate and ammonium ions were quantified using a Lachat FIA auto-analyzer (Zellweger Analytics, Lachat Instrument Division, 6645 W. Mill Road, Milwaukee, WI, 53218).[1]

2. Organic C, Total N, Loss-on-Ignition OM, and POM

Total soil C and N contents of air-dried soil (<2mm), which was ground to pass a 0.5-mm sieve, were determined by the Dumas dry combustion technique using an automated Carlo Erba NA 100 CN analyzer (CE Elantech, Inc., 170 Oberlin Ave. North, Lakewood, NJ, 08701)[1] as described by Schepers et al. (1989). Total C and N results were converted from air-dry to an oven-dry basis. Soil organic C contents were calculated by subtracting carbonate C contents from total C values when soil pH values exceeded 7.2, which indicates the presence of carbonates. Soil organic matter (SOM) by weight loss-on-ignition (LOI) was done as described by Schulte (1988) but modified for a higher combustion temperature as discussed by Combs and Nathan (1998). Briefly, 10 g of air-dried soil (<2 mm) was placed in a pre-ignited (450°C) aluminum weighing pan and heated in an oven at 105°C overnight. The sample was cooled in a desiccator, weighed to three decimal places and ignited in a muffle furnace at 450°C for 4 h. After ignition, samples were cooled in a desiccator containing dried ascarite and then weighed to three decimal places. The weight loss-on-ignition (LOI) SOM content was calculated as the difference in weight between the initial dry weight of soil and weight after ignition for 4 h at 450°C divided by the oven-dry (105°C) weight of soil. An advantage of LOI procedures for organic matter is that samples do not require adjustment for carbonates which are not lost by heating at temperatures below 500°C (Combs and Nathan, 1998).

Particulate Organic Matter (POM) was measured using two methods: (i) that originally described by Cambardella and Elliott (1992), which utilizes a chemical dispersion of soil followed by a physical separation based upon particle size and density to isolate the POM fraction from soil and (ii) POM-LOI method, a modification of the Cambardella and Elliott (1992) procedure to facilitate organic matter analyses by weight loss-on-ignition (LOI) as detailed by Schulte (1988) and described previously. The POM-LOI procedure was modified to use 30 g of air-dried soil dispersed in 90 ml of 5 g L^{-1} sodium hexametaphosphate by shaking for 15 h on a reciprocal shaker. Dispersed soils were passed through 500- and 53-μm nested sieves to isolate two fractions (0.053 to 0.5 mm and 0.5 to 2.0 mm). Each sieved fraction was rinsed several times with tap water until the rinsate was clear. Material retained on sieves was quantitatively transferred into pre-weighed and pre-conditioned (450°C) aluminum pans, oven dried at 50 to 55°C, and weighed. Separated material was placed in a muffle furnace at 450°C for 4 h. After ignition, samples were cooled in a desiccator containing ascarite and weighed. The POM fraction was calculated as the difference of sample weight before and after ignition as described earlier for SOM by LOI. Sand content can also be determined by this procedure by weighing the ash free material retained on the 53- and 500-μm sieves. Cambardella et al. (1999) reported that POM measured by LOI method at 450°C was a sensitive indicator of soil quality.

---

[1] Mention of a trademark, proprietary product, or vendor does not constitute a guarantee or warranty of the product by USDA nor imply its approval to the exclusion of other products that may be suitable.

3. Potentially Mineralizable N

Potentially mineralizable N was estimated using two methods: (i) Ammonium-N produced under anaerobic incubation (AI) as described by Waring and Bremner (1964) and modified by Keeney (1982), and (ii) a hot KCl extraction method (HKCl) as described originally by Gianello and Bremner (1986).

Samples for the AI method were prepared by adding 5 g of fresh soil (<2 mm) and 20.0 ml of distilled water to a 16 × 150-mm test tube and mixing to remove air pockets. Distilled water (1 to 2 ml) was then added to minimize headspace and tubes and stoppered tubes were incubated at 40°C for 7 days. After 7 days, samples were quantitatively transferred into tared 113 ml (4 oz) extraction vials using three 5- to 10-ml portions of 3.33 M KCl and the total volume of solution brought to 50.0 ml with 3.33 M KCl by weighing (soil + 2 M KCL = 59.5 g). The resultant 2 M KCl suspensions were shaken for 30 min, filtered, and analyzed for NH_4-N and NO_3-N as described earlier.

Samples for the hot KCl extraction method (HKCl) were prepared by mixing 3.0 g fresh soil (oven-dry equivalent) and 20 ml of 2.0 M KCl in 100-ml Pyrex digestion tubes. Digestion tubes were stoppered and placed in a block digestor at 100°C for 4 h. Three tubes with 20.0 ml of 2.0 M KCl only were used as reagent controls. After digestion tubes were removed from the block digestor and cooled to room temperature. Initial NH_4- and NO_3-N were determined by preparing as above and incubating them at room temperature for 4 h with occasional shaking. When samples had cooled to room temperature the samples were filtered through Whatman No.1 filter paper which had been washed three times (twice with deionized water and once with 2.0 M KCl) to remove absorbed NO_3 and NH_4. Concentration of NO_3- and NH_4-N in extracts was determined as described earlier.

4. Soil Microbial Biomass and Mineralizable C and N

Soil microbial biomass C was determined using a modification of the chloroform-fumigation incubation (CFI) technique (Jenkinson and Powlson, 1976). The procedure was modified to use 50-g portions of fresh soil in 100-ml glass beakers. Both fumigated and nonfumigated control samples were inoculated with 1.0 g moist nonfumigated soil composited from all treatments at each location. Inoculated soil samples were gently tamped to a bulk density of 1.1 g cm^{-3}, adjusted to 55% water-filled pore space (WFPS) as described by Linn and Doran (1984a), placed in 1.9-L glass canning jars, and sealed with lids fitted with rubber sampling ports. The samples were incubated for 10 days at 25°C at which time 1-ml samples were removed from the jar headspace for determination of CO_2 by gas chromatography as described by Linn and Doran (1984b). Incubation jars for nonfumigated controls were flushed with air to remove accumulated CO_2, resealed, and incubated for another 10 days at 25°C, at which time headspace CO_2 was again measured. Microbial biomass C was calculated by dividing the flush of CO_2-C (CO_2-C from fumigated samples during 0 to 10 days minus 10- to 20-day CO_2-C from nonfumigated controls) resulting from chloroform fumigation by a conversion factor (K_C) of 0.45 as described by Jenkinson (1988).

Soil respiration (C carbon mineralization) and N mineralization for all soils were determined using control samples from the microbial biomass procedure which were incubated at 25 °C over a 20-day time period. These data represent a mineralization potential for disturbed soils (mixed) under optimal moisture (55% WFPS) conditions. Soil NO_3- and NH_4-N levels at 10 and 20 days were compared to initial values using analysis procedures described earlier.

Table 2. Soil texture and bulk density for alternative and conventional management systems at five sites in the Great Plains

Site	Soil depth	Sand		Silt		Clay		Soil bulk density	
		ALT	CONV	ALT	CONV	ALT	CONV	ALT	CONV
	(cm)	———%, g 100g^{-1}———						——g cm^{-3}——	
Mandan, ND									
	0 – 7.5	22	23	51	49	27	28	1.37	1.30
	7.5 – 30	22	21	48	47	30	31	1.31	1.40[a]
	0 – 30	22	22	49	48	29	30	1.32	1.38
Brookings, SD									
	0 – 7.5	46	47	26	26	28	27	1.24	1.25
	7.5 – 30	45	46	27	26	29	28	1.63	1.60
	0 – 30	45	46	27	26	29	28	1.53	1.51
Mead, NE									
	0 – 7.5	11	11	52	49	37	39	1.13	1.24[a]
	7.5 – 30	11	12	51	48	38	40	1.39	1.48[a]
	0 – 30	11	12	51	48	38	40	1.33	1.42[b]
Akron, CO									
	0 – 7.5	37	39	41	39	22	23	1.39	1.31
	7.5 – 30	29	28	39	36	32	36	1.45	1.44
	0 – 30	31	31	40	37	30	33	1.43	1.40
Sidney, NE									
	0 – 7.5	22	41	45	34	33	25	1.12	1.26[b]
	7.5 – 30	46	45	34	34	20	21	1.23	1.30
	0 – 30	40	44	37	34	23	22	1.20	1.29[a]

Means, within location and depth, beween alternative and conventional management differ significantly at [a]P<0.20 and [b]P<0.10.

B. Statistical Analyses

Statistical analysis of all soil parameters was done using SAS (1997), proc mixed. The analysis of variance (ANOVA) was conducted as a multilocation trial, and was performed by sampling depth. Comparisons for the 0- to 30-cm depth range were made using averages or sums, within replicates, for the 0- to 7.5-cm and 7.5- to 30-cm depths. The experimental model for all sites except Sidney NE was a randomized complete block design with three replicates and two treatments. At the Sidney NE location, the experimental model was a randomized complete block design with two replicates and two treatments. Exact procedures followed are described in Littell (1996).

Table 3. Soil pH, electrical conductivity (EC), and nitrate and ammonium N contents for alternative and conventional management systems at five sites in the Great Plains

Site	Soil depth	pH (1:1) ALT	CONV	EC (1:1) ALT	CONV	NO₃-N ALT	CONV	NH₄N ALT	CONV
	(cm)			— dS m⁻¹ —				kg ha⁻¹	
Mandan, ND									
	0 – 7.5	5.67	5.89	0.44	0.17^d	13	3^d	17	0.4^d
	7.5 – 30	5.98	6.27	0.29	0.16^d	10	3^c	2	0.4
	0 – 30	5.82	6.18	0.33	0.16^d	23	6^c	19	1
Brookings, SD									
	0 – 7.5	5.91	6.20	0.36	0.40	22	14^d	1	16^d
	7.5 – 30	6.98	6.47^a	0.25	0.28	34	38	2	33^b
	0 – 30	6.71	6.41	0.28	0.31	56	52	3	49^c
Mead, NE									
	0 – 7.5	6.49	5.60^b	0.29	0.26	8	8	3	2
	7.5 – 30	6.64	6.44	0.19	0.22	8	9	4	6
	0 – 30	6.61	6.23	0.21	0.23	16	18	7	9
Akron, CO									
	0 – 7.5	5.59	6.12	0.31	0.29	12	13	1	1
	7.5 – 30	6.46	7.42^c	0.18	0.33^d	21	37^b	3	13
	0 – 30	6.24	7.08^c	0.21	0.32^d	33	50^a	4	14
Sidney, NE									
	0 – 7.5	6.09	6.44	0.30	0.18^c	12	6^b	1	0.6
	7.5 – 30	6.57	6.50	0.15	0.16	18	19	5	5
	0 – 30	6.45	6.48	0.19	0.16	30	25	6	6

Means, within location and depth, beween alternative and conventional management differ significantly at [a]P<0.20, [b]P<0.10, [c]P<0.05, and [d]P<0.01.

III. Results and Discussion

A. Physical and Chemical Properties

Soil particle size analyses for alternative and conventional management systems at the five locations differed little, except for surface soil (0 to 7.5 cm) in wheat/fallow at Sidney, NE where the sand content was higher and silt content lower for the conventional moldboard plow treatment as compared with no-tillage (Table 2). This difference has been explained by the mixing of coarser materials from depth and the greater loss of fines due to erosion with moldboard plow tillage (O'Halloran et al., 1987). Soil bulk densities differed by 5 to 13%, mainly at locations where tillage management practices varied between management comparisons. At Mandan, ND and Akron, CO surface soil bulk density was 5 to 6% higher for no-tillage management; however, at Sidney, NE density was 13% lower for no-tillage, apparently due to a significantly higher soil organic matter (SOM) content and a lower sand content (Doran et al., 1998). Pikul and Allmaras (1986) demonstrated an inverse relationship between soil bulk density and organic C content in varying crop residue management

Mineral N vs. EC (1:1), 0 to 7.5 cm

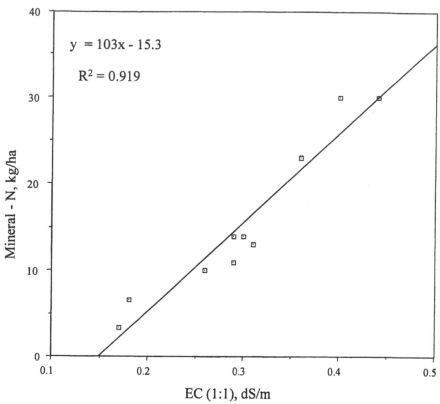

Figure 1. Relationship between EC (1:1) and mineral N (NH_4- and NO_3-N) in the surface soil (0 to 7.5 cm) for alternative and conventional management practices at five sites in the Great Plains.

systems in the semi-arid Pacific Northwest. Thus, for accurate assessment of soil quality and the effects of management on soil nutrient concentrations and carbon storage, soil analyses must be converted from a gravimetric to a volumetric basis using the soil bulk density at time of sampling. In the present study, failure to adjust analytical results to a volumetric basis would have resulted in a relative error in comparison of management systems, at four of five sites, of from 5 to 13%.

In general, soil pH (1:1) values across sites remained within the optimum range for crop growth of 6 to 7.5 except at Mandan and Mead where the pH in the in 0 to 7.5 cm depth soil under alternative and conventional management, respectively, dropped to 5.7 and 5.6 (Table 3). Soil pH differences between management systems likely resulted from differential application and nitrification of mineral N fertilizers and inherent soil properties, such as depth to calcium carbonates (pH>7.2) in soil. However, as discussed by Smith and Doran (1996), soil acidification is an indicator of inefficient use of fertilizer N, leaching of NO_3 from the rooting zone, and of C loss to the atmosphere as CO_2 in calcareous soils. Soil EC(1:1) values ranged from 0.15 to 0.44 dS m^{-1} across all sites, which is within the range considered non-saline and optimal for growth and activity of microorganisms and most crop plants (Table 3). Soil EC differed significantly between the management systems at all sites except Brookings, SD and Mead, NE. Differences in NO_3-N and NH_4-N concentrations between alternative

Table 4. Soil (<2 mm) organic C, total N and organic matter by loss on ignition (450°C) for alternative and conventional management systems at five sites in the Great Plains

Site	Soil depth	Organic C		Total-N		OM (LOI)	
		ALT	CONV	ALT	CONV	ALT	CONV
	(cm)			—Mg ha-depth^{-1} —			
Mandan, ND							
	0 – 7.5	25.9	19.1^d	2.37	1.73^d	60.4	48.4^d
	7.5 – 30	55.8	48.5	5.18	4.56^a	144.5	137.1
	0 – 30	81.7	67.6^b	7.55	6.29^d	204.9	185.5^c
Brookings, SD							
	0 – 7.5	15.9	18.4^c	1.59	1.76^c	45.1	45.7
	7.5 – 30	34.7	48.2^c	3.82	4.67^c	143.2	151.2
	0 – 30	50.6	66.6^c	5.41	6.43^c	188.3	196.9
Mead, NE							
	0 – 7.5	16.4	16.2	1.59	1.58	44.0	44.6
	7.5 – 30	39.2	35.3	3.89	3.68	127.0	124.5
	0 – 30	55.6	51.5	5.48	5.26	171.0	169.1
Akron, CO							
	0 – 7.5	9.9	7.2^c	1.04	0.86^c	31.2	24.7^d
	7.5 – 30	22.4	19.1	2.72	2.68	93.6	94.4
	0 – 30	32.3	26.4	3.76	3.54	124.8	119.1
Sidney, NE							
	0 – 7.5	17.7	11.7^d	1.66	1.12^d	45.6	33.7^d
	7.5 – 30	27.7	31.1	2.77	2.97	85.2	98.4^a
	0 – 30	45.4	42.8	4.43	4.09	130.8	132.1

Means, within location and depth, beween alternative and conventional management differ significantly at ^{a}P<0.20, ^{b}P<0.10, ^{c}P<0.05, and ^{d}P<0.01.

and conventional management systems at all sites were similar to those observed for soil EC. As illustrated in Figure 1, there was a strong positive correlation ($r^2 = 0.92$) between total available mineral N (NO_3-N + NH_4-N) and soil EC for surface soil (0 to 7.5 cm) across all sites. A smaller but significant correlation ($r^2 = 0.55$) was observed for the 0 to 30 cm layer of soil. These results demonstrate the utility of using an easily measured property such as soil EC as an estimator of plant available N as discussed by Smith and Doran (1996). At the Brookings site, NH_4-N contents with conventional management were 16 times greater than those under alternative management at both soil depths. At Mandan, NH_4-N concentrations in the 0 to 7.5 cm layer with alternative management was over 40 times higher than that under conventional management. Examination of the management records for these sites suggests that these differences likely resulted from differential tillage practices and the amount, form, and timing of mineral N fertilizer applications.

B. Soil Organic C, N, and Organic Matter Pools

In most cases we measured greater total organic C and N contents in surface soil with alternative management (Table 4). Soil organic C and total N levels were 9 to 54% greater with alternative other

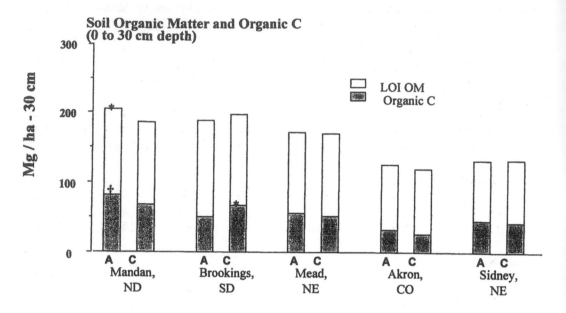

Figure 2. Organic matter by loss on ignition and organic C in the surface 30 cm of soil as related to alternative (A) and conventional (C) management practices at five sites in the Great Plains. Values between alternative and conventional management differ significantly at †P < 0.10 and *P < 0.05.

than conventional management at the Mandan, Akron, and Sidney locations where no tillage was the alternative tillage management practice. The greatest difference (16 to 54%) in C and N levels between alternative and conventional management comparisons occurred in the top 0 to 7.5 cm of soil. Quite different results were obtained from experimental fields at Brookings and Mead, where organic C and total N levels for conventional management were slightly higher (16 to 31%) or not different from these under alternative management. The differences observed between management systems at all locations were more pronounced for the 0 to 7.5 cm layer of soil than for the 0 to 30 cm layer. The C:N ratio was similar for alternative and conventional management systems and ranged from 7.5 to 10.8. The SOM contents of the five sites, as estimated by LOI, followed trends similar to those observed for organic C and total N. For the three locations where no-tillage was included under alternative management (Mandan, Akron, and Sidney), SOM levels in the 0 to 7.5 cm layer of soil averaged 25 to 45% greater than those for conventional management. No differences were observed at the other two sites. For the 0 to 30 cm sampling depth a statistical difference in SOM between management systems was only observed at Mandan although, as illustrated in Figure 2, the trends were similar to those observed for organic C. The use of less involved analytical procedures such as SOM by LOI may be helpful in initial pre-screening of agricultural management systems for their long-term effects on soil quality and soil C storage. It is important to note, however, that failure to express analysis results on a volumetric basis or to sample to below the depth of tillage can result in erroneous conclusions about the effectiveness of varying crop and soil management systems on soil carbon storage. Our general observations from the present survey, however, are similar to those of other researchers: that degree of soil tillage and cropping management practices that alter C inputs to, and losses from, soil are the major factors determining SOM levels and C storage in soil (Campbell and Zenter, 1997; Lesoing and Doran, 1997; and Paustian et al., 1997).

Table 5. Soil (<2 mm) particulate organic matter (POM)[1] fractions and % of loss-on-ignition OM for alternative and conventional management systems at five sites in the Great Plains

Site	Soil depth	POM size fractions						POM as % OM (LOI-450°C)	
		0.05 – 0.5 mm		0.5 – 2.0 mm		0.05 – 2.0 mm			
		ALT	CONV	ALT	CONV	ALT	CONV	ALT	CONV
	(cm)				Mg ha-depth[-1]				
Mandan, ND									
	0 – 7.5	7.7	5.2[d]	4.1	1.7[d]	11.8	6.9[d]	19.6	14.2[b]
	7.5 – 30	10.9	8.2	3.2	1.6	14.1	9.8	9.8	7.1
	0 – 30	18.6	13.3[a]	7.3	3.3[a]	25.9	16.6[c]	12.6	9.0[a]
Brookings, SD									
	0 – 7.5	3.8	3.0[a]	2.0	1.6	5.8	4.5[a]	12.7	9.9[a]
	7.5 – 30	18.5	8.9[a]	2.2	7.0[b]	20.7	15.9	14.4	10.5[a]
	0 – 30	22.2	11.9[a]	4.2	8.5[a]	26.4	20.4	14.0	10.4[a]
Mead, NE									
	0 – 7.5	5.9	4.6	2.2	2.5	8.1	7.1[a]	18.4	15.8[a]
	7.5 – 30	8.5	7.2	0.7	0.8	9.2	8.0	7.2	6.4
	0 – 30	14.4	11.8	2.9	3.3	17.3	15.1	10.1	8.9
Akron, CO									
	0 – 7.5	4.5	3.3[b]	2.2	0.9[d]	6.7	4.2[c]	21.4	16.9[a]
	7.5 – 30	5.8	6.1	0.6	0.6	6.4	6.7	6.8	7.1
	0 – 30	10.3	9.4	2.8	1.5[a]	13.1	10.9	10.5	9.1
Sidney, NE									
	0 – 7.5	7.0	4.5[c]	2.2	0.5[d]	9.2	5.0[d]	20.1	14.8[c]
	7.5 – 30	13.3	13.2	1.0	2.0	14.3	15.2	16.8	15.4
	0 – 30	20.3	17.7	3.2	2.5	23.5	20.2	17.9	15.3

[1]POM by loss-on-ignition method at 450°C for 4 h.
Means, within location and depth, beween alternative and conventional management differ significantly at [a]$P<0.20$, [b]$P<0.10$, [c]$P<0.05$, and [d]$P<0.01$.

The POM fraction (0.05 to 2.0 mm) in the 0 to 7.5 cm sampling layer averaged 1.5 times greater under alternative than under conventional management and was significantly greater at all sites, although trends were less pronounced at Brookings, SD and Mead, NE (Table 5). The greatest consistent differences in POM (0.5 to 2.0 and 0.053 to 0.5 mm fractions) were observed at the Mandan, Akron, and Sidney sites for the 0 to 7.5 cm soil layer. In both management systems POM levels at all sites decreased 2- to 3-fold with depth, and to a greater degree than soil OM which averaged 1.2- to 1.4-fold lower with depth. For surface soil (0 to 7.5 cm), POM generally comprised a significantly greater percentage of the SOM under alternative management compared to conventional management (Table 5). Particulate organic matter content in the 0.053 to 0.5 mm fraction in the 0 to 7.5 cm soil layer averaged 2.2 to 2.9 times greater than POM content in the 0.5 to 2.0 mm POM fraction under alternative and conventional management, respectively. The overall POM fraction (0.05 to 2.0 mm) appeared to be a sensitive indicator of differences in tillage and cropping management systems at the five study sites and although the differences were significant for

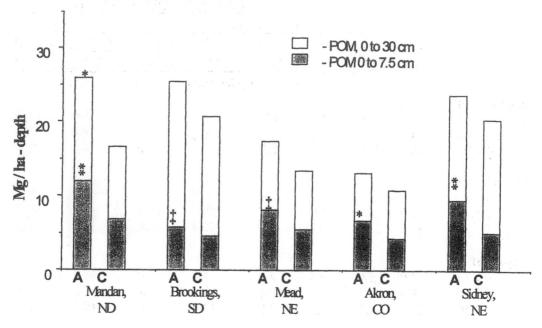

Figure 3. Particulate organic matter (0.5 to 2.0 mm) by loss on ignition in 0 to 7.5 and 0 to 30 cm soil depths as related to alternative (A) and conventional (C) management practices at five sites in the Great Plains. Values between alternative and conventional management differ significantly at $\ddagger P<0.20$, *$P<0.05$, and **$P<0.01$.

only one site in the 0-30 cm soil layer the trends observed were very similar to those observed for the 0 to 7.5 cm layer of surface soil, four of which were statistically significant (Figure 3). Our findings are in general agreement with those of Allmaras et al. (1988) who found that distributions of coarse organic matter fractions (0.5 to 2.0 mm) are sensitive indicators of crop residue incorporation as influenced by cropping and tillage management practices. Bolinder et al. (1999) also concur that macro-organic matter, the size fraction from 0.05 to 2.0 mm, is one of the most sensitive SOM attributes to conservation management practices in eastern Canada soils.

C. Soil Biomass and Mineralizable C and N Pools

Microbial biomass C levels in the 0 to 7.5 cm layer averaged 42% greater in alternative than in conventional management systems across all locations (Table 6). Differences in biomass C between management systems were less pronounced at the 7.5 to 30 cm sampling depth, except at Akron where the biomass C level under conventional management tended to be higher than that under alternative no-till management. These greater microbial biomass levels in alternative plots probably resulted from the larger production and accumulation of crop biomass with reduced tillage and crop rotations. Soil microbial biomass C concentrations at all locations and management systems decreased with depth, as was observed earlier for SOM and POM.

Table 6. Measures of biomass and potentially mineralizable N for alternative and conventional management systems at five sites in the Great Plains

Site	Soil depth	Microbial biomass C (N)[1]		Anaerobic NH$_4$-N release		Hot KCl mineral N release		Lab - 20d mineralizable N[2]	
		ALT	CONV	ALT	CONV	ALT	CONV	ALT	CONV
	(cm)			———kg ha-depth^{-1}———					
Mandan, ND									
	0 – 7.5	491(58)	292(34)^d	56	31^d	17	11^d	52	9
	7.5 – 30	732(86)	557(66)	58	29^d	25	22	24	13
	0 – 30	1223(144)	848(100)^b	113	60^d	42	34^b	76	21
Brookings, SD									
	0 – 7.5	430(51)	295(35)^c	34	21^a	12	12	10	5
	7.5 – 30	999(118)	828(97)	47	53	37	40	28	10
	0 – 30	1429(168)	1123(132)^a	81	74	49	53	38	15
Mead, NE									
	0 – 7.5	689(81)	294(35)^d	78	29^d	12	11	16	7
	7.5 – 30	703(83)	598(70)	35	20^b	19	16	18	12
	0 – 30	1392(164)	892(105)^c	113	49^d	30	28	34	19
Akron, CO									
	0 – 7.5	240(28)	168(20)^a	26	24	7	5^c	24	7
	7.5 – 30	444(52)	525(62)	27	8^c	13	10	21	13
	0 – 30	684(80)	693(82)	53	32^b	20	15	45	20
Sidney, NE									
	0 – 7.5	445(50)	217(26)^d	31	14^b	12	6^d	17	3
	7.5 – 30	649(76)	632(74)	23	36^a	18	21	8	19
	0 – 30	1094(129)	849(100)	56	51	31	27	25	22

[1]Biomass N in parenthesis; estimated from biomass C using a C:N ratio of 8.5:1.

[2]N mineralized in 20d = [(NO$_3^-$ + NH$_4$) 20 - (NO$_3$ + NH$_4$) initial].

Means, within location and depth, beween alternative and conventional management differ significantly at [a]P<0.20, [b]P<0.10, [c]P<0.05, and [d]P<0.01.

Changes in soil mineralizable-N levels are more dramatically influenced by cultural practices than are total C or N, thus differences are more easily detectable (Janzen, 1987; Campbell et al., 1991). In our study, soil potentially mineralizable N levels were significantly influenced by management practice at all sites (Table 6). Microbial biomass N in the 0 to 7.5 cm sampling depth ranged from 28 to 81 kg N ha-depth^{-1} in alternative and 20 to 35 kg N ha-depth^{-1} in conventional management across locations and averaged 60% greater in alternative than in conventional management. Biomass N contents in the 7.5 to 30 cm layer did not differ significantly at any sites. Potentially mineralizable N estimated using the anaerobic incubation method was 1.8 to 6.5 times higher for the 0 to 7.5 cm depth and up to 2.3 times higher for the 7.5-30 cm depth than that estimated by hot KCl extraction in both management systems. Potentially mineralizable N by anaerobic incubation for the 0 to 7.5 cm and 0 to 30 cm soil depths averaged 1.9 and 1.65 times greater, respectively, in alternative management systems than in conventional management systems. Similar, but larger differences, were

Biomass N versus Anaerobic N for Surface Soil (0 to 7.5 cm)

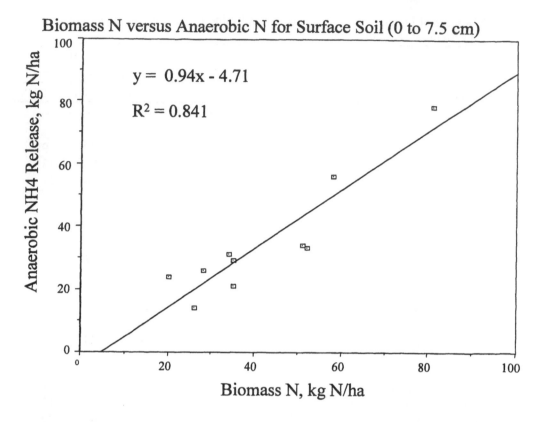

$y = 0.94x - 4.71$

$R^2 = 0.841$

Figure 4. Relationship between NH_4-N released on anaerobic incubation and microbial biomass N in surface soil (0 to 7.5 cm) for alternative and conventional management practices at five sites in the Great Plains.

observed for disturbed soil samples incubated under optimal conditions in the lab for 20 days where N mineralization with alternative management averaged 3.8 and 2.3 times greater than conventional management for the 0 to 7.5 cm and 0 to 30 cm soil layers. In this study, the anaerobic incubation method was a useful tool for assessing potentially mineralizable N levels in soil as influenced by tillage and cropping management systems. Not only does this method appear to be a sensitive indicator of tillage and cropping management but the analysis is less involved and less costly than other measures and is highly correlated with levels of microbial biomass N in surface soil (Figure 4).

As occurred for N, C mineralization potentials for conventional management were much lower than those for alternative and no-till management, likely due to greater crop residue, root mass, and OM accumulation in alternative management. Surface (0 to 7.5 cm) soil respiration, as measured in the laboratory, averaged 1.8 times greater for soil under alternative management than for that under conventional management and decreased in both magnitude and variation with depth (Table 7). Few or no differences in potential C mineralization occurred between management systems at the 7.5 to 30 cm soil depth. Levels of C mineralized from surface soil (0 to 7.5 cm) over 20 days averaged 1.7 times greater for alternative management than for conventional management. Differences were less pronounced at the 7.5 to 30 cm depth which resulted in a lower average increase of mineralized C for alternative versus conventional management for the 0 to 30 cm depth of 1.25-fold. The ratio of C

Table 7. Laboratory measured soil respiration rate, 20d mineralizable carbon, and mineralizable carbon to mineralizable nitrogen ratio for alternative and conventional management systems at five sites in the Great Plains

Site	Soil depth	Laboratory soil respiration rate[1]		Laboratory soil mineralizable C		Mineralizable C / mineralizable N[2]	
		ALT	CONV	ALT	CONV	ALT	CONV
	(cm)	kg CO_2-C/ha 2.25cm/d		kg CO_2-C/ha-depth/20d			
Mandan, ND							
	0 – 7.5	3.0 (4.3)	(3.0) 3.2	300	237	6	25
	7.5 – 30	1.4 (1.5)	(1.0 (3.2)	349	299	14	23
	0 – 30	1.8	(1.5)	649	536	12	24
Brookings, SD							
	0 – 7.5	4.5 (5.6)	2.9 (3.8)	371	247	37	34
	7.5 – 30	1.5 (2.5)	1.7 (2.0)	592	548	21	55
	0 – 30	2.3	2.0	963	795	25	50
Mead, NE							
	0 – 7.5	5.1 (8.7)	3.4 (4.5)	464	305	29	41
	7.5 – 30	1.1 (1.5)	0.8 (1.1)	343	264	19	23
	0 – 30	2.1	1.5	807	569	22	28
Akron, CO		2.9 (3.8)					
	0 – 7.5	0.7 (1.0)	2.1 (2.6)	279	195	11	27
	7.5 – 30	1.3	0.7 (1.1)	221	231	11	18
	0 – 30		1.1	500	426	11	20
Sidney, NE		3.7 (5.5)					
	0 – 7.5	1.0 (1.1)	1.4 (1.7)	352	121	21	40
	7.5 – 30	1.7	1.4 (1.6)	231	351	30	19
	0 – 30		1.4	583	472	28	24

[1]Measured over 10- to 20-day period of incubation, value in parentheses represents the 0- to 10-day period.
[2]Data for N mineralized after 20 days in the laboratory are in Table 6.

mineralized to N mineralized in 20 days was consistently lower for alternative as compared to conventional management averaging 20 and 30, respectively, across all sites and soil depths. Given that, in general, the organic C and N pools were larger in alternative versus conventional management, these results would imply that the chemical nature and susceptibility of these pools to turnover and loss with soil disturbance also varies with management and may be an indicator of long-term soil stability and C storage. As discussed by Parkin et al. (1996), the real value of such laboratory observations lies in the ability to demonstrate the mechanisms controlling C and N cycling in the field.

Table 8. System descriptions and coefficients of cropping intensity/cover, tillage, and carbon storage for tillage and crop rotation systems at five experimental sites in the Great Plains

Site/ Mgmt.	Crop system/ rotation[1]	Mos. in crop (nontilled fallow)	Crop intensity/cover coefficient	Tillage regime	Tillage coeff.	Carbon storage coefficient
			C		T	C × T
Mandan, ND						
ALT	SpW/WW/Sun	21/36(31/36)[2]	0.58(0.97)	No-till	1.0	0.58(0.97)[2]
CONV	SpW/fallow	5/24	0.21	Conventional till[3]	0.50	0.11
Brookings, SD						
ALT	Corn/Soy/SpW/Alf	30/48	0.63	Conventional till[4]	0.4[5]	0.25
CONV	Continuous corn	6/12	0.50	Conventional till[4]	0.50	0.25
MEAD, NE						
ALT	Corn/Soy/Srg/Oat-cl	18/36(31/36)	0.5(0.86)	Reduced till[6]	0.5[5]	0.25(0.29)
CONV	Continuous corn	6/12(10/12)	0.5(0.83)	Disk[7]	0.50	0.25(0.41)
Akron, CO						
ALT	WW/Corn/Millet	20/36(36/36)	0.56(1.0)	No-till	1.0	0.56(1.0)
CONV	WW/fallow	10.5/24(20/24)	0.44(0.83)	Sweep plow (4-5 as needed)	0.70	0.31(0.58)
Sidney, NE						
ALT	WW/fallow	10/24(19.5/24)	0.41(0.82)	No-till	1.0	0.41(0.82)
CONV	WW/fallow	10/24	0.41	Plow[8]	0.40	0.17

[1]Cropping designations; SpW, spring wheat; WW, winter wheat; Sun, sunflower; Soy, soybean; Srg, sorghum; Alf, alfalfa; Oat-cl, oats-clover.
[2]Values in parantheses = (months in crop plus same year untilled fallow)/total period.
[3]Sweep and chisel plow, fall and spring disk.
[4]Fall chisel plow, spring disk and cultivation.
[5]Tillage factor dropped by 0.1 units to account for legume by tillage interaction where legume tilled into soil.
[6]Disked twice in spring plus 1 to 2 cultivations plus 1 year oats-clover.
[7]Disked twice in spring plus 1 to 2 cultivations.
[8]Spring moldboard plow, disk or field cultivation (2), rotary rodweeder.

IV. Conclusions

Many soil quality attributes, which are useful indicators of sustainable management, were significantly influenced by tillage and cropping management practices in this study. The utility of soil EC as an indicator of available mineral N levels in surface soil suggests its value as a tool in determining supplemental early season crop fertilizer N needs or as a post harvest indicator of potential leaching losses and environmental degradation. Declines in soil pH at several sites suggested inefficiency of fertilizer N use and leaching losses of nitrate-N from the rooting zone or a potential loss of C to the atmosphere as CO_2 where calcium carbonates are present in soil. Soil pH was also an indicator of the need to adjust total C determined by high temperature combustion for carbonate content to obtain accurate measurements of soil organic C content. In this study, failure to convert gravimetric analysis results to a volumetric basis, using soil bulk density values at time of sampling, would have resulted in a 5 to 13% relative error in comparing management systems at 4 of 5 sites. For three sites where tillage management varied most, surface soil (0 to 7.5cm) organic C and mineralizable N reserves (by anaerobic incubation) with alternative management averaged 43 and 73% greater, respectively, than those with conventional management. At two sites with similar tillage management, soil organic C levels with conventional management were the same or greater than alternative management. Across all sites, however, surface soil mineralizable N levels were highly correlated with microbial biomass N levels ($r^2 = 0.841$), indicating the utility of this less involved test in assessing biologically active N reserves with varying tillage and cropping management. Sensitivity of soil microbial biomass and organic C and N mineralization to tillage management were dependent on cropping and residue management as was observed in other studies by Doran (1987).

Evaluation of soil C storage potential with varying management was assessed by determining the mass of organic C contained in the surface 0 to 30 cm soil layer, which extends below the depth of tillage at these locations. From 3.7 to 14.1 Mg C ha^{-1} more organic C was contained in soil with alternative management than in soil with conventional management at the three locations where no-tillage management was practiced. These values correspond to an annual difference in soil C content between alternative and conventional management of from 0.1 to 1.0 Mg C ha^{-1} yr^{-1} based on the length of time that management comparisons had been conducted at these sites (Table 1). At the two sites where tillage was similar between management systems soil organic C contents did not vary between management practices or were greater with conventional management (16 Mg C ha^{-1} at Brookings). These results should be interpreted with caution, however, because although different relative amounts of C are retained under varying management practices the overall long-term change in soil C storage may be negative. For example, at the Sidney, NE site, where management comparisons for a wheat/fallow system were initiated in native sod, annual losses of soil organic C over 25 years for no-tillage and conventional tillage management averaged 390 and 490 kg C ha^{-1} yr^{-1}, respectively (Doran et al.,1998). In such situations, the decline in soil organic C and associated decline in soil quality will likely only be slowed by increasing C inputs to soil through use of a more intensive cropping system which increases the time of cropping and reduces the time in fallow.

The utility of weight loss-on-ignition procedures for determination of SOM and POM in assessing the effects of cropping and tillage management on the quantity and quality of soil organic pools was verified in this study. Trends in SOM as affected by tillage and cropping were similar to those observed for soil organic carbon and would lead to similar conclusions of the effects of soil management practices on soil organic C pools. The observation that POM (0.05 to 2.0 mm) comprised a greater proportion of total SOM in alternative management (18%) as compared to conventional management (13%) suggests the utility of this measure as a sensitive indicator of management practices and perhaps an early indicator of long-term trends in soil organic matter. The reliability of LOI measures of soil organic matter which are less costly to perform (time, complexity, and money) warrants their use in initial evaluations of the effects of management practices on soil quality and organic matter dynamics.

Results of this survey generally supported the hypothesis that under alternative management, where tillage is reduced or cropping is intensified or diversified, soil organic matter and mineralizable N levels are greater than with conventional management. Rigorous testing of this hypothesis was complicated by the variation in cropping and tillage management practices at the five locations. However, as illustrated in Table 8, comparisons of quantitative estimates of cropping intensity, tillage effects, and carbon storage coefficients for alternative and conventional management practices were comparable to the differences in soil carbon content observed in the surface 30 cm of soil at each location. For example, the carbon storage coefficients at the Mead and Brookings sites were similar or the same between alternative and conventional management, largely due to similar tillage management between systems and the tillage of legume crops into soil. These were also the only sites where soil C contents for conventional management were the same or higher than those for alternative management. The carbon storage coefficients at Mandan, Akron and Sidney were all considerably higher with alternative (no-tillage) management (81 to 427%) as were levels of SOM and POM (29 to 72%) in surface soil.

Acknowledgments

The authors wish to thank staff at the five field sites for their management of long-term field plots, assistance in obtaining soil samples, and for providing records of management, climate, and crop yields. Specific thanks are extended to Randy Anderson and staff at Akron, CO; Mary Kay Tokach in Mandan, ND; Gary Varvel, Drew Lyon, and Ken Hubbard in Nebraska; and Max Pravecek and David Harris in Brookings, SD. This project was supported in part by a Visiting Scientist Fellowship from the Office of Economic and Cooperative Development (OECD).

References

Allarmaras, R.R., J.L. Pikul Jr., J.M. Kraft and D.E. Wilkins. 1988. A method for measuring incorporated crop residue and associated soil properties. *Soil Sci. Soc. Am. J.* 52:1128-1133.

Anderson, T.H. and K.H. Domsch. 1989. Ratios of microbial biomass carbon to total organic carbon in arable soils. *Soil Biol. Biochem.* 21:471-479.

Bolinder, M.A., D.A. Angers, E.G. Gregorich and M.R. Carter. 1999. The response of soil quality indicators to conservation management. *Can. J. Soil Sci.* 79:37-45.

Cambardella, C.A. and E.T. Elliott. 1992. Particulate soil organic-matter changes across a grassland cultivation sequence. *Soil Sci. Soc. Am. J.* 56:777-783.

Cambardella, C.A., A.M. Gajda, J.W. Doran, B.J. Wienhold and T.A. Kettler. 2001. Estimation of particulate and total organic matter by weight loss-on-ignition. p. 349-359. In: R. Lal, J.F. Kimble, R.F. Follett and B.A. Stewart (eds.), *Assessment Methods for Soil Carbon.* Lewis Publishers, Boca Raton, FL.

Campbell, C.A. and R.P. Zenter. 1997. Crop production and soil organic matter in long-term crop rotations in the semi-arid Northern Great Plains of Canada. p. 317-334. In: E.A. Paul, E.T. Elliott, K. Paustian and C.V. Cole (eds.), *Soil Organic Matter in Temperate Agroecosystems: Long-term Experiments in North America.* Lewis Publishers, Boca Raton, FL.

Campbell C.A., LaFond, G.P., Leyson A.J., Zentner R.P. and H.H. Janzen. 1991. Effect of cropping practices on the initial potential rate of N mineralization in a thin Black Chernozem. *Canadian J. Soil Sci.* 71:43-53.

Combs, S.M. and M.V. Nathan. 1998. Soil organic matter. p. 53-58. In: *Recommended Chemical Soil Test Procedure for the North Central Region.* North Cen. Reg. Pub. No. 221 (revised). Station Bull. 1001, Missouri Agr. Exp. Stn., University of Missouri, Columbia.

Dahnke W.C. and D.A. Whitney. 1988. Measurement of soil salinity. p. 32-34. In: *Recommended Chemical Soil Test Procedure for the North Central Region.* North Cen. Reg. Pub. No. 221 (revised). Bull. No. 499 (rev.) North Dakota Ag. Exp. Stn., North Dakota State University, Fargo.

Dick, R.P., D.R. Thomas, and J.J. Halvorson. 1996. Standardized methods, sampling, and sample pretreatment. p. 107-121. In: J.W. Doran and A.J. Jones (eds.), *Methods for Assessing Soil Quality.* Soil Sci. Soc. Am. Spec. Publ. 49. Soil Science Society of America, Madison, WI.

Doran, J.W. 1987. Microbial biomass and mineralizable nitrogen distributions in no-tillage and plowed soils. *Biol. Fertil. Soils* 5:68-75.

Doran, J.W. and T.B. Parkin. 1996. Quantitative indicators of soil quality: A minimum data set. p. 25-37. In: J.W. Doran and A.J. Jones (eds.), *Methods for Assessing Soil Quality*. Soil Sci. Soc. Am. Spec. Publ. 49. Soil Science Society of America, Madison, WI.

Doran, J.W., E.T. Elliott and K. Paustian. 1998. Soil microbial activity, nitrogen cycling, and long-term changes in organic carbon pools as related to fallow tillage management. *Soil Tillage Res.* 49: 3-18.

Doran, J.W., M. Sarrantonio and M.A. Liebig. 1996. Soil health and sustainability. p. 1-54. In: D.L. Sparks (ed.), Advances in Agronomy Vol. 56. Academic Press, San Diego, CA.

Eckert, D.J. 1988. Recommended pH and lime requirement tests. p. 6-8. In: *Recommended Chemical Soil Test Procedure for the North Central Region*. North Cen. Reg. Pub. No. 221 (revised). Bull. No. 499 (rev.) North Dakota Ag. Exp. Stn., North Dakota State University, Fargo.

Gee, G.W. and J.W. Bauder. 1986. Particle-size analysis. p.404-409. In: A. Klute (ed.), *Methods of Soil Analysis, Part 1. Physical and Mineralogical Methods*. 2nd ed. Agron. Monograph 9. American Society of Agronomy and Soil Science Society of America, Madison, WI.

Gianello, C. and J.M. Bremner. 1986. A simple method of assessing potentially available organic nitrogen in soil. *Commun. Soil Sci. Plant Anal.* 17:195-214.

Gupta, V.V.S.R., M.M. Roper, J.A. Kirkegaard and J.F. Angus. 1994. Changes in microbial biomass and organic matter levels during the first year of modified tillage and stubble management practices on Red Earth. *Australian J. Soil Res.* 32:1339-1354.

Janzen, H.H. 1987. Soil organic matter characteristics after long-term cropping to various spring wheat rotations. *Canadian J. Soil Sci.* 67:845-856.

Jenkinson, D.S. 1988. Determination of microbial biomass carbon and nitrogen in soil. p. 368-386. In: J.R. Wilson (ed.), *Advances in Nitrogen Cycling in Agricultural Ecosystems*. C.A.B. International, Oxon, U.K.

Jenkinson, D.S. and J.N. Ladd. 1981. Microbial biomass in soil: measurement and turnover. p. 415-471. In: E.A. Paul and J.N. Ladd (eds.), *Soil Biochemistry*, Vol. 5, Marcel Dekker, New York.

Jenkinson, D.S. and D.S. Powlson. 1976. The effects of biocidal treatments on metabolism in soil V: A method for measuring soil biomass. *Soil Biol. Biochem.* 8:209-213.

Kaiser, E.A. 1994. Significance of microbial biomass for carbon and nitrogen mineralization in soil. *Z. Pflanzenernh. Bodenk.* 157:271-278.

Keeney, D.R. 1982. Nitrogen-availability indices. p. 727-728. In: A.L. Page., R.H. Miller and D.R. Keeney (eds.), *Methods of Soil Analysis, Part 2. Chemical and Microbiological Properties*, 2nd ed. American Society of Agronomy, Madison, WI.

Keeney, D.R. and D.W. Nelson. 1982. Nitrogen-inorganic forms. p. 674-682. In: A.L. Page, R.H. Miller and D.R. Keeney (eds.), *Methods of Soil Analysis, Part 2. Chemical and Microbiological Properties*. Agronomy Monogr. No. 9, 2nd ed. ASA, CSSA and SSSA, Madison, WI.

Lesoing, G.W. and J.W. Doran. 1997. Crop rotation , manure, and agricultural chemical effects on dryland crop yield and SOM over 16 years in Eastern Nebraska. p. 197-204. In: E.A. Paul, E.T. Elliott, K. Paustian, and C.V. Cole (eds.), *Soil Organic Matter in Temperate Agroecosystems: Long-term Experiments in North America*. Lewis Publishers, Boca Raton, FL.

Linn, D.M. and J.W. Doran. 1984a. Aerobic and anaerobic microbial populations in no-till and plowed soils. *Soil Sci. Soc. Am. J.* 48:794-799.

Linn, D.M. and J.W. Doran. 1984b. Effect of water-filled pore space on CO_2 and N_2O production in tilled and non-tilled soils. *Soil Sci. Soc. Am. J.* 48:1167-1172.

Littell, R.C., G.A. Milliken, W.W. Stroup and R.D. Wolfinger. 1996. Common mixed models, Multilocation trial. p. 75-85. In: *SAS System for Mixed Models*. SAS Institute Inc., Cary, NC.

Lupwayi, N.Z., W.A. Rice and G.W. Clayton. 1998. Soil microbial diversity and community structure under wheat as influenced by tillage and crop rotation. *Soil Biol. Biochem.* 30:1733-1741.

O'Halloran, L.P., L.P. Stewart and E. De Jong. 1987. Changes in P forms and availability as influenced by management practices. *Plant Soil* 100:113-126.

Parkin, T.B., J.W. Doran and E. Franco-Vizcaino. 1996. Field and laboratory tests of soil respiration. p. 231-245. In: J.W. Doran and A.J. Jones (eds.), *Method for Assessing Soil Quality.* Soil Sci. Soc. Am. Spec. Publ. 49. Soil Science Society of America, Madison, WI.

Paustian, K.H., P. Collins and E.A. Paul. 1997. Management controls on soil carbon. p. 15-49. In: E.A. Paul, E.T. Elliott, K. Paustian and C.V. Cole (eds.), *Soil Organic Matter in Temperate Agroecosystems: Long-term Experiments in North America.* Lewis Publishers, Boca Raton, FL.

Pikul, J. L. and R. R. Allmaras. 1986. Physical and chemical properties of a Haploxeroll after fifty years of residue management. *Soil Sci. Soc. Am. J.* 50:214-219.

SAS Institute Inc. 1997. SAS/STAT Software: *Changes and Enhancements through Release 6.12.* p. 571-702. SAS Institute Inc., Cary, NC. 1167 pp.

Sarrantonio, M., J.W. Doran, M.A. Liebig and J.J. Halvorson. 1996. On-farm assessment of soil quality and health. p. 83-105. In: J.W. Doran and A. J. Jones (eds.), *Methods for Assessing Soil Quality.* Soil Sci. Soc. Am. Spec. Publ. 49. Soil Science Society of America, Madison, WI.

Schepers, J.S., D.D. Francis and M.T. Thompson. 1989. Simultaneous determination of total C, total N, and ^{15}N on soil and plant material. *Commun. Soil Sci. Plant Anal.* 20: 949-959.

Schulte, E.E. 1988. Recommended soil organic matter tests. p. 29-32. In: *Recommended Chemical Soil Test Procedure for the North Central Region.* North Cen. Reg. Pub. No. 221 (revised). Bull. No. 499 (rev.) North Dakota Ag. Exp. Stn., North Dakota State University, Fargo.

Smith, J. L. and J. W. Doran. 1996. Measurement and use of pH and electrical conductivity for soil quality analysis. p. 169-185. In: J.W. Doran and A.J. Jones (eds.), *Methods for Assessing Soil Quality.* Soil Sci. Soc. Amer. Spec. Publ. 49. Soil Science Society of America, Madison, WI.

Sparling, G. 1992. Ratio of microbial biomass to soil organic carbon as a sensitive indicator of changes in soil organic matter. *Australian J. Soil Res.* 30:195-207.

Waring, S.A. and J. M. Bremner. 1964. Ammonium production in soil under waterlogged conditions as an index of nitrogen availability. *Nature* (London) 201: 951.

Wienhold, B. and A.D. Halvorson. 1998. Cropping system influences on several soil quality attributes in northern Great Plains. *J. Soil and Water Conserv.* 53:254-258.

Section V

Soil Erosion and Sedimentation

^{137}Cesium for Measuring Soil Erosion and Redeposition: Application for Understanding Soil Carbon

J.C. Ritchie

I. Introduction

Soil erosion is a natural process caused by water, wind, and ice that has affected the earth's surface since the beginning of time. Man's activities often accelerate soil erosion (Sombroek, 1995; Walling, 1983). Soil erosion and its off-site, downstream damages are major concerns around the world (Lal, 1994) causing losses in soil productivity, degradation of landscape, degradation of water quality, and loss of soil organic carbon (Walling, 1983; Brown and Wolf, 1984; Walling, 1987; Morgan, 1995; Lal et al., 1998). Brown (1991) estimated soil loss for the world's cropland to be 23 billion tons. Economic damages from the degradation of soils, water quality, and reservoirs by eroded soil particles are significant (Pimentel et al., 1995; Colacicco et al., 1989).

Current techniques for assessing soil erosion are (1) long-term soil erosion plot monitoring, (2) field surveys, and (3) soil erosion models (Evans, 1995). Each of these techniques has strengths and weaknesses, but measurements of spatially distributed soil erosion patterns on the landscape using these techniques are difficult, time consuming, and expensive (Mutchler et al., 1994). Models such as the Universal Soil Loss Equation (USLE) (Wischmeier and Smith, 1965), the Revised Universal Soil Loss Equation (RUSLE)(Renard et al., 1997), Soil Loss Estimation Model for Southern Africa (SLEMSA) (Elwell, 1990), and many other models have been developed and widely used. These efforts to model soil erosion (Foster, 1991) have had varying degrees of success and applications in management and research. Current techniques to measure or model soil erosion do not produce the spatial patterns of soil erosion and especially the redeposition of eroded particles in the field needed to understand soil loss and changes in soil carbon. A great need still exists for timely and good quality data on soil erosion (Stocking, 1993) especially quantitative data at point, field, and reconnaissance scale for assessment of the soil erosion problem (Walling and Quine, 1993).

Over the last 30 years, research has shown the potential of using radioactive fallout ^{137}Cs (^{137}Cesium) to provide timely and quantitative estimates of soil erosion and redeposition at point, field, and reconnaissance scales. Applications of ^{137}Cs to provide independent measurements of soil erosion rates, patterns, and redeposition are well documented (Ritchie and McHenry, 1990; Walling and Quine, 1990; Ritchie and Ritchie, 1998). The purpose of this chapter is to provide an overview of the application of the ^{137}Cs technique for estimating soil erosion and redeposition and to provide examples of its potential for helping to understand soil carbon.

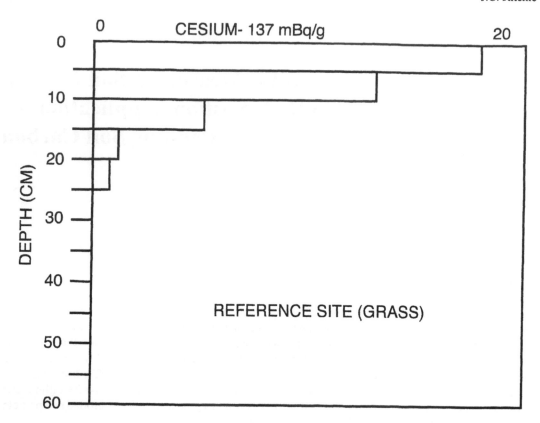

Figure 1. Depth distribution of ^{137}Cs in an undisturbed soil profile.

II. Background

High-yield thermonuclear weapons testing beginning in November 1952 (Cochran et al., 1995) injected ^{137}Cs into the stratosphere where it circulated globally before slowly returning to the atmosphere (Longmore, 1982). Once in the atmosphere, regional patterns and rates of ^{137}Cs fallout to the land surface from the atmosphere are related to rainfall rates and patterns (Davis, 1963). Deposition of fallout ^{137}Cs has decreased since the maximum in the early 1960s with fallout since the mid 1980s being below detection levels (Cambray et al., 1989). However, unique events, such as the Chernobyl accident in April 1986, can cause regional dispersal of measurable ^{137}Cs (Playford et al., 1992; Volchok and Chieco, 1986).

^{137}Cs is rapidly and strongly adsorbed by clay particles in the surface soil and is essentially nonexchangeable once adsorbed to these clay surfaces (Tamura, 1964; Cremers et al., 1988; Livens and Rimmer, 1988). Distribution of ^{137}Cs in soil profiles at undisturbed sites shows an exponential decrease (Figure 1) with depth (Beck, 1966; Ritchie et al., 1970; Campbell et al., 1982) while plowed soils show uniform mixing (Figure 2) of ^{137}Cs in the plowed layer (Bachhuber et al., 1986; Loughran et al., 1987). Depositional profiles show the distribution of ^{137}Cs below the plow depth (See Figure 2). Less than 1% of the ^{137}Cs moves from a catchment in solution immediately after fallout deposition, and generally less than 0.1% moves in solution per year after this initial flush (Eakins et al., 1984;

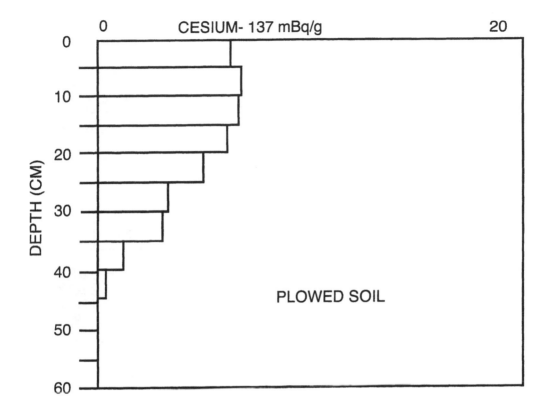

Figure 2. Depth distribution of ^{137}Cs in a plowed soil profile.

Helton et al., 1985). Although biological and chemical processes can move very limited amounts ^{137}Cs in unique environments such as flooded soils, the dominant factors moving ^{137}Cs between and within landscape compartments are the physical processes of water and wind erosion. The movement of ^{137}Cs across the landscape is due to physical processes that redistribute soil particles and are the same physical processes that cause soil erosion. Thus man's activities related to the development of nuclear energy have distributed a radioactive tracer across the landscape surface in discernible patterns that can be used to trace soil erosion and redeposition processes.

Methods for measuring the strong gamma-ray of ^{137}Cs in soil samples using gamma-ray spectrometry are well established and considered routine (Ritchie and McHenry, 1973; Walling and Quine, 1993). Measurements of the redistribution of ^{137}Cs between and within landscape elements provide unique information on soil erosion rates and patterns (Figure 3). The challenge in soil erosion studies is to develop sampling strategies to measure the changing patterns of distribution of ^{137}Cs tagged soil particles on the landscape. Measurements of ^{137}Cs distribution on the landscape provide an estimate of medium-term average (35+ year averages) soil losses and gains. Estimates are site specific and can be made from samples from a single soil core. Multiple cores allow spatial patterns of soil erosion or deposition to be determined.

BASIS OF THE CESIUM-137 TECHNIQUE

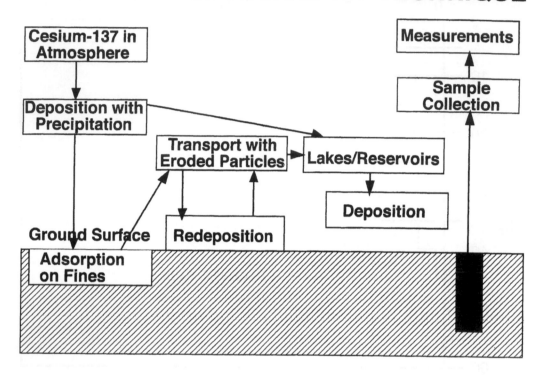

Figure 3. Diagrammatic view of the ^{137}Cs cycle on the landscape.

The unique advantages of the ^{137}Cs technique to study soil erosion rates and patterns are that it (a) requires only one trip to the field; (b) provides results quickly; (c) allows retrospective assessment of soil erosion rates; (d) provides average losses for 35- to 40-year period thus is less influenced by extreme events; (e) represents the total soil erosion process; (f) provides estimates of soil erosion rates, deposition rates, and export rates; and (g) allows a sampling strategy to provide the spatial resolution required.

Determination of areas of soil erosion or deposition on the landscape is based on a comparison of measured ^{137}Cs in a soil profile with the ^{137}Cs originally input to the site. Local measurements of input to a site are usually not available, therefore, a local reference site must be chosen where little disturbance or soil erosion has occurred since the 1950s. Local variation of fallout ^{137}Cs on the landscape can be significant and thus several reference profiles should be measured near each study site (Kiss et al., 1988; Nolin et al., 1993; Wallbrink et al., 1994). A comparison between the sampled soil site and the reference site provides information on gains or losses of ^{137}Cs. Sites with a loss of ^{137}Cs are eroding while sites with a gain of ^{137}Cs are aggrading. Walling and Quine (1993) have provided a handbook for the application of ^{137}Cs to measure soil erosion and deposition.

III. Method

A sampling strategy is designed to provide the spatial resolution required to understand the soil erosion patterns for the landscape element being studied. Samples of the soil profile containing ^{137}Cs are collected. Sampling depth is site dependent but must be deep enough to collect all ^{137}Cs in the profile. The entire soil profile may be sampled as a single sample or the profile may be collected by depth increments. The samples are analyzed to provide an inventory of ^{137}Cs in the total soil profile. This provides a concentration of ^{137}Cs in the sampled profile. Multiple cores allow spatial patterns to be determined.

Sample preparation consists of drying, then sieving through a 2-mm screen to ensure uniform aggregate size for analyses. The sample is weighed into a container and the container is placed on a gamma-ray detector and the gamma emissions are measured using a gamma-ray spectrometer which measures gamma emissions by energy along a spectrum. ^{137}Cs is detected using its 0.662 Mev emission. Measurement time depends on sample size and ^{137}Cs concentration. Measurement times of 30,000 seconds provide measurement precision of ± 4 to 5% for most 1000 g samples. The gamma-ray spectrometer system is calibrated, and efficiency for measuring ^{137}Cs is determined using a known radionuclide standard.

IV. Calculating Soil Erosion Rates from ^{137}Cs Measurements

Estimating soil loss or deposition rates using ^{137}Cs is based on the assumption that a relationship can be developed between increase or decrease of ^{137}Cs soil inventory at a sample site when compared with that of a reference site and the depth of soil loss and deposition. Walling and Quine (1990) discussed the strengths and weaknesses of the various assumptions and approaches. Empirical relationships and theoretical models have been developed to calculate soil erosion from ^{137}Cs data. Empirical relationships are based on soil erosion rates and ^{137}Cs loss measured from experimental plots. Empirical relationships have limited applications and are valid only for the conditions and time period for which they were measured. Empirical studies are most often used to develop an understanding of the processes involved in the movement of ^{137}Cs tagged soil particles. Theoretical models and accounting procedures that simulate the effects of soil processes and ^{137}Cs redistribution are most often used to estimate soil redistribution.

Walling and Quine (1990) reviewed different approaches of using ^{137}Cs to estimate soil erosion and gave an in-depth discussion of strengths and weaknesses of the different modelling approaches. Walling and He (1997) developed a series of quantitative models to estimate soil loss/deposition based on measurements of ^{137}Cs in the soil and have made the software available through the International Atomic Energy Agency, Vienna, Austria. These models range from a simple proportional model to a mass balance model that includes soil movement by tillage. A description of two simple models follows. For more in-depth description of these and other models the reader is referred to the papers by Walling and He (1997, 1998) and Walling and Quine (1990).

A. Proportional Model

Proportional models are the simplest approach. It assumes that ^{137}Cs is uniformly mixed in the plowed layer and that losses and gains of ^{137}Cs from a soil profile are proportional to the loss/gain of soil at a site. The model provides estimates of mean annual soil loss/gain (t ha^{-1} yr^{-1}) based on data on percent loss/gain of ^{137}Cs, soil bulk density, depth of the plow layer, date of sample collection (yr), and the ratio of particle surface area at the sampling site to that of the original soil. In its simplest form the model is:

$$Y = 10 \frac{BDX}{100TP}$$ (1)

where Y is the mean annual soil loss (t ha^{-1} yr^{-1});
      X is the percent loss of ^{137}Cs at sampling site compared to reference site;
      B is the bulk density of soil (kg m^{-3});
      D is depth of the plowed layer (m);
      T is the time since the start of ^{137}Cs deposition (yr);
and   P is a correction factor for particle surface area.

The date of the sample collection is used to calculate the total time of soil erosion. The particle surface area factor is used to correct for the differential erosion of fine particles on which ^{137}Cs is most likely adsorbed (He and Walling, 1996). The model requires a minimum of easily obtained data making it easy to implement. Variations of this model have been widely used for estimating soil erosion and redeposition (i.e., Mitchell et al., 1980; de Jong et al., 1983; Martz and de Jong, 1987; Walling and Quine, 1990). This model is based on a number of simplifying assumptions such as that ^{137}Cs deposition occurred at time T was instantaneously mixed into the soil rather than accumulating over a period of years as it really happened. Also, it does not account for incorporation of soil from below the original plow depth as soil erosion occurs on the surface. This model probably underestimates soil erosion/deposition rates under most conditions.

B. Simplified Mass Balance Model

Mass balance models attempt to take into account all inputs and outputs of ^{137}Cs from the landscape. A Simplified Mass Balance Model, proposed by Zhang et al. (1990), assumed that all ^{137}Cs inputs occurred in 1963 rather than the total period of fallout (1952 to 1970s). The model form is:

$$Y = \frac{10BD}{P} [1 - (1 - \frac{X}{100})^{\frac{1}{T - 1963}}]$$ (2)

where Y is the mean annual soil loss (t ha^{-1} yr^{-1});
      X is the percent loss of ^{137}Cs at sampling site compared to reference site;
      B is the bulk density of soil (kg m^{-3});
      D is the depth of the plowed layer (m);
      T is the year of sample collection;
and   P is a correction factor for particle surface area.

Again eroding sites have lost ^{137}Cs while depositional sites have gained ^{137}Cs relative to the reference site. This model improves on the proportional model by accounting for the incorporation of soil from below the original plow depth. This model, like the proportional model, is easy to use.

While it is an improvement over the proportional model, it also assumes instantaneous mixing of ^{137}Cs in the plow layer and that all deposition occurred in 1963.

More comprehensive mass balance models based on Eq. 2 have been developed that incorporate time variant inputs of ^{137}Cs fallout, fate of newly deposited fallout before it is incorporated into the plowed layer, radioactive decay, and other factors. Mass balance models have been widely used to estimate soil erosion and deposition patterns (i.e., Kachanoski and de Jong, 1984; Quine, 1989; He and Walling, 1997). These models have been found to give realistic estimates of soil erosion.

He and Walling (1997) have further developed the mass balance models to incorporate soil movement by tillage and have developed models to estimate soil loss from uncultivated soils. Thus, a wide range of theoretical models is available to estimate soil loss based on measured patterns of ^{137}Cs. All these models, their applications, advantages, and limitation have been discussed by Walling and He (1997) and He and Walling (1997).

V. Applications to Understanding Soil Carbon

Since most soil organic carbon (SOC) is in the soil layers most vulnerable to soil erosion, large amounts of SOC are redistributed along with the eroding soil particles while other SOC is exposed to oxidation process at the surface. ^{137}Cesium can be used to measure the erosion and redeposition of soil particles of the vulnerable layer. Figure 4 shows the erosion and redeposition of soil particles on five parallel transects upslope from a conservation practice site on a contoured field in Maryland. The redistribution pattern shows both soil erosion and deposition in an area of 40 × 25 m. Net accumulation of soil particles has occurred in this small area. If we assume that SOC was uniformly distributed across the area then a net accumulation rate of SOC has occurred in the area also. However, since the redistribution of the soil particles in the area could expose SOC to a more oxidizing environment there could also be a differential loss between loss of soil particles and SOC in the area. Another study for an agricultural field in Wisconsin (Figure 5) showed the pattern of soil redistribution within a field along a slope profile (Mitchell et al., 1980). In this example, a net loss of soil particles occurred along the slope profile. While no soil carbon data are available in these two examples, they do show how ^{137}Cs can be used to show patterns of soil (and probably SOC) redistribution within a field. The fundamental question to be answered is the relationship between soil particle redistribution and SOC redistribution. Does SOC move uniformly with the soil particles or is there a differential SOC loss due to (a) lighter SOC particles being carried off the field in the transporting water and (b) increased oxidation of SOC due to exposure to a more oxidizing environment?

There is generally a positive relationship between ^{137}Cs content and other soil attributes such as SOC (Figure 6), total carbon, nitrogen and other soil fertility attributes. Using ^{137}Cs to estimate soil erosion rates, we usually find a relationship between soil erosion and soil carbon loss (Figure 7). The data in Figures 6 and 7 were collected as part of a study of the effect of CRPs on soil carbon. Other studies have shown a clear relationship between soil ^{137}Cs content and SOC, total soil carbon, and nitrogen (Pennock, 1998; Ritchie et al., 1975; McHenry and Ritchie, 1977). Studies (de Jong and Kachanoski 1988; Gregorich and Anderson 1985) have shown that long-term carbon losses in prairie soil were due to soil erosion. Generally a relationship exists between soil loss and SOC loss.

However, a study at the Columbia Basin Agricultural Center near Pendleton, Oregon on long-term cropping experimental plots found little evidence of soil erosion based on estimates using ^{137}Cs (Ritchie and Rasmussen, 2000), RUSLE or EPIC yet these plots show steady declines in SOC over the past 50 years. Since there is little evidence of soil movement from the experimental plots to explain the steady decline of SOC, Rasmussen and Albrecht (1998) concluded that the major contributing factor for the declining soil organic matter in these plots must be biological oxidation.

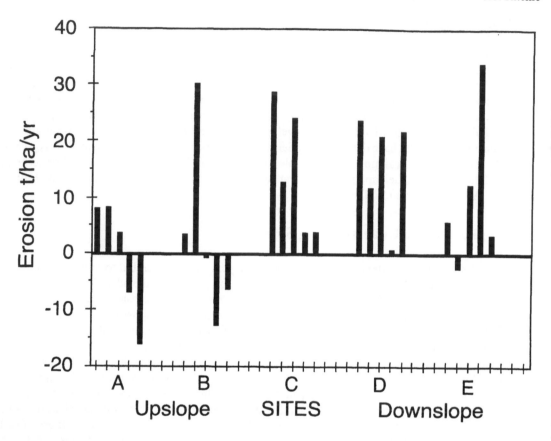

Figure 4. Soil redistribution rates upslope (bottom [E] to the top [A]) from a stiff grass hedge conservation practice at Beltsville, MD estimated based ^{137}Cs content using the proportional model (Equation 1). Each grouping (A-E) represents samples along the contour.

This study indicated that under certain conditions biological oxidation is the most important factor for soil carbon loss; thus we cannot assume that since no soil erosion is occurring then there is no soil carbon loss. While there is great potential to sequester carbon in soils (Lal et al., 1998), it is necessary to understand the mechanism of carbon loss from the soil to determine its fate in the environment.

Accelerated soil erosion is a key factor in moving SOC on landscape units. ^{137}Cs can give information on the redistribution of soil particles within the landscape unit and provide information on the loss of soil particles and SOC from the landscape unit. Once off the field, deposition of the material in reservoirs and lakes can lead to sequestration of carbon. Ritchie (1988) in a study of 58 small reservoirs across the United States estimated sediment accumulation rates in these reservoir using ^{137}Cs. Based on these sediment accumulation rates, organic carbon accumulation rates ranged from 26 to 3700 gC $^{-2}$ yr^{-1} with an average 657 ± 739 gC $^{-2}$ yr^{-1}. Carbon concentration in the reservoir sediments was similar to the carbon concentration in the surface soil (0 to 10 cm), suggesting that carbon was moving with the eroding material. In a similar study in navigation pools along the upper Mississippi River using sediment cores dated using ^{137}Cs, Ritchie (1988) found that organic carbon in the 1964 to 1977 sediment profile was less than the organic carbon was in the 1954 to 1964

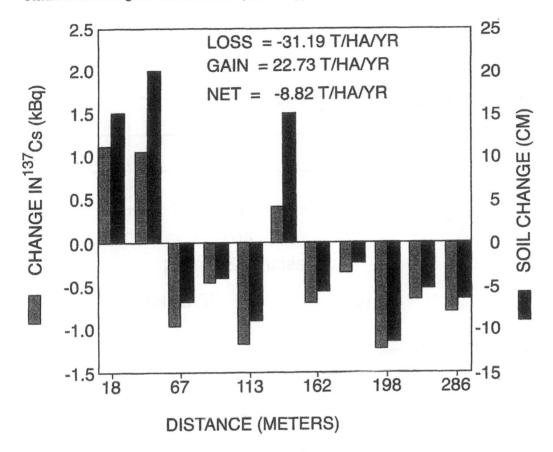

Figure 5. Soil redistribution from the bottom (18 m) to top (286 m) of a slope from an agriculture field in Wisconsin. (Adapted from Mitchell et al., 1980.)

sediment profile, which was less than organic carbon in the sediment deposited before 1954. Rates in these Mississippi River navigation pools were similar to the rates measured in the reservoir study. These studies showed that carbon sequestration can be significant in depositional areas.

VI. Conclusion

The large size of the SOC pool makes it a key factor for understanding global carbon cycles (Rasmussen et al., 1998). Some studies have shown that long-term carbon loss in soil was due to soil erosion while other studies indicate that biological oxidation is the most important factor for soil carbon loss. While there is great potential to sequester carbon in soil, it is necessary to know the mechanism of carbon loss from the soil to determine its fate in the environment. There is little doubt that better estimates of soil redistribution on the landscape would improve our understanding of the SOC pool. Measurement of ^{137}Cs redistribution across the landscape has been shown to provide realistic estimates of soil redistribution on the landscape and thus should be a useful tool for helping understand the SOC pools.

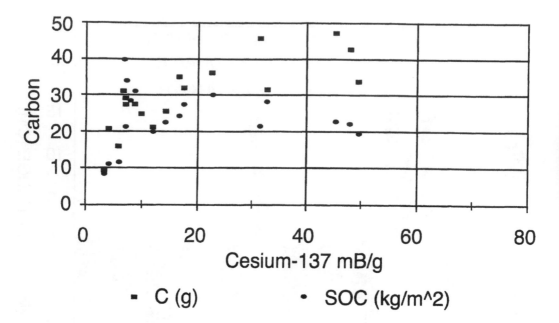

Figure 6. Relationship between ^{137}Cs concentration and soil carbon.

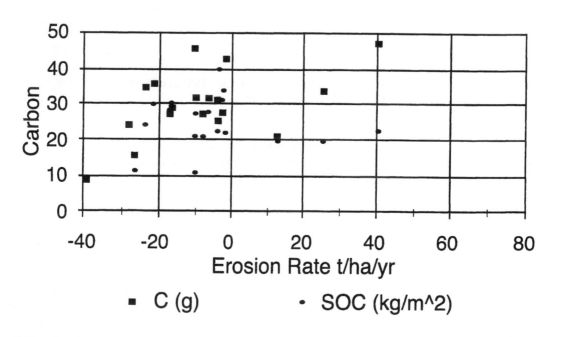

Figure 7. Relationship between soil erosion rates estimated by ^{137}Cs and soil carbon content.

References

Bachhuber, H., K. Bunzl and W. Schimmack. 1986. Spatial variability of distribution coefficients of ^{137}Cs, [65]Zn, [85]Sr, [57]Co, [109]Cd, [141]Ce, [103]Ru, [95m]Tc, and ^{131}I in cultivated soil. *Nucl. Tech.* 72:359-371.

Beck, H.L. 1966. Environmental gamma radiation from deposited fission products 1960-1964. *Health Phys.* 12:313-322.

Brown, L.R. 1991. The global competition for land. *J. Soil Water Conserv.* 46:394-397.

Brown, L.R. and E.C. Wolf. 1984. Soil Erosion: Quiet Crisis in the World Economy. Worldwatch Paper 60, Washington, D.C. 51 pp.

Cambray, R.S., K. Playford, G.N.J. Lewis and R.C. Carpenter. 1989. Radioactive Fallout in Air and Rain: Results to the End of 1988. AERE-R-13575. U.K. Atomic Energy Authority Report, Harwell, U.K.

Campbell, B.L., R.J. Loughran and G.L. Elliott. 1982. Caesium-137 as an indicator of geomorphic processes in a drainage system. *Aust. Geographical Studies* 20:49-64.

Cochran, T.B., W.M. Arkin, R.S. Norris and J.S. Sands. 1995. Nuclear Weapons Data Book. Harper & Row, New York.

Colacicco, D., T. Osborn and K. Alt. 1989. Economic damage from soil erosion. *J. Soil Water Conserv.* 44:35-39.

Cremers, A., A. Elsen, P. De Preter and A. Maes. 1988. Quantitative analysis of radiocaesium retention in soils. *Nature* 335:247-249.

Davis, J.J. 1963. Cesium and its relationship to potassium in ecology. p. 539-556. In: V. Schultz and A.W. Klement, Jr. (eds.) *Radioecology.* Reinhold, NY.

de Jong, E., C.B.M. Begg and R.G. Kachanoski. 1983. Estimates of soil erosion and deposition from some Saskatchewan soils. *Can. J. Soil Sci.* 63:607-617.

de Jong, E. and R.G. Kachanoski. 1988. The importance of erosion in the carbon balance of prairie soils. *Can. J. Soil Sci.* 68:111-119.

Eakins, J.D., R.S. Cambray, K.C. Chambers and A.E. Lally. 1984. The transfer of natural and artificial radionuclides to Brotherswater from its catchment. p. 125-144 In: E.Y. Haworth and J.W.G. Lund (eds.), *Lake Sediment and Environmental History.* University of Minnesota Press, Minneapolis.

Elwell, H.A. 1990. The development, calibration, and field testing of a soil loss and runoff model derived from a small scale physical simulation of the erosion environment on arable land in Zimbabwe. *J. Soil Sci.* 41:239-253.

Evans, R. 1995. Some methods of directly assessing water erosion of cultivated land - a comparison of measurements made on plots and in fields. *Prog. Physical Geogr.* 19:115-129.

Foster, G.R. 1991. Advances in wind and water soil erosion prediction. *J. Soil Water Conserv.* 46: 27-29.

Gregorich, E.G. and D.W. Anderson. 1985. Effects of cultivation and erosion on soils of four toposequences in the Canadian Prairie. *Geoderma* 36:343-345.

He, Q. and D.E. Walling. 1996. Interpreting particle size effects in the absorption of ^{137}Cs and unsupported ^{210}Pb by mineral soils and sediments. *J. Environ. Radioact.* 30:117-137.

He, Q. and D.E. Walling. 1997. The distribution of fallout ^{137}Cs and ^{210}Pb in undisturbed and cultivated soils. *Applied Radiat. Isot.* 48:677-690.

Helton, J.C., A.B. Muller and A. Bayer. 1985. Contamination of surface-water bodies after reactor accidents by the erosion of atmospherically deposited radionuclides. *Health Phys.* 48:757-771.

Kachanoski, R.G. and E. de Jong. 1984. Predicting the temporal relationship between cesium-137 and erosion rate. *J. Environ. Qual.* 13:301-304.

Kiss, J.J., E. de Jong and L.W. Martz. 1988. The distribution of fallout cesium-137 in southern Saskatchewan, Canada. *J. Environ. Qual.* 17:445-452.

Lal, R. (ed.). 1994. Soil Erosion Research Methods. Soil and Water Conservation Society, Ankeny, IA. 340 pp.

Lal, R., J.M. Kimble, R.F. Follett and C.V. Cole. 1998. *The Potential of U.S. Cropland to Sequester Carbon and Mitigate the Greenhouse Effect*. Ann Arbor Press, Ann Arbor, MI. 128 pp.

Livens, F.R. and D. Rimmer. 1988. Physico-chemical controls on artificial radionuclides in soil. *Soil Use Manage*. 4:63-69.

Longmore, M.E. 1982. The caesium-137 dating technique and associated applications in Australia— A review. p. 310-321. In: W. Ambrose and P. Duerden (eds.), *Archaeometry: An Australasian Perspective*. Australian National University Press, Canberra, Australia.

Loughran, R.J., B.L. Campbell and D.E. Walling. 1987. Soil erosion and sedimentation indicated by caesium 137: Jackmoor Brooke catchment Devon, England. *Catena* 14:201-212.

Martz, L.W. and E. de Jong. 1987. Using cesium-137 to assess the variability of net soil erosion and its association with topography in a Canadian prairie landscape. *Catena* 14:439-451.

McHenry, J.R. and J.C. Ritchie. 1977. Distribution of Cs-137 in arid watersheds. *Water Resour. Res.* 13:923-925.

Mitchell, J.K., G.D. Bubenzer, J.R. McHenry, and J.C. Ritchie. 1980. Soil loss estimation from fallout cesium-137 measurements. p. 393-401. In: M. DeBoodt and D. Gabriels (eds.), *Assessment of Erosion*. John Wiley & Sons, London.

Morgan, R.P.C. 1995. Soil Erosion and Conservation. Longman Scientific & Technical, Harlow, Essex, England.

Mutchler, C.K., C.E. Murphree and K.C. McGregor. 1994. Laboratory and field plots for soil erosion research. p. 11-37. In: Lal, R. (ed.), *Soil Erosion*. Soil and Water Conservation Society, Ankeny, IA.

Nolin, M.C., Y.Z. Cao, D.R. Coote and C. Wang. 1993. Short-range variability of fallout ^{137}Cs in an uneroded forest soil. *Can. J. Soil Sci.* 73: 381-385.

Pennock, D. 1998. Cesium-137 measures soil redistribution as an indicator of soil quality. Report on the Second Research Co-ordination Meeting of the Co-ordinated Research Projects D1-RC-629.2 and F3-RC-6442. IAEA/FAO Joint Division. Report and abstracts of presentation at Bucharest, 25-29 May 1998, p. 49.

Pimentel, D., C. Harvey, P. Resosudarmo, K. Sinclair, D. Kurz, M. McNair, S. Crist, L. Shipritz, L. Fitton, R. Saffouri and R. Blair. 1995. Environmental and economic cost of soil erosion and conservation benefit. *Science* 267:1117-1123.

Playford, K., G.N.J. Lewis and R.C. Carpenter. 1992. Radioactive Fallout in Air and Rain: Results to the End of 1990. AEA-EE-0362. United Kingdom Atomic Energy Authority, Harwell, U.K. 27 pp.

Quine, T.A. 1989. Use of a simple model to estimate rates of soil erosion from caesium-137 data. *J. Water Resour.* 8:54-81.

Rasmussen, P.E. and S.L. Albrecht. 1998. Crop management effects on organic carbon in semi-arid Pacific Northwest soils. p. 209-219. In: R. Lal, J.M. Kimble, R.F. Follett and B.A. Stewart (eds.), *Management of Carbon Sequestration in Soil*. Lewis Publishers, Boca Raton, FL.

Rasmussen, P.E., K.W.T. Goulding, J.R. Brown, P.R. Grace, H.H. Janzen, and M. Körschen. 1998. Long-term agroecosystem experiments: Assessing agricultural sustainability and global change. *Science* 282:893-896.

Renard, K.G., G.R. Foster, G.A. Weesies, D.K. McCool and D.C. Yoder. 1997. Predicting Soil Erosion by Water: A Guide to Conservation Planning with the Revised Universal Soil Loss Equation (RUSLE). USDA Agriculture Handbook No. 537, Washington, D.C. 384 pp.

Ritchie, J.C. 1988. Organic matter content in sediments of three navigation pools along the upper Mississippi River. *J. Freshwater Ecol.* 4:343-349.

Ritchie, J.C. 1989. Carbon content of sediments in agricultural reservoirs. *Water Resour. Bull.* 25: 301-308.

Ritchie, J.C., E.E.C. Clebsch and W.K. Rudolph. 1970. Distribution of fallout and natural gamma radionuclides in litter, humus, and surface mineral soils under natural vegetation in the Great Smoky Mountains, North Carolina-Tennessee. *Health Phys.* 18:479-491.

Ritchie, J.C., A.C. Gill and J.R. McHenry. 1975. A comparison of nitrogen, phosphorus, and carbon in sediments and soils of cultivated and noncultivated watershed of the North Central States. *J. Environ. Qual.* 4: 339-341.

Ritchie, J.C. and J.R. McHenry. 1973. Determination of fallout Cs-137 and natural gamma-ray emitting radionuclides in sediments. *Internat. J. Appl. Radiat. Isot.* 24:575-578.

Ritchie, J.C. and J.R. McHenry. 1990. Application of radioactive fallout cesium-137 for measuring soil erosion and sediment accumulation rates and patterns: a review. *J. Environ. Qual.* 19:215-233.

Ritchie, J.C. and P.E. Rasmussen. 2000. Application of Cesium-137 to estimate erosion rates for understanding soil carbon loss on long-term experiments at Pendleton, Oregon. *Land Rehabilitation and Development* 11:75-81.

Ritchie, J.C. and C.A. Ritchie. 1998. Bibliography of publications of ^{137}Cs studies related to soil erosion and sediment deposition. p. 63-116. In: IAEA. Use of ^{137}Cs in the Study of Soil Erosion and Sedimentation. IAEA-TECDOC-1028, Vienna, Austria.

Sombroek, W. 1995. Soil degradation and contamination: A global perspective. *Soil and Environ.* 5:3-13.

Stocking, M. 1993. Soil erosion in developing countries or where geomorphology fears to tread! Paper presented at memorial symposium for Prof. J. De Ploey "Experimental Geomorphology and Landscape Ecosystem change," March 1993, Leuven, Belgium.

Tamura, T. 1964. Consequences of activity release: selective sorption reactions of cesium with soil minerals. *Nucl. Safety* 5:262-268.

Volchok, H.L. and N. Chieco (eds.). 1986. A Compendium of the Environmental Measurement Laboratory's Research Projects Related to Chernobyl Nuclear Accident. USDOE Rep. EML-460, Environmental Monitoring Laboratory, New York.

Wallbrink, P.J., J.M. Olley and A.S. Murray. 1994. Measuring soil movement using [137]Cs: Implications of reference site variability. *IAHS Publ.* 224:95-105.

Walling, D.E. 1983. The sediment delivery problem. *J. Hydrol.* 65:209-237.

Walling, D.E. 1987. Rainfall, runoff, and soil erosion of the land: a global review. p. 89-117. In: Gregory, K.J. (ed.), Energetics of the Physical Environment. John Wiley & Sons, Ltd., Chichester, England.

Walling, D. and Q. He. 1997. Models for Converting [137]Cs Measurements to Estimates of Soil Redistribution Rates on Cultivated and Uncultivated Soils (Including software for model implementation). A contribution to the IAEA Coordinated Research Programmes on Soil Erosion (D1.50.05) and Sedimentation (F3.10.01), Department of Geography, University of Exeter, U.K., 29 pp.

Walling, D.E. and Q. He. 1998. Use of fallout ^{137}Cs measurements for validating and calibrating soil erosion and sediment delivery models. *IAHS Publ.* 249:267-278.

Walling, D.E. and T.A. Quine. 1990. Calibration of caesium-137 measurements to provide quantitative erosion rate data. *Land Degradation and Rehabilitation* 2:161-175.

Walling, D.E. and T.A. Quine. 1993. Use of Caesium-137 as a Tracer of Erosion and Sedimentation: Handbook for the Application of the Caesium-137 Technique. U.K Overseas Development Administration Research Scheme R4579, Department of Geography, University of Exeter, Exeter, U.K. 196 pp.

Wischmeier, W.H. and D.D. Smith. 1978. Predicting rainfall soil erosion losses. Agr. Handbook No. 537. U.S. Department of Agriculture, Washington, D.C. 58 pp.

Zhang, X.B., D.L. Higgitt and D.E. Walling. 1990. A preliminary assessment of the potential for using caesium-137 to estimate rates of soil erosion in the Loess Plateau of China. *Hydrol. Sci. J.* 35:243-251.

CHAPTER 29

Assessing the Impact of Erosion on Soil Organic Carbon Pools and Fluxes

G.C. Starr, R. Lal, J.M. Kimble and L. Owens

I. Introduction

Energy from rainfall, runoff, and wind disintegrate near surface soil structure (Yoder, 1936; Ellison, 1944; Le Bissonnais and Arrouays, 1997). As soil is transported from eroded to depositional positions across watersheds, aggregate breakdown exposes physically protected soil organic carbon (SOC) pools to mineralization, leading to increased CO_2, CH_4, and N_2O emissions to the atmosphere (Lal, 1995). Some transported material (including SOC, nutrients, and minerals) moves into fluvial waterways (Lal, 1995; Stallard, 1998). In aquatic ecosystems, these materials contribute to eutrophication (Vezjak, 1998; Forsberg, 1998; Frielinghaus et al., 1998; Lowery, 1998), anoxia (Vezjak et al., 1998; Lowery, 1998), turbidity (Wass et al., 1997; Riley, 1998), and greenhouse gas (GHG) emissions (Lal, 1995). The amount of SOC eventually stored in sediments (Stallard, 1998) of aquatic ecosystems is only a small fraction of transported SOC (Lal, 1995).

The process of sequestering SOC results in improved soil quality (Lal, 1997), greater aggregate stability (Tisdall and Oades, 1982; Elliott, 1986; Edgerton et al., 1995; Kay, 1998), reduced erodibility (Lal, 1997; Piccolo et al., 1997), and decreased GHG emissions to the atmosphere (Lal, 1997). Accelerated erosion has a serious detrimental effect on SOC levels in eroded landscapes (Varoney et al., 1981; Tiessen et al., 1982; Gregorich and Anderson, 1985; De Jong and Kochanoski, 1988; Kreznor et al., 1989; Genge and Coote, 1991; Changere and Lal, 1995; Mitchell et al., 1998; Gregorich et al., 1998). However, the fate of transported SOC and the implications of erosion for GHG emissions are not well known. For instance, Hedges et al. (1997) argue that most of the 0.25 Pg yr $^{-1}$ of dissolved organic carbon (DOC) and 0.15 Pg Yr^{-1} of particulate organic carbon (POC) transported by rivers to the ocean is ultimately oxidized. Soil deposited on landscapes after erosion events goes through stages of rapid oxidation due to loss of structural SOC protection (Lal, 1995). The loss in productivity caused by erosion is related to a reduction in sequestered carbon (Gregorich et al., 1998). However, Stallard (1998) argues that erosion, coupled with carbon burial in aquatic sediments, may account for an atmospheric CO_2 sink.

When erosional processes cause significant soil loss, they counteract C sequestration processes in the soil pool as is evidenced by both experimental observations and modeling. Water transported C flux is positively correlated with soil loss by erosion (Massey and Jackson, 1952; Starr et al., 2000). Also, reduced levels of SOC have been consistently observed in eroded landscapes (Tiessen et al., 1982; Gregorich and Anderson, 1985; Genge and Coote, 1991; Gregorich et al., 1998). These observations are consistent with modeling results showing the importance of erosional processes for SOC dynamics. One model predicted that erosion, when appreciable, dominates SOC equilibrium

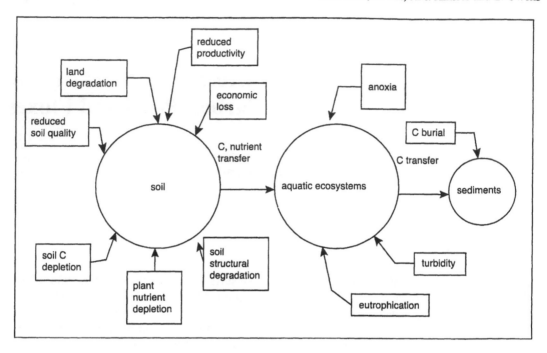

Figure 1. Erosion pressures on environmental quality in soil and aquatic ecosystems.

levels in the long term (Varoney et al., 1981). Another model (Mitchell et al., 1998) showed water and wind erosion were the most significant factors effecting SOC balance in the north central US.

Loss of topsoil by erosion can cause primary productivity reductions of 50% or more in severely impacted regions (Lal, 1998), leading to dramatic reductions in the SOC inputs of litter and dead roots. Although reduced biomass pool and lower input fluxes to the SOC pools are indirect effects of erosion, their influence on the net carbon stored in the soil-plant biome may represent the largest loss of sequestered carbon pool (Gregorich et al., 1998). Clearly, erosion exerts many environmental pressures on soil and aquatic ecosystems (Figure 1). The objective of this chapter is to review recent and relevant literature focusing on the SOC pools and fluxes and methods of assessment of these pools and fluxes in relation to erosional and depositional processes and fluxes of GHGs.

II. Fate of Eroded Carbon

The fate of eroded carbon is understood conceptually in terms of the pools of carbon impacted and the magnitude of the fluxes between pools. Data on global pool assessments and transported C flux are crudely estimated at this time and are the subject of ongoing debate. The problem is somewhat confounded by the interdisciplinary nature of transport problem from the soil/plant biome to the aquatic/sedimentary biome and between both these biomes and the atmosphere (Hutjes et al., 1998). Further complicating matters are methodological difficulties in measuring and scaling the interrelated pools and fluxes.

A. Sediment-borne SOC

Based on the limited available data on the flux of SOC directly transported by runoff waters, the annual rate of SOC translocation by water erosion has been estimated at 5.7 Pg where approximately 90% of translocated SOC is redistributed across the landscape (Lal, 1995). Stallard (1998) hypothesized that aquatic sediments may account for an atmospheric CO_2 sink assuming that the sediment-borne SOC is replaced in the landscape from atmospheric CO_2. This assumption may or may not be a valid one, however, and several studies present data that appear to conflict with it. Tiessen et al. (1982) showed that SOC content following cultivation of a Canadian grassland did not reach equilibrium, while continuing to decline over 60 to 90 years as a result of erosion, and this conclusion agrees with the model prediction by Varoney et al. (1981). These reductions are not easily or quickly reversible, as evidenced by Geng and Coote (1991), who showed that erosion induced reductions in SOC were not replenished even after 17 years of continuous grass cover in Canada. However, they were replenished after 27 years of continuous grass cover in a fine textured soil.

B. Particle Size of Eroded Material and C Pool

The sediments carried in runoff and rivers are concentrated or enriched in clay-sized particles (Young and Onstadt, 1978; Ongley et al., 1981; Pert and Walling, 1982), POC (Ittekkot and Laane, 1991), and DOC (Spitzy, 1991; Lal, 1995) relative to the soils they came from. A size maximum of 0.06 mm diameter is commonly used for suspended river sediments (Pert and Walling, 1982). From 60 to 90% of suspended sediments were <0.002 mm in size in Wilton Creek, Ontario (Ongley et al., 1981). Similarly, microaggregate (<0.06 mm) enrichment has been observed in plot runoff waters (Sutherland and Ziegler, 1996; Wan and El-Swaify, 1997). The most stable SOC size pools in soil (Anderson et al., 1981; Paul et al., 1995) are the fine silt and clay domain because of the physical protection afforded in soil aggregates and because humic materials in this size range are recalcitrant (Kay, 1998).

C. SOC in Depositional Sites

Little is known of the properties and fate of SOC that is translocated from erosion to deposition points on the landscape. The physics of particle settling would tend to suggest that the extent of erosional translocation is negatively correlated with the size and density of detached soil aggregates and primary particles. Thus, soil erosion is a highly selective process that preferentially removes the smallest and lowest density components of soil and transports them great distances (Lal, 1995).

Several researchers have observed increases in SOC content at depositional sites compared with eroded land (Mermut et al., 1983; De Jong and Kochanoski, 1988; Pennoch et al., 1994; Fahnestock et al., 1995). Pennoch et al. (1994) observed that some portions of the foot-slopes initially were depositional with increases in SOC and soil quality but later were eroded and lost SOC. Other portions of the foot-slopes (occupying only 15% of the landscape) were consistently depositional and increased in SOC and soil quality.

Freshwater aquatic sediments play a small but significant role in the global carbon cycle. Mulholland and Elwood (1982) estimated about 0.06 Pg C yr^{-1} and 0.2 Pg C yr^{-1} accumulated in lakes and reservoirs, respectively. These authors found that direct inputs of SOC are small compared with the estimated 0.5 Pg C yr^{-1} attributed to aquatic primary production in inland waters. Thus, the input of nutrients in runoff waters feeding aquatic primary productivity and eutrophication may play a more important role than erosion does in aquatic carbon dynamics. Some authors (Spitzy, 1991; Hedges et al., 1997) propose that the majority of DOC transported by inland waters comes from soil and land plants and is oxidized in the ocean. By contrast, the majority of POC transferred from soil to aquatic

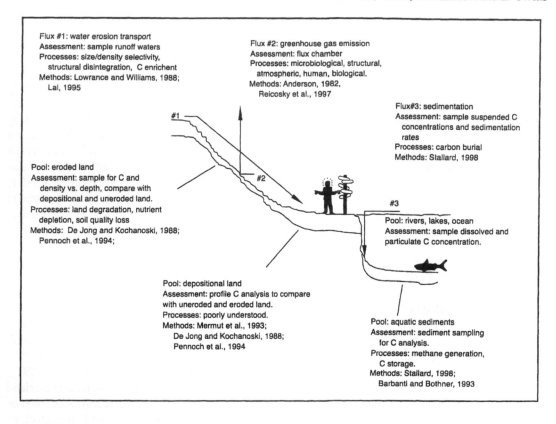

Figure 2. Pools and fluxes of C related to erosion.

ecosystems is either oxidized or deposited as inland aquatic sediments (Ittekkot, 1991). The sink of C accumulating in reservoirs may increase as more rivers are dammed, but may also decrease if the reservoirs are dredged (Mulholland and Ellwood, 1982).

III. Assessment Methods of C Pool over the Landscape

There is a complex interplay of soil, plant, atmospheric, microbiological, and surficial processes (Figure 2) that has a bearing on the net impact of erosion on SOC dynamics and no single method can unravel or elucidate all these interactions. However, a few authors have attempted to synthesize results from many different types of observations (Lal, 1995; Gregorich et al., 1998). Fluxes of importance include direct erosional transport, emission of GHGs, and sedimentation, whereas pools include eroded and depositional land, aquatic sediments and dissolved and particulate suspended sediments C in aquatic ecosystems (Table 1). Each of these has been assessed to some extent in the past, but the interrelationships and processes are not well understood and the data on magnitude and rates of C movement are sketchy.

One branch of assessment methods involves direct measurement of SOC and nutrient fluxes in water erosion events (Rogers, 1941; Massey and Jackson, 1952; Lowrance and Williams, 1999; Owens and Edwards, 1993; Zobish et al., 1995). Another approach is to conduct landscape analysis of soil profiles (Mermut et al., 1983; Gregorich and Anderson, 1985; DeJong and Kochanoski, 1988; Kreznor et al., 1989; Genge and Coote, 1991; Pennoch et al., 1994; Fahnestock et al., 1995) with varying degrees of erosion and deposition. Methods of monitoring carbon dioxide fluxes from the soil (Anderson, 1982) include diffusion gradient methods and static and dynamic chambers (Reicoski et al., 1997). Modeling of erosion and SOC dynamics has also been applied to soils (Van Veen et al., 1981; Varoney et al., 1981; Lal, 1995; Mitchell et al., 1998) and eroded sediments (Lal, 1995; Ludwig et al., 1996; Stallard, 1998).

IV. New Approaches to Assessment

Methodology for assessing SOC losses by erosion becomes increasingly difficult as the scale of the study area increases. Hypotheses drawn from the limited data that are available are sometimes in conflict (Lal, 1995; Stallard, 1998), indicating that our knowledge of how erosion factors in the global carbon cycle is incomplete. Thus, it would be helpful to develop new approaches to the problem, at a range of scales, to improve our understanding of SOC dynamics. This was the rationale of some recent studies that aim to develop alternative approaches to the problem. A brief overview and discussion of a rainfall-driven aggregate fractionation system and a method for assessing erosional SOC transport from watersheds in individual events will be given in the following subsections. These overviews are intended as introductory and will be expanded upon in a later report.

A. Rain Fractionator

Raindrops and runoff impact and detach soil aggregates along with the SOC they contain, and transport colloids in runoff and surface water. Understanding the mechanisms and modes of transport of SOC is necessary for assessing the impact of erosional processes on the global C cycle. Thus, a rainfall simulator method for separating aggregate fractions was tested to realistically assess the pools of SOC available for transport by water erosion. A 50-g sample of dry aggregates placed in a nest of sieves, with a cylindrical splash guard above and a collector below, was subjected to 50 mm/hr simulated rain for 30 min. The rain simulator was described by Chowdhary et al (1997). Aggregates, 5 to 8 mm diameter, were taken from the surface of the same Ohio soil (fine loamy mixed mesic typic Hapludult) under four management practices, and were equilibrated at 100% relative humidity. This method of sample preparation is described in Collis-George and Lal (1971). Soil was fractionated into seven pools: floatable organic matter (FOM, 0.25 to 8 mm), and mineral associated SOC in aggregates in the size ranges of 5 to 8, 2 to 5, 1 to 2, 0.5 to 1, 0.25 to 0.5, and 0 to 0.25 mm.

Repeated testing of this system revealed that mass recovery averaged 98.5% and repeated tests had a standard deviation for individual pool assessment averaging 2.5% of total mass. Forest and pasture soil was consistently more stable under raindrop impact than cropped plateau no-till continuous corn, and conventional till continuous corn had the least stable aggregates. Mean weight diameters were 3.7, 4.0, 1.6, and 1.2 mm (where C % was 3.4, 3.3, 1.8, and 1.0) for forest, pasture, no-till, and conventional till, respectively. A plot of the percentage of total SOC contained in the various pools (Figure 3) suggests a trend that as tillage intensity increases there is a shift of aggregates and mineral associated SOC into the finer aggregate pools, but the FOM pool trends toward a lower percentage of the total. The pools that are expected to be most transportable by erosion are the FOM pool, because of its low density, and the 0 to 0.25 mm pool, because of its small size. Runoff waters, provided the depth of overland flow is comparable to size of material, easily carry the transportable pools.

One advantage of the new approach is the efficient use of water that results in ~400 ml of concentrated micro-aggregates and clay domain SOC. Compared with a traditional wet sieve (Yoder, 1936) method, the proposed technique simulates the impact of natural rainfall conditions on large

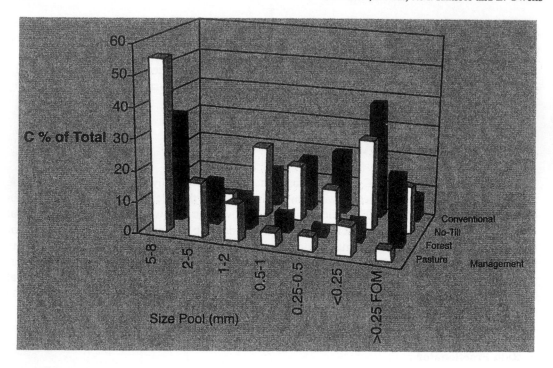

Figure 3. Percent of total soil C in size pools under various management practices.

aggregates. It also has greater potential to analyze the fine size pools that are critical to assessing the impact of water erosion on the pools and fluxes of SOC. Traditional wet sieving yields about 9 L of dilute microaggregates and it is cumbersome to perform analysis on this pool (Kemper and Rosenau, 1986).

B. Flux Assessment by Indirect Method

The loss of SOC in runoff and eroded sediments from watersheds is difficult to assess and methods of evaluating this component of the carbon cycle need to be developed. Because SOC is transported along with sediments, its loss may be highly correlated with runoff, soil, and nutrient losses by erosion. If readily quantifiable, the derived regressions might be used for indirectly calculating SOC transport flux by water erosion from more readily obtained data. This hypothesis was tested using 13 years of runoff and soil erosion data from two conservation tillage watersheds of the North Appalachian Experimental Watersheds in Coshocton, OH (no-till watershed 118 and chisel-till watershed 123 described in Kelley et al., 1975). The corn (*Zea mays*) – Soybean (*Glycine max*) rotation of these watersheds is described in Shipitalo and Edwards (1998).

Flumes, equipped with Coshocton wheel samplers, were used to collect flow-weighted subsamples of runoff that were analyzed for total SOC, N, P, K, and soil losses in rainstorm events. The SOC transport, ranging from 0 to 357 kg ha[-1], depended on the severity of the storm and was highly correlated (R^2 ranging from 0.74 to 0.94) in power law relationships with N, P, K, soil loss, and runoff. No-till and chisel-till watersheds exhibited nearly identical regression models.

The strongest correlation was SOC vs. K or SOC vs. runoff, both of which had a power near 1 indicating direct proportionality. These data suggest that it may be possible to develop empirical relationships by which SOC losses by erosion could be assessed for specific soil and management systems given certain measured or modeled variables such as soil loss, runoff, and nutrient balances. Understanding transport mechanisms may lead to development of conceptual models for predicting C dynamics in agriculture.

V. Summary

There is an array of interrelated pools, fluxes, and processes that have some bearing on the problem of assessing the impact of erosion on SOC dynamics. The fluxes of direct erosional transport, GHG emission, and sedimentation are interlinked with pools of C in eroded and depositional landscapes, buried aquatic sediments, and dissolved and particulate suspended sediment C in rivers, lakes, and oceans. Assessment methods focus on specific pools and fluxes. As a result, data are disjointed and a comprehensive understanding of interconnected pools and fluxes remains elusive. Quantitative inferences drawn from the limited and sketchy data sets that exist are, by necessity, hypothetical when larger (i.e., continental and global) scales are considered or the net impact of erosion on GHG dynamics are estimated. Based on the literature discussed in this chapter, erosion leads to losses of SOC from the soil pool and reductions in above and below ground biomass in eroded landscapes while transferring some C to sediments and depositional landscapes that accumulate C.

VI. Future Directions

Assessment of pools and fluxes, methodologies for their assessment, and our understanding of the underlying processes are improving, but the task is by no means completed. A more complete description would be obtained if we could improve our assessment of:

1. The pools of SOC available for transport.
2. The transformation of material and landscapes by processes such as structural disintegration and selective transport.
3. The fate of transported material in aquatic ecosystems.

References

Anderson, D.W., S. Saggar, J.R. Bettany and J.W.B. Stewart. 1981. Particle size fractions and their use in studies of soil organic matter: I. The nature and distribution of forms of carbon, nitrogen and sulfur. *Soil Sci. Soc. Am. J.* 45:767-772.

Anderson, E.A. 1982. Soil respiration. p. 831-832. In: A.L. Page, R.H. Miller and D.R. Keeney (eds.), *Methods of Soil Analysis. Part 2. Chemical and Microbiological Properties.* American Society of Agronomy, Madison, WI.

Barbanti, A. and M.H. Bothner. 1993. A procedure for partitioning bulk sediments into distinct grain-size fractions for geochemical analysis. *Environmental Geol.* 21:3-13.

Collis-George, N. and R. Lal. 1971. Infiltration and structural changes as influenced by initial moisture content. *Aust. J. Soil Res.* 11:93-105.

Chengere, A. and R. Lal. 1995. Soil degradation by erosion of a Typic Hapludalf in central Ohio and its rehabilitation. *Land Degradation and Rehabilitation* 6:223-238.

Chowdhary, M.A., R. Lal and W.A. Dick. 1997. Long-term tillage effects on runoff and soil erosion under simulated rainfall for a central Ohio soil. *Soil Till. Res.* 42:175-184.

De Jong, E. and R.G. Kochanoski. 1988. The importance of erosion in the carbon balance of prairie soils. *Can. J. Soil Sci.* 68:111-119.

Edgerton, D.L., J.A. Harris, P. Birch and P. Bullock. 1995. Linear relationship between aggregate stability and microbial biomass in three restored soils. *Soil Biol. Biochem.* 27:1499-1501.

Elliott, E. T. 1986. Aggregate structure and carbon, nitrogen, and phosphorus in native and cultivated soils. *Soil Sci. Soc. Am. J.* 50:627-633.

Ellison, W.D. 1944. Studies of raindrop erosion. *Agric. Exp.* 25:131-136,181-182.

Fahnestock, P., R. Lal and G.F. Hall. 1995. Land use and erosional effects on two Ohio alfisols. *J. Sustain. Ag.* 7(2/3): 1995.

Forsberg, C. 1998. Which policies can stop large scale eutrophication? *Wat. Sci. Tech.* 37:193-200.

Frielinghaus, M. and W.-G. Vahrson. 1998. Soil translocation by water erosion from agricultural cropland into wet depressions (morainic kettle holes). *Soil Till. Res.* 46:23-30.

Geng, G.-Q. and D.R. Coote. 1991. The residual effect of soil loss on the chemical and physical quality of three soils. *Geoderma* 48:415-429.

Gregorich, E.G. and D.W. Anderson. 1985. Effects of cultivation and erosion on soils of four toposequences in the Canadian prairies. *Geoderma* 36:343-354.

Gregorich, E.G., K.J. Greer, D.W. Anderson and B.C.Liang. 1998. Carbon distribution and losses: erosion and deposition effects. *Soil Till. Res.* 47:291-302.

Hedges, J.J., R.G. Keil and R. Benner. 1997. What happens to terrestrial organic matter in the ocean? *Org. Geochem.* 27:195-212.

Hutjes, R.W.A., P. Kabat, S.W. Running, W.J. Shuttleworth, C. Field, B. Bass, M.A. F.da Silva Dias, R. Avissar, A. Becker, M. Claussen, A.J. Dolman, R.A. Feddes, M. Fosberg, Y. Fukushima, J.H.C. Gash, L. Guenni, H. Hoff, P.G. Jarvis, I. Kayane, A.N. Krenke, Changming Liu, M. Maybeck, C.A. Nobre, L. Oyebande, A. Pitman, R.A. Pielke Sr., M. Raupach, B. Saugier, E.D. Shultze, P.J. Sellers, J.D. Tenhunen, R. Valentini, R.L. Victoria and C.J. Vorosmarty. 1998. Biospheric aspects of the hydrological cycle. *J. Hydrology* 212-213:1-21.

Ittekkot, V. and R.W.P.M. Laane. 1991. Fate of riverine particulate organic matter. p. 233-243. In: E.T. Degens, S. Kempe and J.E. Richey (eds.), *Biogeochemistry of Major World Rivers.* John Wiley, New York.

Kay, B.D. 1998. Soil structure and organic carbon: a review. p. 169-198. In: R. Lal, J.M. Kimble, R.F. Follett and B.A. Stewart (eds.), *Soil Processes and the Carbon Cycle.* Advances in Soil Science, Lewis Publishers, Boca Raton, FL.

Kemper, W.D. and R.C. Rosenau. 1986. Aggregate stability and size distribution. p. 425-444. In: A. Klute (ed.), *Methods of Soil Analysis. Part I. Physical and Mineralogical Methods.* American Society of Agronomy, Madison, WI.

Kreznor, W.R., K.R. Olson, W.L. Banwart, and D.L. Johnson. 1989. Soil, landscape, and erosion relationships in a northwest Illinois watershed. *Soil Sci. Soc. Am. J.* 53:1763-1771.

Lal, R. 1995. Global soil erosion by water and carbon dynamics. In: R. Lal, J. Kimble, E. Levine and B.A. Stewart (eds.), *Soil Management and Greenhouse Effect.* Advances in Soil Science, Lewis Publishers, Boca Raton, FL.

Lal, R. 1997. Residue management, conservation tillage and soil restoration for mitigating greenhouse effect by CO_2-enrichment. *Soil Till. Res.* 43:81-107.

Lal, R. 1998. Soil erosion impact on agronomic productivity and environment quality. *Critical Rev. Plant Sci.* 17:319-464.

Le Bissonnais, Y. and D. Arrouays. 1997. Aggregate stability and assessment of soil crustability and erodibility: II. Application to humic loamy soils with various organic carbon contents. *European J. Soil Sci.* 48:39-48.

Lowery, T.A. 1998. Modelling estuarine eutrophication in the context of hypoxia, nitrogen loadings, stratification and nutrient ratios. *J. Environ. Management* 52:289-305.

Lowrance, R. and R.G. Williams. 1988. Carbon movement in runoff and erosion under simulated rainfall conditions. *Soil Sci. Soc. Am. J.* 52:1445-1448.

Ludwig, W., J.L. Probst and S. Kempe. 1996. Predicting the oceanic input of organic carbon in continental erosion. *Global Biogeochemical Cycles* 10:23-41.

Massey, H. F. and M.L. Jackson. 1952. Selective erosion of soil fertility constituents. *Soil Sci. Soc. Am. Proc.* 16:353-356.

Mermut, A.R., D.F. Action and W.E. Eilers. 1983. Estimation of soil erosion and deposition by a landscape analysis technique on clay soils in southern Saskatchewan. *Can. J. Soil Sci.* 63:727-739.

Mitchell, P.D., P.G. Lakshminarayan, T. Otake and B.A. Babcock. 1998. The impact of soil conservation policies on carbon sequestration in agricultural soils of the central United States. In: R. Lal, J.M. Kimble, R.F. Follett and B.A. Stewart (eds.), *Management of Carbon Sequestration in Soil*. Advances in Soil Science, CRC Press, Boca Raton, FL.

Mulholland, P.J. and J.W. Elwood. 1982. The role of lake and reservoir sediments as sinks in the perturbed global carbon cycle. *Tellus* 34:490-499.

Ongley, E.D., M.C. Bynoe and J.B. Percival. 1981. Physical and geochemical characteristics of suspended solids, Wilton Creek, Ontario. *Can. J. Earth Sci.* 18:1365-1379.

Owens, L.B. and W.M. Edwards. 1993. Tillage studies with a corn-soybean rotation: surface runoff chemistry. *Soil Sci. Soc. Am. J.* 57:1055-1060.

Paul, E.A., W.R. Horwath, D. Harris, R. Follett, S.W. Leavitt, B.A. Kimball and K. Pregitzer. 1995. Establishing the pool sizes and fluxes in CO_2 emissions from soil organic matter turnover. p. 297-306. In: R. Lal, J.M. Kimble, E. Levine and B.A. Stewart (eds.), *Soils and Global Change*. Advances in Soil Science, CRC Press, Boca Raton, FL.

Pennock, D.J., D.W. Anderson and E. de Jong. 1994. Landscape-scale changes in indicators of soil quality due to cultivation in Saskatchewan, Canada. *Geoderma* 64:1-19.

Pert, M. R. and D.E. Walling. 1982. Particle size characteristics of fluvial suspended sediment. p. 397-407. In: *Recent Developments in the Explanation and Prediction of Erosion and Sediment Yield* (Proceedings of the Exeter Symposium, July 1982). (IAHS Publ. no. 137).

Piccolo, A, G. Pietramellara, and J.S.C. Mbagwu. 1997. Reduction in soil loss from erosion-susceptible soils amended with humic substances from oxidized coal. *Soil Technology* 10:235-245.

Reicosky, D. C., W.A. Dugas, and H.A. Torbert. 1997. Tillage-induced soil carbon dioxide loss from different cropping systems. *Soil Till. Res.* 41:105-118.

Riley, S.J. 1998. The sediment concentration-turbidity relation: its value in monitoring at Ranger Uranium Mine, Northern Territory, Australia. *Catena* 32:1-14.

Ritchie, J.C. 1989. Carbon content of sediments of small reservoirs. *Water Resour. Bull.* 25:301-308.

Rogers, H. T. 1941. Plant nutrient losses by erosion from a corn, wheat, clover rotation on Dunmore silt loam. *Soil Sci. Am. Proc.* 263-270.

Shipitalo, M.J. and W.M. Edwards. 1998. Runoff and erosion control with conservation tillage and reduced-input practices on cropped watersheds. *Soil Till. Res.* 46:1-12.

Spitzy, A. 1991. Dissolved organic carbon in rivers. p. 213-232. In: E.T. Degens, S. Kempe and J.E. Richey (eds.), *Biogeochemistry of Major World Rivers*. John Wiley & Sons, New York.

Stallard, R.F. 1998. Terrestrial sedimentation and the carbon cycle: coupling weathering and erosion to carbon burial. *Global Biogeochemical Cycles* 12:231-257.

Sutherland, R.A. and A.D. Ziegler. 1996. Aggregate enrichment ratios for splash and wash transported sediment from an Oxisol. *Catena* 26:187-208.

Tiessen, H.J.W., B. Stewart and J.R. Bettany. 1982. Cultivation effects on the amounts and concentration of carbon, nitrogen, and phosphorus in grassland soils. *Agron. J.* 74:831-885.

Tisdall, J. M. and Oades, J. M. 1982. Organic matter and water-stable aggregates in soils. *J. Soil Sci.* 33:141-163.

Van Veen, J.A. and E.A. Paul. 1981. Organic carbon dynamics in grassland soils. I. Background information and mathematical simulation. *Can. J. Soil Sci.* 61:185-201.

Varoney, R.P., J.A.Van Veen and E.A. Paul. 1981. Organic C dynamics in grassland soils. 2. model validation and simulation of long term effects of cultivation and rainfall erosion. *Can. J. Soil Sci.* 61:211-224.

Vezjak, M., T. Savsek, and E.A. Stuhler. 1998. System dynamics of eutrophication processes in lakes. *European J. Operations Res.* 109:442-451.

Wan, Y. and S.A. El-Swaify. 1997. Flow-induced transport and enrichment of erosional sediment from a well-aggregated and uniformly-textured Oxisol. *Geoderma* 75:251-265.

Wass, P.D., S.D. Marks, J.W. Finch, G.J.L. Leeks and J.K. Ingram. 1997. Monitoring and preliminary interpretation of in-river turbidity and remote sensed imagery for suspended sediment transport studies in the Humber catchment. *Sci. Tot. Environ.* 194-195:263-283.

Yoder, R.E. 1936. A direct method of aggregate analysis and a study of the physical nature of erosion losses. *J. Am. Soc. Agron.* 28:337-351.

Young, R.A. and Onstad, C.A. 1978. Characteristics of rill and interrill eroded soil. *Trans. ASAE* 21:1126-1130.

Zobisch, M.A., C. Richter, B. Heiligtag and R. Schlott. 1995. Nutrient losses from cropland in the central highlands of Kenya due to surface runoff and soil erosion. *Soil Till. Res.* 33:109-116.

CHAPTER 30

Assessing Water Erosion Impacts on Soil Carbon Pools and Fluxes

P.A. Jacinthe, R. Lal and J.M. Kimble

I. Introduction

Concerns about the rising carbon dioxide (CO_2) concentration in the atmosphere (0.5% yr^{-1}) and the projected warming of world climate have hastened studies of carbon (C) dynamics within known C reservoirs and the fluxes of CO_2 from these reservoirs to the atmosphere. During the last three decades, a wealth of data regarding CO_2 evolution in various ecosystems has been collected and several global C budgets have been constructed. A constant observation of these budgets is an apparent imbalance of the global C cycle, giving rise to the so-called "missing C," the amount of C (1.2 to 1.8 Pg C) needed to balance the cycle. A 1992 report to the Intergovernmental Panel on Climate Change (IPCC) states that estimates of C fluxes to the atmosphere from land-use changes remain among the most uncertain figures in the global C cycle (Watson et al., 1992). It has been suggested that temperate region forests may represent an additional sink not fully accounted for in global C budgets. The capacity of this sink has been estimated at 1.37 Pg C yr^{-1} (Sedjo, 1992). There are indications that this sink can be enhanced due to forest regrowth, CO_2 fertilization and atmospheric N deposition effects (Hudson et al., 1994; Thornley et al., 1991). Conversely, cyclical events like fires can greatly reduce the role of forests as C sink. Nabuurs et al. (1997) estimated that about half of the C sink of European forests (0.101 Pg C yr^{-1}) can be lost due to fire-related emission of CO_2 (0.06 Pg C yr^{-1}).

Next only to the ocean (38×10^3 Pg C) and geological pool (5×10^3 Pg C), soil is the largest C reservoir in the biosphere. Reported global soil organic C pools vary between 1500 and 1760 Pg C (Post et al., 1982; Eswaran et al., 1993) which is about two and three times the amount of C in the atmosphere (750 Pg C) and in the terrestrial biota (570 Pg), respectively (Post et al., 1982). Because of the size of the soil C reservoir, even minor disturbances could have measurable impact on atmospheric CO_2 level. Reinforcing this possibility is the recognition that two-thirds of actively cycling organic carbon in the biosphere is on land, as opposed to only one-third in the ocean (Hedges et al., 1997).

In agricultural soils, soil organic carbon (SOC) reserves change with time reflecting the combined effects of crop residues inputs and processes leading to SOC losses. Mechanisms of SOC losses include physical (eg. erosion, leaching) and biological (soil respiration) processes. However, in many studies, temporal decline in SOC levels is often ascribed only to soil respiration. Soil erosion is an SOC depleting process that can be of significance in some context. It involves the breakdown of soil aggregates, the release of physically protected C and SOC translocation. Released SOC can be displaced to other parts of the landscape, undergo microbial decomposition or be buried in depositional sites. The relative importance of each of these possible fates remains speculative at the present time.

This review recognizes the lack of data relating soil erosion and emission of CO_2 from eroded soils and attempts to address this information gap. This is a serious oversight given the potential economical and environmental consequences of these two global and interrelated environmental issues. In this chapter, we first discuss SOC distribution within the soil matrix with particular attention to the restriction on SOC decomposition imposed by soil architecture and a possible increase in SOC decomposition when this architecture is disturbed by water erosion. Then, we derive estimates of the magnitude of erosion-induced CO_2 fluxes, based on current understanding of SOC dynamics and published data on SOC pools.

II. Organic Carbon and the Soil Architecture

A. Ecological Importance and Sources of SOC

The roles of SOC as plant nutrient storage, a source of energy for soil microbes, and binding agent in development and stabilization of soil structure have been extensively studied (Allison, 1973). Decomposition of SOC by soil microbes results in CO_2 and CH_4 release into the atmosphere depending on the oxidation status. The extent of the decomposition process is controlled by SOC quality (labile vs. recalcitrant), accessibility and activity of the microbial community.

B. Characterization of SOC

Researchers have long been interested in SOC characterization in an attempt not only to understand its composition but, more importantly, to be able to relate composition to ecologically meaningful attributes. There is broad agreement among scientists that SOC can be partitioned into several hypothetical and kinetically defined pools based on resistance to microbial degradation: labile, partly labile and passive pools (Jenkinson and Rayner, 1977; Parton et al., 1994). The labile C pool has a short turnover time (< 5 yr), and includes the rapidly decomposable SOC fractions. The passive C pool has a turnover time of several hundred years and is comprised of microbial-resistant complex macromolecules such as the humic substances (Nicolardot et al., 1994). Chemical and biochemical approaches have been used to estimate the size of these pools. In Table 1, we summarize data regarding the size of the different SOC fractions associated with labile SOC.

Dissolved organic carbon (DOC) is often used as an index of SOC availability given its correlation with heterotrophic microbial activity (Cook and Allan, 1992), soil respiration (Burford and Bremner, 1975) and CO_2 emission (Rochette and Gregorich, 1998). Wet-dry (Zsolnay and Gorlitz, 1994) and freeze-thaw (Wang and Bettany, 1993) cycles generally increase DOC concentration. This increase is ascribed to death of microbes, disruption of soil structure and release of previously protected carbon. Except during these events, and in samples from wetlands (Tao and Lin, 2000), DOC concentrations in soils typically represent < 1 to 2% of the total SOC (Table 1).

Densimetric methods allow separation of SOC into light, intermediate and heavy fractions relative to a solvent of known density (Strickland and Sollins, 1987). The light fraction (LF), composed of unrecognized and incompletely decomposed plant and microbial debris (Janzen et al., 1992; Gregorich et al., 1998), is often taken as the labile organic C pool and is correlated with microbial biomass C and basal soil respiration. Mineralization of the LF carbon proceeds at rates 1.5 to 5 times faster that the medium and heavy fractions, respectively (Alvarez et al.,1998). Curtin et al. (1998) observed that the amount of CO_2-C emitted from soils was similar to the amount by which the LF C had decreased. This observation was taken as an indication that the labile C, represented by the LF C, was a major source

Table 1. Size of labile soil organic carbon pools reported in studies from various ecological regions and in soils of different land use history

Literature reference	Location and soil type	Cropping system	Total SOC	Biomass C	Density fractions[1]			Mineralized C[2]	DOC
					LF	MF	HF		
			g C kg⁻¹		———— mg C kg⁻¹ ————				
		Temperate region							
Lundquist et al., 1999	Sacramento Valley, CA, USA Typic Xerothents	Organic tomato	7.8	104					117
	Salinas Valley, Pachic Haploxeroll	Conventional tomato	7.6	67					14
		Lettuce-celery	15.8	83					19
Hu et al., 1999	Davis, CA, USA Mollic Xerofluvent	Fallow for 2 yr., then oat-vetch during the year prior to sampling	10.8	80 - 310				600	30 -80
Ajwa et al., 1998	Konza prairie, Kansas, USA Mesic pachic Argiudoll	Tallgrass prairie, grazing	22.1					1792 (2561)	27
		Continuous corn (1989-1991), wheat (1992-1994)	12.5						
Larney et al., 1997	Alberta, Canada Typic Cryoboroll	Wheat - fallow, MT	13.8		603			347	
		Wheat - fallow, NT	13.9		653			367	
Biederbeck et al., 1996	Saskatchewan, Canada Brown Chernozem	Wheat - barley, 10 yr (0 - 7.5 cm)							
		Wheat - barley, 10 yr (7.5 - 15 cm)							

Table 1. (continued)

Literature reference	Location and soil type	Cropping system	Total SOC	Biomass C	Density fractions[1]			Mineralized	
					LF	MF	HF	C[2]	DOC
			g C kg⁻¹		mg C kg⁻¹				
Gregorich et al., 1995	Winchester, Canada Humic Gleysol	Forest	139		3475				
		Forage (60 yr)	21		756				
		Corn (15 yr)							
Haider et al., 1991	Bavaria, Germany	Bare fallow	7.7	70 - 110					
		Potato	10	200 - 220					
		Grassland	17.6	970 - 1130					
		Barley-potato-wheat (manure)	11.5	330 - 370					
		Barley-potato-wheat (fertilizer)	12.9	420 - 680				1792 (2561)	27
				Tropical region					
Alvarez et al., 1998	Pergamino, Argentina Typic Argiudoll	Corn-soybean, CT	15.8	157	379[3]	2314	12853	160 (1300)	
		Corn-soybean, NT	16.5	219	778	2804	12963	200 (1867)	
		Pasture (15 yr)	22.4	669	1832	4102	16516	467 (3300)	
Guggenberg and Zech, 1999	Nuerta Norte, Costa Rica Typic Humitropepts	Primary forest	72		7222				
		Pasture	58		2320				
		Secondary f. (3.5 yr)	71		4575				
		Secondary f. (12.5 yr)	75		5460				
		Secondary f. (18.5 yr)	81		6292				

Reference	Location	Soil	Treatment					
Haynes, 1999	Canterbury, New Zealand	Udic Dystrochrept	Barley with burning of straw	29	350	500	300	250[4]
			Ryegrass -white clover	42	800	600	450	370
				42	780	580	450	380
			Continuous ryegrass + 150 kg N ha^{-1} yr^{-1}	32	370	650	320	260
			Ryegrass, plowed under in summer	37	600	700	380	350
			Ryegrass, no-till	63	1200	1700	560	460
			Pasture (20 yr)					
Barrios et al., 1996	Machakos, Kenya	Kandic Rhodustalf	Continuous corn	9.6		331		
			Cowpea - corn	9.6		507		
			Pigeon pea - cowpea intercropped with corn	9.6		371		
				9.6		400		
			Pigeon pea - corn intercrop	9.6		443		
			Corn + prunnings of Glircidia	9.6		358		
			Corn - Glicidia intercrop	9.6		385		
			Gliricidia stand					

[1]LF, MF, HF means light, medium and heavy fractions, respectively. Density fractions were determined using various solvents. For example, the data inserted in the table from the study by Barrios et al. (1996) are for LF carbon using sodium iodide. In the same study, LF carbon extracted with sodium polytungstate was generally twice as much, whereas LF carbon was about twice less when ludox was used.
(continued next page)

Table 1. (continued)

[2]Mineralized C is typically obtained during a 10 - 30 day incubation. Ajwa et al. (1998) and Alvarez et al. (1998) used a 40- and 24-wk incubation, respectively. Values in parentheses are potentially mineralizable C, which is derived by fitting experimental data to a first order model of SOC decomposition.

[3]Alvarez et al. (1998) reported that 58, 37 and 10% of the LF, MF and HF carbon, respectively, was mineralized during a 160-day incubation in the laboratory.

[4]In this study, DOC was extracted with 0.5 M K_2SO_4.

Abbreviations: MT = minimum tillage; CT = conventional tillage; NT = no-till.

of respired C (Curtin et al., 1998). Research data summarized in Table 1 shows that the LF C represents between 2 and 10% of the total SOC.

Compared to chemical methods, biochemical methods which rely on the determination of SOC decomposition products (CO_2, CH_4) are thought to be more integrative, better mimic *in situ* conditions and provide more valuable information regarding the fate of organic carbon in soils. Various biochemical indicators including microbial biomass and mineralizable C (Jenkinson and Powlson, 1976; Jenkinson and Ladd, 1981) have been used to assess the size of the labile C pool. Mineralizable C is typically determined by measuring the amount of CO_2 evolved during a 10-day incubation. In some cases (Ajwa et al., 1998; Alvarez et al., 1998), incubation is carried out for a longer period of time (100 to 300 days) and the potentially mineralizable C is computed by fitting experimental data to decomposition models.

Regardless of the pools considered and the approach used to estimate their size, the active C fraction represents, at best, 4 to 15% of the total SOC (Table 1). This estimate considers the labile SOC pool to include the DOC, part of the microbial biomass and LF C. It is consistent with kinetically derived estimates of potentially mineralizable C (8 to 14% of the total SOC) listed in Table 1. Although representing a small fraction of the total SOC, the labile C pool has great ecological significance in that it has a short turnover time, is more responsive to management practices (Powlson et al., 1987; Carter, 1991) and therefore can be used to trace short-term change in SOC dynamics. Available data suggest that the most biologically active SOC fractions are the most prone to loss when soils are cultivated. In a study of labile C in soils of different land-use histories, Woods and Schuman (1988) observed that after 1 year of cultivation, the size of the active C pool dropped sharply to levels comparable to soils that had been cultivated for 25 years. Data from Campbell et al. (1999) showed that depletion of the LF C occurs at rates four to five times faster than the total SOC.

C. Distribution of Organic Carbon within the Soil Matrix

The quality and amount of SOC associated with soil aggregates and particles vary with the size of these structural units. It is often observed that C in micro-aggregates are older and less labile than C in macro-aggregates (Elliott, 1996; Jastrow et al., 1996). Bonde et al. (1992) noted that, in general, more than half of the total SOC is associated with the clay (< 2 μm) fraction, probably due to the protective effect of clay on SOC mineralization (Saggar et al., 1996). The degree of protection provided by clay appears to be dependent on the type of clays with the highly porous (eg. allophane), expanding [and high surface-charged clays (eg., montmorillonite) (Zech et al., 1997)] offering more protection than kaolinite (Hassink, 1994; Franzluebbers et al., 1996).

D. Soil Microorganisms within the Soil Architecture

The spatial distribution of microbial life in soils is largely dictated by the heterogeneous nature of soils and the variable environmental conditions encountered even at the micro-scale level. The capillary pores (2 to 6 μm) are viewed as the most favorable micro-habitats for bacteria given the long-term availability of water in these pores, and the reduced possibility for these microbes to be grazed by other soil organisms such as flagellates, amoebae and ciliates (Hattori and Hattori, 1976). Thus, the soil bacterial population is more abundant in the inner than in the outer region of soil aggregates, while the reverse is true for soil fungi (Gupta and Germida, 1988).

Soil microbes are essentially immobile and dispersed within the soil mass. It follows that substrate quality is a partial determinant of *in situ* decomposition of SOC. The physico-chemical interactions between SOC and soil minerals, the spatial separation between substrate and microbes, and the limited diffusion of degraders within the soil matrix are strong controllers of the decomposition process. Soil disturbances such as water erosion could alter the soil architecture, dislodge protected SOC, and stimulate mineralization of previously occluded SOC.

III. Water Erosion

A. Extent of the Problem

Soil erosion is one of the most pervasive soil degradation processes affecting an estimated 1100 Mha by water erosion and 550 Mha by wind erosion worldwide (Oldeman, 1994), and resulting in the transport of 20 to 25 × 10^9 Mg of soil to the ocean per year (Brown, 1984; Milliman and Syvitski, 1992). Pimentel et al. (1995) reported that 10^7 ha of agricultural lands are degraded every year due to erosion which thereby contributes to the continuous shrinking of the soil resource-base on which the growing world population depends for food.

While natural erosion rates typically range between 0.1 and 7 Mg ha^{-1} yr^{-1} (Lal, 1990), human-induced erosion rates can be several orders of magnitude higher. Worldwide rates of soil erosion on cropland average 30 Mg ha^{-1} yr^{-1} with a range of 0.5 to 500 Mg ha^{-1} yr^{-1} (Pimentel et al., 1995). Average soil loss rates in Europe and the United States are in the 10 to 20 Mg ha^{-1} yr^{-1} range (Pimentel, 1993), whereas in cultivated steepland in some parts of the tropics, erosion rates can be one or two orders of magnitude higher (Lal, 1986; Pimentel et al., 1987).

B. The Physics of Water Erosion

Water erosion is initiated by the breaking of soil aggregates impacted by raindrops. Detached particles are further reduced in size and transported away in runoff waters. Soil erosion is a self-reinforcing process; detached soil particles clog pore spaces, reduce soil water infiltration and further increase erosion intensity.

Soil erosion is considered the most complete form of land degradation resulting in declined soil productivity (Lal, 1990) and diminished ecosystem stability (Pimentel et al., 1995). Soil erosion affects soil physical and chemical properties and undermines the biological processes supporting nutrient cycling in soils, and thereby impairs the capacity of these resources to sustain long-term production. Erosion removes the most fertile and the most biologically active portion of the soil profile and exposes the subsoil which often does not provide physical and chemical conditions favorable for plant growth. The low density organic matter, concentrated in the surface layers of soils, is readily removed during erosional events. Van Noordwijk et al. (1997) stated that the organic carbon content of eroded material is at least twice that of the average topsoil. Several other studies have also documented this C enrichment (E_R) effect. Palis et al. (1997) reported E_R between 1 and 3 in runoff sediments, with the greatest enrichment in the 0.25 to 1 mm size fraction and in sediments collected during the first 5 minutes following the onset of runoff. In experiments conducted on Alfisols in western Nigeria, Lal (1976) reported E_R for organic C averaging 2.2 to 2.7. Starr et al. (2000) reported an average carbon E_R of 2.1 for run-off sediments collected during the period 1984–1996 at the North Appalachian Experimental Watershed, Coshocton, Ohio.

C. Water Erosion and Soil Organic Carbon Reserves

Following conversion of natural ecosystems into cropland, significant decreases in SOC levels are often recorded. Decline in SOC can be attributed to the biophysical changes in the soil environment resulting from physical disturbances such as land clearing, burning and tillage. In addition to mixing the topsoil with the SOC-poor subsoil, cultivation increases soil aeration, alters soil moisture and temperature patterns, and ultimately leads to accelerated rates of SOC decomposition. Soil carbon stocks in cultivated soils typically drop to 30 to 50% of their original levels after 50 to 100 years of cultivation (Campbell et al., 1967; Schlesinger, 1986). The rates of SOC depletion vary with the intensity of cropping and management practices. Detwiller and Hall (1988) estimated that conversion of forest soils to permanent agriculture results in an average SOC decline of 0.8% yr^{-1}, whereas decline can be in the order of 9 to 13% yr^{-1} if forested land is converted to shifting cultivation.

Soil erosion and off-site transport of organic C are significant contributors to SOC decline in cultivated soils, but only a limited number of studies (De Jong and Kachanoski, 1987; Pennock and al., 1994) have attempted to quantify its contribution. Cultivation increases soil erosion risks by reducing the amount of residues on soil surface and rendering soil aggregates less resistant to breakdown (Woods and Schuman, 1988; Fuller et al., 1995).

IV. Linking Erosion and Carbon Dynamics

A. General

The interrelationship between SOC and erosion is well established. Both the content and the distribution of organic matter within the bulk soil (Powlson, 1996) have important implications with regard to susceptibility of soils to erosion. In fact, SOC content is one of the master variables in determining the soil erodibility factor (K) used in the universal soil loss equation (USLE) (Wischmeier and Smith, 1978). Maintenance of plant residues on the soil surface through such management practices as no-till and mulching has been found to substantially reduce soil erosion rates (Paningbatan et al.,1995; Roose and Ndayizigiye, 1997). While the protective role of SOC against water erosion is well known, there is limited data to document the changes in SOC dynamics associated with erosional events, suggesting that the second aspect of the interrelationship deserves additional research.

Cultivation generally results in loss of SOC through increased soil respiration and erosion. Long-term research has shown that, in the years immediately following conversion of natural ecosystems, SOC contents decline sharply and then decrease at much moderate rates afterwards to stabilize at around 30 to 50% of the original level depending on intensity of cultivation and management practices (Bohn, 1978; Monreal et al., 1997).

By comparing SOC levels in native and cultivated soils, rates of SOC depletion have been computed and these rates are often used to determine the contribution of land use change to atmospheric CO_2 (Houghton et al., 1983; Smith et al., 1997; Mullen et al., 1999). This approach assumes that soil respiration is the major cause of SOC loss and can yield erroneous results if processes such as leaching and erosion are significant.

The contribution of erosion to CO_2 fluxes into the atmosphere has not been taken into consideration in most studies documenting SOC decline in agricultural soils. Several researchers (Elliott et al., 1996; Lee et al., 1996; Gregorich et al., 1998) have alluded to soil erosion as a perturbing factor of the global C cycle but quantitative estimates of soil erosion contribution have not been provided. Schlesinger (1986) discounted erosional C loss as an important factor explaining SOC decline following land-use conversion. Schlesinger (1984) stated that erosion-induced SOC losses are small compared to soil respiration. While this may be correct for the years immediately following forest conversion to agriculture, this statement may not hold beyond the initial flush of decomposition, specially for cultivated steeplands. Erosional C losses can be substantial in these regions where agricultural soils are eroding at rates in excess of 50 Mg ha^{-1} yr^{-1} (Lal, 1986; Pimentel et al., 1987). Slater and Carleton (1938) monitored SOC evolution in continuous fallow plots in Missouri and estimated that rates of SOC decline due to erosion were 18 times those due to oxidation. Gregorich and Anderson (1985) estimated that 20 to 70% of organic C losses in Chernozemic soils in Canada is attributable to erosion. Using the Century model, Gregorich et al. (1998) predicted that, after 20 to 71 years of cultivation, 67% of the total SOC in Chernozemic soils would be lost when erosion is included in the simulation as opposed to a 20% SOC loss without erosion. These estimates compare well with data from de Jong and Kachanoski (1988) who, based on change in ^{137}Cs concentration profiles, reported erosional C losses averaging 27 g C m^{-2} yr^{-1} in the Canadian prairies. In a study investigating SOC evolution in different landscape elements, Pennock et al. (1994) attributed 70% of the observed SOC decline to export from upper to lower landscape positions.

Erosion is a multi-component process involving detachment, translocation and deposition of particles. Erosion selectively removes the lighter and smaller soil particles which are deposited in depressions and in lower landscape positions. If the biologically active SOC fractions are associated with the most easily eroded materials (Voroney et al., 1981; Tiessen and Stewart, 1983), one should therefore expect C in depositional areas to be more active than residual C in the upslope positions (Gregorich et al., 1998). Gregorich (1996) found that mineralizable C was lowest at the mid-slope positions and greatest (2 to 2.4 times) at the lower slopes, suggesting translocation of the labile fractions of SOC to depositional sites. Anderson et al. (1986) reported soil accretion in lower slope positions in cultivated soil toposequences indicating transport of eroded soil. A lower C content at depositional sites than in the native soils was interpreted as an indication of active C mineralization in depression areas.

Current knowledge about soil erosion processes allows a good qualitative description of possible pathways for eroded SOC. Essentially, macro-aggregates are disrupted by raindrop impact and the particles released are further reduced in size by the shearing force of runoff waters. Abrasion of soil particles could also result in the liberation of SOC coating mineral grains. The combined net effect of these processes is the release of finer soil particles and of organic C occluded within the aggregates. The organic C liberated is transported as dissolved and particle-associated C to varying distances over the eroding landscape depending on weight and size of particles and carrying capacity of run-off. Some of the mobilized C is deposited in terrestrial deposits while the remainder is carried to streams and rivers.

In quantitative terms, however, the fate of SOC mobilized during erosional events has not been adequately assessed. This information gap contrasts sharply with the wealth of published data regarding the fate of terrestrial C in aquatic systems which, in recent years, has resulted in the publication of numerous review papers regarding riverine C and the global C budget (Degens et al., 1991; Hope et al., 1994; Benner et al., 1995; Ludwig and Probst, 1996; Ludwig et al., 1997).

B. The Various Schools of Thought

The magnitude of soil erosion impact on atmospheric CO_2 loading is largely unknown. Yet, the question has been raised in several reviews and articles (Lowrance and Williams, 1988; Beyer, 1993; Lal, 1995; Lee et al., 1996) which, upon analysis, reveal significant differences in the assumptions put forth by various investigators.

A series of climate/CO_2 scenarios for the U.S. corn belt developed by Lee et al. (1996) predicted that, over a 100-year period, soils in this region would lose 4.8 Mg C ha^{-1}, about half of which would be transported off site, presumably by erosion. Lee et al. (1996) stated that some of the mobilized SOC will be oxidized to CO_2 but provided no estimates of probable CO_2 emission to the atmosphere. Using CPMAS ^{13}C-NMR spectroscopy, Beyer et al. (1993) compared functional groups in SOC extracted from eroded and depositional sites in Germany. From the differences among sites observed in the NMR data, Beyer et al. (1993) inferred that 70% of SOC in eroded soil is decomposed during transport and deposition. Lal (1995) assumed that 20% of C contained in eroded soils can be mineralized and, using world's data on soil erosion, estimated that erosion could contribute to an annual flux of 1.14 Pg C into the atmosphere. In a rainfall simulation study, Lowrance and Williams (1988) measured erosion-induced SOC loss between 5.8 to 15.2 g C m^{-2} (3.5 to 7.5 kg C ha^{-1} mm^{-1} rainfall), approximating about 1% of the total SOC. They also reported that 90 to 95% of the total C in runoff was in the particulate form (POC) but presented no data on degradability of the exported SOC.

In contrast to investigators who assumed a partial decomposition of eroded SOC, van Noordwijk et al. (1997) suggested that relocation of SOC into swamp and aquatic sediments could actually contribute to C sequestration, given the low rates of C mineralization in these environments (Schimel et al., 1985). In the same vein, Stallard (1998) proposed that eroded C can be sequestered and removed from the global C cycle. His assertion stems from the fact that, instead of being transported to streams and rivers, most eroded soil remains trapped in terrestrial deposits and suggested that this process could lead to sequestration of 0.6 to 1.5 Pg C yr^{-1} accounting for most of the so-called "missing carbon."

There are, however, two fundamental issues which have not been adequately evaluated in formulating the hypothesis of eroded SOC sequestration in terrestrial deposits. One is that historical evidence strongly suggests that soil organic matter and productivity of land-based ecosystems are linked, and that accelerated soil erosion reduces SOC levels and affects productivity. Data from Lal (1993) indicated a linear to exponential decrease in corn and cowpea yields with cumulative topsoil loss. Dazhong (1993) estimated that 3% of the potential annual grain production in China is lost due to severe soil erosion. Studies from various regions of the world and with several crops (Olson and Nizeyimana, 1988; Mokma and Sietz, 1992) suggested a 12 to 65% decline in crop productivity due to erosion. Diminished productivity also means a progressive reduction in the amount of crop residues returned to the soil to maintain organic matter levels. Bauer and Black (1994) estimated that each Mg ha^{-1} of organic matter contributes to the yearly production of 35 kg ha^{-1} of wheat biomass. Erosion-induced decline in crop productivity can be attributed to a series of interacting factors including nutrients depletion, degradation of soil structure and reduced soil water holding capacity. Monreal et al. (1997) showed that when erosion rates exceed 14 Mg ha^{-1} yr^{-1}, there is a net loss of SOC even with fertilizer application. Therefore, the assumption that growing biomass will replace translocated SOC at the eroded site does not seem tenable in the long run. A second issue needing further clarification in the erosion-sequestration of SOC hypothesis is that eroded SOC decomposes slowly at some depositional sites due to limited oxygen availability in these environments. This scenario must be weighted against the potential for the production of nitrous oxide (N_2O) and methane (CH_4) which, owing to their greater global warming potential relative to CO_2, could also have serious impact on global climate.

The lack of research data on the fate of erosion-derived SOC and global CO_2 flux has been recognized in the IPCC (1996) manual for national greenhouse gas inventory, which states that the net effect of erosion on CO_2 evolution is unclear at the present time and that individual countries must decide whether and how to account for soil erosion on their national C inventories. In the meantime, the IPCC (1996) methodology suggests that SOC depletion over time be equated with soil contribution to atmospheric CO_2. Although not stated, an implicit assumption in this approach is that SOC mobilized with eroded soil will be mineralized to CO_2. This assumption can be problematic, since in reality only a fraction of the C removed is likely to be mineralized, and currently the size of this fraction is unknown.

C. Estimating Mineralizable Carbon in Eroded Soils and Derivation of CO_2 Fluxes

Run-off from eroded fields can be viewed as a complex mixture of water, microbial cells, plant debris, freely-dissolved organic C (DOC) and particle-associated organic C (POC). Thus, we will consider the fate of mobilized organic C in both the aqueous (DOC) and the solid phases (POC).

Filtrates of DOC provide an environment in which there is minimal restriction on SOC decomposition due to SOC-mineral associations, spatial separation between substrate and microbes, and limited diffusion of substrates. Mineralization in this case depends primarily on substrate quality. Research has shown that DOC encompasses numerous classes of organic compounds ranging from simple sugars and amino acids to more complex constituents such as the humic acids (Thurman, 1985). It is often suggested that the more aromatic humic acids are refractory while the hydrophilic moieties (proteins, polysaccharides) are more amenable to microbial decomposition (Spitzy and Leenheer, 1991). Martin-Mousset et al. (1997) studied the biodegradability of DOC in rivers and reservoirs from various regions in France and observed that between 8 and 23% of the DOC was degraded during 8 days of incubation. Mogren et al. (1990) reported a 12% degradation rate for DOC in the Ohio River. Qualls and Haines (1992) reported that 14 to 33% of soil, forest and stream DOC was decomposed during a 130-day incubation. Cropland DOC was found to be equally partitioned into refractory and labile C fractions; 85% of the latter is degradable (Zsolnay and Steindl, 1991). Boyer and Groffman (1996) reported that 20 to 25% of DOC extracted from forest and cultivated soils is degradable. In that study, it was also observed that cropland DOC was more labile than forest DOC and this observation was attributed to better quality of litter from fertilized soils compared to forest litter (Boyer and Groffman, 1996). Overall, the aforementioned studies suggest that only a fraction (15 to 30%) of DOC is decomposable.

During erosional events, some of the SOC mobilized remains associated with soil particles. Based on runoff plots data from Elliott et al. (1989), Starr et al. (2000) projected that 73% of the material mobilized during erosion are > 0.062 mm in size and are likely to be redeposited at other points over the eroding landscape. Materials in this size class include soil organisms (excluding virus), plant debris, and the coarse and the very fine POC (Thurman, 1985; Hope et al., 1994). Colonization of the SOC released and, consequently, an enhancement of CO_2 production during transport of the eroded soil over the landscape are expected. The data listed in Table 1 suggest that the labile C pool in most soils represents 4 to 15% of the total SOC. Runoff plot (Lal, 1976; Palis et al., 1997; Starr et al., 2000) and toposequence studies (Gregorich, 1996; Gregorich et al.,1998) have shown that total SOC in sediments is typically twice that of the original soil. We assume that a similar enrichment occurs with regard to the labile C fractions in translocated soils during erosional events, and predict that 8 to 30% of the eroded C associated with the solid phase is labile. This proposed value is consistent with Ittekot's (1988) estimates that at least 35% of POC in rivers is labile and decomposable. Our assumption is also supported by Schreiber and McGregor's (1979) finding that sediment and aqueous phase SOC are no different in terms of their biodegradability as measured by biological oxygen demand (BOD_5) assay.

Taking the mid-range values of C mineralization in the two phases (22% for DOC, and 19% for POC), and assuming a DOC/POC ratio of 1, we project that 20% of the SOC mobilized by water erosion could be mineralized. This value is similar to mineralization rate estimates proposed earlier by Lal (1995). Even in actively eroding landscapes, such as plowed soils (Schreiber and McGregor, 1979; Lowrance and Williams, 1988), in which more than 90% of the eroded SOC is in the particulate form, the weighted-average mineralization rate is still 19%.

To gain some perspective of the impact of erosion on global CO_2 fluxes, one needs to be reminded that current fluxes of sediments to the world's oceans are in the order of 25×10^9 Mg yr^{-1} (Brown, 1984; Milliman and Syvitski, 1992). The amount of sediments delivered to the oceans typically represents 13 to 20% (Walling, 1983) of the total soil mass mobilized by water erosion (100 to 200 $\times 10^9$ Mg yr^{-1}). Assuming an average C content of 3%, the potential CO_2 flux would range between 0.3 and 1.8 Pg C yr^{-1}. If we consider only the arable lands (1.51×10^9 ha; FAO, 1998) and assuming an average erosion rate of 30 Mg ha^{-1} yr^{-1} (Pimentel et al., 1995), and sediment C content of 3%, the erosion-induced CO_2 flux would be 0.27 Pg C yr^{-1}. Table 2 shows several global estimates of erosion-

Table 2. Estimates of global carbon dioxide fluxes linked to water erosion

Reference	Scale	Estimates, Pg C yr^{-1}†	Comments
Reported in:			
Lal (1995)	Global	1.14	Assuming annual global soil loss, 190 Pg; average C content of sediment, 3%; mineralization of eroded C, 20%.
This review	Global	0.3 - 1.8	All eroding landscapes included, and assuming: soil displaced annually by water erosion, 100 - 200 Pg; average C content of eroded soil, 3%; mineralization of eroded C, 20%.
This review	Cropland	0.3	Cultivated lands only, with the following assumptions: average annual erosion rate, 30 Mg ha^{-1}; average C content of sediment, 3%; mineralization of eroded C, 20%; and total cropland area: 1.51×10^9 ha (FAO, 1998).
Calculated from:			
Stallard (1998)	Global	0.6	The following was provided in the referenced paper: - C delivered to world ocean: 0.53 Pg C yr^{-1}; - C retained in terrestrial deposits: 0.6 - 1.5 Pg C yr^{-1}; - Total sediment in motion: 30 - 100 Pg yr^{-1}; We then took the mid-point of the data listed above, and assumed a C content of sediment of 3% to compute CO_2-C evolution during transport and deposition of eroded soil as: (70 Pg × 3%) - (0.53 + 1.01) = 0.56 Pg C yr^{-1}.
Several sources	Cropland	0.24	Using: - global estimate of C loss in cultivated soil: 100 to 300 g C m^{-2} yr^{-1} (Schlesinger, 1986), and - erosion contribution to soil C depletion: 20 to 70% (Gregorich and Anderson, 1985), 47% (Gregorich et al.,1998) erosion-induced CO_2-C flux from cropland was computed as: (200 g C m^{-2} yr^{-1})(40%)(20%)(1.51×10^{13} m^{-2})

† Other sources of atmospheric CO_2: fossil fuel burning (5.4 Pg C yr^{-1}); deforestation in the tropics and land-use (1.6 Pg C yr^{-1}). Data from Lal (1995).

Table 3. A hierarchical framework for studies linking soil erosion and the global carbon cycle

Approach	Brief description of proposed approach	Expected results	Limitations
Analogy	Comparison with disturbances such as tillage and freeze-thaw events which, like water erosion, cause breakdown of soil aggregates and are often followed by increased rates of CO_2 production.	These disturbances could provide an indication of the magnitude of expected CO_2 fluxes during erosional events.	- Compared to water erosion, these biophysical processes are less selective with regard to mobilization of the most biologically-active soil C fractions. - Although tillage erosion causes some soil movement, transport processes are of lesser magnitude than during water erosion.
Laboratory studies	Release of SOC from soil aggregates subjected to various levels of disturbances can be studied in the laboratory. Disturbed aggregates can then be incubated to monitor mineralization of the SOC released. Examples of disturbances include: - forcing soil aggregates through sieves of different mesh sizes; - submit soil aggregates to different energy levels within the range of kinetic energy typically recorded for rainfall in a particular area. Ultrasonic dispersion can be used for that purpose.	Bench-top research leads to: - understanding of labile C distribution within soil aggregates; - quantifying of labile C released as a function of intensity of disturbances; - inference with regard to mineralization of eroded SOC.	Although laboratory studies provide much insight on relevant processes (namely SOC release, SOC quality and degradability), it is usually problematic to extrapolate laboratory study data to field, regional and global scales.

| Field studies | These studies can be carried out using natural or simulated rainfall. They can also be carried out at different scales using soil blocks, runoff plots, or at watershed level. Eroded material can be trapped and incubated. Data to be collected include:
- particle size distribution in eroded materials;
- organic carbon distribution among the various particle sizes and biodegradability of OC associated with each size;
- mineralization rate of organic C in the aqueous phase;
- measurement of CO_2 emission rates.

Conducting this type of research at the scale of a watershed may require installing of sediment traps at different locations within the basin. As the timing of natural rainfall is difficult to predict and rainfall events often occur overnight, sediment traps may need to be somewhat automated. | Field studies allow:
- assessment of amount and quality of SOC release by specific physical processes (peeling, crushing, abrasion...) through variation of (simulated) rainfall characteristics;
- direct measurement of erosion-induced CO_2 fluxes;
- extrapolation to other ecosystems;
- integration of knowledge gained during detailed scales studies;
- capture of ecological, morphological and management differences among sites;
- better description of processes difficult to ascertain at finer scales. This is specially the case for sediment transfer (deposition-resuspension) processes which must be studied at scale of watershed or higher to yield meaningful results. | - Require commitment of important financial and technical resources;
- Interpretation of data requires information obtained during laboratory studies;
- As the scale of study gets larger, expertise from various disciplines (soil science, agricultural engineering, hydrology...) is often needed. |

Table 3. (continued)

Approach	Brief description of proposed approach	Expected results	Limitations
Modeling	Erosion impact on CO_2 flux can be evaluated through various modeling approaches depending on availability of data, understanding of processes, and level of details in the expected output. In general, the modeling exercise encompasses physical and biological components. Two examples are provided below.	Modeling approaches provide an indirect way to estimate erosion-induced CO_2 emissions during water erosion. This approach is supported by available data.	- Estimates are based on aggregation of data obtained at finer scales and often using non-standardized methods. Propagation of errors is a legitimate concern; - Modeling output often lacks site and management-specific details; - Laboratory and field studies are still needed to refine model output.

A. Up scaling approach
- Collate data on mineralizable C content of a wide range of soils;
- Collect data on erosion rates, sediment delivery ratios, and OC content of sediments from various eco-regions;
- Compute potential erosion-induced CO_2 emission using sediment flux and eroded C mineralization data.

B. Mass balance approach
- Collect historical data on SOC stocks in agro-ecosystems under different managements;
- Determine rate of SOC depletion through comparison of managed and native ecosystems;
- Based on crop residue inputs and humification rates, determine the replenishment rate of soil humus;
- Compute expected rates of SOC stocks depletion due only to mineralization of humus;
- SOC export is taken as depletion of SOC stocks in excess of humus respiration.
- After discounting for other fates (leaching, burial in terrestrial deposits, delivery into aquatic systems), rate of CO_2 evolution during transport and deposition of eroded SOC can be computed.

induced CO_2 fluxes. These estimates indicate that soil erosion can be a significant source of CO_2 in the atmosphere with a contribution ranging between 20 and 70% that of tropical deforestation. Mullen et al. (1999) and Schlesinger (1995) estimate that agricultural soils added between 0.7 and 0.8 Pg CO_2-C yr^{-1} into the atmosphere during the last century. These estimates were based on the assumption that mineralization of soil humus was the leading cause of SOC decline in cultivated soils. Our findings suggest a downward revision of these estimates to account for erosional SOC losses through emission as CO_2 into the atmosphere, entrapment in terrestrial deposits and exports to aquatic ecosystems.

We must stress, however, that estimates such as the ones produced in this chapter are reasonable only if the underlying assumptions are justified. There are several areas of uncertainty in our estimates. One such area is the partitioning of the labile C between the solid and aqueous phases. We also need to investigate the time course of CO_2 evolution following an erosional event and how post-erosion processes such as particle settling will affect mineralization of eroded SOC as re-aggregation of the soil mass is initiated. Thus, much research remains to be done in this area.

In Table 3 we outline a series of research approaches for improving estimates of erosion-induced CO_2 fluxes. The information presented is an attempt to delineate a couple of research protocols to address this fundamental question: How much CO_2 is emitted into the atmosphere during transport and deposition of eroded soils? The proposed approaches are broadly described and are listed with the main purpose of stimulating reflection and discussion among interested investigators.

Estimates of erosion-induced CO_2 fluxes presented in this chapter are based on the up-scaling approach described in Table 2. In another publication, the mass balance approach will be used with SOC data from various eco-regions and land management practices to further refine our estimates of CO_2 fluxes during erosional events. Finally, laboratory as well as field experiments will be carried out to validate these estimates.

V. Summary and Conclusions

Water erosion, depletion of SOC reserves and atmospheric CO_2 loading are closely linked. Despite the recognition of such linkages, quantitative data documenting water erosion and CO_2 fluxes are scarce. This chapter provides a discussion on possible fates of mobilized SOC during erosional events. Analysis of available data suggests that 20% of the SOC in eroded materials could be mineralized during transport and deposition. A global erosion-induced CO_2 flux of up to 1.3 Pg C yr^{-1} was computed, 0.3 Pg C yr^{-1} of which originates from croplands. The magnitude of the erosion-induced CO_2 flux depends on the total SOC mobilized by water erosion. Adoption of management practices such as conservation tillage will both increase SOC storage and reduce the contribution of water erosion to atmospheric CO_2 loading.

References

Ajwa, H.A., C.W. Rice and D. Sotomayor. 1998. Carbon and nitrogen mineralization in tallgrass prairie and agricultural soil profiles. *Soil Sci. Soc. Am. J.* 62: 942-951.

Allison, F.E. 1973. *Soil Organic Matter and Its Role in Crop Production*. Development in Soil Science 3. Elsevier, Amsterdam.

Alvarez, C.R., R. Alvarez, M.S. Grigera and R.S. Lavado. 1998. Associations between organic matter fractions and the active soil microbial biomass. *Soil Biol. Biochem.* 30:767-773.

Anderson, D.W., E. de Jong, G.E. Verity and G.E. Gregorich. 1986. The effects of cultivation on the organic matter of soils of the Canadian prairies. *Trans. XIII Cong. Int. Soc. Soil Sci.* (Hamburg) 7:1344-1345.

Barrios, E., R.J. Buresh and J.I. Sprent. 1996. Organic matter in soil particle size and density fractions from maize and legume cropping systems. *Soil Biol. Biochem.* 28:185-193.

Bauer, A. and A.L. Black. 1994. Quantification of the effect of soil organic matter content on soil productivity. *Soil Sci. Soc. Am. J.* 58:185-193.

Benner, R., S. Opsahl, G. ChinLeo, J.E. Richey and B.R. Forsberg. 1995. Bacterial carbon metabolism in the Amazon River system. *Limnology and Oceanography* 40:1262-1270.

Beyer, L., R. Frund, U. Schleuss and C. Wachendorf. 1993. Colluvisols under cultivation in Schleswig-Holstein. 2. Carbon distribution and soil organic matter composition. *J. Plant Nutr. and Soil Sci.* 156:213-217.

Biederbeck, V.O., C.A. Campbell, H. Ukrainetz, D. Curtin and O.T. Bouman. 1996. Soil microbial and biochemical properties after ten years of fertilization with urea and anhydrous ammonia. *Can. J. Soil Sci.* 76:7-14.

Bohn, H.L. 1978. On organic soil carbon and CO_2. *Tellus* 30:472-475.

Bonde, T.A., B.T. Christensen and C.C. Cerri. 1992. Dynamics of soil organic matter as reflected by natural C-13 abundance in particle size fractions of forested and cultivated Oxisols. *Soil Biol. and Biochem.* 24:275-277.

Boyer, J.N. and P.M. Groffman. 1996. Bioavailability of water extractable organic carbon fractions in forest and agricultural soil profiles. *Soil Biol. and Biochem.* 28:783-790.

Brown, L.R. 1984. The global loss of topsoil. *J. Soil Water Conserv.* 39:162-165.

Burford, J.R and J.M. Bremner. 1975. Relationships between the denitrification capacities of soils and total water-soluble and readily decomposable soil organic matter. *Soil Biol. and Biochem.* 7:389-394.

Campbell, C.A., E.A. Paul, D.A. Rennie and K.J. MaCallum. 1967. Applicability of the carbon-dating method of analysis to soil humus studies. *Soil Sci.* 104:217-224.

Campbell, C.A., V.O. Biederbeck, G. Wen, R.P. Zentner, J. Schoenau and D. Hahn. 1999. Seasonal trends in selected soil biochemical attributes: Effects of crop rotation in the semiarid prairie. *Can. J. Soil Sci.* 79:73-84.

Carter, M.R. 1991. The influence of tillage on the proportion of organic carbon and nitrogen in the microbial biomass of medium-textured soils in a humid climate. *Biol. and Fert. Soils* 11:135-139.

Cook, B.D. and D.L. Allan. 1992. Dissolved organic carbon in old field soils: total amounts as a measure of available resources for soil mineralization. *Soil Biol. and Biochem.* 24:585-594.

Curtin, D., F. Selles, H. Wang, C.A. Campbell and V.O. Biederbeck. 1998. Carbon dioxide emissions and transformation of soil carbon and nitrogen during wheat straw decomposition. *Soil Sci. Soc. Am. J.* 62:1035-1041.

Dazhong, W. 1993. Soil erosion and conservation in China. p. 63-85. In: D. Pimentel (ed.), *World Soil Erosion and Conservation.* Cambridge University Press, Cambridge.

de Jong, E. and R.G. Kachanoski. 1987. The importance of erosion in the carbon balance of prairie soils. *Can. J. Soil Sci.* 68:111-119.

Degens, E.T., S. Kempe and J.E. Richey. 1991. Summary: Biogeochemistry of major world rivers. p. 323-347. In: E.T. Degens, S. Kempe and J.E. Richey (eds.), *Biogeochemistry of Major World Rivers*, SCOPE 42. John Wiley & Sons, New York.

Detwiller, R.P. and C.A.S. Hall. 1988. Tropical forests and the global carbon cycle. *Science* 239:42-47.

Elliott, E.T., K. Paustian and S.D. Frey. 1996. Modeling the measurable or measuring the modelable: A hierarchical approach to isolating meaningful soil organic matter fractionations. p. 161-179. In: D.S. Powlson, P. Smith and J.U. Smith (eds.), *Evaluation of Soil Organic Matter Models*. Springer-Verlag, Berlin.

Elliott, W.J., A.M. Liebenow, J.M. Laflen and K.D. Kohl. 1989. A compendium of soil erodibility data from WEPP cropland soil field erodibility experiments, USDA Agricultural Research Service, NSERL Report No. 3, U.S. Department of Agriculture, Washington, D.C.

Eswaran, H., E. Vandenberg and P. Reich. 1993. Organic carbon in soils of the world. *Soil Sci. Soc. Am. J.* 57:192-194.

FAO. 1998. *Production Yearbook*. Volume 52. Food and Agricultural Organization of the United Nations, Rome, Italy.

Franzluebbers, A.J., R.L. Haney, F.M. Hons and D.A. Zuberer. 1996. Active fractions of organic matter in soils with different texture. *Soil Biol. and Biochem.* 28:1367-1372.

Fuller, L.G., T.B. Goh and D.W. Oscarson. 1995. Cultivation effects on dispersible clay of soil aggregates. *Can. J. Soil Sci.* 75:101-107.

Gregorich, E.G., B.H. Ellert and C.M. Monreal. 1995. Turnover of soil organic matter and storage of corn residue carbon estimated from natural ^{13}C abundance. *Can. J. Soil Sci.* 75:161-167.

Gregorich, E.G. 1996. Soil quality: A Canadian perspective. p. 40-52. In: K.C. Cameron, I.S. Cornforth, R.G. McLaren, M.H. Beare, L.R. Basher, A.K. Metherell and L.E. Kerr (eds.), *Soil Quality Indicators for Sustainable Agriculture in New Zealand*. Proceedings of a workshop. Lincoln University, Christchurch, New Zealand.

Gregorich, E.G. and D.W. Anderson. 1985. Effects of cultivation and erosion on soils on four toposequences in the Canadian Prairies. *Geoderma* 36:343-354.

Gregorich, E., R. Kachanoski and R.P. Voroney. 1988. Ultrasonic dispersion of aggregates: distribution of organic matter in size fractions. *Can. J. Soil Sci.* 68:395-403.

Gregorich, E.G., K.J. Greer, D.W. Anderson and B.C. Liang. 1998. Carbon distribution and losses: erosion and deposition effects. *Soil Tillage Res.* 47:291-302.

Gupta, V.V.S.R. and J.J. Germida. 1988. Distribution of microbial biomass and its activity in different soil aggregate size classes as affected by cultivation. *Soil Biol. and Biochem.* 20:777-786.

Guggenberger, G. and W. Zech. 1999. Soil organic matter composition under primary forest, pasture, and secondary forest succession, Región Huerta Norte, Costa Rica. *Forest Ecol. Management* 124: 93-104.

Haider, K., F.F. Groblinghoff, T. Beck, H.R. Schulten, R. Hempfling and H.D. Ludermann. 1991. Influence of soil management practices on the organic matter structure and the biochemical turnover of plant residues. p. 79-91. In: W.S. Wilson (ed.), *Advances in Soil Organic Matter Research: The Impact on Agriculture and the Environment*. The Royal Society of Chemistry, Cambridge, U.K.

Hassink, J. 1994. Effects of soil texture and grassland management on soil organic C and N mineralization. *Soil Biol. and Biochem.* 26:1221-1231.

Hattori, T. and R. Hattori. 1976. The physical environment in soil microbiologyan attempt to extend principles of microbiology to soil microorganisms. *CRC Critical Reviews in Microbiology* 4:423-461.

Haynes, R.J. 1999. Labile organic matter as an indicator of organic matter quality in arable and pastoral soils in New Zealand. *Soil Biol. and Biochem.* 32:211-219.

Hedges, J.I., R.G. Keil and R. Benner. 1997. What happens to terrestrial organic matter in the ocean? *Organic Geochem.* 27:195-212.

Hope, D., M.F. Billet and M.S. Cresser. 1994. A review of the export of carbon in river water: fluxes and processes. *Environ. Pollut.* 84:301-324.

Houghton, R.A., J.E. Hobbie, J.M. Melillo, B. Moore, B.J. Peterson, G.R. Shaver and G.M. Woodwell. 1983. Changes in the carbon content of terrestrial biota and soils between 1860 and 1980: A net release of CO_2 to the atmosphere. *Ecolog. Monograph* 53:235-262.

Hu, S.J., A.H.C. van Bruggen and N.J. Grunwald. 1999. Dynamics of bacterial populations in relation to carbon availability in a residue-amended soil. *Appl. Soil Ecol.* 13:21-30.

Hudson, R.J.M., S.A. Gherini and R.A. Goldstein. 1994. Modeling the global carbon cycle: Nitrogen fertilization of the terrestrial biosphere and the "missing" CO_2 sink. *Global Biogeochem. Cycles* 8:307-333.

IPCC (Intergovernmental Panel on Climate Change). 1996. *Revised Methodology for Greenhouse Gas Inventory.* Intergovernmental Panel on Climate Change, Zurich.

Ittekot, V. 1988. Global trends in the nature of organic matter in river suspensions. *Nature* 332:436-438.

Janzen, H.H., C.A. Campbell, S.A. Brandt, G.P. Lafond and L. Townley-Smith. 1992. Light-fraction organic matter in soils from long-term crop rotations. *Soil Sci. Soc. Am. J.* 56:1799-1806.

Jastrow, J.D., T.W. Boutton and R.W. Miller. 1996. Carbon dynamics of aggregate-associated organic matter estimated by carbon-13 natural abundance. *Soil Sci. Soc. Am. J.* 60:801-807.

Jenkinson, D.S. and J.N. Ladd. 1981. Microbial biomass in soil: measurement and turnover. *Soil Biol. and Biochem.* 5:415-471.

Jenkinson, D.S. and D.S. Powlson. 1976. The effects of biocidal treatments on metabolism in soil. V. A method for measuring soil biomass. *Soil Biol. and Biochem.* 8:209-213.

Jenkinson, D.S. and J.H. Rayner. 1977. The turnover of soil organic matter in some of the Rothamsted classical experiments. *Soil Sci.* 123:298-305.

Lal, R. 1976. *Soil Erosion Problems on an Alfisol in Western Nigeria and Their Control.* ITA Monograph No. 1, Ibadan, Nigeria.

Lal, R. 1986. Soil surface management in the tropics for intensive land use and high, sustained production. *Advances Soil Sci.* 5:51-105.

Lal, R. 1990. Soil erosion and land degradation: the global risks. p. 129-172. In: R. Lal and B.A. Stewart (eds.), *Soil Degradation*, Volume 11, Advances in Soil Science. Springer-Verlag, New York.

Lal, R. 1993. Soil erosion and conservation in West Africa. p. 7-25. In: D. Pimentel (ed.), *World Soil Erosion and Conservation.* Cambridge University Press.

Lal, R. 1995. Global soil erosion by water and carbon dynamics. p. 131-142. In: R. Lal, J. Kimble, E. Levine and B. Stewart (eds.), *Soil and Global Change.* Advances in Soil Science. Lewis Publishers, Boca Raton, FL.

Larney, F.J., E. Bremer, H. H. Janzen, A.M. Johnston and C. W. Lindwall. 1997. Changes in total, mineralizable and light fraction soil organic matter with cropping and tillage intensities in semiarid southern Alberta, Canada. *Soil and Tillage Res.* 42:229-240.

Lee, J.J., D.L. Phillips and R.F. Dodson. 1996. Sensitivity of the US corn belt to climate change and elevated CO_2: II. Soil erosion and organic carbon. *Agricultural Systems* 52:503-521.

Lowrance, R. and R. Williams. 1988. Carbon movement in runoff and erosion under simulated rainfall conditions. *Soil Sci. Soc. Am. J.* 52:1445-1448.

Ludwig, W. and J.L. Probst. 1996. Predicting the oceanic input of organic carbon by continental erosion. *Global Biogeochem. Cycles* 10:23-41.

Ludwig W., P. Amiotte-Suchet and J.L. Probst. 1996. River discharges of carbon to the world's oceans: Determining local inputs of alkalinity and of dissolved and particulate organic carbon. *Comptes Rendus de l'Académie des Sciences, Série II., Sciences de la Terre et des Planètes* 323: 1007-1014.

Lundquist, E.J., L.E. Jackson and K.M. Scow. 1999. Wet–dry cycles affect dissolved organic carbon in two California agricultural soils. *Soil Biol. Biochem.* 31:1031-1038.

Martin-Mousset, B., J.P. Croue, E. Lefebvre and B. Legube. 1997. Distribution and characterization of dissolved organic matter of surface waters. *Water Research* 31:541-553.

Milliman, J.D. and J.P.M. Syvitski. 1992. Geomorphic / tectonic control of sediment discharge to the ocean: The importance of small mountainous rivers. *J. Geol.* 100:325-344.

Mogren, E.M., P. Scarpino and R.S. Summers. 1990. Measurement of biodegradable dissolved organic carbon in drinking water. Proceedings of the Annual American Water Works Association Conference, Cincinnati, Ohio (USA), 18 - 21 June.

Mokma, D.L. and M.A. Sietz. 1992. Effects of soil erosion on corn yields on Marlette soils in south-central Michigan. *J. Soil Water Conserv.* 47:325-327.

Monreal, C.M., R.P. Zentner and J.A. Robertson. 1997. An analysis of soil organic matter dynamics in relation to management, erosion and yield of wheat in long-term crop rotation plots. *Can. J. Soil Sci.* 77:553-563.

Mullen, R.W., W.E. Thomason and W.R. Raun. 1999. Estimated increase in atmospheric carbon dioxide due to worldwide decrease in soil organic matter. *Commun. Soil Sci. and Plant Analy.* 30: 1713-1719.

Nabuurs, G. J., R. Päivinen, R. Sikkema and G.M.J. Mohren. 1997. The role of European forests in the global carbon cycle–a review. *Biomass and Bioenergy* 13:345-358.

Nicolardot, B., J.A.E. Molina and M.R. Allard. 1994. C and N fluxes between pools of soil organic matter: model calibration with long-term incubation data. *Soil Biol. and Biochem.* 26:235-243.

Oldeman, L.R. 1994. The global extent of soil degradation. p. 99-118. In: D.J. Greenland and I. Szabolcs (eds.), *Soil Resilience and Sustainable Land Use*. CAB International, Wallingford, U.K.

Olson, K.R. and E. Nizeyimana. 1988. Effects of soil erosion on corn yields of seven Illinois soils. *J. Prod. Agricult.* 1:13-19.

Palis, R.G, H. Ghandiri, C.W. Rose and P.G. Saffigna. 1997. Soil erosion and nutrient loss .3. Changes in the enrichment ratio of total nitrogen and organic carbon under rainfall detachment and entrainment. *Australian J. Soil Res.* 35:891-905.

Paningbatan, E.P., C.A. Ciesiolka, K.J. Coughlan and C.W. Rose. 1995. Alley cropping for managing soil erosion of hilly lands in the Philippines. *Soil Technology* 8:193-204.

Parton, W.J., D.S. Ojima, C.V. Cole and D.S. Schimel. 1994. A general model for soil organic matter dynamics: sensitivity to litter chemistry, texture and management. p. 147-167. In: Soil Sci. Soc. Am. Special Publication No. 39. Soil Science Society of America, Madison, WI.

Pennock, D.J, D.W. Anderson and E. de Jong. 1994. Landscape-scale changes in indicators of soil quality due to cultivation in Saskatchewan, Canada. *Geoderma* 64:1-19.

Pimentel, D. 1993. Overview. p. 1-5. In: D. Pimentel (ed.), *World Soil Erosion and Conservation*. Cambridge University Press, Cambridge.

Pimentel, D., J. Allen, A. Beers, L. Guinand, R. Linder, P. McLaughlin, B. Meer, D. Musonda, D. Perdue, S. Poisson, S. Siebert, K. Stoner, R. Salazar and A. Hawkins. 1987. World agriculture and soil erosion. *Bioscience* 37:277-283.

Pimentel, D., C. Harvey, P. Resosudarmo, K. Sinclair, D. Kurz, M. McNair, S. Crist, L. Shpritz, L. Fitton, R. Saffouri and R. Blair. 1995. Environmental and economic costs of soil erosion and conservation benefits. *Science* 267:1117-1123.

Post, W.M., W.R. Emmanuel, P.J. Zinke and A.G. Stangenberger. 1982. Soil carbon pools and world life zones. *Nature* 298:156-159.

Powlson, D.S. 1996. Why evaluate soil organic matter models. In: D.S. Powlson, P. Smith and J.U. Smith (eds.), *Evaluation of Soil Organic Matter Models*. Springer-Verlag, New York.

Powlson, D.S., P.C. Brookes and B.T. Christensen. 1987. Measurement of soil microbial biomass provides an early indication of changes in total soil organic matter due to straw incorporation. *Soil Biol. and Biochem.* 19:159-164.

Qualls, R.G. and B.L. Haines. 1992. Biodegradability of dissolved organic matter in forest throughfall, soil solution, and stream water. *Soil Sci. Soc. Am. J.* 56:578-586.

Rochette, P. and G. Gregorich. 1998. Dynamics of soil microbial biomass C, soluble organic C and CO_2 evolution after three years of manure application. *Can. J. Soil Sci.* 78:283-290.

Roose, E. and F. Ndayizigiye. 1997. Agroforestry, water and soil fertility management to fight erosion in tropical mountains of Rwanda. *Soil Technol.* 11:109-119.

Saggar, S., A. Parshotam, G.P. Sparling, C.W. Feltham and P.B.S. Hart. 1996. ^{14}C-labeled ryegrass turnover and residence times in soils varying in clay content and mineralogy. *Soil Biol. and Biochem.* 28:1677-1686.

Schimel, D.S., E.F. Kelly, C. Yonker, R. Aguilar and R. Heil. 1985. Effects of erosional processes on nutrient cycling in semiarid landscapes. p. 571-580. In: D.E. Caldwell, J.A. Brierley and C.L. Brierley (eds.), *Planetary Ecology*. Van Nostrand Reinhold, New York.

Schlesinger, W.H. 1984. Soil organic matter: A source of atmospheric CO_2. p. 111-127. In: G.M. Woodwell (ed.), *The Role of Terrestrial Vegetation in the Global Carbon Cycle: Measurement by Remote Sensing*. SCOPE 23, John Wiley & Sons, New York.

Schlesinger, W.H. 1986. Changes in soil carbon storage and associated properties with disturbance and recovery. p. 194-220. In: J.R. Trabalka and D.E. Reichle (eds.), *The Changing Carbon Cycle: A Global Analysis*. Springer-Verlag, New York.

Schlesinger, W.H. 1990. Evidence from chronosequence studies of a low carbon-storage potential of soils. *Nature* 348:232-234.

Schlesinger, W.H. 1995. An overview of the carbon cycle. p: 9-25. In: R. Lal, J. Kimble, E. Levine and B. Stewart (eds.), *Soils and Global Change*. Advances in Soil Science. Lewis Publishers, Boca Raton, FL.

Schreiber, J.D. and K.C. McGregor. 1979. The transport and oxygen demand of organic carbon released to runoff from crop residues. *Progress Water Technol.* 11:253-261.

Sedjo, R.A. 1992. Temperate forest ecosystems in the global carbon cycle. *Ambio* 21:274-277.

Slater, C.S. and E.A. Carleton. 1938. The effect of erosion on losses of soil organic matter. *Soil Sci. Soc. Am. Proc.* 3:123-128.

Smith, W.N., P. Rochette, C. Monreal, R.L. Desjardins, E. Pattey and A. Jaques. 1997. The rate of carbon change in agricultural soils in Canada at the landscape level. *Can. J. Soil Sci.* 77:219-229.

Spitzy, A. and J. Leenheer. 1991. Dissolved organic carbon in rivers. p. 214-232. In: E.T. Degens, S. Kempe and J.E. Richey (eds.), *Biogeochemistry of Major World Rivers*, SCOPE, John Wiley & Sons, New York.

Stallard, R.F. 1998. Terrestrial sedimentation and the carbon cycle: Coupling weathering and erosion to carbon burial. *Global Biogeochem. Cycles* 12:231-257.

Starr, G.C., R. Lal, R. Malone, D. Hothem, L. Owens and J. Kimble. 2000. Modeling soil carbon transported by water erosion processes. *Land Degradation and Develop.* 11:83-91.

Strickland, T.C. and P. Sollins. 1987. Improved method for separating light and heavy fraction of organic material from soil. *Soil Sci. Soc. Am. J.* 51:1390-1393.

Tao, S. and B. Lin. 2000. Water soluble organic carbon and its measurement in soil and sediment. *Water Res.* 34:1751-1755.

Thornley, J.H.M., D. Fowler and G.R. Cannell. 1991. Terrestrial carbon storage resulting from CO_2 and nitrogen fertilization in temperature grasslands. *Plant, Cell and Environ.* 14:1007-1011.

Thurman, E.M. 1985. *Developments in Biochemistry: Organic Geochemistry of Natural Waters*. M. Nijhoff, Dr. W. Junk Publishers, Dordrecht.

Tiessen, H. and J.W.B. Stewart. 1983. Particle-size fractions and their uses in studies of soil organic matter: II. Cultivation effects on organic matter composition in size fractions. *Soil Sci. Soc. Am. J.* 47:509-514.

Van Noordwijk, M., C. Cerri, P.L. Woomer, K. Nugroho and M. Bernoux. 1997. Soil carbon dynamics in the humid tropical forest zone. *Geoderma* 79:187-225.

Voroney, R.P., J.A. van Veen and E.A. Paul. 1981. Organic carbon dynamics in grassland soils. 2. Model validation and simulation of long-term effects of cultivation and rainfall erosion. *Can. J. Soil Sci.* 61:211-224.

Walling, D.E. 1983. The sediment delivery ratio problem. *J. Hydrol.* 65:209-237.

Wang, F.L. and J.R. Bettany. 1993. Influence of freeze-thaw and flooding on the loss of soluble organic carbon and carbon dioxide from soil. *J. Environmental Qual.* 22:709-714.

Watson, R.T., L.G. Meira-Filho, E. Sanhueza and A. Janetos. 1992. Greenhouse gases: Sources and sinks. p. 25-46 In: J.T. Houghton, B.A. Callander and S.K. Varney (eds.), *Climate Change 1992*. The Supplementary Report to the IPCC Scientific Assessment, Cambridge University Press, Cambridge.

Wischmeier, W.H. and D.D. Smith. 1978. *Predicting Rainfall Erosion Losses: A Guide to Conservation Planning*. U.S. Department of Agriculture, Agricultural Handbook No. 537.

Woods, L.E. and G.E. Schuman. 1988. Cultivation and slope position effects on soil organic matter. *Soil Sci. Soc. Am. J.* 52:1371-1376.

Zech, W., N. Senesi, G. Guggenberger, K. Kaiser, J. Lehmann, T.M. Miano, A. Miltner and G. Schroth. 1997. Factors controlling humification and mineralization of soil organic matter in the tropics. *Geoderma* 79:117-161.

Zsolnay, A. and H. Gorlitz. 1994. Water-extractable organic carbon in arable soils: Effects of drought and long-term fertilization. *Soil Biol. and Biochem.* 26:1257-1261.

Zsolnay, A. and H. Steindl. 1991. Geovariability and biodegradability of the water-extractable organic material in an agricultural soil. *Soil Biol. and Biochem.* 23:1077-1082.

CHAPTER 31

Soil Organic Carbon Erosion Assessment by Cesium-137

Y. Hao, R. Lal, L.B. Owens and R.C. Izaurralde

I. Introduction

Soil organic carbon (SOC) is a major pool that impacts the global carbon cycle (Lal, 1999). Increasing SOC pool is desirable because of its favorable effects on improving soil fertility, decreasing water and air pollution, and mitigating the greenhouse effect caused by various energy utilization activities such as fossil fuel combustion. The amount of SOC depends on kinetic competition between various input and output processes. The input processes include plant growth (plant residue, root excretion, and organic matter through-fall), addition of organic material (manure, sewage sludge, and other organic wastes) through soil management, and deposition through soil erosion. The output processes comprise decomposition into gases, leaching into groundwater, and removal through soil erosion. Assessment of these processes is one of the steps toward adopting the strategy of increasing SOC content.

The SOC erosion implies its translocation, redistribution over the landscape, and removal from the ecosystem by natural and anthropogenic forces such as rain, wind, gravity, and tillage. It is both an input and output process, depending on the soil boundaries involved. Because SOC is concentrated in the surface layers, its erosion dramatically affects and controls SOC pool (Slater and Carleton, 1938; Rogers, 1941; Logan et al., 1991; Ruppenthal et al., 1997; Gregorich et al., 1998; Lal, 1998). The SOC erosion can be assessed directly by quantifying the eroded SOC mass, and indirectly by evaluating other SOC processes determining its mass balance. Direct assessment of SOC erosion can facilitate quantification of other processes such as root excretion, plant organic matter through-fall, organic matter mineralization, and leaching, some of which are even more difficult to measure than SOC erosion.

Direct assessment of SOC erosion usually involves the techniques used in assessing soil erosion and sediment SOC content. Soil erosion can be assessed by several methods such as sediment collection (Mutchler, 1994), Universal Soil Loss Equation (USLE) or its revised version (RUSLE) prediction (USDA, 1997), soil ^{137}Cs radioactivity loss estimation (Ritchie and McHenry, 1990; Walling and He, 1999), soil fly ash content change measurement (Jones and Olson, 1990; Hussain et al., 1998), and soil horizon depth and property change assessment (Kreznor et al., 1992). The sediment SOC content can be determined either directly by analyzing the sediment or indirectly by characterizing *in-situ* field soil. For the latter case, the uncertainties involved in assessing the enrichment ratio (Er [a]) of SOC can be a source of error.

Soil ^{137}Cs method is based on the principle of tenacious binding of atmospheric fallout [137]Cs (major fallout occurring from 1956 to 1965) with soil particles and thus using it as a soil tracer (Ritchie and McHenry, 1990). The method has been developed and used extensively during the past two decades to assess soil erosion, especially for the cultivated sites characterized by drastic impact of agricultural activities on soil erosion. The method has several advantages over the other methods for assessing soil erosion, and thus is an unique tool for assessing SOC erosion.

The objective of this chapter is to examine the principles and application of ^{137}Cs method in assessing SOC erosion for a cultivated site, and to outline its pros and cons in comparison with other SOC erosion assessment methods. To accomplish the above objective, the chapter describes the study conducted in 1998 at the North Appalachian Experimental Watershed Research Station in Coshocton, Ohio. The study assessed the historic SOC erosion since 1950s at a small (0.79 ha) cultivated watershed with the three methods: sediment collection, RUSLE prediction, and soil ^{137}Cs loss estimation. Comparison of the results obtained with different methods elucidates the important issues in application of ^{137}Cs method for SOC erosion assessment.

II. Principles of [137]Cs Method

Soil ^{137}Cs originates from fallout of atmospheric ^{137}Cs produced in nuclear testing. The fallout ^{137}Cs tenaciously binds with mineral particles especially with illite and vermiculite (Nishita et al., 1956; Schulz et al., 1959; Lomenick and Tamura, 1965; Schulz, 1965; Nishita and Essington, 1967). It is not absorbed by plants. Therefore, soil is essentially labeled with ^{137}Cs whose change can only be accounted for by natural decay and horizontal soil movement into and out of a soil profile.

The cumulative amount of ^{137}Cs fallout during 1950s and 1960s is normally higher than 2,000 Bq[b] m^{-2} (Ritchie et al., 1970; Ritchie and McHenry, 1978; Walling and Quine, 1991). Due to the limited vertical movement, ^{137}Cs is confined in the surface 10 to 30 cm of soil with a concentration normally higher than 2 to 3 mBq g^{-1} (assuming soil bulk density of 1.5 Mg m^{-3}). With a strong gamma-ray emission (662 keV) and a long half-life (30.2 years), ^{137}Cs activity can be detected in soil for a long time.

As a soil tracer, ^{137}Cs concentration at a landscape position indicates net soil erosion (erosion and deposition) induced by water, wind, gravity, and tillage (Walling and Quine, 1991; Lobb et al., 1995). Since the fallout started in the 1950s, ^{137}Cs concentration can indicate the amount of historic soil erosion retrospectively back to the 1950s (Ritchie and McHenry, 1990). For a specific time and location, the input of fallout ^{137}Cs can be inferred from its concentration in an undisturbed forest or grassland site where there is no soil erosion or deposition. Furthermore, the soil ^{137}Cs determined at many sites around the world during the past two decades provides the initial ^{137}Cs values for a long-term study (Kachanoski and de Jong, 1984; Kachanoski, 1987). If the ^{137}Cs values were determined after 1965 when the fallout had ended, the long-term study by resampling the soil ^{137}Cs at later times can provide more reliable assessment of soil erosion than the historic study since the 1950s which involves the confounding factor of temporal- and spatial-dependent ^{137}Cs fallout rate.

Several models are used to convert soil ^{137}Cs change into soil loss or gain (Ritchie and McHenry, 1990; Walling and He, 1999). These models are either based on theoretical analysis such as

---

[a] Er is the ratio of soil or SOC contents in sediment to field soil.

[b] Bq (becquerel) is one nuclear transition per second.

proportional and mass balance equation or empirical model such as an exponential equation. For a cultivated site, a proportional equation (Equation 1) is often used:

$$S = 10^4 * \rho_b * d_p * (A_1 - A_2) / (A_1 * Yr) \qquad \text{(Eq. 1)}$$

where S is annual soil loss or gain (Mg ha^{-1} yr^{-1}), ρ_b is soil bulk density (Mg m^{-3}), d_p is plow depth (m), A_1 is reference-site value or study-site initial (after corrected for natural decay) value of ^{137}Cs (Bq m^{-2}), A_2 is study-site end value of ^{137}Cs (Bq m^{-2}), and Yr is number of years. When used for historic assessment of soil erosion with an undisturbed forest or grassland ^{137}Cs as a reference value, this equation is based on five assumptions: (1) the input of fallout ^{137}Cs is equal for the reference and study sites; (2) the entire input during the fallout period is mixed into the plow layer at the start of the fallout; (3) tillage dilution of ^{137}Cs in the plow layer with removal of surface soil is negligible; (4) the mixing in the plow layer is homogeneous; and (5) the ^{137}Cs Er is one. These assumptions are oversimplifications of reality, and can be major sources of error.

The proportional equation causes less error when used for a long-term study using the study-site ^{137}Cs value determined after 1965 as an initial value. In this situation, the above assumptions involve the last three for the plow till site and the last two for the no-till site. Furthermore, the mass balance equations capable of reducing the errors produced by the assumptions have been developed, which give a more accurate measurement of soil erosion than the simple proportional equation. Readers are referred to the description of these equations by Kachanoski and de Jong (1984) and Walling and He (1999).

An exponential equation was developed initially by Ritchie et al. (1974) and later modified by Ritchie and McHenry (1990) (Equation 2):

$$S = 0.87 * [100 * (A_1 - A_2) / A_1]^{1.18} \qquad \text{(Eq. 2)}$$

The equation was based on the correlation of the sediment and USLE data with ^{137}Cs, ^{90}Sr, and ^{85}Sr data published in several studies and had a regression coefficient of 0.95. In Equation 2, the ^{137}Cs loss is cumulative instead of annual, and the soil loss is annual instead of cumulative. However, conceptually, the soil loss assessed by Equation 2 is the total loss instead of annual loss. Therefore, we modified Equation 2 by dividing it with the number of years in the study period to obtain Equation 3 which computes the annual soil loss:

$$S = 0.87 * [100 * (A_1 - A_2) / A_1]^{1.18} / Yr \qquad \text{(Eq. 3)}$$

This equation is referred to as the revised exponential equation in this chapter.

To assess SOC erosion, the relationship between sediment (mainly minerals) and eroded SOC has to be established. The relationship can be determined by measuring the SOC content in the field soil (excluding plant residue at the soil surface) and estimating the SOC-Er value. The techniques for determining SOC content include dry combustion and dichromate oxidation (Nelson and Sommers, 1996).

Depending on the surface roughness of the field soil, extent of soil erosion, and sampling frequency, the depth of top sampling layer should be as shallow as possible and normally <1 cm because soil erosion is a surface phenomenon. This is especially important for the no-till site where SOC content changes dramatically within the top few centimeters (Dick et al., 1997). The number of sampling points depends on tillage, cropping methods, watershed area and shape, and slope length. Plow till and no till create less heterogeneous SOC content than other tillage methods such as ridge till. Sampling points at upper, middle, and lower slope positions are normally required. Due to a slow temporal change in SOC content (see discussion below), sampling frequency of once per year can be sufficient for the accuracy demanded in SOC erosion assessment.

For historic assessment of SOC erosion, the SOC content for the previous years back to the 1950s have to be estimated if it is not available. The following guidelines are helpful in such estimation and are explained by Monreal et al. (1997), Paustian et al. (1997), Rasmussen et al. (1998), and Brady and Weil (1999):

(1) The annual rate of SOC change in content averaged over a long time is low and normally $< 0.05\%$ per year or < 1 Mg ha^{-1} yr^{-1} in a furrow slice (assuming 2000 Mg in one ha furrow slice). The SOC usually decreases when the forest and grasslands are converted to agricultural use and increases when the plow till is replaced by no till. In the former case, if the conversion occurred around 1950s, the SOC content in the forest or grasslands close to the converted area needs to be determined and is often equal to the SOC content in the agricultural soil at the time of conversion.

(2) The equilibrium SOC content is attained normally within 25 to 50 years after a new land use is adopted (Buyanovsky and Wagner, 1997). This information is useful for estimating the previous SOC content for the areas where the agricultural land use has been subject to none or minor changes for a long time, and

(3) The SOC content and its temporal dynamics in an ecosystem can be inferred from the ecosystems with similar characteristics including climate, soil type, agricultural management history, and topography. The SOC content normally increases with increase in precipitation, decrease in temperature, increase in clay content, decrease in tillage, and increase in deposition in a depressional site. The SOC content data for the early 20th century are available for a few long-term agriculture research sites in Europe and North America. The temporal dynamics of SOC content in these regions has also been modeled starting from the 1900s (Patwardhan et al., 1997). These data on SOC content and its temporal dynamic can be extrapolated to similar soils and ecoregions.

According to the assumption of a unit ^{137}Cs-Er underlining the proportional equation, and because soil erosion is enriched with fine particles and both SOC and ^{137}Cs preferentially bind with fine particles (Kreznor et al., 1992; Higgitt, 1995; Hird et al., 1995; He and Walling, 1996), SOC-Er should not be included in calculations using the proportional equation. For the mass balance equations accounting for enrichment process, both Er values have to be determined and included in the calculation. Because the exponential equation is empirical, only SOC-Er can be used in the calculation.

The Er values of SOC and ^{137}Cs have been reported in several studies (Barrows and Kilmer, 1963; Lal, 1979; Kachanoski and de Jong, 1984; Palis et al., 1997; Wan and El-Swaify, 1997; Bernard et al., 1998), and depend on many factors such as soil texture, rainfall intensity, topography, and the amount of soil erosion. Both Er values are generally in the range of 1 to 5. For historic assessment of SOC erosion, the Er values cannot be estimated precisely, and can be a source of error.

III. Application of ^{137}Cs Method to Assess SOC Erosion

A. Site Description

The watershed was located in the North Appalachian Experimental Watershed research station (40° 22′ N, 81° 48′ W, within elevation of 300 to 600 m) near Coshocton, OH, USA. The watershed had an area of 0.79 ha, an average slope length of 132 m, an average slope steepness of 10%, and a shape of an approximately equilateral triangle. The soil type is Coshocton silt loam (fine loamy, mixed, mesic Aquultic Hapludalf), containing 24% illite and 31% vermiculite in the clay fraction, and developed from the sandstone and shale bedrock (Kelley et al., 1975).

The land uses were plow till corn (*Zea mays* L.)-wheat (*Triticum aestivum* L., winter wheat, soft red type)-meadow-meadow rotation from 1943 to 1970, plow-till corn from 1971 to 1975 except for the no-till corn in 1974, meadow from 1976 to 1983, and no-till corn-soybean (*Glycine max* L. Merr) rotation since 1984. All farming operations were done on the contour since 1943. Corn stover was chopped and left on the soil surface which was then lightly disked. Timothy (*Phleum pratense* L.) was seeded along with wheat in early October and timothy, red clover (*Trifolium pratense* L.), and alsike clover (*Trifolium hybridum* L.) were seeded in March. The vegetative cover was clipped in late July to discourage weed growth. In the two meadow years, hay was harvested in June and August. Rye (*Secale cereale*) was aerially seeded into the soybean prior to leaf drop, and later killed with herbicides in April/May prior to sowing corn in the spring.

The undisturbed forest (mixed Oaks, *Quacus* spp.) and long-term grassland plots (Kentucky blue grass, *Poa pratensis*, and orchard grass, *Dactylis glomerata*) were within 100 m distance from the experimental watershed and had the same soil type as the watershed. The close proximity of the experimental and reference site is important to validate the assumption that difference in ^{137}Cs fallout between the two sites is minimal. Before 1971, the grassland site had the same land use as the watershed except that the crop rotation started 1 year earlier.

B. Data Collection and Analyses

Soil samples were collected at the watershed in 1970, 1984, 1990, 1995, and 1998. Soil bulk density was measured by the core method for the post-1970 samples (Blake and Hartge, 1986) and had an average of 1.33 Mg m[-3]. The SOC content was measured with the dichromate method for 1970 samples and with dry combustion method after removal of inorganic carbon for the post-1970 samples. The average SOC content was 1.51% (ranging from 1.31 to 1.80%). Soil ^{137}Cs activity was measured for 1998 samples. Surface runoff water and sediment were collected starting from 1945 for all rainfall events. Runoff volumes were measured using H-flume and water stage recorder and calculated from hydrographs. Sediment concentration in runoff water was determined by filtration of 100 cm[3] runoff water collected with a Coshocton wheel sampler (Brakensiek et al., 1979). In events that generated high suspended load, the sediment was also collected by continuous-flow centrifugation and SOC contents in sediments were determined. The mean annual SOC Er from 1987 to 1995 ranged from 1.60 to 1.82, with an average of 1.71.

Soil samples taken in December, 1998 for ^{137}Cs analysis were from upper, middle, and lower landscape positions for the forest, grassland, and watershed sites. Three replicates, one meter apart, were taken at each landscape position and at 0.1 m depth increment up to 0.5 m depth. Soil samples were air-dried and ground to pass 2-mm sieve. Soil ^{137}Cs was measured in USDA-ARS Hydrology Laboratory, Beltsville, MD, according to the technique described by Ritchie and Rasmussen (2000), with measurement precision of ± 4 to 6% on most samples. Estimates of the soil ^{137}Cs radioactivity (Bq kg[-1]) were made using Canberra Genie-2000 software and transformed into the area-based ^{137}Cs radioactivity (Bq m[-2]) using the average soil bulk density of 1.33 (Mg m[-3]). The area-based ^{137}Cs radioactivity for the whole soil profile was computed by obtaining the sum of the area-based ^{137}Cs radioactivity for each 0.1 m sampling layer.

Assuming that the atmospheric ^{137}Cs fallout started in 1951 (Ritchie and McHenry, 1990) and the soil ^{137}Cs activity was equal in the forest and watershed sites in 1951 and for the grass and watershed sites in 1970, the amount of soil ^{137}Cs loss from the watershed can be calculated for three periods: from 1951 to 1970, from 1971 to 1998, and from 1951 to 1998. Because there were no significant differences in soil ^{137}Cs between the upper, middle, and lower slope positions at the three sites (see discussion below), the averages of ^{137}Cs values of the three landscape positions were used to calculate ^{137}Cs loss. Using Equations 1, 2, and 3 and assuming plow depth of 0.2 m, the annual amounts of soil erosion can be calculated from the soil ^{137}Cs loss. For sediment method, the annual soil erosion rates for the three periods were calculated by adding the sediment mass values for each year in the periods and then dividing the sum by the numbers of the years.

According to the cropping history and agricultural management, the parameters for RUSLE were: $R = 115$ hundreds of ft $\times$ ton $\times$ in $\times$ acre$^{-1} \times$ h$^{-1} \times$ yr^{-1}, $K = 0.35$ h $\times$ hundreds of ft$^{-1} \times$ in^{-1}, LS = 1.60 for the meadow period and 1.80 for the other periods, P = 1.0 for the meadow period and 0.89 for the other periods, and C = 0.050, 0.154, 0.002, and 0.009 for the four consecutive land use periods, respectively (USDA, 1997). Then, the RUSLE predicted annual soil loss expressed as Mg ha^{-1} yr^{-1} is equal to 2.24 * R * K * LS * C * P. To obtain the average annual soil loss for the periods from 1971 to 1998 and from 1951 to 1998, we divided the sums of soil erosion by 28 for the former period and by 48 for the latter period.

The annual amounts of SOC erosion was equal to the annual amounts of soil erosion multiplied by 1.51% (the average SOC content) for Equation 1, and further by 1.71 (the average SOC-Er) for Equation 2, Equation 3, sediment mass measurement, and RUSLE prediction.

C. Results and Discussion

The data in Table 1 indicate that the ^{137}Cs content decreased with depth for the forest site and below 0.2 m for the grassland and watershed sites. The ^{137}Cs concentration often reached an undetectable level at 0.4 to 0.5 m depth, indicating a slow vertical movement and tenacious binding with clay (Schulz et al., 1959; Lomenick and Tamura, 1965; Schulz, 1965). However, significant amounts of ^{137}Cs at 0.4 m depth was detected, as was also reported by some other studies in Ohio (Bajracharya et al., 1998). Therefore, sampling depth of at least 0.4 m is necessary to obtain a complete inventory of soil ^{137}Cs. The average of ^{137}Cs activity (kBq m^{-2}) over nine replicates (nine soil cores) was 5.41 at the forest site, 4.30 at the grassland site, and 3.03 at the watershed site (Table 1). These values are higher than many published soil ^{137}Cs values.

The coefficients of variations were >25% for the unplowed layers, <25% for the plowed layers, and increased with depth indicating heterogeneous vertical movement of ^{137}Cs into deeper soil profile through macropore, clay illuviation, and pedoturbation (Southard and Graham, 1992). With nine replicates at each site, the CV was 35% for the forest site, 18% for the grassland site, and 21% for the watershed site (Table 1). The spatial variation of ^{137}Cs for the forest sites was high (CV > 20%, Wallbrink et al., 1994) due to preferential transfer of ^{137}Cs intercepted by forest canopy to the soils through stem flow (Franklin et al., 1967; Gersper, 1970). Based on normal distribution of ^{137}Cs in forest soil (Nolin et al., 1993), an allowable error at 95% confidence level with nine replicates was 23%, 12%, and 14% for the forest, grassland, and watershed sites, respectively.

The slope length (132 m) and steepness (10%) of the watershed would indicate occurrence of concentrated flow and thus soil deposition at the lower slope position where the soil ^{137}Cs would be higher than the upper and/or middle slope positions. However, least significant test indicated insignificant differences between the three landscape positions, probably due to the high spatial variation of ^{137}Cs in the soils. Because the relationship between ^{137}Cs loss and soil and SOC loss follows linearity as shown in Equation 1 or does not significantly deviate from linearity as shown in Equations 2 and 3, there were also no significant differences in soil and SOC erosion between the three landscape positions.

The annual soil and SOC erosion rates calculated with the proportional equation were much higher than those by the sediment method and RUSLE prediction (Table 2). The proportional equation estimated the values which were approximately 20 to 55 times the sediment values and 4 to 6 times the RUSLE values for soil erosion, and 10 to 30 times the sediment values and 2 to 3.5 times the RUSLE values for SOC erosion. The exponential equation (Equation 2) estimated even higher values than the proportional equation (data not shown). However, the revised exponential equation (Equation 3) estimated the values similar to (1.2 to 1.6 times) the sediment values except for the period from 1951 to 1970 (3 times the sediment values) (Table 2). The values obtained with the revised

Table 1. Soil ^{137}Cs radioactivity (kBq m^{-2}) in depth (cm) profiles for upper, middle, lower, and total slope areas of the forest, grassland, and watershed sites and its coefficient variation (CV%) shown in parentheses

Depth	Upper	Middle	Lower	Slope
		Forest		
0-10	4.20 (54)	3.19 (19)	2.03 (59)	3.14 (52)
10-20	1.42 (38)	1.34 (56)	2.20 (25)	1.66 (41)
20-30	0.47 (32)	0.35 (27)	0.57 (82)	0.46 (59)
30-40	0.25 (98)	NS [a]	0.20 (92)	0.23 (87)
40-50	NS	NS	0.00	NS
0-50	6.34 (48)	4.88 (7.1)	5.01 (35)	5.41 (35)
		Grassland		
0-10	1.62 (15)	1.64 (13)	1.45 (12)	1.57 (13)
10-20	1.82 (14)	2.20 (24)	1.58 (6.2)	1.87 (21)
20-30	0.95 (30)	0.81 (65)	0.65 (34)	0.80 (43)
30-40	0.04 (173)	0.13 (87)	0.00	0.06 (155)
40-50	NS	NS	NS	NS
0-50	4.43 (11)	4.78 (23)	3.68 (2.4)	4.30 (18)
		Watershed		
0-10	1.05 (7.6)	1.07 (13)	1.26 (8.0)	1.13 (12)
10-20	1.15 (8.9)	1.24 (12)	1.24 (26)	1.21 (16)
20-30	0.73 (62)	0.56 (54)	0.42 (107)	0.57 (66)
30-40	0.29 (173)	0.00	0.09 (173)	0.12 (233)
40-50	0.00	0.00	0.00	0.00
0-50	3.22 (29)	2.86 (5.3)	3.01 (27)	3.03 (21)

[a] = NS not sampled.

exponential equation were lower than (0.21 to 0.35 times) the RUSLE values for all three periods (Table 2).

When we used the revised exponential equation to calculate the amounts of soil erosion from the previously published data, the equation also gives a close approximation to the sediment values except for the experiment by Kachanoski (1987) (Table 3). The experiment by Kachanoski is not a historic assessment of soil erosion but a long-term study based on sampling soil ^{137}Cs both at the start and the end of the experiment for the period since 1965. The data showed that the proportional equation gave a much closer approximation to the sediment value than the two exponential equations. Further, Lance et al. (1986) used Equation 2 with annual soil ^{137}Cs loss instead of the total loss and gave an estimation close to the sediment data (Table 3).

The study results obtained by Mitchell et al. (1980), Soileau et al. (1990), and Montgomery et al. (1997) indicated good match in annual soil erosion rate estimation between the proportional equation and USLE or RUSLE methods (Table 3). In contrast, our study did not show such close match. Although RUSLE was developed with its large proportion of data collected during the 1940s to 1970s from the midwestern U.S. (Yoder and Lown, 1995), the RUSLE predicted annual soil erosion rate for the Ohio watershed was 3 to 14 times the sediment values. This indicates that the RUSLE, as a soil conservation tool, may overestimate the soil erosion risks.

Table 2. Annual rate (Mg ha[-1] yr[-1]) of soil and SOC erosion from the watershed (0.79 ha) as determined by [137]Cs[a], sediment, and RUSLE methods

Period	Land use[b]	Proportional		Exponential		Sediment		RUSLE	
		Soil	SOC	Soil	SOC	Soil	SOC	Soil	SOC
1951 - 1970	CWMM PT	27.3	0.413	1.54	0.040	0.50	0.013	7.22	0.186
1971 - 1998	CC-M-CS PT-NT	28.1	0.424	1.69	0.044	1.39	0.036	4.75	0.123
1951 - 1998		24.4	0.368	1.58	0.041	1.02	0.026	5.78	0.149

[a]Using Equation 1 (proportional equation) and Equation 3 (exponential equation) to calculate the amounts of soil erosion and multiplying Equation 1 by 1.51% and Equation 3 by 1.51% and 1.71% to calculate the amounts of SOC erosion. The amounts of ^{137}Cs radioactivity in Equations 1 and 3 are the averages of those at the upper, middle, and lower slope positions.
[b]CWMM = corn-wheat-meadow-meadow rotation, C-M-CS = 5 years of corn followed by 8 years of meadow followed by 15 years of corn-soybean rotation. PT = plow till, and NT = no till.

The error involved in the sediment mass measurement using H-flume and Coshocton wheel sampler is expected to be less than that of the manual sampling which may be as much as 83% (Zöbisch et al., 1996). If the error in sediment measurement in this study is assumed to be 50%, the accuracy of the revised exponential equation in estimating historic soil and SOC erosion is high for all three study periods. The accuracy of RUSLE prediction is high only for the period from 1971 to 1998 but low from 1951 to 1998 and poor from 1951 to 1970. Whereas the accuracy of the estimation with the proportional equation is poor for all three periods (Table 2).

Walling and He (1999) and VandenBygaart et al. (2000) identified numerous sources of error based on the assumptions underlining the use of the proportional equation. With regard to this study, the errors involved in using the proportional equation are mainly caused by the infrequent tillage in the 4-year rotation which could lead to significant ^{137}Cs loss with runoff water, POM, and hay harvest (the tillage dilution and the high spatial variability of ^{137}Cs in soils can also cause errors). Natural vegetation can retain some radionuclides for several weeks (Aleksakhin and Tikhomirov, 1973). The amounts of organic matter in runoff water collected from 1984 to 1998 were generally much higher than those in sediments (data not shown), indicating the potential loss of ^{137}Cs with POM prior to 1965. Frere and Roberts (1963) found that at the Coshocton research watersheds there was no correlation between the ^{90}Sr loss and the recorded water runoff or recorded soil loss for the years 1954 to 1960. Menzel (1960) reported that the ^{90}Sr concentration in the runoff was 10 to 30 times that in the field soils. Bremer et al. (1995) indicated a potential significant loss of ^{137}Cs before it was incorporated into the soil managed with the stubble-mulch system.

Furthermore, the discrepancy in soil and SOC erosion assessed by the ^{137}Cs method and sediment measurement for the period from 1971 to 1998 could indicate that the ^{137}Cs concentrations in soil were significantly different by 1970 for the grassland and watershed sites, which invalidated the assumption for using the proportional equation. The difference could be caused by the different vegetation covers at the two sites for most of the years during the period from 1951 to 1970 when ^{137}Cs fallout occurred because the two sites 50 m away from each other had similar topography, the same soil type, and the same crop rotation sequence except that the rotation started 1 year earlier for the grassland than for the watershed.

Table 3. Annual rate (Mg ha^{-1} yr^{-1}) of soil erosion determined by ^{137}Cs[a], sediment, and USLE methods in the previous studies

Study	Site	Period	Proportional	Exponential	Sediment	USLE
Menzel, 1960[b]	Plots in La Crosse, WI	1957 5 months	NA[g]	Corn: 5.09 Oats: 5.09 Clover: 0.25	Corn: 2.24 Oats: 2.24 Clover: 0.067	NA
	Plots in Tifton, GA			Corn: 1.28 Oats: 0.34 Clover: 1.34	Corn: 1.34 Oats: 0.67 Clover: 1.57	
Rogowski and Tanura, 1970[c]	Plots in Oak Ridge, TN	NA 2 years	NA	Bare: 37 (74) Poor meadow: 14 (28) Good meadow: 4.3 (8.6)	Bare: 27 (54) Poor meadow: 1.3 (2.5) Good meadow: 0.45 (0.09)	NA
Mitchell et al., 1980	Watershed transect A, C, and D in Soawano County, WI Corn-corn-oat-meadow-meadow rotation.	1955-1976 22 years	A: 1.6 C: 2.8 D: 16	NA	NA	A: 3.6 C: 3.6 D: 1.3
Lance et al., 1986[d]	Watershed in El Reno, OK Wheat-sorghum rotation	NA 8 years	NA	1.8 (1.3)	2.2	NA
Kachanoski, 1987[e]	Plots in Guelph, Ontario	1965-1976 10 years	31	2.2	34	NA
Soileau et al., 1990	Watershed transects in Colbert County, AL Corn, cotton, pasture, and hay	1954-1987 34 years	A: 45 B: 33 C: 28	A:3.9 B: 2.7 C: 2.2	NA	A:25 B:26 C:26
		1984-1988 5 years	NA	NA	0.9-3.9	8.9-52

Table 3. (continued)

Study	Site	Period	Proportional	Exponential	Sediment	USLE
Montgomery et al., 1997[f]	Watershed in Whitman County, WA Wheat, barley, dry pea	1963-1990 27 years	46	NA	NA	44

[a]Using Equation 1 for the proportional calculation and Equation 3 for the the exponential calculation. The values for the exponential method were calculated in this study except for the values reported in Menzel, 1960 and in parentheses.

[b]Strontium-90 was used by spraying onto the soil surface and the site was not plowed. The values were for 5 months.

[c]The values in parentheses were reported in the original paper.

[d]The value in parentheses was reported in the original paper using annual ^{137}Cs loss percentage.

[e]The ^{137}Cs activity at the study site was measured at the start and the end of the experiment to determine ^{137}Cs loss without using a reference value at a forest site. The value was the average of the values obtained from 10 plots.

[f]RUSLE instead of USLE was used.

[g]NA = data not available.

Table 4. Comparison of ^{137}Cs method with sediment and RUSLE methods in assessing soil and SOC erosion

Comparison	^{137}Cs	Sediment	RUSLE
Erosion type	Erosion by water, wind, gravity, and tillage	Water erosion	Water erosion excluding concentrated flow
Temporal range	Long term (>10 years); retrospective back to 1950s	Short and long terms	Long term (>10 years)
Spatial range	Point, plot, and watershed scales; comparative study for different topographic positions; world wide	Plot and watershed scales; world wide	Plot and watershed scales; mainly in U.S.A.
Accuracy	Low in retrospective determination of soil and SOC erosion from 1950s; higher accuracy in repeat sampling after 1965	High; SOC erosion can be directly determined with sediment	Low, and risks of overestimation
SOC Er	Er cannot be determined but should not be used with proportional equation and mass balance equations unless ^{137}Cs Er is also known	Er can be determined	Er cannot be determined
Cost	Low	High	Low

Based on the results of the present and other studies, it can be concluded that satisfactory assessment of soil and SOC erosion can be obtained by using the empirical revised exponential equation for historic assessment back to the 1950s and by using the proportional equation for a long-term study with soil ^{137}Cs determined at the start of the experiment. In addition, the potential ^{137}Cs loss with runoff, POM, and fine mineral particles is a key issue in improving the accuracy of using the proportional equation for historic and long-term soil and SOC erosion assessment, which needs to be studied in more detail.

IV. Comparison of ^{137}Cs Method with Sediment and RUSLE Methods

Table 4 shows that the main advantage of ^{137}Cs technique compared to the sediment and RUSLE methods is its capability to assess total soil and SOC erosion by water, wind, gravity, and tillage, whereas the other two methods cannot. In comparison with the sediment method, the ^{137}Cs method is more rapid and less costly for assessing historic, comparative, and long-term (> 10 years) soil and SOC erosion. The procedure of the ^{137}Cs method includes sampling, air-drying, grinding, and counting soil samples once or twice as compared with labor intensive long-term sampling with sediment method. Although the cost of gamma-ray counting equipment is high, there is a saving in labor and cost of sediment collection equipment and its maintenance. In comparison with the RUSLE method, the main advantage of ^{137}Cs method is its worldwide applicability.

The main disadvantage of the ^{137}Cs method is that it cannot directly provide SOC content and Er values whereas the sediment method can, and that it is more expensive than the RUSLE method.

V. Conclusion

The SOC dynamics is an important issue in increasing soil fertility and improving water and air quality. The SOC erosion is one of the processes depleting the SOC pool. The methods for assessment of SOC erosion can elucidate the effects of management on SOC dynamics by erosion, oxidation, and leaching. Soil ^{137}Cs method is a unique tool for assessing SOC erosion in that it can assess total erosion by water, wind, gravity, and tillage. It is a simple, rapid, and relatively low cost method for historic, comparative, and long-term erosion study for point, plot, and watershed scales. The analyses of our data and those of others indicate that the revised exponential equation and proportional equation can provide a satisfactory assessment for historic and long-term soil and SOC erosion, respectively. The accuracy of the ^{137}Cs method can be further improved by obtaining information about the potential loss of ^{137}Cs with runoff, POM, and fine mineral particles.

Acknowledgments

This research was funded by the United States Department of Energy. Soil sampling for ^{137}Cs was conducted by Gordon Starr and Mahmoud Ahmadi. Soil ^{137}Cs radioactivity was analyzed in Jerry Ritchie's USDA-ARS Hydrology Laboratory. Historical data of sediment, SOC, and crop management at the studied sites were provided by the technical staff at the North Appalachien Experimental Watershed Research Station. Norman L. Widman of the Natural Resources Conservation Service, the United States Department of Agriculture, Columbus, Ohio, assisted with selection of parameters for RUSLE.

References

Aleksakhin, R.M. and F.A. Tikhomirov. 1973. Radionuclide migration in forest biogeocenosis. p. 126-140. In: V.M. Klechkovskii, G.G. Polikarpov and R.M. Aleksakhin (eds.), *Radioecology*. John Wiley & Sons, New York.

Bajracharya, R.M., R. Lal and J.M. Kimble. 1998. Use of radioactive cesium-137 to estimate soil erosion on three farms in west central Ohio. *Soil Sci.* 163:133-142.

Barrows, H.L. and V.J. Kilmer. 1963. Plant nutrient losses from soils by water erosion. *Adv. Agron.* 15:303-316.

Bernard, C., L. Mabit, S. Wicherek and M.R. Laverdiere. 1998. Long-term soil redistribution in a small French watershed as estimated from cesium-137 data. *J. Environ. Qual.* 27:1178-1183.

Blake, G.R. and K.H. Hartge. 1986. Bulk density. p. 363-376. In: A. Klute (ed.), *Methods of Soil Analysis. Part 1.* Agron. Monogr. 9. American Society of Agronomy and Soil Science Society of America, Madison, WI.

Brady, N.C. and R.R. Weil. 1999. *The Nature and Properties of Soils.* 12th edition. Prentice-Hall, Upper Saddle River, NJ. 881 pp.

Brakensiek, D.L., H.B. Osborn and W.J. Rawls (coordinators). 1979. *Field Manual for Research in Agricultural Hydrology.* USDA Agriculture Handbook 224. 550 pp.

Bremer, E., E. de Jong and H.H. Janzen. 1995. Difficulties in using ^{137}Cs to measure erosion in stubble-mulched soil. *Can. J. Soil Sci.* 75:357-359.

Buyanovsky, G.A. and G.H. Wagner. 1997. Crop residue input to soil organic matter on Sanborn field. p. 73-84. In: E.A. Paul, K. Paustian, E.T. Elliott and C.V. Cole (eds.), *Soil Organic Matter in Temperate Agroecosystems.* CRC Press, Boca Raton, FL.

Dick, W.A., W.M. Edwards and E.L. McCoy. 1997. Continuous application of no-tillage to Ohio soils: changes in crop yields and organic matter-related soil properties. p. 171-182. In: E.A. Paul, K. Paustian, E.T. Elliott and C.V. Cole (eds.), *Soil Organic Matter in Temperate Agroecosystems.* CRC Press, Boca Raton, FL.

Frere, M.H. and H. Roberts, Jr. 1963. The loss of strontium-90 from small cultivated watersheds. *Soil Sci. Soc. Am. Proc.* 27:82-83.

Franklin, R.E., P.L. Gersper and N. Holowaychuk. 1967. Analysis of gamma-ray spectra from soils and plants: II. Effect of trees on the distribution of fallout. *Soil Sci. Soc. Am. Proc.* 31:43-45.

Gersper, P.L. 1970. Effect of American beech trees on the gamma radioactivity of soils. *Soil Sci. Soc. Am. Proc.* 34:318-323.

Gregorich, E.G., K.J. Greer, D.W. Anderson and B.C. Liang. 1998. Carbon distribution and losses: erosion and deposition effects. *Soil Tillage Res.* 47:291-302.

He, Q. and D.E. Walling. 1996. Interpreting particle size effects in the adsorption of ^{137}Cs and unsupported ^{210}Pb by mineral soils and sediments. *J. Environ. Radioact.* 30:117-137.

Higgitt, D.L. 1995. The development and application of caesium-137 measurements in erosion investigations. p. 287-305. In: I.D.L. Foster, A.M. Gurnell and B.W. Webb (eds.), *Sediment and Water Quality in River Catchments.* John Wiley & Sons, New York.

Hird, A.B. and D.L. Rimmer. 1995. Total caesium-fixing potentials of acid organic soils. *J. Environ. Radioactivity.* 26:103-118.

Hussain, I., K.R. Olson and R.L. Jones. 1998. Erosion patterns on cultivated and uncultivated hillslopes determined by soil fly ash contents. *Soil Sci.* 163:726-738.

Jones, R.L. and K.R. Olson. 1990. Fly ash use as a time marker in sedimentation studies. *Soil Sci. Soc. Am. J.* 54:855-859.

Kachanoski, R.G. 1987. Comparison of measured soil 137-cesium losses and erosion rates. *Can. J. Soil Sci.* 67:199-203.

Kachanoski, R.G. and E. de Jong. 1984. Predicting the temporal relationship between soil cesium-137 and erosion rate. *J. Environ. Qual.* 13:301-304.

Kelley, G.E., W.M. Edwards, L.L. Harrold and J.L. McGuinness. 1975. *Soils of the North Appalachian Experimental Watershed.* USDA-ARS Misc. Publ. 1296. U.S. Government Printing Office, Washington, D.C.

Kreznor, W.R., K.R. Olson and D.L. Johnson. 1992. Field evaluation of methods to estimate soil erosion. *Soil Sci.* 153:69-81.

Lal, R. 1979. Soil erosion problems on an Alfisol in western Nigeria and their control. International Institute of Tropical Agriculture (IITA) Monograph No.1. Ibadan, Nigeria. 208 pp.

Lal, R. 1998. Soil erosion impact on agronomic productivity and environmental quality. *Crit. Rev. Plant Sci.* 17:319-464.

Lal, R. 1999. Soil management and restoration for C sequestration to mitigate the accelerated greenhouse effect. *Prog. Environ. Sci.* 1:307-326.

Lance, J.C., S.C. McIntyre, J.W. Naney, and S.S. Rousseva. 1986. Measuring sediment movement at low erosion rates using cesium-137. *Soil Sci. Soc. Am. J.* 50:1303-1307.

Lobb, D.A., R.G. Kachanoski and M.H. Miller. 1995. Tillage translocation and tillage erosion on shoulder slope landscape positions measured using ^{137}Cs as a tracer. *Can. J. Soil Sci.* 75:211-218.

Logan, T.J., R. Lal and W.A. Dick. 1991. Tillage systems and soil properties in North America. *Soil Tillage Res.* 20:241-270.

Lomenick, T.F. and T. Tamura. 1965. Naturally occurring fixation of cesium-137 on sediments of lacustrine origin. *Soil Sci. Soc. Am. Proc.* 29:383-386.

Menzel, R.G. 1960. Transport of strontium[90] in runoff. *Science.* 131:499-450.

Mitchell, J.K., G.D. Bubenzer, J.R. McHenry, and J.C. Ritchie. 1980. Soil loss estimation from fallout cesium-137 measurements. p. 393-401. In: M. de Boodt and D. Gabriels (eds.), *Assessment of Erosion*. John Wiley & Sons, New York.

Monreal, C.M., R.P. Zentner and J.A. Robertson. 1997. An analysis of soil organic matter dynamics in relation to management, erosion and yield of wheat in long-term crop rotation plots. *Can. J. Soil Sci.* 77:553-563.

Montgomery, J.A., A.J. Busacca, B.E. Frazier and D.K. McCool. 1997. Evaluating soil movement using cesium-137 and the revised universal soil loss equation. *Soil Sci. Soc. Am. J.* 61:571-579.

Mutchler, C.K., C.E. Murphree and K.C. McGregor. 1994. Laboratory and field plots for erosion research. p. 11-37. In: R. Lal (ed.), *Soil Erosion Research Methods*. Soil and Water Conservation Society, Ankeny, IA.

Nelson, D.W. and L.E. Sommers. 1996. Total carbon, organic carbon, and organic matter. p. 961-1010. In: A.L. Page et al. (eds.), *Methods of Soil Analysis. Part 3*. American Society of Agronomy and Soil Science Society of America, Madison, WI.

Nishita, H., B.W. Kowalewsky, A.J. Steen and K.H. Larson. 1956. Fixation and extractability of fission products contaminating various soils and clays: I. Sr90, Y91, Ru106, Cs137, and Ce144. *Soil Sci.* 81:314-326.

Nishita, H. and E.H. Essington. 1967. Effect of chelating agents on the movement of fission products in soils. *Soil Sci.* 103:168-176.

Nolin, M.C., Y.Z. Cao, D.R. Coote and C. Wang. 1993. Short-range variability of fallout ^{137}Cs in an uneroded forest soil. *Can. J. Soil Sci.* 73:381-385.

Palis, R.G., H. Ghandiri, C.W. Rose and P.G. Saffigna. 1997. Soil erosion and nutrient loss. 3. Changes in the enrichment ratio of total nitrogen and organic carbon under rainfall detachment and entrainment. *Aust. J. Soil Res.* 35:891-905.

Patwardhan, A.S., A.S. Donigian, Jr., R.V. Chinnaswamy, and T.O. Barnwell. 1997. A retrospective modeling assessment of historical changes in soil carbon and impacts of agricultural development in central U.S.A., 1900 to 1990. p. 485-498. In: R. Lal, J.M. Kimble, R.F. Follett and B.A. Stewart (eds.), *Soil Processes and the Carbon Cycle*. Lewis Publishers, Boca Raton, FL.

Paustian, K., H.P. Collins, and E.A. Paul. 1997. Management controls on soil carbon. p. 15-50. In: E.A. Paul, K. Paustian, E.T. Elliott and C.V. Cole (eds.), *Soil Organic Matter in Temperate Agroecosystems*. Lewis Publishers, Boca Raton, FL.

Rasmussen, P.E., S.L. Albrecht and R.W. Smiley. 1998. Soil C and N changes under tillage and cropping systems in semi-arid Pacific Northwest agriculture. *Soil Tillage Res.* 47:197-205.

Ritchie, J.C., E.E.C. Clebsch and W.K. Rudolph. 1970. Distribution of fallout and natural gamma radionuclides in litter, humus and surface mineral soil layers under natural vegetation in the Great Smoky Mountains, North Carolina-Tennessee. *Health Physics* 18:479-489.

Ritchie, J.C., J.A. Spraberry and J.R. McHenry. 1974. Estimating soil erosion from the redistribution of fallout [137]Cs. *Soil Sci. Soc. Am. Proc.* 38:137-139.

Ritchie, J.C. and J.R. McHenry. 1978. Fallout cesium-137 in cultivated and noncultivated North Central United States watersheds. *J. Environ. Qual.* 7:40-44.

Ritchie, J.C. and J.R. McHenry. 1990. Application of radioactive fallout cesium-137 for measuring soil erosion and sediment accumulation rates and patterns: a review. *J. Environ. Qual.* 19:215-233.

Ritchie, J.C. and P.E. Rasmussen. 2000. Application of [137]ceasium to estimate erosion rates for understanding soil carbon loss on long-term experiments at Pendleton, Oregon. *Land Degrad. Develop.* 11:75-81.

Rogers, H.T. 1941. Plant nutrient losses by erosion from a corn, wheat, clover rotation on Dunmore silt loam. *Soil Sci. Soc. Am. Proc.* 6:263-271.

Rogowski, A.S. and T. Tamura. 1970. Environmental mobility of cesium-137. *Radiat. Bot.* 10:35-45.

Ruppenthal, M., D.E. Leihner, N. Steinmuller and M.A. El-Sharkawy. 1997. Losses of organic matter and nutrients by water erosion in cassava-based cropping systems. *Expl. Agric.* 33:487-498.

Schulz, R.K. 1965. Soil chemistry of radionuclides. *Health Physics* 11:1317-1324.

Schulz, R.K., R. Overstreet and I. Barshad. 1959. On the soil chemistry of cesium 137. *Soil Sci.* 89:16-27.

Slater, C.S. and E.A. Carleton. 1938. The effect of erosion on losses of soil organic matter. *Soil Sci. Soc. Am. Proc.* 3:123-128.

Soileau, J.M., B.F. Hajek and J.T. Touchton. 1990. Soil erosion and deposition evidence in a small watershed using fallout cesium-137. *Soil Sci. Soc. Am. J.* 54:1712-1719.

Southard, R.J. and R.C. Graham. 1992. Cesium-137 distribution in a California Pelloxeret: evidence of pedoturbation. *Soil Sci. Soc. Am. J.* 56:202-207.

United States Department of Agriculture. 1997. Predicting soil erosion by water: a guide to conservation planning with the revised universal soil loss equation (RUSLE). Agriculture Handbook 703. 384 pp.

VandenBygaart, A.J., D.J. King, P.H. Groenevelt and R. Protz. 2000. Cautionary notes on the assumptions made in erosion studies using fallout ^{137}Cs as a marker. *Can. J. Soil Sci.* 80:395-397.

Wallbrink, P.J., J.M. Olley and A.S. Murray. 1994. Measuring soil movement using [137]Cs: Implications of reference site variability. *International Assoc. for Hydrolog. Publ.* 224:95-105.

Walling, D.E. and Q. He. 1999. Improved models for estimating soil erosion rates from cesium-137 measurements. *J. Environ. Qual.* 28:611-622.

Walling, D.E. and T.A. Quine. 1991. Use of ^{137}Cs measurements to investigate soil erosion on arable fields in the U.K.: potential applications and limitations. *J. Soil Sci.* 42:147-165.

Wan, Y. and S.A. El-Swaify. 1997. Flow-induced transport and enrichment of erosional sediment from a well-aggregated and uniformly-textured Oxisol. *Geoderma* 75:251-265.

Yoder, D. and J. Lown. 1995. The future of RUSLE: inside the new Revised Universal Soil Loss Equation. *J. Soil Water Cons.* 50:484-489.

Zöbisch, M.A., P. Klingspor and A.R. Oduor. 1996. The accuracy of manual runoff and sediment sampling from erosion plots. *J. Soil Water Cons.* 51:231-233.

Section VI

Modeling and Scaling Procedures

A Simple Method to Estimate Soil Carbon Dynamics at the BOREAS Northern Study Area, Manitoba, Canada

G. Rapalee

I. Introduction

High latitude ecosystems would be expected to experience the largest temperature changes if the Earth warms. Boreal forest soils, where most soil carbon accumulates in the moss layer and organic and humic horizons, are both an important source of and sink for atmospheric carbon (C). Therefore, responses of boreal forest soils to warming could have important effects on the interactions between climate change and terrestrial carbon storage. Understanding the present patterns of soil carbon distribution and the soil properties that covary with soil carbon is essential for projecting future changes.

Scaling plot-level measurements to landscape-level patterns is a difficult and important task. A model was developed that estimates soil carbon stocks and fluxes of soil profile data from a study area in the boreal forest of Manitoba, Canada. The model is simple enough to be assembled as a series of linked files in a spreadsheet program on a personal computer and yet detailed enough to describe this complex landscape.

II. Background

Predicting how the vast stores of organic carbon in the boreal forest biome will be affected by possible future global warming requires an understanding of the factors controlling the production, decomposition, and storage of organic C in boreal ecosystems. Inverse models that calculate the latitudinal distribution of CO_2 sources and sinks from observed CO_2 distributions suggest that boreal regions may be significant C sinks (Tans et al., 1990; Ciais et al., 1995). However, forest inventories and land use studies do not explain where all the carbon is sequestered, perhaps because these approaches do not fully account for the soils, which may store as much as two-thirds of the missing carbon (D. Schimel, in Kaiser, 1998).

An alternative approach to determining C fluxes at the landscape or regional scale is to develop models that examine the processes of C dynamics and to use those models, together with a knowledge of the spatial distribution of important controlling factors (such as precipitation, temperature, and soil drainage), to extrapolate C fluxes from the plot-scale of observations to larger regions. Because it is process based, this latter approach may also be used to predict the response of the landscape to changes in controlling parameters.

For example, models developed in each of the following four studies examined the processes of C dynamics in the soil, moss, and trees. In a study of boreal forest soils in Finland, Liski and Westman modeled the effects of a temperature gradient (1997a) and site productivity (1997b) to estimate and determine regional patterns of soil carbon storage. Burke et al. (1989) examined the effects of the major controls over soil organic matter content and predicted regional patterns of C storage in grasslands by modeling relationships between soil C and soil texture and differing precipitation and temperature conditions. Bonan and Korzuhin (1989) examined the ecological significance of interactions among site conditions, tree growth, and moss layer development in boreal forests of interior Alaska by simulating different climatic and forest canopy conditions.

Responses of boreal forest soils to warming, changes in drainage, or changes in fire frequency have all been proposed to be important for terrestrial C storage (Bonan, 1993; Moore and Knowles, 1990; Gorham, 1991; Kasischke et al., 1995; Kurz and Apps, 1995). To confirm these inferences based on atmospheric CO_2 concentrations and to better understand the geophysical and biophysical factors, a better understanding of the surface biogeochemical processes expected in carbon cycling is needed.

To examine these processes in the boreal forest soils, field studies of C input, storage, and turnover in the northern region of the boreal forest were used (Trumbore and Harden, 1997; Harden et al., 1997). These data were collected as part of the BOReal Ecosystem-Atmosphere Study (BOREAS),[1] to develop simple models linking soil C storage and rates of accumulation to soil drainage class and the time since last fire. These models were previously combined with a soil drainage map for a 120-year-old black spruce (*Picea mariana*) stand at the BOREAS Northern Study Area (NSA) Old Black Spruce (OBS) site (1 km²), to compare the soil component of net C storage with tower-based eddy-covariance measurements of net ecosystem production (Harden et al., 1997). This study expands this scaling approach at the 1 km² site to estimate the soil C storage and flux for a 733 km² area within the BOREAS NSA, in which two soil surveys were undertaken during the 1994 and 1996 field seasons (Veldhuis, 1998; Veldhuis and Knapp, 1998; Trumbore et al., 1998).

III. Model Description

The model incorporates area, depth, and time in estimating soil carbon fluxes in the boreal forest. The model has three inputs (Table 1): soil drainage, forest stand age (or time in years since the last disturbance, which is most often fire), and soil carbon stocks. Output (Table 2) shows annual carbon flux derived from the three inputs.

Drainage and incidence of fire are the two factors thought to be most important in controlling annual accumulation rates of soil carbon in the boreal forest. Drainage affects the severity of fires, the kind of vegetation, and the rate of regeneration. Periodic fires characterize the northern boreal region, controlling variations in C storage across the landscape. Using data from a soil survey for the NSA (Veldhuis and Knapp, 1998), the study area was stratified by drainage class (Table 3). Drainage class also corresponds to vegetation type. The age of each of the forest stands in the study area was determined from fire scars detected on satellite imagery and fire history maps, and from forest inventory and tree core data (as described in Rapalee et al., 1998b).

The upland soil profiles were divided into two layers that are distinctly different in their C dynamics: (i) a surface layer that includes moss and soil that is recognizable as organic material and (ii) a deep layer consisting of more highly decomposed organic matter (humic material), charred material, and the mineral A horizon where minor amounts of organic matter are incorporated. The

---

[1] BOREAS, a study under NASA's Earth Science Enterprise.

Table 1. Model Parameters — Input

Parameter	Units	Description	Equation	Reference/Source
D	—	Soil drainage class (See Table 3, this chapter.)	—	Veldhuis and Knapp, 1998
t	yrs	Forest stand age (Time since last disturbance)	—	Rapalee et al., 1998a
mp	km^2	Mapped soil polygon	—	Veldhuis and Knapp, 1998
F	%	Fraction of mapped soil polygon	—	Veldhuis and Knapp, 1998
Surface Soil C Stock				
$C_{t,D}$	kg C m^{-2}	Average of surface horizons by drainage class (D) and stand age (t)	$(I_{surface,D}/k_{surface,D}) \times (1 - e^{-kt})$	(1) Harden et al., 1992, 1997
$C_{surface,F}$	kg C m^{-2}	Area-weighted average of map fraction (F)	$C_{t,D} \times F_{t,D}$	(2) Davidson and Lefebvre, 1993
$C_{surface,mp}$	kg C m^{-2}	Area-weighted average of map polygon (mp)	$\sum_{mp} C_{surface,F,mp}$	(3) Davidson and Lefebvre, 1993
$I_{surface}$	kg C m^{-2} yr^{-1}	Soil carbon input rate, surface soil layers	—	Trumbore and Harden, 1997 (See Table 3, this chapter.)
$k_{surface}$	yr^{-1}	Decomposition constant, surface soil layers	—	Trumbore and Harden, 1997 (See Table 3, this chapter.)
Deep Soil C Stock				
C_{SS}	kg C m^{-2}	Average of deep horizons by soil series (SS)	$\sum_h BD_h \times \%C_h \times T_h$	(4) Davidson and Lefebvre, 1993
$C_{deep,F}$	kg C m^{-2}	Area-weighted average of map fraction (F)	$C_{SS} \times F_{SS,t,D}$	(5) Davidson and Lefebvre, 1993
$C_{deep,mp}$	kg C m^{-2}	Area-weighted average of map polygon (mp)	$\sum_{mp} C_{deep,F,mp}$	(6) Davidson and Lefebvre, 1993

Table 1. Model Parameters — Input (cont.)

Parameter	Units	Description	Equation		Reference/Source
BD_{obs}	g cm^{-3}	Bulk density (observed in soil surveys)	—		Veldhuis, 1998 Trumbore et al., 1998
BD_{est}	g cm^{-3}	Bulk density (estimated from regression equations)	$\ln(BD_{est}) = 0.271 - (0.066 \times \%C)$ $\ln(BD_{est}) = 0.132 - (0.072 \times \%C)$	(7) (8)	Rapalee et al., 1998b from Veldhuis, 1998 from Trumbore et al., 1998
$\%C$	%	Organic carbon	—		Veldhuis, 1998 Trumbore et al., 1998
h	—	Soil horizon	—		Veldhuis, 1998 Trumbore et al., 1998
T	cm	Soil horizon thickness	—		Veldhuis, 1998 Trumbore et al., 1998
SS	—	Soil series	—		Veldhuis and Knapp, 1998 Trumbore et al., 1998
I_{deep}	kg C m^{-2} yr^{-1}	Soil carbon input rate, deep soil layers	—		Trumbore and Harden, 1997 (See also Table 3, this paper.)
k_{deep}	yr^{-1}	Decomposition constant, deep soil layers	—		Trumbore and Harden, 1997 (See also Table 3, this paper.)

Total Soil C Stock

Surface + Deep Carbon Stock

Parameter	Units	Description	Equation		Reference/Source
$C_{total,F}$	kg C m^{-2}	Area-weighted average of map fraction (Eq. 2 + Eq. 5)	$C_{surface,F} + C_{deep,F}$	(9)	Rapalee et al., 1998a,b
$C_{total,mp}$	kg C m^{-2}	Area-weighted average of map polygon (Eq. 3 + Eq. 6)	$C_{surface,mp} + C_{deep,mp}$	(10)	Rapalee et al., 1998a,b

Table 2. Model Output

Parameter	Units	Description	Equation		Reference/Source
Surface Soil C Flux					
			$I - kC$		Harden et al., 1992, 1997
$dC_{t,D}/dt$	g C m^{-2} yr^{-1}	Average of surface horizons by C stocks (C), drainage class (D), and stand age (t)	$I_{surface,D} - (k_{surface,D} \times C_{t,D})$	(11)	Rapalee et al., 1998a,b
$dC_{surface,F}/dt$	g C m^{-2} yr^{-1}	Area-weighted average of map fraction (F)	$(dC_{t,D}/dt) \times F_{t,D}$	(12)	Rapalee et al., 1998a,b
$dC_{surface,mp}/dt$	g C m^{-2} yr^{-1}	Area-weighted average of map polygon (mp)	$\sum_{mp} dC_{surface,F,mp}/dt$	(13)	Rapalee et al., 1998a,b
Deep Soil C Flux					
			$I - kC$		Harden et al., 1992, 1997
dC_{SS}/dt	g C m^{-2} yr^{-1}	Average deep horizons by soil series (SS) and drainage (D) class	$I_{deep,D} - (k_{deep,D} \times C_{SS})$	(14)	Rapalee et al., 1998a,b
$dC_{deep,F}/dt$	g C m^{-2} yr^{-1}	Area-weighted average of map fraction (F)	$(dC_{SS}/dt) \times F_{SS,t,D}$	(15)	Rapalee et al., 1998a,b
$dC_{deep,mp}/dt$	g C m^{-2} yr^{-1}	Area-weighted average of map polygon (mp)	$\sum_{mp} dC_{deep,F,mp}/dt$	(16)	Rapalee et al., 1998a,b
Net Soil C Flux		Surface + Deep Carbon Flux			
$dC_{net,F}/dt$	g C m^{-2} yr^{-1}	Area-weighted average of map fraction (Eq. 12 + Eq. 15)	$(dC_{surface,F}/dt) + (dC_{deep,F}/dt)$	(17)	Rapalee et al., 1998a,b
$dC_{net,mp}/dt$	g C m^{-2} yr^{-1}	Area-weighted average of map polygon (Eq. 13 + Eq. 16)	$(dC_{surface,mp}/dt) + (dC_{deep,mp}/dt)$	(18)	Rapalee et al., 1998a,b

mineral B horizon was also included in this deep layer. Surface layers accumulate carbon between fires, and turnover times are about 10 times shorter than for deep layers in which C accumulates slowly, integrating over many fire cycles (Trumbore and Harden, 1997).

Because C dynamics are different in surface and deep soil layers, total carbon stocks were estimated using a different method for each layer. C stocks for surface layers were estimated based on a time-dependent model of moss growth after fire. For the deep soil layers, carbon stocks were estimated by common soil series based on soil survey observations and data from individual soil profiles which had been analyzed using <2 mm particle size.

For the output (Table 2), surface and deep soil carbon flux were estimated for 1994, the year in which most of the field studies were conducted, based on C stocks and a simple model of C turnover derived from Trumbore and Harden's (1997) radiocarbon field studies. The models for surface C flux based on fire frequency and drainage class were derived from the work of Harden et al. (1997).

The spatial base for these analyses is a soil polygon map from Veldhuis and Knapp (1998). Spatial variation within soil polygons $(mp)^2$ is accounted for by fractional components (F) with similar attributes within each polygon. Each fractional component of each soil polygon is described by a list of attributes, including drainage class (D) and soil series (SS). As model input for each fractional component, D and SS were determined from the soil survey, and age since disturbance (t) from fire scars and tree core data.

From the two sets of field data, Veldhuis (1998) and Trumbore et al. (1998), average C stock in the deep layers was calculated for each soil series, or group of similar soil series, as identified in Veldhuis and Knapp (1998).

As outlined in Tables 1 and 2, the model estimates soil C stock and flux per unit area at three scales: (i) by computing the total or average at a site in a particular drainage and age class for surface stocks and surface and deep fluxes, and a particular soil series for deep C stocks; (ii) by computing C stock and flux for each fractional component (F) by multiplying by the percentage of the total polygon area the component occupies; and (iii) by computing the C stock and flux for each soil polygon (mp) by summing respective stocks and flux from (ii). (ii) and (iii) yield area-weighted averages of each soil map fraction and polygon, respectively, accounting for spatial variation within the landscape of the study area. Area-weighted averages in (iii) include fractional components identified in the soil survey as non-soil areas (rock, water, and lake).

C mass (stock and flux) of fractional components and soil polygons was calculated by multiplying C stock or flux by respective area. Summing the mass of each polygon yielded total carbon stock and flux for the area.

IV. Model Input

The model has several inputs (Table 1) and employs independent intermediate steps that lead to estimating C stocks in the surface and deep soil layers, which then were added together to estimate total C stock for the entire soil profile at the three scales described in Section III. Although the methods differ for estimating surface and deep C stocks, the model estimates all combinations of drainage class (D) and stand age (t) for each of the surface layer input (I) rates and decomposition (k) constants (Table 3) and deep layer soil series (SS).

Input (I) rates and decomposition (k) constants in Table 3 are from two related studies and were determined using two approaches. First, surveys of the C inventory in organic matter above the most

---

[2] See Table 1 for description of model parameters listed throughout this chapter.

recent char layer of the deep soil horizons were conducted across a series of sites that differed in time since fire (Harden et al., 1997). A second approach used vertical accumulation rates obtained by radiocarbon analyses to determine the age of accumulated carbon (Trumbore and Harden, 1997). For modeling C dynamics in surface soil (including mosses) for moderately well, imperfectly, and poorly drained upland sites, estimates of I and k based on Harden et al.'s (1997) chronosequence studies were used. Where no chronosequence data were available (well-drained sites) and for the wetlands, Trumbore and Harden's (1997) radiocarbon-derived I and k values were used.

A. Surface C Stocks

Soil C stocks in the surface layers ($C_{t,D}$) between fires were modeled using the following equation from Harden et al. (1992, 1997):

$$(I_{surface,D}/k_{surface,D}) \times (1 - e^{-kt}) \qquad \text{Eq. (1)}$$

Equation 1 was derived from the following equation (Harden et al., 1992, 1997), which calculates a simple balance of C inputs and first-order decomposition:

$$dC/dt = I - (k \times C) \qquad \text{Eq. (1a)}$$

Surface soil carbon for the map fraction ($C_{surface,F}$) equals the product of $C_{t,D}$ and the percent area the fraction occupies:

$$C_{t,D} \times F_{t,D} \qquad \text{Eq. (2)}$$

The area-weighted average of surface soil carbon of the map polygon ($C_{surface,mp}$) was calculated by summing Equation 2:

$$\sum_{mp} C_{surface,F,mp} \qquad \text{Eq. (3)}$$

B. Deep C Stocks

In the deep soil layers, C stocks were estimated by common soil series (SS) identified in Veldhuis and Knapp (1998). Carbon stocks (C_{SS}) of each deep horizon (h) of each profile were calculated by multiplying bulk density (BD) by percent C (%C) and thickness (T), and the products were summed for each deep soil series profile, including B horizons, when present (Rapalee et al., 1998a,b, as outlined in Davidson and Lefebvre, 1993):

$$\sum BD_h \times \%C_h \times T_h \qquad \text{Eq. (4)}$$

Deep soil carbon for the map fraction ($C_{deep,F}$) equals the product of C_{SS} and the percent area the fraction occupies:

$$C_{SS} \times F_{SS,t,D} \qquad \text{Eq. (5)}$$

Finally, the area-weighted average of deep soil carbon of the map polygon ($C_{deep,mp}$) was calculated by summing Equation 5:

$$\sum_{mp} C_{deep,F,mp} \qquad\qquad \text{Eq. (6)}$$

Non-linear regression equations were developed using both field data sets to estimate bulk density for profiles for which bulk density measurements were not available.

$$\ln(BD_{est}) = 0.271 - (0.066 \times \%C) \quad \text{Veldhuis, 1998} \qquad\qquad \text{Eq. (7)}$$

$$\ln(BD_{est}) = 0.132 - (0.072 \times \%C) \quad \text{Trumbore et al., 1998} \qquad\qquad \text{Eq. (8)}$$

C. Total C Stocks

C stock for the total profile was calculated by adding surface plus deep C stocks:

$$C_{total,F} = C_{surface,F} + C_{deep,F} \quad (\text{Eq. 2} + \text{Eq. 5}) \qquad\qquad \text{Eq. (9)}$$

$$C_{total,mp} = C_{surface,mp} + C_{deep,mp} \quad (\text{Eq. 3} + \text{Eq. 6}) \qquad\qquad \text{Eq. (10)}$$

V. Model Output

A. Surface C Flux

Surface C fluxes ($dC_{t,D}/dt$), representing net annual storage into upland soils, were calculated using Equation 1a with I and k values (Table 3) and C inventory ($C_{t,D}$):

$$I_{surface,D} - (k_{surface,D} \times C_{t,D}) \qquad\qquad \text{Eq. (11)}$$

Surface soil flux for the map fraction ($C_{deep,F}$) equals the product of $dC_{t,D}/dt$ and the percent area the fraction occupies:

$$(dC_{t,D}/dt) \times F_{t,D} \qquad\qquad \text{Eq. (12)}$$

The area-weighted average of surface soil carbon flux of the map polygon ($dC_{surface,mp}/dt$) was calculated by summing Equation 12:

$$\sum_{mp} dC_{surface,F,mp}/dt \qquad\qquad \text{Eq. (13)}$$

B. Deep C Flux

Deep C fluxes represent net annual losses from upland soils and net annual storage in the wetlands. Deep C fluxes (dC_{SS}/dt) were calculated using I and k values (Table 3) and C inventory (C_{SS}):

$$I_{deep,D} - (k_{deep,D} \times C_{SS}) \qquad \text{Eq. (14)}$$

Deep soil flux for the map fraction ($C_{deep,F}$) equals the product of dC_{SS}/dt and the percent area the fraction occupies:

$$(dC_{SS}/dt) \times F_{SS,t,D} \qquad \text{Eq. (15)}$$

Finally, the area-weighted average of deep soil carbon flux of the map polygon ($dC_{deep,mp}/dt$) was calculated by summing Equation 15:

$$\sum_{mp} dC_{deep,F,mp}/dt \qquad \text{Eq. (16)}$$

C. Net C Flux

C flux for the total profile was calculated by adding surface plus deep C flux:

$$dC_{net,F}/dt = (dC_{surface,F}/dt) + (dC_{deep,F}/dt) \quad \text{(Eq. 12 + Eq. 15)} \qquad \text{Eq. (17)}$$

$$dC_{net,mp}/dt = (dC_{surface,mp}/dt) + (dC_{deep,mp}/dt) \quad \text{(Eq. 13 + Eq. 16)} \qquad \text{Eq. (18)}$$

In summary, the approach to modeling and scaling up to the region was to: (i) combine soil survey data of drainage class and the record of time since fire to calculate the stocks and 1994 fluxes of C into surface moss layers using I and k values determined for the six drainage classes (Table 3); (ii) use data collected in soil surveys on C inventory to determine the deep C inventory for each polygon of the soil map; and (iii) determine fluxes of C lost from the deep C pool using the deep C soil stock estimates and I and k values (Table 3).

VI. Results

The results, as reported in Rapalee et al. (1998b), indicate that soil carbon stocks are tied to drainage, with the largest C stocks occurring in very poorly drained fens and bogs and those poorly drained upland peat sites where humic material and a charred layer derived from fire residues lie at the base of the moss layer. Calculations of net C flux are sensitive to the decomposition rate for this large pool of deep soil, which can range from a small sink to a source. Depending on both drainage class and time since last fire, net flux estimates indicate the entire profile is a small C sink, with annual accumulations highest in the most recently burned areas and in the oldest, more poorly drained sites.

Figures 1 and 2 are examples of the output, modeled as area-weighted average annual flux in the surface and deep layers and net flux for the entire study area. Figure 1 graphs the results of summing

Table 3. Input (I) rates and decomposition (k) constants for surface and deep soil layers

Soil horizon	Drainage class	I, kg C m^{-2} yr^{-1}	k, yr^{-1}
Surface	Well	0.06	0.07
	Moderately well	0.08	0.013
	Imperfect	0.07	0.0105
	Poor		
	Sphagnum moss	0.06	0.008
	Palsa	0.08	0.013
	Very poor	0.0324[a]	0.02[a]
Deep	Well	0.015	0.01
	Moderately well	0	0.003
	Imperfect	0	0.002
	Poor		
	Sphagnum moss	0	0.0007
	Palsa	0	0
	Very poor	0.064	0.0004

Information in this table is from Trumbore and Harden (1997).

[a] These values are fixed for a C inventory of 13 kg C m^{-2}, so as to give 0.064 kg C m^{-2} yr^{-1}, the input to the deep layers.

Table is from Rapalee et al. (1998b), used with permission of the American Geophysical Union.

the modeled fractional components (F) by each combination of drainage class (D) and time since fire (t), so that (A) is the flux in the surface layers (i.e., summing Equation 12), and (B) the deep soil fluxes (i.e., summing Equation 15). Figure 1 clearly shows that soil C flux levels depend on both time since fire and drainage class, and illustrates the difference between fluxes in the two layers. In the surface layers (A), where C is stored in the soil in years after fire, flux is highest in the most recently burned sites and progressively declines in the older stands. The deep layers (B), however, are releasing C to the atmosphere, with greatest losses to the atmosphere in the moderately well and imperfectly drained sites.

Figure 2 represents area-weighted net annual soil carbon flux of the map polygons from Veldhuis and Knapp (1998). Net flux values were calculated using Equation 18. The map depicts the spatial variation in the study area. Greatest annual C accumulation is in the area of the recent burn, while losses (release of C to the atmosphere) occur in the well-drained sites.

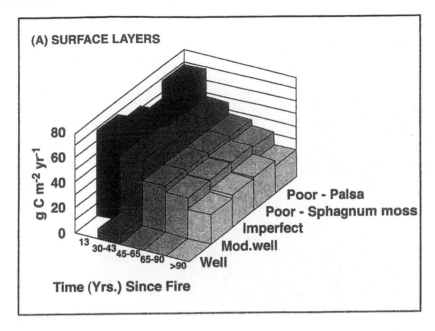

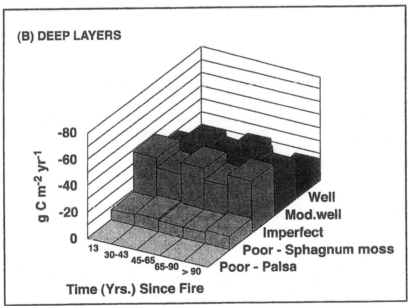

Figure 1. (A) Rates of C accumulation (positive values) in surface layers (including moss) for drainage class and by time since fire (stand age). Values are derived from applying Equation 12 (Table 2) over the entire 733 km^2 study area. (B) Rates of C loss (negative values) from deep organic layers (below moss) and mineral soil as a function of soil drainage and time since fire. Fens and collapse scar bogs (not shown) are gaining an estimated 29 and 12 g C m^{-2} yr^{-1}, respectively, in the deep layers. Values are derived from applying Equation 15 (Table 2) over the entire study area. Figure is from Rapalee et al. (1998b), used with permission of the American Geophysical Union.

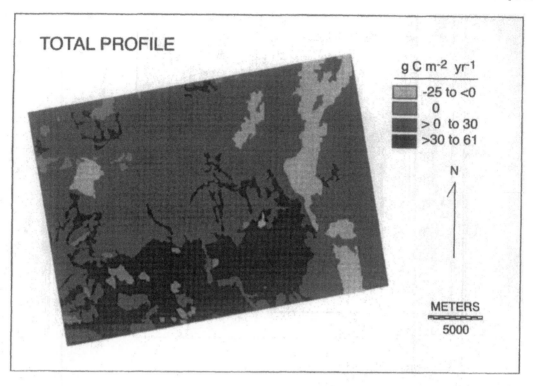

Figure 2. Rates of soil carbon accumulation or loss in the 733 km^2 study area. The map represents area-weighted averages of map polygons of the total soil profile. Values are derived from applying Equation 18 (Table 2) over the entire study area. Areas with zero C accumulation or loss are those polygons mapped as lake or rock. Area-weighted averages of the soil polygons shown on this map include fractional components (F) mapped as water and rock. Figure is from Rapalee et al. (1998b), used with permission of the American Geophysical Union.

VII. Model Validation

The modeled results are close to those of other studies in the northern latitudes. For instance, the estimate of total C stock (42 kg C m^{-2}, from Rapalee et al., 1998b), averaged over the entire 733 km^2 study area, is within the range (40-70 kg C m^{-2}) reported in the Lacelle et al. (1998) soil organic carbon map of North America. The model estimates for annual C flux in the fens (28.8 g C m^{-2} yr^{-1}) are in accord with Gorham's (1991) estimates (29 g C m^{-2} yr^{-1}) derived from synthesizing results of several studies of undrained and unmined boreal and subarctic peatlands. Finally, flux estimates for deep soils in the older black spruce sites (40 g C m^{-2} yr^{-1}; Figure 1B) are within the range of Goulden et al.'s (1998) findings, averaged over 3 years, of an implied loss of 80 ± 50 g C m^{-2} yr^{-1} at a 120-year-old site within the study area. At the same site Harden et al.'s (1997) estimates that the moss and surface soil layer accumulate 20–40 g C m^{-2} yr^{-1} compare well with the modeled estimate of this study of 25 g C m^{-2} yr^{-1} in imperfectly drained sites >90 years old (Figure 1A).

VIII. Summary

This simple and portable model is capable of generating area-weighted spatially referenced estimates of soil carbon stocks and fluxes in the northern region of the boreal forest. The model accounts for spatial variation within the soil polygons at finer scales than were previously practical. Modeled values are within the ranges reported in other field studies of the northern latitudes. The methodology also can be applied to other geographic areas.

Acknowledgments

This material is based on work supported by NASA under prime grant NAG5-2306, The Regents of the University of California, and the USGS Global Change Program. I acknowledge additional support from grants NAG5-31752 and NAG5-7301. I thank K.F. Huemmrich for useful comments about presenting this paper, E.A. Davidson for help in preparing Tables 1 and 2, and E.R. Levine for help with final revisions.

References

Bonan, G.B. 1993. Physiological controls of the carbon balance of boreal forest ecosystems. *Canadian Journal of Forestry Research* 23:1453-1471.

Bonan, G.B. and M.D. Korzuhin. 1989. Simulation of moss and tree dynamics in the boreal forests of interior Alaska. *Vegetation* 84:31-44.

Burke, I.C., C.M. Yonker, W.J. Parton, C.V. Cole, K. Flach and D.S. Schimel. 1989. Texture, climate, and cultivation effects on soil organic matter in U.S. grassland soils. *Soil Science Society of America Journal* 53:800-805.

Ciais, P., P.P. Tans, M. Trolier, J.W.C. White and J.R. Francey. 1995. A large northern hemisphere CO_2 sink indicated by $^{13}C/^{12}C$ of atmospheric CO_2. *Science* 269:1098-1102.

Davidson, E.A. and P.A. Lefebvre. 1993. Estimating regional carbon stocks and spatially covarying edaphic factors using soil maps at three scales. *Biogeochemistry* 22:107-131.

Gorham, E. 1991. Northern peatlands: Role in the carbon cycle and probable responses to climate warming. *Ecological Applications* 1:182-195.

Goulden, M.L., S.C. Wofsy, J.W. Harden, S.E. Trumbore, P.M. Crill, S.T. Gower, T. Fries, B.C. Daube, S.-M. Fan, D.J. Sutton, A. Bazzaz and J.W. Munger. 1998. Sensitivity of boreal forest carbon balance to soil thaw. *Science* 279:214-217.

Harden, J.W., E.T. Sundquist, R.F. Stallard and R.K. Mark. 1992. Dynamics of soil carbon during deglaciation of the Laurentide ice sheet. *Science* 258:1921-1924.

Harden, J.W., K.P. O'Neill, S.E. Trumbore, H. Veldhuis, and B.J. Stocks. 1997. Moss and soil contributions to the annual net flux of a maturing boreal forest. *Journal of Geophysical Research* 102 (D24):28,805-28,816.

Kaiser, J. 1998. New network aims to take the world's CO_2 pulse. *Science* 281:506-507.

Kasischke, E.S., N.L. Christensen and B.J. Stocks. 1995. Fire, global warming, and the carbon balance of boreal forests. *Ecological Applications* 5:437-451.

Kurz, W.A. and M.J. Apps. 1995. An analysis of future carbon budgets of Canadian boreal forests. *Water, Air, and Soil Pollution* 82:321-331.

Lacelle, B., C. Tarnocai, S. Waltman, J. Kimble, F. Orozco-Chavez and B. Jakobsen. 1998. *Soil Organic Carbon Map of North America (Provisional).* Scale 1:10,000,000. Agriculture and Agri-Food Canada, U.S. Department of Agriculture, Instituto Nacional de Estadística Geografía e Informática (INEGI), México, and the Institute of Geography, University of Copenhagen, Denmark.

Liski, J. and C.J. Westman. 1997a. Carbon storage in forest soil of Finland 1. Effect of thermo-climate. *Biogeochemistry* 36:239-260.

Liski, J. and C.J. Westman. 1997b. Carbon storage in forest soil of Finland 2. Size and regional patterns. *Biogeochemistry* 36:261-274.

Moore, T.R. and R. Knowles. 1990. Methane emissions from fen, bog, and swamp peatlands in Quebec. *Biogeochemistry* 11:45-61.

Rapalee, G., E. A. Davidson, J. W. Harden, S. E. Trumbore, H. Veldhuis and D. Knapp. 1998a. BOREAS TGB-12 soil carbon and flux data of NSA-MSA in raster format (part of "Input, Accumulation, and Turnover of Carbon in Boreal Forest Soils: Integrating ^{14}C Isotopic Analyses with Ecosystem Dynamics"). In *Collected Data of the Boreal Ecosystem-Atmosphere Study,* Volumes 1-12. Edited by J. Newcomer, D. Landis, S. Conrad, S. Curd, K. Huemmrich, D. Knapp, A. Morrell, J. Nickeson, A. Papagno, D. Rinker, R. Strub, T. Twine, F. Hall and P. Sellers. Volume 4: HTML_PAGES/GROUPS/TGB/TGB12/TGB12_SOIL_CARBON_MAP.HTML. Published on CD-ROM by NASA, 2000. Also on-line at http://www-eosdis.ornl.gov/, Distributed Active Archive Center, Oak Ridge National Laboratory, Oak Ridge, TN.

Rapalee, G., S.E. Trumbore, E.A. Davidson, J.W. Harden and H. Veldhuis. 1998b. Soil carbon stocks and their rates of accumulation and loss in a boreal forest landscape. *Global Biogeochemical Cycles* 12:687-701.

Tans, P.P., I.Y. Fung and T. Takahashi. 1990. Observational constraints on the global atmospheric CO_2 budget. *Science* 247:1431-1438.

Trumbore, S.E. and J.W. Harden. 1997. Input, accumulation, and turnover of carbon in organic and mineral soils of the BOREAS northern study area. *Journal of Geophysical Research* 102 (D24): 28,817-28,830.

Trumbore, S. E., J. W. Harden, E. T. Sundquist and G. C. Winston. 1998. BOREAS TGB-12 soil carbon data over the NSA (part of "Input, Accumulation, and Turnover of Carbon in Boreal Forest Soils: Integrating ^{14}C Isotopic Analyses with Ecosystem Dynamics"). In *Collected Data of the Boreal Ecosystem-Atmosphere Study,* Volumes 1-12. Edited by J. Newcomer, D. Landis, S. Conrad, S. Curd, K. Huemmrich, D. Knapp, A. Morrell, J. Nickeson, A. Papagno, D. Rinker, R. Strub, T. Twine, F. Hall and P. Sellers. Volume 1: HTML_PAGES/GROUPS/-TGB/TGB12/TGB12_SOIL_CARBON.HTML. Published on CD-ROM by NASA, 2000. Also on-line at http://www-eosdis.ornl.gov/, Distributed Active Archive Center, Oak Ridge National Laboratory, Oak Ridge, TN.

Veldhuis, H. 1998. BOREAS TE-20 NSA soil lab data (part of *Soils of Tower Sites and Super Site, Northern Study Area, Thompson, Manitoba, Canada*). In: *Collected Data of the Boreal Ecosystem-Atmosphere Study,* Volumes 1-12. Edited by J. Newcomer, D. Landis, S. Conrad, S. Curd, K. Huemmrich, D. Knapp, A. Morrell, J. Nickeson, A. Papagno, D. Rinker, R. Strub, T. Twine, F. Hall, and P. Sellers. Volume 1: HTML_PAGES/GROUPS/TE/TE20/TE20_-NSA_SOIL_LAB.HTML. Published on CD-ROM by NASA, 2000. Also on-line at http://www-eosdis.ornl.gov/, Distributed Active Archive Center, Oak Ridge National Laboratory, Oak Ridge, TN.

Veldhuis, H. and D. Knapp. 1998. BOREAS TE-20 soils data over the NSA-MSA and tower sites in raster format (part of *Soils of Tower Sites and Super Site, Northern Study Area, Thompson, Manitoba, Canada*). In: *Collected Data of the Boreal Ecosystem-Atmosphere Study*, Volumes 1-12. Edited by J. Newcomer, D. Landis, S. Conrad, S. Curd, K. Huemmrich, D. Knapp, A. Morrell, J. Nickeson, A. Papagno, D. Rinker, R. Strub, T. Twine, F. Hall and P. Sellers. Volume 4: HTML_PAGES/GROUPS/TE/TE20/TE20_NSA_SOILS_RASTER.HTML. Published on CD-ROM by NASA, 2000. Also on-line at http://www-eosdis.ornl.gov/, Distributed Active Archive Center, Oak Ridge National Laboratory, Oak Ridge, TN.

CHAPTER 33

Methods Used to Create the North American Soil Organic Carbon Digital Database

B. Lacelle, S. Waltman, N. Bliss and F. Orozco-Chavez

I. Introduction

The North American Soil Organic Carbon Digital Database was created to determine carbon stocks for North America and to investigate the variations in existing estimates. Prior to this project, soil organic carbon was investigated by Agriculture and Agri-Food Canada (Tarnocai, 1994) and the U.S. Department of Agriculture (USDA) (Bliss et al., 1995). The methods to create these two soil organic carbon digital databases were documented (Lacelle, 1998; Bliss et al., 1995). An evaluation of the techniques used to compile the source data for both countries revealed almost identical methodology, so it was expected that the process of combining and mapping these two data sets would be relatively easy. The purpose of this chapter is to discuss the differences in procedure that alter the results or skew perception of the data. After we mapped the United States and Canadian soil organic carbon contents together, a distinct break at the border was apparent. We describe the data characteristics that caused the break and the solutions that were devised to integrate the United States and Canadian soil carbon data. We also mention the general process for calculating soil organic carbon for North America, the data sources, the database model, methods for adding Mexico and Greenland to the database, and map generation.

II. Standardize Process to Calculate Soil Organic Carbon Contents

Both the United States State Soil Geographic (STATSGO) data and the Canadian soil organic carbon data are vector format coverages stored in ARC/INFO[1] geographic information systems. Related tables describe the characteristics of each map unit, component, and soil layer. For the purposes of this chapter, specific procedures (modified from Bliss et al., 1995 and Lacelle, 1998) are presented for each region to show the aggregation of soil organic carbon contents from the individual soil layers to the map unit level.

---

[1]Any use of trade, product, or firm names is for descriptive purposes only and does not imply endorsement by the U.S. Government.

A. Soil Organic Carbon Content of a Soil Layer

The soil organic carbon (SOC) content for each layer of each soil was calculated by multiplying the thickness of the soil by the bulk density (BD) and then by the organic carbon (OC):

$$\text{SOC content}_{layer} \ (kg \ m^{-2}) = \text{Thickness}_{layer}(cm) \times \text{Bd}_{layer}(g \ cm^{-3}) \times \text{Oc}_{layer}(\%)/10 \times Cf$$

Cf is a factor related to the amount of coarse fragments in the soil. Please note that the procedures established by the United States correct for coarse fragments for each soil layer, as indicated in the equation above, and the Canadian procedures correct for coarse fragments at the soil component level.

B. Soil Organic Carbon Content of a Soil

The soil organic carbon content of each soil was calculated by summing the soil organic carbon contents of each layer:

$$\text{SOC content}_{soil} \ (kg \ m^{-2}) = (\sum \text{SOC content}_{layer} \ (kg \ m^{-2}))$$

C. Soil Organic Carbon Content of a Spatial Polygon

The soil organic carbon content of the polygon was calculated by summing the soil organic carbon contents of all the soils in the polygon after they had been adjusted by the percentage that they occupy in the polygon.

$$\text{SOC content}_{polygon} \ (kg \ m^{-2}) = \sum (\text{SOC content}_{soil}(kg \ m^{-2}) \times \text{Percentage}_{soil})$$

where N is the number of soils in the polygon. The total mass of soil organic carbon for a polygon was calculated by multiplying the soil organic carbon content of the polygon by the area of the polygon.

III. Sources

The following data sources were used or are being used to develop the North American Soil Organic Carbon digital database. Source data were compiled from existing soil maps and legends or a combination of Landsat imagery and ground truth site information. Data missing or inaccurate at the source were upgraded at the source.

United States:
1. 1:250,000- and 1:1,000,000-scale STATSGO Database for the conterminous United States and Alaska, respectively (U.S. Soil Survey Staff, 1997b)
2. Forest cover images at 1-km resolution (Powell et al., 1992; Zhu and Evans, 1994; U.S. Forest Service, 1992)
3. Laboratory Soil Characterization (pedon) data (U.S. Soil Survey Staff, 1997a)

Canada:
1. 1:1,000,000-scale Soil Organic Carbon Database (Tarnocai and Lacelle, 1996)

Mexico:
1. 1:1,000,000- and 1:250,000-scale Soil Maps of Mexico (INEGI, 1982)
2. Pedon data (INEGI, 1982)

Greenland:
1. 1:2,500,000-scale Soil Map of Greenland (Jakobsen and Eiby, 1997)

IV. Database Model and Map

A simple database model was designed with links to the original source data. Physically joining the individual areas into one cover was not feasible and of no benefit. Each coverage was projected into the same projection parameters and a database with identical structures was created (see Table 1). Each map unit will be rated for its reliability, and the rating will be added to the database. A strategy for rating reliability has not yet been developed.

Using the coverages and the attribute data for soil organic carbon, we prepared a preliminary map (Figure 1) of the soil organic carbon for North America (Lacelle et al., 1998). The next section describes some of the problems we encountered in creating the database and making the map and also the solutions that we are applying.

Table 1. Database variables

Variable name
Area of map unit (m^2)
Map unit ID
SOC content for 30-cm depth (kg m^{-2})
SOC content for 100-cm depth (kg m^{-2})
SOC content for total depth recorded (kg m^{-2})
SOC mass for 30-cm depth (kg $\times 10^6$)
SOC mass for 100-cm depth (kg $\times 10^6$)
SOC mass for total depth recorded (kg $\times 10^6$)
Land area of map unit (km^2)
Water area of map unit (km^2)
Ice area of map unit (km^2)
Miscellaneous area of map unit (km^2)
Reliability of SOC estimates

V. Problems and Solutions

As a test, the calculated soil organic carbon contents for the United States and Canada were mapped together so that the fit could be viewed. United States soil organic carbon contents were distinctly lower than those for Canada for similar soils. After investigating the source data, we discovered differences related to the surface organic layer, missing bulk density data, depth inconsistencies, and international border correlation.

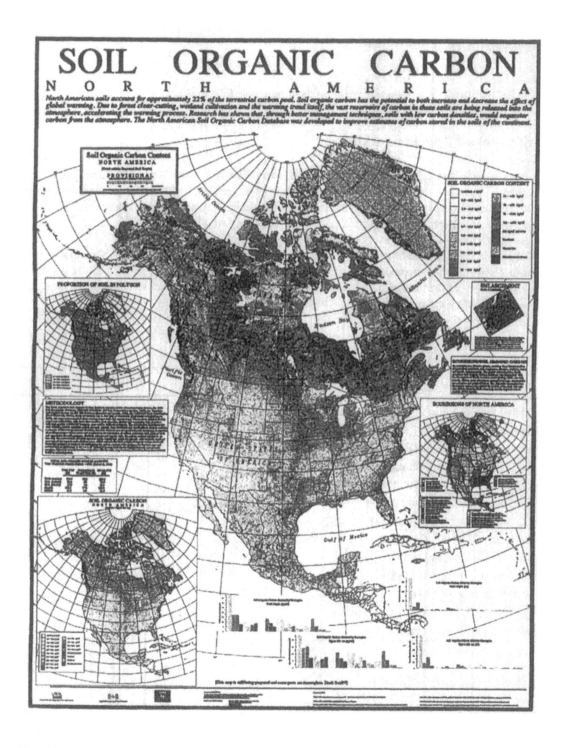

Figure 1. Soil organic carbon of North America map (actual dimensions are 36″ x 47″).

A. Surface Organic Layer Data

1. Problem

Surface organic layer data were not recorded in the United States STATSGO database, but the Canadian soil organic carbon database does contain values for the soil organic carbon content of surface layers. Calculating how much carbon is stored in Canada for undecomposed/partially decomposed litter and organic materials deposited under saturated conditions gave us an indication of the importance of this layer when trying to assess the total carbon pool. The masses of these layers were summed and found to hold approximately 19 gigatons (18%) of the total soil organic carbon pool for mineral soils for Canada (derived from Canadian soil organic carbon database) (1 gigaton = 1 petagram = 10^{15} grams). Therefore it is necessary to obtain values for this layer for the United States data.

2. Solution

A method to add appropriate soil organic carbon litter values from mapped United States cover types (U.S. Forest Service, 1992) to STATSGO map units is being developed. We are combining the STATSGO vector data with the 1-km gridded forest covers to obtain percentages of forest types within map units. We will research published literature to find measured litter soil organic carbon values for each forest type and then recalculate soil organic carbon contents for each polygon.

B. Missing Bulk Density Data

1. Problem

A large proportion of the STATSGO data have missing bulk density data and/or missing organic matter data, resulting in low soil organic carbon contents for those map units.

2. Solution

The USDA National Soil Survey Laboratory (NSSL) pedon database (U.S. Soil Survey Staff, 1997a) will serve as a source to populate the STATSGO database with measured values. The pedon data, however, also have many missing values for bulk density. Elissa Levine and Dan Kimes from NASA were approached to see if neural net techniques could be used to determine missing bulk density values for the 20,000 NSSL pedon records. The basic concept is outlined below.

First method: Bulk density values were filled in with a bulk density value from an upper or lower horizon of the same pedon if the horizon had similar clay content and organic matter.

Second method: Neural net relationships were developed for six horizon categories: amorphous, organic, cemented or fragipan, high $CaCO_3$, frozen, and all other horizons.

The pedon database is being updated by these methods; the first method is used if possible. For the 'all other horizons' category, the missing bulk density values can be predicted reliably by this process (r^2 is approximately .70). Once the pedon database is ready, the STATSGO database will be updated. The methods to populate the STATSGO database with the measured or estimated pedon data have not been developed yet, but we will consider taxonomic classification and geographic occurrence of the pedon sample.

C. Depth Inconsistencies

1. Problem

Peat deposits store large amounts of soil carbon, so missing layer data in areas of peat result in extremely low estimates of soil organic carbon content. Peat data in the STATSGO database are recorded to a maximum depth of 2 m except in Alaska, which only has data to 1 m. The peat deposits in the Canadian carbon database are recorded to the depth of the peat deposit.

In the Canadian database, data on mineral soils are recorded to a depth of 1 m or to the depth of a lithic contact that is less than 1 m. One meter is adequate for Canadian soils because glaciation stripped most of the soil away, leaving shallow soils. The United States data for mineral soils are typically recorded to a depth of 1.5 m or to a lithic or paralithic contact, except in Alaska, where many of the early data were sampled only for the active layer in permafrost-affected soils.

2. Solution

USDA Natural Resources Conservation Service (NRCS) plans to upgrade the Alaskan STATSGO database with pedon data. The variations in depth are an artifact of the existing data. The USDA-NRCS is in the process of using ground penetration radar to map the actual depth of peat and organic soils.

D. International Border Correlation

1. Problem

Different sources of data almost always have inconsistencies in definitions or methods. The conceptual framework for each source was nearly the same, but the border regions needed to be examined and reassessed by the national soil correlators of each country.

2. Solution

All international borders for North America have been correlated by this procedure. The first step was to combine the original source data from each region in a series of maps. Using these maps, correlators from each country worked together to revise the landscape boundaries (Figure 2). The lines were then adjusted digitally in the geographic information system to accommodate similar landscape features. Polygons on either side of the border that are conceptually the same were linked by a number system and mapped to ensure the correct joins (Figure 3).

Algorithms that were written and tested performed a weighted average of the associated soil organic carbon contents on the basis of the percentage of area they occupy in the joined polygon. The new weighted average soil organic carbon content was coded as an attribute for the polygons on each side of the border.

$$\text{SOC content}_{join} = [(\text{SOC content}_{Canada} \times \text{Percentage}_{Canada}) + (\text{SOCcontent}_{us} \times \text{Percentage}_{us})]$$

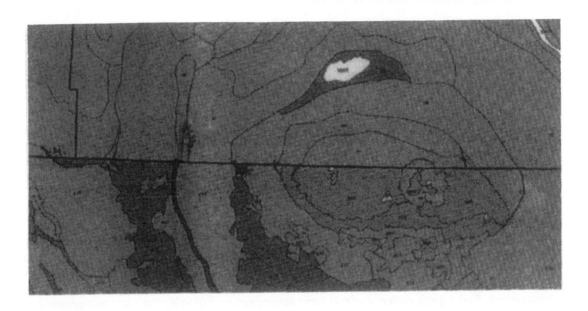

Figure 2. Manual correlation of Turtle Mountain (discrepancies in soil organic carbon contents before the correlation exercise).

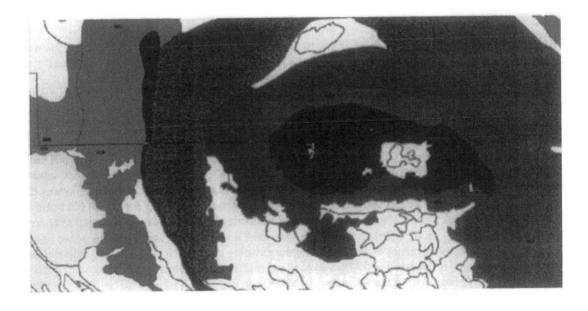

Figure 3. Plot verifying border correlated soil organic carbon contents.

VI. Adding Mexico and Greenland

Completing the North American digital database required obtaining soil organic carbon contents for Mexico and Greenland. These areas did not have digital soil data available.

A. Mexico

Source materials and guidance for the Mexican part of the database were provided by the Instituto Nacional de Estadistica, Geografia e Informatica (INEGI). The 1:1,000,000-scale Soil Map of Mexico (INEGI, 1982) was digitized by the U.S. Geological Survey's EROS Data Center. The map legend is based on the classification developed by the Food and Agriculture Organization of the United Nations (FAO). Attribute data were assembled by the USDA from pedon data provided with the 1:250,000-scale Soil Maps of Mexico (INEGI, 1982). The pedon data were entered for nearly 2,000 sites and averaged for each mapped soil class to obtain organic matter and bulk density values. Missing bulk density values were determined by using particle size distribution and cation exchange capacity data; the methods of Baumer (1992) were used. The exception was Andisols, which were assigned a bulk density of 0.9 g cc^{-1}. Layer data goes to a depth of 2 m. Pedons sharing the same FAO soil designation were averaged for soil organic carbon and assigned to the appropriate 1:1,000,000-scale soil map components. Soil organic carbon contents were determined by following the same procedures used with the United States data (Bliss et al, 1995). The international border between Mexico and the United States was correlated using the same procedure established for the Canadian and United States border.

B. Greenland

Soil organic carbon contents and a hardcopy base soil map for Greenland were provided by Bjarne Jakobsen, University of Copenhagen (Jakobsen and Eiby, 1997). This soil base data was digitized and the digital database prepared by Agriculture and Agri-Food Canada.

VII. Mapping Standards

Misinterpreting data does not only occur when statistics are calculated or data are evaluated. It can also happen when the data are presented. The determination of soil organic carbon contents for Canada was based on the soil area of the polygon, while for the United States it was based on the land area of the polygon. We discovered this when analyzing one particular area at the border where 90% of the surface area was rock and 10% was soil. Canada had a fairly high soil organic carbon content coded for a very large polygon. The United States data showed a low value for the extension of the same feature. We solved this problem by standardizing the mapping criteria. Now the soil organic carbon contents are mapped on the basis of the land area, and a small inset map portrays the percentage of soil in the polygon. This presents a clearer picture to users.

Mapping Criteria:
1. Map soil organic carbon content on the basis of land area.
2. Include an index map indicating the proportion of soil in the polygon.
3. Use natural breaks of soil organic carbon content.
4. Include a box of text describing the methodology.
5. Include an inset map showing the soil organic carbon for total depth recorded in the database.

VIII. Conclusions

This project has enabled us to assemble a database of soil organic carbon information for North America. It was necessary to update the source data in each country to enable a consistent calculation of the soil organic carbon contents. Because of budget limitations, these problems had to be dealt with in timely, cost-effective ways. We used our knowledge of the basic soil sampling and analytical techniques to guide our methods for aggregating and mapping the data. The methods were suitable for the generalized nature of the source maps. Existing databases were assessed for data compatibility and integrity. Standards vary between organizations, and documentation was not always available. Procedures to calculate soil organic carbon contents and mapping techniques were defined. The original data sources have gone through many changes in the past few years, and these calculations should be repeated in the future to incorporate new information on soil organic carbon contents and other soil properties. Work is in progress to integrate updated values from soil pedon and forest litter data sources. Results of future research, including evaluations of the soil organic carbon content at other depths, will be linked to the source data.

The data structure makes it easy to compare soil organic carbon contents to other thematic data. For example, soil organic carbon contents have been evaluated by ecological regions (Bailey, 1995). These results and other facts are part of the map titled "Soil Organic Carbon of North America" (Lacelle et al., 1998; figure 1).

The database on soil organic carbon for North America will be useful for a variety of applications, including ecological studies and research on the global carbon cycle. The carbon in soils may act as a source or sink of greenhouse gases. Changes in soil carbon can both contribute to and be affected by climate change. This database provides a baseline calculation of soil carbon stocks.

Acknowledgments

We thank the national soil correlators in the United States and Canada who participated in updating the international border polygons and to the many soil scientists who participated in compiling the source data used to calculate soil organic carbon contents. John Kimble (USDA-NRCS, Lincoln, NE) and Charles Tarnocai (Agriculture and Agri-Food Canada) instigated this joint project and obtained funding through their agencies. We extend our great appreciation to Elissa Levine and Dan Kimes (NASA, MD, USA) for helping overcome a very serious problem with missing data.

References

Bailey, R. 1995. *Ecoregions of the Continents.* U.S. Department of Agriculture, Forest Service, Washington, D.C.

Baumer, O.W. 1992. Prediction of soil hydraulic parameters. p. 341-354. In: *International Workshop on Indirect Methods for Estimating the Hydraulic Properties of Unsaturated Soils.* U.S. Department of Agriculture, Agricultural Research Service, U.S. Salinity Laboratory, Department of Soil and Environmental Sciences, University of California-Riverside.

Bliss, N., S. Waltman and G. Petersen. 1995. Preparing a soil carbon inventory for the United States using geographic information systems. p. 275-290 In: R. Lal, J.M. Kimble, E. Levine and B.A. Stewart. *Soils and Global Change.* Advances in Soil Science. Lewis Publishers, Boca Raton, FL.

INEGI. 1982. *1:1,000,000 and 1:250,000 Soil Maps of Mexico.* Instituto Nacional de Estadistica, Geografia e Informatica, Depto. de Edafologia, Aguascalienties, Mexico.

Jakobsen, B. and A. Eiby. 1997. *A Soil Map of Greenland.* II International Conference on Cryopedology, Syktyvkar, Russia.

Lacelle, B. 1998. Canada's soil organic carbon database. p. 93-101. In: R. Lal, J.M. Kimble, R.F. Follett and B.A. Stewart. *Soil Processes and the Carbon Cycle.* Advances in Soil Science. Lewis Publishers, Boca Raton, FL.

Lacelle, B., C. Tarnocai, J. Kimble, S. Waltman, F. Orozco-Chavez and B. Jakobsen. 1998. *Soil Organic Carbon of North America Map (Provisional).* Agriculture and Agri-Food Canada, Research Branch, Ottawa.

Powell, D.S., J.L. Faulkner, D. Darr, Z. Zhu and D.W. MacCleery. 1992. *Forest Resources of the United States. General Technical Report.* U.S. Department of Agriculture, Forest Service, Rocky Mountain Forest and Range Experiment Station, Fort Collins, CO.

Tarnocai, C. 1994. Amount of organic carbon in Canadian soils. p. 67-74. In: *Proceedings of the 15th World Congress of Soil Science*, International Society of Soil Science, Acapulco, Mexico.

Tarnocai, C. and B. Lacelle. 1996. *The Soil Organic Carbon Digital Database of Canada.* Agriculture and Agri-Food Canada, Research Branch, Ottawa, Canada.

U.S. Forest Service. 1992. *Forest Maps of the United States. Forest Type Groups and Forest Density From Satellite Data.* RPA program. CD-ROM media. U.S. Department of Agriculture, Forest Service, Southern Forest Experiment Station, Starkville, MS.

U.S. Soil Survey Staff. 1997a. *National Characterization Data.* U.S. Department of Agriculture, Natural Resources Conservation Service, Soil Survey Laboratory, Lincoln, NE.

U.S. Soil Survey Staff. 1997b. *The State Soil Geographic Database, STATSGO. National Collection. Digital Soil Maps and Associated Attribute Tables.* U.S. Department of Agriculture, Natural Resources Conservation Service, National Soil Survey Center, Lincoln, NE.

Zhu, Z. and L.D. Evans. 1994. U.S. forest types and predicted percent forest cover from AVHRR data. *Photogrammetric Engineering and Remote Sensing* 60:525-531.

CHAPTER 34

Basic Principles for Soil Carbon Sequestration and Calculating Dynamic Country-Level Balances Including Future Scenarios

O. Andrén and T. Kätterer

I. Introduction

The potential for sequestering carbon in soils needs to be rapidly assessed, since the current political/economical decisions (Kyoto Protocol, possible trading with "carbon shares," etc.) need a firm scientific base (Lal et al., 1997). For example, it has been recommended (Kyoto, Dec. 1997) that all nations compare carbon emissions between 1990 and projections for 2010. However, reliable measurements of soil carbon stocks and emissions on a national scale are scarce, and high-precision estimates of, e.g., annual carbon inputs to the soil are even scarcer.

On the other hand, the main factors affecting soil carbon stocks and fluxes are reasonably well known, and the influence of these factors is mathematically described in models (Paustian et al., 1997b; Smith et al., 1998). These models, almost exclusively based on first-order assumptions, may look trivial when presented as box-and-arrow diagrams, but the first-order assumptions (which we think are close to the truth) have some consequences that are not always recognized. For example, sequestering carbon only by increasing input to the soil will actually *increase* the CO_2 emission from the m^2 in question – but sequestering carbon through reduced cultivation will *decrease* CO_2 emission, at least for a limited time. It may be relevant to discuss if both strategies are of equal value with respect to carbon sequestration – and if they in the end should receive identical subsidies.

What kind of information is available for making national carbon budgets? Usually, there are regional data available on land use, and detailed information for agricultural land on cropping systems, crop yield, fertilizer levels, etc. in most cases can be found at the regional level. Often there are also some results from local agricultural field trials available, e.g., comparisons of topsoil carbon concentrations after 10 years of wheat vs. 10 years of grass crops for hay production. Climatic data, such as air temperature, precipitation and evapotransporation are also available for most regions, and if it is possible to compensate for differences in climate, soils, etc., generally applicable results and principles from quite different regions can be used, and this is where mathematical models are necessary.

In this study we use a simple, analytically solved, soil carbon balance model (ICBM; Andrén and Kätterer, 1997; *http://www.mv.slu.se/vaxtnaring/olle/ICBM.html*) for two purposes:

1. To illustrate some properties of modeled (and hopefully real) soil carbon dynamics, particularly the effects of different strategies for carbon sequestration. This is an attempt to narrow the gap between modelers and ordinary mortals. Those with an influence on legislation and/or subsidy

policies must begin to understand the very basics of soil carbon dynamics – otherwise the remedies may well be worse than the illness.

2. To show how to integrate mixed-quality information on carbon pools, land use, climate, and soils into a national soil carbon model – and how to use this for predictions of consequences of, e.g., climate change or changes in land use.

II. Mathematical Principles for Carbon Sequestration

A. The Simplest Possible Soil Carbon Model

Imagine the soil carbon as a homogeneous substrate (Figure 1, top). Assume that carbon input to the soil is constant at a rate of i kg year^{-1}. Assume also that a certain fraction of the soil carbon is respired as CO_2-C each year. Further, assume that this fraction is constant, e.g., $k_1 = 0.01$, so if the soil carbon mass is two times as high, the respiration is two times as high, since the *fraction* respired is constant (See, e.g., Hénin and Dupuis, 1945). This is the simple first-order assumption that is used in most carbon models, but it has some non-trivial consequences that need to be understood for meaningful carbon sequestration discussions.

Written as a differential equation, the model looks like this:

$$dC/dt = i - k_1C.$$

In words: For each infinitesimally small time step (t) the change in soil carbon (C) is the input (i) minus the fraction parameter (k_1) times the carbon mass present in the soil (C).

We can immediately see that at equilibrium (**steady-state** in soil carbon) $dC/dt = 0$. The conditions for equilibrium are thus $i = k_1C$. In words: When the input (i) is equal to the respiration output (k_1C) there is no change in the amount of soil carbon, even if the inputs and outputs are large. The amount of C present in the soil at steady-state thus becomes $C_{SS} = i/k_1$

One can also see that if C is large, i must be large to maintain C, unless k_1 is very small. In other words, if we have increased C, we must maintain a high i to maintain the high C mass. Also, if we reduce i to zero, soil carbon will decrease to zero, but at a decreasing rate. This is the case, since (k_1) is a constant and C is decreasing, and so is then the output (k_1C). If C is plotted against time, we get the familiar "exponential decay curve" which becomes less and less steep with time.

Which assumptions are then necessary to make the carbon mass dynamics linear, as most non-modelers intuitively think is true? This would be equal to assuming zero-order kinetics, i.e., a constant *mass*, not a constant *fraction* is lost per time unit. A consequence of zero-order kinetics is that regardless of whether you put 1 or 2 kg wheat straw into the soil, you would lose, e.g., 0.1 kg year^{-1}. This is hardly ever true, and the first-order assumption that regardless of the amount presented to the soil organisms, they will digest and respire a certain fraction, is almost always superior.

What will then happen in the first-order model if we start from equilibrium, and want to sequester carbon by doubling the annual input? Note that we continue to add the double input every year, otherwise the soil C will more or less slowly revert back to the earlier equilibrium.

Well, as shown above, the twice as high i will eventually result in a twice as high C pool at steady-state, but the C pool in most soils is large compared to i, so this is a slow process.

Finally, just as the exponential decay curve starts out steep and then gradually flattens out, the increase towards the steady-state after a doubling of input is steepest when far from steady-state, and

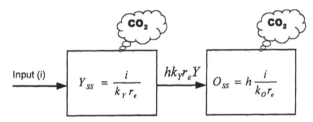

The simplest model

The ICBM model

Figure 1. Models describing soil carbon balances: i = (annual) input, C = pool of soil carbon, k_I = fraction of C that decomposes (per year), Y = young soil carbon, O = old soil carbon, k_Y = fraction of Y that decomposes (per year), k_O = fraction of O that decomposes (per year), h = humification coefficient, r_e = external influence coefficient. The index "SS" denotes the equation for calculating the steady-state value for that pool.

as stated above, the change becomes zero (mass plotted against time is a horizontal line) at steady-state.

The final insight from this exercise with the simplest possible soil C model is then that: After an increase in input (or decrease in kI) the soil C mass will gradually move towards a new steady-state and flatten out there. This limit in C increase is due the balance of inputs and outputs, and it is not necessary to assume any heterogeneity or 'limited protection capacity' of the soil to explain this – it is simply a consequence of the basic assumptions.

If the properties of this simple model are clear, the more complex soil C models are quite easy to understand, and we suggest that you already have taken the giant leap, and only have a small step left before becoming a modeler.

B. ICBM, the Introductory Carbon Balance Model

The Introductory Carbon Balance Model (ICBM), was developed as a minimum approach for calculating soil carbon balances in a 30-year perspective (Andrén and Kätterer, 1997). The model has

Table 1. The parameters of the ICBM model, their typical dimensions and the effect on total soil carbon mass of an increase in parameter value

Parameter	Symbol	Typical dimension	Effect on soil C mass of increase
Input	i	kg year^{-1}	Positive
Decomposition rate constant for Y	k_y	year^{-1}	Negative
Humification coefficient	h	dimensionless	Positive
Decomposition rate constant for O	k_o	year^{-1}	Negative
External influence on k_y and k_o	r_e	dimensionless	Negative

two state variables or pools, "Young" (Y) and "Old" (O) soil carbon. Two pools were considered a minimum, since the model was intended to handle inputs of different qualities, such as wheat straw vs. farmyard manure. ICBM has five parameters: i, k_y, h, k_o, and r_e. (Table 1, Figure 1, bottom).

The "humification coefficient" (h) controls the fraction of Y that enters O and ($1 - h$) then represents the fraction of the outflow from Y that immediately becomes CO_2–C. The parameter r_e summarizes all external influence on the decomposition rates of Y and O. Note that r_e only affects decomposition rates; r_e does not influence i or h.

The model is analytically solved, i.e., simulation techniques are not necessary, model properties can be mathematically analyzed and the model can be run and optimized in an ordinary spreadsheet program (Excel, etc.). There are also, in analogy with the minimum model above, equations for steady-state conditions, i.e., when the pools are constant and the inputs and outputs balance out. The steady-state equation for Y is: $Y_{ss} = i/k_y r_e$.

The corresponding equation for O (when Y is in steady-state) is: $O_{ss} = ih/k_o r_e$.

C. Dynamic Properties of Soil Carbon Illustrated by the ICBM

In the following we give four examples of results from different parameter settings of the ICBM (Figure 2). The boxes to the left indicate total topsoil carbon ($Y+O$), in this case initially 4.33 kg m^{-2}, (0 to 25 cm depth), and within the boxes the cumulated change in total soil carbon during 30 years (DC_{30}) is given. The arrows pointing to the boxes indicate annual carbon inputs through roots, crop residues, manure etc., and the arrows pointing upwards indicate total annual carbon loss (as carbon dioxide respiration from soil organisms) to the atmosphere. The diagrams to the right indicate the dynamics of these annual losses during a 30-year period (see figure legend).

The graphs indicate annual CO_2-C from topsoil at steady-state (kg m^{-2} year^{-1}) during 30 years after the change (if any). The upper line is total emission, and the lower indicates emission from the young pool.

Initial parameter setting: $i = 0.2$, $k_y = 0.8$, $h = 0.125$, $k_o = 0.00605$, $r_e = 1$. Changes relative to top figure: i doubled, r_e halved and h doubled, respectively. All figures start from the same initial topsoil C pool sizes: Young = 0.25, Old = 4.13 kg m^{-2}.

The top figure shows a parameter setting at steady-state, and obviously 0.2 kg centers each year and 0.2 kg leaves, but the size of the soil carbon pool is constant. Most of the carbon lost comes from the young pool, in spite of the fact that it only constitutes 0.25/4.38 = 5.7% of total soil C. This is a

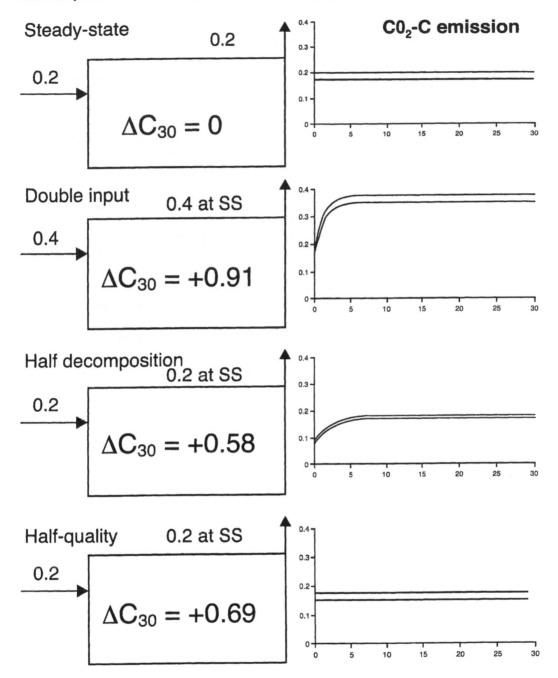

Figure 2. Different strategies for (temporarily) sequestering carbon in soils. Values at arrows indicate annual input to topsoil (0-25 cm) and emission of CO_2-C from topsoil at steady-state (kg m^{-2} year^{-1}). ΔC_{30} indicates 30-year changes in the total topsoil carbon pool (kg m^{-2} 30 years^{-1}). Note that when steady-state is reached, ΔC by definition equals zero.

model prediction due to the relative sizes of parameters k_y and k_O, but the parameter values used here are based on actual field experimental data (Andrén and Kätterer, 1997).

The second figure shows what will happen if we want to sequester carbon only by *doubling the input*. From the steady-state equations given above we can immediately see that that the new steady-state will be $2 \times 4.38 = 8.76$ kg. However, after 30 years of double input we will only have gained 0.91 kg, and we still have a very long way to go. This is due to the fact that usually about 50% of the crop residues will be lost in respiration each year (as indicated by $k_Y = 0.8$; Annual mass loss: $1 - e^{-0.8} = 55\%$ year^{-1}; you do not have to understand, just accept, this). Only a small fraction (in this case h = 0.125) will enter the 'humus,' 'slow,' old' 'SOM' pool – or whatever we want to call it. Results of this type have been observed in long-term field experiments (see, e.g., Kätterer and Andrén, 1999), so we are quite confident that they exist outside cyberspace.

So, by doubling the input we have actually *increased the CO_2 emission to the atmosphere* (right diagram), and when we finally reach steady-state we will have doubled the annual CO_2 emission (vertical arrow). In the 30-year perspective, we have increased the input with $30 \times 0.2 = 6$ kg, but we have only increased the soil C pool (actually mainly the 'Young,' rapidly decomposing pool) with 0.91 kg.

Note also that we have to maintain the high input to maintain the soil C at the higher level; in other words, when you have a bank loan, you have to keep on paying the annual interest just to maintain the debt at a constant level.

For a global budget, the alternative use of the crop residues is of central importance. If you burned the straw in the field, CO_2 emissions would be immediate, and you would not even get the 0.91 kg into the soil. On the other hand, if you used the straw as building material it would be out of circulation for maybe 50 years, so global CO_2 emissions would decrease. Finally, if you burn it instead of fossil fuel for energy, that the fossil fuel can stay in the ground and will not contribute to global change.

If we instead try to *lower the decomposition rates* in the soil, we have more of a win-win situation. Assume that we reduce the decomposition rates of both pools to half its present value (in this case we set $r_e = 0.5$). Actually, the well-known beneficial effects on soil carbon of pastures as opposed to row crops are to a great extent due to the fact that the grass transpires more than row crops and that soil cultivation is less frequent (Paustian et al., 1990). Both the drier soils and the reduced cultivation will decrease actual decomposition rates (or, in the model, decrease r_e).

So if we reduce r_e to 0.5 (Figure 1; 'Half decomposition') we will immediately halve the CO_2 emissions and start to build up a larger soil C pool. Unfortunately, as the young pool rapidly increases, so will the emission (diagram), but at least we will after 30 years have increased the soil C pool with 0.58 kg without any change in input and with a reduced emission. At steady-state, in a distant future, we will have doubled the soil carbon and increased the soil C with 4.38 kg, and will be back to 0.2 kg emission.

Another win-win strategy may be to *reduce the quality of the input*. In decomposition terminology 'quality' is reversed – high quality means that the material decomposes rapidly ("Volvo is a high-quality car" would mean it decomposes faster than its competitors). In ICBM terms: if the humification coefficient (h) is increased, the old pool, which is slowly decomposing, will increase. If we double h to 0.25 (Figure 1; 'Half-quality input'), we will reduce emissions, until we finally reach steady-state again. We will also increase total soil C by 0.69 kg over a 30-year period, and all of this increase will be in the slow pool.

Actually, this may be an idea worth pursuing. If, e.g., plant breeders could come up with a corn with a higher stem and root lignin content (increases h), the North American corn belt could reduce carbon emissions for a considerable time even without any other changes (Eldor Paul, personal communication).

In conclusion, it is not a straightforward process to sequester carbon in the soil, and we have to be aware of the full consequences of what we do, also outside the soil itself. Just because carbon sequestration neatly ties into 'conservation tillage,' 'no-till,' etc. does not mean it is the only, or even always the best, way to reduce CO_2 concentrations in the atmosphere. At worst, we could create a 'carbon bomb' in the soil – having a large and mobile soil pool that needs huge inputs to maintain steady-state, and the bomb may 'explode' whenever we, e.g., increase cultivation again. However, the added benefits of increasing soil carbon contents (erosion control, water holding capacity, structure, overall fertility leading to more photosynthesis) may well outweigh the negative consequences (Paustian et al., 1997a; Lal et al., 1997).

III. Integration of Information into Country-Level Dynamic Carbon Budgets and Analysis of Possible/Impossible Scenarios

A. The Data Set – Swedish Agricultural Land

We will use agricultural land in Sweden as the basis for the calculations in the following (Figure 3). Note, however, that this is not a report of the status of Swedish agricultural soil carbon – there may be severe errors or at least extremely low precision in the data – the intention is just to show how dynamic descriptions and predictions can be made. The information immediately available concerning Swedish agricultural land is fairly typical, and it can serve as a general example of the mixed-quality data we usually will have to work with.

Swedish agricultural land is annually classified according to usage (Anonymous, 1996). Using these values, the total area of agricultural land (arable + grazing land) in Sweden in 1994 (including also small farms < 2ha) can be calculated to about 3.6 million ha. The total area of Sweden is 41.1 million ha, excluding lakes and rivers. Thus, about 9% of the land was used for agriculture in 1994 (reindeer grazing land excluded).

One can also calculate that the total area of *cultivated* land was about 3.2 million ha. The "natural," never cultivated, grazing land, about 400 000 ha, was not included, so in the following we will use 3.2 million ha as reference point.

The classification into Central, North and South is illustrated in Figure 3. The classification was based on the existing divisions for agricultural statistics, which were bulked into three zones mainly according to climate (growing season, annual mean temperature, etc.). The fraction of total arable land occupied by each climatic zone is 30%, 10% and 60% for Central, North and South, respectively. From Figure 3 it should be obvious that the arable land in the North only occupies a very minor part of the land area; actually, agriculture is more or less limited to a strip along the coast.

We bulked the crops into three main groups (categories in agricultural statistics given within brackets): *Annuals* (grain crops, legumes, oleiferous plants, potatoes, sugar beet); *grass* for hay, silage, grazing (improved grassland, other crops, not used grass, not used arable, cultivated grazing land); *bare fallow* (arable land kept without vegetation). Thus the areas for 1994 became (in thousands of hectares): Annuals, 1,505; Grass, 1,624; and Bare fallow, 57. We are aware of the existence of crop rotations, i.e., that a given hectare one year can be under grass, the next year under winter wheat, etc. However, this does not severely affect the calculations here – we are only interested in regional and national levels – not the fate of an individual field.

Three soil types (topsoil only, 0 to 25 cm depth).were considered: *Heavy* (> 20% clay), *Light* (<20% clay) and *Organic* (> 12% C). Using a synthesis of experimental data from agricultural

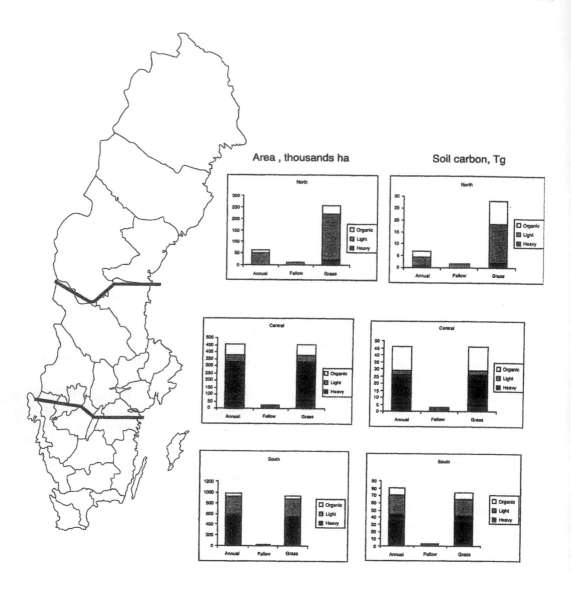

Figure 3. Map of Sweden with the three climatic zones indicated. The bar graphs at the right indicate present agricultural area and topsoil carbon mass. Annual = annual crops, Fallow = black fallow, Grass = improved grassland for grazing, hay, silage. Organic, light, heavy indicate soil types.

field trials (Mattsson, 1996), and a recent national soil sampling (Eriksson et al., 1997) we estimated the soil type distribution in each of the three geographic areas. Assuming no particular preference for each crop type to soil type, we then could calculate the area a certain crop type on a certain soil type would occupy in a given region. The total distribution of Swedish agricultural soils according to this classification is approximately: Heavy, 59%; Light, 31%; and Organic, 10%.

Since we have three values for each of the factors Zone, Soil and Crop, we have $3 \times 3 \times 3 = 27$ possible permutations, which we call 'virtual regions.' The resulting areal distribution into the 'virtual regions' is shown in Figure 3.

The present stocks of carbon in Swedish agricultural soils were estimated from available sources in a three-step procedure: (1) Obtaining an overall mean for Swedish agricultural soils, (2) modifying this according to zone (North to South), and finally (3) according to soil type.

The overall mean was calculated from a comprehensive and representative sampling of Swedish agricultural topsoils, where more than 3 000 points were sampled and humus concentrations (= carbon concentration/0.58) are reported. The overall mean carbon concentration was 3.65% (organic carbon) and the median value was 2.38%. The reason for the great difference between mean and median values is the 10% organic soils with extremely high carbon content >12 % C that increase the mean, and the median value is more representative for mineral soils (Eriksson et al., 1997).

To calculate carbon mass per unit area, topsoil depth and dry bulk density are needed. First, we calculated the carbon mass in mineral soils, i.e., Heavy + Light. Average bulk density for "heavy" (high clay content) soils is commonly set to 1.25, and the corresponding value for "light" (low clay content) soils is 1.35 (A. Andersson, personal communication). Using the areal distribution given above we can calculate a weighted average of 1.28. Assuming an average topsoil thickness of 25 cm (Eriksson et al., 1997) and a density of 1.28 we can calculate a mean mass of topsoil carbon for mineral soils as $0.0238 \times 0.25 \times 1280 = 7.6$ kg m^{-2}.

Second, organic (peat, gyttja etc.) soils have to be treated quite differently due to subsidence, i.e., the loss of carbon from the topsoil to the atmosphere lowers the surface of the field, and ploughing and continuously improved drainage replenishes the topsoil carbon store from below. Thus, one can assume that the topsoil carbon is fairly close to steady-state – increased losses through, e.g., intensive cultivation induces rapid subsidence and consequentially a higher input from below. Assuming a carbon content of 45% and a bulk density of 0.2 g cm^{-3}, the average carbon mass in Swedish organic soils can be calculated as $0.45 \times 0.25 \times 200 = 22.5$ kg m^{-2}. We do not distinguish here between different types of organic soils, e.g., peat, gyttja, mineral-mixed peat soils, but simply assume that those with lower C concentration have a correspondingly higher bulk density. Since the organic soils constitute about 10% of the total area we can calculate a weighted mean carbon mass for all arable topsoil of 9.0 kg m^{-2}, corresponding to a total stock for Sweden of about 290 Tg (million ton).

Differences in mean carbon content along the North-South gradient are due to both differences in soil types and in climatic factors. For a given soil type we assume that if Central Sweden (29.2% of the area) is set to 1, Northern Sweden (10.4%) is set to 1.1 and Southern Sweden (60.4%) is set to 0.98. These assumptions are based on calculations on subsets of the data set by Eriksson et al. (1997).

Redistributing the overall mean carbon mass to the different soil types is quite simple, since we already have C mass in organic soils (22.5 kg m^{-2}) and the average for the mineral soils (7.6 kg m^{-2}). There is usually a lower carbon concentration in light (sandy) soils than in heavy (clay) soils, but the average bulk density is higher in light soils, so we let these two factors cancel each other out, setting 7.6 kg m^{-2} as the topsoil carbon mass for all mineral soils. See Figure 3 for the resulting soil carbon mass for each virtual region.

Annual crop grain production (weighted average for all cereals) is 4 ton ha^{-1} for Sweden as a whole, which equals about 0.2 kg m^{-2} of carbon. The straw production is about the same, 0.2 kg C m^{-2}. The below-ground input (roots) to soil in cereals was approximated to about 28% of total above-ground production (Paustian et al., 1990 and papers cited therein), i.e., 0.11 kg m^{-2}. If straw is exported at harvest, the stubble left is about 35% of the straw, i.e., 0.07 kg, and the total C input becomes $0.11 + 0.07 = 0.18$ kg. If straw is left in the field, the input becomes $0.11 + 0.2 = 0.31$ kg C.

Table 2. Example of basic parameter settings and multipliers according to zone, crop, and soil. Y_o and O_o indicate starting values, i.e., the initial amount of carbon (kg m^{-2}) in the Young and Old pool, respectively

	i	k_y	k_o	h	r_e	Y_o	O_o
Base	0.318	0.800	0.006	0.177	1.000	0.500	8.200
South	1.10	1.00	1.00	1.00	1.20	0.98	0.98
Central	1.00	1.00	1.00	1.00	1.00	1.00	1.00
North	0.70	1.00	1.00	1.00	0.90	1.10	1.10
Annual	1.00	1.00	1.00	1.00	1.00	1.00	1.00
Grass	1.10	1.00	1.00	0.75	0.73	1.00	1.00
Fallow	0.00	1.00	1.00	1.00	1.33	1.00	1.00
Light	0.90	1.00	1.00	0.75	0.90	1.00	1.00
Heavy	1.00	1.00	1.00	1.00	1.00	1.00	1.00
Organic	1.00	1.00	1.00	0.75	1.00	2.00	3.00

The fraction of the area cultivated with annuals where straw is left in the field was estimated at about 50%, so the average input of plant remains in annuals becomes 0.245 kg m^{-2}.

Farmyard manure (FYM) poses a special problem, since it is a partly decomposed product of hay from perennials, other feed such as grain, as well as residues of straw from annuals used as floor cover in stables, etc. FYM production in Sweden is 18 Tg with a dry matter fraction of 0.12 and a C fraction of 0.5 (Steineck, 1991), which corresponds to an annual carbon input to the soil of 1.1 million tons. Since most of the manure is applied to annual crops, we assumed that that all manure was spread there. On average, the amount of manure carbon per ha of annual crops then becomes 1.1 million ton/1.5 million ha = 0.73 ton ha^{-1} year^{-1}, or 0.073 kg m^{-2} year. This amount should be added to the input for annuals given above, 0.245 + 0.073 = 0.318. We also have to change the humification quotient, h, since this differs for straw and the partly decomposed manure (0.125 and 0.31, respectively; Andrén and Kätterer, 1997). The weighted average of h then becomes: (0.245 × 0.125 + 0.073 × 0.31)/0.318 = 0.167.

B. Parameterization and Calibration of ICBM – Regions to Country

First, we set up a benchmark parameter set from the "virtual region" we know best: Central, Annual, Heavy. Based on all available information (e.g., Andrén et al., 1990; Andrén and Kätterer, 1997; Anonymous, 1997; Eriksson et al., 1997; Paustian et al., 1990) we used the following parameter set, which results in more or less steady-state when $Y_0 = 0.5$ and $O_0 = 7.1$: $i = 0.318$, $k_Y = 0.8$, $h = 0.167$, $k_O = 0.00605$, $r_e = 1.0$.

Second, we set up a matrix of parameter modifiers for each 'virtual region.' The parameter value for each 'virtual region' was modified by the product of the three modifiers for each climatic zone, crop and soil type, respectively (Table 2). For example, the i parameter for South, Annual, Light becomes 0.318 × 1.1 × 1 × 0.9 = 0.315 kg m^{-2} year^{-1}. Note that the baseline (Central, Annual, Heavy) is multiplied by 1 only. Also note that only i, h and r_e are changed for mineral soils – we recommend using the k parameters as "global constants" (Andrén and Kätterer, 1997). However, in the Excel program described below we allow for changes also in these.

Third, we let the computer calculate the parameter values for each one of the 27 'virtual regions' and use these parameters to model the changes in soil carbon during, e.g., 30 years. The actual reasoning behind each multiplier value would be too complex to present here, but a few examples can be given. Differences in i between climatic zones are based on crop yield statistics, and the fact that grasslands usually both have a higher input and creates a lower soil moisture is reflected in the "grass factors" for i and r_e, respectively. The increased "humification quotient" or "protective capacity" of clay soils is reflected in the reduced h for non-clay soils etc. Organic soils need special attention due to subsidence, and we simply added a front-end model that calculates the total input and weighted h from subsidence and normal input.

Fourth, the results for each 'virtual region' in kg m^{-2} were scaled in relation to the actual proportion of Swedish arable land occupied. The balances were summed according to zone, etc. and national balances calculated.

C. Using the Icbm27 Program for Balance Calculations Based on Different Scenarios

All steps are performed within an Excel program (ICBM27.xls) that can be interactively run over the Internet, or downloaded (see Introduction for www-address). The basic logic is as described above, using an overall mean and starting from a baseline parameter combination.

All changes are propagated through the chain of calculations; e.g., if one of the baseline parameter values is changed, all results from each virtual region (and the total sums) will be immediately changed. If, on the other hand, one wants to change, e.g., the effect of high/low clay content on h, one changes the 'Heavy' multiplicator in the table above.

If one wants to predict changes in land use, one can either change only the total area of arable land or change the areal distribution between the 27 'virtual regions.' General climate change (or changes in cultivation intensity) is mimicked by changing r_e, affecting soil carbon decomposition rates. If one believes that the change also will affect crop production and, consequently, the amount (and quality) of the input to the soil, one simply also changes i (and h).

1. Business as Usual

Using the tentative parameter settings and the 27 zone/crop/soil permutations for the virtual regions, we came up with the following overall results (Figure 4). The mineral soils are fairly close to steady-state, with a pool of 218.6 Tg (286.3 Tg including organic soils). However, although the organic soils are more or less in steady-state in the topsoil, the subsidence creates a CO_2-C emission of about 1 Tg year^{-1}.

The loss through subsidence is also illustrated by the difference in accumulated inputs, where the cumulative inputs with subsidence after 30 years are about 30 Tg higher than when subsidence is not included. Note that the cumulative input is equal to the soil pool after about 25 years, but this does not mean that the total soil pool is replaced after this time, since most of the input never reaches the old pool (cf. II. Mathematical principles for carbon sequestration).

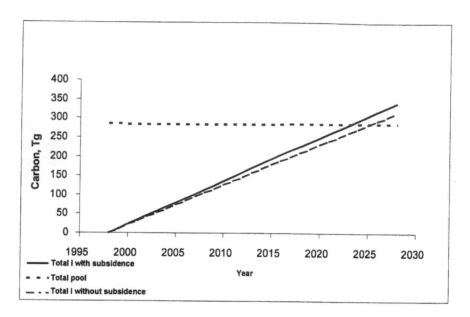

Figure 4. Topsoil carbon and cumulative inputs during the next 30 years of Swedish agriculture, assuming "Business as usual," continuation of present agricultural area and practices and assuming no change in climate.

2. Climate Change

As can be seen from the steady-state equations in Figure 1, a change in climate that affects decomposition rates (r_e) can be counterbalanced by an increase in i. If a warmer and moister climate increases r_e by 20%, and crop production and consequently i also increases by 20% the steady-state value will be the same and no change will occur (See Kätterer and Andrén, 1999).

However, let us assume a radical change in climate and that we cannot increase our i – maybe we will have plant disease problems or social unrest due to lack of skiing opportunities. We increase r_e to 5.36 for central Sweden, assuming a hot and humid rainforest climate (Andrén and Kätterer, 1997). The consequences of this not very probable change are illustrated in Figure 5. Note that even this radical climate change "only" would decrease the soil carbon pool by 134 Tg in 30 years. About 153 Tg would still remain in the soil.

However, the subsidence and consequential CO_2 emission from the organic soils would increase considerably, from about 1 to almost 6 Tg year^{-1}.

A more probable scenario could be along these lines: Climate change will increase the summer temperatures and increase rainfall, which will increase both i and r_e.

However, an increased variability in the weather patterns (violent rainstorms, spring droughts, occasional summer frosts) will prevent us from increasing i by more than 10%. However, the higher summer temperature and moisture cause a 20% increase in r_e (Figure 6). For comparison, a 10 °C

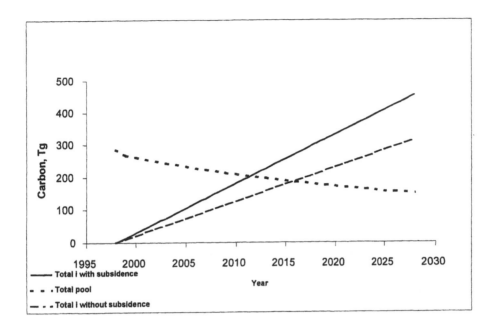

Figure 5. Topsoil carbon and cumulative inputs during the next 30 years of Swedish agriculture. Here we assume all parameters are the same as in Figure 4, but r_e has been changed to 5.36, which would mean a radical climate change to a hot and humid climate.

increase in soil temperature would double r_e, using the common assumption that $Q_{10} = 2$ (Lomander et al., 1998).

The changes in soil carbon pools will not be very dramatic, but after 30 years the pool will have decreased from 296 to 280 Tg. Again, the increase in r will increase losses by subsidence.

3. Land Use Change

If we change land use radically, e.g., plant forest on arable land or convert the land into parking lots, we just reduce the total area of arable land with the corresponding area. However, this is not very helpful from a national carbon budget view, but there is nothing that stops us from having a differently parameterized model for all forest land – or parking lots, where i and r_e probably will be low – and just increase the forest area and decrease the arable area when forests are planted on arable land.

However, let us instead make a scenario where we convert all annual cropland to grass production – which can be incinerated in heating plants for energy purposes. We simply replace all annuals and black fallow with grass parameter settings in the model and the results for soil carbon would be as shown in Figure 7. Note that we now have to put the manure on the grasslands, and if we spread it evenly on the area we will have a manure input of 1.1 Tg/3.2 Mha = 0.034 kg m^{-2} year. The total input

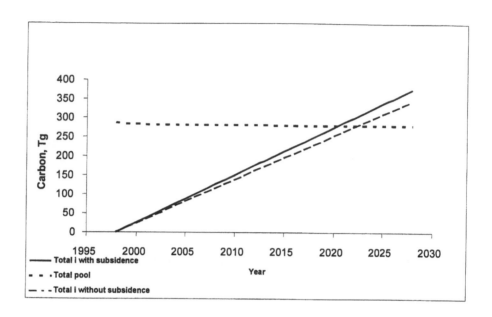

Figure 6. Topsoil carbon and cumulative inputs during the next 30 years of Swedish agriculture. Here we assume all parameters are the same as in Figure 4, but i has been increased by 10% and r_e has been increased by 20%, both due to a mild climate change towards a more variable, warm and humid climate.

will therefore be (cf. Table 2) $1.1 \times 0.318 + 0.034 = 0.384$ kg m^{-2} year. The humification coefficient will then be the weighted average: $(1.1 \times 0.318 \times 0.125 + 0.034 \times 0.31)/0.384 = 0.141$. Finally, r_e will be the one used for grass, 0.73. Also, we have to set the subsidence input for all organic soils to the value for grass, 0.20 kg m^{-2} year^{-1}.

This would increase soil carbon from 286 to 298 Tg after 30 years, and it would considerably decrease the emissions due to subsidence from organic soils (the two sloping lines in Figure 7 become almost parallel). The reason for this is the assumed low subsidence input for organic soils under grass; annuals = 0.40, grass = 0.20, and fallow = 0.64 kg m^{-2} year^{-1}.

IV. Concluding Remarks

Most humans think linearly, but first-order kinetics gives non-linear results. Therefore models are necessary, and fortunately, if the model is simple enough, most of us can understand the results and even how they were generated. The approach here, using a very simple core model, and modifying parameters by multiplication should be transparent to most users – and the projected results not too far from those we will be able to measure in the future.

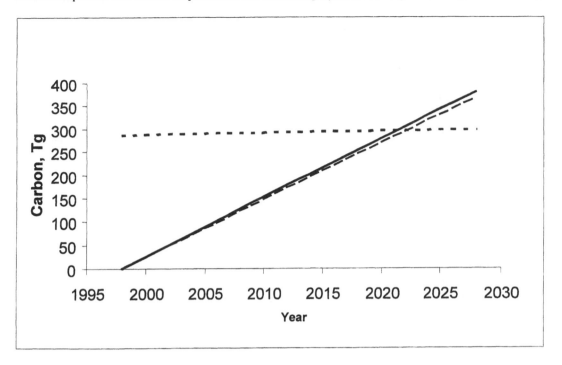

Figure 7. Topsoil carbon and cumulative inputs during the next 30 years of Swedish agriculture. Here we assume all arable land has been turned to grassland, which affects i, r_e, h and subsidence. See text for details.

Perhaps the most important insight is that sequestering carbon into soils in most cases is a pretty wasteful strategy, compared with storing the crop residues under dry conditions in barns or abandoned mines – or burning them instead of fossil fuels. However, increasing soil carbon contents – or at least not reducing them – usually is beneficial for crop growth (more carbon is fixed from the atmosphere), erosion control etc. – so when all factors are added up for a given case soil carbon sequestration still may be the best present option.

The actual spreadsheet makes 'changing the future' easy – converting the entire Swedish agriculture to grass/bioenergy production takes 5 minutes of parameter changes – and to predict the results for 30 years into the future takes less than 1 s of processing time. We believe the weakest link in the predictions is in the data available (but this is naturally what modelers usually claim). However, the simple, interactive approach also is of help here – the minute we get, e.g., improved estimates of the amount and quality of root input we can change the values for parameters i and h.

The simple approach also has the advantage that we only need very basic information (although obtaining this is admittedly not a trivial task). In other words, the model does not add any parameters that warrant special investigations or guesses, such as 'the influence of input quality on microbial efficiency.'

Finally, we are not claiming that this is the only way to go – a high-resolution GIS map with high-precision data for each hectare would be an excellent structure for 3.2 million simultaneous applications of, e.g., the ICBM model. But this approach also stands and falls with the quality of the data and assumptions – even if we can create colorful and visually striking maps from fairly rough guesses massaged by advanced interpolation routines.

Acknowledgments

This work contributes to the GCTE Core Research Programme, Category 1, which is a part of the IGBP. Financial support was received from the Swedish Environmental Protection Agency.

References

Andrén, O. and T. Kätterer. 1997. ICBM - the Introductory Carbon Balance Model for exploration of soil carbon balances. *Ecol. Appl.* 7:1226-1236.

Andrén O. T. Lindberg, U. Boström, M. Clarholm, A.-C. Hansson, G. Johansson, J. Lagerlöf, K. Paustian, J. Persson, R. Pettersson, J. Schnürer, B. Sohlenius and M. Wivstad. 1990. Organic carbon and nitrogen flows. In: O. Andrén, T. Lindberg, K. Paustian and T. Rosswall (eds.), Ecology of Arable Land – Organisms, Carbon and Nitrogen Cycling. *Ecological Bulletins* (Copenhagen) 40:85-126.

Anonymous. 1997. *Yearbook of Agricultural Statistics.* Statistiska Centralbyrån, Örebro, Sweden (in Swedish).

Eriksson, J., A. Andersson and R. Andersson. 1997. Tillståndet i svensk åkermark. *NV Rapport 4778,* Stockholm.

Hénin, S. and M. Dupuis.1945. Essai de bilan de la matière organique du sol. *Annales Agronomiques* 15:17-29.

Kätterer, T. and O. Andrén. 1999. Long-term agricultural field experiments in Northern Europe: Analysis of the influence of management on soil carbon stocks using the ICBM model. *Agric. Ecosys. Environ.* 72:165-179. Erratum: 75:145-146.

Lal, R., J.M. Kimble, R.F. Follett and B.A. Stewart. (eds.). 1997. *Management of Carbon Sequestration in Soil.* Advances in Soil Science. Lewis Publishers, Boca Raton, FL.

Lomander, A., T. Kätterer and O. Andrén. 1998. Modelling the effects of temperature and moisture on CO_2 evolution from top- and subsoil using a multi-compartment approach. *Soil Biol. Biochem.* 30:2023-2030.

Mattsson, L. 1996. *Markbördighet och jordart i svensk åkermark.* Naturvårdsverket (Swedish Environmental Protection Agency) Report 4533, Stockholm (in Swedish, with English summaries).

Paustian, K., O. Andrén, M. Clarholm, A.-C. Hansson, G. Johansson, J. Lagerlöf, T. Lindberg, R. Pettersson and B. Sohlenius. 1990. Carbon and nitrogen budgets of four agro-ecosystems with annual and perennial crops, with and without N fertilization. *J. Appl. Ecol.* 27:60-84.

Paustian, K., O. Andrén, H.H. Janzen, R. Lal, P. Smith, G. Tian, H. Tiessen, M. van Noordwijk and P. Woomer. 1997a. Agricultural soils as a sink to mitigate CO_2 emissions. *Soil Use Manag.* 13:230-244.

Paustian, K., E. Levine, W.M. Post and I.M. Ryzhova. 1997b. The use of models to integrate information and understanding of soil C at the regional scale. *Geoderma* 79:227-260.

Paustian, K., E.T. Elliott and M.R. Carter. 1998. Tillage and crop management impacts on soil C storage: Use of long-term experimental data. *Soil Tillage Res*. 47:7-12.

Steineck, S., L. Djurberg and J. Ericsson. 1991. Stallgödsel. Swedish Univ. Agric. Sci., *Speciella skrifter*, 43 (in Swedish).

Smith, P., O. Andrén, L. Brussaard, M. Dangerfield, K. Ekschmitt, P. Lavelle, M. Van Noordwijk and K. Tate. 1998. Soil biota and global change at the ecosystem level: The role of soil biota in mathematical models. *Global Change Biol*. 4:773-784.

CHAPTER 35

Examining the Carbon Stocks of Boreal Forest Ecosystems at Stand and Regional Scales

J.S. Bhatti, M.J. Apps and H. Jiang

I. Introduction

Terrestrial ecosystems of the Northern Hemisphere contain large pools of biospheric carbon (C), and are known to play a very dynamic role in the global carbon cycle (Apps et al., 1993). Canadian forests account for approximately 25% of the C in the boreal zone and 10% of the world's forested area. The net budget of C-fluxes between Canadian forests and the atmosphere is therefore an important component of the global C-cycle (Schindler, 1998). The boreal forest is the most extensive forest biome in Canada and is estimated to contain 40% of the total biotic C in Canada (Price and Apps, 1993). Short growing seasons, low temperatures, and high moisture content are important factors that limit the decomposition of organic matter in these ecosystems (Kimmins, 1996). Consequently, the soils of Canada's boreal region contain significant C reservoirs that have accumulated over thousands of years and play an active role in the source/sink relationships for terrestrial C (Apps et al., 1999a). An increase in large-scale, stand-replacing disturbances since ca. 1970 has been reported for Canadian forests (Kurz et al., 1995b). This has resulted in transitory decreases in net ecosystem productivity and increases in the pools of decomposing organic matter, causing the Canadian boreal forest to become a small net C-source rather than a C-sink (Kurz and Apps, 1996). The change in disturbance regimes for these northern ecosystems may be a response to the larger scale phenomenon of global change, which results from human-induced changes in the physical climate system, land-use changes, and atmospheric pollution (IGBP, 1998; Woodwell et al., 1998). Changes in the disturbance regime, and the resulting forest response, are the consequence of both direct (e.g., forestry operations) and indirect (e.g., climate change) effects of human activity. As global change alters weather patterns, the frequency and severity of natural disturbances such as wildfire and insect outbreaks are affected, leading to changes in ecosystem dynamics (Figure 1). Under present projections of climate change, the rate of natural disturbance will likely increase, resulting in an increased proportion of younger age stands in the forests (Kurz and Apps, 1999). Other vegetative and soil processes are also expected to change in response to changes in temperature, precipitation, ambient CO_2 concentration, and other global climatic factors. The responses may also be influenced by changes in land-use practice and atmospheric deposition or pollution which also vary strongly over time and with local conditions. Understanding the complex relationship between global change and forest ecosystem processes is necessary to predict both the feedbacks to global change and the future forest resource. To achieve this understanding the appropriate use of spatial and temporal scale is essential (Houghton et al., 1998; Apps, 1993; Holling, 1992).

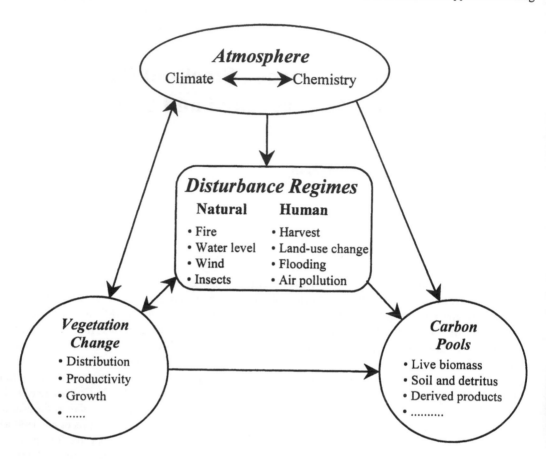

Figure 1. Relationship between forest ecosystem components, their interactions, and atmospheric-climate feedbacks. (Modified from Apps, 1993.)

Feedback mechanisms associated with interactions of altered disturbance regimes, vegetation structure and function, and biospheric C pools all contribute to the linkages with the atmosphere that are presented in Figure 1. Climatic variables and atmospheric deposition of nutrients are important ecosystem driving variables that affect processes such as plant growth and detrital decomposition. Site-specific conditions and small-scale processes determine the relative importance of these variables. These processes in turn modify the net atmospheric fluxes of energy, momentum and matter (especially water and greenhouse gases) and affect the physical climate system (Ojima et al., 1992). Of particular importance are the changes in (1) net C fluxes via plant production and decomposition (Harden et al., 1997); (2) trace gas exchange, largely controlled by nutrient cycling (Bhatti et al., 1998); (3) water and energy fluxes altered by precipitation, evapotranspiration and run-off (Kimball et al., 1997); and (4) C allocation within plants and ecosystem components. The processes controlling these factors are highly sensitive to environmental change and response at leaf, tree and stand levels occur within days to years (Shugart and Smith, 1996).

Mean annual changes in regional precipitation and temperature patterns are not good indicators for predicting plant productivity and decomposition rates, as these process changes take place unevenly throughout the annual cycle or geographic region. Ecosystem processes at the leaf and tree level are more sensitive to climate characteristics such as first frost occurrence and the durations of the wet and dry seasons. Such characteristics are not well represented by average values of climatic parameters (Hogg, 1999), but constrain critically important processes such as germination, growth initiation, and mortality.

Global climate change also alters processes that affect ecosystem structure and function at a larger spatial and temporal scale. Alterations in resource availability and its partitioning amongst biotic components, changes in forest structure and changes in disturbance regimes are three ways by which changes at the forest or biome level occur. These changes may not be immediately observed and instead may occur over an extended period of time—from years to decades or centuries. The importance of changes in regional nutrient inputs through atmospheric deposition has been widely noted (Schindler and Bayley, 1993; Townsend et al., 1996). Nutrient cycling, regeneration, species composition, and succession patterns at the landscape scale are also strongly influenced by climate change (Haxeltine et al., 1996; Thomson et al., 1996; Woodward et al., 1995; Neilson and Marks, 1994). Changes in forest structure, both in terms of age-class structure and spatial distribution, can modify the local, regional, and global scale climate through alteration of albedo, humidity, and ground-level wind patterns (Bhatti et al., 1999; Jarvis et al., 1997).

A key factor affecting vegetation structure and function is the natural or anthropogenic disturbance regime (Larsen and MacDonald, 1998). Apps and Kurz (1993) have shown that stand-replacing disturbances play a complex but very important role in determining both the annual exchange of CO_2 with the atmosphere and the transfer of C within Canadian forest ecosystems. An expected impact of global change is a significant alteration to the frequency and timing of disturbance events in northern forests (Weber and Flannigan, 1997; Weber and Stocks, 1998). Overpeck et al. (1997) demonstrated that frequent large-scale forest disturbances such as wildfires could accelerate forest biome adaptation to changing climate conditions. Kurz et al. (1995a) showed that changes in the disturbance regime will be both an important mechanism by which northern ecosystems respond to climate change and an important feedback mechanism to that change.

From a carbon budget perspective, forest ecosystem carbon can be divided into three major pools — biomass, detritus, and soil (Apps et al., 1999b). Understanding the influence of disturbances on the size of these pools is essential for evaluating the forest C cycle at any scale. This chapter discusses the influence of disturbance on C pools of forested ecosystems in the boreal forests of Central Canada at the point, stand and regional scale using the CENTURY model (Peng et al., 1998) and a simplified version of the Carbon Budget Model of the Canadian Forest Sector (CBM-CFS2; Kurz and Apps, 1999). The changes in the relationship between vegetation biomass, litter, and soil C pools in response to variation in disturbances are also examined.

II. Materials and Methods

A. The CENTURY Model and Its Parameterization

CENTURY, a point-level biogeochemistry model, simulates the long-term (100 to 10,000 years) dynamics of C, nitrogen (N), phosphorus (P), and sulfur (S) for a variety of plant–soil ecosystems. Parton et al. (1987; 1993) and Metherell et al. (1993) have described the model in detail while Peng et al. (1998) and Peng and Apps (1998) reported testing and application of the forest version (CENTURY 4.0) in the boreal forests of central Canada. CENTURY's forest production module partitions biomass into several compartments: foliage, fine and coarse roots, fine branches, and large wood. C and N are allocated to the different plant parts using a fixed allocation scheme (Peng et al., 1998). Gross primary productivity (GPP) is calculated as a function of maximum gross productivity, moisture, soil temperature, and live leaf area index (LAI). Litterfall and decomposition in a series of dead organic matter pools are simulated as functions of soil texture, monthly temperature, and precipitation.

Table 1. Site parameters used in CENTURY model for location near Prince Albert

Parameters	Location
Latitude	53.37N
Longitude	101.08W
Mean monthly minimum (°C)	−25.3
Mean monthly maximum (°C)	24.5
Annual precipitation (mm)	398
Dominant vegetation	*Pinus banksiana*
Soil type	Developing Spodosols
Soil texture	
Clay (%)	13
Sand (%)	73

The model operates on a monthly time step. The major input variables for the model include both biotic and abiotic site factors such as monthly mean maximum and minimum air temperature, monthly precipitation, soil texture, atmospheric and soil N inputs, plant lignin content, and initial values for soil nutrients (C, N, P, and S). Given soil texture, monthly temperature, and precipitation data, other input variables can be estimated internally by the model. For the present application, site-specific parameters and initial conditions, such as soil texture (clay, silt and sand content), bulk density, soil pH, soil C content for the 0 to 20 cm layer, soil C content for the 0 to 100 cm layer and drainage characteristics of soil were obtained from field data (Siltanen et al., 1997) (Table 1). Mean maximum and minimum monthly temperatures and monthly precipitation were calculated by CENTURY 4.0 using the 30-year normals (1950-1980) for a nearby climate station (AES, 1983).

CENTURY 4.0 simulations were run for about 6000 years using random disturbance sequences (Apps et al., 1999a). These were used to simulate changes in boreal forest ecosystem C pools at a site along the southern limits of the boreal forest near Prince Albert, Saskatchewan. Simulations were performed using six different random disturbance sequences. For each such sequence, the average fire return interval was 100 years over the 4000-year analysis period. Simulations were also carried using uniform disturbance sequences with strictly periodic return intervals of 25, 50, 75, 100, 150, 200, 500, and 1000 years. To avoid initialization artifacts, the average C pool size for biomass, litter, and soil were calculated using only the last 4000 years of simulation. A typical random sequence was used to examine the variations in the biomass, litter, and soil C pool sizes over a 1200-year record having intervals of low, medium, and high disturbance.

B. CBM-CFS2 Model

CBM-CFS2 is a spatially distributed simulation model that accounts for C pools and fluxes in forest ecosystems whose dynamics are primarily driven by stand-replacing disturbances (Kurz and Apps, 1999). In the model, the simulation area is divided into spatial units having broadly similar vegetation characteristics (Apps and Kurz, 1993). Within each of these spatial units, the model simulates the dynamics of groups of stands (State Variable Objects, or SVOs) having similar species, productivity, stocking, and age-class characteristics. Biomass growth curves that describe the accumulation of biomass (C) as a function of stand age are derived from forest inventory data (Kurz et al., 1992) and associated with each SVO to simulate changes in both the above and below ground biomass (Kurz and Apps 1999; Kurz et al., 1996; Apps and Kurz, 1993; Kurz et al., 1992). In each SVO, forest vegetation is represented by 12 pools consisting of four aboveground (foliage, submerchantable, merchantable, and other) and two below ground (coarse and fine roots) biomass pools for both a hardwood and a softwood group of species. Soil and litter dynamics are represented by four soil/detritus C pools (designated as very fast, fast, medium, and slow). These pools have different decomposition rates that are distinguished by vegetation type, origin and size of detrital material, mean annual temperature, stand conditions that are associated with each SVO, and the time since last disturbance (stand age). The slow soil C pool represents humified organic matter and receives C from the three other pools (very fast, fast, and medium). Litterfall and mortality are derived from the growth curves and disturbance data (Kurz et al., 1992). These inputs of dead organic matter are then used with a simple soil decomposition model (Kurz and Apps, 1999; Apps and Kurz, 1993) to account for changes in litter and soil pools between disturbances.

The number of SVOs, and the area associated with each, changes during the simulation as disturbances are applied to the region. During a disturbance event, transfers of biomass C to litter pools (including coarse woody debris), to the forest product sector (in the case of harvesting) and from both biomass and litter to the atmosphere (in the case of fire) are specified as proportions of the undisturbed SVO component. In this way, the amount of C in the biomass and soil pools simulated in the model is an explicit function of ecosystem type and past disturbance history. This chapter discusses only the dynamics of the slow C pool.

In Saskatchewan's boreal forest, the major tree species are jack pine *(Pinus banksiana)*, white spruce *(Picea glauca* (Moench) Voss), black spruce *(Picea mariana* (Mill.) B.S.P.), and trembling aspen *(Populus tremuloides* Michx.). Of the 457 growth curves (GC) developed from the national forest inventory by Kurz and Apps (1999), the major proportion of the Saskatchewan boreal forest area (24.5 M ha) is associated with 6 different GC, as discussed in Apps et al. (1999a). In this analysis, GC 4 was not used, as the area associated with it is less than 0.005% of the total area and it is almost identical to growth curve 3. The five growth equations used for analysis are shown in Figure 2. GC 1, 5, and 6 depict high, medium, and low productivity coniferous stands, respectively, GC 2 is for mixed wood, and GC 3 represents deciduous stands. The simplified model based on these five growth curves does not use the spatial disturbance database from the forest inventory of Saskatchewan's boreal forest but does represent the broad range of possible conditions for the region.

Fifteen different random disturbance sequences, each having an average return disturbance interval of 100 years over the period, were generated and used for model simulations as discussed in Apps et al. (1999a). Simulations were also carried out with a uniform disturbance regime with a 100-year return cycle. Both random and uniform disturbance sequences were applied to all the growth curves. To avoid initialization artefacts, simulation results were examined for periods between 2001 and 6000. The total Saskatchewan boreal forest biomass C, litterfall C, and soil C pools were then estimated using an area-weighted sum over all five growth equations.

A typical random sequence (the same sequence as used in the CENTURY simulation) was selected to estimate the variations in biomass, litterfall rate, and soil C contents over the 1200-year period for each growth curve for each forest type. The average C pool size for biomass, litter production, and soil was calculated for each apparent period of low, medium, and high disturbance.

C. Soil C in Relation to Biomass C

Organic C in forest soil and litter pools is derived entirely from shed, dead, and decaying vegetation biomass. Changes in the carbon content of dead organic matter pools (litter and soil) of a forest ecosystem are a simple balance between inputs of fresh organic matter from living vegetation (through litterfall, root exudates and turnover, and mortality) and losses (through combustion, translocation, and decomposition) (Martin, 1998). The inputs and losses are both functions of the ecosystem conditions, the environmental conditions, and the disturbance regime. The latter factor was examined by considering the relationship between biomass, dead organic matter, and disturbances.

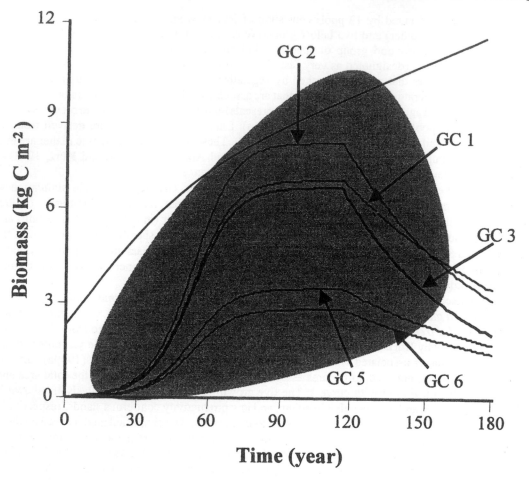

Figure 2. A typical CENTURY growth curve (unlabeled) and the five growth curves (GC 1 to GC 6) used in the simplified CBM-CFS2 model to simulate forest growth. The shaded area represents the range of observed biomass in Saskatchewan's boreal fores. (From Halliwell et al., 1995.)

Disturbances influence the dead organic matter C pools in three ways (Apps, 1993). First, there is a direct input of organic matter from the killed biomass. These inputs are highest for insect- or disease-induced mortality, somewhat lower (by the amount that is released directly through combustion) for fire, and least for harvesting (by the amount of woody material removed from the site). The inputs may have both a pulsed component (at the time of the disturbance) and a continuing, but declining, component associated with post-disturbance mortality. Second, the disturbance event may directly remove or release existing dead organic matter carbon from the site. This is most important for fire, where as much as 85% of the organic matter of the forest floor can be lost through combustion (Weber and Stocks, 1998; Auclair and Carter, 1993). Third, the disturbance event generally alters the microsite conditions dramatically through loss of canopy, removal of the duff layer, alteration of microsite hydrology, and similar impacts (Bhatti et al., 1999). Such alterations may significantly influence subsequent decomposition rates by changing both the decomposer population and their habitat—changes whose effect (post disturbance losses) is presumed to dissipate over time as the vegetation recovers from the disturbance.

The amount of fresh material added to the forest floor by a disturbance event depends both on the event itself and on the amount of biomass present in the undisturbed ecosystem. Even in the absence of disturbances, the input of organic matter to the forest floor through litterfall and individual tree mortality is highly correlated to the vegetation biomass present and, as a general rule, increases with stand age (Apps et al., 1999a). Stands with younger age-class trees have lower litterfall rates compared to older age-class trees. Thus the input of C to the dead organic matter pools at a given site is directly related to the biomass at that site and the time that has elapsed since the last disturbance.

The losses of C from litter and soil pools due to decomposition are generally assumed to be proportional to the amount of C in those pools (C_i)

$$dC_i(t) = -kC_i(t)dt \qquad \text{Eq. 1}$$

where k is the first order decomposition rate parameter (yr^{-1}) specific to that pool. In reality the parameter k depends on a number of biotic and abiotic variables (including temperature) that change over time following a disturbance, and with other changes in the environmental conditions. Indeed, process-based models like CENTURY are designed to simulate the effects of such changes.

Equation 1 implies that fresh C input $I(t_1)$ to a given dead organic matter pool C_i at time t_1 plays a declining role in the pool at a later time t_2. Specifically at time t_2, the amount remaining is reduced by the factor e^{-kt} where $t = t_2 - t_1$ is the elapsed time. This declining contribution applies both to the pulsed inputs associated with disturbances and the smoothly changing inputs associated with litter production in normal stand growth processes. Integrating (1) gives:

$$C_i(t) = \int_{-\infty}^{t} I(t')e^{-kt'}dt' \qquad \text{Eq. 2}$$

This relationship states that the C_i in the organic pool at a time t is determined by the site history of inputs of C to that pool.

The final step in our argument is to note that the inputs change in synchrony with the site biomass—i.e., the litter and disturbance inputs are proportional to the amount of biomass on site and therefore follow a related course through time. Thus it can be postulated that:

$$C_i(t) \propto \int_{-\infty}^{t} B(t')e^{-kt'}dt' \qquad \text{Eq. 3}$$

where $B(t)$ is the vegetation biomass on the site at time t. The proportionality symbol ($\propto$) is used rather than an equality to indicate that the trends over time should be similar, although numerically very different. To make equation 3 into an equality, terms that include the direct disturbance-induced emissions and translocation of C from the pool C_i must be included (Apps, 1997).

To test this hypothesis, the right-hand side of Equation 3 was calculated from the simulated biomass using typical litter decomposition rates and compared to the more detailed soil C pool calculations performed in the CENTURY and CBM-CFS2 simulations. Estimated soil C from Equation 3 and the simulated values from CENTURY and the simplified CBM-CFS2 model were compared using regression analysis (Zar, 1984).

Table 2. Average C in the aboveground biomass, litter pool, and slow soil pools of the upper 20 cm of soil as simulated by CENTURY 4.0 for a series of random disturbance sequences (each having 100-year average return period) and uniform disturbances (having return periods between 25 and 1000 years)

Disturbance sequence	Biomass	Litter pool	Slow soil
		$kg\ m^{-2}$	
Random			
100 (average)	7.3 – 9.2	4.0 – 4.1	4.9 – 5.2
Uniform			
25	2.70	3.06	4.02
50	4.23	3.62	4.61
75	5.65	3.80	4.78
100	6.68	3.91	4.88
150	8.25	4.05	5.03
200	9.53	4.16	5.15
500	12.9	4.49	5.62
1000	14.7	4.68	5.91

III. Results and Discussion

A. Point-Scale Simulations with CENTURY

Simulation results for biomass C, litter pool C, and slow soil pool C under random and uniform disturbance regimes are presented in Table 2. The simulation shows that higher biomass, litter, and soil C pools occur under the random disturbance regime with an average return interval of 100 years than under the uniform disturbance regime having the same return interval.

Biomass C simulated by the CENTURY model increases as the disturbance return interval increases (Table 2). The biomass C simulated by CENTURY with a 100-year return interval appears to overestimate biomass when compared to that estimated from the forest inventory data by Bonnor (1985). One reason for the discrepancy could be an artifact of CENTURY's failure to represent regeneration delay (Figure 2). In CENTURY, the growth curve is assumed to start exponentially limited growth immediately following disturbance. This results in higher C accumulation rate in the early stages of recovery from disturbance than are generally found in natural boreal ecosystems.

The litter pool C simulated with CENTURY was between 4.0 and 4.1 kg C m^{-2} under random disturbance regimes and varied between 3.1 and 4.7 kg C m^{-2} under different uniform disturbance sequences (Table 2). Both simulations compare well with the value of 3.7 kg C m^{-2} for the boreal forest litter pool reported by Matthews (1997). With a random disturbance cycle, the simulated soil C content ranged from 4.9 to 5.2 kg m^{-2} – within 5% of the observed values reported by Siltanen et al. (1997). As the return interval under uniform regime was increased from 25 to 1000 years, a substantial increase in soil C content was found. These results show that CENTURY simulations using uniform disturbance regimes underestimate the C stocks relative to the simulations with random disturbance regimes.

The actual disturbance rate has a significant influence on the C dynamics of biomass, litter, and soil pools. This is demonstrated in Figure 3, which shows a 1200-year record for a typical random disturbance sequence. For example, the simulated average biomass for the site over the entire 1200 years was 7.2 kg C m^{-2} but varies between 0.4 to 16.2 kg C m^{-2}, 0.6 to 8.4 kg C m^{-2}, and 0.2 to 6.5 kg C m^{-2} during periods of apparent high, medium, and low disturbance.

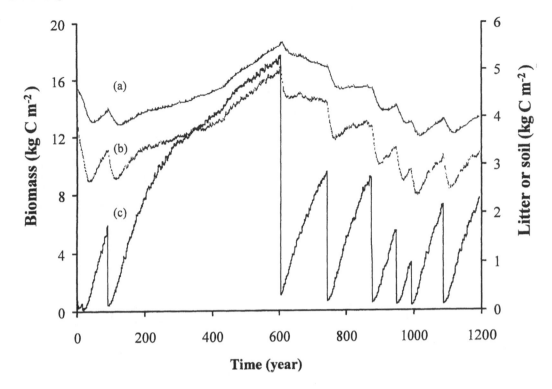

Figure 3. CENTURY simulations of biomass (c), litter (b), and soil C (a) (0–20 cm depth) pools over 1200-year interval using a typical random disturbance sequence for a boreal forest stand at a site near Prince Albert, Saskatchewan.

Simulated biomass, litter and soil C pools with CENTURY, in the absence of disturbance or with very low disturbance, appear to increase over time without limit in an unrealistic manner (see Figure 2 and simulation years 100 to 300 in Figure 3). In reality, for the even-aged stand development typical of Canadian forests following disturbance, biomass follows a sigmoidal growth pattern (as for example, in the Chapman-Richards representation, Avery and Burkhart 1983). Inventory data indicates that biomass accumulation slows or ceases after maturity is reached, typically 100 to 200 years after disturbance (Kurz et al., 1992). Even if not disturbed stand biomass often declines in later stages of stand development (Venevsky and Shvidenko, 1997; Alban and Perala, 1992; Kurz et al., 1992) as older individual trees die and are replaced by younger cohorts. In addition, as stand age increases, vigor and productivity decrease, the stand becomes more susceptible to insect or disease, and becomes more prone to non-stand replacing disturbances (Kurz and Apps, 1994). In Figure 3, litter and soil C pools follow the same trend as biomass C, but with some lag period. With increases in disturbance frequency, there is a substantial decrease in litter and soil C pools.

B. Stand-Scale Simulations with CBM-CFS2

There were substantial differences in the biomass, litter production, and soil C accumulation for different stands in Saskatchewan's boreal forest (Table 4). Using a typical random disturbance sequence (average 100-year return period over a 1200-year sequence), the average biomass C simulated by the simplified CBM-CFS2 model for Saskatchewan boreal forests was 3.42 kg C m^{-2},

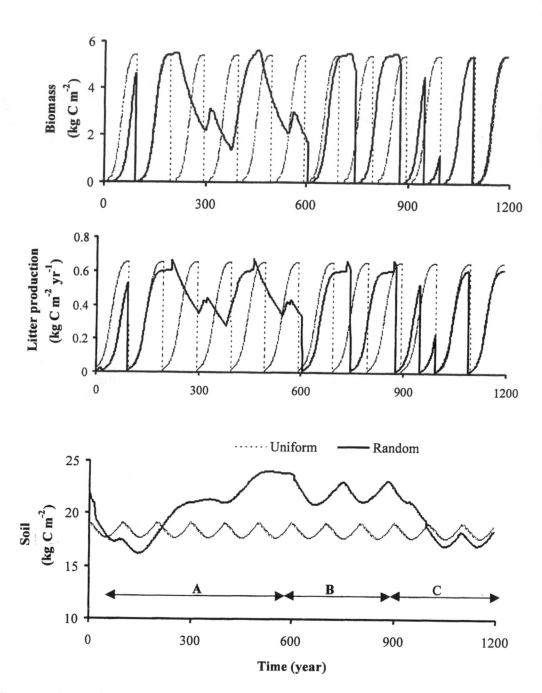

Figure 4. Simulated biomass, litter fall, and soil C pools using a uniform (100-year return period) and a typical random disturbance sequence (average 100-year return period) in a boreal forest stand using the simplified CBM-CFS2 model over a 1200-year interval. Intervals A, B, and C represent intervals of a random sequence that have periods of apparent low, medium, and high disturbance, respectively.

with a range of 1.75 to 4.93 kg C m^{-2} depending upon the productivity and type of vegetation. Biomass C simulated for the same stands under the uniform disturbance varied from 1.58 to 4.46 kg C m^{-2}. The biomass C of low productivity coniferous stands was 1.74 kg C m^{-2}, less than half that of high productivity stands (4.18 kg C m^{-2}). These were within the range of 0.52 to 5.38 kg C m^{-2} (average 2.41 kg C m^{-2}) reported by Simpson et al. (1993) from direct measurements for western Canada's boreal forest.

The amount of C stored in biomass at a specific site depends on the successional stage of stand development and, hence, on the actual disturbance history of that site (Figure 4). During periods of apparent low disturbance, many stands reach their mature/over mature growth phase with relatively even-aged forests of 100 to 200 years and older. The rate of C sequestration in the biomass of these forests is relatively low, but the amount of C stored is large. Depending upon the forest type, stand biomass C pools under an apparent period of low disturbances were 20 to 21% higher than the average over the period (Table 3). During periods of intermediate disturbances, many forest stands, being in the young to mature stage, accumulate C at near maximum rates. During periods of high disturbance, many stands are in the regeneration and immature growth phase and few reach the age of 50. In this period, biomass C accumulation rates were relatively high but C stocks were about 44% lower (Table 3) than the temporal average for the entire record.

Simulated litter production for boreal forest stands varied from 0.28 to 0.52 kg C m^{-2} yr^{-1} under the random disturbance regime as compared to 0.24 to 0.45 kg C m^{-2} yr^{-1} under the uniform disturbance regime. These were within the range of litter production of $\leq$ 0.27 kg C m^{-2} yr^{-1} reported by Matthews (1997), 0.46 kg C m^{-2} yr^{-1} for jack pine, and 0.26 kg C m^{-2} yr^{-1} for black spruce stands in Ontario observed by N.W. Foster (personal communication, 1998). As a consequence, the litter production varied considerably amongst the stands making up the boreal forests in Saskatchewan. Stands in the younger age classes, during periods of high disturbance, had lower litter production than those in periods of low disturbance (Figure 4). Moreover, the physical size of individual litter components also differs across the forest age-class distribution. Both these factors (litterfall rates and size of litter components) influence the quantity (mass) and the quality (decomposability) of litter input to the detritus and soil pools.

Large amounts of C are present in the soils of the boreal forests (Figure 4). The average soil C simulated under random disturbance regimes for boreal forest stands in Saskatchewan was 23 kg C m^{-2} with a range from 16 to 29 kg C m^{-2}. Soil C content, under a uniform disturbance regime with the same average return interval (100 years), varied from 14 to 27 kg C m^{-2} — lower than that obtained under random disturbance regimes. These values were a little higher than the soil C content of 14 to 20 kg C m^{-2} reported by Pastor and Post (1988), but much lower than the 50 kg C m^{-2} estimated by Tarnocai (1998) from interpreted soil C maps. These results further indicate that assumptions about the stochasticity of the disturbance regime affect the estimations and predictions of C storage in boreal forest stands. They also suggest that most models presently in use underestimate the C stocks because they employ a uniform disturbance regime assumption.

C. Regional-Scale Simulations with CBM-CFS2

Due to variation in their disturbance history, different forest stands will have different stocks of C (biomass, litter, and soil pool) resulting from differences in the age class structure. On the regional scale, under an unchanging disturbance regime, the distribution of stand age classes does not change over time. Under such conditions the variation in stand histories balance each other out so that over the region the area disturbed per year is relatively constant. The net ecosystem C flux (or net biome productivity, NBP) tends to zero and C stocks approach a steady-state value at the landscape level.

In such circumstances, the average C stock for all the stands can be used to calculate regional C pools. For the same average disturbance return interval (100 years), the average biomass C for Saskatchewan boreal forest simulated by the simplified CBM-CFS2 model varied from 2.67 to 2.70 kg C m^{-2} under random disturbance regimes, and was 2.40 kg C m^{-2} under a uniform disturbance regime (Table 4). These values of biomass C were close to the biomass C measurements for boreal forest of 2.41 kg C m^{-2} reported by Simpson et al. (1993) through direct measurements. Litter production over the 4000-year period (2001 to 6000) were 0.35 to 0.38 kg C m^{-2} yr^{-1} under random disturbance regimes compared to 0.31 kg C m^{-2} yr^{-1} under a uniform disturbance regime and was within the range of litter production of $\leq$ 0.27 kg C m^{-2} yr^{-1} reported by Matthews (1997).

Under random disturbance regimes, the average soil C estimated by the simplified CBM-CFS2 model varied between 18.8 and 20.1 kg C m^{-2}. Under the uniform disturbance regime, the estimated soil C was 18.3 kg C m^{-2} (Table 4). These simulated soil C values were close to the soil C content of 13.5 to 19.5 kg C m^{-2} (to a depth of 100 cm) reported by Pastor and Post (1988) for the boreal forest of North America. The simulated soil C content using CBM-CFS2 was 60% higher than that reported by Bhatti and Apps (1999) for the upland boreal forest using field data reported by Siltanen et al. (1997), but much lower than the 50 kg C m^{-2} estimated from interpreted soil carbon maps by Tarnocai (1998).

There are several explanations for the higher CBM-CFS2 estimates relative to the Bhatti and Apps (1999) estimates, which are believed to be the most accurate. First, the CBM-CFS2 soil module contains a relatively simple representation of the processes governing soil organic C dynamics, including a simplistic parameterization of the partitioning of litter decomposition products between soil organic matter and the atmosphere (Kurz et al., 1992). Second, both the nature of the historical disturbances (fire) and the average frequency (100 years) used in the present simulations are clearly simplifications of a much more complex history. As shown in an earlier part of the chapter, stands experiencing different disturbance periodicity can have considerably different amounts of soil C (Table 3).

Changes in the disturbance regime alter the age-class structure. With frequent disturbance, the proportion of younger age-class stands increases (Kurz et al., 1995a), while under less frequent disturbance, more of the forest stands will reach older age classes. In the boreal forest, stand age influences the C accumulation rate more than any other site parameters (Rapallee et al., 1998). Simulations show that during periods of apparent low to medium disturbances, the soil C content of forest stand remains the same or achieves even higher values (Table 3). With higher forest biomass C accumulation and higher litter production there is a higher C input to the soil C. The balance between decomposition losses and litter inputs is achieved at a higher soil C value. Therefore, soil appears to act as a C sink during a transition from a period of high to lower disturbances—a result previously noted by Kurz and Apps (1995).

In contrast, during periods of high disturbance, there was about 10% decrease in soil C content: the soil appears to act as a source of atmospheric C. This phenomenon arises in the simulations (and, we suggest, in reality) for three reasons (i) during times of higher disturbance, there is a higher proportion of younger age stands and, hence lower litter transfers (as described previously); (ii) there is decreased input of coarse woody debris (with low turnover rates) due to the decrease in the proportion of older age stands (Harmon et al., 1990); and (iii) there is an increased rate of decomposition of detritus and soil C pool due to changed micro-environmental conditions associated with the younger stands (Bhatti et al., 1999). These situations might occur when forest biomass productivity is limited by climate conditions (for example), when input from disturbance is low because of severe fire loss or through harvesting, or when the decomposition rate is high. Therefore, the soil of any individual stand may act as a sink or a source of C depending upon the actual disturbance history of that site.

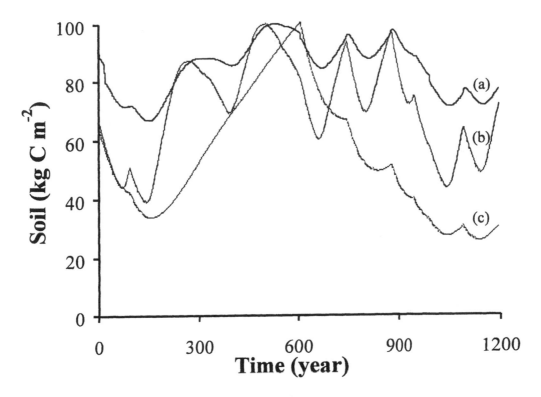

Figure 5. Soil C content estimated using the analytical relationship (see text) and biomass C data simulated by CENTURY (c) and simplifies CBM-CFRS2 (b). For comparison, curve (a) shows the more detailed simulation of soil C with CBM-CFS2 for a stand in Saskatchewan's boreal forest.

Liski et al. (1998) observed similar results for the soils of boreal forests in western Finland. He reported that with frequent fires, the amount of soil C could be 25% less than that estimated for a no-fire scenario. At any particular time, if a site were in a period of apparent low disturbance, the soil C content would be higher than it would have been in a period of high disturbance.

D. Soil C and Litterfall C in Relation to Biomass C and Disturbances

Soil C content estimated using the analytical relationship (3) with biomass input derived from CENTURY and the simplified CBM-CFS2 simulations are presented in Figure 5. The postulated relationship appears to capture the soil C dynamics related to changes in biomass C inputs resulting from variations in disturbances. Soil C estimated from (3) and modelled values (CENTURY and CBM-CFS2) follow similar trends over the 1200-year period. Soil C content estimated using the analytical relationship (3) were more closely correlated with soil C simulated by CBM-CFS2 ($r^2 = 0.71$) than with the CENTURY simulation values ($r^2 = 0.39$).

Table 3. Simulated biomass, litter production and soil C content for different stands using the simplified CBM-CFS2 model under uniform and random disturbance regimes having 100-year average return interval over the 1200-year simulation record; within the random sequences, estimates are also shown for periods of apparent high, medium and low disturbances

| | Coniferous[a] | | | | |
	High productivity	medium productivity	Low productivity	Mixed wood[b]	Deciduous[c]
Average biomass C (kg m^{-2})					
Uniform	3.80	1.96	1.58	4.46	3.69
Random average	4.18	2.16	1.75	4.93	4.07
Low period	5.02	2.60	2.10	5.95	4.92
Medium period	4.99	2.57	2.08	5.82	4.79
High period	2.34	1.20	0.97	2.77	2.29
Average litter production C (kg m^{-2} y^{-1})					
Uniform	0.45	0.28	0.24	0.37	0.32
Random average	0.52	0.32	0.28	0.39	0.41
Low period	0.63	0.41	0.35	0.46	0.51
Medium period	0.58	0.35	0.31	0.39	0.43
High period	0.31	0.20	0.17	0.28	0.26
Average soil C (kg m^{-2})					
Uniform	27	16	14	27	20
Random average	30	19	16	29	24
Low period	30	19	17	30	24
Medium period	31	20	17	31	26
High period	28	18	15	28	23

[a]High productivity conifers use GC 1, medium productivity conifers, GC 5, low productivity conifers, GC 6; [b]mixed wood = GC 2; [c]deciduous = GC 3.

The simplicity of the relationship (3) and the correlation with more detailed simulations shown in Figure 5 further emphasizes the importance of past biomass dynamics and prior disturbance history for present soil carbon stocks and changes. The relative weight of prior disturbance event is determined by the product of the decomposition rate parameter (k in equation 5) and the time (t) elapsed since that event. For boreal systems, the values of k are large relative to other forest ecosystems, increasing the importance of the historical variation in disturbances. A more detailed analysis is under way (Apps, unpublished manuscript) that takes into account, for example, the influence of disturbances on decomposition rate and accounts for cascading carbon flows between multiple soil compartments. The importance of the present analysis is that it shows that the projection of likely trends under conditions of changing disturbance regimes may be inferred in a simple way.

E. Role of Disturbance Frequency in C Sequestration under a Changing Climate

One of the significant factors influencing the response of boreal forests to climate change is the potential for significant changes in disturbance regimes (Weber and Stocks, 1998; Weber and

Table 4. Biomass C, litterfall C, and slow soil pool C simulated with the simplified CBM-CFS2 model for Saskatchewan boreal forests under uniform and random disturbance regimes having the same average return interval (100 years); (the range reported is the variation between different 100-year random series)

C pool	Uniform	Random
Biomass (kg C m^{-2})	2.40	2.67 – 2.70
Litter fall (kg C m^{-2} y^{-1})	0.31	0.35 – 0.38
Slow soil (kg C m^{-2})	18.3	18.8 – 20.1

Flannigan, 1997; Apps, 1993). The variations in climatic conditions in central Canada over the last 20 years appear to have resulted in higher disturbances (Kurz et al., 1995b) at a national scale. Regionally the changes may be larger – the mean fire return interval for some sites in Alberta and Saskatchewan was reported to be as low as 34 years (Larsen and MacDonald, 1998). As indicated in Table 3, the simulated C pools in biomass, litter, and soils during periods of high disturbance are substantially lowered. With more frequent disturbances, more of the stands are in younger age classes and release greater amounts of C to the atmosphere as their elevated detritus and soil pools decay (equation 2 and 3).

Simulation results, using the simplified version of CBM-CFS2, show that during periods of high disturbance, Saskatchewan's boreal forests lost about 27g C m^{-2} y^{-1} (on average) to the atmosphere (Table 4). This is not a firm estimate of reality since the actual loss depends upon the initial C stocks and the previous disturbance history, but it is consistent with other estimates. Kurz et al. (1992) reported that a three-fold increase in Canadian wildfires between a high-fire year (1989) and a reference year (1986) would result in an estimated 86% reduction in the net ecosystem C sink at a national scale. Subsequent and more detailed analyses (Kurz and Apps 1996, 1999) suggest that central Canada's boreal forests may have become a source of atmospheric C as a result of such changes in the natural disturbance regime.

This situation is expected to worsen as climatic change proceeds and there is a shortening of fire return intervals (Weber and Flannigan, 1997). With an increase in fire disturbance, the present analysis suggests a potential rapid increase in CO_2 release to the atmosphere rather than a net increase in terrestrial C storage (Smith and Shugart, 1993). There are a number of possible offsetting factors that have not been explicitly included in the present analyses, including increases in biomass productivity (B in relationship 3) associated both with natural responses to the changing environment (e.g., Peng and Apps, 1998; Peng et al., 1998) and with response to management intervention (e.g., Binkley et al., 1997).

Management options include reducing the regeneration delay through seeding and plantations, enhancing forest productivity, changing the forest rotation length through control and suppression of disturbance by fire, pest and disease and the judicious use of forest products (Apps et al., 1999b). A crude calculation using the simulation results reported here suggests that with 100% fire prevention, average soil C content could be increased by 2.4 g C m^{-2} yr^{-1}. This rate is only a potential increase, however: it may not be feasibly accomplished and it certainly could not be maintained. It is beyond the scope of this chapter to examine these effects in detail. It should be noted, however, that a protected forest acts as sink only during the period of transition from a higher to a lower period of disturbance. Can such protection be maintained if the changing climatic conditions favor increased disturbances? Moreover, if the protection is removed, it must be stressed that the same areas will be a source as the disturbance regime relaxes back to a period of higher disturbance.

IV. Conclusions

Using CENTURY and a simplified version of CBM-CFS2, the average C stocks in simulated biomass, litter, and organic matter in slow turnover soil pools were shown to be higher under a random disturbance regime than under a uniform disturbance regime having the same average return interval. This relationship was found to hold at the point, stand, and regional scales. The actual C content biomass, litter, and soil pools at the stand level also depend on productivity and species composition, and at the regional level on the distribution of forest types (of different productivity and composition) within the region.

For a given site and forest type, the time since last disturbance was a primary determinant of the biomass C pool and the litter input to dead organic matter on the forest floor. The two models — CENTURY (having a point-based representation of decomposition processes) and a simplified version of CBM-CFS2 (having a simple regionally parameterized representation) — both indicate that sustained periods of low, intermediate, and high disturbance rates are associated with very different levels of C stocks in biomass, litter and soil. Both models suggest that under sustained periods of low disturbance, C stocks for a given forest type (at the point, stand or regional level) are higher than under sustained (uniform) high disturbance. Differences in C stocks simulated by the two models become evident at low disturbance rates where the simulation of biomass of mature and over-mature stands become increasingly important (neither model explicitly considers changes in species associated with succession).

Changes in the timing of disturbance events were found to have a strong and lasting influence on the dynamics of the C stocks, and a direct effect on the net atmospheric fluxes with both models. At point and stand scales, changes over time in site-specific C content of dead organic matter (litter and soil) were strongly related to the time since last disturbance, but also to the entire prior disturbance history. Thus at the site scale (point and stand), the present dynamics of the C stocks, as well as the level of these stocks, are strongly related to the disturbance history of that site. This relationship begins with the individual disturbance event itself: during or shortly after the event, large pulses of detritus (including dead roots) are transferred to the forest floor and litter input from living biomass changes abruptly. Litter fall (both above and below ground) continues to vary with the stage of stand development until a subsequent disturbance repeats the cycle of regrowth. Thus, the net C pools in dead organic matter change sharply following a disturbance event and continue to change gradually over time with the alteration both in inputs from above- and belowground biomass and in losses through decomposition, translocation and combustion.

Characteristically, in the absence of other external changes, the net organic matter pools following a disturbance relax back towards a steady state condition (inputs = losses) that in Canadian forest stands is rarely reached before a subsequent disturbance event. In reality, variations in succession and site productivity caused by alterations in other environmental factors, tend to make this steady state an elusive concept. In addition, a given site retains memory of its past disturbance history in the structure and chemical makeup of its soils. For soil carbon, the relative contribution of past biomass C declines over time as it is decomposed.

A simple analytical equation was derived to express the declining contribution to present soil C content of past biomass inputs. This relationship was tested with results from both CENTURY and the simplified CBM-CFS2 model and was shown to exhibit the same temporal dynamics. At the point and stand level this analytical estimation of soil C content compared well with that simulated by the much more complex CENTURY and CBM-CFS2 models.

With changing disturbance regimes, whether at the stand or landscape scale, soil C was strongly affected by the previous disturbance pattern (which determines the present age-class structure). During periods of more frequent disturbance, soil C content decreases as a result of both decreased fresh litter input and changes in the micro-environment of the disturbed areas. During such periods, the forest likely acts as a source of C until the regrowth inputs overtake the increased decomposition losses. With effective forest protection, it may be possible to temporarily increase the C sequestration capacity of the boreal forests. Such measures can delay the loss of C back to atmosphere only as long as the management intervention is effectively maintained.

Acknowledgments

We thank D. Halliwell for carrying out the modeling simulations with CBM-CFS2 and for informative discussions. Funding for this study was provided in part by the Energy from the Forest (ENFOR) program of the Federal Panel on Energy Research and Development (PERD) and by the Sustainable Forest Management Network of Centres of Excellence.

References

Alban, D.H. and D.A. Perala. 1992. Carbon storage in Lake States aspen ecosystem. *Can. J. For. Res.* 22:1107-1110.

Auclair, A.N.D. and T.B. Carter. 1993. Forest wildfire as a recent source of CO_2 at northern latitudes. *Can. J. For. Res.* 23:1530-1536.

Apps, M.J. 1997. Biomass Burning: Accounting for Fire in the IPCC Guidelines. Background invited paper for IPCC Workshop on Biomass Burning, Land-Use Change and Forestry, Rockhampton, Australia, September 1997. Unpublished manuscript, 34 pp.

Apps, M.J. 1993. NBIOME: A biome-level study of biospheric response and feedback to potential climate changes. *World Res. Rev.* 5:41-65.

Apps, M.J. and W.A. Kurz. 1993. The role of Canadian forests in the global carbon balance. p.14-28. In: Carbon Balance of World's Forested Ecosystems: Towards a Global Assessment. Publ. Acad. Finland No. 3/1993 Helsinki, Finland.

Apps, M.J., J.S. Bhatti, D. Halliwell and H. Jiang. 1999a. Influence of uniform versus random disturbance regimes on carbon dynamics in the boreal forest of central Canada. In: R. Lal, J. Kimble, H. Eswaran and B.A. Stewart (eds.), *Global Climate Change and Cold Ecosystems*. Lewis Publishers, Boca Raton, FL.

Apps, M.J., W.A. Kurz, S.J. Beukema and J.S. Bhatti. 1999b. Carbon budget of the Canadian forest product sector. *J. Environ. Sci. Policy* 2:25-41

Apps, M.J., W.A. Kurz, R.J. Luxmoore, L.O. Nilsson, R.A. Sedjo, R. Schmidt, L.G. Simpson and T.S. Vinson. 1993. The changing role of circumpolar boreal forests and tundra in the global carbon cycle. *Water, Air and Soil Pollution* 70:39-53.

Atmospheric Environmental Service (AES). 1983. Canadian climate normals, 1951–1980, temperature, precipitation. Prairie Provinces. Environmental Canada, Downsview, Ontario.

Avery, T.E. and H.E. Burkhart. 1983. *Forest Measurements* (3rd ed.). McGraw-Hill, New York.

Bhatti, J.S., and M. J. Apps. 1999. Carbon and nitrogen storage in upland boreal forests. In: Lal, R., J. Kimble, H. Eswaran and B.A. Stewart (eds.), *Global Climate Change and Cold Ecosystems*. Lewis Publishers, Boca Raton, FL.

Bhatti, J.S., R.L. Fleming, P.A. Arp and N.W. Foster. 2000. Modelling pre- and post-harvesting fluctuations in soil temperature, soil moisture, and snowpack in boreal forests using ForHYMIII. *Forest Ecology and Management* (in press).

Bhatti, J.S., N.W. Foster, T. Oja, M.H. Moayeri and P.A. Arp. 1998. Modelling potential sustainable biomass productivity in jack pine forest stands. *Can. J. Soil Sci.* 78:105-113.

Binkley, C.S., M.J. Apps, R.K. Dixon, P. Kauppi and L.O. Nilsson. 1997. Sequestering carbon in natural forests. *Critical Reviews in Environmental Science and Technology* 27:S23-45.

Bonnor, G.M. 1985. Inventory of Forest Biomass in Canada. Canadian Forest Service, Petawawa National Forestry Institute, Chalk River, ON. 62 pp.

Halliwell, D.H., M.J. Apps and D.T. Price. 1995. A survey of the forest site characteristics in a transect through the central Canadian boreal forest. *Water, Air, Soil Pollut.* 82:257-270.

Harmon, M.E., W.K. Ferrell and J.F. Franklin. 1990. Effects of carbon storage of conversion of old-growth forests to young forests. *Science* 247:699-702.

Harden, J.W., K.P. O'Neill, S.E. Trumbore, H. Veldhuis and B.J. Stocks. 1997. Moss and soil contribution to the annual net carbon flux of a maturing boreal forests. *J. Geophy. Res.* 102:28805-28816.

Haxeltine, A., I.C. Prentice and I. D. Cresswell. 1996. A coupled carbon and water flux model to predict vegetation structure. *J. of Veg. Sci.* 7:651-666.

Hogg, E.H. 1999. Simulation of interannual response of trembling aspen stands to climatic variation and insect defoliation in western Canada. *Ecol. Modelling* 114:175-193.

Holling, C.S. 1992. Cross-scale morphology, geometry, and dynamics of ecosystems. *Ecol. Monogr.* 62:447-502.

Houghton, R.A., E.A. Davidson and G.M. Woodwell. 1998. Missing sink, feedbacks, and understanding the role of terrestrial ecosystems in global carbon balance. *Global Biogeoch. Cycle* 12:25-34.

IGBP Terrestrial Carbon Working Group. 1998. The terrestrial carbon cycle: Implications for the Kyoto Protocol. *Science* 280:1393-1397.

Jarvis, P.G., J.M. Massheder, S.E. Hale, J.B. Moncrieff, M. Rayment and S.L. Scott. 1997. Seasonal variation of carbon dioxide, water vapor, and energy exchanges of a boreal black spruce forest. *J. Geophy. Res.* 102:28953-28966.

Kimmins, J.P. 1996. Importance of soil and role of ecosystem disturbance for sustained productivity of cool temperate and boreal forest. *Soil Sci. Soc. Am. J.* 60:1643-1654.

Kimball, J.S., M.A. White and S.W. Running. 1997. BIOME-BGC simulations of stand hydrological processes for BOREAS. *J. Geophy. Res.* 102:29043-29051.

Kurz, W.A. and M.J. Apps. 1999. A 70-year retrospective analysis of carbon fluxes in the Canadian forest sector. *Ecol. Appl.* 9:526-547.

Kurz, W.A. and M.J. Apps. 1996. Retrospective assessment of carbon flows in Canadian boreal forests. p. 173-182. In: M.J. Apps and D.T. Price (eds.), *Forest Ecosystems, Forest Management and Global Carbon Cycle*. Springer-Verlag, Berlin.

Kurz, W.A. and M.J. Apps. 1995. An analysis of future carbon budgets of Canadian boreal forests. *Water, Air, Soil Pollut.* 82:321-331.

Kurz, W.A. and M.J. Apps. 1994. The carbon budget of Canadian forests: A sensitivity analysis of changes in disturbance regimes, growth rates and decomposition rates. *Environ. Pollut.* 83:55-61.

Kurz, W.A., M.J. Apps, B.J. Stocks and W.J.A. Volney. 1995a. Global climate change: Disturbance regimes and biospheric feedbacks of temperate and boreal forests. p. 119-133. In: G.M. Woodwell and F.T. Mackenzie (eds.), *Biotic Feedbacks in the Global Climatic System*. Oxford University Press, New York.

Kurz, W.A., M.J. Apps, S.J. Beukema and T. Lekstrum. 1995b. Twentieth century carbon budget of Canadian forests. *Tellus* 47B:170-177.

Kurz, W.A., M.J. Apps, T.M. Webb and P.J. McNemee. 1992. The carbon budget of the Canadian Forest Sector: Phase I. Information report NOR-X-326. Forestry Canada, Northwest Region, Northern Forestry Centre, Edmonton, Alberta. 93 pp.

Kurz, W.A., S.J. Beukema and M.J. Apps. 1996. Estimation of root biomass and dynamics for the Carbon Budget Model of the Canadian Forest Sector. *Can. J. For. Res.* 26:1973-1979.

Liski, J., H. Ilvensiemi, A. Mäkelä and M. Starr. 1998. Model analysis of the effects of soil age, fires and harvesting on the carbon storage of boreal forest soils. *European J. Soil Sci.* 49:407-416.

Larsen, C.P.S. and G.M. MacDonald. 1998. Fire and vegetation dynamics in a jack pine and black spruce forest reconstruction using fossil pollen and charcoal. *J. Ecology* 86:815-828.

Martin, P.H. 1998. Soil carbon and climate perturbations: using the analytical biogeochemical cycling (ABC) scheme. *J. Environ. Sci. and Policy* 1:87-97.

Matthews, E. 1997. Global litter production, pools and turnover times: estimates from measurement data and regression models. *J. Geophy. Res.* 102:18771-18880.

Metherell, A.K., L.A. Harding, C.V. Cole and W.J. Parton. 1993. CENTURY soil organic matter model environment. Technical documentation, Agroecosystem Version 4.0. Great Plains System Research Unit, Technical Report No 4. USDA-ARS, Fort Collins, CO.

Neilson, R.P. and D. Marks. 1994. A global perspective of regional vegetation and hydrologic sensitivities from climatic change. *J. Veg. Sci.* 5:715-730.

Ojima, D.S., T.G.F. Kittel, T. Rosswall and B.H. Walker. 1992. Critical issues for understanding global change effects on terrestrial ecosystems. *Ecol. Appl.* 1:316-325.

Overpeck, K., D. Hughen, D. Hardy, R. Bradley, R. Case, M. Douglas, B. Finney, K. Gajewski, G. Jacoby, A. Jennings, S. Lamoureux, A. Lasca, G. MacDonald, J. Moore, M. Retelle, S. Smith, A. Wolfe and G. Zielinski. 1997. Arctic environmental change of the last four centuries. *Science* 278:1251-1257.

Parton, W.J., J.M.O. Scurlock, D.S. Ojima, T.G. Gilmanov, R.J. Scholes, D.S. Schimel, T. Kirchner, J.C. Menaut, T. Seastedt, E. Garcia Moya, A. Kamnalrut and J.I. Kinyamario. 1993. Observations and modeling of biomass and soil organic matter dynamics for the grassland biome worldwide. *Global Biogeochemical Cycles* 7:785-809.

Parton, W.J., D.S. Schimel, C.V. Cole and D.S. Ojima. 1987. Analysis of factor controlling soil organic levels in Great Plains grasslands. *Soil Sci. Soc. Am. J.* 51:1173-1179.

Pastor, J. and W.M. Post. 1988. Response of northern forests to CO_2-induced climate change. *Nature* 334:55-58.

Peng, C.H. and M.J. Apps. 1998. Simulating carbon dynamics the Boreal Forest Transect Case Study (BFTCS) in central Canada: 2. Sensitivity to climate change. *Global Biogeochemical Cycles* 12:393-402.

Peng, C.H., M.J. Apps, D.T. Price, I.A. Nalder and D.H. Halliwell. 1998. Simulating carbon dynamics along the Boreal Forest Transect Case Study (BFTCS) in central Canada 1. Model testing. *Global Biogeochemical Cycles* 12:381-392.

Price, D.T. and M.J. Apps. 1993. Integration of boreal ecosystem-process models within a prognostic carbon budget model for Canada. *World Resour. Rev.* 5:15-31.

Rapaleee, G., S.E. Trumbore, E.A. Davidson, J.W. Harden and H. Veldhuis. 1998. Soil carbon stocks and their rate of accumulation and loss in a boreal forest landscape. *Global Biogeochemical Cycle* 12:687-701.

Schindler, D.W. 1998. A dim future for boreal waters and landscapes. *BioScience* 48:157-164.

Schindler, D.W. and S.E. Bayley. 1993. The biosphere as an increasing sink for atmospheric carbon: estimates from increased nitrogen deposition. *Global Biogeochemical Cycles* 7:717-733.

Siltanen, R.M., M.J. Apps, S.C. Zoltai, R.M. Mair and W.L. Strong. 1997. A soil profile and organic carbon database for Canadian forest and tundra mineral soils. Natural Resources Canada, Canadian Forest Service, Northern Forest Center, Edmonton, Alberta. Inf. Rep. Fo42-271/1997E. 37 pp.

Simpson, L.G., D.B. Botkin and R.A. Nisbet. 1993. The potential aboveground carbon storage of North American forests. *Water, Air, Soil Pollut.* 70:197-205.

Shugart, H.H. and T.M. Smith. 1996. A review of forest patch models and their application to global research. *Climatic Change* 34:131-154.

Smith, T.M. and H.H. Shugart. 1993. The transient response of terrestrial carbon storage to a perturbed climate. *Nature* 361:523-526.

Tarnocai, C. 1998. The amount of organic carbon in various soil order and ecological provinces in Canada. p. 81-92. In: R. Lal, J. Kimble and B.A. Stewart (eds.), *Soils and Global Change*. CRC Lewis Publishers, Boca Raton, FL.

Thompson, V.M, J.T. Randerson, C.M. Malmstrom and C.B. Field. 1996. Change in net primary production and heterotrophic respiration: How much is necessary to sustain the terrestrial carbon sink? *Global Biogeochmical Cycles* 10:711-726.

Townsend, A.R., B.H. Braswell, E.A. Holland and J.E. Penner. 1996. Spatial and temporal patterns in terrestrial carbon storage due to deposition of fossil fuel nitrogen. *Ecol. Appl.* 6:806-814.

Venevsky, S. and A. Shvidenko. 1997. Modeling of stand growth dynamics with a destructive stage. p. 28-32. Proceedings of the 7th Annual Conference of the International Boreal Forest Research Association, St. Petersburg, Russia.

Weber, M.G. and M.D. Flanningan. 1997. Canadian boreal forest ecosystem structure and function in a changing climate: impact on fire regimes. *Environmental Reviews* 5:145-166.

Weber, M.G. and B.J. Stock. 1998. Forest fires in the boreal forests of Canada. p. 215-233. In: J.M. Moreno (ed.), *Large Forest Fires*. Backbuys Publishers, Leiden, The Netherlands.

Woodward, F.I., T.M. Smith and W.R. Emanuel. 1995. A global primary productivity and phytogeography model. *Global Biogeochemical Cycles* 9:471-490.

Woodwell, G.M., F.T. Mackensie, R.A. Houghton, M. Apps, B. Gosham and E. Davidson. 1998. Biotic feedbacks in the warming of the Earth. *Climatic Change* 40:495-518.

Zar, J. H. 1984. *Biostatistical Analysis*. Prentice-Hall, Englewood Cliffs, NJ.

CHAPTER 36

Predicting Broad-Scale Carbon Stores of Woody Detritus from Plot-Level Data

M.E. Harmon, O.N. Krankina, M. Yatskov and E. Matthews

I. Introduction

Woody detritus in the form of dead tree parts such as boles, stumps, branches, and coarse roots is an important store of carbon in forest ecosystems (Harmon and Chen, 1992). Not only does this material represent a large and frequently overlooked pool (Harmon et al., 1986), but also it is a crucial component of heterotrophic respiration (Turner et al., 1996). In recent years methods to study the size and dynamics of these detritus pools have been developed and applied to various ecosystems (Harmon et al., 1999). Despite an increase in these plot-level efforts, however, no reliable inventory-based estimates exist at the regional, national, or global scales (Turner et al., 1996; Kurtz et al., 1992).

In this chapter we review plot level methods and then present a series of complementary methods that can be used to estimate potential steady-state and actual stores of woody detritus at regional to global scales. These include (1) correction and conversion factors for incomplete inventories; (2) dead:live wood expansion factors; (3) predictions of steady-state stores from input:decomposition rate-constant ratios; and (4) adjustments to include disturbance regimes.

II. Types of Woody Detritus

Woody detritus or debris takes many forms in forested ecosystems (Figure 1). The most useful distinctions are based on the size (length and diameter) and position (standing, downed, buried in soil) of the detritus. Other systems, such as those used in fire fuels, depend upon the moisture time lag, but are operationally divided by diameter (Fosburg, 1971). Unfortunately the use of these terms to describe woody detritus has been quite unstandardized, leading to major problems in comparisons. The most important distinction concerning size is between the coarse and fine fractions. This distinction often depends upon the ecosystem being examined, but a frequently used size "break point" is 10 cm diameter at the large end of a piece (Harmon and Sexton, 1996). Given the increasing number of pieces per unit area as diameter decreases, another useful size breakpoint is at 1-cm diameter. Woody branches, twigs, and bark pieces less than this diameter can be exceedingly numerous and are best treated as a special case of fine litter (in fact they usually are included in litterfall studies). In the case of dead roots, however, the cut-off between fine and coarse woody detritus is often at 1-cm diameter.

The term woody detritus or woody debris should be used to include all the forms of dead woody material above- and belowground. Aboveground woody detritus can be divided into coarse- or fine-

Figure 1. Examples of woody detritus in forest ecosystems: A. coarse, downed wood in an old-growth, *Pinus contorata*, forest; B. standing dead wood following wildfire in *Pinus contorta forest*; C. remnant, charred wood after intense wildfire (note that woody detritus was the only aboveground component remaining); D. woody slash left after timber harvest in Pseudosuga/Tsuga forests in 1930s; and E. stumps and other slash left after pasture conversion in Mexico tropical forests. (From Rosenberg et al., 1999. With permission.)

fractions. The minimum dimensions for coarse woody detritus are usually 10-cm diameter at the large end and 1.5 m in length. Pieces smaller than these size limits are usually considered fine woody detritus. Coarse fractions can in turn be divided into snags (or standing dead) and logs (or dead and downed). The separation of snags from logs is usually a 45° angle. In addition to snags, we also recognize stumps in managed settings. Vertical pieces resulting from natural processes should always be called snags. The term stump is used for short, vertical pieces that were created by cutting. Fine fractions can also be divided into suspended or downed fractions. In the case of suspended fine wood, one must distinguish between that attached to living woody plants and that attached to dead woody plants. Belowground woody detritus has been rarely studied (McFee and Stone, 1966), but we recommend this be divided into buried wood (very decayed material in the mineral soil or forest floor) and dead coarse roots. The distinction between these two types of belowground wood is primarily based on whether you can tell its origin (similar to the distinction between O1 and O2 layers in the organic soil layers).

III. Plot Level Measurements

Harmon et al. (1999), Harmon and Sexton (1996), and Harmon et al. (1986) provide thorough reviews of plot-level methods including the relative merits of different approaches — we will therefore provide only a brief review of plot level protocols here.

A. Mortality or Input Rates

Inputs of woody detritus take several forms: (1) that associated with tree mortality and (2) that associated with pruning of branches and roots. The majority of coarse wood inputs are associated with tree mortality. Tree mortality rates are best measured using permanent plots, although various reconstruction methods have also been used to estimate long-term rates (McCune et al., 1988). The only parameters required to estimate woody detritus inputs are the diameter at breast height and species of the dead tree at the time of plot remeasurement. Biomass equations are then used to compute the amount added. Coarse woody detritus inputs are highly variable in both time and space. As a first approximation, data should not be dependent upon the inclusion or exclusion of a single tree. If one desires to be within 10% of the live stem mortality rate (defined here as the percentage of tree stems dying each year), then at least 10 trees have to die within some combination of area and time. On the basis of observed rates of mortality 5 hectare-years would appear to be the minimum sample required to generate reliable data for temperate and tropical hardwood forests. In contrast, the minimum sample for temperate conifer forests would appear to be 10 hectare-years.

Although one could measure the input of fine woody detritus directly, it is difficult given that many trees die standing. Therefore inputs of fine woody detritus can be indirectly estimated using allometric equations. These estimates can be based on the diameters of the trees that died in the mortality surveys (Sollins 1982). A relatively small fraction of the downed fine woody detritus is input from branches that have snapped off by windstorms or ice and snow damage. These inputs can be measured on fixed area plots (generally <4 m² has been used) as long as one separates freshly killed branches from decayed branches that have fallen from snags or from dead branches attached to live trees (Swift et al., 1976). Neither of the latter forms are actually new inputs of woody detritus (i.e., a change from live to dead wood); rather, these represent a change in the form of woody detritus (i.e., suspended to downed). Finally, it is also possible to use the allometric approach to estimate the mass input of dead coarse roots, although no one has tested the accuracy of this method.

B. Decomposition Rates

There are several methods to determine rates that woody detritus decomposes, forms soil organic matter, and immobilizes or releases nutrients. Two frequently used approaches are: (1) chrono-sequences which give a short-term snap shot of processes and (2) a time series approach which is a long-term effort yielding excellent resolution of temporal patterns and processes. In a chronosequence one ages as many pieces as possible in various states of decay and examines how a parameter such as density changes through time. Dates can be determined from fall scars, seedlings, living stumps, and records of disturbance (e.g., fire, insect outbreak, windstorm, thinning). This approach has been used extensively for coarse woody detritus (e.g., Graham, 1982; Grier, 1978; Harmon et al., 1987; Means et al., 1987; Sollins et al., 1987), but also can be used for downed fine woody detritus (Erickson et al., 1985) and dead coarse roots (Fahey et al., 1988). The interpretation of chronosequence data varies depending upon its use: (1) conversion of volume measurements into mass or nutrient stores; and (2) to determine the rate mass is lost or nutrients are immobilized or released. If the aim is to use a decay chronosequence to estimate rates of mass loss or nutrient release, then the data must be adjusted for past fragmentation losses to estimate these rates correctly (Harmon and Sexton, 1996). Although chronosequences produce results quickly, there are serious temporal resolution problems caused by errors in dating and estimates of initial conditions of the dead trees. A time series circumvents these problems by examining how a cohort of pieces progresses through time, thereby avoiding the substitution of space for time. Although the method requires substantial investments in effort and time, it lends itself nicely to process studies. In addition to examining a chronosequence of pieces, one can also indirectly estimate decomposition rates of woody detritus from a chronosequence of different aged stands (e.g., Gore and William, 1986; Spies et al., 1987) by assuming each stand-creating disturbance left a similar amount of material. In many cases this assumption is not justified and can lead to significant uncertainty concerning decomposition rates. Finally, the ratio of input to stores can be used to indirectly estimate decomposition rate constants of woody detritus assuming the pool is in steady-state (Sollins, 1982). This is subject to the same errors as for fine litter, compounded by the fact both inputs and stores of woody detritus are highly variable.

C. Fine Woody Detritus Stores

Several alternative methods exist for measuring fine woody detritus stores (Harmon et al., 1999; Harmon and Sexton, 1996). The most straightforward is harvesting and weighing material within small plots (<4 m^2). Downed fine woody detritus can also be estimated using planar transects in which the number of pieces in size classes is recorded and then converted to stores using the mean diameter and bulk density of the size class. Unfortunately, the latter two parameters are rarely measured or reported (see Harmon and Sexton, 1996); therefore, stores estimated by this method have questionable accuracy. Moreover, although the planar transect estimates the volume of downed, surface wood, it does not measure other important forms of woody detritus including dead branches or dead coarse roots.

D. Coarse Woody Detritus Stores

For coarse woody detritus (<10-cm diameter and < 1 m long) it is impractical to remove and weigh pieces. Therefore, for downed logs, standing dead trees, and stumps it is more usual to record piece dimensions within fixed area plots (Harmon et al., 1987) or along planar transects (Warren and Olsen, 1964; Van Wagner, 1968; Brown, 1974) to estimate volume. This parameter is then converted to mass

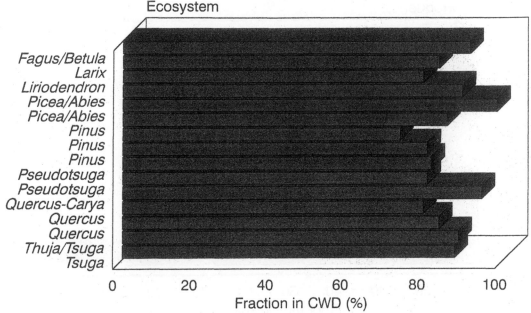

Figure 2. Fraction of total aboveground woody detritus in the coarse fraction. (Modified from Harmon et al., 1986.)

and/or carbon using decay class specific bulk density and carbon concentration values. If planar transects are used, one must measure the volume of standing dead trees and stumps with a plot-based system.

IV. Correction, Conversion, and Expansion Factors

Measurements are rarely taken or reported in the forms required for broad-scale analysis. It is therefore necessary to either correct the data for missing components (e.g., coarse wood to total woody detritus), to convert one set of units to another (e.g., volume to mass), or to base estimates of woody detritus on other related pools. We term each of these correction, conversion, and expansion factors, respectively.

A. Correction Factors

Inventories of woody detritus are rarely complete and therefore corrections to total stores are usually required. While making corrections, it is important to bear in mind that some forms of woody detritus are highly correlated while others are not. For example, size classes of woody detritus are often highly correlated (Figure 2). This is probably because as trees die, the proportions of size classes reflect the proportions of size classes of living tree parts (Leopold, 1971). Thus it is not unusual to find that fine downed wood comprises ≈20% of the total downed woody detritus (Harmon et al., 1986) just as it makes up ≈20% of the live biomass. Similar proportions might be expected for suspended fine wood versus snags. There is also probably a good correlation between aboveground woody detritus and dead coarse roots as this is largely controlled by the structure of living trees. Perhaps the worst correlation between forms of woody detritus is for standing or suspended material and downed

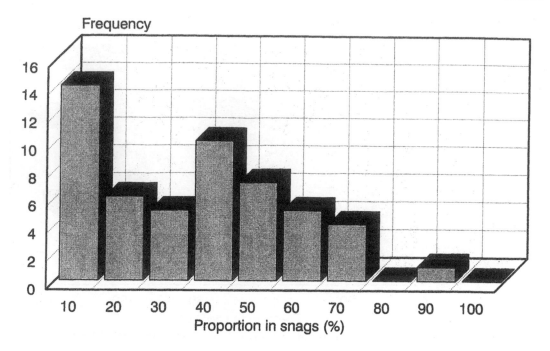

Figure 3. Frequency distribution of the fraction of coarse woody detritus in snags (i.e., standing dead trees) for *Pseudotsuga/Tsuga* forests of the Pacific Northwest. (Based on Wright, 1998.)

material. Even within a single forest type, such as the frequently studied *Pseudotsuga/Tsuga* forest of the Pacific Northwest, it is not unusual to find that snags can make up as little as 2% to as much as 98% of the coarse dead wood biomass (Figure 3). Another correction that is difficult to make is to account for missing decay classes. Many woody detritus inventories are focused on estimating merchantable volume and thus only consider the less decomposed material. Unfortunately the fraction of undecayed wood can range from being high immediately following a disturbance to very low in young forests (Figure 4).

B. Conversion Factors

The most commonly used conversion factor is to convert volume to mass or carbon. The volume of large woody detritus estimated from the dimensional data or along transects can be converted to mass and nutrient stores using decay class specific bulk density data for various stages of decay (Table 1). These values can be taken from the literature, although potential errors exist by not using site-specific values. Using an existing or establishing a new decay class system requires that physical characteristics be used to distinguish between decay classes. These include: presence of leaves, twigs, branches, bark cover on branches and boles, sloughing of wood, collapsing and spreading of log (indicating the transition from round to elliptic form), degree of soil contact, friability or crushability of wood, color of wood, and if the branch stubs can be moved.

Mortality is another parameter often requiring conversion factors as it is not unusual for mortality to be reported as stems, basal area, volume, or mass lost per unit time. Fortunately the latter three variables are equivalent when expressed in relative terms as they are all correlated (i.e., basal area is

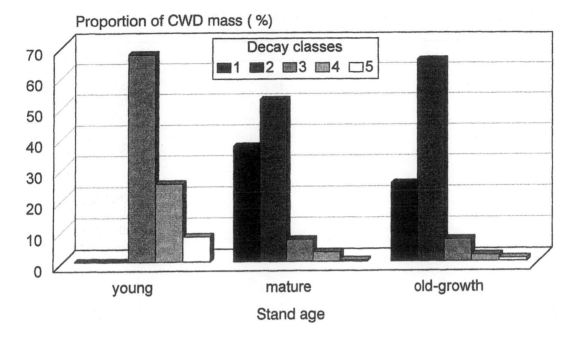

Figure 4. Fraction of coarse woody detritus in 5 decay classes for *Picea* forests in Northwestern Russia. Class 1 is the least and class 5 is the most decomposed woody material. Based on data from Krankina and Harmon (1999).

often used as a proxy for volume and mass is volume times wood density). The number of stems is not, however, well correlated to either volume or mass. Unfortunately the bulk of mortality numbers are expressed as relative stems lost per unit time. We assume here a conversion factor of 1; exact conversion factors from stem to mass input awaits more studies that report mortality in both forms.

C. Expansion Factors

The most commonly used expansion factor used is the ratio of dead to live wood (Harmon and Chen, 1992; Harmon et al., 1993; Matthews, 1997). These ratios are usually developed using field measurements of both dead and live wood mass (Krankina and Harmon, 1995). The basic logic behind these ratios is that given our much greater knowledge of live wood stores, this ratio allows one to make an estimate from that database. The major problem with these ratios is that they are best used for older stands as the ratio ranges from near infinity for recently disturbed forests (i.e., where little live wood exists) to a fairly stable number less than 1 in old-growth forests to close to 0 in forests where timber salvage and thinning are practiced (Krankina and Harmon, 1995). Although convenient, these ratios have not been estimated for more that a handful of ecosystems. Based on a sampling of 50 stands in the Pacific Northwest (Wright, 1998), it would appear that dead:live ratios are distributed as negative exponential to log-normal distributions (Figure 5).

Table 1. Density of woody detritus by decay classes and species (mean and standard error in parentheses) for Northwest Russia; class 1 is the least and class 5 the most decomposed.

Species	Decay class	Density (Mg m^{-3})
Birch		
	1	0.480 (0.010)
	2	0.442 (0.016)
	3	0.235 (0.023)
	4	0.125 (0.020)
	5	0.094 (0.003)
Spruce		
	1	0.347 (0.017)
	2	0.309 (0.015)
	3	0.207 (0.018)
	4	0.110 (0.023)
Pine		
	1	0.384 (0.018)
	2	0.311 (0.019)
	3	0.236 (0.012)
	4	0.111 (0.018)
	5	0.108 (0.004)

(After Krankina and Harmon, 1999.)

In addition to being a convenient number, use of the dead:live ratio has a theoretical basis. If it is assumed that the forest is approaching a steady-state then the inputs equal the outputs (Olson, 1963). The outputs are defined as:

$$\text{Output} = k \, \text{Dead}_{ss} \tag{Eq. 1}$$

where k is the decomposition rate-constant and Dead_{ss} is the mass of dead material are steady-state. Therefore:

$$\text{Dead}_{ss} = \text{Input}/k \tag{Eq. 2}$$

If we relativize the inputs by assuming the steady-state live biomass (Live_{ss}) is equal to 1 then:

$$\text{Dead}_{ss} = m/k \tag{Eq. 3}$$

where m is the mortality rate-constant. As Live_{ss} is equal to 1, equation 3 is equivalent to the dead:live ratio. This allows one to take advantage of the great number of mortality and decomposition rate-constants that have been reported.

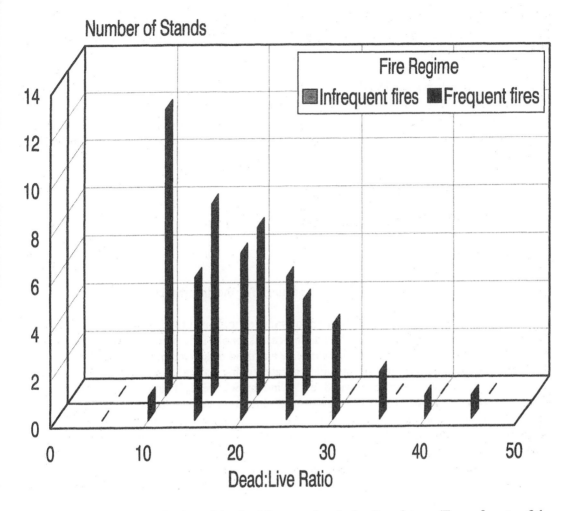

Figure 5. Frequency distribution of the dead:live wood ratio for *Pseudotsuga/Tsuga* forests of the Pacific Northwest. (Adapted from Wright, 1998.)

V. Estimating the Effects of Disturbance

Although the steady-state is a useful concept, it is rare that ecosystems reach this condition. At a regional scale a more realistic condition includes areas in various ages since the time of last catastrophic disturbance. By modeling the relative change of live and dead wood stores over succession one can estimate how to adjust the steady-state stores.

As with using mortality and decomposition rate-constants to estimate the steady-state dead:live wood ratio, we can simplify the calculations by setting $Live_{ss}$ equal to 1.0. The live biomass ($Live_t$) at any time (t) can be described by a Chapman-Richards function:

$$Live_t = (1 - exp[-B_0 t])^{B_1} \qquad \text{Eq. (4)}$$

where B_0 and B_1 are parameters describing the rate of live biomass accumulation. Changes in dead wood mass relative to the steady-state live mass can be calculated from the input (I) versus the decomposition loss (k):

$$\Delta \, dead_t = I - k \, Dead_{t-1} \hspace{4cm} \text{Eq. (5)}$$

Inputs to the dead wood pool occur either as "normal" mortality associated with competition, wind, an other non-catastrophic forms of death or because of disturbances that kill the entire forest. In the former case inputs are:

$$I = m \, Live_t \hspace{6cm} \text{Eq. (6)}$$

where m is the mortality rate constant. In the latter case inputs are:

$$I = (1 - d) \, Live_t \hspace{5cm} \text{Eq. (7)}$$

where d is the proportion of the live mass remaining after the disturbance.

This relatively simple model can be used to estimate the dynamics of dead and live wood after a catastrophic disturbance (Figure 6). The simplest case is for old-field succession where both live and dead mass start at 0 (Figure 6a). In this case live and dead mass accumulation parallel each other. A more complicated situation occurs after a catastrophic natural disturbance. Assuming the disturbance removes a minimum of wood (e.g., wind throw), $Dead_0$ is equal to the sum of 1 and the dead:live ratio (Figure 6a). This is followed by a monotonic decline to the steady-state dead:live ratio. When the disturbance removes a fraction of the woody mass (e.g., timber harvest) $Dead_0$ can range anywhere between 0 and 1+dead:live. Current timber harvest, for example, might remove 70% of the live biomass leading to a value of $Dead_0$ of 0.5 (Figure 6b). Dead mass declines and in some cases may drop below the dead:live ratio at steady-state. This is because the replacement of dead wood lags behind decomposition in the middle stages of succession (Harmon et al. 1986, Spies et al., 1987). Perhaps the most complicated case is when forests are converted to intensive, short-rotation forestry. Here the live mass does not recover to the steady-state level and a large fraction of the mortality is removed as intermediate timber harvest in thinning and salvage. This leads to a decrease in the dead:live ratio to a value much lower than the steady-state value. The final successional pattern, not shown, is found after forest clearing for agriculture. In this case dead mass continues to decline as non-woody plants dominate the ecosystem.

Using these successional patterns to adjust steady-state estimates requires information about the age structure of patches created by catastrophic disturbances. Although it would be best if the specific age structure was used, in many regions this is not known. Therefore one may assume that the proportion of each age class is equal to the inverse of the mean interval between disturbances in a region. The mean dead:live ratio for each region with natural disturbance is calculated as:

$$Dead:Live_{mean} = \mathbf{D'A} \hspace{5cm} \text{Eq. (8)}$$

where $\mathbf{D}$ is a vector describing the dead:live ratio of each age class and $\mathbf{A}$ is a vector describing the proportion of each disturbance age class. It is also possible to use this approach to estimate the carbon stores of live woody biomass with natural disturbance regimes by substituting $\mathbf{L}$, a vector describing the relative live carbon store for each age class, for vector $\mathbf{D}$.

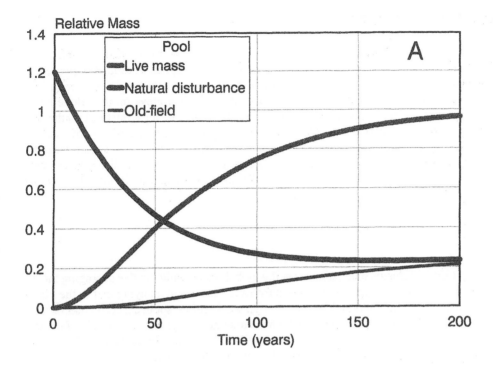

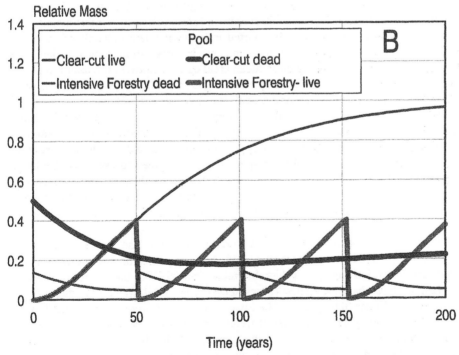

Figure 6. Change in woody detritus stores relative to live woody stores following catastrophic disturbances. A. Temporal patterns during old-field succession, or following a catastrophic natural disturbance or B. various types of timber harvest.

VI. Application to Broad-Scales

Having described how broad-scale calculations can be made we will demonstrate their use by estimating global level stores of woody detritus. We present a progression of estimates, proceeding from steady-state estimates based on stores averages to those based on observed dead:live ratios to those based on mortality and decomposition rate-constants. We then move on to include natural disturbance regimes considered typical of large regions and finish by making reductions for the forest area historically removed for agriculture. Our calculations are based on a compilation of data from the literature and from unpublished values from our research (http://www.fsl.orst.edu/lter/pubs/-globalcwd.htm). For biomes without reported data, we assume values from similar biomes. Other assumptions are presented with the actual calculations. The biome areas and undisturbed live biomass used in this analysis follow those of Matthews (1997).

A. Steady-State Estimates

The most straightforward steady-state estimates at the global scale would involve averaging the old-growth or primary ecosystem woody detritus mass for each biome and applying that to the biomes total area. As most studies only report downed material, we therefore applied a correction factor of 25% to all biomes except evergreen forests in which we assumed snags comprised 40% of the total coarse woody detritus. We also increased all numbers by 20% to account for the fact that fine woody detritus, especially dead coarse roots, are not usually reported. After these correction factors were applied, mean carbon stores in all forms of woody detritus range from as little as 1.5 Mg C ha^{-1} in evergreen woodland to as much as 55 Mg C ha^{-1} in evergreen forests (Table 2). Unfortunately, there are three biomes in which numbers have not been reported and several others with only a few studies. Nonetheless, the generally accepted pattern that stores increase in forests from tropical to deciduous to evergreen forests is supported. Applying these mean areal carbon stores to the total biome area yields as undisturbed forest steady-state global total of 157 Pg C (Table 3). This is in comparison to a total of a total of 552 Pg C of live biomass or a global dead:live ratio of 0.28.

Our second estimate of steady-state woody detritus stores is based on observed dead/live ratios. Unfortunately, the database for these calculations is even scantier than for stores, with a total of 35 observations. From this limited sample it would appear that tropical and deciduous forests are relatively similar with values of 0.14 and 0.15, respectively. Evergreen forests have the largest sample, with a mean of 0.25; evergreen woodlands appear to be even higher, although there are too few samples to have confidence in this value. Applying these number to the undisturbed live woody carbon stores gives an undisturbed forest carbon store in woody detritus of 114 Pg C which indicates a mean dead:live ratio of 0.21.

Our final estimate of steady-state woody detritus stores is based on dead:live ratios calculated from the mortality:decomposition rate-constant ratio. The first component of this calculation is the mortality rate-constant for which there is considerably more data (N=108) than stores or dead:live ratios (Table 2). Tropical forest have the highest values (0.0167 year^{-1}) followed by deciduous (0.012 year^{-1}) and then evergreen forests (0.01 year^{-1}). The lowest value is for evergreen woodland, with a mortality rate-constant of 0.004 year^{-1}. The overall trend in mortality rate-constants appears to be an increase with productivity. We therefore used this general relationship to "guestimate" values for the biomes without data. Decomposition rate-constants, the second component of this calculation also decreased from tropical (0.176 year^{-1}) to deciduous (0.080 year^{-1}) to evergreen forests (0.032 year^{-1}). Deciduous shrubland appears to have the highest decomposition rate-constant, possibly due to the presence of termites, which rapidly remove woody material. Although tropical forests have the highest decomposition rate-constants of any major biome, the distribution of values appears bimodal

Table 2. Area, mean undisturbed woody biomass, woody detritus, dead:live ration, mortality, and decomposition rates for biomes dominated by woody plants (numbers in bold are from literature)

Biome[a]	Area (ha × 10^5)	Average undisturbed live biomass (Mg C ha^{-1})	Woody detritus observed (Mg C ha^{-1})	Dead:live ratio	Mortality rate-constant (year^{-1})	Decomposition rate-constant (year^{-1})
Tropical forest	1323	116.4	12 (1.5)[b] 19	0.14 (0.08) 3	0.0167 (0.001) 40	0.176 (0.050) 16
Evergreen forest	1631	109.7	55 (8.4) 40	0.25 (0.05) 14	0.010 (0.001) 44	0.032 (0.003) 33
Deciduous forest	1722	106.5	21.9 (3.5) 33	0.15 (0.03) 16	0.012 (0.002) 22	0.080 (0.013) 23
Evergreen woodland	497	61.2	1.5 (NA)[c] 2	0.36 (NA) 2	0.004 (NA) 2	0.01 (NA) 0
Deciduous woodland	735	45.6	2 (NA) 0	0.36 (NA) 0	0.004 (NA) 0	0.02 (NA) 1
Evergreen shrubland	226	27.0	2.8 (NA) 0	0.50 (NA) 0	0.005 (NA) 0	0.01 (NA) 0
Deciduous shrubland	145	28.0	2.8 (NA) 1	0.25 (NA) 0	(NA) 0	1.06 (NA) 1
Xeromorphic formations	3595	22.9	2.8 (NA) 0	0.25 (NA) 0	0.010 (NA) 0	0.03 (NA) 0

[a]Definitions of biomes follow Matthews (1983); [b]Mean (standard error), N, if N = 0 then the value is estimated; [c]NA is not applicable.

Table 3. Undisturbed steady-state estimates of total woody detritus carbon stores for biomes dominated by woody plants

Biome[a]	Observed woody detritus (Pg C)	Observed dead:live ratio (Pg C)	Mortality:decomposition-rate constant ratio (Pg C)	Adjusted for natural disturbances (Pg C)
Typical forest	15.9	20.5	13.9	13.3–14.0
Evergreen forest	89.7	42.5	53.2	50.1–61.7
Deciduous forest	37.7	22.1	22.0	22.0–21.3
Evergreen woodland	0.7	8.7	9.7	13.0–15.7
Deciduous woodland	1.5	9.6	10.7	14.3–17.3
Evergreen shrubland	0.6	2.4	2.4	2.8–3.3
Deciduous shrubland	0.4	0.8	0.3	0.4–0.5
Xeromorphic formations	10.1	7.5	10.0	10.1–13.4
Global total	**156.6**	**114.2**	**122.3**	**131.1–148.4**

[a]Definitions of biomes follow Matthews, 1983.

(Figure 7), with a peak at <0.04 year^{-1} and another at >0.12 year^{-1}. This may be a reflection of two groups of species, one containing compounds toxic to fungi and insects in their heartwood and a second group that has little decay-resistance. Evergreen and deciduous ecosystems appear to have unimodal distributions of decomposition rate-constants.

Dead:live ratios modeled from mortality and decomposition rate-constants appear to be generally similar to observed values (Figure 8). Given the lack of a rigorous global database it is difficult to determine which is correct and whether the slightly lower ratio for tropical forests or the higher ratio for evergreen forests estimated by rate-constants is correct or not. The total woody detritus carbon store using this method is estimated to be 122 Pg C and the mean dead:live ratio is 0.22. As this estimate is based on the largest database we will use this in all subsequent calculations that adjust for disturbance effects.

B. Natural Disturbance Effects

We used equations 4 to 8 to estimate the influence of natural catastrophic disturbances on woody detritus stores. We used a range of mean intervals between these disturbances with evergreen forests having the most frequent and tropical forests the least frequent disturbances (Table 4). We also assumed that tropical forests would recover their biomass the fastest, while woodlands and evergreen shrublands were the slowest. We also assumed that the lag required to start producing biomass increased as site severity increased (i.e., parameter B_1). Adjusting for these disturbance effects gener-

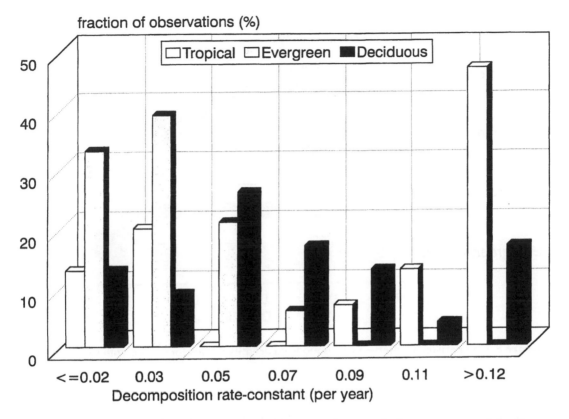

Figure 7. Frequency distribution of decomposition rates from tropical, evergreen, and deciduous forests.

ally increases the dead:live ratio, however, for tropical and deciduous forest the effect is minimal. This is probably due to the long interval between disturbances which allows woody detritus to stay near the steady-state value for much of succession. In contrast, in biomes where decomposition is slow and the interval between disturbances is short the disturbance adjusted dead:live ratio can be much higher than the non-disturbance ratio. Using these adjusted ratios yields a global store of woody detritus ranging from 131 to 148 Pg C, for the longest and shortest intervals between disturbance, respectively. Applying the same logic to live biomass yields a range of 344 to 420 Pg C. Disturbance therefore creates dead wood at the expense of live wood (Figure 9).

C. Agricultural Removals

Assuming that woody detritus in agricultural lands cleared prior to the 1950s has completely decomposed, we can adjust for this effect by reducing the land area for each biome that is now in agricultural land use. This would indicate that there was 113 to 129 Pg C in woody detritus in the middle of this century. For live biomass a range of 295 to 362 Pg C is indicated. We have not calculated the effects of post 1950s agricultural clearing, but is has probably influenced live stores more than woody detritus stores.

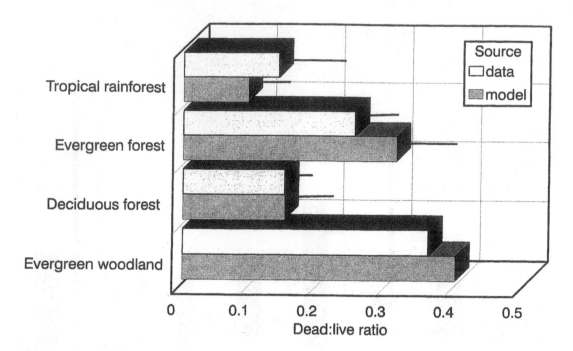

Figure 8. Comparison of observed versus modeled dead:live ratios for biomes with values reported.

Table 4. Parameters used to model the effects of natural disturbance on woody detritus stores (see Equations 4 to 8)

Biome[a]	Disturbance interval[b] (years)	B_0 (year^{-1})	B_1 (dimensionless)	Steady-state dead:live ratio	Disturbance adjusted dead:live ratio
Tropical forest	500–1000	0.033	2	0.095	0.095–0.096
Evergreen forest	150–300	0.015	2	0.313	0.321–0.363
Deciduous forest	500–750	0.015	2	0.150	0.150–0.152
Evergreen woodland	200–300	0.01	3	0.400	0.535–0.646
Deciduous woodland	200–300	0.01	3	0.400	0.535–0.646
Evergreen shrubland	200–300	0.01	4	0.500	0.566–0.683
Deciduous shrubland	100–200	0.033	2	0.100	0.127–0.157
Xeromorphic formations	100–200	0.015	3	0.333	0.335–0.445

[a]Definitions of biomes follow Matthews, 1983; [b]after Harmon et al., 1993.

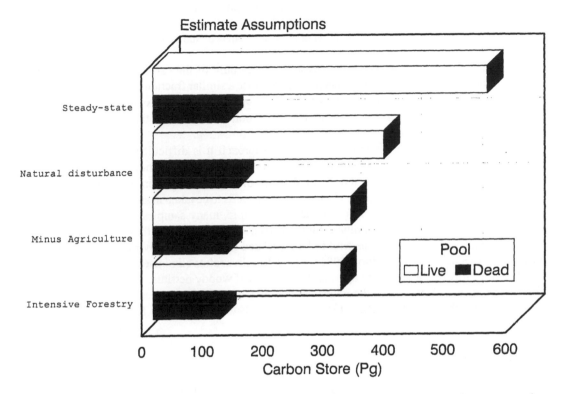

Figure 9. Comparison of live and woody detritus carbon stores for various assumptions concerning disturbance.

D. Intensive Forestry

Intensive forestry practices of short-rotations, thinning, and dead wood removal are estimated to be occurring in 10% of the worlds forest land (Dixon et al., 1994). Assuming the majority of this is taking place in evergreen forests would mean that 36% of those forest are currently under intensive management. We modeled this effect using a 50-year rotation length and assuming 50% of the mortality is harvested, and 50% of the live carbon is removed at the time of harvest. This indicates a dead:live ratio of 0.075 relative to the steady-state live mass of 1.0. That is instead of storing an equivalent of ≈30% of the maximum live carbon, the intensively managed forest stores only 7.5%. Moreover, the average live mass is reduced to 11% of the maximum. Applying these value to evergreen forests indicates the current global woody detritus store is 104 to 118 Pg C. Using a similar approach for live biomass yields a total of 283 to 340 Pg C when intensive forestry is taken into account. Using completely different methods Dixon et al. (1994) estimated a current live wood store of 359 Pg C globally. Unfortunately, they do not give a woody detritus estimate.

VII. Discussion

Despite the wide range of estimates that we calculated (113 to 156 Pg C) they are considerably narrower than those made by Harmon and Chen (1992) and Harmon et al. (1993) which spanned an order of magnitude (50 to 500 Pg C). They are all higher than the 66 to 80 Pg C estimated by Matthews (1997). These earlier estimates applied many of the same methods, but the difference is that new estimates of mortality and decomposition rate-constants have been published in the intervening years.

Regardless of the method all estimates indicate the bulk of the carbon is stored in evergreen forests. Together tropical and deciduous forests comprise a similar fraction to evergreen forests. All other biomes together appear to contribute ≈10 to 25% of the total global woody detritus store. Given the scarcity of data of any sort for these biomes, some future investment in these areas might improve global estimates significantly.

Given small number of samples used in each approach it is difficult to rate them as to their reliability. It is very likely, however, that the direct biomass approach is more prone to overestimate than the ratio approaches. First, the reported stores of woody detritus in ecosystems are neither random nor systematic. As with many fields, sampling tends to be taken in areas where large amounts of woody detritus occurs. In conifer forests, for example, many samples have been taken in the Pacific Northwest. We tried to eliminate this effect by averaging studies within regions first. Second, direct estimates of woody detritus tend to be correlated to the mass of living wood and hence the productivity (Krankina and Harmon, 1995). Using dead:live ratios eliminates some of this bias so that small stature forests contain more appropriate amounts of woody detritus. Our method of adjusting for disturbance effects makes theoretical sense, and increased stores as expected. Its primary shortcoming is the lack of specific information on the disturbance regime for many biomes.

VIII. Conclusions

In recent years, methods to study the size and dynamics of these detritus pools have been developed and applied to various ecosystems. Despite an increase in plot-level efforts, no reliable inventory-based estimates exist at the regional, national, or global scales. We presented a series of complementary methods that can be used to estimate potential steady-state and actual stores at regional to global scales. These include: (1) correction and conversion factors for incomplete inventories, (2) dead:live wood expansion factors, (3) predictions of steady-state stores from input:decomposition rate ratios, and (4) adjustments to include disturbance regimes. These methods indicate that global stores of carbon in woody detritus could be 114 to 157 Pg in the absence of disturbance, 131 to 148 Pg with natural disturbance, and 114 to 129 Pg after pre-1950s deforestation is considered.

The estimates of woody detritus stores we present are quite crude, but they strongly indicate several points. Considerable woody detritus is present at the global level, with the bulk of it stored in evergreen forests. Globally these estimates exceed those for surface litter (50 to 60 Pg C) and are roughly half of that estimated for peat. Although smaller than the commonly accepted store for soils (≈1500 Pg C), woody detritus is highly sensitive to natural disturbance and management by humans. This analysis indicates better inventories of dead woody detritus and measurement of its dynamics would considerably improve the overall understanding the role of carbon stores in the terrestrial biota.

Acknowledgments

This research was sponsored by the National Science Foundation grants DEB-9632929 and DEB-9632921 and in part by USDA-CSRS-NRICGP (Contract #95-37109-2181). We also thank the Oregon State University Forest Science Data Bank (FSDB) and Gody Spycher for assistance in managing and processing the data used in this paper.

References

Brown, J.K. 1974. Handbook for inventorying downed woody material. USDA Forest Service General Technical Report INT-16.

Dixon, R.K., S. Brown, R.A. Houghton, A.M. Solomon, M.C. Trexler and J. Wisniewski. 1994. Carbon pools and flux of global forest ecosystems. *Science* 263:185-190.

Erickson, H.E., R.L. Edmonds and C.E. Peterson. 1985. Decomposition of logging residue in Douglas-fir, western hemlock, Pacific silver fir, and ponderosa pine ecosystems. *Can. J. For. Res.* 15:914-921.

Fahey, T.J., J.W. Hughes, P. Mou and M.A. Authur. 1988. Root decomposition and nutrient flux following whole-tree harvest in Northern hardwood forest. *For. Sci.* 34:744-768.

Fosburg, M.A. 1971. Climatological influences on moisture characteristics of dead fuel: theoretical analysis. *For. Sci.* 17:64-72.

Gore, J.A. and A.P. William. 1986. Mass of downed wood in north hardwood forests in New Hampshire: potential effects in forest management. *Can. J. For. Res.* 16:335-339.

Graham, R.L. 1982. Biomass dynamics of dead Douglas-fir and western hemlock fir boles in mid-elevation forests of Cascade Range. Ph.D. dissertation. Oregon State University, Corvallis.

Grier, C.C. 1978. A *Tsuga heterophylla-Picea sitchensis* ecosystem of coastal Oregon: Decomposition and nutrient balance of fallen logs. *Can. J. For. Res.* 8:198-206.

Harmon, M.E., K.J. Nadelhoffer and J.M. Blair. 1999. Measuring decomposition, nutrient turnover, and stores in plant litter. p. 202-240. In: G.P. Robertson, C.S. Bledsoe, D.C. Coleman and P. Sollins (eds.), *Standard Soil Methods for Long Term Ecological Research*. Oxford University Press, New York.

Harmon, M.E., S. Brown and S.T. Gower. 1993. Consequences of tree mortality to the global carbon cycle. p. 167-178. In: T.S. Vinson and T.P. Kolchugina (eds.), *Carbon Cycling in Boreal and Subarctic Ecosystems*. Corvallis, OR. EPA Rept No. 600-93/084.

Harmon, M.E. and H. Chen. 1992. Coarse woody debris dynamics in two old-growth ecosystems. *Bioscience* 41:604-610.

Harmon, M.E., K. Cromack, Jr. and B.G. Smith. 1987. Coarse Woody Debris in mixed-conifer forests, Sequoia National Park, California. *Can. J. For. Res.* 17:1265-1272

Harmon, M.E., J.F. Franklin, F. Swanson, P. Sollins, S.V. Gregory, J.D. Lattin, N.H. Anderson, S.P. Cline, N.G. Aumen, J.R. Sedell, G.W. Lienkaemper, K. Cromack, Jr. and K.W. Cummins. 1986. Ecology of Coarse Woody Debris in Temperate Ecosystem. *Adv. Ecol. Res.* 15:133-302.

Harmon, M.E. and J. Sexton. 1996. Guidelines for measurements of woody detritus in forest ecosystems. LTER Network Pub. No. 20. University of Washington, Seattle. 91 pp.

Krankina, O. N. and M. E. Harmon. 1999. Nutrient stores and dynamics of woody detritus in a boreal forest: modeling potential implications at the stand level. *Can. J. For. Res.* 29:20-32.

Krankina, O.N. and M.E. Harmon. 1995. Dynamics of the dead wood carbon pool in northwestern Russian boreal forests. *Water Air Soil Pollut.* 82:227-238.

Kurz, W.A., M.J. Apps, T.M. Webb and P.J. McNamee. 1992. The carbon budget of the Canadian forest sector: Phase I. Info. Rept. NOR-X-326. Forestry Canada, Edmonton, Alberta, Canada.

Leopold, L.B. 1971. Trees and streams: the efficiency of branching patterns. *J. Theor. Biol.* 31:339-354.

Matthews, E. 1997. Global litter production, pools, and turnover times: Estimates from measurement data and regression models. *J. Geophys. Res.* 102:771-800.

Matthews, E. 1983. Global vegetation and land use: new high-resolution data bases for climate studies. *J. Clim. and App. Meteorol.* 22:474-487.

McFee, W.W. and E.L. Stone. 1966. The persistence of decaying wood in humus layers of northern forests. *Soil Sci. Soc. Am. J.* 30:513-516.

McCune, B., C.L. Cloonan and T.V. Armentano. 1988. Tree mortality and vegetation dynamics in Hemmer Woods, Indiana. *Am. Mid. Nat.* 120:416-431.

Means, J.E., K. Cromack, Jr. and P.C. Macmillan. 1987. Comparison of decomposition models using wood density of Douglas-fir logs. *Can. J. For. Res.* 15:1092-1098.

Olson, J.S. 1963. Energy stores and the balance of producers and decomposers in ecological systems. *Ecology* 44:322-331.

Rosenberg, N.J., R.C. Izaurralde and E.L. Malone. 1999. *Carbon Sequestration in Soils: Science, Monitoring and Beyond.* Batelle Press, Columbus, OH. 201 pp.

Sollins, P. 1982. Input and decay of coarse woody debris in coniferous stands in Western Oregon and Washington. *Can. J. For. Res.* 12:18-28.

Sollins, P., S.P. Cline, T. Verhoeven, D. Sachs and G. Spycher. 1987. Patterns of log decay in old-growth Douglas-fir forest. *Can. J. For. Res.* 17:1585-1595.

Spies, T.A., J.F. Franklin and T.B. Thomas. 1988. Coarse woody debris in Douglas-fir forests of western Oregon and Washington. *Ecology* 69:1689-1702.

Swift, M.J., I.N. Healey, J.K. Hibbard, J.M. Sykes, V. Bampoe and M.E. Nesbitt. 1976. The decomposition of branch-wood in the canopy and floor of a mixed deciduous woodland. *Oecologia* 26:139-149.

Turner, D.P., G.J. Koerper, M.E. Harmon and J.J. Lee. 1995. A carbon budget for forests of the conterminous United States. *Ecol. App.* 5:421-436.

Van Wagner, C.E. 1968. The line intercept method in forest fuel sampling. *For. Sci.* 14:20-26.

Warren, W.G. and P.F. Olsen. 1964. A line transect technique for assessing logging waste. *For. Sci.* 10:267-276.

Wright, P. 1998. Coarse woody debris in two fire regimes of the central Oregon Cascades. Master's thesis. Oregon State University, Corvallis.

CHAPTER 37

Soil C Dynamics: Measurement, Simulation and Site-to-Region Scale-Up

R.C. Izaurralde, K.H. Haugen-Kozyra, D.C. Jans, W.B. McGill,
R.F. Grant and J.C. Hiley

I. Introduction

Soil organic matter (SOC) has long been recognized as a primary soil property with an essential role in soil conservation and sustainable agriculture (Johnston, 1994). SOC participates prominently in the global carbon cycle by serving as a repository that regulates the amounts that transfer annually among land, atmosphere and oceans. The degree of this regulation, however, is subject to management. Soils have acted as net sources of atmospheric CO_2 during the conversion of forests and grasslands to agriculture. The Intergovernmental Panel on Climate Change (IPCC) (Cole et al., 1996) estimates losses from cultivated soils to have been 55 Pg (including 11 Pg from wetland soils). Similar estimates were reported by Scharpenseel (1993) and Houghton (1995). There is increasing scientific evidence, however, that soils can serve as net sinks of atmospheric CO_2 when carbon-sequestering practices are applied to agricultural soils or when native vegetation is allowed to return in agricultural land of marginal quality.

Long-term field experiments have been instrumental for the quantitative and mechanistic study of SOC. Recent workshops and symposia on long-term experiments and simulation models have served as points of convergence for demonstrating how purposeful management can make soils sequester carbon (Powlson et al., 1996; Lal et al., 1998a; Lal et al., 1998b; Paul et al., 1997). Management for soil carbon sequestration includes practices that conform to principles of sustainable agriculture (e.g., reduced tillage, erosion control, diverse cropping systems, improved soil fertility).

The case may be made for the global use of agricultural soils for carbon storage, yet the Kyoto Protocol signed in December 1997 did not recognize soils as valid carbon sinks due to perceived uncertainties in its determination and monitoring. Izaurralde et al. (1998) concluded that although there is a "compelling scientific basis for believing that a substantial Canadian C sequestering activity is occurring, the science and supporting measurement protocols have not been d eveloped to the point that this can be adequately quantified at the farm or regional scale and accepted for market purposes." They noted the importance of scale, sampling, analyses, modeling, and marketing for acceptance of soils as a C sink. In this chapter, we review aspects of soil organic C (SOC) measurements and present a methodology for simulating and projecting SOC changes at regional scales.

II. Carbon Sequestration in Soils: Assessing the Potential to Mitigate Global Warming

In chapter 23 of the second assessment of the IPCC, Cole et al. (1996) estimated the range of global potential rate of C sequestration at 0.4 to 0.6 Pg y^{-1} with a duration of 50 to 100 years. Soil C gain from restoration of degraded lands were estimated to provide an *additional* 0.0 to 0.2 Pg y^{-1}. Total potential C gain from better management and restoration of degraded lands is about 0.4 to 0.8 Pg y^{-1}. Lal et al. (1998c) concluded that the U.S. potential for C sequestration in agricultural soils ranged from 0.08 to 0.2 Pg y^{-1}. The realization of this soil-C sequestration potential and its ultimate impact in attenuating the rate and extent of global warming would largely depend on the wide adoption of recommended practices and the implementation of incentives.

Whether soil C sequestration is widely adopted or remains only a potential will depend on its value to producers relative to other options. The role of soil C sequestration was explored within the context of other C capture and sequestration technologies using the global energy model Mini-CAM98.3 (Edmonds et al., 1996a,b; 1999) and presented in Rosenberg et al. (1999) (Figure 1). Besides soil C sequestration the analysis also included other capturing measures such as C capture at power plants and fuel transformation to H$_2$. Further for the model runs, it was assumed that energy efficiency could increase at an annual rate of 1%, and conventional fossil fuel and power generation to improve annually by 1%. Improvements also extended to non-fossil power sources and biomass energy. Consistent with international projections, world population was set to stabilize in the model runs at 11.2 billion and to benefit from increments in the Gross National Product. Soils were set to sequester C from the atmosphere at rates ranging between 0.4 and 0.8 Pg y^{-1}, with sequestration rates per unit area during the early years being twice that of later years. The two divergent C emission trajectories in Figure 1 contrast a "business-as-usual" future (upper trajectory surpassing 18,000 Tg y^{-1} in 2095) with one in which emissions stabilize at equivalent CO$_2$ concentration of 550 ppm (lower trajectory –WRE 550– with emissions just above 6,000 Tg y^{-1} in 2095). The results suggest that soil C sequestration alone would contribute substantially toward the reduction in C emissions required to stay within the stabilization path especially during the early years of its global deployment. Such deployment of C-sequestering practices might allow time for the development and application of other cost-effective technologies later in the century.

III. Soil C Dynamics: Aspects of Its Measurement

In anticipation of soil C sequestration as a mitigation tool to counter global warming a group of energy companies in Canada examined the requirements to implement a C trading system. A major requirement for the recognition of soil C sequestration as a valid mitigation activity lies in the development of practical, cost-efficient methods to measure soil C changes at the field level.

A program using only direct measurements of soil C to detect management-induced changes would be most desirable but not cost effective (Izaurralde et al., 1998). Three issues that relate to the measurement of soil C changes in time and space will be briefly discussed here: (a) calculation of soil C changes; (b) selection of appropriate controls; and (c) need for fast and cost-effective instrumentation.

The organic C concentration of a soil can be determined with relative accuracy and precision using dry combustion methods on dry, sieved, homogeneous and carbonate-free samples. Reporting SOC as mass per unit volume requires measurement of soil bulk density. This calculation is necessary to compute temporal changes of SOC induced by management. Soil management, however, can induce changes in both SOC concentration and soil bulk density. Changes in the latter suggest that soil samples taken at equal depth increments between two periods or from various treatments may not

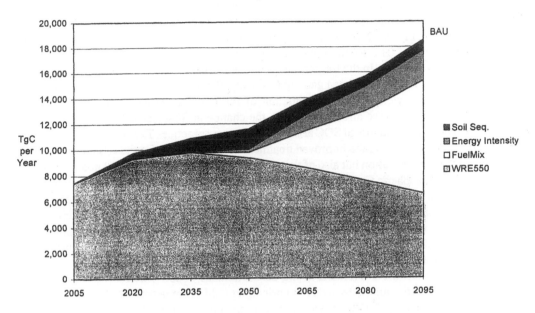

Figure 1. Modeled path for reducing global C emissions from the "business-as-usual" to the WRE 550 ppm concentration pathway, under a scenario in which credit for soil C sequestration is allowed, obtained with MiniCAM (Edmonds et al., 1996a,b; 1999). The modeled results suggest that soil C sequestration alone could help achieve the necessary reductions in net C emissions, especially early in the 21st century. From the middle of the century, additional reductions in emission should arise from changes in the energy system – such as fuel switching and reductions in total energy consumptions (Rosenberg et al., 1999. With permission.)

contain the same amount of soil. A classical example is that of the increase in bulk density often measured when soils are converted from conventional to no-tillage agriculture. Soil sampling using equal-depth increments is not likely to produce valid comparisons between treatments or periods. To analyze this problem, Ellert and Bettany (1995) recalculated C mass from 15 previously reported experiments. This recalculation consisted in adjusting the depth of each layer in the soil profile so that the soil mass of treatments being compared was equal. In their view, this procedure allowed them to make fair treatment comparisons. The results were sobering. In 12 of the 15 experiments, the traditional method indicated a loss of C under cultivation. When the equivalent-soil mass method was used, gains were detected in two experiments but there was only one in which the results of the two methods coincided. In the experiments detecting loss of C, the equivalent-soil mass method enhanced the losses by an average 9% with respect to the control treatment. In six no-tillage experiments, where SOC increased, the increases were smaller when calculated with the equivalent-soil mass method. We conclude that the equivalent mass method should be used when comparison of volumetric mass of SOC is to be made between soils with different bulk densities.

The second issue concerns the selection of appropriate controls for determining temporal changes in soil organic C. Determination of short-term soil organic C changes induced by management is challenging because these changes are small (e.g., 1 to 4 Mg C ha^{-1}) compared to the large soil C stocks (e.g., 20 to 80 Mg C ha^{-1}). Thus, several years of continued management is needed to accumulate changes in C that are large enough to be measured accurately. The options for selecting the control measure when determining temporal changes in SOC, however, are sensitive to whether SOC content is at steady state at time zero of the measurement interval (McGill et al., 1996). The first option is a time-zero measurement followed by another at time t. This method has been used in many long-term experiments. The second option is the comparison at time t of the new treatment with respect to the conventional. In the former method, C sequestration is truly measured only if the soil was at steady state. Had the soil been gaining C already, the difference in C content between time t and zero would have yielded an overestimation of the C gains arising from the new practice. Alternatively, if the soil were already losing C, the change in SOC between time t and time zero would not detect the avoided loss of SOC arising from the new practice. The side-by-side comparison between the conventional and the improved treatment at time t allows for the estimation not only of the C gains due to sequestration but also of the avoided loss. McGill et al. (1996) noted that from the point of view of atmospheric C emissions both forms of sequestration are important. Availability of data for both types of controls appears to be the best – albeit complex – solution to account for the impact of erosion and non-steady state conditions on soil C sequestration determinations.

The third issue relates to the need – including a plea for innovation – for the development of fast and reliable methods to measure and monitor SOC on a volumetric basis. Although we currently have reasonably accurate methods to take soil samples and determine soil C, we lack the field instrumentation necessary to make fast and reliable *in situ* determinations of soil C across varying landscapes. Instruments that map soil salinity by induction principles (EM38) or determine soil water content by time domain reflectometry (TDR) provide excellent examples of the advantages from developing *in situ* technologies.

IV. Simulation of Soil C Dynamics and Verification of Soil C Models

Efforts to develop models that explain SOC turnover have been ongoing for more than five decades (Jenny, 1941). Most models in use today divide SOC into various compartments – pools – and apply principles of first-order kinetics to explain its turnover (McGill, 1996). Although these models have been primarily used as research tools to describe how management and environmental variables influence SOC, there is great stimulus to use them to predict SOC stocks and associated C fluxes across large regions. Extending their use from sites to regions is considered essential to make projections of the global impact of carbon sequestration as a mitigation option. Inter-model comparisons and analyses of their performance against long-term data sets are essential steps to identify whether or not a model works under a set of conditions and select those that suit objectives best (Powlson et al., 1996).

In this section, we summarize results from an internal report with limited distribution by Izaurralde et al. (1996). The objective of the work reported was to select a model and an aggregating technique with which to project SOC changes and simulated C flux from localities to regions across two ecodistricts in Alberta, Canada. First, candidate SOC models and long-term experimental data sets were selected. The next step entailed simulation of the long-term SOC values and the subsequent statistical analysis of model performance. Finally, the model that best met the scientific, statistical and operational criteria was selected and used to scale up C fluxes across large regions.

A. Description of the Models

Six models were selected for the test: CENTURY (Parton et al., 1987), DNDC (DeNitrification DeComposition model; Li et al., 1994), *ecosys* (Grant, 1996a,b), EPIC (Erosion Productivity Impact Calculator; Williams, 1995), RothC (Jenkinson and Coleman, 1994) and SOCRATES (Soil Organic Carbon Reserves And Transformations in agro-EcoSystems; Grace and Ladd, 1995). Models were selected for their explicit representation of soil C dynamics, past testing, performance and accessibility.

CENTURY was specifically developed for long term – years to centuries – simulation of SOC dynamics, plant productivity and nutrient cycling (N, P and S). SOC is divided into three compartments: active, slow and passive. Active SOC comprises material of rapid turnover (0.5 y), including biomass C. Slow SOC is interpreted as physically protected material – rich in plant lignin – with turnover times of at least a decade. Passive SOC – the largest compartment – contains physical and chemical protection against microbial access and turns over in periods of 500 years or more. Rate constants are affected by soil water content, soil temperature and soil texture. The model runs on weekly time steps.

In DNDC, soil organic C is partitioned into four compartments: litter, microbial biomass, active humus and passive humus. The first three compartments are further divided into sub-compartments according to their resistance to decomposition. DNDC also applies principles of first-order kinetics to describe decomposition and adjusts rate constants according to soil water content, soil temperature and soil texture. Production of soluble C is required for denitrification to occur. The model computes decomposition processes on a daily time step but uses hourly steps to quantify denitrification.

The agroecosystem model *ecosys* divides SOC into four substrate-microbe complexes: animal manure, plant residues, active SOC and passive SOC. Each complex is further resolved into various substrate and microbial components. Microbial communities within each component are identified by their functional type. Of the models discussed here, *ecosys* is the one with the most detailed treatment of decomposition and transformation processes. While plant residues are made up of proteins, carbohydrates, cellulose and lignin; microbial populations are separated into labile, resistant, and storage components. Active microbial biomass, substrate concentration and environmental variables (e.g., soil temperature, soil water content and oxygen availability) regulate substrate decomposition. Running on hourly time steps, *ecosys* has the shortest time step and the highest resolution and computational requirements.

EPIC is designed to simulate wind and water erosion processes in agroecosystems and their impact on soil productivity. Treatment of soil C in EPIC is connected to nutrient cycling (e.g., N and P) and transport processes (e.g., wind and water erosion). SOC occurs as fresh organic matter – including microbial residues – and as stable organic matter. Fresh organic matter decomposes by first-order kinetics as a function of the C:N and C:P ratios of the soil layer, crop residue composition, temperature and soil water. The rate of decomposition is changed in a step-wise manner as the original residue is transformed: the first 20% decomposes easily, the next 70% has an intermediate rate, and the last 10% decomposes slowly. Microbial biomass is assumed to immobilize 40% of C in fresh crop residue; immobilization may be limited by shortage of N or P. Stable organic matter decomposes very slowly.

In the RothC model, SOC exists as decomposable plant material, resistant plant material, microbial biomass, humified organic matter or inert organic matter. Plant residues decompose by first-order kinetics to CO_2, microbial biomass or humified organic matter. Both microbial biomass and humified organic matter decompose by first-order kinetics to give more CO_2, microbial biomass and humified organic matter. The inert organic component is resistant to microbial transformation. Decomposition is regulated by temperature, moisture and clay content. Because RothC does not simulate crop growth, it requires instead user-defined values for the annual input of plant C into soil.

Soil organic C in SOCRATES is separated into five conceptual compartments. Plant residues consist of decomposable and resistant organic matter. Native SOC contains two microbial biomass fractions: protected and unprotected. The unprotected biomass has a rapid turnover (rate constant of 0.95 per week under optimal conditions) and is present only during initial stages of plant residue decomposition. Approximately 98% of total organic C is assumed to exist as stable humus. All fractions decompose by first-order kinetics as a function of temperature and moisture.

B. Long Term Experiment Datasets

Experimental data for the model selection test were obtained from the University of Alberta research plots at Breton (Alberta) and from Agriculture and Agri-Food Canada Research Station plots at Swift Current (Saskatchewan). Soils at Breton are dominantly Gray Luvisols complexed with Dark Gray Luvisols. Soil C data from six long-term experiments ranging from 15 to 27 years were used to evaluate model performance. The cropping systems included (1) non-fertilized wheat (*Triticum aestivum*)-fallow; (2) fertilized wheat-fallow; (3) non-fertilized 5-year rotation of wheat-oat (*Avena sativa*) – barley (*Hordeum vulgare*) -hay-hay; (4) fertilized 5-year rotation; (5) conventional-tillage continuous barley and (6) no-till continuous barley. Soils at Swift Current are Orthic Brown Chernozemic. The 28 years of soil C data used in the test corresponded to (1) continuous wheat and (2) wheat-fallow management systems. Management and soil characteristics have been described in detail elsewhere (Izaurralde et al., 1996; Nyborg et al., 1995; Campbell and Zentner, 1993).

C. Statistical Analyses

All models were evaluated in their ability to mimic long-term changes in soil C storage data. Simulated values were compared to experimental data for discrete changes in SOC between sampling dates. Comparing discrete changes in soil organic C versus cumulative changes ensures the data are independent and prevents the violation of assumptions of independence in least squares analysis techniques (Hess and Schmidt, 1995). The statistical tests included correlation coefficient (r), mean error (ME), mean-square error (MSE), chi-square (c^2), paired t-test and – depending on availability of replicated data – goodness of fit.

The coefficient of correlation was used to reveal how close the simulated data agreed with the experimental data (Ott, 1984). The correlation coefficient ranges from +1 to –1 where +1 indicates perfect positive correlation, zero no correlation and –1 perfect negative correlation. The ME test is used to detect the occurrence of systematic bias in the simulations. MSE indicates the scatter around a 1:1 line in a plot of simulated versus observed values (Gabrielle et al., 1995). The test c^2 statistic was used to rank model performance (Molina et al., 1990). A low c^2 value suggested a good fit between model and experimental data. The paired t-test was used to analyze paired data (experimental and simulated changes in SOC over a time interval) under the null hypothesis that the mean difference was zero (Ott, 1984). Finally, the goodness of fit method was used when replicated experimental data were available (Whitmore, 1991). This method divides the sum of squares of differences between measured and simulated values into two components. The first estimates the size of the difference between simulated values and the mean of replicate measurements (lack-of-fit). The second component calculates the variance within each set of replicate measurements (pure error). The lower the ratio of mean square due to lack of fit / mean square due to error the better the fit between model and experimental data.

Table 1. Salient features of six models with capability to simulate soil organic carbon dynamics; CENTURY, DNDC, *ecosys*, EPIC, RothC and SOCRATES

	CENTURY	DNDC	*ecosys*	EPIC 5125	RothC	SOCRATES
Spatial scale	Plot to field	Plot	Plot	Plot to basin	Plot to field	Plot
Time-step	Week	Hour and day	Hour	Day	Month	Week
Period for SOC simulation	Months to centuries	Days to centuries	Hours to years	Years to centuries	Years to centuries	Years to centuries
Soil profile	Uniform – 20 cm depth	Surface layers – 50 cm depth	Soil layers – up to 10	Soil layers – up to 10	Uniform – 23 cm depth	Uniform – 10 cm depth
SOC pools	Three	Three	Three	Two	Three	Three
Texture effect	Yes	Yes	Yes	No	Yes	Yes
Weather data	Precipitation and temperature	Precipitation and temperature	Radiation, precipitation, temperature, humidity, wind speed	Radiation, precipitation, temperature, humidity, wind speed	Precipitation, temperature, open pan evaporation	Precipitation and temperature
Plant growth simulation	Cereals, legumes, grasses, trees	Cereals, legumes, grasses	Cereals, oil-seeds, grass-land, legumes, forests	Cereals, legumes, oil-seeds, grasses, trees	No	Cereals, legumes, oil-seeds, grasses
Rotations	Multiple crops per year, unlimited rotations	Multiple crops per year, rotations limited to 20 years	Multiple crops, unlimited rotations	Multiple crops per year, unlimited rotations	None	One crop per year, limited to 20 years
Nutrients and fertilizers	N,P,S	N,P	N,P	N,P	None	N
Tillage	Many options	Many options	Many options	Many options	None	
Erosion	No	No	No	Wind and water	No	No
CO_2 effect	Yes	No	Yes	Yes	No	No

D. Model Evaluation and Selection

Only four of the six models met specific criteria for inclusion in site-specific simulations: CENTURY, *ecosys*, EPIC and SOCRATES (Table 1). The DNDC and RothC models were excluded in the first selection phase, based on their ease of use, input requirements and run time. The selected models were further evaluated for their ability to reproduce long-term SOC data from research sites at Breton (Alberta) and Swift Current (Saskatchewan). The evaluation comprised statistical comparisons between predicted and observed values of SOC (Table 2). The final model selection was based not only on the statistical evaluation but also on practical objectives such as ease of use, input requirements, output format and run-time.

1. Breton Cereal-Fallow Rotation (Fertilized and Nonfertilized)

This rotation was sampled for SOC determination in 1968, 1972, 1979 and 1990. All models predicted the SOC decline observed in the non-fertilized wheat-fallow rotation, albeit not to the same degree. In the fertilized cereal-fallow rotation SOC content declined less steeply to about 1.1% in

Table 2. Model performance statistics of discrete changes in SOC for four rotation/tillage experiments

Statistic	Century	*ecoyss*	EPIC	SOCRATES
Wheat - fallow rotation (Breton), n = 4[1]				
Nonfertilized				
Correlation	-1.00	0.97	0.79	1.00
Mean error	-0.04	-0.06	-0.05	-0.03
Mean square error	0.01	0.01	0.01	0.01
Chi-square	0.86	0.75	0.74	0.52
Paired t-test	-0.65	-0.96	-0.87	-0.67
Fertilized				
Correlation	-0.54	-0.75	0.15	0.91
Mean error	-0.06	-0.07	-0.04	-0.03
Mean square error	0.01	0.02	0.01	0.01
Chi-square	1.13	1.49	0.83	0.56
Paired t-test	-0.93	-0.91	-0.74	-0.69
Wheat - oat - barley - hay - hay rotation (Breton), n = 4				
Nonfertilized				
Correlation	0.09	–	0.06	-0.44
Mean error	-0.06	–	0.04	0.03
Mean square error	0.01	–	0.00	0.00
Chi-square	4.53	–	1.52	1.29
Paired t-test	-0.97	–	1.22	1.02
Fertilized				
Correlation	-0.79	–	1.00	-0.75
Mean error	-0.07	–	0.03	0.02
Mean square error	0.04	–	0.00	0.01
Chi-square	6.76	–	0.61	1.03
Paired t-test	-0.54	–	0.71	0.32
Continuous barley (Breton), n = 2				
No-tillage				
Correlation	–	–	–	–
Mean error	0.33	0.48	0.43	0.39
Mean square error	0.11	0.23	0.18	0.15
Chi-square	–	–	–	–
Paired t-test	–	–	–	–
Conventional tillage				
Correlation	–	–	–	–
Mean error	0.14	0.27	0.24	0.17
Mean square error	0.02	0.02	0.06	0.03
Chi-square	–	–	–	–
Paired t-test	–	–	–	–

Table 2. (continued)

Statistic	Century	*ecoyss*	EPIC	SOCRATES
		**Cereal rotation (Swift Current), n = 6**		
		Continuous wheat		
Correlation	0.06	−0.55	-0.64	0.85
Mean error	0.01	0.04	0.07	−0.02
Mean square error	0.03	0.05	0.06	0.03
Chi-square	0.80	1.12	1.52	0.69
Paired t-test	0.14	0.34	0.55	−0.25
MSLOFIT/MSE Ratio	1.36	1.88	2.59	1.17
		Wheat - Fallow		
Correlation	−0.65	−0.54	−0.23	−0.71
Mean error	0.01	−0.01	0.07	−0.03
Mean square error	0.02	0.04	0.04	0.02
Chi-square	1.10	2.31	2.21	1.10
Paired t-test	0.12	−0.05	0.71	−0.51
MSLOFIT/MSE Ratio	0.41	0.86	0.82	0.41

[1]Number of SOC sampling dates.

1990. A relative ranking of performance statistics revealed SOCRATES as the best model with CENTURY being a close second despite negative correlation coefficients.

2. Breton 5-Year Rotation (Fertilized and Nonfertilized)

The Breton 5-year rotation (wheat-oat-barley-hay-hay) has been in practice for more than half a century. The hay component is a mixture of alfalfa and brome grass. Sampling dates were the same as for the previous rotation. EPIC and SOCRATES mimicked best the gradual increase in SOC observed over a 27-year period. CENTURY over-predicted measured SOC values.

3. Breton Continuous Barley (Conventional and No-Tillage)

This continuous barley system has been conducted under contrasting tillage management for over a decade (Nyborg et al., 1995). The selected treatments retained all crop residues and received 56 kg N ha^{-1} y^{-1}. SOC was determined in 1980 and 1990. Initial SOC was approximately 1.4% (0 to 15 cm) and reached 1.8% under no tillage. Although all models tested under estimated SOC, most reproduce a gradual SOC increase. The sparse sampling schedule resulted in only one comparable discrete change in SOC over the 10-year period restricting statistical analysis to the mean square error and mean error terms. CENTURY had the least spread and statistical bias of the four models, with SOCRATES a close second.

4. Swift Current Continuous Wheat and Wheat-Fallow Rotations

These rotations were established in 1966 (Campbell and Zentner, 1993). The first rotation has received on average N at 27.5 and P at 9.6 kg ha^{-1} annually. The wheat-fallow rotation has received

Table 3. Performance and operational ratings of the CENTURY, *ecosys*, EPIC and SOCRATES models

Criteria	Model			
	CENTURY	*ecosys*	EPIC	SOCRATES
Soil C simulation performace	Good to excellent	Good	Good to excellent	Excellent
Input requirements	Fair (expertise required)	Complex (expertise required)	Complex (expertise required)	Easy
Ease of use	Easy - fair	Complex	Fair - good	Easy
Output performance	Excellent	Fair (modeler only)	Good	Unsatisfactory
Run time	<10 s	2 h	<60 s	<5 s

on average N at 3.2 and P at 6.4 kg ha^{-1} (annual equivalent). The initial SOC was 1.96% (0 to 15 cm) and the sampling dates were 1976, 1981, 1984, 1990 and 1993. SOC concentrations have either been maintained or decreased slightly. The decrease has been attributed to the impact of drought on yields over a 9-year period. Predicted SOC reproduced the observed trends. Since the Swift Current data were replicated, the statistical evaluations also included the goodness of fit method. Overall, SOCRATES ranked first in all statistical tests while CENTURY placed second.

5. Summary of Model Selection

The final selection of the simulation model for scale-up analysis was based not only on how the models reproduced long-term SOC changes but also on several operational criteria (Table 3). Of the four models included in the final comparison, SOCRATES was the best at reproducing observed data on soil C dynamics. CENTURY, EPIC, and *ecosys* had a relatively close performance. Analysis of the various statistics in Table 2 revealed that the order of performance changed from data set to data set. Our selection of SOCRATES as the model for the scale-up work, therefore, is justified not only by its relative accuracy at simulating soil C changes but also by its relative simplicity. This was important since the C-flux analysis required the inclusion of entire populations of soils and management within ecodistricts (e.g., an ecodistrict with 10 soil series and 10 management systems makes 100 possible soil-management combinations). The selection of SOCRATES was also favored by the availability of crop yield data from census information aggregated at ecodistrict level, which made the task of estimating crop productivity redundant. Model selection, however, depends heavily on model objectives. For example, SOCRATES does not treat either erosion or tillage systems explicitly. Had the intention been to account for erosion impacts on C balance, then probably EPIC would have been selected. The rigorous manner in which these models were examined provides confidence in the direct comparability of their outputs.

Figure 2. Ecodistrict map of Alberta showing the locations of ecodistricts 727 and 828.

V. Scaling C Changes and Fluxes at the Regional Level

Ecodistricts were selected for simulating and scaling soil C changes and C fluxes at the regional level. Ecodistricts are biophysical landscape units with common climatic, landform, topographic, edaphic and agricultural features (Ecological Stratification Working Group, 1996). Hiley et al. (1989), Hiley and Wehrhahn (1991) and Izaurralde et al. (1992) have demonstrated that resource areas at this scale provide meaningful results when assessing the impact of land use and agricultural practices on resource quality. Two ecodistricts in Alberta (727 and 828) having contrasting climate, soils, and management were selected to test the scaling procedures (Figure 2).

A. Ecodistrict 727

The Leduc Plain in central Alberta occupies 850,000 hectares of undulating landscape with inclusions of level and hummocky topography. Soils are predominantly Black Chernozemic, complexed with Dark Gray Chernozemic and Solonetzic. There is a slight to moderate heat limitation for crop production in the region (Pettapiece, 1989). Mixed grain, cattle and dairy operations are common, with barley, oilseeds and feeds/forage being typical crop types.

B. Ecodistrict 828

The Foremost Plain occupies over 1 million hectares of mixed grass prairie in southern Alberta. The landform is largely undulating and the soils are mainly Brown Chernozemic (Pettapiece 1989). Moisture deficits severely limit crop production and determine wheat-fallow to be an extensive cropping system in the region. Irrigation water is used for producing hay and cash crops.

C. Representing Populations of Climate, Soil and Management Combinations (CMSC)

In order to identify populations of significant climate, soil and management combinations we examined several datasets with varying degrees of geographic precision. The datasets described features (e.g., proportion of soils and associated production systems) within a region but were 'partially spatial' (Jackson et al., 1995) because their location and distribution were unknown. The climate data were at the coarsest level of resolution, in that data from a representative Environment Canada weather station within the ecodistrict were selected to provide the climate data for the entire ecodistrict. The soils and management datasets were initially defined on a Soil Landscape of Canada (SLC) unit, and then were spatially weighted and aggregated up to ecodistrict level (Table 4). Therefore, the product of the number of soil types multiplied by the number of management systems identified in an ecodistrict determined the ultimate population of CSMCs for the region.

Management information was derived from published databases. The first database linked 1991 Statistics Canada Agricultural Census data to soil-landscape polygons (SLC) weighted by census enumeration areas (Hiley and Huffman, 1994). The database was further processed so that every farm within an SLC was categorized based on total sales (Statistics Canada, Census of Agricultural Division, N. Hillary, personal communication). This economic stratification aggregated farms by major production systems (Table 5). The census data also contained information to segregate land into three categories: cultivated, improved range and unimproved range.

Crop rotations were derived for each production system from census data by first adding the total area of each crop and then calculating each as a percentage of the total cultivated area. The number of occurrences of a crop within a 10-year rotation was calculated by dividing the crop percentages by 10. The percentage of cultivated land in the total agricultural area was 76% for ecodistrict 727 and 68% for ecodistrict 828. Crop sequences were reviewed (Agriculture Canada 1993a,b; 1994a,b,c; Hiley et al., 1989; Izaurralde et al., 1992) to determine the most appropriate combinations.

Crop yields were not modeled but specified as input to the model using a variety of sources and expert opinion. The remainder of the area in each ecodistrict that it was not under cultivation was assigned as either improved or unimproved range. For ecodistrict 727 this portion represented 24% of the total area while for ecodistrict 828 it was 32%. Separate simulations were conducted for rangeland conditions. The resulting crop-soil-management (CSM) combinations for the two ecodistricts are presented in Table 6.

Table 4. Characteristics of soil series found in Ecodistrict 828 and 727

Ecodistrict (Agricultural area in ha)	Soil series	Name	Subgroup	Area[a] (%) (rangelands)		Clay content (%)	Soil organic C (%)
727	AGS	Angus Ridge[b]	Evluviated Black	26.3	(25.1)	30	4.0
(653457 ha)	CMO	Canmore	Black Solodized Solonetz	3.4	(3.3)	25	4.0
	DUG	Daugh	Black Solonetz	3.3	(3.1)	50	4.0
	FLU	Falun	Dark Gray Chernozem	6.8	(6.6)	21	4.0
	KVG	Kavanagh	Black Solodized Solonetz	1.1	(1.1)	15	4.0
	MCO	Mico	Orthic Dark Gray	2.4	(2.3)	40	3.0
	MDR	Mundare	Orthic Black	3.2	(3.0)	6	4.0
	MMO	Malmo	Eluviated Black	16.4	(15.7)	38	4.0
	NVR	Navarre	Gleyed Black	6.9	(6.6)	40	4.0
	PHS	Peace Hills	Orthic Black	2.6	(2.5)	16	3.0
	POK	Ponoka	Eluviated Black	18.0	(17.2)	24	3.5
	TBY	Thorsby	Dark Gray Solod	1.7	(1.6)	20	3.0
	WKN	Wetaskiwin	Black Solodized Solonetz	2.9	(2.8)	30	4.0
	WTB	Winterburn	Orthic Dark Gray	4.8	(4.6)	25	3.0
	BRK[c]	Bright Bank	Dark Gray Luvisol	–	(1.3)	6	3.0
	ELP[c]	Elk Point	Dark Gray Luvisol	–	(0.7)	9	2.0
	PRM[c]	Primula	Eluviated Eutric Brunisol	–	(2.3)	2	0.4
828	CFD	Cranford	Orthic Brown	24.2	(22.9)	23	2.1
(747657 ha)	CHN	Chin	Orthic Brown	5.1	(4.8)	21	3.3
	FMT	Foremost	Orthic Brown	6.3	(6.0)	20	2.0
	HDY	Halliday	Brown Solod	3.0	(2.8)	25	2.0
	MAB	Maleb	Orthic Brown	47.3	(44.8)	28	1.8
	MSN	Masinasan	Orthic Brown	7.9	(7.5)	20	2.0
	ROL	Ronaline	Solonetzic Brown	6.4	(6.0)	30	2.0
	CVD[c]	Cavendish	Orthic Brown	–	(2.5)	8	1.5
	VST[c]	Vendisant	Rego Brown	–	(0.8)	4	1.0
	MKR[c]	Milk River	Cumulic Regosol	–	(1.0)	7	0.8
	PHN[c]	Pinhorm	Orthic Brown	–	(1.0)	10	3.1

[a]Percent of the agricultural area in which the particular soil series occurs; bracketed values are for rangeland, where more soil series are included in the analysis.
[b]Data for the Angus Ridge soil taken from Larney et al., 1995; all other soils data derived from the Alberta Soil Layer File.
[c]Soils of poor quality and used primarily for range.

D. Evaluation of Scaling-Up Options

We evaluated three methods for scaling-up or aggregating soil C storage and change to the ecodistrict level. The first method (CSM) was the most complex because it required the use of all climate-soil-management combinations to calculate soil C storage at the regional level. The second method used one dominant soil type to model all production systems (PSS). The last method (TAS or Total Area Scale-up) was the simplest because it modeled soil C changes using one average production system

Table 5. relevant production systems identified from the 1991 Alberta Census for ecodistricts 727 and 828.

Eco-district	Production system	Definition	Rotation	Area (%)
727	DAIRY	Farms on which sales of dairy products exceed 51% of total sales[a]	B-B-O-B-B-H-H-H-H-H[e]	5.7
	CATTLE	Farms on which the sales of cattle and calves make up 51% total sales[b]	W-O-B-F-B-B-H-H-H-H	29.2
	PIGS	Farms on which the sales of hogs make up 51% of total sales	W-B-B-F-OS-W-B-B-B-H	1.8
	WHEAT	Farms on which the sales of spring wheat, durum wheat, and winter wheat make up 51% of total sales	W-W-W-B-F-W-W-B-W-W	1.3
	OILSEED	Farms on which the sales of canola, flax-seed, soybeans, mustard seed, sunflower and safflower make up 51% of total sales	OS-OS-B-OS-OS-W-OS-OS-W-F	1.0
	GRAIN	Farms on which the sales of small grains[c] make up 51% of total sales	W-B-B-F-OS-W-B-OS-H-H-H	35.0
	FIELD CROPS	Farms on which the sales of corn (silage), hay, other field crops cut for hay and silage, potatoes, tobacco, forage seed, other unspecified field crop items and honey make up 51% of total sales	H-H-H-H-B-F-OS-H-H-H	2.0
TAS[d]			W-O-B-F-OS-B-B-H-H-H	76.4
828	CATTLE	See above	W-B-F-W-F-W-W-F-H-H	8.5
	WHEAT	See above	W-W-F-W-F-W-F-W-W-F	49.7
	GRAIN	See above	W-W-F-W-Bn-W-Bn-H-H	6.6
	FIELD CROPS	See above	W-F-W-Bn-B-Bn-W-Bn-H-H	2.6
TAS			W-W-F-W-B-F-W-F-W-F	67.8

[a]Alternate DAIRY definition – farms on which sales of dairy products make up 40 to 50.9% of total sales, provided that the value of dairy sales plus the value of the sales of cattle, calves and hay make up 51% or more of total value of sales.

[b]Alternate CATTLE definition – farms on which sales of cattle and calves make up 40 to 50.9% of total sales, provided that the value of the sales of cattle and calves plus the value of sales of hay make up 51% or more of the total value of sales.

[c]Small grains include wheat, oats, barley, mixed grains, corn (*Zea mays*) for grain, autumn rye (*Secale cereale*), spring rye, canola (*Brassica* spp.), flaxseed (*Linum usitatisimum*), soybeans (*Glycine max*), mustard seed (*B. juncea, B. nigra*), dry field peas (*Pisum sativum*), lentils (*Lens culinaris*), dry white beans (*Phaseolus* spp.), faba beans (*Vicia faba*), other dry beans, sunflowers (*Helianthus annuus*), buckwheat (*Fagopyrum esculentum*), canary seed, millet (*Setaria italica*), triticale, safflower (*Carthamus tintorius*), and caraway seed (*Carum carvi*).

[d]Total area scale-up – total area for each crop summed for the entire ecodistrict, and expressed as a percentage of the total cultivated area in the ecodistrict.

[e]Crop name abbreviations: B-barley; W-wheat; OS-oilseeds; F-fallow; O-oats; H-alfalfa/grass hay; Bn-beans.

Table 6. Area of each ecodistrict occupied by soils (%), production systems (%) and resulting soil × management combinations (ha)

		DAIRY	CATTLE	PIGS	WHEAT	OILSDS	GRAIN	FLDCRPS	
					Ecodistrict 727				
Soil	Area, %	5.72	29.17	1.84	1.30	0.97	35.01	2.05	
			Area (ha) of soil × management combinations						Sum, ha
AGS	26.28	9823	50093	3160	2232	1666	60122	3520	130617
CMO	3.45	1290	6576	415	293	219	7893	462	17147
DUG	3.27	1222	6233	393	278	207	7481	438	16253
FLU	6.85	2560	13057	824	582	434	15671	918	34046
KVG	1.14	426	2173	137	97	72	2608	153	5666
MCO	2.40	897	4575	289	204	152	5491	322	11928
MDR	3.16	1181	6023	380	268	200	7229	423	15706
MMO	16.41	6134	31280	1973	1394	1040	37542	2198	81561
NVR	6.87	2568	13095	826	584	435	15717	920	34145
PHS	2.64	987	5032	317	224	167	6040	354	13121
POK	18.03	6739	34368	2168	1532	1143	41248	2415	89613
TBY	1.71	639	3259	206	145	108	3912	229	8499
WKN	2.94	1099	5604	353	250	186	6726	394	14612
WTB	4.85	1813	9245	583	412	307	11096	650	24105
		37378	190613	12024	8495	6339	228775	13396	497019

					Ecodistrict 828				
Soil	Area, %		8.50		49.74		6.63	2.59	
			Area (ha) of soil × management combinations						Sum, ha
CFD	24.17	–	15357	–	89867	–	11979	4679	121882
CHN	5.08	–	3228	–	18892	–	2518	984	25622
FMT	6.30	–	4002	–	23419	–	3122	1219	31762
HDY	2.96	–	1880	–	11004	–	1467	573	14924
MAB	47.26	–	30033	–	175747	–	23426	9151	238357
MSN	7.89	–	5012	–	29331	–	3910	1527	39780
ROL	6.35	–	4037	–	23625	–	3149	1230	32041
	Sum	–	63551	–	371885	–	49570	19364	504369

on a dominant soil type. We hypothesized that the CSM method would render the most detailed simulation of soil C fluxes because it contained all the information about the system modeled. Annual changes in soil C were calculated as concentration, mass per unit area and mass per soil-management combination. Here we restrict the presentation and discussion of results largely to the last calculation method when applied to cultivated lands (Izaurralde et al., 1996).

1. CSM Method Scale-Up

Ecodistrict 727: Cultivated lands were predicted to gain an average of 61 kg C ha^{-1} y^{-1}, while improved rangelands were predicted to gain only 35. Unimproved rangelands were predicted to lose an average of 5 kg C ha^{-1} y^{-1}. As expected from the model structure, coarse-textured soils gained less C than soils with high clay content. The trends were evident in both cultivated and rangelands. Predicted mean annual SOC change on cultivated lands indicated that dairy farms, with their high proportion of cereals and hay crops would have greater potential for C sequestration than other farm types. The inclusion of fallow or oilseeds in the rotation would reduce C storage in grain and oilseed farms likely because of reduced C inputs. Despite the partially spatial nature of the datasets (Jackson et al., 1995), we consider the spatially weighted C storage values (Table 7) to be valid estimates of C sequestration potential because all possible combinations were considered in its calculation.

In all, cultivated lands in ecodistrict 727 were predicted to sequester C at a rate of 32,625 t C y^{-1}. Improved rangelands would gain less than cultivated lands (2,631 t C y^{-1}) while the change in unimproved rangelands would be -238 t C y^{-1}. The annual change in C storage predicted for all land uses in ecodistrict 727 is 35,018 t C y^{-1}. Donigian et al. (1995) reported similar results using CENTURY and DNDC to model C storage and greenhouse-gas emissions in 23 states of the central U.S. Using an annual increase of 1.5% in crop yields over a 40-year period, these researchers consistently modeled SOC gains in cultivated lands especially in rotations that retained significant amounts of crop residues.

Ecodistrict 828: On average, cultivated lands in this ecodistrict were predicted to sequester C at 82 kg C ha^{-1} y^{-1} while rangelands were predicted to lose 22 kg C ha^{-1} y^{-1}. The predicted trends of increased C sequestration on clay-rich soils were less clear in this ecodistrict than in No. 727 because of reduced textural variability found among soil series (Table 4). Production systems with high proportion of irrigated hay and proportionately less fallow years had greater C sequestration potential than other types of management. The inclusion of 4 years of fallow in the 10-year management plan of wheat farms reduced C sequestration potential because of substantial decreases in C inputs. The C gains induced by 2 years of irrigated hay in cattle production systems more than offset the C losses during the fallow years. The annual change in C storage in cultivated lands of ecodistrict 828 was 6,170 t C y^{-1} while in range lands it was $-5,172$ t C y^{-1} (Table 7). The net C storage of 998 t C y^{-1} for this ecodistrict is much lower than that predicted for ecodistrict 727.

2. PSS Method Scale-Up

Ecodistrict 727: The dominant soil used for this ecodistrict was the Angus Ridge (Table 4). The relative ranking of C sequestration potential by production system was similar to that derived using the CSM method (Table 8). Changes in C storage in this ecodistrict were predicted at 29,456 t C y^{-1} in cultivated lands and 2,499 t C y^{-1} in improved rangelands. Similar to the CSM calculation, the PSS method predicted a loss of 1,119 t C y^{-1} in unimproved rangelands. Thus, the PSS method yielded a net C storage of 30,836 t C y^{-1} across the entire ecodistrict. Cattle and grain farms, dominant in the area, made up more than 75% of the regional C storage.

Ecodistrict 828: The Maleb series was the dominant soil used in the simulations (Table 4). As in ecodistrict 727 the relative ranking of C sequestration potential by production system calculated with this method was similar to the CSM method (Table 8). C storage in cultivated lands of ecodistrict 828 was predicted to be 1,119 t C y^{-1}. For rangelands, the prediction was -5,347 t C y^{-1}. This resulted in a total net loss of C storage of 4,228 t C y^{-1}. Changes in soil C storage on wheat farms contributed significantly to these results.

Table 7. Spatially weighted C storage values in cultivated lands simulated with the CSM method

		DAIRY	CATTLE	PIGS	WHEAT	OILSDS	GRAIN	FLDCRPS	
					Ecodistrict 727				
Soil	Area, %	5.72	29.17	1.84	1.30	0.97	35.01	2.05	
					C storage (t C y⁻¹)				Sum
AGS	26.28	1277	3451	177	145	−71	2579	183	7741
CMO	3.45	139	342	16	14	−4	236	18	762
DUG	3.27	264	972	56	43	12	943	55	2346
FLU	6.85	233	509	21	20	−11	265	24	1061
KVG	1.14	29	48	1	1	−3	0	1	78
MCO	2.40	175	613	36	27	8	592	35	1486
MDR	3.16	51	0	−5	−2	−11	−160	−4	−132
MMO	16.41	1005	3131	154	132	18	2782	171	7393
NVR	6.87	444	1413	78	61	10	1287	80	3373
PHS	2.64	94	262	14	11	−1	212	14	605
POK	18.03	762	2100	104	86	−4	1609	116	4772
TBY	1.71	69	212	11	9	0	168	11	480
WKN	2.94	143	386	20	16	0	324	20	909
WTB	4.85	236	721	40	32	4	678	40	1751
	Sum	4920	14160	724	596	−55	11515	765	32625

					Ecodistrict 828				
Soil	Area, %		8.50		49.74		6.63	2.59	
					C storage (t C y⁻¹)				Sum
CFD	24.17	–	1657	–	−3154	–	934	833	270
CHN	5.08	–	168	–	−1645	–	75	132	−1271
FMT	6.30	–	401	–	−700	–	231	201	133
HDY	2.96	–	220	–	−286	–	128	107	168
MAB	47.26	–	4021	–	−2285	–	2345	1903	5985
MSN	7.89	–	502	–	−877	–	290	252	167
ROL	6.35	–	541	–	−399	–	315	261	717
	Sum	–	7510	–	−9346	–	4318	3689	6170

Table 8. Spatially weighted C storage values in cultivated lands simulated with the PSS method

Ecodistrict	DAIRY	CATTLE	PIGS	WHEAT	OILSDS	GRAIN	FLDCRPS	
				C storage (t C y⁻¹)				Sum
727	4589	13133	672	552	−272	9814	697	29456
828	–	6857	–	−13505	–	3866	3449	1119

3. TAS Method Scale-Up

This method was the simplest aggregation technique to implement because it required only the use of a single rotation applied to a dominant soil type (same as the PSS method). Application of this method to an entire ecodistrict required making significant abstraction of the management information. Predicted changes in C storage were 29,374 t C y^{-1} for ecodistrict 727 and –11,909 t C y^{-1} for ecodistrict 828. The TAS rotation derived for ecodistrict 727 (Table 5) was similar to the other rotations used with the PSS and CSM methods. Instead, the TAS rotation derived for ecodistrict 828 was similar only to wheat production systems and therefore influenced the regional outcome of C sequestration potential. Practically, it was like dedicating the entire cultivated area in ecodistrict 828 to a wheat-fallow production system with the consequent negative impact on soil C storage.

4. Evaluation of the Regional Simulations

The three scale-up methods discussed here ranged from very complex with high spatial resolution (CSM) to simple with coarse spatial resolution (TAS). The PSS method fell between the two in terms of simplicity and resolution. We examined each of the scale-up methods based on their accuracy and their sensitivity to non-normal distribution of soil-management combinations. The accuracy of the PSS and TAS methods was evaluated by comparing regional net C storage to that of the CSM method, which was considered to represent the true regional net C storage. Favorable comparisons of the alternative methods (PSS and TAS) with the CSM method would suggest the possibility of using simplified approaches for aggregating C fluxes at the regional level, with the consequent reduction in time and cost. For ecodistrict 727, the PSS and TAS methods produced estimates that were similar to the true simulated net C storage for each type of agricultural land (Table 9). Use of a dominant soil in ecodistrict 727 resulted in estimates that were within 12% (PSS) and 17% (TAS) of the CSM method. This would imply that the use of simpler methods to scale up regional C fluxes in ecodistricts similar to No. 727 would not jeopardize the accuracy of the predictions.

The results from Ecodistrict 828 were not encouraging, however. There were wide discrepancies in the regional net C storage values predicted by the three methods. The discrepancies in estimates grew larger as the method became simpler and with coarser resolution (Table 9). While the CSM method predicted C sequestration rates greater than 1,000 t C y^{-1} the TAS scale-up predicted net C releases ten times greater than the CSM procedure. The use of simplified scale-up methods to calculate regional C storage in ecodistricts similar to No. 828 should not be recommended. A frequency analysis of changes in soil C storage revealed that both ecodistricts had similar degree of skewness and kurtosis (Izaurralde et al., 1996). However, the coefficient of variation found in ecodistrict 828 was much greater (413%) than that found in ecodistrict 727 (230%) and caused, in our view, the divergent results in C storage. The exact nature of this mechanism, however, remains unknown.

5. Summary of Model Selection and Scale-Up

SOCRATES was objectively selected for this conceptual study because it mimicked long-term C change well and it was easy to operate. However, the extended use of SOCRATES for modeling regional C storage may be limited by lack of management options. While stubble management and grazing were modeled explicitly, other practices – wheat on fallow versus wheat on stubble – were represented implicitly through yield inputs. In view of the importance of zero tillage on soil C seques-

Table 9. Comparison of C storage among three scale-up methods (CSM, PSS and TAS) in ecodistricts 727 and 828

Scale-up option	Cultivated lands	Improved range	Unimproved range	Totals
	C storage (t C y^{-1})			
Ecodistrict 727				
CSM	32, 625	2,631	−238	35,325
PSS	29,456	2,499	−1,119	30,836
TAS	27,908	2,615	−1,149	29,374
Ecodistrict 828				
CSM	6,170	−	−5,172	1,021
PSS	1,119	−	−5,347	−4,228
TAS	−6,593	−	−5,316	−11,909

tration, it might be argued that other models with a more complex representation of management options (e.g., CENTURY or *ecosys*) would have been better suited for predicting long-term soil C trends. EPIC could have been the model of choice for simulating erosion impacts on net C storage as well as for simulating C-fertilization effects due to climate change.

Our evaluation suggests that scale-up of soil C change may not require the use of complex models. Such models are indispensable in helping us understand the processes within the C cycle. However, some of their complexity may be redundant, even excessive, in scaling-up exercises which, by definition, have to make many simplifying assumptions. As shown by the good performance of SOCRATES, comparatively simple models might furnish reliable estimates with fewer demands on data inputs and computing resources.

The selection of a model for predicting changes of ecosystem properties at the regional level will ultimately depend on research and application objectives. In multi-objective projects, more than one model might be needed to make these projections. Thus, model-to-data and model-to-model comparisons are deemed essential methodological steps toward building robust procedures to estimate C fluxes at regional or national levels.

We believe that the selection of accurate procedures for regional scaling should be approached not only with knowledge of the regional distribution of climate, soils and management but also with an understanding of a model's sensitivity to these combinations. Soil-landscape models (MacMillan, 2000) and remotely sensed data hold promise for improving the resolution of spatial data and the precision of C-flux estimates.

VI. Summary and Conclusions

We have reviewed the importance of C sequestration in soils in the context of greenhouse warming, examined aspects of soil C dynamics, and discussed results of model selection and scale-up procedures to project C changes from sites to regions. The analysis by Edmonds et al. (1999) demonstrated that the role of soils is during the early stages in any concerted action against global warming. Their results show that soil C sequestration alone would contribute significantly toward reduction in C emissions required to stay within a stabilization path of atmospheric CO_2 concentrations. If soil C sequestration gains acceptance as a temporary but valid technology to contain global warming, then methods and protocols will be needed to verify and monitor soil C changes over large regions. In this context, we have discussed the need to address methodological aspects of soil C measurement at the field level. Finally, we have shown the need for homogeneous groupings of climate-soil-management combinations if aggregation techniques are not to become unwieldy on one hand or unreliable on the other. Further work on measurement and projections of soil C dynamics is needed to advance the science of C sequestration and its application on large scale.

Acknowledgments

Financial support for this work was provided by the Agriculture and Agri-Food Canada Greenhouse Gas Initiative. The technical advice of H.H. Janzen and N.J. Rosenberg is gratefully acknowledged. R.C. Izaurralde thanks the Office of Science of the U.S. Department of Energy and the U.S. Environmental Protection Agency for support to complete this manuscript.

References

Agriculture Canada. 1993a. Agricultural policies and soil degradation in western Canada: An agroecological economic assessment (Report 1: Conceptual Framework). Technical Report 2/93. Agriculture Canada, Policy Branch, Ottawa. 52 pp.

Agriculture Canada. 1993b. Agricultural policies and soil degradation in western Canada: An agroecological economic assessment (Report 2: The Environmental Modeling System). Technical Report 5/93. Agriculture Canada, Policy Branch, Ottawa. 110 pp.

Agriculture Canada. 1994a. Agricultural policies and soil degradation in western Canada: An agroecological economic assessment (Report 3: The Integration of the Environmental and Economic Components). Agriculture Canada, Policy Branch, Ottawa. 130 pp.

Agriculture Canada. 1994b. Agricultural policies and soil degradation in western Canada: An agroecological economic assessment (Report 4: Modifications to CRAM and Policy Evaluation Results). Agriculture Canada, Policy Branch, Ottawa. 131 pp.

Agriculture Canada. 1994c. Agricultural policies and soil degradation in western Canada: An agroecological economic assessment (Report 5: Project Summary). Agriculture Canada, Policy Branch, Ottawa. 31 pp.

Campbell, C.A. and Zentner, R.P. 1993. Soil organic matter as influenced by crop rotations and fertilization. *Soil Sci. Soc. Am. J.* 57:1034-1040.

Cole, V., C. Cerri, K. Minami, A. Mosier, N. Rosenberg and D. Sauerbeck. 1996. Agricultural options for mitigation of greenhouse gas emissions. p. 744-771. In: R.T. Watson, M.C. Zinyowera and R.H. Moss (eds.), Climate Change 1995: *Impacts, Adaptations, and Mitigation of Climate Change: Scientific-Technical Analyses. Contribution of Working Group II to the Second Assessment Report of the Intergovernmental Panel on Climate Change.* Cambridge University Press, Cambridge. 880 pp.

Coleman, K. and D.S. Jenkinson. 1996. RothC-26.3 — A model for the turnover of carbon in soil. p. 237-246 In: D.S. Powlson, P. Smith and J.U. Smith (eds.), Evaluation of soil organic matter models using existing long-term datasets. NATO ASI Series I, Vol. 38, Springer-Verlag, Heidelberg.

Donigian, A.S., A.S. Patwardhan, R.B. Jackson, T.O. Barnwell, K.B. Weinrich and A.L. Rowell. 1995. Modeling the impacts of agricultural management impacts on soil carbon in the central U.S. p. 121-135. In: R. Lal, R., J. Kimble, E. Levine and B.A. Stewart (eds.), *Soil Management and Greenhouse Effect*. Advances in Soil Science. Lewis Publishers, Boca Raton, FL.

Ecological Stratification Working Group. 1996. A National Ecological Framework for Canada. Agriculture and Agri-Food Canada, Research Branch, Centre for Land and Biological Resources Research and Environment Canada, State of the Environmental Directorate, Ecozone Analysis Branch, Ottawa/Hull. Report and national map at 1:7 500 000 scale.

Edmonds, J., M. Wise, H. Pitcher, R. Richels, T. Wigley and C. MacCracken. 1996a. An integrated assessment of climate change and the accelerated introduction of advanced energy technologies: an application of MiniCAM 1.0. *Mitigation and Adaptation Strategies for Global Change* 1:311-339.

Edmonds, J., M. Wise, R. Sands, R. Brown and H. Kheshgi. 1996b. Agriculture, Land-Use, and Commercial Biomass Energy: A Preliminary Integrated Analysis of the Potential Role of Biomass Energy for Reducing Future Greenhouse Related Emissions. PNNL-11155. Pacific Northwest National Laboratory, Richland, WA.

Edmonds, J., J. Dooley and S.H. Kim. 1999. Long-term energy technology: needs and opportunities for stabilizing atmospheric CO_2 concentrations. p. 81-107 In: C.E. Walker, M.A. Bloomfield and M. Thorning (eds.), *Climate Change Policy: Practical Strategies to Promote Economic Growth and Environmental Quality*. Monograph Series on Tax, Trade, and Environmental Policies and U.S. Economic Growth. American Council for Capital Formation, Center for Policy Research, Washington, D.C.

Ellert, B.H. and J.R. Bettany. 1995. Calculation of organic matter and nutrients stored in soils under contrasting management regimes. *Can. J. Soil Sci.* 75:529-538.

Gabrielle, B., S. Menasseri and S. Houot. 1995. Analysis and field evaluation of the CERES model water balance component. *Soil Sci. Soc. Am. J.* 59:1403-1412.

Grace, P.R. and J.N. Ladd. 1995. SOCRATES v2.00 User Manual. Cooperative Research Centre for Soil and Land Management. PMB 2 Glen Osmond 5064, South Australia.

Grant, R.F. 1996a. Ecosys. p. 65-74. In: *Global Change and Terrestrial Ecosystems Focus 3: Wheat Network. Model and Experimental Metadata*. 2nd Edition. GCTE Focus 3 Office, NERC Centre for Ecology and Hydrology, Wallingford, Oxon, U.K.

Grant, R.F. 1996b. Ecosys. p. 19-24. In: *Global Change and Terrestrial Ecosystems Task 3.3.1: Soil Organic Matter Network (SOMNET): 1996 Model and Experimental Metadata*. GCTE Focus 3 Office, NERC Centre for Ecology and Hydrology, Wallingford, Oxon, U.K.

Hess, T.F. and S.K. Schmidt. 1995. Improved procedure for obtaining statistically valid parameter estimates from soil respiration data. *Soil Biol. Biochem.* 27:1-7.

Hiley, J.C. and E.C. Huffman. 1994. Land use analysis and monitoring system to assess soil quality. p. 4-1, 4-9. In: D.F. Acton (ed.), *A Program to Assess and Monitor Soil Quality in Canada*. Soil Quality Evaluation Program Summary Report. Centre for Land and Biological Resources Research, Research Branch, Agriculture and Agri-Food Canada. Ottawa, ON.

Hiley, J.C. and R.L. Wehrhahn. 1991. Evaluation of the sustainability of extensive annual cultivation within selected agroecological resource areas of Alberta. Agriculture Canada. Edmonton. LRCC Contribution No. 91-18. 88 pp.

Hiley, J.C., K.E. Toogood, W.W. Pettapiece and R.L. Wehrhahn. 1989. Description of agricultural production in Alberta using an integrated land resource/land use database (CRITICALE-ALBERTA, 1981). Alberta Institute of Pedology No. M-89-1. University of Alberta, Edmonton. 109 pp.

Houghton, R.A. 1995. Changes in the storage of terrestrial carbon since 1850. p. 45-65. In: R. Lal, J. Kimble, E. Levine and B.A. Stewart (eds.), *Soils and Global Change*. Advances in Soil Science. Lewis Publishers, Boca Raton, FL.

Izaurralde, R.C., J. Tajek, F. Larney and P. Dzikowski. 1992. Evaluation of the suitability of the EPIC model as a tool to estimate erosion from selected landscapes in Alberta. University of Alberta, Edmonton, Alberta, Agriculture Canada and Alberta Agriculture.

Izaurralde, R.C., Y. Feng, J.A. Robertson, W.B. McGill, N.G. Juma and B.M. Olson. 1995. Long-term influence of cropping systems, tillage methods, and N sources on nitrate leaching. *Can. J. Soil Sci.* 75:497-505.

Izaurralde, R.C., W.B. McGill, D.C. Jans-Hammermeister, K.L. Haugen-Kozyra, R.F. Grant and J.C. Hiley. 1996. Development of a technique to calculate carbon fluxes in agricultural soils at the ecodistrict level using simulation models and various aggregation methods. Final Report, Agriculture and Agri-Food Canada Greenhouse Gas Initiative. University of Alberta, Edmonton. 67 pp.

Izaurralde, R.C., W.B. McGill, A. Bryden, S. Graham, M. Ward and P. Dickey. 1998. Scientific challenges in developing a plan to predict and verify carbon storage in Canadian Prairie soils. p. 433-446. In R. Lal, J. Kimble, R. Follett and B.A. Stewart (eds.), *Management of Carbon Sequestration in Soil*. Advances in Soil Science. Lewis Publishers. Boca Raton, FL.

Jackson IV, R.B., A.L. Rowell and K.B. Weinrich. 1995. Spatial modeling using partially spatial data. p. 105-115. In: R. Lal, J. Kimble, E. Levine, and B.A. Stewart (eds.), *Soils and Global Change*. Advances in Soil Science. Lewis Publishers, Boca Raton, FL.

Jenkinson, D.S. and K. Coleman. 1994. Calculating the annual input of organic matter to soil from measurements of total organic carbon and radiocarbon. *European Journal of Soil Science* 45:167-174.

Jenny, H. 1941. *Factors of Soil Formation*. McGraw-Hill, New York.

Johnston, A.E. 1994. The Rothamsted Classical Experiments. p. 9-37. In: R.A. Leigh and A.E. Johnston (eds.), *Long Term Experiments in Agricultural and Ecological Sciences*. CAB International, Oxon, U.K. 428 pp.

Lal, R., J. Kimble, R. Follett and B.A. Stewart (eds.). 1998a. *Management of Carbon Sequestration in Soil*. Advances in Soil Science. Lewis Publishers, Boca Raton, FL.

Lal, R., J. Kimble, R. Follett and B.A. Stewart (eds.). 1998b. *Soil Processes and the Carbon Cycle*. Advances in Soil Science. Lewis Publishers, Boca Raton, FL.

Lal, R., J. Kimble, R. Follett and C.V. Cole (eds.). 1998c. *The Potential of U.S. Cropland to Sequester Carbon and Mitigate the Greenhouse Effect*. Ann Arbor Press, Chelsea, MI. 126 pp.

Larney, F.J., R.C. Izaurralde, H.H. Janzen, B.M. Olson, E.D. Solberg, C.W. Lindwall and M. Nyborg. 1995. Soil erosion-crop productivity relationships for six Alberta soils. *J. Soil Water Cons.* 50:87-91.

Li, C., S. Frolking and R. Harriss. 1994. Modeling carbon biogeochemistry in agricultural soils. *Global Biogeochemical Cycles* 8:237-254.

MacMillan, R.A., W.W. Pettapiece, S.C. Nolan and T.W. Goddard. 2000. A generic procedure for automatically segmenting landforms. *J. Fuzzy Sets Systems* 113:81-109.

McGill, W.B. 1996. Evaluation of the soil organic matter models. p. 111-132 In: D.S. Powlson, P. Smith and J.U. Smith (eds.), *Evaluation of Soil Organic Matter Models Using Existing Long-Term Datasets*. NATO ASI Series I, Vol. 38, Springer-Verlag, Heidelberg.

McGill, W.B., Y.S. Feng and R.C. Izaurralde. 1996. Soil organic matter dynamics: from past frustrations to future expectations. Soil Biology Symposium, Solo/Suelo 1996 XIII Congresso Latino Americano de Ciéncia do Solo, August 4-8 1996, Águas de Lindóia, São Paulo, Brazil.

Molina, J.A.E., A. Hadas and C.E. Clapp. 1990. Computer simulation of nitrogen turnover in soil and priming effect. *Soil Biol. Biochem.* 22:349-353.

Nyborg, M., E.D. Solberg, R.C. Izaurralde, S.S. Malhi and M. Molina-Ayala. 1995. Influence of long-term tillage, straw and N fertilizer on barley yield, plant N uptake and soil-N balance. *Soil Tillage Res.* 36:165-174.

Ott, L. 1984. *An Introduction to Statistical Methods and Data Analysis.* PWS Publishers, Boston.

Parton, W.J. and P.E. Rasmussen. 1994. Long-term effects of crop management in wheat-fallow: II. CENTURY model simulations. *Soil Sci. Soc. Am. J.* 58:530-536.

Paul, E.A., K. Paustian, E.T. Elliott and C.V. Cole (eds.). 1997. *Soil Organic Matter in Temperate Agroecosystems: Long-Term Experiments in North America.* Lewis Publishers, Boca Raton, FL.

Pettapiece, W.W. 1989. Agroecological resource areas of Alberta. Map, scale 1:2M. Agriculture Canada, Alberta Land Resource Unit, Edmonton.

Powlson, D.S., P. Smith and J.U. Smith (eds.). 1996. Evaluation of soil organic matter models using existing long-term datasets. NATO ASI Series I, Vol. 38, Springer-Verlag, Heidelberg.

Rosenberg, N.J., R.C. Izaurralde and E.L. Malone (eds.). 1999. *Carbon Sequestration in Soils: Science, Monitoring and Beyond.* Battelle Press, Columbus, OH. 201 pp.

Scharpenseel, H.W. 1993. Major carbon reservoirs of the pedosphere; source-sink relations; potential of D ^{14}C and δ ^{13}C as supporting methodologies. *Water Air Soil Pollution* 70:431-442.

Seligman, N.G. and van Keulen, H. 1981. PAPRAN: A simulation model of annual pasture production limited by rainfall and nitrogen. p. 192-221 In: M.J. Frissel and J.A. van Veen (eds.), *Simulation of Nitrogen Behaviour of Soil-Plant Systems.* Pudoc Centre for Agricultural Publishing and Documentation, Wageningen, The Netherlands.

Soil Science Working Group. 1993. Soil Names File - Generation 2, Users Handbook, L.J. Knapik and J.A. Brierley (eds.). Alberta Land Resource Unit, Agriculture and Agri-Food Canada.

Whitmore, A.P. 1991. A method for assessing the goodness of computer simulation of soil processes. *J. Soil Sci.* 42:289-299.

Williams, J.R., 1995. The EPIC model. p. 909–1000. In: V.P. Singh (ed.), *Computer Models of Watershed Hydrology.* Water Resources Publications, Highlands Ranch, CO.

CHAPTER 38

Some Factors Affecting the Distribution of Carbon in Soils of a Dryland Agricultural System in Southwestern Australia

R.J. Harper and R.J. Gilkes

I. Introduction

Soil carbon is not randomly distributed across landscapes, being affected by the factors which affect pedogenesis generally (Birkeland, 1984) as well as by soil management practices and erosion (Lal et al., 1998). An understanding of the influence of these factors on soil carbon contents is essential as it will form the basis for scaling up point observations of carbon contents and fluxes to project, regional and national scales and allow the development of efficient sampling schemes. This is a key issue both for developing reliable national estimates of carbon inventories as part of the Kyoto Protocol (United Nations, 1997) and also as a means of devising efficient monitoring protocols for carbon emissions trading.

For example, as part of Australia's commitments to the Kyoto Protocol an inventory of Australia's soil carbon stocks is being developed under the National Carbon Accounting System. Similarly, an estimate will have to be made of the changes in soil carbon in the 2008-2012 compliance period, as part of an assessment of whether Australia's carbon emission targets have been met (Australian Greenhouse Office, 1999). Soil carbon may also play a role in carbon emissions trading, based on changes in land-use, such as afforestation and reforestation of farmland (Australian Greenhouse Office, 1998) or changing agricultural practice (Lal et al., 1998).

Estimates of the effects of afforestation and reforestation on soil carbon may also be required. Farmland afforestation not only has the potential to sequester large amounts of carbon, but will also produce other significant benefits (Shea, 1998). In much of southern Australia, for example, there is a strong need to establish trees to control degradation of agricultural land due to salinization, waterlogging and wind-erosion. In southwestern Australia alone it is estimated that 3 Mha out of 18 Mha of farmland will require afforestation to reverse salinity (Government of Western Australia, 1996). The rate of landholder adoption of afforestation in the absence of commercial returns has been poor; however, payment for sequestered carbon, both in the trees and soil, through a carbon emissions trading system (Shea, 1998), could increase the rate of afforestation.

Unlike other developed countries (Hodges, 1991) Australia does not have a national soil survey, and that map coverage that does exist is uneven in quality (McKenzie, 1991). Most soil mapping of the agricultural areas of Western Australia is at scales of 1:100,000 with the remainder at 1:2,000,000 (Northcote et al., 1967), with little mapping at larger scales. Many surveys provide limited soil analytical data; a survey of 336,000 ha (Overheu et al., 1993), for example, only characterized ten pedons. Where pedon data are available this is often from virgin soils (e.g., McArthur, 1991). Gifford (1998) describes some general limitations of Australian soil carbon data including (a) poor quantification of soil carbon contents below 10 or 30 cm, (b) few bulk density data and (c) poor characterization of soil carbonates in terms of amounts, distribution and composition.

Consequently, our knowledge of the distribution of carbon in the soils of southwestern Australia is rudimentary. One approach to provide this knowledge is to undertake extensive and costly soil sampling. We suggest another approach in that significant prior knowledge about the factors that can affect the distribution of soil properties across landscapes, in the form of soil-landscape models, is available (Mulcahy, 1960; Bettenay and Hingston, 1964; McArthur, 1991). It may be possible to use this knowledge to predict soil carbon distribution, despite the soil-landscape models often not explicitly describing carbon distribution.

The purpose of this chapter is to test the above hypothesis. It describes the systematic variations in soil carbon in a landscape in semi-arid south-western Australia, where a strong understanding has been developed of the factors which control the distribution of soil properties (Harper, 1994). It will be demonstrated that variations in soil carbon can be related to geomorphological and pedological processes and can be perturbed by land management practices. It should be stressed that the intensity of sampling and soil carbon analyses available for this study area is atypical for this region. Thus, the emphasis of this chapter will be on portraying trends in soil carbon, to provide a framework for future sampling schemes, rather than developing an expert system for managing data, which are unlikely to be routinely available.

II. Methods

A. Study Area

An area of 5 000 ha located 400 km south–east of Perth, Western Australia (118°45'—118°51' E; 33°43'—33°45' S), was selected as having soils and landforms representative of the general area (Figure 1). This area has a semi-arid Mediterranean climate, with a seasonal drought from November to April, a mean annual rainfall of 350 mm and mean annual evaporation of ~1600 mm. Mean minimum and maximum temperatures are 12.4°C and 21.7°C, respectively. This area comprised parts of four farms that were mostly developed from mallee woodland (*Eucalyptus* spp.) from 1964 to 1968. Farming involves annual rotations of cereal (*Triticum aestivum, Hordeum vulgare*) or legume (*Lupinus angustifolius*) crops with improved annual legume (*Trifolium subterraneum*) and grass (*Lolium rigidum*) pastures (Harper, 1994).

B. Geomorphological and Soil Survey

A general-purpose soil survey, using the free survey technique (Dent and Young, 1981), was undertaken at a scale of 1:12 500, at a mean observation density of 0.3 inspections ha^{-1}. Inspection holes mostly extended through sandy topsoils to a depth of 10 cm into underlying clayey subsoil horizons, to a maximum total depth of 300 cm. The soils were described according to McDonald and Isbell (1984). The soils and geomorphology of the study area are described in detail elsewhere (Harper, 1994). Six Geomorphic Units, comprising a total of 15 Soil Series, were identified.

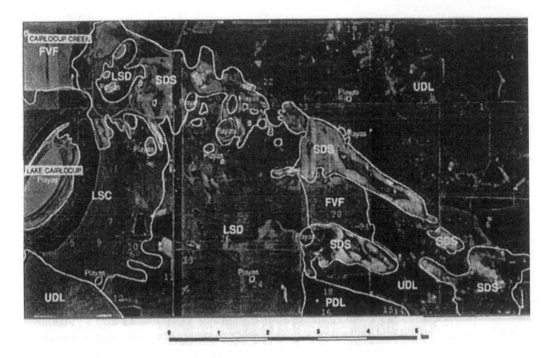

Figure 1. Vertical aerial photograph of the Cairlocup Study Area, showing the major Geomorphic Units as described in the text and pedon locations. Major landscape features include (a) playas, (b) single or multiple arrays of lunettes southeast of the playas and (c) sand dunes and sheets that occur southeast of Cairlocup Creek. Results of severe wind erosion of 1980–1981 are readily apparent.

C. Sampling

Surface soils: Samples were selected randomly from previously cultivated areas of up to 0.25 ha from each of the Soil Series. Each of the 219 samples were composites of 20 to 30 cores taken from a depth of 0 to 10 cm, the depth of sampling routinely used for soil testing in Western Australia. This also corresponds to the mean depth of the Ap horizon. For 133 texture contrast soils with a sandy E horizon paired samples were taken from 0 to 10 and 10 to 20 cm.

Pedons: 27 pedons, representing modal soils, were excavated to depths of at least 100 cm, described and sampled by horizon. Several were augered to depths of up to 4 m. Samples from each horizon were ~1 kg in weight, apart from those from horizons containing pedogenic segregations which ranged up to 6.5 kg.

D. Soil Analysis

Samples were air dried, gently crushed, and passed through a 2-mm sieve. The >2 mm fraction was washed, oven dried and quartz, calcrete and ferricrete segregations separated and weighed. Particle size analysis was performed by the pipette method (Gee and Bauder, 1986) following pre-treatments to remove organic carbon, carbonates and salts as necessary, and the clay (<2 μm), silt (2 to 20 μm), fine sand (20 to 200 μm) and coarse sand (200 to 2000 μm) fractions determined. Mean sand size was calculated from the fractions derived from sieving sand samples (45 to 2000 μm) at 0.5Φ intervals. Exchangeable cations (Ca, Mg, Na, K) were determined by the method of Gillman and Sumpter (1986). This technique is unbuffered and as dissolution of carbonates apparently occurred in alkaline soils, results are only reported for soils with pH <7. pH and electrical conductivity were determined on 1:5 soil: water extracts (Rayment and Higginson, 1992).

Soil organic carbon (SOC) content was determined by spectrophotometer (Surface samples) or by titration (Pedon samples) following wet oxidization by a dichromate-sulfuric acid mixture (Nelson and Sommers, 1982). Organic carbon contents (%w/w) were converted to an areal basis (t ha^{-1}) using the sampling depth, the mass of the >2 mm fraction, and an assumed bulk density of 1.5 t m^{-3}, a value representative of uncompacted sandy surfaced Western Australian soils (Tennant et al., 1992). Charcoal, a major feature of some Australian soils, was not quantified.

Total carbonate content in the <2 mm fraction was determined for a sub-set of 76 surface soil samples and all profile samples by the Calcium Carbonate Equivalent (CCE) method of Moore et al. (1987). This method is based on the measurement of the equilibrium pH of a dilute acetic acid solution following shaking with a soil sample for several hours. The CCE content was derived from the pH values using a calibration curve obtained for standard calcium carbonate and this expressed in terms of elemental carbon.

X-ray diffraction (XRD) was used to identify carbonate minerals. Samples were scanned at 0.5° min^{-1} using a Cu Kα radiation source, with a Philips PW1140 X-ray generator and a PW1050 diffractometer, operated at 50 kV and 20 ma. The diffractometer was equipped with a 1° divergence slit, a 0.3 mm receiving slit and a 4° anti-scatter slit. Soils containing abundant soft carbonate were scanned between 25 to 40° 2θ, and the calcrete nodules and concretions between 4 to 65° 2θ.

For the pedon samples, total carbon in the surface meter was estimated from the CCE analysis (carbonates in the <2 mm fraction), the mass of calcrete (carbonate segregations >2 mm) and the SOC carbon content. For each horizon these values were expressed in terms of the whole soil, taking into account the thickness of each horizon. It was assumed that the calcretes contained 70% carbonate. Bulk density data were absent and a value of 1.5 g m^{-3} was again assumed.

E. Erosion Assessment

A remote sensing analysis of wind erosion patterns in 1980 and 1981 was used to indicate the maximum extent of erosion. In this analysis (Carter and Houghton, 1984), Landsat-3 Multispectral Scanner (MSS) images were classified into eroded and non-eroded areas, based on reflectance values. Although erosion may have occurred in other years, on other soils, it is assumed that the extensive nature of the erosion in 1980 and 1981 will indicate the relative erodibility of the different soils. Sampling sites from the present study were classified as eroded or non-eroded on the basis of this classification (Harper and Gilkes, 1994a). ^{137}Cs specific activity (mBq g^{-1}) was determined on a sub-set of 142 sites (Harper and Gilkes, 1994a). ^{137}Cs activities (mBq g^{-1}) were converted to an areal basis (Bq m^{-2}), using the same procedure as for SOC.

F. Data Analysis

Soil attributes were log transformed to make data conform to the assumptions of the General Linear Model. The success of the classification, for each attribute, was assessed with a one-way analysis of variance (ANOVA), with the proportion of the variance accounted for related to the intra-class correlation (Webster and Oliver, 1990).

III. Results and Discussion

A. Geomorphic Units and Associated Soils

The study area comprises a valley floor (290 m a.s.l) approximately 4 km wide bounded by gently undulating ridges with up to 40 m relative relief. The ridges are comprised almost entirely of deeply weathered granitic rocks (Harper, 1994). The landscape was described with a framework of six Geomorphic Units and hence six parent materials, which form a repeating pattern in the landscape (Figure 1). These units contain fifteen distinct Soil Series and could form a basis for presenting soil information at scales of 1:50,000 or smaller.

The six Geomorphic Units can be considered in terms of the following three groups:

1. Deeply weathered granitic ridges and slopes

Two Geomorphic Units were defined in the deeply weathered granitic terrain, based on the degree of landscape dissection. Here the soil parent materials have been deeply weathered (lateritised), and subsequently dissected to various extents, with the soil parent materials comprising the various horizons of the previous lateritic profiles (Mulcahy, 1960). Unit UDL occurs where dissection appears to have been minimal and the soils have formed on ferricrete horizons, and associated slope deposits. Sandy surfaced, texture contrast soils occur within these areas (Plinthoxeralfs, Natrixeralfs (Soil Survey Staff, 1987); Dy 3.81, 3.84, 5.86; Gn 2.21 (Northcote, 1979)).

Unit PDL occurs where the pallid zone, a deeper horizon of the former lateritic profile has been exposed. Sandy surfaced texture contrast soils also occur within these areas (Natrixeralfs; Dy 2.53, 2.83, 4.86), often containing calcrete nodules or concretions.

2. Valley floor

The valley floors are poorly drained and comprise sequences of Quaternary sediments (Thom et al., 1984) here dominated by Lake Cairlocup, a 200 ha hypersaline playa. This is bounded by a series of source bordering lunettes (clay or parna dunes) which extend 5 km to the southeast (Harper, 1994). Playas with similar features are described in Texas (Reeves, 1965). These clay dunes were formed from materials eroded by wind from the playa bed during former dry periods (Bowler, 1973) and represent a chronosequence, with the youngest materials closest to the playa shore (Harper, 1994). Pedogenesis of the clayey sediments has produced an array of soils.

Those areas of the valley floor with lunettes and associated swales have been separated into two Geomorphic Units — those comprised of loamy soils (Xerochrepts; Dy 2.13, Gc 1.12, Gn 3.93), which occur immediately adjacent (0-2 km) to Lake Cairlocup (Unit LSC), and those with sandy texture contrast soil profiles (Natrixeralfs; Dy 2.83, 3.43, 3.83, 5.83), some distance (2 to 5 km) from Lake Cairlocup (Unit LSD). A third, small, Geomorphic Unit (Unit FVF) was defined for those areas of the valley floor, without playas or lunettes. This contained texture contrast soils with carbonate nodules at depth (Natrixeralfs; Dy 2.23).

3. Sandy eolian deposits

Unit SDS occurs as a discontinuous, NW-SE oriented, strip ~10 km long and 2 km wide, directly southeast of the ephemeral Cairlocup Creek. Deep (>100 cm) quartzose sands (Typic Quartzipsamments; Uc 2.21, 2.23) overlay a range of substrates and landscape positions. This feature most likely represents former quartzose eolian deposits, derived from an ephemeral drainage line.

Although illitic and smectitic clays occur in the more clayey soils adjacent to playas, the clay fraction in the sandy surfaced soils is comprised predominantly of kaolin and iron oxides (Singh, 1991).

B. Spatial Pattern of Soil Carbon

1. Between Geomorphic Units

There is a clear spatial pattern in soil carbon in this landscape, with marked differences in the amounts of both organic (SOC) and inorganic carbon (SIC) between the Geomorphic Units.

Deep sandy soils, which occur on source-bordering sand deposits (SDS), had a median SOC content (0 to 10 cm) of 6.8 t ha^{-1} compared to values of 10.8 t ha^{-1} for both sandy lunettes and swales (LSD) and undissected laterites (UDL) (Table 1). Partially dissected laterites (PDL) (18.2 t ha^{-1}) and clayey lunettes and swales (LSC) (33.2 t ha^{-1}) contained significantly larger amounts of SOC.

Soil samples for fertilizer management in Western Australia are routinely taken from the surface 10 cm (Duncan, 1992). In this study area this generally corresponds to the median depth of the Ap horizon of 9 cm ($n = 689$). For 133 sites with sandy, texture contrast profiles, paired samples from 0 to 10 and 10 to 20 cm had mean SOC contents of 10.1 and 3.5 t ha^{-1}, respectively, with no difference in proportion between Geomorphic Units.

SOC, however, also occurs in the soils at depths >20 cm, albeit in amounts of up to 0.3%, thus SOC was determined in the top meter for 27 pedons (Table 2). The total SOC of the different Geomorphic Units varies between 22±8 t ha^{-1} (±SE) for Unit SDS and 71±9 t ha^{-1} for Unit LSC. The 0 to 10 cm and 0 to 50 cm depth increments contain 52 and 82%, respectively, of the SOC in the surface meter. Again there is no significant difference in these proportions between the Geomorphic Units, suggesting that an estimate of SOC in the top meter can be obtained by doubling the 0 to 10 cm measurements.

The amounts and forms of SIC also differ between the different Geomorphic Units (Table 2). For the pedons total SIC in the surface meter was estimated. Again, while SIC was absent in the SDS and UDL Geomorphic Units, mean values of SIC of 267±45, 208±81, 84±46 and 23±1 t ha^{-1} occurred in each of the Units LSC, LSD, PDL and FVF, respectively (Table 2). Thus, total carbon in the Geomorphic Units in this area varied from 22±8 t ha^{-1} in Unit SDS to 331±49 t ha^{-1} in Unit LSC. The form of the SIC also varied between Geomorphic Units. SIC mainly occurred in the <2 mm soil (measured as CCE) in Unit LSC, this occurring throughout the soil profiles. In contrast, the SIC in the soils of Units LSD and PDL mainly occurred as calcrete nodules or concretions that occurred in discrete horizons. SIC also occurred in several soils >1 m deep.

Soil inorganic carbon was composed of calcite and/or dolomite. Where soft carbonates occurred throughout a soil profile the proportion of dolomite increased with depth. Pedon 9 in Unit LSC is a typical example. This soil contained 3% CCE as calcite in the surface (0 to 7 cm) horizon, 24% CCE composed of calcite and dolomite (ratio 3.1:1) between 25 to 40 cm and 12% CCE calcite and dolomite (ratio 1.0:1) at 180 to 190 cm. There was no clear spatial pattern in the composition of the nodular calcretes; of eleven examined seven were composed only of calcite, two were dolomite and two calcite and dolomite. pH values of up to 10.4 in deeper horizons of some soils in Unit LSC suggest that bicarbonates are also an important component in these soils.

The pedon soil carbon data are far more variable than those for the surface soils. The coefficients of variation for the untransformed SOC data ranged from 24% (Unit PDL) to 55% (Unit LSC). Coefficients of variation, for each Geomorphic Unit, for the pedon CCE ranged from 57 to 158% and for calcrete from 78 to 172%. Despite this variability the geomorphic classification has identified gross differences in both carbon amount and type across the landscape, with the total amounts varying 15-fold between Unit SDS and Unit LSC.

Table 1. Median values for major properties of the Ap (0 to 10 cm) horizons for each of the Geomorphic Units; n is the number of sampler from each class

Geomorphic Unit	n	Organic carbon (%)	Organic carbon (t ha^{-1})	Particle size distribution (<2 mm) Clay (%)	Silt (%)	FS (%)	CS (%)	Exchangeable cations Ca (cmol kg^{-1})	Mg (cmol kg^{-1})	Na (cmol kg^{-1})	K (cmol kg^{-1})	EC (S m^{-1})	pH
UDL	59	0.72	10.8	2.6	1.2	28.7	67.0	1.24	1.20	0.07	0.09	0.01	5.8
PDL	20	1.30	18.2	6.0	3.2	32.4	57.0	6.05	3.20	0.48	0.22	0.01	6.4
LSC	42	2.22	33.2	22.1	20.4	26.0	40.0	n.d.	n.d.	1.22	2.08	0.03	8.0
LSD	73	0.72	10.8	3.0	2.1	28.7	65.5	3.00	2.10	0.08	0.17	0.01	6.0
SDS	25	0.46	6.8	1.2	0.6	25.5	72.7	1.20	0.60	0.03	0.04	0.00	6.0

PSD: clay <2, silt 2 to 20, FS 20 to 200, CS 200 to 2000 μm; EC: electrical conductivity.

Table 2. Mean contents (t ha^{-1}) of carbon as soil organic carbon (SOC) and inorganic carbon (SIC) in the top meter of pedons from within each Geiomorphic Unit; P is probability of a significant difference between units derived from a one-way ANOVA; n.s. not significant

Geomorphic Unit	n	Organic carbon						Inorganic carbon				Total carbon	
		0 to 10 cm			10 to 50 cm		50 to 100 cm	<2 mm[a]		>2 mm[b]			
		Mean	SE	%[c]	Mean	SE	Mean	SE	Mean	SE	Mean	SE	Mean	SE
							— t ha^{-1} —							
UDL	8	**14**	1	56	**8**	2	**5**	1	**0**	0	**0**	0	27	4
PDL	4	**21**	2	52	**16**	4	**5**	1	**28**	23	**56**	40	127	38
LSC	6	**36**	5	52	**24**	4	**11**	2	**214**	47	**53**	35	331	49
LSD	5	**19**	2	47	**13**	3	**8**	1	**64**	22	**144**	74	248	83
SDS	2	**9**	3	50	**7**	7	**6**	4	**0**	0	**0**	0	22	8
FVF	2	**16**	3	48	**10**	1	**7**	0	**15**	3	**8**	2	57	1
		0.0001		n.s.	0.01		n.s.		0.0003		n.s.		0.0005	

[a]Estimated from calcium carbonate equivalent (CCE).
[b]Estimated from weight of calcrete segregations and assumed 70% CCE.
[c]As proportion of top meter.

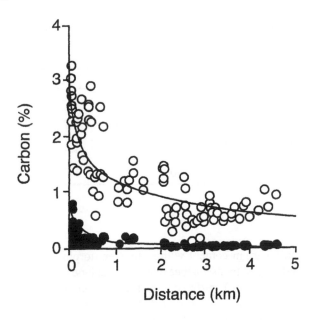

Figure 2. Changes in the content (%) of SOC (○) and <2mm SIC (●) with distance (km) from a large playa, Lake Cairlocup, aligned with the prevailing wind (NW-SE). Samples from 0 to 10 cm. SOC (%) = −0.95 log distance (km) + 1.21, r^2 = 0.69; SIC (%) = −0.19 log distance (km) + 0.11, r^2 = 0.65).

2. Within Geomorphic Units

a. Lunette Sequence

The systematic decrease in SOC and SIC contents along the sequence of Geomorphic Units increasingly distant from the playa (LSC-LSD-PDL), indicated in Tables 1 and 2, is confirmed by examination of the surface samples southeast (down-wind) of Lake Cairlocup. There are logarithmic decreases in the values of both SOC and SIC with distance from the playa explaining 69% of the variation in soil SOC content and 65% of that in <2 mm SIC (Figure 2). The contents of clay, silt and exchangeable cations also decrease in a similar fashion with distance (Harper, 1994).

Carbonates, as recorded from field observations during the initial soil survey, occur in a discrete plume southeast of Lake Cairlocup (Harper, 1994), not only in the lunettes and swales, but also in the deeply weathered Unit PDL. Here ferricrete, which is generally indicative of a prior lateritic deep weathering environment, is cemented by calcrete. Where calcrete nodules occur in lunettes (of Unit LSD) they are unequivocally pedogenic as they are beyond the size of material that can be carried by saltation, which was the mode of transport of the clayey parent materials.

The lunette/swale array is most likely a chronosequence, overlaid with a plume of eolian dust (desert loess) derived from the playa, which contained both carbonates and illitic clay minerals. These suspension sediments, coeval with the clay dunes that are eolian saltation deposits, were subsequently pedogenically modified (Harper, 1994). This model does not therefore preclude the localized redistribution of calcretes by groundwater. The eolian transport of clay materials, and carbonate, from playas has been previously reported in this environment by Bettenay (1962).

Table 3. Median soil properties for a toposequence on undissected lateritic terrain (Geomorphic Unit UDL); (0 to 10 cm sample)

Geomorphic Unit	A+E depth		Organic carbon		Particle-size distribution (<2 mm)				pH
					Clay	Silt	FS	CS	
	$\underline{n}$	(cm)	(%)	(t ha^{-1})	—————— % ——————				
Crest	11	30	0.78	11.7	4.6	1.7	28.7	63.9	5.6
Midslope	17	35	0.67	10.0	2.3	1.4	27.7	67.8	6.0
Footslope	12	50	0.60	9.0	2.0	1.0	26.6	70.2	6.0

b. Laterite Toposequence

Variations in soil organic carbon content within the Geomorphic Unit UDL were examined by grouping samples according to a toposequence of crest, midslope or foot-slope (Table 3). There was a small, but insignificant ($P<0.06$), difference in the SOC content with slope position, despite systematic, and highly significant ($P<0.0001$) differences in other soil attributes (clay and silt contents, pH, depth of sand horizon) with slope position.

Thus, better definition within this Geomorphic Unit does not provide a better indication of carbon distribution. This suggests that in some environments the use of soil-landscape models, which rely on assumptions of slope mediated distribution of soils, will not provide any useful information about carbon distribution.

C. Relationship of Soil Carbon to Other Soil Attributes

Although each of the geomorphic surfaces contains an array of soils, increasing SOC values are generally related to increasing soil clay contents (Table 1). Non-parametric Spearman rank correlations were determined between different soil attributes (Table 4) with SOC being strongly related to several other soil attributes, such as clay and silt contents. Many of these differences are related to differences in clay content, with a bivariate relationship between log clay and log SOC explaining 66% of the variation. In this study area clay can be regarded as a surrogate property that indicates both the chemical and physical fertility of the soils.

D. An Alternative Classification Using Surface Horizon Properties

The Geomorphic Units may thus have been able to portray differences in soil SOC content due partly to differences in clay contents between Units and the subsequent strong relationship between SOC and clay content. Given that the derivation of the Geomorphic Units is relatively subjective, and these Units contain a range of soils, the question arises as to whether a classification based on clay content, or its surrogate field texture, may be more reliable and furthermore be more transferable to other areas.

To examine this hypothesis, the 15 identified Soil Series, which contained a total of 27 phases, were combined into five classes (FC I–V), derived on the basis of differences in field texture and consistence of the surface (A+E) horizons as described elsewhere (Harper and Gilkes, 1994c). These classes encompass several pedological classes, and have a clear spatial distribution. Such a classifi-

Table 4. Matrix of Spearman rank correlation coefficients (p) between selected soil attributes (0 to 10 cm sample); (n = 219)

	OC (%)	Clay (%)	Silt (%)	pH	EC (S m^{-1})	Mean sand size (Φ)
OC (%)	1.00					
Clay (%)	0.87	1.00				
Silt (%)	0.83	0.87	1.00			
pH	0.62	0.56	0.62	1.00		
EC (S m^{-1})	0.84	0.79	0.82	0.66	1.00	
Mean sand size (Φ)	0.35	0.41	0.33	0.28	0.31	1.00

cation has been useful for describing the distribution of water repellency (Harper and Gilkes, 1994c), hardsetting (Harper and Gilkes, 1994b) and wind erodibility (Harper and Gilkes, 1994a) in the same field area. This is consistent with the approach of Bouma (1989), who considered that soils could be grouped in terms of functional rather than pedological differences.

Field texture and consistence are broadly related to clay content. Clay contents increased systematically with successive Field Classes (Table 5). The differences between successive Field Classes are highly significant ($P<0.0001$), when mean log–transformed values were compared, although these differences are often only small. There is a similar systematic increase in SOC, with increasing Field Class.

For each attribute the overall ANOVA model was highly significant ($P<0.0001$). The success of the classification in explaining variation, however, varies between attributes. For those attributes strongly related to the field properties used to set up the classification (texture, consistence) such as clay content, the classification explained between 89% of the total variation. Where there is a reasonable underlying relationship between clay content and other soil attributes (such as SOC) the classification provides a reasonable separation between classes (53%). For attributes unrelated to the classifying criteria (e.g. mean sand size) the classification is poor (21%). We will return to the implications of this for soil carbon prediction using soil or geomorphic surveys in the General Discussion.

IV. Effects of Wind Erosion on Soil Carbon Contents

In some cases SOC content may be affected by factors which were not used to derive the field classification. The effects of wind erosion of soil SOC contents were estimated from a remote sensing study and cesium-137 analysis, with results being previously reported by Harper and Gilkes (1994a). Wind erosion was confined to soils with <5% clay. A highly significant difference ($P<0.0001$) in ^{137}Cs contents existed between eroded and non-eroded sandy soils (<5% clay) with mean activity values of 243±17 Bq m^{-2} and 386±13 Bq m^{-2}, respectively. Non-eroded soils with higher clay contents had a mean ^{137}Cs content of 421±26 Bq m^{-2}. There were highly significant ($P<0.0001$) differences in the mean SOC contents (7.2±0.4 vs 11.3±0.5 t ha^{-1}) of the surface 20 cm between eroded and non-eroded soils. The relationship between soil SOC and ^{137}Cs contents for eroded soils was highly significant ($P<0.0001$) but weak ($r^2 = 0.25$).

An assumption of many ^{137}Cs studies is that uniform labeling has occurred across the landscape, and that erosion can be determined by comparing small numbers of samples from purportedly stable "reference" sites and those that are thought to be eroded. There is, however, considerable variability

Table 5. Median values for major properties of the Ap (0 to 10 cm) horizons for each of the field classes (FC I to V)

Field Class	n	Organic carbon (%)	Organic carbon (t ha^{-1})	Particle size distribution (<2 mm) Clay (%)	Silt (%)	FS (%)	CS (%)	Exchangeable cations Ca (cmol kg^{-1})	Mg	Na	K	EC (S m^{-1})	Mean sand size (ϕ)
I	45	0.48	7.1	1.5	0.9	26.1	71.2	1.23	0.24	0.04	0.05	0.004	1.72
II	50	0.66	9.9	2.5	1.6	28.0	67.3	1.54	0.35	0.08	0.14	0.007	1.68
III	61	0.92	13.4	4.1	2.1	30.6	62.6	1.82	0.51	0.09	0.14	0.007	1.78
IV	20	1.41	21.0	8.1	6.6	31.9	53.3	4.78	1.91	0.46	0.48	0.015	1.85
V	42	2.21	33.2	22.1	20.4	26.0	40.0	n.d.	n.d.	1.21	2.08	0.031	1.91

in the ^{137}Cs values, for both eroded and non-eroded areas, as previously reported in other areas (Lance et al., 1986; Sutherland, 1991). Consequently, although ^{137}Cs could discriminate between eroded and non-eroded areas, it was estimated that from 27 to 96 replicate samples were required to provide statistically valid estimates of ^{137}Cs loss in this study area. With respect to variability and the need for adequate sampling ^{137}Cs is therefore no different from other soil attributes, but significantly more costly to analyze.

V. General Discussion

This chapter has demonstrated that the distribution of soil carbon across the landscape is not random, and does vary with other factors including soil parent material, soil properties and past erosion. It is similarly likely to vary with soil management. Soil-landscape studies may thus provide a broad framework for scaling up individual point observations to the farm and catchment scale; however, there is obviously a need to apply the correct soil-landscape model. Thus a model of eolian deposition explained strong differences in both soil SOC and SIC, whereas a model relying on differences in slope position showed no difference in soil SOC.

A well-recognized feature of soil surveys is that if the attributes of interest are not correlated with those used to develop the field classification, no amount of grouping in one set will affect groups in the other (Butler, 1980). This provides a general indication of the usefulness of soil or geomorphic surveys for predicting the distribution of carbon. In this study area the soil survey was able to provide information about SOC distribution because of the underlying relationships between field texture, clay and SOC contents. Such relationships, however, do not hold in all areas (Webster and Butler, 1976), thus soil surveys will not be expected to be a useful predictor of SOC in all areas. Where strong relationships do occur, these relationships may be localised, indicating the need to develop and calibrate pedo-transfer functions at a local level.

Poor success from soil and landscape classifications is also expected where SOC has been substantially affected by past soil management practices, particularly where these practices are applied across different soils. Thus, there was a strong influence from wind erosion on soil SOC, which was resolved using a previous remote sensing analysis and ^{137}Cs as an indicator of erosion. Losses of SOC in this area could be estimated because of a prior remote sensing analysis; however, such analyses will not be available in most areas. This, and the influence of other factors that affect soil organic matter content (such as tillage practice) which are not considered here, will further degrade the usefulness of both the geomorphic and soil mapping approaches in estimating soil carbon.

This chapter has only been concerned with quantities of soil carbon, whereas a crucial component of carbon accounting will be in determining fluxes of carbon, particularly in response to changes in land management. Although the SIC are a major pool of carbon in these soils fluxes in response to soil management are not clear. Given the acidifying nature of southwestern Australian farming systems (Porter, 1992), it is feasible that some SIC and in particular the soft calcite segregations could dissolve with agricultural soil management, releasing carbon. Of more immediate importance, however, and before further quantification of different pools of carbon and their landscape distribution is undertaken, is to quantify the changes in SOC from changes in land management activities such as afforestation and changing tillage practice. This is a priority for future research.

References

Australian Greenhouse Office. 1998. Greenhouse Challenge Vegetation Sinks Workbook. Quantifying Carbon Sequestration in Vegetation Management Projects. Version 1.0. Australian Greenhouse Office, Canberra.

Australian Greenhouse Office. 1999. Strategic Plan for a National Carbon Accounting System for Land Based Sources and Sinks 1999-2001. Commonwealth of Australia, Canberra.

Bettenay, E. 1962. The salt lake systems and their associated aeolian features in the semi-arid areas of Western Australia. *J. Soil Sci.* 13:10–17.

Bettenay, E. and F.J. Hingston, 1964. Development and distribution of soils in the Merredin area, Western Australia. *Aust. J. Soil Res.* 2:173-186.

Birkeland, P.W. 1984. *Soils and Geomorphology*. Oxford University Press, New York.

Bouma, J. 1989. Using soil survey data for quantitative land evaluation. *Adv. Soil Sci.* 9:177–213.

Bowler, J.M. 1973. Clay dunes: their occurrence, formation and environmental significance. *Earth–Science Reviews* 9:315–338.

Butler, B.E. 1980. *Soil Classification for Soil Survey*. Clarendon Press, Oxford.

Carter, D.J. and H.J. Houghton. 1984. Integration of Landsat MSS and auxiliary data for resource management – Lake Magenta area, Western Australia. p. 207-215. In: E. Walker (ed.), LANDSAT 84. Third Australian Remote Sensing Conference, Brisbane.

Dent, D. and A. Young. 1981. *Soil Survey and Land Evaluation*. Allen and Unwin, London.

Duncan, I. 1992. Understanding and interpreting a CSBP soil analysis service report. *Productivity Focus* 10: 1-6.

Gee, G.W. and J.W. Bauder. 1986. Particle-size analysis. p. 383-411. In: A. Klute (ed.), *Methods of Soil Analysis, Part 1. Physical and Mineralogical Methods*. Agronomy Monograph N° 9. American Society of Agronomy—Soil Science Society of America, Madison, WI.

Gifford, R., 1998. Soil carbon fluxes: uncertainty and methodology development. p. 245-262. In: National Carbon Accounting System Expert Workshop Report. Australian Greenhouse Office, Canberra.

Gillman, G.P. and E.A. Sumpter. 1986. Modification to the compulsive exchange method for measuring exchange characteristics of soils. *Aust. J. Soil Res.* 24:61-66.

Government of Western Australia. 1996. *Western Australian Salinity Action Plan*. Government of Western Australia, Perth.

Harper, R.J. 1994. The nature and origin of the soils of the Cairlocup area, Western Australia, as related to contemporary land degradation. Ph.D. dissertation, The University of Western Australia.

Harper, R.J. and R.J. Gilkes. 1994a. Evaluation of the ^{137}Cs technique for estimating wind erosion losses for some sandy Western Australian soils. *Aust. J. Soil Res.* 32:1369-1387.

Harper, R.J. and R.J. Gilkes. 1994b. Hardsetting in the surface horizons of sandy soils and its implications for soil classification and management. *Aust. J. Soil Res.* 32:603-619.

Harper, R.J. and R.J. Gilkes. 1994c. Soil attributes related to water repellency and the utility of soil survey for predicting its occurrence. *Aust. J. Soil Res.* 32:1109-1124.

Hodges, J.M. (ed.). 1991. Soil survey - a basis for European soil protection. Soil and Groundwater Research Report 1. Commission of the European Communities, Luxembourg.

Lal, R., J.M. Kimble, R.F. Follett and C.V. Cole. 1998. *The Potential of U.S. Cropland to Sequester Carbon and Mitigate the Greenhouse Effect*. Ann Arbor Press, Chelsea, MI.

Lance, J.C., S.C. McIntyre, J.W. Naney and S.S. Rousseva. 1986. Measuring sediment movement at low erosion rates using cesium-137. *Soil Sci. Soc. Am. J.* 50:1303-1309.

McArthur, W.M. 1991. *Reference Soils of South-western Australia*. Australian Society of Soil Science Inc. (WA Branch), Perth.

McDonald, R.C. and R.F. Isbell. 1984. Soil profile. p. 83-126. In: R.C. McDonald, R.F. Isbell, J.G. Speight, J. Walker and M.S. Hopkins (eds.), *Australian Soil and Land Survey Field Handbook*. Inkata Press, Melbourne.

McKenzie, N.J. 1991. A strategy for coordinating soil survey and land evaluation in Australia. Divisional Report 114, CSIRO Australia Division of Soils.

Moore, T.J., R.H. Loeppert, L.T. West and C.T. Hallmark. 1987. Routine method for calcium carbonate equivalent of soils. *Commun. Soil Sci. Plant Anal.* 18:265-277.

Mulcahy, M.J. 1960. Laterites and lateritic soils in south-western Australia. *J. Soil Sci.* 11:206–225.

Nelson, D.W. and L.E. Sommers. 1982. Total carbon, organic carbon and organic matter. p. 539-579. In: A.L. Page, R.H. Miller and D.R. Keeney (eds.), *Methods of Soil Analysis. Part 2, Chemical and Microbiological Methods*. Agronomy Monograph N° 9. American Society of Agronomy–Soil Science Society of America, Madison, WI.

Northcote, K.H. 1979. *A Factual Key for the Recognition of Australian Soils*. Rellim Technical Publishers, Adelaide.

Northcote, K.H., E. Bettenay, H.M. Churchward and W.M. McArthur. 1967. *Atlas of Australian Soils. Explanatory Data for Sheet 5. Perth—Albany—Esperance Area*. CSIRO Australia and Melbourne University Press, Melbourne.

Overheu, T.D., P.G. Muller, S.T. Gee and G.A. Moore. 1993. Esperance land resource survey. Land Resources Series 8, Western Australian Department of Agriculture.

Porter, W.M. 1992. Managing soil acidity for stable cropping systems. In: G.J. Hamilton, K.M. Howes and R. Attwater (eds.), 5th Australian Soil Conservation Conference. Proceedings of the 5th Australian Soil Conservation Conference. Department of Agriculture 4:74-82.

Rayment, G.E. and F.R. Higginson. 1992. *Australian Laboratory Handbook of Soil and Water Chemical Methods*. Australian Soil and Land Survey Handbook. Inkata Press, Melbourne.

Reeves, C.C. 1965. Chronology of west Texas pluvial lake dunes. *J. Geol.* 73:504-508.

Shea, S. 1998. Farming carbon. *Landscope* 14:17-22.

Singh, B. 1991. Mineralogical and chemical characteristics of soils from south-western Australia. Ph.D. dissertation, The University of Western Australia.

Soil Survey Staff. 1987. Keys to Soil Taxonomy. SMSS Technical Monograph N° 6. Cornell University Press, Ithaca, NY.

Sutherland, R.A. 1991. Examination of caesium-137 areal activities in control (uneroded) locations. *Soil Technology* 4:33-50.

Tennant, D., G. Scholz, J. Dixon and B. Purdie. 1992. Physical and chemical characteristics of duplex soils and their distribution in the south-west of Western Australia. *Aust. J. Exp. Agric.* 32:827-843.

Thom, R., R.J. Chin and A.H. Hickman. 1984. Newdegate, Western Australia. 1:250 000 Geological Series—Explanatory Notes. Geological Survey of Western Australia, Perth.

United Nations. 1997. Kyoto Protocol to the United Nations Framework Convention on Climate Change.

Webster, R. and B.E. Butler. 1976. Soil classification and survey studies at Ginninderra. *Aust. J. Soil Res.* 14:1–24.

Webster, R. and M.A. Oliver. 1990. *Statistical Methods in Soil and Land Resource Survey*. Oxford University Press, Oxford.

CHAPTER 39

A National Inventory of Changes in Soil Carbon from National Resources Inventory Data

M.D. Eve, K. Paustian, R. Follett and E.T. Elliott

I. Introduction

Changes in land use and agricultural land management have been and are currently occurring in the United States (Kellogg et al., 1994; CTIC, 1998; Paustian et al., 1998). These changes can dramatically impact soil carbon stocks (Bruce et al., 1999; Buyanovsky and Wagner, 1998; Houghton et al., 1999; Lal et al., 1999; Paustian et al., 1997a; Paustian et al., 1997b). A means of estimating changes in terrestrial carbon storage at a national level is required to fulfill U.S. obligations under the United Nations Framework Convention on Climate Change (UNFCCC; UNFCCC, 1995). Furthermore, an accurate and defensible estimate of carbon storage in agricultural cropland soils is critical for the development of effective U.S. agricultural and environmental policies and strategies. The Intergovernmental Panel on Climate Change (IPCC) has developed a standardized inventory approach for monitoring greenhouse gas emissions (IPCC, 1997a). The approach was developed to be usable by any of the signatory nations of the UNFCCC. These include many developing countries in which data on land use and management and soil resources are limited.

In the U.S., detailed data are available that are well suited to application of the IPCC approach. Every 5 years, the U.S. Department of Agriculture Natural Resources Conservation Service (USDA-NRCS) collects detailed land use, land cover, and other natural resource related data at an extensive network of permanent sampling sites across the U.S. as part of the National Resources Inventory (NRI). Each year, the Conservation Technology Information Center (CTIC) Crop Residue Management Survey tracks changes in tillage management including shifts from conventional tillage practices to conservation tillage or no-till (CTIC, 1998). Through analysis of these data sets, an estimate of changes in soil C stocks for the U.S. can be derived.

Our research focuses on estimating net changes in soil carbon storage in U.S. croplands for the UNFCCC baseline year (1990). Estimates are derived utilizing application of the balance sheet approach to national greenhouse gas inventories that was developed by the IPCC (IPCC, 1997a, b, c). Results were previously reported that applied the IPCC inventory approach on data aggregated to six climate regions of the conterminous U.S. (Eve et al., 2000). The current work analyzes changes in soil C stocks resulting from land use change at each of the over 800,000 NRI sites individually. Evaluating C changes prior to data aggregation should allow for more precise measurement of changes in soil C. In addition to the NRI data, we used input data from sources including the Crop Residue Management Survey and the PRISM (Parameter-elevation Regressions on Independent Slopes Model) climate mapping program (Daly et al., 1994; Daly et al., 1998). These data sources provide information on land use, soils, crops, tillage practices, and climate, which we then applied in a series of spreadsheet models that estimate changes in C storage in cropland soils.

II. Materials and Methods

A. IPCC Inventory Approach

The IPCC was established in 1988 by the World Meteorological Organization (WMO) and the United Nations Environment Programme (UNEP). Its role is to assess the scientific, technical and socio-economic information relevant for understanding human-induced climate change risk, making this information available to WMO and UNEP members and the world scientific community. The IPCC provides assessments, reports, guidelines, and methodologies in support of the 1990 UNFCCC. More specifically, the IPCC has been responsible for developing and refining an internationally agreed upon technique that IPCC member countries and Parties to the UNFCCC can use in conducting and reporting greenhouse gas inventories (IPCC, 1997a).

The methods developed by IPCC for inventorying greenhouse gas emissions are comprehensive, covering the full range of anthropogenic influences on sources and sinks of greenhouse gases, such as agricultural, industrial, energy, waste, and forestry and land use activities. The section dealing with land use change accounts for changes in terrestrial carbon storage in plant biomass as well as soils. Since CO_2 is exchanged between plant/soil systems and the atmosphere through the processes of photosynthesis and respiration, net changes in plant/soil C stocks can be equated to net changes in CO_2 emissions. The research reported here focuses on changes in soil carbon stocks related to changes in land use and/or agricultural management practices. Land use change includes such activities as rangeland being converted into cropland, cropland being converted to grass or trees through enrollment in the Conservation Reserve Program (CRP), agricultural land being converted to urban development, or agricultural land being planted back into forest. Changes in agricultural management include changing cropping systems or tillage management practices. Changes in cropping systems would include activities such as shifting from a corn/soybean rotation to a corn/hay rotation or shifting from a wheat/fallow rotation to a continuous wheat rotation. Changes in tillage management would include changes in tillage practices like shifting from conventional tillage to a no-till strategy where the soil surface is left undisturbed from harvest until planting the next crop. Documentation related to the inventory methods for land use and management change can be found in IPCC Workbook Module 5 (Land-Use Change and Forestry; IPCC, 1997b) and Reference Manual Chapter 5 (Land-Use Change and Forestry; IPCC, 1997c).

The method was developed as a practical first order approach using simple assumptions about the effects of land use change on carbon stocks, and then applying those assumptions in order to estimate changes in carbon stocks due to land use change over the past 20 years. Changes in soil carbon stocks as a function of land use and land management practices are estimated using a series of coefficients based on climate, soil type, disturbance history, tillage intensity, productivity and residue management (C input rate) (IPCC, 1997b). The default method applies to the upper 30 cm of the soil profile only. The IPCC method is very general in order to facilitate broad application by Parties to the UNFCCC, but flexible enough that more detailed analysis can be conducted if data are available. Information is laid out in a series of worksheets, each related to a different source of carbon flux. The worksheets contain the formulas necessary to compute soil carbon storage (IPCC, 1997b). The authors of the approach searched the literature to establish default values for native carbon levels and changes in carbon stocks under different land use change scenarios. For the most basic application of the inventory, the investigator needs only the estimated area under each land use/management system at the beginning and end of the inventory period. If more detailed information is available for the country being inventoried, carbon values and change factors can be adjusted to make the inventory as accurate as possible.

Table 1. Description of the IPCC climate categories that occur in the conterminous U.S.

Climate zone	Zone code	Annual average temperature (°C)	Average annual precipitation (mm)	Length of dry season (months)
Cold temperate, dry	CTD	<10	<PET[a]	NA
Cold temperate, moist	CTM	<10	≥PET	NA
Warm temperate, dry	WTD	10 - 20	<PET	NA
Warm temperate, moist	WTM	10 - 20	≥PET	NA
Sub-tropical, dry	STD	>20	<1,000	Usually long
Sub-tropical, moist (w/short dry season)	STM	>20	1000 - 2000	<5

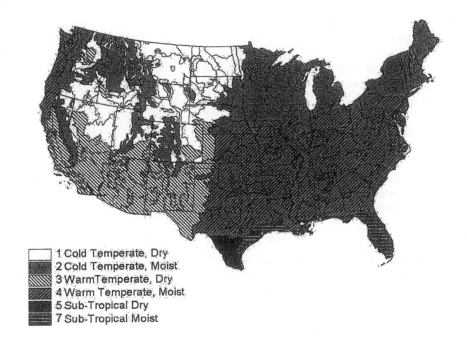

1 Cold Temperate, Dry
2 Cold Temperate, Moist
3 Warm Temperate, Dry
4 Warm Temperate, Moist
5 Sub-Tropical Dry
7 Sub-Tropical Moist

[a]PET is Potential Evapotranspiration.

Figure 1. Climatic regions of the conterminous U.S. delineated using PRISM average precipitation and temperature summarized by Major Land Resource Area (MLRA; USDA-NRCS) according to the IPCC climatic region definitions; the lines overlaying the climate zones are the state boundaries, the thin lines in the background are the MLRA boundaries. (From Eve et al., in press. With permission.)

B. Input Data Sources

The primary data requirements for the IPCC method deal with the land use and land management changes over time. However, information is also needed to stratify these changes according to climate and soil type. Under the IPCC approach, climate is divided into eight distinct categories based upon

Table 2. IPCC inventory soil categories (and associated taxonomic soil orders) and the IPCC default estimates of soil organic C that would exist under native vegetation

| IPCC inventory soil categories | USDA taxanomic soil orders | Approximate quantity of soil organic C under native vegetation by climatic region (MT C ha[-1], 0-30 cm depth) | | | | | |
		Cold temperate dry	Cold temperate moist	Warm temperate dry	Warm temperate moist	Sub-tropical dry	Sub-tropical moist
High clay activity mineral soils	Vertisols, Mollisols, Inceptisols, Aridisols and high base status Alfisols	50	80	70	110	60	140
Low clay activity mineral soils	Ultisols, Oxisols, acidic Alfisols, and many Entisols	40	80	60	70	40	60
Sandy soils	Sand and loamy sand textured Entisols	10	20	15	25	4	7
Volcanic soils	Andisols, Spodosols[a]	20	70	70	130	50	100
Aquic soils	Soils denoted as Hydric (excluding Histosols)	70	180	120	230	60	140
Organic soils	Histosols	NA[b]	NA	NA	NA	NA	NA

[a]Spodosols do not fit well within IPCC categories but were lumped with Andisols as the group having the most similar C levels across climate categories (J.M. Kimble, personal communication).
[b]Changes in soil organic C are figured separately for organic soils.
(Adapted from IPCC, 1997c.)

average annual temperature, average annual precipitation, and the length of the dry season (IPCC, 1997c). Six of these climatic regions occur in the conterminous U.S. The climatic regions are delineated primarily by temperature and precipitation (Table 1; Figure 1).

Soils are grouped by merging soil orders into one of six classes based upon texture, morphology, and ability to store organic matter (Table 2; IPCC, 1997a and 1997c).
1. High clay activity mineral soils are those defined as having a large proportion of high activity clays (i.e. expandable clays such as montmorillonite) that are effective in long-term stabilization of soil organic matter.
2. Low clay activity mineral soils are those dominated by low activity, relatively non-expandable clays (such as kaolinite or gibbsite) which have less ability to stabilize soil organic matter.
3. Sandy soils have less than 8 percent clay and greater than 70 percent sand, have poor structural stability, and poor C stabilization ability.
4. Volcanic soils include the andisols, which generally have allophane as the primary colloidal mineral and are rich in carbon and highly fertile. Spodosols are also included in this category because

their carbon content, fertility, and ability to stabilize soil organic carbon are most similar to the andisols (J.M. Kimble, personal communication[1]).

5. Aquic soils are mineral soils that have developed in wet sites with poor drainage. Under native conditions, these soils typically have high organic matter content and a reduced rate of decomposition.

6. Organic soils form under water-saturated conditions with greatly reduced decomposition and have very high organic matter content.

The IPCC method provides default estimates of C contents for each of the five mineral soil classes under native (i.e., pre-agricultural) conditions (Table 2).

The types of land use and/or land management change are defined specifically for the country being inventoried. It is important only that the systems identified capture the changes over the previous 20 years (Cannell et al., 1999; IPCC, 1997c). Land use categories used in our research are presented in Table 3.

For this type of inventory, the U.S. has better data available than many other parts of the world. We have utilized much data and conducted a detailed inventory. We started with the Major Land Resource Areas (MLRA; SCS, 1981) as our basic spatial unit. MLRAs were originally delineated by the USDA-NRCS (formerly the Soil Conservation Service; USDA-SCS) in the 1960s as a tool to assist land managers and land use planners; and have undergone occasional revisions since then. Each MLRA represents a geographic unit with relatively similar soils, climate, water resources, and land uses (SCS, 1981). The input data required for implementation of the IPCC inventory were derived as follows:

1. Climate

Climate in the U.S. is monitored through an extensive network of National Weather Service (NWS) cooperative weather stations. Other national agencies also maintain specific climate databases such as the USDA-NRCS Snotel network and the Global Gridded Upper Air Statistics database maintained by the National Climatic Data Center (NCDC). The PRISM Climate Mapping Program has combined the 1961–1990 averages from each of these stations (point data) and sources with topographic information derived from digital elevation models (DEM; grid data) to generate gridded (4-km grid cells) estimates of temperature and precipitation for the U.S. (Daly et al., 1994; Daly et al., 1998). Average annual precipitation and average annual temperature were derived for each MLRA from PRISM model outputs. These averages were used to aggregate the nearly 180 MLRAs that make up the conterminous U.S. into the six prescribed IPCC climatic zones represented within the U.S. (Figure 1).

2. Land Use

Land use and change information was derived from the NRI database developed by the USDA-NRCS. The NRI is a stratified two-stage area sample of over 800,000 points across the United States (Nusser and Goebel, 1997). Each point in the survey is assigned an area weight (i.e., expansion factor) based on other known areas and land use information so that each point has a statistically assigned area that it represents (Nusser and Goebel, 1997). Based on these area weights, the NRI data show a land base of approximately 782.5 Mha in the conterminous U.S. (Figure 2). An extensive amount of soils, land use, and land management data are collected each time all, or nearly all, sites are visited every 5 years

---

[1] Dr. John M. Kimble is a Research Soil Scientist at the USDA Natural Resources Conservation Service, National Soil Survey Laboratory in Lincoln, Nebraska.

Table 3. Land use and management categories grouped from the Natural Resources Inventory data

General land use categories	Code	Specific management related sub-categories
Agricultural land	IRR	Irrigated crops
	CRC	Continuous row crops
	CSG	Continuous small grains
	CRS	Continuous row crops and small grains
	RCF	Row crop / fallow rotation
	SGF	Small grains / fallow rotation
	SSF	Small grains, small grains then fallow rotation
	RSF	Row crops and small grains with fallow rotations
	RCH	Row crops in rotation w/ hay and/or pasture
	SGH	Small grains in rotation w/ hay and/or pasture
	RSH	Row crops and small grains in rotation w/ hay and/or pasture
	VIR	Vegetable crops (continuous or in a rotation with other crops)
	RIR	Rice (continuous or in a rotation with other crops)
	LRA	Low residue annual crops (such as cotton or tobacco)
	CPH	Continuous perennial and/or horticultural crops
	CHA	Continuous hay
	CPA	Continuous pasture
	CRP	Conservation Reserve Program (CRP)
	MIS	Other miscellaneous crop rotations
Range land	RNG	
Forest	FST	
Urban land	URB	
Water	WAT	
Federal land[a]	FED	
Miscellaneous non-cropland[b]	MSC	

[a]NRI sample points falling on U.S. federally owned land are noted as such, and no further data are collected at those sites.
[b]Miscellaneous non-cropland includes such things as roads and highways and barren areas such as mines or beaches.

(Nusser et al., 1998). NRI was designed as a tool to assess conditions and trends for soil, water, and related natural resources primarily on non-federal lands of the U.S. (Kellogg et al., 1994; Nusser and Goebel, 1997). Specifically, NRI is intended to provide monitoring and status information in support of natural resource conservation policy development and program implementation (Nusser et al., 1998). Because the data points and the information collected have remained fairly constant since 1982, NRI is a useful source for much of the data required for the U.S. inventory. Land use and land management information for 1982 and changes from 1982 to 1992, enrollment in the Conservation Reserve Program (CRP), dominant soil order and other soil characteristics were obtained from the NRI (NRCS, 1994). For the purposes of our research, the land use information in NRI was merged into a combination of land use and management systems as listed in Table 3. Each NRI point was assigned to one of these systems based upon the land use data collected in the 1982 and 1992 surveys as well as the cropping history data recorded for the 3 years prior to each survey (NRCS, 1994). Each of the

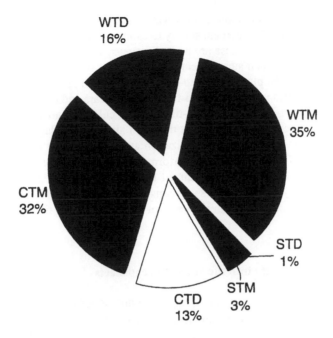

Figure 2. The amount of the conterminous U.S. that lies within each of the IPCC defined climatic regions; based on average temperature and precipitation computed using PRISM for each NRCS MLRA. By NRI estimates, the conterminous U.S. comprises approximately 782.5 Mha.

over 800,000 NRI points was assigned an aerial extent based upon the weighted expansion factors discussed earlier.

3. Soils

The dominant taxonomic soil order was obtained for each point by querying the soils database that accompanies the NRI data which is linked through a common pointer (NRCS, 1994). A relatively small percentage of the points did not have an associated soil order record in the soils database. These points were mostly in water or urban areas and did not create a problem. For the remaining NRI sites where soil information was needed and not available, the site was assigned to the low activity mineral soils IPCC category because that category would minimize potential over- or underestimation of C changes at that site.

4. Tillage Management

Reliable estimates of the adoption of improved tillage practices such as conservation tillage or no-till are critical for our inventory because much of the gain in C storage in U.S. cropland is projected to come through the adoption of these practices (Bruce et al., 1999; Lal et al., 1999; Paustian et al., 1997b). The Conservation Technology Information Center (CTIC), a nonprofit information and data transfer facility that promotes environmentally and economically beneficial natural resource systems, conducts annual surveys of the adoption of conservation tillage management (CTIC, 1998). Each year they conduct a Crop Residue Management survey to estimate the portion of cropland being managed under the various tillage systems. These are annual surveys, and indicate hectares of a specific crop planted by tillage system for each survey year. However, the survey does not account for rotations where one year's crop may be planted into no-till and the following crop planted using tillage. Nor does it account for producers who may try no-till for a few years and then revert back to some other tillage system. Hence, the areas reported in no-till for a particular year are greater than the areas under continuous no-till. Field experiments suggest that with alternating tillage and no-till management, most of the benefits of no-till for soil C stabilization are lost (Pierce et al., 1994; Reicosky, 1997; Stockfisch et al., 1999). Thus, supplemental information from CTIC was required to estimate the adoption of more continuous no-till and reduced tillage systems by region for the U.S. For our research, use of a specific system for 5 years was considered adoption. CTIC provided these estimates for both 1982 and 1992 to coincide with the years of NRI data being utilized (Table 4; D. Towery, personal communication[2]).

C. Integration of Existing Data into the IPCC Spreadsheets

Changes in soil C were estimated for each of the over 800,000 NRI points in the conterminous U.S. To do this, each point was assigned to a climatic region (based on Figure 1). The dominant soil order for each point was determined from the soils database associated with the NRI data set. Each point was assumed to represent the area (number of hectares as estimated by the NRI expansion factor) assigned to that point in the NRI data. The extent of each cultivated cropland category was further separated into relative proportions of conventional tillage, reduced tillage, and no-till hectares by utilizing the information provided by CTIC. Based upon the soil order, land use, agricultural crop rotation, and tillage management in place in 1982, the soil C stock for each point was determined using the IPCC inventory approach and default factor values. Carbon stocks for non-cropland areas were computed as:

$$C_n = ha_n * SCUN_n * BF_n *_{Ifn} \tag{1a}$$

and C stocks for cropland sites were computed as:

$$C_n = (haNT_n * SCUN_n * BF_n * IF_n * TF_{NT}) + (haRT_n * SCUN_n * BF_n * IF_n * TF_{RT})$$

$$+ (haCT_n * SCUN_n * BF_n * IF_n * TF_{CT}) \tag{1b}$$

where: C_n is the estimated 1982 C stock for the area represented by NRI point (n)

ha_n is the area (in hectares) associated with point (n) in 1982

$SCUN_n$ is the IPCC default estimate of soil C under native vegetation at point (n) given the climatic zone and soil type at that point

BF_n is the IPCC base factor, or the relative percentage of soil C that has been lost historically by point (n) being used a particular way

IF_n is the IPCC input factor for point (n), which adjusts soil C levels based upon the level of inputs such as irrigation, or utilizing hay in the rotation

---

[2] Dan Towery is Natural Resource Specialist and Certified Crop Advisor at the Conservation Technology Information Center, West Lafayette, Indiana.

haNT$_n$ is the number of hectares related to point (n) that represent no-tillage management systems in 1982

haRT$_n$ is the number of hectares related to point (n) that represent reduced tillage management systems in 1982

haCT$_n$ is the number of hectares related to point (n) that represent conventional tillage management systems in 1982

TF$_{NT}$ is the IPCC tillage factor for a no-till system

TF$_{RT}$ is the IPCC tillage factor for a reduced tillage system

TF$_{CT}$ is the IPCC tillage factor for a conventional tillage system

Table 4. Tillage system adoption by percent of cropland type for the beginning and end of the inventory period (as estimated by CTIC, D. Towery, personal communication)

	1982			1992		
System	No-till (%)	Reduced tillage (%)	Conv. tillage (%)	No-till (%)	Reduced tillage (%)	Conv. tillage (%)
Sub-tropical dry region (STD)						
Row crop rotations	0	2	98	0	4	96
Small grain rotations	0	0	100	0	2	98
Vegetables in rotation[a]	0	2	98	0	4	96
Low residue agriculture	0	2	98	0	4	96
Sub-tropical moist region (STM)						
Row crop rotations	0	0	100	0	20	80
Small grain rotations	0	0	100	0	10	90
Vegetables in rotation[a]	0	2	98	0	3	97
Low residue agriculture	0	2	98	0	3	97
Warm temperate dry region (STD)						
Row crop rotations	0	0	100	0	10	90
Small grain rotations	0	2	98	0	15	85
Vegetables in rotation[a]	0	2	98	0	1	99
Low residue agriculture	0	2	98	0	1	99
Warm temperate moist region (WTM)						
Row crop rotations	0	5	95	10	30	60
Small grain rotations	0	5	95	5	30	65
Vegetables in rotation[a]	0	8	92	1	10	89
Low residue crops	0	8	92	1	10	89
Cool temperate dry region (CTD)						
Row crop rotations	0	2	98	2	25	73
Small grain rotations	0	5	95	4	25	71
Vegetables in rotation[a]	0	0	100	1	2	97
Low residue crops[b]	0	0	100	1	2	97
Cool temperate moist region (CTM)						
Row crop rotations	0	10	90	5	30	65
Small grain rotations	0	10	90	5	30	65
Vegetables in rotation[a]	0	0	100	1	2	97
Low residue crops[b]	0	0	100	1	2	97

[a]Vegetables in rotation was assumed to follow an adoption pattern similar to the low residue crops category.

[b]Low residue crops (primarily cotton) in the CTM and CTD region make up very few acres.

The sum of the hectares (haNT), (haRT) and (haCT) in Eq. 1b for each NRI point will sum to the area related to that point (based on the NRI expansion factor). Carbon stocks for 1992 were computed the same way utilizing the 1992 data on soil order, land use, agricultural crop rotation, and tillage management related to each NRI point.

Once the soil C stocks for the 1982 inventory and the 1992 inventory were completed for each NRI point, the change in C stocks at that point was simply computed as the 1992 stock minus the 1982 stock. The IPCC defaults are designed to estimate changes over a 20-year period. The available NRI data, however, only span 10 years. In order to derive an estimate of change over a 10-year inventory period (rather than the IPCC default 20-year inventory) we scaled the amount of change by half.

Finally, the computed change in C stocks for each point was converted to an annual average change over the 10-year period. The annual average change was aggregated by type of land use change, and summed as million metric tons of C per year (MMT C yr^{-1}) for each climatic region.

Organic soils are handled differently in the IPCC approach. Organic soils that are under native vegetation are excluded from the inventory under the assumption that they are not significantly affected by human activity. Organic soils that are intensively managed are assigned a default rate of C loss based on land use system and the climatic region where they are located (IPCC, 1997c). Only two types of managed systems are considered: cropland and introduced pasture/forest (IPCC, 1997b). Estimated C loss from croplands established on organic soils is four times greater than loss from pasture and forest plantations on organic soils in the same climatic region (IPCC, 1997b).

III. Results and Discussion

A. Changes in Land Use and Management

The NRI data show that changes in land use are occurring in the United States (Kellogg et al., 1994). One of the most notable changes in land use is the enrollment of cropland in the Conservation Reserve Program (CRP) portrayed in Figure 3. CRP is a federal program of the 1985 U.S. Food Securities Act intended to take highly erodible cropped land out of agricultural production by planting it back to grass or trees for a 10-year period. As of the 1992 NRI survey, nearly 14 Mha (of the 782.5 Mha conterminous U.S. land base; Figure 2) had been enrolled in CRP (Kellogg et al., 1994). Another notable land use shift between 1982 and 1992 is the loss of native range land (Figure 4). Over 4.5 Mha was broken out of range and converted to crop and urban land uses. Possibly the most striking land use change is the over 5 Mha increase in urban area in the U.S., representing a 17% increase in the area of urban lands from 1982 to 1992 (Figure 5). Urban development often occurs on high quality farmland, and often involves removal and sale of the topsoil. As neighborhoods are established, high-input lawns and trees are established. The fate of soil C during the process of urban establishment and the decades subsequent to development are not well researched. However, results of Groffman et al. (1995) suggest that urban forests (defined to include residential lawns, parks and golf courses) can accumulate soil C stocks which exceed those of rural forests.

Management of agricultural lands changed somewhat during the period 1982 to 1992 as well. With implementation of the 1985 farm bill, more producers shifted to reduced tillage and no-till systems (Table 4). No-till was rare in 1982. By 1992, in the warmer areas of the Corn Belt (warm temperate, moist climatic region), about 10% of the area in continuous row crops was being managed under a continuous no-till system. In other crops and other areas, adoption has been lower (Table 4). Systems that reduce the intensity of tillage, however, have received more widespread adoption (Reduced Tillage, Table 4).

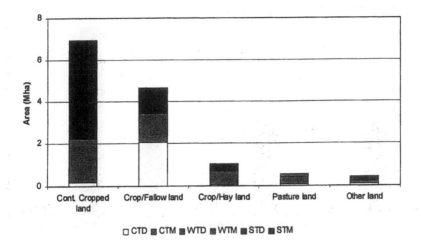

Figure 3. Land use categories from 1982 that were converted to Conservation Reserve Program by 1992 for each climatic region in the conterminous U.S. based on USDA-NRCS National Resources Inventory data. The climatic region codes relate to the region names in Table 1.

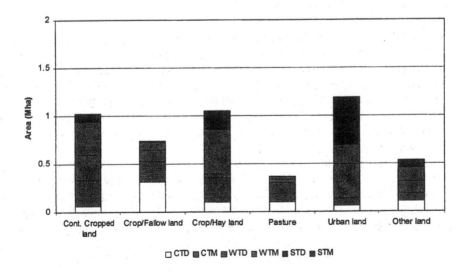

Figure 4. Lands reported as rangeland in the 1982 National Resources Inventory and converted to another land use by the 1992 inventory for each climatic region in the conterminous U.S. The climatic region codes relate to the region names in Table 1.

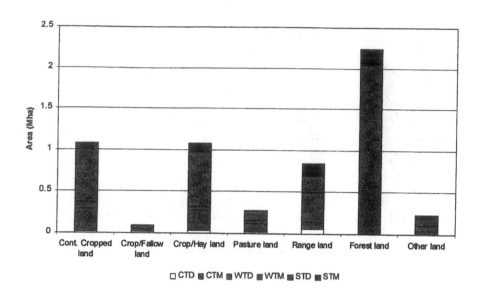

Figure 5. Land use categories from 1982 that were converted to urban land by 1992 for each climatic region in the conterminous U.S. based on USDA-NRCS National Resources Inventory data. The climatic region codes relate to the region names in Table 1.

B. Changes in Carbon Stocks in the U.S.

Our estimate indicates that changes in land use and management during the 10-year period 1982 to 1992 in the U.S. have resulted in a net increase in soil C in mineral soils for every climatic region except the sub-tropical, moist region (Table 5). The region contributing the most to the C storage is the warm temperate, moist region (Figure 6). This region is the largest of the regions (Figure 2), with soils and climate that produce high-yielding crops. It is also the area where (as noted earlier) the rate of adoption of no-till is the highest in the country. Overall, changes in land use and agricultural management on mineral soils during the period of 1982 to 1992 have produced an annual increase in soil C of 14.02 MMT C yr^{-1} (Figure 6, Table 5).

Organic soils, especially the large areas of intensively managed organic soils in the southeast U.S. (sub-tropical, moist region), continue to lose soil C at an annual rate of about 6.01 MMT C yr^{-1} (Figure 6, Table 5). In every climatic region except the sub-tropical moist region, the mineral soils are storing enough C to compensate for the loss of C from the organic soils. Combining the emissions from organic soils with the sink in mineral soils, we estimate a net sink related to changes in land use and agricultural management in the U.S. of 8.01 MMT C yr^{-1} (Figure 7, Table 5).

Our results are a conservative estimate of actual changes in soil C storage resulting from documented changes in land use and agricultural management. Lal et al. (1998; 1999) estimate the potential of agricultural soil to sequester C at between 75 and 208 MMT C yr^{-1}. Bruce et al. (1999) estimate this potential at 75.2 MMT C yr^{-1}. These projections of potential sequestration account for widespread adoption of conservation tillage techniques, intensification of agricultural cropping (increasing biomass inputs), improved irrigation management, and accelerated land conversion through CRP, buffer strips, wetland restoration, etc. Our estimate includes what actually happened in most of

Table 5. IPCC inventory estimate of annual carbon emissions due to changes in land use and agricultural management practices during the period 1982 to 1992. Based on point-to-point assessment of the USDA-NRCS National Resources Inventory (NRI). Negative values indicate a net sink of atmospheric C in the soil.

Climatic region	Estimated soil C emissions		
	Mineral soils (MMT C yr^{-1})	Organic soils (MMT C yr^{-1})	Total emissions (MMT C yr^{-1})
Cool temperate, dry	−1.0041	0.0000	−1.004
Cool temperate, moist	−3.8866	0.3907	−3.496
Warm temperate, dry	−0.8746	0.4676	−0.407
Warm temperate, moist	−9.0851	1.0580	−8.027
Sub-tropical, dry	−0.0754	0.0000	−0.075
Sub-tropical, moist	0.9035	4.0954	4.998
Total	−14.0224	6.0116	−8.010

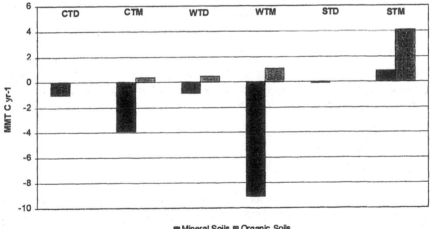

Figure 6. Estimated annual carbon emissions for mineral and organic soils resulting from land use and management change between 1982 and 1992. Emissions are shown by climatic region. Positive values represent C emissions to the atmosphere and negative values represent a C sink rate. Estimates are based upon the IPCC inventory approach and NRCS National Resources Inventory data.

these categories during the period 1982 to 1992, which is far short of the potential soil C storage ability. Furthermore, Lal et al. (1998; 1999) include the potential development of bio-fuel technologies, improved erosion control, and other soil management strategies that our research does not consider.

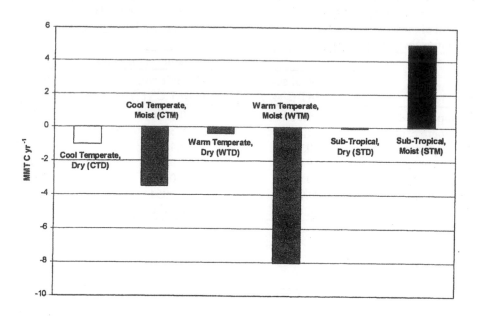

Figure 7. Net annual soil carbon emissions by climatic region combining both the mineral and organic soils. This would be the total net emissions resulting from land use and agricultural management changes between 1982 and 1992, based upon the IPCC inventory approach and NRCS National Resources Inventory data.

C. Limitations of the IPCC Approach

The IPCC inventory approach accounts for changes in C stocks resulting from changes in land use. It also takes into account agricultural land management such as cropping practices and tillage systems, and is sensitive to changes in these management practices. The approach was designed to address only "anthropogenic" sinks and sources – those occurring as a result of human management or impact on the landscape. The approach does not consider any changes in C stocks that result from natural processes or from change that occurred more than 20 years prior (10 years prior in our application). For example, agricultural land that was converted from annual crops to hay prior to the 1982 inventory period could still be accumulating soil C. This approach, however, would indicate no change in soil C because the site remained in hay throughout the inventory periods. As another example, some research indicates that native organic soils may continue to sequester C almost indefinitely (Armentano, 1980; Armentano and Menges, 1986). IPCC inventory methods do not account for this type of increase in soil C. Also, for mineral soils, the approach assumes that native portions of the landscape (including federal lands, forest, rangeland, etc.) have arrived at a long-term equilibrium of soil C and are in a steady state, neither gaining nor losing C stocks.

In computing change in C stocks for organic soils, the IPCC inventory assigns a constant emission rate for all agriculturally managed lands without regard to tillage practices. If there is a reduction in emissions resulting from less intensive disturbance (e.g., no-till management) of organic soils, the IPCC approach does not capture this change. The result would be an overestimation of C emissions from organic soils.

Detailed land use change data tend to be available only for the relatively recent past, making it difficult to incorporate more long-term trends in land use and land management. Finally, the IPCC inventory does not account for changes in soil C due to erosion and sedimentation. There is no estimate made in the method for C lost to the atmosphere during an erosion event or for C buried in sediments.

D. Uncertainty

As with any analysis, there is a level of uncertainty inherent in the estimates. Each of the input data sets has an associated level of uncertainty that gets passed through the analysis and results in uncertainty in the final estimates. It is extremely difficult, therefore, to objectively quantify the level of uncertainty in this type of analysis (Cannell et al., 1999; Houghton et al., 1999). In order to minimize uncertainty, we utilized data from tightly controlled data sets with rigid data collection procedures in place from highly respected sources. Ongoing research is aimed at identifying the sources and magnitude of uncertainty, and refining the approach to minimize that uncertainty. Some sources of uncertainty currently being addressed will be discussed below.

For our estimate of carbon stocks under native vegetation, as well as the base, tillage, and input factors, we started with the default values contained in the IPCC inventory documentation (IPCC, 1997b). The IPCC authors determined these defaults after searching the relevant literature and experimental data. The defaults are set according to soil type and climatic region, and were intended for application anywhere in the world where better information is not readily available. The default values carry with them a high degree of uncertainty. Many of the important default values are not well established or are highly variable from region to region (IPCC, 1997a). This uncertainty can be reduced substantially for the U.S. by using field experiment data to adjust the default values used in the analysis. For this preliminary work, the default values have been evaluated and determined to be consistent with the available field data. More comparison with available field data and further refinements to the values may result in decreased uncertainty. Cannell et al. (1999) note that estimating soil carbon flux is much more difficult, data intensive, and therefore uncertain than estimating changes in biomass C. More field research and new experimental data will result in better inventory inputs and reduced uncertainty.

Another possible source of uncertainty is in the spatial scaling of the data. The NRI data that we used was designed for regional and national analysis of natural resources, and implemented and tested primarily at the county level (Nusser and Goeble, 1997). The sampling framework may not be intensive enough to allow reliable estimation of parameters at more localized scales. The effects of analyzing the NRI and other data at various scales, from county level to the larger climatic regions, remains uncertain, and is part of our ongoing investigation. This will be as much a test of the sensitivity of the approach as it is a test of uncertainty.

IV. Conclusions

Estimating changes in soil C using specific land use change at each NRI point should be more accurate than estimating the change from the same data aggregated to a regional level prior to analysis. In the aggregated approach (Eve et al., 2000), an area of specific land use change within a region may be offset by another area where the exact opposite change occurred. For example, if a given climatic region had 100 Mha of land that changed from forest to cultivated land and 50 Mha that were converted from cultivated land back into forest, only the net (50 Mha) change of forest to agriculture would be accounted for in the aggregated approach. Using the point-by-point method reported here, the full amount of actual (gross) change is accounted for. However, the two approaches resulted in similar numbers for total annual changes in soil C stocks – 8.01 MMT reported here versus 5.29 using aggregated data (Eve et al., 2000) – but uncertainty of the estimate is expected to be reduced using the point-by-point analysis.

There are several explanations for a C sink in agricultural soils. The most logical include:

- most upland agricultural soils have been in production long enough that they are no longer a net source of atmospheric C (Cole et al., 1993),
- crop yields (and crop biomass) have increased substantially since the 1940s (Allmaras et al., 1998; Buyanovsky and Wagner, 1998),
- the adoption of conservation tillage practices has increased (Kern and Johnson, 1993; Lal and Kimble, 1997), and
- cropland has been converted to grass and trees through the CRP (Gebhart et al., 1994; Lal et al., 1999).

The IPCC inventory approach for estimating changes in soil C stocks is efficient and can be accomplished with reasonable land use, soils and climate data inputs. The use of the IPCC inventory approach involves limitations and uncertainty. However, many of the limitations can be overcome with minor modifications to the approach, or addressed with other modeling techniques. Through application of more detailed data such as presented here, uncertainty can be minimized. Techniques need to be sought out and applied in order to accurately quantify the level of uncertainty. There is a continuing need for experimental research and information to refine and improve the values used as input factors to compute changes in C stocks. Also, additional factors may be needed that can accommodate new or improved agricultural technologies for enhancing soil C stocks. Research is also ongoing to apply the CENTURY ecosystem model to derive regional and national estimates of changes in C stocks. Use of a simulation model approach will require more detailed input data, but may facilitate more rigorous quantification of uncertainty.

Acknowledgments

This work was funded by the USDA-ARS, with additional support from USEPA and the Colorado State University, Natural Resource Ecology Laboratory (CSU-NREL). Appreciation is expressed to Dr. John Kimble and Sharon Waltman (USDA-NRCS National Soil Survey Laboratory) and Dan Towery (Conservation Technology Information Center) for their input. Author K. Paustian was Co-Chair for the soils methodology group within the Land Use Change and Forestry section of the IPCC Guidelines for National Greenhouse Gas Inventories.

References

Allmaras, R.R., D.E. Wilkins, O.C. Burnside and D.J. Mulla. 1998. Agricultural technology and adoption of conservation practices. p. 99-158 In: F.J. Pierce, and W.W. Frye (eds.), *Advances in Soil and Water Conservation*. Sleeping Bear Press, Chelsea, MI.

Armentano, T.V. 1980. Drainage of organic soils as a factor in the world carbon cycle. *BioScience* 30:825-830.

Armentano, T.V. and E.S. Menges. 1986. Patterns of change in the carbon balance of organic soil – wetlands of the temperate zone. *J. Ecology* 74:755-774.

Bruce, J.P., M. Frome, E. Haites, H. Janzen, R. Lal and K. Paustian. 1999. Carbon sequestration in soils. *J. Soil and Water Conserv.* 54:382-389.

Buyanovsky, G.A. and G.H. Wagner. 1998. Carbon cycling in cultivated land and its global significance. *Global Change Biology* 4:131-141.

Cannell, M.G.R., R. Milne, K.J. Hargreaves, T.A.W. Brown, M.M. Cruickshank, R.I. Bradley, T. Spencer, D. Hope, M.F. Billett, W.N. Adger and S. Subak, 1999. National inventories of terrestrial carbon sources and sinks: The U.K. experience. *Climatic Change* 42:505-530.

Cole, C.V., K. Flach, J. Lee, D. Sauerbeck, and B. Stewart. 1993. Agricultural sources and sinks of carbon. *Water, Air, and Soil Pollution* 70:111-122.

CTIC. 1998. 1998 Crop residue management executive summary. Conservation Technology Information Center, West Lafayette, IN.

Daly, C., G.H. Taylor, W.P. Gibson, T. Parzybok, G.L. Johnson and P.A. Pasteris. 1998. Development of high-quality spatial datasets for the United States. p. I-512 – I-519. In: *Proc. 1st International Conference on Geospatial Information in Agriculture and Forestry*. Lake Buena Vista, FL. June 1–3, 1998.

Daly, C., R.P. Neilson and D.L. Phillips. 1994. A statistical-topographic model for mapping climatological precipitation over mountainous terrain. *J. Applied Meteorology* 33:140-158.

Eve, M.D., K. Paustian, R. Follett and E.T. Elliott. 2000. *A Preliminary Inventory of Carbon Emissions and Sequestration in U.S. Cropland Soils. Soil Science Society of America Journal*, Special Publication (in press).

Eve, M.D., K. Paustian, R. Follett and E.T. Elliott. An inventory of carbon emissions and sequestration in U.S. cropland soils. In: R. Lal and K. McSweeney (eds.), *Soil Management for Enhancing Carbon Sequestration*. Special Publication. Soil Science Society of America, Madison, WI, in press.

Gebhart, D.L., H.B. Johnson, H.S. Mayeux and H.W. Polley. 1994. The CRP increases soil organic carbon. *J. Soil and Water Conserv.* 49:488-492.

Groffman, P.M., R.V. Pouyat, M.J. McDonnell, S.T.A. Pickett and W.C. Zipperer. 1995. Carbon pools and trace gas fluxes in urban forest soils. p. 147-158 In: R. Lal, J. Kimble, E. Levine and B.A. Stewart (eds.), *Soil Management and the Greenhouse Effect*. Advances in Soil Science. Lewis Publishers, Boca Raton, FL.

Houghton, R.A., J.L. Hackler and K.T. Lawrence. 1999. The U.S. carbon budget: Contributions from land-use change. *Science* 285:574-578.

IPCC. 1997a. Revised 1996 IPCC Guidelines for National Greenhouse Gas Inventories Reporting Instructions (Volume 1). J.T. Houghton, L.G. Meira Filho, B. Lim, K. Treanton, I. Mamaty, Y. Bonduki, D.J. Griggs and B.A. Callender (eds.). Intergovernmental Panel on Climate Change.

IPCC. 1997b. Revised 1996 IPCC Guidelines for National Greenhouse Gas Inventories Workbook (Volume 2). J.T. Houghton, L.G. Meira Filho, B. Lim, K. Treanton, I. Mamaty, Y. Bonduki, D.J. Griggs and B.A. Callender (eds.). Intergovernmental Panel on Climate Change.

IPCC. 1997c. Revised 1996 IPCC Guidelines for National Greenhouse Gas Inventories Reference Manual (Volume 3). J.T. Houghton, L.G. Meira Filho, B. Lim, K. Treanton, I. Mamaty, Y. Bonduki, D.J. Griggs and B.A. Callender (eds.). Intergovernmental Panel on Climate Change.

Kellogg, R.L., G.W. TeSelle and J.J. Goebel. 1994. Highlights from the 1992 National Resources Inventory. *J. Soil and Water Conserv.* 49:521-527.

Kern, J.S. and M.G. Johnson. 1993. Conservation tillage impacts on national soils and atmospheric carbon levels. *Soil Sci. Soc. Amer. J.* 57:200-210.

Lal, R., R.F. Follett, J. Kimble and C.V. Cole. 1999. Managing U.S. cropland to sequester carbon in soil. *J. Soil Water Conserv.* 54:374-381.

Lal, R., J. Kimble, R.F. Follett and C.V. Cole. 1998. *The Potential of U.S. Cropland to Sequester Carbon and Mitigate the Greenhouse Effect*. Sleeping Bear Press, Chelsea, MI. 128 pp.

Lal, R. and J.M. Kimble. 1997. Conservation tillage for carbon sequestration. *Nutrient Cycling Agroecosystems* 49:243-253.

NRCS. 1994. 1992 National Resources Inventory Digital Data. United States Department of Agriculture, Natural Resources Conservation Service (formerly Soil Conservation Service), Washington, D.C.

Nusser, S.M. and J.J. Goebel. 1997. The National Resources Inventory: a long-term multi-resource monitoring programme. *Environmental Ecological Statistics* 4:181-204.

Nusser, S.M., F.J. Breidt and W.A. Fuller. 1998. Design and estimation for investigating the dynamics of natural resources. *Ecological Applications* 8:234-245.

Paustian, K., O. Andren, H.H. Janzen, R. Lal, P. Smith, G. Tian, H. Tiessen, M. Van Noordwijk and P.L. Woomer. 1997a. Agricultural soils as a sink to mitigate CO_2 emissions. *Soil Use and Management* 13:230-244.

Paustian, K., H.P. Collins and E.A. Paul. 1997b. Management controls on soil carbon. p. 15-49. In: E.A. Paul, K. Paustian, E.T. Elliott and C.V. Cole (eds.), *Soil Organic Matter in Temperate Agroecosystems: Long-Term Experiments in North America*. Lewis Publishers, Boca Raton, FL.

Paustian, K., C.V. Cole, D. Sauerbeck and N. Sampson. 1998. CO_2 mitigation by agriculture: An overview. *Climate Change* 40:135-162.

Pierce, F.J., M.C. Fortin, and M.J. Staton. 1994. Periodic plowing effects on soil properties in a no-till farming system. *Soil Sci. Soc. Amer. J.* 58:1782-1787.

Reicosky, D.C. 1997. Tillage methods and carbon dioxide loss: Fall versus spring tillage. p. 99-111. In: R. Lal, J.M. Kimble, R.F. Follett, and B.A. Stewart (eds.), *Management of Carbon Sequestration in Soil*. Advances in Soil Science. Lewis Publishers, Boca Raton, FL.

SCS. 1981. *Land Resource Regions and Major Land Resource Areas of the United States*. USDA Agriculture Handbook 296. United States Department of Agriculture, Natural Resources Conservation Service (formerly Soil Conservation Service), National Soil Survey Center, Lincoln, NE. 156 pp.

Stockfisch, N., T. Forstreuter, and W. Ehlers, 1999. Ploughing effects on soil organic matter after twenty years of conservation tillage in Lower Saxony, Germany. *Soil and Tillage Research* 52:91-101.

UNFCCC, 1995. *Executive Summary of the National Communication of the United States of America*. FCCC/NC/7. 13 pp.

Section VII

Economics and Policy Issues

CHAPTER 40

Assessing the Economics of Carbon Sequestration in Agriculture

L. Tweeten, B. Sohngen and J. Hopkins

I. Introduction

Most industrial countries have signed the United Nation's Framework Convention on Climate Change to reduce anthropogenic contributions to climate change. Towards that end, international negotiators proposed the Kyoto Protocol in December 1997, setting targets for emissions reductions for individual countries. For the United States, the Protocol calls for reduction in atmospheric carbon emissions by 7% from the 1990 level by years 2008–2012. This amounts to a challenging, 25 to 30% reduction from "business-as-usual" trend levels (Bruce et al., 1998). This Protocol, though not yet ratified by Congress, has been signed by the President and will continue to be a benchmark for debate.

The Protocol permits a combination of carbon sinks and emission reductions to reach targets. As such, agriculture becomes a potential major sink. The global soil organic carbon pool, 1550 Pg (Pg = petagram = 10^{15}g), is twice the atmospheric pool and three times the biotic pool (Lal, 1997). Currently, most agricultural soils in the United States and Canada are nearly neutral with respect to emissions – they are neither major sources of carbon dioxide nor sinks of carbon (Bruce et al., 1998).

North American soils, accounting for about 22% of the terrestrial global carbon pool, have lost 5.5 Pg of carbon since land opened for agriculture but could regain some of that pool if economic and other incentives were favorable (Bruce et al., 1998). Many private and public decisions regarding measures to address (or not address) global warming will be based on benefits relative to costs, hence on economics. The objective of this chapter is to examine the economics of expanding carbon sequestration in agricultural soils.

Alternative agricultural tillage, crop rotation, livestock waste disposal, and other practices influence the level of carbon in farm soils. The growing popularity of conservation tillage, including no-till, which adds to carbon in the soil has heightened the interest in the role agriculture might play in sequestering carbon to diminish greenhouse gases and global warming. To ascertain the potential role of agriculture in sequestering carbon to alleviate global warming, several issues are addressed in this chapter:

1. The economic benefits of carbon sequestration to society.

2. The cost of controlling the level of atmospheric carbon through sequestration in agricultural soils. In addition to cropping, alternatives considered include fossil fuel energy conservation and growing of trees to sequester carbon.

3. Complementarities to enhance economic feasibility of soil carbon sequestration. Examples include combining goals of soil conservation and water quality with carbon sequestration.

4. Public policies to increase carbon stocks in soils.

II. Benefits of Sequestering Carbon

Economic benefits to society of carbon sequestration include (a) lower cost of adjustments, dislocations, and other impacts of averted global warming from greenhouse gases in the atmosphere and (b) less soil erosion, higher water quality, and favorable nutrient release associated with greater organic matter retained in soils. Table 1 contains only benefits from (1). Benefits from (2) are discussed later.

Nordhaus and Yang (1996, p. 745) cautioned that "Estimates of the economic impacts or damages from climate change are sparse at this stage" partly because of complex systems determining climate. According to Nordhaus (1993, p. 28), major greenhouse gases have an atmospheric residence time of over a century. Because of the great thermal inertia of the oceans, the climate has a lag of several decades behind the change in greenhouse gases (Nordhaus, 1993).

Nordhaus (1993, p.43) calculated that global real economic output would fall 0.6% from 1989 levels by 2105 due to inefficiencies caused by policies to reduce greenhouse gas emissions by 20%. However, a more nearly optimal emissions policy could slightly increase global real economic output. As shown in Table 1, the optimal tax equal to the benefit per ton of atmospheric carbon reduction ranged from approximately $6 in year 2000 to $21 in year 2100 to address the presumed 3°C "business as usual" warming expected in the 21st century in the absence of ameliorative public policy (Nordhaus, 1993; Nordhaus and Yang, 1996). Nordhaus (1993) reported five previous studies which "...find conclusions that are roughly similar to those reported here."

Frankhauser and Pearce (1994) estimated that each ton of carbon emitted into the atmosphere causes some $20 in damages (Table 1). Estimates of the global economic benefits of reducing carbon dioxide summarized by Brown (1998) ranged from zero to $300 per ton of carbon. Brown placed the most likely range of global benefits at $8 to $59 per ton of carbon.

Based on the results of a systematic set of new studies, Mendelsohn and Neumann (1998) examined implications for the United States economy resulting from 1.5, 2.5, and 5.0°C increases in temperature and 0, 7, and 15% increases in precipitation over base levels by year 2060. Scenarios depict the impact of the sea level rising 33 to 66 cm by year 2100. Under eight of the nine scenarios presented by Mendelsohn and Neumann (1998), U.S. gross domestic product (GDP) was projected to increase from global warming. However, the increases (and one decrease for 5° Celsius warming, no precipitation change) in GDP were so small they can be viewed as not statistically different from zero (Table 1). Some sectors such a coastal areas, energy, and water are worse off while agriculture and timber industries increase their real output due to carbon fertilization. The study did not examine adjustment costs or nonmarket impacts such as health effects.

Other U.S. analyses (see Schimmelpfenning, 1997) like that of Mendelsohn and Neumann (1998) report no major net impact on agricultural output, although costs could be sizable from regional dislocations and adjustments as warming shifts output north. Global analysis suggests a range of potential impacts, with developing countries near the equator experiencing the largest losses (for a review see Watson et al., 1998; Reilly, 1996).

The estimates in Table 1 permit a first approximation of the benefits of carbon sequestration in soils. No-till on average sequesters 3 tons per ha based on 27 studies primarily of the United States summarized by Paustian et al. (1997). Estimates of carbon gains ranged from -4 to +10 tons per ha. If benefits of sequestered carbon currently are $6 per ton and each ha of no-till cropland under continuous no-till production methods can sequester 3 tons, then the public can afford to pay $18 per ha once-for-all for perpetual sequestration, or $.90 annual rent per ha in perpetuity (assuming a discount rate of 5%).

Now consider an upper estimate of the benefit of carbon sequestration to reduce global warming. Under proper conditions, conservation tillage will build soil carbon stocks at a rate of 0.5 tons per ha per year. Carbon stocks might be built to over 10 tons per ha with time depending on soil, climate, production practices, and incentives (Lal et al., 1998). Assuming 10 tons of carbon per ha and an

Table 1. Economic benefit per ton of less carbon emissions or of more carbon sequestration

Analysts	Benefit per ton of atmospheric carbon reduction	Comments
Frankhauser and Pierce, 1994	$20/ton (Global)	Simplified model; covered range of carbon levels
Nordhaus, 1993; Nordhaus and Yang, 1996	$6/ton in year 2000 to $21/ton in 2100 (Global)	Measured by tax justified to bring marginal cost in line with marginal benefit in efficient solution
Brown, 1998	$8 to $59 (Global)	Summary from eight studies. Actual range was from zero to $300. Numbers to left most likely range.
Mendelsohn and Neumann, 1998	Zero (U.S.)	Gains to farming, timber, and recreation industries; loss to coastal, energy, and water sectors.

upper value of $60 per ton carbon (see Brown, 1998, in Table 1), the higher value of sequestered carbon is $600 once-for-all per ha or $30 per ha per year based on a 5% discount rate. Expressed in annual terms, the value of sequestering carbon from $.90 to $30 per ha per year is not large relative to typical earnings (rent) of $250 per ha for producing crops in the Midwest. The $18 to $600 value in perpetuity of a ha of cropland for the purpose of carbon sequestration is well below a typical value of cropland, $5,000 per ha, in Midwest cash grain farms.

As noted later, however, modest values for carbon sequestration do not preclude economic feasibility when complementarities of soil organic matter with soil and water conservation and other benefits are recognized. It is also noted that opportunities to sequester carbon tend to be greatest in cool, moist climates and least in hot, dry climates.

The conclusion from the above carbon studies is that analysts have not reached consensus on the benefits of carbon sequestration to avoid global warming — even if the latter is occurring. The incremental cost per ton of atmospheric carbon accumulation rises with higher greenhouse gas accumulation and global warming. Thus economists' estimates of the value of sequestering carbon vary in part because scientists lack consensus on the extent of future global warming as well as the consequences of a given amount of warming. Although the benefits of sequestering carbon are unclear, the Kyoto Protocol suggests that policy may move forward anyhow. It is therefore important to address the costs of controlling greenhouse gases. The costs of carbon sequestration in cropland soils must be balanced against the cost of alternatives such as fossil fuel conservation or carbon sequestration through afforestation.

III. Costs of Sequestering Carbon

Carbon in the atmosphere is accumulating in part from utilizing fossil fuels to produce energy, fertilizer, and pesticides central to our food supply and standard of living. Massive reduction in carbon emissions called for by the Kyoto Protocol will reduce living standards and food supplies unless cutbacks are achieved in a cost-effective manner through a combination of carbon emissions reduction and sequestration.

This section mainly addresses carbon sequestration but recognizes that an overall strategy for reducing greenhouse gases also would feature measures to conserve energy and fossil fuel consumption through taxes, subsidies, regulation, research, and education. We focus on carbon because carbon dioxide is the most abundant greenhouse gas and because agriculture can sequester carbon.

Fossil fuels producing energy (while releasing carbon dioxide) are the lifeblood of industrial economies. Reduction of greenhouse gases by 50% mainly through greenhouse gas emission reduction would require major cutbacks in carbon dioxide and could cost about 1% of global output (Nordhaus, 1993, p. 36). The average productivity of fossil fuels is so large that cutbacks have severe and pervasive ill effects as apparent in the energy crises of the 1970s. Marginal, carefully selected cutbacks can be made, however, with a gain to society as illustrated by the example of the Green Light energy saving initiative launched by the U.S. Environmental Protection Agency in January 1991 (Table 2). Green Light featured incentives for firms and agencies to adopt energy-efficient fluorescent bulbs, occupancy sensors, and a number of other measures to reduce lighting costs which account for one-fourth of energy use in the United States. Decanio (1995) estimated an internal rate of return of 45% for investment in Green Light based on a 3-year payback – near the average for 3,673 Green Light projects up to 1995.

Green Light is but one illustration of opportunities to reduce carbon dioxide emissions while raising real national income. Another option is a tax on fossil fuels. While potentially costly, fossil fuel taxes might have many benefits in addition to reducing carbon dioxide emissions: less traffic congestion, urban sprawl, smog, and scarcity from depleting energy reserves.

A. Sequestration with Crops and Trees

While achieving the goals outlined at Kyoto would require less carbon emissions, an efficient policy is likely to entail a wide range of activities. At some point, carbon sequestration in cropland soils or trees may offer greater economic payoffs than additional emission reductions to cut greenhouse gases. Trees and cropland carbon sequestration opportunities are examined here. Trees sequestering carbon can be grown on forestland or on former cropland.

There are several issues to consider with using forests to sequester carbon. First, prior use of land is important. Some trees store more terrestrial carbon above the ground than some plants but some grasses efficiently build large stores of soil organic matter below the ground. Net carbon storage is likely to occur only if land is converted from cultivated crop use to forest. Second, the ultimate fate of the wood products must be considered. Carbon is sequestered for long periods of time if the forest is never harvested or burned. Even if the forest is harvested, however, sequestration can be long lived if slow-growing older trees are processed to lumber used in construction of houses that last for decades. Alternatively, sequestration is short lived if trees are burned or processed into paper that is recycled into animal bedding and later spread with manure on fields. Current estimates suggest that 35% of carbon removed from U.S. forests is stored immediately, 30% returns to the atmosphere through decay, and 35% is burned for energy (Heath et al., 1996).

One of the first estimates of the costs of planting trees investigated how much land would be needed to sequester all carbon emissions (Sedjo, 1989). With temperate forest plantations sequestering 6.24 tons per ha per year, Sedjo estimated that some 465 million ha of temperate zone forest plantations could sequester an additional 2.9 Pg of carbon emissions per year for 30 to 50 years. Assuming land rental costs are zero, Sedjo estimated that it would cost $3.50 per ton of carbon stored.

Adams et al. (1993) estimated the marginal cost of sequestering carbon by converting some U.S. agricultural land to tree plantations in regions with over 18 inches of rainfall per year (Table 2). Adams et al. (1993) estimated incremental cost per ton of carbon sequestered ranged from $7.40 for 132 million tons of carbon (2.5% of U.S. annual carbon emissions and taking approximately 5 million

Table 2. Summary of costs for reducing carbon emissions or sequestering carbon

Analysts	Ecosystem	Cost/ton carbon	Comments
Lower emissions Decanio, 1995	U.S. businesses	Negative	45% internal rate of return on costs from measures to increase energy lighting efficiency
Carbon sequestration: forest ecosystems and tree plantations			
Moulton and Richards, 1990	U.S. forest plantations on marginal agricultural land	$8.50 to $38.29	Overall cost $700 million to $19.5 billion yr^{-1}
Adams et al., 1993	U.S., conversion of agricultural land to tree plantations	$7.40 to $60.75	Assumes timber harvest with lower cost for 32 million metric tons, and higher cost for 635 million metric tons without timber harvest
Parks and Hardie, 1995	U.S. marginal agricultural land	$9 to $10	8.9 million ha costing $3.7 billion to sequester 4.7 tons ha^{-1} yr^{-1}
Sedjo, 1989	Temperate zone forests	$3.50	465 million ha required to offset 2.9 Pg carbon accumulation per year @ 6.24 tons of carbon ha^{-1} could drive down lumber prices
Hoen and Solberg, 1994	Norwegian forests	$79	Examined different management strategies
Stavins, 1999	37 U.S. countries in the South	Less than $66	Marginal cost nearly linear to 9 million tons carbon storage yr^{-1}
Sohngen et al., 1998	Global, emphasizing subtropical forest plantation strategy	Not available	Tree plantation in subtropical region most promising
Richards and Stokes, 1995	Review of studies over wide range of ecosystems	$1 to $187	Cost depends on region, growing factors, and other conditions
Carbon sequestration: cropland			
Babcock, 1998 (preliminary)	12-state U.S. Midwest region, all major crops	$0 to 11 million tons, $200 at 19 million tons, $400 at 22 million tons	One-fourth of U.S. crop production, conservation tillage only; marginal cost of 1 ton of cumulative carbon buildup per year

ha of agricultural land and assuming timber harvest) to $60.65 for 635 million tons of carbon sequestered (50% of total U.S. annual carbon emissions and taking 111 million ha of agricultural land and assuming no timber harvest).

Using data from Conservation Reserve Program contracts, Parks and Hardie (1995) estimate a cost of $9 to $10 per ton of carbon sequestered by trees on 8.9 million ha of marginal cropland. This study, however, did not address the indirect effects of purchasing land. Afforestation of extensive areas of cropland would raise food and land values – a factor accounting for the higher costs shown by Adams et al. (1993) compared to the estimates of Moulton and Richards (1990) who assumed constant land costs as tree plantations expand on croplands (Table 2).

Stavins (1999) examined how carbon sequestration policies would affect land markets and land use in a 36-county region of the southern U.S. He found that 7 to 8 million tons of carbon could be sequestered each year for less than $66 per ton. Alig et al. (1997, p. S108) used a U.S. timber and agriculture model to predict that for each ha of agricultural land converted to forest plantations, a ha of forestland somewhere else would be converted back to agriculture, resulting in no net carbon storage.

Sohngen et al. (1998) taking a global perspective noted that deliberate forest expansion in some regions to sequester carbon would lower world timber prices, thus discouraging economic timber plantations elsewhere. Fewer trees planted elsewhere would offset some of the intended benefits of sequestration. On the other hand, lower timber prices diminish incentives to harvest the boreal forest of the North and rainforests of the tropics, thus protecting wildlife and biological diversity while sequestering substantial carbon.

Richards and Stokes (1995), in a comprehensive review of literature provided a wide range of cost estimates – from $1 to $187 per ton of carbon sequestered. In general, costs were lower for small efforts and expanded as more carbon was sequestered.

Notably absent from the Richards-Stokes review were estimates of sequestration costs on cropland used for crops. Bruce Babcock (1998, private correspondence) of Iowa State University provided preliminary estimates of carbon sequestration with conservation tillage on cropland in a 12-state Midwest region (Table 2). In 1998, this region accounted for one fourth of U.S. crop production. Several conclusions follow:

- Approximately 11 million metric tons of carbon are sequestered at private expense and at no direct cost to the public because it is complementary with other economic crop production practices.

- The marginal incentive required for adding another ton of carbon cumulatively per year increases from near zero dollars to $200 per ton at 19 million metric tons of soil carbon in the 12-state region.

- Marginal private costs for sequestration rise sharply above 19 million tons and reach $400 for adding another ton of carbon cumulatively when soil carbon stocks reach 22 million metric tons.

- Per ha payments required to induce conservation tillage range from zero dollars (no charge) to entice up to 40% of crop area participation, rise at a nearly linear rate from zero to $86 per ha up to 95% participation, and become very high to entice further participation. These numbers may be compared with the average annual cost of $120 per ha for the Conservation Reserve Program of general land retirement. The conclusion is that carbon sequestration is economically most feasible within the range where its complementarities with best management crop production practices are strongest.

B. Sequestration by Restoring Degraded Soils

Attention now turns to estimates of costs and benefits of sequestering carbon on degraded lands currently or historically used for agriculture. Lal (1997) calculated that soil restoration by growing trees and vigorous cover crops "...could lead to carbon sequestration at the rate of 3 Pg per year....", a rate near the annual increase in carbon in the atmosphere (Lal et al., 1995). Lal (1997) cautioned that "...this rate of increase in humus content may be difficult to achieve in arid and semi-arid tropics, and if so only for a limited period of time."

Dregne and Chou (1992) classified 216 million ha or 47% of global rainfed cropland as degraded (moderate to very severe desertification), 43 million ha or 30% of irrigated land as degraded, and 3,333 million ha or 73% of rangeland as degraded. The worldwide total, 3,592 million ha, exceeds estimates by Oldeman et al. (1990) that nearly 2 billion ha of soil are degraded worldwide, but includes arid and semi-arid tropical rangeland with limited capacity to sequester carbon.

Progress in restoring degraded soils will depend on economic incentives. In this context, it is notable that Dregne and Chou (1992) classified 276 million ha or 85% of U.S. rangeland as moderately or more severely desertified. Low rainfall precludes sizable economic buildup of soil organic carbon in much of this rangeland. Restoration of many degraded lands can be profitable according to calculation by Dregne and Chou shown in Table 3. The cost-benefit ratio for rehabilitating land ranges from 3.5 for rangeland to 1.9 for rainfed cropland, and averages 2.6 over all land uses.

Several considerations suggest viewing the favorable cost-benefit ratios in Table 3 with caution. Many assumptions underlying the estimates are not stated but the cost-benefit ratios may more nearly depict what is judged to be socially beneficial rather than privately profitable.

Although following practices that avoid degradation is cheaper than reclaiming degraded land, allowing soils to degrade may be cheaper for producers than paying the costs of conservation. This is because annual loss of productivity is judged by farmers and ranchers to be too small to justify conservation measures. Practices that degrade land continue also because "downstream" costs of desertification are not charged to operators or owners of eroding soil. The lack of rehabilitation of degraded land suggests that rehabilitation is not privately profitable. Thus, rehabilitation of degraded lands will not be as rapid as implied by the favorable cost-benefit ratios in Table 3.

Because of high private discount rates and because for atmospheric carbon "downstream" is global, the social benefits of soil, water, and air improvements exceed private benefits. It follows that public monetary inducements for land conservation and rehabilitation could be effective and economically justified in poor countries using revenues from industrial countries purchasing carbon emission permits.

IV. Complementarities Enhancing Economic Feasibility of Carbon Sequestration

A conclusion from the foregoing analysis is that raising crops and trees for carbon sequestration becomes economic only through complementarity of cropland with other private and social benefits derived from soil, water, and air quality conservation. Carbon sequestration is best viewed as one component of a package of environmental benefits economists classify as *externalities* — benefits *or* costs that accrue to society but not to the parties engaged in market transactions. Potential externalities complementary to carbon sequestration include conservation of soil nutrients, fossil fuels, water quality, wildlife habitat, and biodiversity. The magnitudes of these externalities remain points of controversy.

Table 3. Income foregone over 20-year period and rehabilitation cost on desertified land that would produce positive cost-benefit ratio when rehabilitated

Land use	Area to be rehabilitated	Benefit: added income if land is rehabilitated	Cost of rehabilitation	Cost-benefit ratio
	(million ha)	———— (million $) ————		(B/C)
Irrigated land	43.1 (100%)	215,500	86,200	2.5
Rainfed cropland	150.9 (70%)	114,684	60,360	1.9
Rangeland	1666.8 (50%)	233,352	66,672	3.5
Total	1860.8 (52%)	563,536	213,232	2.6

(Adapted from Dregne and Chou, 1992.)

Table 4. Soil properties and yields by soil: results of a 10-year study in Indiana

Soil series	Crop yield		Phosphorus	Plant available water	Clay content	Organic matter
	Corn	Soybeans				
	Percentage change, severe versus slight erosion					
Corwin	−9	−20	−34	−49	11	−39
Miami	−18	−17	−29	−70	44	−20
Morley	−14	−24	−38	−51	53	−16

(Adapted from Weesies et al., 1994.)

Sanders, Southgate, and Lee (1995) cite four studies indicating that U.S. on-farm costs of soil erosion range from $500 million to $1.2 billion per year. This contrasts with off-site *externality* damage costs of $2.2 billion (Clark et al., 1985) to $7 billion (Ribaudo et al., 1989). Off-site costs include sediment with associated nitrogen, phosphorus, and pesticides deposited in rivers and urban reservoirs. Even the on-site costs of soil erosion and nutrient loss may not be properly controlled by markets if producers do not know the magnitude of costs or if the costs are considered too small each year to trigger the operator's remedial action. Depletion of fossil fuels and phosphate through erosion could entail externalities to the extent that the market under-prices reserves of phosphate and fossil fuels. Fossil fuels are feedstocks used to produce pesticides and nitrogen fertilizer that sometimes contaminate water.

The complementarity among yields, nutrients, and organic matter carbon is illustrated by Table 4. Soil organic matter (SOM) loss is both a cause and effect of erosion. As the SOM declined by 39% on Corwin soils, SOM that averaged 3.03% on slightly eroded Corwin soils fell to 1.86% on severely eroded soils. Sound conservation management practices that raise SOM from that found in severely eroded soil to that found in moderately eroded soil would raise SOM by 0.31 percentage points (19%). Carbon is sequestered in that SOM.

Severe erosion attending low SOM reduced corn yields on Corwin soils in Indiana by 9%, Miami soils 18%, and Morley soils 14%. Soybean yields fell proportionately more. Phosphorus fell 29 to 38% and plant available water fell 49 to 70% due to severe erosion on these Indiana soils. Thus, more commercial fertilizer would be required to maintain yield after erosion on the Indiana soils. Relative clay content increased as larger size mineral particles eroded more rapidly than clay.

A. Marginal Cropland and Carbon Sequestration

Because of the large and inelastic demand for global food, trees will not compete with crops for prime farmland. Trees can compete with crops for more marginal land, however. Shakya and Hitzhusen (1997, p. 24) compared net present value per ha for South-central Ohio land in the Conservation Reserve Program (CRP) used for (1) cropland, (2) white pine tree plantation, and (3) continued in CRP. The analysis contrasted incentives for the landowner considering (1) only private costs and returns and (2) social costs and returns. Social costs included environmental effects such as soil erosion and water quality. Ignoring the unique and special case of a contract to supply pulpwood to a local paper company, the highest *private* return was from cropping the land. The highest *social* return was from growing white pine. Using a social discount rate of 4% and considering environmental impacts, the white pine plantation offered a net present social return of $1376 per ha compared to a *negative* net present social return of $2134 for cropping the land.

The value of carbon sequestration was not considered, but if wood were "sequestered" as lumber in houses for many decades the social return could be $60 higher (a 4% gain over $1376) based on an estimate from the previous section. The conclusion is that full accounting for externalities would call for conversion of many additional ha of marginal, environmentally sensitive cropland in the eastern United States to timber plantations. In 1998 many of those ha were in the 14 million ha Conservation Reserve Program.

The complementarity between carbon sequestration and cropping practices is especially apparent from a recent study of North America as a carbon sink. The study (Fan et al., 1998) found that the continent takes up 1.7 Pg of carbon annually, a large portion indeed of the 1.0 to 2.2 Pg global terrestrial annual uptake of carbon. The authors listed "...regrowth on abandoned farmland and previously logged forest" as one source of the sink. Marginal land has become unprofitable for crops because science and technology have raised productivity, lowering crop prices. These sink numbers, three-fifths of the net carbon added each year to the atmosphere, are tentative. But they reveal how agricultural science and technology investments having a high economic payoff in greater crop and livestock productivity alone have a potentially higher payoff after accounting for cropland released from food production to grow trees and preserve wildlife.

B. Conservation and Carbon Sequestration

Much of the interest in carbon sequestration in agriculture stems from major opportunities to build soil organic matter through conservation tillage. Adoption of conservation tillage was encouraged by advances in planting equipment and chemical weed control coupled with enhanced knowledge of benefits of conservation tillage for soil conservation and water quality. Other contributing factors included conservation education and the 1985 and 1996 farm programs requiring conservation compliance in order to receive direct payments from the government.

Producers appear to have exploited most opportunities for profitable conservation tillage in the United States; conservation tillage area was plateauing in the country by the mid-1990s (Table 5). Conservation tillage was practiced on 36% of cropland planted in 1996, up only 1%age point since 1993 and near the mid-1980s level. No-till, one form of conservation tillage, disturbs the soil least, preserves the most surface crop residue, and sequesters the most carbon. It increased from 5% to 15% of all "tillage" from 1989 to 1996. Considering only conventional tillage, the share classified as "reduced tillage" (15 to 30% residue) remained at one fourth of all cropland while the share with less than 15% residue fell from 49% in 1989 to 38% in 1996.

Much of the decline in low residue conventional tillage is accounted for by less use of moldboard plows. In the 1950s nearly all U.S. corn** was tilled with a moldboard plow. By 1995, only 8%

Table 5. U.S. crop production tillage practices

Item	1989	1993	1996
	(Percent of total)		
Total cropland planted			
Conventional tillage (<30% residue)	74.4	65.1	64.2
Reduced tillage (15 to 30% residue)	25.3	26.3	25.8
Conventional tillage (<15% residue)	49.1	38.8	38.4
Conservation tillage (>30% residue)	25.6	34.9	35.8

(Adapted from U.S. Department of Agriculture, 1997.)

of all U.S. corn and indeed of all crops on average was tilled with a moldboard plow (U.S. Department of Agriculture, 1997, p. 162). Although moldboard plow tillage is not dominant for any major U.S. crop, conventional tillage with less than 30% residue remains the practice of choice on 98% of cotton acres, about three fourths of wheat acres, and over two thirds of southern soybean acreage (U.S. Department of Agriculture, 1997, p. 162). Thus, the potential is great for further conservation tillage – given sufficient inducements.

Farmers tend to adopt conservation tillage if it is privately profitable. The numerous studies of the profitability of conservation tillage reveal that in some circumstances net returns per ha for conservation tillage are higher and in other circumstances are lower than returns for conservation tillage — depending on soil type, drainage, rainfall, crop, and other factors (Doster et al., 1983; Klemme 1985; Hopkins et al., 1996; Stephen et al., 1987). Yield is frequently less in the short run with conservation than with conventional tillage. But production costs tend to be less for conservation tillage, hence it is often more profitable than conventional tillage even when conventional tillage yields are higher.

Results of a recent comprehensive study by Day et al. (1998) are shown in Table 6. Net returns for com on all soil productivity classes are higher for conservation mulch tillage (at least 30% residue) than for any of the other three tillage methods considered. Conventional moldboard plow tillage was least profitable on low and medium productivity soils and nearly tied with no-till for least profitability on high productivity soils. In addition to soil conservation, advantages of conservation tillage include less power and labor requirements and less overall production costs. Disadvantages of conservation tillage include learning and time required to bring yield and profitability to conventional tillage levels, poor stands or late germination (causing yield loss) because of a cool and wet seedbed on tight soils shaded by mulch during cool and wet springs, and potential herbicide contamination of water. According to Day et al. (1998), economic returns tend to vary more with conservation than with conventional tillage.

Recent innovations include fall strip tillage leaving considerable mulch along with a lightly cultivated band of cleared soil. This strip provides a warm seedbed for early spring germination. Other adaptations also will expand conservation tillage, but economic incentives currently seem inadequate to reach the goal of up to 75% adoption in the United States projected by Lal (1997).

Conservation tillage may provide even less opportunity to raise soil organic carbon (SOC) in the tropics. Conservation tillage leads to high SOC near the surface compared to conventional tillage (Lal, 1997). Surface SOC is especially vulnerable to volatilization from high heat and humidity. Conversion to conservation tillage also may have little effect on SOC for soils with coarse texture in arid climates that have been cultivated for many years (Lal, 1997).

The stalling of conservation tillage in the United States at just over one third of cropland indicates that "business as usual" may not bring satisfactory progress toward soil conservation, improved water quality, or carbon sequestration. A multifaceted approach discussed in the next section could speed

Table 6. Corn net returns over variable costs, by tillage system and soil productivity class, ten states,[a] 1996

| Tillage system | Soil productivity class[b] | | |
	Low $Net return ha^{-1}	Medium $Net return ha^{-1}	High $Net return ha^{-1}
Conventional w/o plow	433	524	607
Conventional with plow	315	418	539
Conservation: mulch	470	558	619
Conservation: no-till	416	555	534

[a]States are IA, IL, IN, MI, MN, MO, NE, OH, SD, and WI.
[b]Productivity classes are based on soil-crop sufficiency conditions for bulk density, water holding capacity, permeability, clay content, pH, soil depth, and soil profile.
(Adapted from Day et al., 1998.)

progress towards environmental goals, recognizing complementarity in reaching targets, including low cost food.

V. Policy Options

Carbon sequestration can be profitable on cropland as a joint product with crop production through practices such as conservation tillage, cover crops, legume forages, grass waterways, filter strips, recycled wastes, and the like. These practices largely pay for themselves even in the absence of carbon sequestration. Recycling of degraded lands also offers promise as noted earlier.

Data presented in the previous sections indicate that carbon sequestration in tree plantations as a complement to production of lumber and other forest products is feasible at relatively low costs. However, costs of carbon sequestration rise as area expands. Although a ha of forest may sequester more carbon than a ha of crops, tree plantations cannot be markedly expanded on cropland or grazing land in temperate zones without generating high costs to society from rising prices for food. The greatest opportunities for expanding carbon sequestration at low cost is on tree plantations in the subtropics (Sohngen et al., 1998) and in the temperate zone on land marginal for crops but where rainfall is adequate.

Free market incentives alone may not reach socially desirable carbon sequestration levels in soils because not all value to society enters the accounts of market participants. If global warming is occurring and creating costs worth avoiding, several public policies potentially can help to alleviate the situation. Global cooperation, a market in carbon, and U.S. farm policy initiatives are addressed below.

A. International Cooperative Agreement

Because air is common property, any one individual, firm, or nation bears only a small share of the cost of the carbon each adds to the atmosphere. Because private costs that motivate decisions fall far short of social costs, too much carbon dioxide accumulates in the atmosphere. The result is the so called "tragedy of the commons." Everyone desires to be a freerider – not taking action but waiting to reap the benefits of carbon control by others. One solution among countries is an international cooperative agreement to limit carbon emissions or to sequester carbon as in the Kyoto Protocol. This Protocol does not provide a mandate to guide actions of individual firms and agencies, however.

B. Trading Carbon Emission and Sequestration Permits in the Market

In theory, a system of property rights can align private marginal costs of releasing carbon dioxide into the atmosphere with social marginal cost. And a system of taxes, subsidies, regulations, education, and other tools can bring marginal social costs for controlling carbon emissions in line with marginal social costs of carbon sequestration — and all in line with marginal social benefits of using fossil fuels to provide energy. Assuming controversies over the need to sequester carbon are settled and global accord is reached, for reducing atmospheric carbon, the task remains to find a system that achieves target carbon levels cost effectively.

Lessons from past successes creating a market in SO_2 emissions from power plants for cost-effective environmental protection might be applied to a carbon market extended to farms and forests. One procedure is to auction carbon emission rights to the highest bidders. The public sector would receive the receipts from the initial auction. An alternative procedure would begin by establishing historic carbon emission and sequestration levels for each firm or agent. Emitters then would be assigned carbon emission reduction quotas based on the target reduction below historic levels.

Regardless of how the issue of initial carbon permit allocations is decided, once established, a private market for carbon would allow trading in permits to promote efficiency — those with higher benefits from adding atmospheric carbon and high cost of cutting back would purchase permits from entities with low benefits from adding atmospheric carbon or low costs of sequestering carbon. The credits purchased by firms permitting them to exceed allowances would be exactly offset by credits sold by firms that cut emissions below allowances or that sequester carbon in excess of requirements.

The market in carbon would create some inequities. Farmers who have diligently built soil organic carbon over the years would have little scope to collect receipts for carbon credits. Another problem with this system of carbon permits is inability of most farmers to sell carbon permits in sufficient volume to hold down administrative, verification, and monitoring costs. Farmers desiring to sell carbon sequestration credits might form cooperative pools to reduce such transaction costs. Officials could spot-check farmer compliance periodically. Compliance might most efficiently be measured by use of best practices rather than measured by actual soil carbon. Also to reduce administrative costs, only farmers who wish to enter the sequestration market would need to have their historic base established.

For sequestors, the sale of a carbon credit would constitute a lien on property to sequester that carbon in perpetuity. A farmland buyer would be committed to maintain the contract level of sequestered carbon or return to the carbon market to buy back credits to sequester carbon.

C. Taxes and Subsidies

Historically, the public has used the "carrot" of payments to farmers rather than the "stick" of regulation to serve the environment. Strong political pressure may continue production flexibility contract transition payments as "green" payments after the 1996 farm bill expires in year 2002. Such green payments could be for conservation compliance extended to practices that encourage carbon sequestration as well as other environmental goals such as conservation tillage, cover crops, grass waterways, and filter strips that benefit society as well as farmers.

Previous discussion by economists has focused on eventually targeting transition or green payments to control soil erosion and water quality (see Tweeten and Zulauf, 1997). A far wider base of political support for programs could emerge if the payments, scheduled to end in year 2002, were targeted in future years to carbon sequestration as well. Nearly every farmer can sequester carbon but not all face soil erosion or water quality problems. In short, helping to correct the negative externalities of greenhouse gas accumulation, soil erosion, and water pollution by using funds with low opportunity cost (otherwise they might go to farmers for doing nothing) would likely have much political appeal.

VI. Summary and Conclusions

Carbon sequestration in soils is less likely to be another cash "crop" than a farm income supplement. Carbon sequestration through soil organic matter improves soil tilth, nutrient release, and moisture holding capacity. Because such benefits add economic returns to farmers, producers can be expected to make decisions that provide such benefits.

Other benefits of soil organic matter are not well allocated by markets because they are public rather than private goods — benefits accrue to society rather than to farm market participants. Traditionally the public interest in organic matter in soils has focused on soil conservation and water quality — externalities that can entail costs to "downstream" entities not part of on-farm costs and returns which determine a farmer's decisions. Growing concerns about global warming add another public interest in private decision making — build-up of organic matter that sequesters carbon to reduce greenhouse gas accumulation and possible attendant global warming.

Alternative agricultural tillage, crop rotations, livestock waste disposal, and other practices influence the degree of organic matter in farm soils. The growing popularity of conservation tillage, including no-till, which adds to carbon in the soil has heightened the interest in the role agriculture might play in sequestering carbon and thereby diminishing greenhouse effects.

To ascertain the potential role of agriculture in sequestering carbon to alleviate global warming, several questions were addressed:

1. What is the value to society of sequestering carbon?
2. What is the cost to agricultural producers of sequestering carbon?
3. How does the cost of carbon sequestration in agricultural soils compare with alternative means of carbon sequestration and fossil fuel energy conservation? An example of the former is growing of trees; an example of the latter is the Green Light energy saving initiative discussed in the text.
4. What are the limits of *economically feasible* carbon sequestration in agriculture? Is it a potential major or minor player in an effort to reduce greenhouse gases?
5. What public policies might most effectively increase carbon sequestration in agriculture?

The current incremental value of carbon sequestration cannot be estimated with precision but some estimate the value to be about $5 per ton or less to the United States. The value may be considerably larger after accounting for nations with larger costs of global warming. Also, the value may rise in the future (see Table 1). The cost to agriculture of sequestering carbon is low or negative (it pays) for up to perhaps 40 million tons, is modest for the next 40 million tons, and becomes high for over 80 million tons (see Table 2). Other means such as afforestation and fossil fuel energy conservation can efficiently reduce atmospheric carbon. However, soil carbon sequestration can compete with other approaches to reduce greenhouse gases over some ranges (see Table 2).

Question 4 above cannot be answered with much confidence. Appropriate public policy decisions regarding global warming are not possible without better measures of the marginal social costs and benefits of soil carbon sequestration. Estimates of costs (and benefits) of carbon sequestration on cropland are primitive, an empirical vacuum waiting to be filled. Proper modeling to find answers could use a mathematical programming model, parametrically raising the price of carbon per ha from low to high levels while observing the profitable increase in carbon sequestration on representative resource situations (typical farms) across the nation. Such analyses would need to account for complementarities of soil organic matter (SOM) with crop yields, water quality, and soil erosion over time – considering on- and off-site costs and benefits. Analysis needs to consider the time sequence of SOM buildup under alternative fertilization, tillage, rotation, and other cropping practices. The total discounted cost of commercial fertilizer would be estimated over time recognizing that more fertilizer may be applied in early years to build SOM, and less applied in later years as SOM conserves soil and nutrients while appropriately releasing them for plant growth. Such measures of incremental social costs of soil carbon sequestration then can be compared with improved measures of incremental social costs of global warming to formulate proper carbon emission control and sequestration levels.

Turning now to question 5 above, three public policy initiatives were proposed to address soil carbon sequestration. One was the need for collaborative international arrangements to avoid the "tragedy of the commons" – the tendency of individuals, firms, and nations to be "free riders" who wait for others to act. The second initiative was to establish a carbon emission and sequestration permit market to minimize costs of meeting established greenhouse gas targets. The third was to target production flexibility contract payments – if they are continued – to "green" payments for soil conservation, water quality, and carbon sequestration.

References

Adams, R., D. Adams, J. Calloway, C. Chang and B. McCarl. 1993. Sequestering carbon on agricultural land: social cost and impacts on timber markets. *Contemporary Social Issues* 12:76.

Alig, R., D. Adams, B. McCarl, J. Callaway and S. Winnett. 1997. Assessing effects of mitigation strategies for global climate change with an international model of the U.S. forest and agricultural sectors. In: R. Sedjo, R. Sampson and J. Wisniewski (eds.), *Economics of Carbon Sequestration in Forestry*. Lewis Publishers, Boca Raton, FL.

Babcock, B. 1998. Unpublished worksheets on carbon sequestration through conservation tillage in 12 Midwest states, Center for Agricultural and Rural Development, Iowa State University, Ames.

Brown, S. 1998. Global warming policy: some economic implications, *Economic Review,* Dallas, TX: Federal Reserve Bank, Fourth Quarter, 26-35.

Bruce, J., M. Frome, E. Haites, H. Janzen, H., R. Lal and K. Paustian. 1998. Carbon sequestration in soils. Soil and Water Conservation Society, Carbon Sequestration in Soils Workshop, May, 1998, Calgary, Alberta.

Clark, E., J. Haverkamp and W. Chapman. 1985. *Eroding Soils: The Off-Farm Impact.* Conservation Foundation, Washington, D.C.

Day, J., C. Sandretto, W. McBride and V. Breneman. 1998. Conservation tillage in U.S. corn production: an economic appraisal. Paper presented at the annual meeting of the American Agricultural Economics Association, August, 1998, Salt Lake City, UT. Economic Research Service, U.S. Department of Agriculture, Washington, D.C.

Decanio, S. 1995. The Energy Paradox: Bureaucratic and Organizational Barriers to Profitable Energy Saving Investments, Working Paper in Economics No. 19-95, Santa Barbara: Department of Economics, University of California.

Doster, D., D. Griffith, J. Mannering and S. Parsons. 1983. Economic returns from alternative corn and soybean tillage systems in Indiana. *J. Soil and Water Conservation* 38:504.

Dregne, H. and N. Chou. 1992. Global desertification dimensions and costs. p. 249-281. In: H.E. Dregne (ed.), *Degradation and Restoration of Arid Lands.* International Center for Arid and Semi-Arid Land Studies, Texas Tech University, Lubbock.

Fan, S., M. Gloor, J. Mahlman, S. Pacala, J. Sarmiento, T. Takahashi and P. Tans. 1998. A large terrestrial carbon sink in North America implied by atmospheric and oceanic carbon dioxide data and models. *Science* 282:442.

Frankhauser, S. and D. Pearce. 1994. The social costs of greenhouse gas emissions. In: *Economics of Climate Change,* Organization for Economic Cooperation and Development, Paris.

Heath, L., R. Birdsey, C. Row and A. Plantinga. 1996. Carbon pools and fluxes in U.S. forest products. *NATO ASI Series* 1(40):271-278.

Hoen, H. and B. Solberg. 1994. Potential and economic efficiency of carbon sequestration in forest biomass through silvicultural management. *Forest Science* 40:429.

Hopkins, J., G. Schnitkey and L. Tweeten. 1996. Impacts of nitrogen control policies on crop and livestock farms at two Ohio farm sites. *Review of Agricultural Economics* 18:311.

Klemme, R. 1985. A stochastic dominance comparison of reduced tillage systems in corn and soybean production under risk. *Amer. J. Agricultural Economics* 67:550.

Lal, R. 1997. Residue management, conservation tillage, and soil restoration for mitigating greenhouse effect by CO_2-enrichment. *Soil and Tillage Research* 43:81.

Lal, R., J. Kimble, E. Levine and C. Whitman. 1993. World soils and greenhouse effect: an overview. p. 1-8. In: R. Lal, J. Kimble, E. Levine and B.A. Stewart (eds.), *Soils and Global Change.* Advances in Soil Science. Lewis Publishers, Boca Raton, FL.

Lal, R., J. Kimble, R. Follett and C. Cole. 1998. *The Potential of US Cropland to Sequester Carbon and Mitigate the Greenhouse Effect.* Sleeping Bear Press, Chelsea, MI.

Mendelsohn, R. and J. Neumann. 1998. *The Market Impact of Climate Change on the U.S. Economy,* Cambridge University Press, Cambridge.

Moulton, R. and K. Richards. 1990. *Costs of Sequestering Carbon Through Tree Planting and Forest Management in the U.S.*, General Technical Report WO-58, Forest Service, USDA, Washington, D.C.

Nordhaus, W. 1993. *Rolling the "Dice": An Optimal Transition Path for Controlling Greenhouse Gases,* Cowles Foundation Paper No. 836, Cowles Foundation for Research and Economics, Yale University, New Haven, CT.

Nordhaus, W. and Z. Yang. 1996. A regional dynamic general-equilibrium model of alternative climate change strategies. *American Economic Review* 6:741.

Oldeman, R., R. Hakkeling and W. Sombroeck. 1990. *World Map of the Status of Human-Induced Soil Degradation: An Explanatory Note.* International Soil Reference and Information Center, United Nations Environmental Program, Nairobi, Kenya.

Parks, P. and I. Hardie. 1995. Least-cost forest carbon reserves: cost-effective subsidies to convert marginal agricultural land to forests. *Land Economics* 71:122.

Paustian, K., O. Andren, H. Janzen, R. Lal, P. Smith, G. Tian, H. Tiessen, M. van Noordwijk and P. Woomer. 1997. Agricultural soil as a C sink to offset CO_2 emissions. *Soil Use and Management* 13:230.

Reilly, J. 1996. Agriculture in a changing climate: impacts and adaptation. In: R. Watson, M. Zinyowem and R. Moss (eds.), *Climate Change 1995-Impacts, Adaptations, and Mitigation of Climate Change: Scientific-Technical Analysis.* Cambridge University Press, Cambridge.

Ribaudo, M., D. Colacicco, A. Barbarika and C. Young. 1989. The economic efficiency of voluntary soil conservation programs. *J. Soil Conservation* 44:40.

Richards, K. and C. Stokes. 1995. National, Regional, and Global Carbon Sequestration Cost Studies: A Review and Critique, Draft Paper, Indiana University.

Sanders, J., D. Southgate and J. Lee. 1995. The economics of soil degradation: technological change and policy alternatives, Technical Monograph Number 22, USAID Soil Management Support Service, Washington, D.C.

Schimmelpfenning, D. 1997. Global climate change: could U.S. agriculture adapt? *Agricultural Outlook,* Economic Research Service, USDA, Washington, D.C.

Sedjo, R. 1989. Forests, a tool to moderate global warming? *Environment* 13:14.

Shakya, B. and F. Hitzhusen. 1997. A benefit-cost analysis of the conservation reserve program in Ohio: are trees part of a sustainable future in the Midwest? *J. Regional Analysis and Policy* 27:13.

Sohngen, B., R. Mendelsohn and R. Sedjo. 1998. The Effectiveness of Forest Carbon Sequestration Strategies with System-Wide Adjustments. Paper presented at 1998 Annual Meeting of American Agricultural Economics Association, July 21, 1998, Salt Lake City, UT.

Stavins, R. 1998. The costs of carbon sequestration: a revealed-preference approach. *American Economic Review* 89:994.

Stephen, J., W. Henderson and D. Stonehouse. 1987. Effects of soil tillage and time of planting on corn yields and farm profits in southern Ontario. *Canadian J. Agricultural Economics* 36:127.

Tweeten, L. and C. Zulauf. 1997. Public policy for agriculture after commodity programs. *Review Agricultural Economics* 19:263.

U.S. Department of Agriculture. 1997. *Agricultural Resources and Environmental Indicators, 1996–1997.* Agricultural Handbook No. 712, ERS, USDA, Washington, D.C.

Watson, R., M. Zinyowera and R. Moss (eds.). 1998. *The Regional Impacts of Climate Change: An Assessment of Vulnerability,* A special report of IPCC WORKING GROUP 11, Cambridge, U.K. Cambridge University Press, Cambridge.

Weesies, G., S. Livingston, W. Hosteter and D. Schertz. 1994. Effects of soil erosion on crop yield in Indiana: results of a 10-year study. *J. Soil and Water Conservation*, p. 598.

CHAPTER 41

Climate Change Policy and the Agricultural Sector

D. Zilberman and D. Sunding

I. Introduction

Climate change has become a major public policy issue and is the subject of a great deal of current economic research. While scientists continue to refine their estimates of the likelihood and dimensions of climate changes, economists continue to assess the impacts of climate change on the economy and design policies to slow climate change while minimizing adverse economic impacts.

This chapter provides an overview of economic research on climate change, with an emphasis on its effects on agriculture. First, general results of studies predicting the impacts of climate change on the agricultural sector are presented. Next, alternative strategies to address the impacts of climate change are discussed. The last section concentrates on policies to delay the process of climate change through mechanisms that will lead to accelerated sequestration of carbon and other greenhouse gases.

II. Economic Impacts of Climate Change on Agriculture

Economic predictions regarding the outcome of climate change depend on scientific knowledge. Since predictions of the physical dimensions of climate change are subject to vast uncertainty, the economic impacts are uncertain as well. The predicted impacts of change also depend on assumptions regarding other phenomena, such as population growth, the rate of technological change, and government policies. Furthermore, there are numerous models of climate change impacts. They vary in their assumptions about risk and uncertainty, the degree of aggregation, and variability and spatial heterogeneity. The major modeling approaches to predict climate change impacts in agriculture are hedonic price models, programming models and Delphi studies.

A. Hedonic Price Models (Mendelsohn et al., 1994)

The two main premises underlying this approach are the following:

(a) Changing Asset Values. The economic impact of changes affecting agriculture is reflected by changes in asset values due to the competitive nature of agriculture and its relatively small impact on the price of other, non-fixed inputs.

(b) Correlation with Temperature. Land prices capitalize the values of the assets' attributes. For example, the price of land can be broken down into components that reflect the value of the attributes of land, including soil quality, location, and weather characteristics. The contribution of one unit of temperature to the price of land is called the hedonic price of temperature. By using existing land price data, Mendelsohn et al. (1994) estimated how changes in temperature and other climatic variables affect land prices throughout the United States. They combined these estimates with prediction of changes in temperature and other indicators of climatic changes to obtain an estimate of the impact of climate change on asset values throughout the country. Through aggregation, they obtained overall estimates on the impacts of climate change on the value of land and agricultural assets in the United States.

Mendelsohn et al. (1994) predict that, under most scenarios, the impact of climate change on U.S. agriculture will be modest and result in less than a 10% change in the value of fixed agricultural assets. The impacts will vary significantly across regions; some regions will gain and others may lose.

The hedonic price approach is clever, but it also relies on restrictive assumptions about competitive behavior of agricultural markets and does not consider explicitly the impacts of consumers' well-being. The approach may be elegant and simple, but its simplicity does not provide as much information as decision-makers would like. There are advantages to alternative, more process-oriented measurement techniques.

B. Programming Models (Adams et al., 1992)

Programming models operate under the assumption that farmers maximize profit. These models require information on land allocation, input use, and, in particular, crop budgets for various locations in the country. They predict land allocations, commodity output and prices, and profits at various locations. The key components of programming models are calibrated to fit current land-use, resource allocation and price patterns. Then the calibrated models are run under various global climate scenarios about the global climate, and the impacts of climate change on yields and costs.[1] These latter impacts are obtained from agronomic models that relate output per acre to temperature, precipitation, and concentration of carbon dioxide in the atmosphere.

The main results of the programming models are consistent with those of the hedonic price models in that they predict climate change will have a modest overall impact on U.S. agriculture. For example, neither output levels nor prices of most commodities will change by more than 10%. Field crop producers will actually gain from climate change, while livestock producers and consumers will lose. The largest impact will be on the regional distribution of economic surplus: some regions may gain while other regions such as the Plains states may lose significantly.

C. Delphi Studies (Doering, 1998)

Doering reported the efforts of extension specialists who interviewed numerous economists and natural scientists in the private and public sectors for their expert assessment on the impact of climate change under various scenarios. Their conclusions include:

(a) Geographic Shift. With climate change, there will be a northward shift of 75 to 150 miles of weather and land-use patterns in most regions of the United States. That may not affect the central areas in most regions but may drastically affect the peripheral locations. Some counties in Kansas and Oklahoma may lose their agricultural viability while some areas in the Dakotas may actually gain. The overall effect of climate change may not be as significant as its impact on specific regions, especially the outlying areas of major agricultural regions.

---

[1]Some of the models may also explicitly include demand equations that relate output price to quantity consumed. The models will compute supply of output for different prices and, using demand and supply relationships, find equilibrium prices the *source* for both prices and quantities of agricultural commodities.

(b) Adjustment Costs. The major costs of adjustment to climate change will result from increased variability and changes in the distribution of temperatures and sunlight over space and time. For example, in California, climate change may lead to faster snowmelt and increased rainfalls, resulting in significant flooding and requiring adjustment in current water containment facilities. In general, ability to develop infrastructure to contain floods will be as important as the ability to deal with droughts. Segerson et al. (1998) found that in some cases the most severe impact of climate change on yield might be related to increased variability in weather conditions during seasons and even 24-hour periods.

(c) Importance of Timing. The pace of climate change is an important determinant of its economic impacts. Climate change will require changes in production practices, design of new pest management strategies, reallocation of resources, and reorganization of the agricultural system. Even in the absence of climate change, agriculture will undergo a natural process of change. Firms are established and then shut down. Old farmers retire, and new ones enter the system. Adjustments to climate change will require acceleration and expansion of these processes of change. The agricultural economy is designed to deal effectively with moderate rates of change. However, rapid changes in weather patterns may pose significant problems for the development of agronomic and technological responses to changing conditions as well as their introduction and implementation.

(d) Research Needs. There is need for expanded capacity of research and resource adjustments. The increased uncertainty that climate change brings to the agricultural system requires further research on alternative responses to drastic change in production and environmental conditions as well as improved physical infrastructure capable of adjusting to change. Governments may need to design institutional arrangements and partnerships between the public and private sectors to coordinate response to swift changes in climate.

Research on the impact of climate change on agriculture emphasizes domestic issues, but the most significant effects are likely to occur internationally. U.S. agriculture has been a major exporter, and one major effect of climate change on agriculture is through its impact on export demand. These types of links and the overall impact of climate change on international agriculture have not been investigated as thoroughly as its impact on U.S. agriculture.

Furthermore, current studies look primarily at the impact of changes in climate based on current agricultural supply and demand relationships. However, the climate change events that we are concerned with will likely occur far into the future—20 to 50 years from now.

Two other dynamic phenomena that will affect future periods are population growth and technological change. The expected increase in both population and income in the next 50 years will result in increased food demand (USDA). Thus, it is important to evaluate the impact of climate change and its disruptive effect on supplies in the context of the growing demand for food and fiber. On the other hand, the impact of population growth may be countered or overcome by technological change in agricultural production systems. Over the last 100 years, the rate of technological change has grown faster than the rate of population growth, which has led to a relative decline in food prices. Technological changes may also improve the capacity to address the impacts of climate change. There is no assurance that past rates of innovations and technological changes will continue forever. Part of the growth in supply and agricultural production was due to over-harvesting of natural resources and the use of unsustainable production techniques. Thus, it is uncertain whether increased knowledge will enable the agricultural sector to develop the technology that will counter the major increase in population, the erosion of agricultural resource base, and the negative impact of the planet's productive capacity due to climate change.

For this reason, it is worthwhile to analyze global resource allocation and production patterns under alternative scenarios combining changes in climate population growth and technological change. More global analyses of the impact of climate change on agriculture will improve predictions. Nevertheless, it is reasonable to assume that the overall impacts on climate change will be modest, relative to other major phenomena such as population growth and anticipated changes in technology.

On the other hand, climate change may have a drastic effect on certain regions and lead to significant political impacts. In particular, climate change may be associated with rising sea levels that cause flooding and destruction of millions of acres of land and displace millions of human beings. Climate change may also result in significant desertification of various regions. These physical impacts may lead to severe levels of economic dislocation, which may result in violence and political upheaval.

III. Components of Climate Change Policy

Due to the high degree of uncertainty regarding the timing and magnitude of climate change, policy responses to address its impacts should be flexible and adaptive. Some analysts are tempted to take a "wait and see" approach and advocate little action until more information is accumulated. However, such an approach is unwise because present decisions affect our capacity to alter the speed of climate change and to address some of its future implications. Decisions made today regarding the use of resources such as land and water may be irreversible or costly to reverse. Thus, in making major irreversible resource use decisions, it is important to consider, within a probabilistic framework, alternative climate change scenarios. This suggests that long-term investments and resource allocation plans that are likely to be unfavorably affected under alternative climate change scenarios will become less desirable. On the other hand, resource allocation choices and investments that may alleviate the negative probable impacts of climate change are more valuable.

Because climate change increases the range of weather phenomena we may encounter, more resources should be allocated to research and development activities that improve our ability to handle extreme weather events such as droughts. For example, we should consider introducing water management schemes that conserve water in areas where climate change will significantly reduce water availability in the future. Because of the uncertainty about climate change, "no regret" policies should be pursued. These are policies that are worthwhile regardless of climate change. Fortunately, policies that will slow climate change and mitigate some of its negative consequences and also enhance other environmental objectives. They may also have a positive effect on reducing air and water pollution, conserving natural resources such as land and water, preserving biodiversity, etc. Because of the positive correlation between climate change mitigation and other environmental objectives, to a large extent climate change policies have become an umbrella to numerous policies that aim to meet a wide range of environmental objectives.

There are two types of climate change policies — those that aim to delay climate changes and those that mitigate some of its implications once it occurs. Much emphasis is given to delay strategies, which will be discussed in the rest of this chapter. They include both activities to reduce human contribution to climate change by reducing carbon dioxide emissions and other climate change gases and sequestration activities that aim actually to sequester carbon from the atmosphere. Policies that aim to change human behavior towards climate change-delaying activities include direct control and economic incentives. Considerations related to the use of these policies are highlighted below.

IV. Incentives and Resource Conservation

There is a wide body of theoretical literature and empirical evidence on the role of incentives as policy tools leading to resource conservation. The basic premise of these policies is that producers are profit seekers and, as prices change, they will alter their production strategies to reduce cost or increase revenues. Farmers may choose environmentally unfriendly activities because they may not pay the full social cost of their action. For example, chemical residues applied by farmers may contaminate bodies of water and cause damage to the fish population. If farmers are not required to pay the cost of the damage to the fish, then they will not modify their behavior. Increasing the cost of the inputs (e.g., water and chemicals) will, however, lead to a change in behavior.

Khanna and Zilberman (1997) provide several examples to illustrate that, as price of the input increases, there is gradual change in behavior. A modest increase in the price of an input may lead to reduction in its use with an existing production technology. A larger increase in an input price may lead to a switch in technology. For example, farmers may switch to integrated pest management in response to an increase in the price of chemical input or, if the price of water increases significantly, farmers may switch from flood irrigation to drip irrigation. As the price of inputs increases drastically, some producers may stop their operation altogether.

Increases in input prices or regulations that limit their use do not only affect the behavior of farmers but also have a strong impact on the behavior of technology manufacturers. The design and introduction of agricultural technology are subject to economic consideration, and it was demonstrated empirically by Hayami and Ruttan (1970) who introduced the hypothesis of induced innovation. According to this hypothesis, societies will adopt technologies that save inputs that are scarce in those societies. The relative scarcity of labor in the United States, compared to the rest of the world, may explain the introduction of labor-saving technology in the United States. The introduction of both the cotton harvester and the tomato harvester has been explained by labor scarcity and inability to obtain cheap domestic and migrant labor to harvest this crop (Schmitz and Seckler, 1978).

The response to the energy crisis is especially relevant as we consider policies to address climate change. The increase in the price of oil and gasoline by 100 percent or more between 1973 and 1975 changed machinery, particularly the characteristics of automobiles that were introduced in the American market. In the 1970s, fuel-efficient cars became popular, and energy-use predictions made before the energy crisis were significantly lowered.

Several studies have documented how scarcity and the increase in effective prices led to changes in behavior and institutions during the recent drought in California (Zilberman et al., 1992). During the 5-year drought of 1987–1991, precipitation was between 40% and 70% of normal. Water inventories provided the means to cope with these shortages in the first 2 years without water cutting supplies to agricultural producers. However, during the later years of the drought, supply was reduced significantly, especially to farmers with junior water rights. The impact on farm income and revenues was not significant. Farmers overcame surface water supply reduction through three means: (1) relying on ground water reservoirs, (2) adopting modern, water-saving irrigation technologies, and (3) fallowing land mainly used for low-value crop. Since 80% of agricultural production was grown with 20% of water, fallowing of low-value crops did not affect income very much. Furthermore, as the drought became more severe, California introduced a water bank, which enabled trading of water. Trading supplied water to producers of high-value crops in the cities. Actually, some of the growers who fallowed land did it to sell their water rights to cities. The drought left California's irrigated agriculture in good shape because of the institutional and technological changes it spawned such as the movement towards increased use of drip, sprinkler, and computerized irrigation systems. While the experiences of the drought and the energy crisis suggest that farmers do in fact respond to incentives, they also imply that the response may take time and that incentives must be quite significant for farmers to alter the way they operate.

The literature on the use of incentives for environmental protection and resource management (Baumol and Oates, 1975) states that producers will increase resource use efficiency and reduce activities that may cause negative environmental price effects if they receive the correct price signals. There are several mechanisms that may increase the cost of resource use or environmental damage. They include taxation, introduction of trading rights in resource use or pollution generation, and subsidies for resource conservation or pollution reduction. While these policies may have the same final outcome in terms of resource allocation, they have significantly different distributional effects. Farmers may oppose bitterly resource or pollution taxation because it may lead to reduction in production and transfer of income to the government. They would be less averse to setting upper limits to total resource use or pollution generated by the industry and introducing a market for emission rights. The industry would prefer this policy to pollution taxation because the tax revenues would remain within the industry rather than be transferred to the government. Actually, resource-using industries most favor subsidies for resource conservation. These subsidies generally make the industry better off relative to no intervention at all, but increased concern for government spending makes this type of solution less achievable politically.

The recent success of tradable emission permits as tools to reduce air pollution (Foster and Hahn, 1995) makes this incentive mechanism more attractive. On the other hand, King argues convincingly that introduction of trading in wetland development rights has not worked as well. The basic idea was to have developers wanting to drain a wetland invest in an activity that would generate a wetland of the same quality elsewhere. The economic volume of these activities has been quite substantial. King (1998) argues that the replacement wetlands are not equivalent to the original ones. Success of the air pollution permit versus the problems of trading in wetlands development rights raises the importance of issues such as monitoring and commodification of environmental goods. Air-polluting gases are well-defined substances, and their emissions can be easily monitored. It is much more difficult to quantify and measure wetlands and, under the current arrangement, the performance approval mechanism requires much subjective evaluation and assessment from government officials. According to King (1998), laxity in defining standards and in allocating of resources for their enforcement led to underperformance of the wetland development trading system.

Several issues impede the introduction and implementation of incentives for environmental performance. These issues include the following:

1. Difficulty in Evaluating Environmental Amenities and Natural Resources

Because of this difficulty, Baumol and Oates (1975) suggested that governments determine target levels of environmental quality improvements and use trading as a mechanism to achieve these objectives. The determination of this target level will imply a value of these resources. Higher reduction levels suggest in most cases a higher evaluation of environmental amenities. While this type of approach may not yield the optimal resource allocation where the economic cost of environmental improvement is equal to the economic benefit at the margin, it is a practical approach that has been taken by policymakers in many arenas including climate change.

2. Monitoring Environmental and Pollution Reduction Activities

The key to having an effective incentive system is measurement of performance. However, it is very difficult to monitor pollution in many cases, for example, emission of carbon dioxide with various tillage activities or seepage of agricultural chemicals into ground water (Segerson, 1998). That will make implementation of direct pollution reduction incentives difficult. An alternative approach is to provide incentives for proxies. For example, instead of taxing water pollution directly, a variable fee based on land use can be levied. This fee can also be adjusted to water application technologies. For example, a farmer who grows oranges with drip irrigation may receive a subsidy or pay a lower fee than a farmer who grows alfalfa with flood irrigation.

3. Adjustment Time and Cost

Adjustment to incentives such as taxes and subsidies take time; therefore, the improvement in environmental quality resulting from the introduction of such incentives is delayed. There is significant heterogeneity among producers, and cost of adjustment for some are much higher than others, so they may pay the extra tax or forego a subsidy in order to continue with old technologies. Eventually, producers may adjust, but policymakers and the public may become impatient with the adjustment process. Stricter policies (direct control) have the advantage of meeting a higher level of compliance in less time. On the other hand, they may entail higher cost in terms of both production and other forms of adjustment.

One form of adjustment cost that has not received much attention in the literature (see Hochman et al., 1977) is insolvency. An environmental regulation that increases the cost of production may reduce producers' profit margin and not avail to them the sufficient resources to pay their debt. The debt equity ratios vary among farmers. Financial viability of some producers is very vulnerable to a relatively small reduction in income, and environmental regulation may lead to their insolvency. In situations of insolvency, one may not observe significant change in resource allocation because the new owners will utilize the resources. But the social cost of the insolvency may be quite substantial, and may provide the main reason for objection to environmental regulation. Hanemann et al. (1987) showed that assessing the impact of drainage control in California revealed that penalties on drainage that may reduce acreage in the Central Valley by 5% may lead to insolvency of farms with about 25% of the land.

V. Implementation of the Kyoto Protocol

The Kyoto Protocol provides a framework to slow and control the buildup of global warming gases in the atmosphere. Each country signing this agreement sets a target level of global warming gases not to exceed the year 2007. These target levels are based on global gas emissions of 1990. The United States aims to reduce its emissions by 6% of the 1990 benchmark. Both the European Union and Japan aim to reduce their emissions by 7%. The Soviet Union and other Eastern Bloc nations have a "black hole." Their target levels are equal to the 1990 emission level but, since their economies and emission levels have declined since 1990, they actually have an added capacity to increase their emission levels.

Countries are allowed to form blocs that will meet the aggregate emission level of the countries forming the bloc. This type of arrangement can result in trading of emission rights. If, for example, the United States and Russia form a bloc, then the United States may pay Russia to use some of the unutilized reduction credit that is available to Russia because its emissions are below the 1990 benchmark level. Furthermore, in some cases it may be worthwhile for utilities in the United States not to invest in reduction of emissions but, rather, to buy emission rights of a utility or a producer in the United States or in countries in the same bloc as the United States.

Each government may use a variety of tools to meet its target level. It may include taxation of carbon dioxide and fuels and direct controls that will require certain technologies to be used in certain activities. For example, it may require that all new power stations use a certain type of energy, say, natural gas, while phasing out inefficient and polluting coal-burning facilities. The Conservation Development Methods (CDM) arrangements enable developed nations to earn emission credit by investing in emission reduction projects in developing nations.

The formal accounting of emission credits is conducted at the national level, so individual countries will have internal accounting. However, what matters at the end will be the national aggregations. For accounting purposes, an international cooperation with units in different countries is not considered one entity. Each plant will be part of the emission accounting of its own country. The Kyoto agreement will be binding once it is signed and verified by 126 nations. Thus far, mostly developed nations have signed the agreement. Developing countries that have signed include Costa Rica, Argentina, and some of the island nations. Some of the major developing countries such as India, China, and Brazil have not yet signed.

These countries have not signed the agreement, not because they oppose setting a global limit on carbon emissions, but because they would like developed nations to pay the cost of reducing emissions. China and India are already among the largest emitters of carbon dioxide, but their per capita energy consumptions are far below that of developed nations. They will continue to grow, and that implies increasing energy consumption. Under current prices, it may mean more than doubling their emissions of global warming gases. The perception in some developing countries is that they did not cause the current mess and should not have to pay for it. One perspective that is supported by some groups in India is to assign nations emission rights based on their population. Of course, such an approach may increase incentives for population growth. It will also obviously benefit populated countries in Asia that can use it as a base for trade. Chakravorty (1998) views the current stance as part of a negotiation process where developing countries would like to make their participation in the carbon emission curtailment agreement subject to partial subsidization by the developed world of the modernization of their energy sector.

While the President of the United States has signed the Kyoto agreement, the treaty has not been ratified by the U.S. Senate. It is not clear, under the current situation, if it has sufficient support to be ratified. Senators of agricultural states control crucial votes, and many in agriculture object to the treaty because reduction in U.S. emissions will lead to increased energy taxes, which may increase the cost of agricultural production. Agriculture is likely to view this treaty more favorably if it provides farmers an opportunity to gain extra income through credit for sequestration of global warming gases.

Agricultural carbon sequestration may be one of the most cost-effective way to slow processes of global warming, but it will be practiced only when such activities are considered part of the global warming gases accounting that are recognized by the Kyoto agreement. The present implementation protocol of the Kyoto agreement recognized forest sequestration as part of the emission-reduction activities of the participating nations. The next section will discuss some of the elements needed to incorporate agricultural carbon sequestration in this process.

VI. The Economics of Trading in Soil Carbon

According to Lal et al. (1998), there is twice as much carbon in the soil as in the atmosphere and, until the later stage of the industrial revolution, tillage activities were the major contributors to the accumulation of carbon in the atmosphere. By modifying soil management practices, for example, adoption of minimal tillage or planting of a wide variety of crops (e.g., legumes) on marginal land, significant levels of carbon can be sequestered. According to some estimates, about 20% of the U.S. annual carbon emissions can be sequestered every year for the next 50 years by changing crop management practices and rebuilding marginal soils. The adoption of these soil management practices has another advantage: they are part of no-regret strategies since they help reduce soil erosion and improve water quality and are consistent with a move to more sustainable and less chemically dependent agriculture.

There is an even larger potential for carbon sequestration by improved soil management if it is done on a global basis (Lal et al., 1998). This type of agricultural management activity can accompany forest expansion and management as part of a strategy to reduce buildup of carbon gases in the atmosphere or even reduce their stock. These activities are especially important during a transition period of 50 years or so when alternative energy sources are being developed.

Two sets of issues are raised when incorporating soil carbon as part of accounting of global warming gases conducted as part of the Kyoto agreement. First, carbon sequestration has to be economical and compete with alternative means of reducing global warming gas emissions. The second, and more difficult, set of issues are modification of soil carbon sequestration activities.

A. The Conditions for Economical Sequestration of Soil Carbons

The economic cost of soil carbon sequestration depends on alternative activities that reduce global warming gas stocks. These alternatives include the following:

1. Improving Efficiency of Existing Power Plants and Switching to Clean Energy Sources

Khanna and Zilberman (1997) showed that improvement in management practices could reduce India's CO_2 emissions by about 10% and increase energy production from existing facilities by about 10%. A more drastic approach involves closing old, inefficient coal-burning facilities and switching to natural gas, wind, or nuclear power.

2. Adopting More Fuel-Efficient Appliances and Improving Energy Management Strategies to Reduce Energy Demand

Increasing energy prices and taxing carbon or fossil fuel can induce these strategies. These taxes may be combined with subsidies for low-income individuals.

3. Expanding Forest Areas and Agroforestry Activities

Managers of forests in Costa Rica have negotiated selling their sequestration rights to a consortium of Canadian utilities. The potential of sequestration by trees has been significantly investigated, and sequestration by trees and use of biofuel are considered by some as an efficient strategy for slowing climate change.

It is difficult to estimate an equilibrium price for sequestered carbon. It ranges from $10 per ton to as much as $150 per ton (Tweeten et al., 1999). We assess that carbon sequestration at below $30 per ton is economically viable and that sequestration between $30 and $50 per ton may be economical under some circumstances. If the cost per ton is above $60, then economic viability is highly improbable.

However, when carbon sequestration activities provide added benefits that can be enumerated economically in terms of reduced soil erosion and water contamination, then benefits have to be taken into account in designing an optimal social resource management plan. When these benefits contribute to the profitability of farm operation, then rational growers will automatically take them into account. However, if these benefits contribute to the societal well-being, but not the well-being of individual producers, then individual decision-makers will take them into account only when appropriate incentives (e.g., subsidies) are established. It should be noted that other policies to control global warming (e.g., carbon taxes) also have beneficial side effects such as reduction of traffic congestion and improvements in the balance of trade.

Future research will aim to provide information for a quantitative analysis of the economics of soil carbon sequestration. It should assess the cost of these activities at each location and provide some measure of their carbon sequestration potential and other benefits. Such research will be conducted consistently on a national or global basis. We have Geographical Information Systems that can identify the area where soil carbon sequestration provides the best economic opportunity, and policies will be designed to target these opportunities. These policies will include both payment for carbon sequestration and compensation for other socially beneficial side effects. Babcock et al. (1996) develop a methodology for targeting environmental conservation activities for the Conservation Reserve Program. Such methodologies should be developed for targeting and pricing carbon sequestration activities.

B. Implementation and Modification of Soil Carbon Sequestration Activities

The establishment of a workable implementation plan to incorporate soil carbon sequestration to a quantitative climate change control strategy requires addressing several challenging problems including:

1. Organizational Structure

The remuneration and compensation of soil carbon sequestration activities may be conducted within a market system or as part of a government program. Figure 1 depicts the main organizational structure needed to facilitate soil carbon sequestration in a market for carbon emission rights. Buyers and sellers of emission rights will interact through an exchange (such as the Chicago Mercantile Exchange) which will manage documentation and transfer of rights and funds. The exchange will also likely be responsible for recording and incorporating various transactions within the appropriate national accounting systems for global gases since national accounts are officially recognized by the Kyoto Protocol. Because of the vast amounts of land and large number of farming units involved in

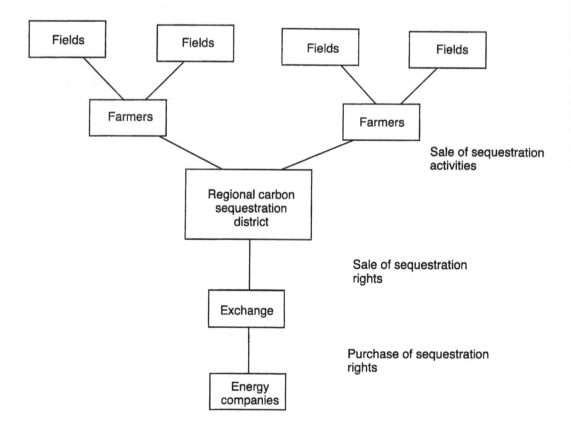

Figure 1. Organizational structure needed to facilitate soil carbon sequestration.

soil sequestration activities and the relative small monetary volume of soil sequestration activities of individual fields or farms, intermediate regional units will emerge. These units that will be referred to as carbon sequestration districts will sell (and sometimes buy) carbon sequestration contracts to the exchange. Utilities and other emitters of global warming gases will purchase these contracts. The sequestration districts will manage the accounting with individual farms and monitor activities at the field level. The exchange may monitor the performance of the sequestration districts.

This type of decomposition is not unusual and is consistent with organizations such as water or pest management districts. The decomposition reduces transition costs and enables establishing different types of contracts for units of different sizes.

Soil carbon sequestration activities may be remunerated through a government program. This may be an expansion of the Conservation Reserve Program (CRP) where the USDA (or other agency) will pay farmers for specified soil management activities through contracts. The government will obtain the rights to the carbon sequestered by these activities, and they will be credited to the national emission reduction effort or be traded. Remuneration through a government program can rely on existing infrastructure for compensation and monitoring of soil management activities. Farmers' receipts from the program may add compensation for extra social benefits generated by carbon sequestering soil management practices. It may reduce transaction costs, since national governments are the entities engaged in the carbon accounting. On the other hand, credibility of carbon sequestration activities will be enhanced if they will be commodified and traded with a commercial market system.

2. Measurement and Monitoring

The remuneration of carbon sequestration will not be based (in most cases) on actual sequestration levels but on estimated sequestration. Therefore, it is crucial to develop procedures, formulas, and software to translate farm management activities in the field to carbon sequestration units. The formulas rely on information on initial situations in the field, agroeconomic activities, and climatic conditions to estimate sequestration levels. Obviously, accurate accounting requires accurate documentation of benchmark conditions and continuous monitoring of agroeconomic activities, and that may result in high transaction costs.

The design of accounting and remuneration procedures is challenged in balancing the gains from increased accuracy with the cost of extra transaction costs. It is expected that software and accounting procedures will be modified and improved as new experience is accumulated and technologies progress. If carbon sequestration credits are traded within markets, the regional sequestration districts are likely to sell and sometimes buy)[2] sequestration contracts (in units of, say, 100 tons of carbon) and buy from farmers long-term contracts (say, 5 to 10 years) that specify agronomic practices. Over time, we may expect that insurance, trading, and accounting arrangements will emerge, so that farmers will be able to buy the rights to discontinue contracted practices, and districts will be able to protect themselves if they fail to deliver a promised contract.

Agricultural practices that will be paid for by sequestration contracts have to be easily observable. They may be specified in terms of crop selection, choice of irrigation and soil management practices, etc. Districts will monitor individual farmer behavior, and it will be desirable to develop a mechanism (e.g., random checkup) to monitor districts' activities. When formulas require quantitative information on yields and actual input use, over time districts may obtain access to accounting data or monitor activities through remote sensing.

---

[2]Sequestration districts may need to buy emission rights from time to time because some farmers may not fulfill their contracts, or changes in conditions over time may make it uneconomical to conduct carbon sequestering activities (actually carbon emitting activities become profitable).

Establishing highly automated, electronic monitoring systems may be challenging, but it is feasible in developed countries. The monitoring procedure may vary by location, and a bigger challenge is to develop effective monitoring procedures and remunerating formulas in developing countries. Research aimed at designing sequestration rights trading mechanisms have to emphasize the development of mechanisms that will facilitate implementing such markets in developing countries.

3. Abuse Potential and Equity Considerations

Ideally, all soil emission and sequestration of carbon should be part of an aggregate national carbon accounting and treated like emissions from major industrial sources. Thus, the percentage (6% in the case of the United States) of reduction in carbon emissions will apply to all sources. The agricultural sector would have gained from such arrangement if emission reductions exceeded the national target and would have had extra costs if emission reductions did not meet the national target.

The across-the-board incorporation of soil carbon emissions and sequestration in the national carbon accounting is not likely to materialize partially because of monitoring and transaction cost consideration and mostly because of lack of political support. This arrangement is likely to eliminate much of the strategic behavior and abuse which may accompany the arrangements that are more likely to occur. These arrangements will apply only to part of the national land base, in particular, to locations with explicit engagement in sequestration activities. The voluntary (at least initially) participation in sequestration activities may motivate landowners to engage in carbon-emitting activities prior to agreeing to a carbon sequestration contract. Such activities (intensive tillage) may increase immediate profits and generate larger potential for sequestration gains. Furthermore, with finite sequestration contracts, farmers may reverse the sequestration gains by changing land management activities when their contracts expire.

To reduce the potential for such abuses, the term of contracts may be very long, but farmers may exit from their contracts and reverse some of their sequestration activities by purchasing emission rights. Furthermore, the benchmark determining the potential for carbon sequestration at a certain location will be decided by the state of the location at a moment in the recent past, say 1998, to eliminate incentives to carbon emission activities behavior before enrollment in sequestration programs.

The introduction of rewards for carbon soil sequestration may negatively affect individuals who have managed their soil soundly and thus have small potential for enhanced sequestration. This equity problem may reward unsound soil stewardship in the past and may divert most of the reward for restoration to regions with marginal and/or bad managed soils. These distributional side effects may have political economic implications where representatives of regions with high-quality, well-preserved lands may not be major supporters of remuneration for soil carbon sequestration. Thus, the legislation to introduce rewards for soil carbon sequestration may be needed to accompany rewards for other environmental stewardship activities that are applicable to regions with lower potential for carbon sequestration.

The implementation difficulties associated with the introduction of rewards for soil carbon sequestration are even more daunting in developing countries. These difficulties may prevent obtaining much international support to inclusion of these activities in the Kyoto Protocol, especially in the long run. Since these soil carbon sequestration activities have the potential to be an important contributor to a global warming remediation strategy and, since the United States may benefit from remuneration of soil carbon sequestration, it may be useful if a soil carbon sequestration program is introduced unilaterally in the United States, so that the implementation issues will be ironed out and the feasibility of the strategy will be examined. The successful implementation of this initial effort will domestically facilitate incorporating soil carbon sequestration in the Kyoto Protocol.

Some of the problems associated with wetlands development rights trading may also promote trading with carbon sequestration rights. It is important to develop mechanisms for measurement and verification to ensure problems of irreversibility are addressed.

The first problem is technical and may be the least difficult. Some basic formulas need to be established so one can translate soil management activities in the field to carbon sequestration levels. If one has the description of the initial situation of a field and certain activities occur, one can compute the levels of carbon sequestration over time. Secondly, activities over time may improve or worsen carbon situations in the atmosphere. Thus, accurate accounting requires some dynamic monitoring. Since changes in behavior affect carbon in the atmosphere, it would have been more accurate if credits and contracts were based on annual performance. However, that may result in high transaction costs, and one may have to consider longer term contracts to specify behavior for set periods, say, 10 years. There will be penalties for deviation from the pre-specified agreements. The cost of implementation largely depends on the size of trading units. Ideally, it will be desirable to monitor behavior on a field-by-field basis, but it may useful to establish larger units that will be responsible for sequestration activities. It may be a farm or a sequestration district. Land-use activities within this district will be monitored to provide accounting of emissions of sequestration, and that may be compared to some prespecified plan. The districts may maintain bookkeeping transactions with some carbon sequestration banks. Then we will handle bookkeeping activities with each individual farmer. This approach of dividing national farming areas into districts that are responsible for all the transactions with the individual units is quite common in land resource management, water delivery and soil conservation, and can be pursued as a proven mechanism of trading soil carbon sequestration.

VII. Conclusions

Climate change presents major uncertainties for the future of agriculture both domestically and globally. Future impacts have to take into account expected developments such as population growth, increase in demand of agricultural products, and changes in technologies that will likely increase productivity. Within this context, the overall impacts of climate change on agriculture are not likely to be drastic, and accommodation and adjustment for climate change have to be incorporated with policies that address these other phenomena.

The distributional effects of climate change are likely to be much more significant than its overall impacts. The economic viability of agriculture in some locations, especially on the periphery of major agricultural production regions, may be drastically altered. Some of these locations may benefit from improved climatic conditions, but production in others may no longer be economically feasible. Rising sea levels may present a major risk for farmers and other inhabitants in coastal zones. Adjusting to these locations and production systems will likely be a major challenge in coping with climate change.

Effective research, extension, and marketing capacities as well as an ability to raise funds for infrastructure and to design and finance appropriate adjustment schemes will determine the capacity to accommodate and adjust to climate change events. The difficulty of adjustment will depend on the speed of these changes; obviously, fast changes in weather patterns will pose a much bigger challenge in terms of adjustment and accommodation. Uncertainty regarding the magnitude and locations of climate change phenomena should not hinder pursuit of economically viable policies that will expand our future options. No-regret policies that attain desirable objectives are especially valuable. Climate change policies should be integrated with other policies, especially those addressing environmental problems in agriculture.

Adjustment costs and the uncertainty of climate change are essential in slowing and reducing the magnitude of this process which is a major feature of the Kyoto agreement. Agriculture maybe somewhat affected by increased energy prices as governments aim to reduce carbon emissions in energy production and transportation. The cost of adjustment to financial incentives and regulations that may result from the need to meet reduced emission targets are likely to be alleviated by the development and adoption of resource-conserving technologies that may be induced by these new policies.

Agriculture may become a major contributor to slowing global climate change if all carbon sequestration activities will be undertaken as part of the carbon emission reduction accounting initiated by the Kyoto Protocol and be compensated accordingly. Agricultural activities that sequester soil carbon are sources of numerous other environmental benefits and are important components of no-regret strategies. However, the implementation of schemes to incorporate soil carbon sequestration within the overall carbon accounting is quite challenging. They require establishment of organizational structure that will facilitate measurement, computation, and monitoring of soil carbon sequestration. It demands development of accurate estimation schemes to accurately quantify the impact of agricultural practices on soil carbon. There are technological and institutional challenges in implementing effective monitoring and enforcement schemes to address possible problems of moral hazards. The implementation difficulties may hamper the enactment of soil carbon sequestration as part of the Kyoto Protocol in the short run. Because of the high potential benefits of these activities, the United States should unilaterally begin experimenting with policies and organizational arrangements to encourage carbon sequestration in agriculture, and should incorporate these arrangements as part of its national agricultural policies.

References

Adams, R., C. Chang, B. McCarl and J. Callaway. 1992. The role of agriculture in climate change: a preliminary evaluation of emission control strategies. In: J.M. Reilly and M. Anderson (eds.), *Economics Issues in Global Climate Change: Agriculture, Forestry, and Natural Resources*. Westview Press, Boulder, CO.

Babcock, B., P.G. Lakshminargyan, J. Wu and D. Zilberman. 1996. The economics of a public fund for environmental amenities: a study of CRP contracts. *Am. J. Agri. Econ.* 78:961-971.

Baumol, W. and W. E. Oates. 1975. *The Theory of Environmental Policy*. Prentice-Hall, Englewood Cliffs, NJ.

Chakravorty, U. 1998. Climate change and carbon trading, an international perspective. Presented at CO_2 Sequestration Schemes and Markets for Carbon Trading in U.S. Agriculture and Energy Sectors, November 20, 1998, Washington, D.C.

Doering, O. 1998. Presented at CO_2 Sequestration Schemes and Markets for Carbon Trading in U.S. Agriculture and Energy Sectors, November 20, 1998, Washington, D.C.

Foster, V. and R. W. Hahn. 1995. Designing more efficient markets: lessons from Los Angeles smog control. *J. Law and Economics* 38:19-48.

Hanemann, M., E. Lichtenberg, D. Zilberman, D. Chapman, L. Dixon, G. Ellis and J. Hukkinen. 1987. Economic Implication of Regulating Agricultural Drainage to the San Joaquin River. Technical Committee Report. Western Consortium for the Health Professions, 1987.

Hayami, Y. and V. W. Ruttan. 1970. Factor prices and technical change in agricultural development: The United States and Japan, 1880-1960. *J. Political Economy* 78.

Hochman, E., D. Zilberman and R. E. Just. 1977. Internalization in a stochastic pollution model. *J. Environmental Economics and Management* 4:25-39.

Khanna, M. and D. Zilberman. 1977. Incentives, precision technology and environmental quality. *Ecological Economics* 23:25-43.

King, D. 1998. Lessons from wetland mitigation and issues in debit/credit criteria. Presented at CO_2 Sequestration Schemes and Markets for Carbon Trading in U.S. Agriculture and Energy Sectors, November 20, 1998, Washington, D.C.

Lal, R., J. Kimble, R. Follett and C. Cole. 1998. *The Potential of U.S. Cropland to Sequester Carbon and Mitigate the Greenhouse Effect.* Sleeping Bear Press, Chelsea, MI.

Mendelsohn, R. and J. Neumann. 1998. *The Market Impact of Climate Change on the U.S. Economy.* Cambridge University Press, Cambridge.

Schmitz, A. and D. Seckler. 1978. Mechanized agriculture and social welfare: the case of the tomato harvester. *American Journal of Agricultural Economics* 52:567-577.

Segerson, K. 1998. The uncertainty and impact of climate change. Presented at CO_2 Sequestration Schemes and Markets for Carbon Trading in U.S. Agriculture and Energy Sectors, November 20, 1998, Washington, D.C.

Tweeten, L., B. Sohngen and J. Hopkins. 1999. Assessing the economics of carbon sequestration in agriculture. Presented at Conference on Assessment Methods for Soil Carbon Pool, November 2–4, 1999, Columbus, Ohio.

U.S. Department of Agriculture. 1998. *Long-Term Demand Forecasts.* World Agricultural Outlook Board, Washington, D.C.

Zilberman, D., A. Dinar, N. MacDougall, M. Khanna, C. Brown and F. Castillo. 1992. How California responded to the drought. Department of Agricultural and Resources Economics, University of California at Berkeley, 1992.

CHAPTER 42

Approaches to Assessing Carbon Credits and Identifying Trading Mechanisms

A. Manale

I. Introduction

Information I am providing is not to be taken as current policy of the United States Government except where I explicitly state it as such. The whole policy arena involving carbon sinks, particularly involving agriculture, is unclear. The Kyoto Protocol (United Nations, 1997) leaves many unanswered questions that must be resolved in subsequent meetings of international bodies. As has been stated by members of the Administration: "The Treaty is a work in progress" (Eizenstat, 1998). Even in the case of forestry, where the Protocol explicitly provides for credits for activities that sequester carbon, uncertainty abounds regarding the definitions of the activities (afforestation, reforestation, and deforestation) and the mechanisms for crediting. Until the United States Senate ratifies the Kyoto Protocol — or some manifestation thereof — nothing in the treaty represents U.S. policy unless there is precedence in existing law. That said, let me try to present the preconditions, from the perspective of government, that would most likely have to be satisfied in order for a market in soil carbon credits to develop.

The United States can clearly set up its own domestic carbon trading schemes regardless of international policies and treaties. But the increase in atmospheric levels of greenhouse gases is a global problem demanding a global solution, as represented by the Kyoto Protocol or some manifestation thereof. In this case, compliance with the Protocol would impose conditions upon trades if the reductions in carbon emissions represented by the trades are to count towards meeting our national cap on total emissions. International concerns about the laxity in the manner in which signatories meet their commitments will dictate the stringency of guidelines for assessing carbon credits for agriculture, or for forestry for that matter. Thus, real or perceived uncertainty associated with measuring soil carbon and assessing carbon by various practices or activities will determine its acceptance internationally.

This is a policy arena in which there is no clear, established path to the ultimate policy goals. There is nothing in economics or law that can be immediately adapted to establish soil carbon markets. Nevertheless, there are some precedents from which we can surmise what can or what cannot work.

This chapter presents the issues that need to be resolved with regard to the assessment and trading of credits. The relative difficulty of addressing these issues is then compared against two existing environmental markets, and the concept of equivalency where the risk of nonperformance may not be quantifiable is introduced. Finally, alternative mechanisms for encouraging trades in soil carbon credits are discussed.

II. Assessment of Credits

The following three sets of issues are critical to the development of a commodity that can be traded. The first, which I call the dimensionality issue, is primarily technical and reflects the state of the science. It encompasses (1) the ease of measurement, (2) the timescale of the measurement, and (3) baseline projections. The second set relates to legal and institutional issues regarding trading in carbon credits. The third, the issue of equivalence, is critically important if the credits are to be used to meet international commitments under a treaty.

Soil carbon is not like bushels of corn, discrete items or set of items that can be clearly and unequivocally measured, given agreement on how measurement will occur. In fact, the carbon in a discrete plot of soil is not likely ever to be actually traded for the purpose of meeting international or even domestic commitments to reduce greenhouse gas emissions. Indeed, what is traded is the promise of the performance of a service whereby carbon is sequestered in soils and maintained there for a given period of time. The service involves using agricultural practices that have been scientifically shown to increase the level of carbon in soil and possibly also in surface biomass. The precision of the measurement that is important to trading relates to the ability to predict the quantity of product, or stock, provided by the service at a given point in time and within a given quantity of land.

Experience in trading in other environmental policy arenas can provide insight into the difficulty of establishing a market in soil carbon credits. Two recent examples of trading schemes serve as the two poles of ease and difficulty of implementation: sulfur oxide (SO_x) emission trading and wetlands mitigation banking. SO_x emission trading is considered by both government and industry to be a success both in terms of reducing the cost of meeting environmental standards and in terms of having reduced the emission of pollutants into the environment. On the other hand, wetland mitigation banking has, up to now, failed to stem the loss of wetlands.

Trading in reductions in SO_x emissions was established with the 1990 amendments to the Clean Air Act (42 U.S.C. §§ 7401 et seq. (1970) and USEPA, 1999). SO_x, a combustion product from coal-fired electrical power plants, is a major air pollutant that contributes to acid rain. Electric utilities are, by regulation, given allotments of SO_x based upon historical emissions of the plants. A reduction in emission over a given time frame over the allotment serves as a credit that can be sold to other energy producers.

Wetlands mitigation banking is a federal program under which landowners needing to "mitigate" or compensate for deleterious impacts to wetlands have the option of purchasing credits from an approved mitigation bank instead of having to mitigate onsite (U.S. Army Corps of Engineers, 1994). Developers thereby save themselves the trouble and expense of restoring or creating wetlands on or near the development site. A wetland bank may be created when a government agency, a corporation, or a nonprofit organization undertakes such activities under a formal agreement with a regulatory agency. The value of a bank is determined by quantifying the wetland functions and values restored or created in terms of "credits." Because full restoration of functions and values may take many years, the credit is tantamount to a service that promises to provide a benefit after a unspecified number of years (King, 1993, 1994, 1997).

A. Dimensionality Issues

Table 1 shows how SO_x trading, wetland mitigation banking, and trading in soil carbon credits compare against dimensionality issues. Measuring SO_x emissions from a smokestack is relative straightforward; the reduction in flow over time is the credit. Because of Clean Air Act regulatory requirements, long- and short-term measurements at individual plant have generally been made over many years before implementation of the trading program. The technology used for monitoring is

Table 1. Dimensionality issues

	SO$_x$	Wetlands mitigation banking	Soil carbon
Ease of measurement	Relatively straightforward	# of acres relatively easy to determine, functionality and values difficult	Difficult, monitoring data would have to calibrate models to estimate amounts sequestered
Timescale	Measured over short and longer term periods with good precision	Functionality measured over years	Multitude of years for a given farm; less for greater spatial scale.
Baseline projections	Relatively straightforward	Comparability difficult to assess	Difficult, need data for many years; moral hazard problem

not controversial. Wetlands mitigation banking, on the other hand, has little historical precedence and the science of restoration or conversion, let alone measurement of values and functions, is still in its infancy (Wainger, 1999). It is not a flow that is measured, but an accounting of stock (King, 1997). Though one can count individual acres of wetlands, determination of equivalency in functions and values is extremely difficult and successful restoration may not be determined for many years. No wonder that there is considerable skepticism on the part of experts as to whether or not wetlands mitigation banking, at least in the near future, can be a low-cost approach to the protection of wetlands.

Measuring soil carbon credits can be as difficult as determining the functions and values of wetlands for wetlands mitigation banks. Soil carbon levels can vary within a field and across different soil types, even more as a consequence of its agricultural history. Because of cost, the number of soil samples that can be taken is limited. Models must be used to extrapolate to larger spatial scales. These models necessarily simplify reality which contributes to uncertainty in the resulting estimates. And, similar to the situation for wetlands mitigation banks, the time dimension of a credit for carbon sequestered in specific fields must necessarily be long. If the time period is particularly long, such as 30 or more years, a multitude of things could occur that could adversely affect soil carbon levels in these fields. The result is that less carbon is sequestered carbon than the amount of the credit that has been traded.

With regard to baseline projections, determining how much carbon there was in the soil preceding the change in practice that increases the amount of carbon in the soil is likely to be at least as difficult as establishing comparability between wetland acres that are to be destroyed and those created in wetland mitigation banks. The amount of soil carbon credits is a function of the change in soil carbon over time over the baseline sequestration level – that is, the amount that would be in the soil were the practice not put in place. For some soils that have been depleted of carbon in relation to their pre-agricultural state, the lower the level of carbon is at the time at which the contract period for the carbon credits begins, the greater will be the potential soil carbon credit. Hence, farmers face the incentive to give erroneous information on tillage history or to manage the soil so as to decrease soil carbon levels before the onset of the contract period – a problem economists call moral hazard. Uncertain baselines make carbon credits extremely difficult to verify. And because larger credits benefit both the farmer and national interests, the incentive definitely exists for all reporting parties to understate initial levels and hence overstate claims.

B. Legal and Institutional Issues

The establishment of property rights necessarily precede markets for certain goods, especially those that are traded but that are never seen by the participants in the trade. Thus, there may potentially be value associated with a soil carbon credit, as would be the case if a regulatory cap is imposed on fossil fuel emissions such that any increase in emissions by industry in general or beyond an allotment given to an emitter would have to be offset by an equivalent sink that constitutes the credit. Where there is uncertainty regarding property rights, trades may still occur, although they may exact a discount that reflects the degree of risk upon the seller or buyer. The greater the risk or lack of trust, the greater the discount. If the risk is too great and hence the discount is too great, there may be too few buyers and sellers and, from the perspective of the public good, there may be too few trades or too few items that are traded.

There are many legal and institutional issues surrounding the definition of the right to the traded property – that is, the credit for soil carbon – that can affect riskiness as well as the overall cost of transacting a trade. Four that are particularly important are: (1) the time period over which the carbon must be maintained in the ground, (2) who is liable for failure to perform the service, (3) the begin date of the service, and (4) who is responsible for monitoring and verification. Whether or not carbon trading is more like SO_x or wetlands mitigation banking on these issues depends in large measure on the relative ease of institutionally addressing the following questions: If a farmer contracts to sequester carbon and sell the credit to the energy industry, what is the minimum period that the carbon must be maintained in the ground through practices that prevent its release? If a contract is with an association of farmers, what would the time frame be for contracts? If the farmer or an association to which a farmer belongs which is party to the contract fails to provide the service or tills the land after only a few years, releasing the carbon that had been stored, who is liable for the carbon that has not been sequestered or the carbon that has been emitted as a result of the purchase of the credit? Does the farmer owe the buyer of the credit or is the reduction in credits entered into the utility's account? If the farmer sells the land, what mechanism can be used to ensure that subsequent owners maintain the carbon levels in the soil? Can a farmer buy a credit from another farmer in order to use the land under contract in a manner inconsistent with maintaining or sequestering carbon in the soil? At what point does the credit for carbon that will be sequestered as a consequence of the use of a practice accrue so that the credit may be sold?

The last of the four legal and institutional issues, monitoring and verification, concerns not just government as a consequence of its responsibilities under international commitments, but also the actual parties to a trade. Reaching international agreement on soil carbon sinks as an offset to greenhouse gas emissions depends to a great extent on treaty signatories' perception that government, and the United States Government in particular, is able to enforce the conditions of a trade and to ensure landowner and farmer compliance. Who is responsible for keeping records on carbon that has been sequestered through a trade? If the credit for x number of years of practices that sequester carbon is given at the beginning of the contract period, who monitors performance since, once the credit has been established, it is in neither the farmer's nor the utilities interest to execute the conditions of the contract? Clearly third-party certification could resolve the problem, but legal requirements would have to be established for it in advance of trades.

Table 2 shows how trading in soil carbon credits compares to SO_x trading and wetland mitigation banking regarding the difficulty in addressing legal and institutional issues. Soil carbon credits are again more like wetlands mitigation banking credits than those for SO_x emission trading. And in

Table 2. Legal and institutional issues

	SO$_x$ trading	Wetlands mitigation banking	Soil carbon credits
Time dimension of practice that leads to credit	Clock starts and stops w/action that establishes the credit	Clock starts with credit, indefinite future with easement	Requires a legal mechanism like an easement or covenant on agricultural lands
Liability	Producer responsibility	No consistent requirements regarding unclear liability; bonding requirements for some banks, not necessarily based upon actual risk of failure	To be statutorily or contractually determined
Begin date of credit	After performance	Before performance since full functionality requires decades	Before performance since soil carbon may have to remain in the soil for decades
Monitoring and verification	Company monitors, government verifies	Private and government monitoring; government verification	For private contracts, unclear (3rd party?); otherwise, government

relation to the time dimension and begin date of credit, trading in soil carbon credits more closely resembles wetland mitigation banking in that the credit is for a service.

For SO$_x$ trading, the credit is established only after the SO$_x$ has not been emitted during the time period specified in the allowance contract. On the other hand, soil carbon credits, similar to wetlands mitigation banking, must necessarily accrue before there is certainty that the conditions of the contract have been fully executed since confirmation that the wetland is fully functional or the concentration of carbon in the soil has increased as much as required may take many years.

Since there may be a strong incentive for the landowner to till the soil so as to lose the carbon and thus reestablish the potential for future carbon gain once a contract has ended, trades in soil carbon credits may require a legal mechanism, such as an easement or a covenant, that restricts how the land is managed into the future. [This may not be an insurmountable problem under a program of comprehensive national accounting which would track the movement of carbon into and out of the atmosphere from all major economic activities. A shift in land use from sink to emitter would presumably be identified in the system with a corresponding increase in the debit account for carbon emissions. The nation or an industrial sector would thus be responsible for reducing its emissions that much more.] Alternatively, he or she may "mine" it for its nutrient content whereby he or she grows crops without replacing the nitrogen or phosphorous that is lost to the crop from the soil. Because trading in carbon credits would necessarily have to involve large numbers of acres – so as to reduce measurement uncertainty, a likely requirement for international acceptance – the geographic and temporal scale of restrictions on activities on private lands may pose a significant political issue.

Regarding the liability issue, trading in soil carbon credits resembles wetland mitigation banking. For SO$_x$ trades, who is responsible for establishing the credit is readily apparent. A major reason that wetland mitigation banking has not been as successful as had been hoped has been the lack of clarity regarding liability associated with many wetland mitigation banks should the restored wetland fail after the limited contractual period. A variety of devices could be legally required to address this problems, such as surety bonding, that accounts for the actual risk of failure well into the future or the establishment of soil carbon banks that serve as insurance against failure to sequester or maintain the sequestration of soil carbon.

Because meeting domestic – or international – agreements require a periodic accounting of the stock of wetlands in banks or carbon in agricultural soils, some entity must be responsible for this monitoring and verification. In fact, under existing climate treaties, the United States and other signatories must develop emission baselines (United Nations Framework Convention on Climate Change, 1992). All sources of emissions must be inventoried, agriculture included, and the inventory serves to establish the baseline amount against which the U.S. commitment would be measured. Government has, however, not yet established baseline levels of agricultural soil carbon for any specific date either nationally or regionally. This initial assessment of agricultural soil carbon stocks or at least current rates of change in soil carbon levels predicates an active program of trading in soil carbon credits.

In the case of wetland mitigation banking, government performs this task. Similarly, government currently conducts surveys of soil carbon for the purpose of evaluating the quality and quantity of natural resource stock. Presumably for a soil carbon credit trading scheme, government, perhaps with compensation by the private sector, could expand its monitoring for specific trades or, alternatively, third-party certifiers which would be licensed by government would monitor with verification by the government.

The information generated from monitoring would have to be conveyed not just to parties involved, but to government as well to show compliance with international agreements. There needs to be an institutional structure to collect and disseminate this information, such as now exists for wetland mitigation banks in which the Corps of Engineers, U.S. Environmental Protection Agency, and the United States Department of Agriculture develop and share databases and make these data available to the public.

C. Equivalence

Equivalence is a measure affecting the degree of uncertainty regarding predictions – that is, the inability to predict with any degree of precision the future state of soil carbon on a particular plot of land. It manifests itself in the relationship between the near certain probability that a ton of carbon that is involved in a trade will be released and the much lower probability that the same amount of carbon will have been sequestered and that it will remain in the soil at least until some point in the future. How much lower depends upon the following risk factors (Wainger, 1999):

- flaws in the underlying science or engineering;
- inadequate planning, funding, or management;
- unexpected events affect performance, such as climate change or weather;
- performance successful, but unexpected events result in loss — floods, distant water, diversion projects, lower water table, etc.;
- performance succeeds but there are changes in the rules;
- performance succeeds, but there is a change in political commitment, reinterpretations, or disregard for commitment;
- economic pressures to use land previously committed to sequestration differently;
- unexpected vandalism, etc.

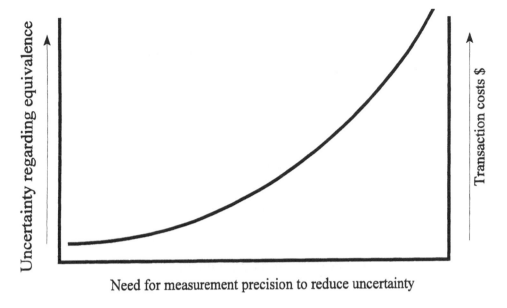

Figure 1. Relationship between uncertainty regarding equivalence and transaction costs.

The greater the time horizon over which the trade commitment or performance is considered, the greater the likelihood that one or more of the above occurs, approaching 100% for a specific acre of land.

Figure 1 illustrates the relationship among equivalence, the need for precision in measurement to reduce uncertainty (as well as improved legal and institutional approaches) and transaction costs. By the latter, I mean not just the administrative costs associated with the actual trades, but also any costs associated with monitoring, acquiring surety bonds, documenting, or otherwise assuring that a given amount of carbon or carbon equivalent is sequestered for a given length of time. The greater the uncertainty regarding the future state of the land, the greater the need for more precision in measuring soil carbon and estimating the amount of carbon in a trade, which in turn leads to higher transaction costs. Improvements in the resolution of dimensionality issues and articulation and development of institutional and legal approaches can, to some extent, offset or reduce problems of equivalence.

For example, the problem of equivalence can in part be addressed by scaling the amount of land involved in a credit trade or by factoring in policy-determined (as opposed to scientifically based) risk adjustment values into the estimate of soil carbon which serves as credit. In the former scenario, the geographic unit for trading becomes the area, the size of a county or preferably an agroecoregion, within which carbon is sequestered in the soil. In the latter, the amount of carbon sequestered in the soil per acre over a specified number of years is divided by a number that reflects the likelihood that the amount of carbon will indeed still be in the soil after a number of decades. In other words, the carbon credit is discounted by a factor reflected by the degree of market risk.

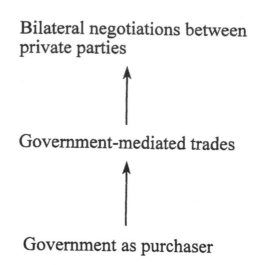

Figure 2. The extremes of trading mechanisms.

III. Trading Mechanisms

By trading mechanisms, I mean the mechanism by which a farmer who can put carbon into the ground, and thus establish a carbon credit relative to baseline emissions, can be compensated for doing so. At the two extremes of compensation mechanisms are, on the one hand, government as purchaser and, at the other, bilateral negotiations between private parties with government's role limited to judicial redress. In-between are, as illustrated in Figure 2, all manner of trades with varying degrees of government mediation.

Government as purchaser pays farmers for practices that serve to sequester carbon. The manner in which such payments could occur would resemble current United States Department of Agriculture programs, like the Environmental Quality Incentive Program or the Conservation Reserve Program, that provide cost-share and direct payments for management practices or land retirement that provides a variety of environmental benefits, such as water quality or wildlife conservation. Alternatively, it could request bids for a quantity of soil carbon wherein legal language would specify the conditions, including protections for the environment, that would have to be satisfied for a contract to be awarded. Farmers, farm organizations or cooperatives, or even brokers would compete for these contracts, which would presumably be awarded to the lowest cost bidders. The observable or documentable adoption of practices that are known to sequester carbon at a given rate under predetermined conditions could substitute for field-level monitoring to lower transaction costs.

Clearly, there are all manner of variations on government-mediated trades. A simple version would have government identify the amount or number of credits created by agriculture. These credits, possibly in the form of emission vouchers, could then be sold to industry. The profits could be redistributed to farmers, possibly through farm programs or through regional councils established for the purpose of administering carbon credit programs, in accordance with such factors as expected crop yields. Farmers would need only keep records and provide documentation that certain practices have been followed. Either the farmer or government would be responsible for periodically assaying carbon levels in fields for the purpose of verification. Figure 3 lists a number of existing United

---

Conservation Reserve Program provides annual rental payments for 10 to 15 years to put sensitive croplands under vegetative cover.

Wetlands Reserve Program provides financial incentives to restore wetlands on agricultural lands.

Environmental Quality Incentive Program provides for 5- to 10-year contracts for technical, financial, and educational assistance to address environmental concerns in designated priority areas.

Conservation Farm Option provides direct payment to producers to implement innovative methods for addressing natural resource concerns.

Conservation Technical Assistance provides technical asssistance to farmers to improve agricultural lands to achieve long-term sustainability of the resource.

Forestry Incentives Program provides cost-share for tree planting and timber stand improvements.

Grazing Lands Conservation Initiative provides technical and educational assistance to improve the management of grazing lands, to conserve energy and other natural resources, to provide habitat for wildlife, to sustain forage and grazing plants, to use plants to sequester greenhouse gases, and to generate biomass for energy and raw industrial material.

Wildlife Habitat Incentives Program provides technical and financial assistance for improving fish and wildlife habitat on private agricultural lands.

Farmland Protection Program provides funding to state, local, or tribal entities with existing farmland protection programs to preserve farm land.

Stewardship Incentive Program provides technical and financial assistance to encourage private forest landowners to maintain the productivity and health of their lands.

National Conservation Buffer Initiative provides assistance to landowners to install buffers that improve soil, air, and water quality; enhance wildlife habitat; restore biodiversity; and create scenic landscapes.

Figure 3. USDA conservation programs that could be used to encourage carbon sequestration or creation of carbon credits.

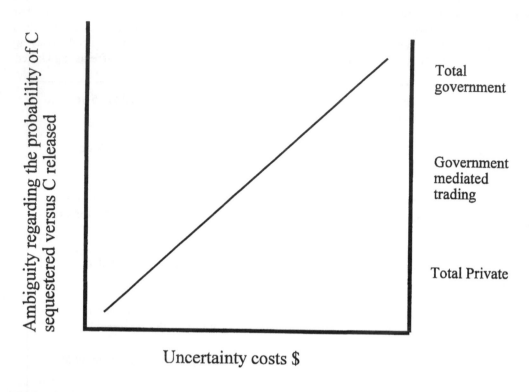

Figure 4: Relationship between ambiguity regarding ratio of C sequestered versus C released and the type of trading mechanism.

States Department of Agriculture programs (16 U.S.C. 3801 et seq.) that could be adapted to encourage sequestration of carbon in agricultural soils.

In totally private trading, two or more private parties negotiate with no government involvement other than judicial redress. Trades could occur between individual firms or consortia of firms from the energy industry or perhaps the transportation sector and either agricultural cooperatives or even individual farms. Given the need, however, for documentation to verify treaty commitments, a limited role would continue for government monitoring of trades.

What manner of trading mechanism is likely to emerge initially is not totally a matter of chance. Problems of dimensionality, the status of resolution of legal and institutional issues and equivalence affect the transaction costs associated with a trade which translate into the discount for an acre of carbon sequestered. On the other hand, the producer faces a certain degree of uncertainty, particularly as a consequence of weather, pests, or natural disasters, that affects his or her costs of production of a unit of carbon. He or she may require a price over and above the cost of production that increases with his or her perception of production risk. This risk-related divergence in price which buyers are willing to pay and at which producers are willing to produce constitute, what I refer to here as the cost of uncertainty. As illustrated in Figure 4, low uncertainty costs lead to a low relative total cost per unit of soil carbon sequestered, whereupon the private sector is more likely to engage in trades. The higher the cost of uncertainty, the higher the total cost per unit of soil carbon sequestered, and hence the fewer the likely buyers. Government can increase the likelihood of trades by, in effect, assuming some portion of this cost – or ensuring a portion of the transactional risk, thus reducing the total cost per unit soil C. At the extreme where uncertainty costs are very high, the only buyer for the good is government.

IV. Relative Advantages and Disadvantages of Government versus Private Trading

Having government as the sole purchaser of carbon credits eliminates much of the transactional cost because government assumes most of the financial risk. Government bears the cost of monitoring and verification and can set the payment to producers at a politically determined price that may or may not have resemblance to the market or world price. The extent of carbon credit is negotiated internationally and need not bear strict relationship to actual amounts sequestered. Thus, there is little need for strict precision with regard to issues of dimensionality. Legal conditions, such as easements on development or conversion of the land to a use unsuitable to maintenance of the carbon sequestered in the soil or covenants on future use of the land may not be necessary since government, as the sole purchaser of credits, needs only an aggregate estimate of sequestered soil carbon over all agricultural land, not how much is sequestered on a particular plot. Contracts, other than existing short-term contracts specifying practices, need not govern transactions. Equivalence is but a minor issue provided that there are no major shifts in land use away from agriculture into a use in compatible with maintenance of soil carbon in the aggregate.

Because transactions involve just two entities, government and the set of any or all farmers, transaction costs are relatively low. The issue of equity, regarding rewards for or lack of reward for early adopters, can be eliminated by dispensing payments equally across all farmers adopting agreed upon practices.

Another area of advantage of government as purchaser is the existing structure for technical assistance and monitoring represented by the Department of Agriculture's Natural Resource Conservation and Extension Services. No new oversight system needs to be put in place. Furthermore, the program can be implemented immediately with Congressional approval and need not require changes in international agreements in the short run. Quantities of carbon sequestered in agricultural soil could, at a future date through international agreement, provide offsets for carbon emissions.

On the negative side, a government program that provides rewards for carbon sequestration generally across agricultural producers creates few incentives for individual producers to improve carbon sequestration or to maintain carbon stocks. Rewards confer across all producers regardless of individual production efficiencies. There is no incentive for the administering agencies to monitor practices or maintain sequestered carbon quantities because of the lack of penalty or reward. In addition, even those increases in carbon stocks that have been achieved over time are likely to accrue at a lower rate than under a system of private trades since there are no incentives for maintaining stocks (unless there is a long-term easement associated with the land) or, as there are likely to be in private transactions, penalties to both parties for failure to comply with contractual conditions. Thus, a totally government system results, over time, in a potentially a less efficient system that fails to develop technologies as rapidly as it potentially could in order to reduce the social cost to society. The overall consequence is less carbon sequestered over time at higher social cost than what could ideally be achieved.

On the other hand, a totally private system initially may produce too little sequestered carbon. High uncertainty costs result in high transaction costs that impede trades or produces too few trades and hence not enough carbon sink activity. The promise of greater future efficiency in soil carbon production may improve over time as technical, legal and institutional issues are resolved.

V. Conclusion

Numerous technical, legal, and institutional issues must be resolved before a system of trading in agricultural soil carbon credits is to become a reality. Because soil carbon credits represent a promise of a service rather than a discrete and measurable good, a market in this commodity resembles wetlands mitigation banking, rather than SO_x emissions trading. Hence, a successful program that encourages the widespread adoption of practices that sequester carbon in the soil will require resolution of numerous complicated policy issues in addition to technical issues pertaining to measurement and prediction of carbon stocks. In the short term, a program involving of government purchase of credits or payment for practices, possibly through modification of existing federal agricultural conservation programs, can provide the incentives to sequester carbon in agricultural soils. Shifting over time to private trading involving minimal government mediation as our ability to predict future carbon levels improves should result in greater economic efficiency and lower social cost of mitigating greenhouse gas emissions.

References

Eizenstat, S. Statement of Stuart Eizenstat, White House Climate Taskforce, before the U.S. Senate Committee on Agriculture, Nutrition, and Forestry March 5, 1998. Http://www.senate.gov/~agriculture/eizenst.htm.

King, D., C. Bohlen and K. Adler. 1993. Watershed management and wetland mitigation: a framework for determining compensation ratios. A report prepared for EPA, Office of Policy Analysis, Washington, D.C., 1994. CEES Contribution # UMCEES-CBL-94-047.

King, D. 1997a. Valuing wetlands for watershed planning. *National Wetlands Newsletter* (May-June): 5-10.

King, D. 1997b. The fungibility of wetlands. *National Wetland Newsletter*, Vol. 19, No. 5, Environmental Law Institute, Washington, D.C.

King, D. 1998. Leading indicators of ecosystem services and values. In: C. Grayson (ed.), *Lost Human Uses of the Environment*. Island Press, Washington, D.C.

United Nations Framework Convention on Climate Change. 1992. http://www.UNFCCC.org/resource/conv/conv_002.html.

United Nations, Framework Convention on Climate Change, Conference of the Parties, Report of the Conference of the Parties on Its Third Session Held at Kyoto from 1 to 11 December 1997, Part II: Action Taken by the Conference of the Parties at Its Third Session, FCCC/CP/1997/Add.1 (18 March 1998).

United States Army Corps of Engineers. 1994. Water Resources Support Center Institute for Water Resources, National Wetland Mitigation Banking Study, Report #92-WMB-1-6.

United States Environmental Protection Agency. 1999. Acid Rain Program SO_2 Emissions Trading Program. http:www.epa.gov/acidrain/trading.html.

Wainger, L.A. and D.M. King. 1999. Managing Risks in Carbon Sequestration Trading: The Use of Spatial and Temporal Variables to Score Carbon Credits. University of Maryland Center for Environmental Sciences, Technical Report Series No. TS-209-99-CBL. Prepared under a cooperative agreement between the University of Maryland, Center for Environmental Science, U.S. Environmental Protection Agency, Office of Policy Analysis, and the U.S. Department of Agriculture, Natural Resource Conservation Service CA-68-7482-8-335.

Section VIII

Synthesis

CHAPTER 43

Methodological Challenges Toward Balancing Soil C Pools and Fluxes

R. Lal, J.M. Kimble and R.F. Follett

I. Introduction

World soils represent the third largest global C pool. Soils contain about 1550 Pg of soil organic carbon (SOC) and 750 Pg of soil inorganic carbon (SIC) (Batjes, 1996). The SOC pool is about 2.2 times the atmospheric pool (720 Pg) and 2.8 times the biotic pool (560 Pg). The geologic pool is estimated at 5000 Pg, comprising 4000 Pg of coal and 500 Pg each of gas and oil (Post et al., 1990). There is a direct linkage between the SOC pool and the atmospheric pool. Change in SOC pool by 1 Pg is equivalent to a 0.47 ppmv change in the atmospheric CO_2 concentration. The atmospheric pool has steadily increased from a pre-industrial concentration of 280 ppmv around 1850 to 365 ppmv in 1995 (IPCC, 1996; Hansen et al., 1998). The atmospheric pool has increased at the expense of the soil, biotic and the geologic pools.

The SOC pool is in dynamic equilibrium with its environment and is readily influenced by anthropogenic activities. Human perturbations with a drastic impact on SOC pool include deforestation, biomass burning, conversion of natural ecosystems to agricultural and other managed ecosystems, tillage, crop cultivation, grazing, applications of fertilizers and manures, etc. Most of these activities have a negative impact on the SOC pool, especially those related to conversion from natural to agricultural ecosystems and the continued cultivation of agricultural lands (Figure 1).

Both the SOC and SIC pools are strongly influenced by the biophysical environments of the pedosphere, hydrosphere, lithosphere and the atmosphere. Therefore, it is important to understand the dynamics of the SOC and SIC pools in relation to anthropogenic perturbations to the pedosphere, hydrosphere and the atmosphere, and their interactions. An important prerequisite to understanding the processes that affect the dynamics of SOC and SIC is quantification of the temporal and spatial changes in the magnitude of these pools. Credible measurements of SOC and SIC pools at different scales (e.g., pedon, soilscape, landscape, watershed, ecoregion, national and global) are needed to understand dynamics of these pools.

There is strong interest in the soil C pools (SOC and SIC) and the associated fluxes because of their significance to the global C cycle and their impacts on greenhouse gases. Quantitative data on C pool and fluxes, at scales ranging from pedon or soil-scape to ecosystem and national and global scales, are needed to assess the magnitude of soil C pool in relation to the biotic and atmospheric pools and to evaluate the contributions (flux) of the soil C pool to the atmospheric pool. There is also an interaction with the biosphere because of the effect of CO_2 fertilization and the long-term effects this can have on biomass production that can lead to changes in the soil pools. Identifying the link between soil and atmospheric processes is necessary to develop and validate land use change and

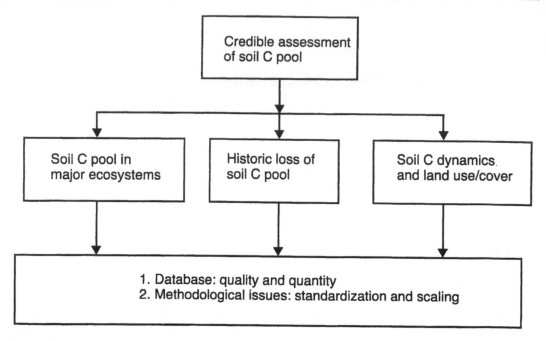

Figure 1. Principal issues in credible assessment of soil C pools.

management practices that might reduce emissions. Once these links are understood, policy options can be identified that will encourage land managers to adopt practices that enhance C sequestration. This chapter summarizes the state-of-the-knowledge related to assessment methods for soil C pools. It also outlines important knowledge gaps and prioritizes researchable issues related to fluxes and the global C cycle.

II. Different Types of Soil C Pools

Soil C pool comprises two principal components: SOC and SIC (Table 1). With regard to the SOC pool, there are at least three principal constituents, shown in Table 1 with their mean residence times. The most dynamic or the labile pool primarily comprises the microbial biomass carbon. The intermediate pool includes remains of plants and animals at various stages of decomposition and humification. The passive pool is a complex material highly resistant to decomposition, often called humus, and is associated with the clay in the soil. The SIC pool, comprising carbonates and biocarbonates, consists of two principal components – primary and secondary carbonates.

Primary carbonates are of geologic origin and contained in the parent material. The secondary carbonates are of pedologic origin, formed by soil-related processes, and may contribute to C sequestration in soil if there is a source of Ca other than from the desolution of primary carbonates. Available data on the C pools and dynamics are sketchy, incomplete, often measured only for the topsoil within the plow layer (10 to 30 cm), and generally without reliable information on soil bulk density for the specific layer. Most available data are difficult to compare because of a wide range of procedures used in laboratories around the world. Large differences in the data about soil C pools can occur due to differences in procedures used for soil sampling, sample aggregation or compositing,

Table 1. Different types of soil C pools

Soil C pool	Specific component	Turnover time (years)
1. Organic	(i) Labile pool	< 10
	(ii) Intermediate pool	$10^1 - 10^2$
	(iii) Passive pool	$10^2 - 10^4$
2. Inorganic	(i) Primary carbonates	——
	(ii) Secondary carbonates	——

analyses, and extrapolations. Further, most data represent points, and procedures for scaling point data to landscape, watershed, or ecological/geographical regions need to be reviewed and standardized (Eswaran et al., 2000).

III. Causes of Variation in Estimates of Soil C Pools

The available research data on global C pools are highly variable. Estimates of the size of the SOC pool range from 1200 to 2200 Pg (Bohn, 1982; Buringh, 1984; Kimble et al., 1990) and are considered low because of inadequate data for soils of the cold regions and peat areas. Similarly, estimates of the size of the SIC pool range widely from 700 to 946 Pg (Batjes, 1996; Eswaran et al., 2000). Some important reasons for the wide variations in estimates include the following:

   (i)   The depth of C assessment is usually limited to the plow layer or rooting zone. For most soils, the depth of C assessment should be 2 or 3 m or to a lithic or paralithic contact. The amount of C in soil below 1-m depth constitutes a significant part of the total C pool in many soils, particularly in humid environments.

  (ii)   The SIC pool is an important component of the total C pool, and yet is neither quantified nor reported.

 (iii)   Methods of C determination vary among laboratories, and data are often not validated against reference samples. The lack of standard methods restricts comparative evaluation of data obtained by different research groups.

 (iv)   Reliable estimates of soil bulk density are not available.

  (v)   Scaling procedures from point source to large scales are not standardized or do not even exist.

IV. Important Issues with Regard to Soil C Pool

Because of its significance in the global C cycle, several important issues with regard to the credible assessment of the soil C pool need to be considered. These issues can be grouped into three categories: (i) assessment of soil C pool in different ecosystems, (ii) historic loss of soil C pool due to anthropogenic activities, and (iii) soil C dynamics in relation to land use/land cover changes. Credible assessment of these three entities depend on two considerations: (i) quantity (extent or magnitude of the database) and (ii) standardization of methods used to assess the pools and fluxes. Improving the database depends on several important issues that need to be addressed.

1. Soil Sampling and Sample Preparation: Errors in soil sampling and sample preparation can result in unusable data. What a sample is to represent determines the type, size, mode, design and numbers of samples to be collected for a given area. Site selection to obtain a representative sample is critical. While the most active (labile C) is in the top 10 cm, determining the total C pool requires sampling to at least 2-m depth. In addition to C, related soil properties must also be determined (e.g., pH, bulk density, texture, aggregation, NPKS, rock fragments, and root biomass). Field variability is important consideration in designing a sampling protocol. The variability may be systemic or random. Sampling protocol should be designed to decrease the coefficient of variation.

2. Soil C Pools: The soil C pool has two principal components: organic (SOC) and inorganic (SIC). The SOC pool comprises plant and animal residue, microbial mass and by-products, and primary and secondary chemical constituents of varying complexities based on a wide range of degradative (decomposition, mineralization) and aggradative (synthesis, sequestration, stabilization, humification, chemical-biochemical-physical transformations) processes. The SIC pool consists of CO_2, carbonates and bicarbonates. The carbonates may be primary or secondary.

3. Methods of C estimation: The methods used for soil C determination depends on the component to be evaluated. The macro-organic matter (MOM) or particulate organic matter (POM) is a very important fraction that cannot be ignored. As shown in Table 1, the SOC pool comprises different fractions that need to be quantified (Table 2). The magnitude of different components depend on the land use history and ecoregional characteristics. Different components may also be classified into light and heavy fractions. Total C can be fractionated into different components by chemical, physical, and biochemical separation techniques. Once fractionated, there are analytical techniques (combustion, titrimetry, chromatography, NMR, IR, and isotopic methods, etc.) to quantify different fractions. The techniques for quantification of different pools range from molecular scale to pedon and landscape scale. There is a need to standardize both fractionation and analytical techniques for the quantification of the different C pools. The choice of a method also depends on the objective and the use of the data related to measuring C.

4. Separating SIC and SOC fractions: High temperature combustion is a useful technique to determine total soil C. However, acidification is required to remove SIC from the sample so it can be determined by the difference in an acidified and non-acidified sample. There is a need to

Table 2. Components of total system C

Pool	Relative size	Turnover time
(a) Above ground		
1. Woody biomass	Large	Rapid
2. Herbaceous biomass	Intermediate	Slow
3. Litter	Small	Rapid
(b) Below ground		
4. Roots	Intermediate	Rapid
5. Soil carbon (SOC and SIC)	Large	Slow (varies as shown in Table 1)
6. Charcoal C	Intermediate	Very slow

remove charcoal C as it is also measured in high temperature combustion, and it may be an important component of SOC in some soils and ecosystems.

5. Soil bulk density and equivalent mass: Soil bulk density is an important property that needs to be determined. Further, the C pool can be quantified on "equivalent mass " basis rather than on "equivalent depth" basis but this requires a measurement of the bulk density. Soil bulk density measurements need to be corrected for the coarse fragment and roots. Differences among field methods of determining soil bulk density depend on methods of measuring soil volume. There is a need to standardize these methods.

6. Charcoal C: This can be an important component, especially in fire-dependent ecosystems. There may be no apparent trend or relationship between charcoal C and total soil C. The relative amount of charcoal C may range from 0% to 50% of the total C. The charcoal C is a relatively resistant or inert material.

7. Importance of different fractions in C sequestration: The labile C pool has a rapid turnover time and the slow (passive C pool) changes at a very slow rate (Table 1). The important fraction in relation to C sequestration, and which responds to land use and management, is the intermediate C pool. The conversion of the labile C pool to the intermediate pool depends on the availability of essential nutrient elements (e.g., N, P, S). It is important to identify the association of the passive C pool with soil constituents, e.g., clay/silt fraction, aggregation, etc.

8. Soil C indices: Land use and management systems may lead to change in soil C and some pools can change more than others. The dynamics of different C pools depend on the characteristics of the pool. Yet, it is important to know the distribution of the C pool in relation to the turnover or the residence time. There are some indices that are useful in identification of pools with different turnover times. Two important indices are: (i) Functionality Index, and (ii) Carbon Management Index (see Chapters 5 and 23, this volume). Development and validation of such indices are important for site-specific situations.

9. Waste disposal hazard: Waste disposal from analytical laboratories is an important consideration in the choice of different methods. Use of isotopic and $KMNO_4$ techniques require a careful waste disposal system. In contrast, optical techniques (NIR, MIR) and sample combustion techniques have little or no waste disposal hazard.

10. Soil erosion and C: The fate of carbon in eroded sediments is not known, and there is little quantitative information about the C budget in eroded sediments based on direct measurements. Important unknowns are what happens to the transported/depositional C, and what is the global significance of buried C (soil sediments, water bodies). An important question is: How much carbon is transported and where is it deposited? An understanding of the effects of erosion on aggregation and the effects of aggregation on rates of decomposition is needed.

11. C sequestration by agricultural practices: Global agricultural productivity has to be enhanced and sustained to meet the needs of a growing world population. It is important, therefore, to identify and encourage widespread adoption of those agricultural systems that lead to C sequestration in soil and terrestrial ecosystems. Avoiding loss of C through adoption of relevant practices is an important consideration, e.g., in soil conservation. It is also important to identify ecosystems or regions with high potential for C sequestration, and a long residence time after it is sequestered. While conservation tillage is useful, it is also important to know what happens to SOC sequestered when a long-term no-till field is tilled?

12. Monitoring and verification of C sequestration: Use of remote sensing and models (e.g., neural network) is inevitable. Relevant models are those that relate quantity and quality of input to site-specific factors, e.g., soil color, moisture and temperature regime, geographical coordinates:

$$C_s = (Input - Output)_{\Theta, T, clay}$$

where C_s is C sequestration, Θ is moisture regime, and T is temperature. Both Θ and T are functions of geographical coordinates, and are also influenced by soil color. Soil C is not randomly distributed over the landscape. It is strongly correlated with the clay content. Thus, is important to integrate monitoring techniques with predictive models for a wide range of soils and ecoregions. The C_s may be a practice-based index validated on some benchmark sites.

13. <u>Dead wood:</u> Dead wood and woody detritus are often ignored in computations of the C budget in forest ecosystems. Estimates of biotic C pool range from 400 to 800 Pg, but without consideration of the dead wood. The latter is estimated at 150 Pg on global basis. The ratio of dead to live wood ranges from 0.10 to 0.50, depending on the ecoregion.

14. <u>Economics of C sequestration:</u> An important strategy is to adopt policies that lead to C sequestration in the soil. The cost of C sequestration is estimated to range from $1/ton to $187/ton and the cost is generally low for low rate of sequestration. Restoration of degraded soils, important to C sequestration as it may be, has not happened because it may not be economical to individual farmers.

15. <u>Carbon credits:</u> Developing tradable permits is an important strategy. Specific land management activities (e.g., no-till, cover crops, rotations) need to be translated into measurable credits. There are three strategies: (i) tax or punish objectionable activities, (ii) subsidize desirable activities or reward good deeds, and (iii) develop transferable permits.

16. <u>Policy issues:</u> Identification of appropriate policies depends on (i) assessment of credits and (ii) development of trading mechanisms. There are three issues involved in assessment of credits: (i) dimensionality issue (scaling), (ii) liability or legality issue, and (iii) equivalency issue. There are at least two relevant models, i.e., SO_2 model and the wetland model, that may provide help in developing a trading mechanism. Trading mechanisms may be based on government or private organizations.

V. Knowledge Gaps

Management of soil organic matter has always been an important issue for agronomists because it has a strong impact on productivity. However, its significance in relation to the global C cycle and the potential greenhouse effect has not yet been fully appreciated. There are numerous knowledge gaps (Figure 2) that restrict our understanding of the processes involved.

1. <u>Standardization of laboratory methods:</u> Laboratory techniques of assessing different pools need to be standardized. Precise measurements of different SOC pools, and secondary carbonates (SIC) remain a major challenge. The quantity and quality of soil C pool need to be correlated with soil attributes, e.g., CEC, clay content, aggregation, etc. There is a need to strengthen the data bank at pedon level, especially for principal soils in tropical ecoregions. The available data are severely constrained by the lack of information on soil bulk density, and the C pool for sub-soil (50-cm to 2-m depth) horizons.

2. <u>Field assessment of C pool:</u> Reliable assessment of C pools depends on the accuracy and precision of the methods used at pedon, farm, ecoregional and geographical levels. Reliable field assessment of C sequestration rate remains a challenge. The field assessment methods may have to be based on (i) quality and quantity of input used, (ii) land use and cultural practices adopted, (iii) assessment of ground cover, residue returned, and soil color through remote sensing, and (iv) ground truth validation on key benchmark sites with soil, geographical coordinates, and soil moisture and temperature regimes. The method must be simple, routine, rapid, and applicable over large scale. There is a need to develop scaling procedures from pedon level to a geograph-

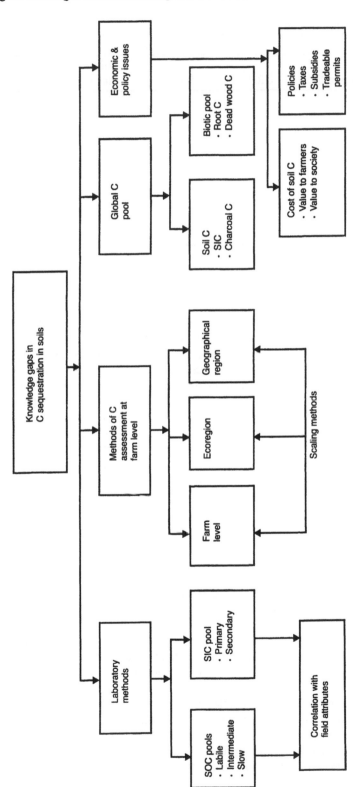

Figure 2. Knowledge gaps in soil C pools.

Table 3. Matrix of C sequestration rate

Ecoregion MLRA/ Soil type	Agricultural practices				
	Conservation tillage	CRP	WRP	Soil fertility management	Land restoration
I A					
B					
C					
D		(C sequestration rate)			
II A					
B					
C					
C					

ical/political region (pedon → soil-scape → farm → MLRA → ecoregion → geographical region). Scaling up may involve use of remote sensing, neural networks, and fractal methods.

3. Global C pool and fluxes: The data on global terrestrial pools (e.g., soil C and biotic C) need to be refined. The soil C pool comprises several constituents that are not adequately assessed e.g. charcoal C, and soil inorganic C including secondary carbonates. Estimates of the biotic C pool need to be improved especially with regard to the dead wood C and root biomass C.

4. Economics and policy issues: There is a strong need to assess the value of soil C on weight basis (per ton) or area basis (per hectare). There are two components of the value: (i) benefits to the farmer (because of enhancement in soil quality and contributions to the nutrient supply and water holding capacity), and (ii) benefits to the society (due to improvements in the environment, e.g., water quality, biodiversity, wildlife habitat, and the greenhouse effect). Once the societal value is defined, farmers and land managers need to be rewarded for good management and taxed for inappropriate practices. A mechanism needs to be established for developing tradable/ transferable credits.

VI. Future Priorities

Assessment and management of soil C pool and fluxes are important global issues. Using the C sequestration potential of world soils to help mitigate the greenhouse effect is based on the following research and development priorities:

1. Identifying regions (soils, ecosystems) with a high C sequestration potential. An important strategy is to identify those regions that have a large sink capacity and long residence time.

2. Estimating C sequestration rates for innovative practices for different ecoregions. Establishing a matrix of rates of C sequestration (Table 3) is an important stop in developing C credits and transferable permits.

3. Developing soil C monitoring and verification system at farm, agro-ecoregion or county level. Such a system should be practice-based and validated against ground truth measurements. The use of remote sensing methods can play an important role in monitoring and verification.

4. Establishing the baseline soil C pool for pre-cultivation and 1990 levels through assessment of the size of the C pool in undisturbed ecosystems. It is important to identify reference points over time and space.

5. Standardizing soil sampling, preparation, and analyses protocols to minimize errors.

6. <u>Assessing</u> the fate of eroded C as it is deposited and redistributed over the landscape remains a high priority.
7. <u>Establishing</u> links between C pools and fluxes for different land uses and soil/crop management systems is a high priority.
8. <u>Evaluating</u> the value of soil C and developing credit trading mechanisms needs to be done and land managers need to be involved in the process of C sequestration.
9. <u>Identifying</u> strategies of realizing the potential of C sequestration in soil. These strategies include: (i) reducing loss of C through better management strategies, (ii) increasing the C size and pool input into the soil, (iii) using biotechnology and plant breeding techniques to alter lignin content and C:N ratio in plants, and (iv) using fossil fuel off-set through production of biofuels.
10. <u>Educating</u> land managers, policy makers, and the public at large about the role of soil C and its relationship to the greenhouse effect is needed.

VII. Conclusions

Reliable assessment of soil C pool depends on five principal considerations:
1. Standardizing procedures for:
   (i)   physical sampling depth in relation to land use and soil management (e.g., tillage methods),
   (ii) sample preparation and handling for different types of C pools (e.g., macro-organic matter, microbial biomass C, labile carbon, intermediate pool and passive pool, SIC, secondary carbonates),
   (iii) measurement of soil bulk density and calculating C pool on equivalent weight rather than equivalent depth, and
   (iv) evaluating different forms of C using stable isotopic techniques or C dating methods.
2. Evaluating total system C and its components.
3. Standardizing scaling procedures from pedon level to global scale.
4. Establishing rationale for assigning monetary value to soil C.
5. Developing methods for establishing C credits.

Standardization of the procedures is critical to improving the database. There is a strong need to improve the database especially with regards to the following:
1. Impact of land use and management on soil C pool in relation to the following agricultural practices:
   (i)   irrigation and drainage on soil C pool (both SOC and SIC components),
   (ii) tillage methods, cover crops, and residue management,
   (iii) fertilizer use and soil-specific or precision farming, and
   (iv) use of manure, compost and biosolids.
2. Effect of soil degradative processes on soil C pool especially with regard to the following:
   (i)   erosional processes due to water and wind,
   (ii) decline of soil structure and physical degradation,
   (iii) soil pollution and contamination,
   (iv) drastic soil disturbance by mining and urbanization,
   (v)   drainage and cultivation of organic soils,
   (vi) intensive use of calciferous soils by irrigation, use of acidifying fertilizers, and
   (vii) exposure of caliche and other calciferous material to climatic elements and anthropogenic activities.
3. Impact of soil restorative measures on soil C pool in relation to the following:
   (i)   erosion control and restoration of eroded soils,

(ii) reclamation of salt-affected soils and growing halomorphic plants by irrigation with saline water,

(iii) mineland reclamation, and

(iv) wetland restoration.

References

Batjes, N.H. 1996. Total C and N in soils of the world. *European J. Soil Sci.* 47:151-163.

Bohn, H.L. 1982. Estimates of organic carbon in world soils. *Soil Sci. Soc. Am. J.* 46:1118-1119.

Buringh, P. 1984. Organic carbon in soils of the world. *SCOPE* 23:91-109.

Eswaran, H., P.F. Reich, J.M. Kimble, F.H. Beinroth, E. Padmanabhan and P. Moncharoen. 2000. Global carbon stocks. p. 15-26. In: R. Lal, J.M. Kimble, H. Eswaran and B.A. Stewart (eds.), *Global Climate Change and Pedogenic Carbonates*. Advances in Soil Science. Lewis Publishers, Boca Raton, FL.

Hansen, J.E., M. Sato, A. Lacis, R. Ruedy, I. Tegen and E. Matthews. 1998. Climate forcing in the industrial era. *Proc. Natl. Acad. Sci.* 95:12753-12758.

IPCC. 1996. *Climate Change 1995.* Working Group 1, Cambridge University Press, Cambridge, U.K.

Kimble, J., T. Cook and H. Eswaran. 1990. Organic matter in soils of the tropics. p. 250-259. In: Symposium on Characterization and Role of Organic Matter in Different Soils. 14th International Congress Soil Science, August 12–18, 1990, Kyoto, Japan.

Post, W.M., T.H. Peng, W.R. Emmanuel, A.W. King, V.H. Dale and D.L. de Angelis. 1990. The global C cycle. *Am. Sci.* 78:310-326.

Index